AF832895

# BIOFUEL PRODUCTION – RECENT DEVELOPMENTS AND PROSPECTS

Edited by **Marco Aurélio dos Santos Bernardes**

**Biofuel Production – Recent Developments and Prospects**
Edited by Marco Aurélio dos Santos Bernardes

**Published by InTech**
Janeza Trdine 9, 51000 Rijeka, Croatia

**Copyright © 2011 InTech**
All chapters are distributed under the Creative Commons Non Commercial Share Alike Attribution 3.0 license, which permits to copy, distribute, transmit, and adapt the work in any medium, so long as the original work is properly cited. After this work has been published by InTech, authors have the right to republish it, in whole or part, in any publication of which they are the author, and to make other personal use of the work. Any republication, referencing or personal use of the work must explicitly identify the original source.

Statements and opinions expressed in the chapters are these of the individual contributors and not necessarily those of the editors or publisher. No responsibility is accepted for the accuracy of information contained in the published articles. The publisher assumes no responsibility for any damage or injury to persons or property arising out of the use of any materials, instructions, methods or ideas contained in the book.

**Publishing Process Manager** Petra Zobic

**Technical Editor** Teodora Smiljanic

**Cover Designer** InTech Design Team

Additional hard copies can be obtained from orders@intechopen.com

Biofuel Production – Recent Developments and Prospects, Edited by Marco Aurélio dos Santos Bernardes
p. cm.
ISBN 978-953-307-478-8

# We are IntechOpen, the world's leading publisher of Open Access books
## Built by scientists, for scientists

4,000+
Open access books available

116,000+
International authors and editors

120M+
Downloads

151
Countries delivered to

Our authors are among the
Top 1%
most cited scientists

12.2%
Contributors from top 500 universities

WEB OF SCIENCE™

Selection of our books indexed in the Book Citation Index in Web of Science™ Core Collection (BKCI)

Interested in publishing with us?
Contact book.department@intechopen.com

Numbers displayed above are based on latest data collected.
For more information visit www.intechopen.com

For EU product safety concerns:
IN TECH d.o.o., Prolaz Marije Krucifikse Kozulić 3, 51000 Rijeka, Croatia,
info@intechopen.com or visit our website at intechopen.com.

# Meet the editor

Dr.-Ing. Marco Aurélio dos Santos Bernardes serves as a postdoc researcher at the Centre de Recherche Public Henri Tudor in Luxembourg. His expertise is in the area of energy analysis, life cycle assessment, renewable energy and biofuels. Dr.-Ing. Bernardes has had 10 papers published in journals such as Solar Energy, International Journal of Life Cycle Assessment, ASME Heat Transfer, a book and book chapters as well. Dr.-Ing. Bernardes' areas of interest include CFD, heat transfer modelling, Solar Chimney Power Plants, thermal processes, thermodynamics. He received his Ph.D. in Mechanical Engineering at Stuttgart University in Germany and conducted a postdoctoral research at the Stellenbosch University in South Africa. He was awarded with the UNEP/SETAC Life Cycle Assessment Award for LCA Projects in Development Countries. Dr.-Ing. Bernardes served as a full professor in the Department of Mechanical Engineering at CEFET-MG in Belo Horizonte for more than 13 years.

Contents

# Preface

Over the past 20 years, there has been a substantial increase in research and development in the area of biofuels. Many researchers around the world have dealt with environmental, economic, policy and technical subjects aspects relating to these studies. In a way, this book aspires to be a comprehensive summary of current biofuels issues and thereby contribute to the understanding of this important topic. Chapters include digests on: the development efforts on biofuels, their implications for the food industry, current and future biofuels crops, the successful Brazilian ethanol program, insights of the first, second, third and fourth biofuel generations, advanced biofuel production techniques, related waste treatment, emissions and environmental impacts, water consumption, produced allergens and toxins.

Relating theoretical and experimental analyses with many important applied purposes of current relevance will make this book extremely useful for researchers, scientists, engineers and graduate students, who can make use of the experimental and theoretical investigations, assessment and enhancement techniques described in this multidisciplinary field. Additionally, the biofuel policy discussion is expected to be continuing in the foreseeable future and the reading of the biofuels features dealt with in this book, are recommended for anyone interested in understanding this diverse and developing theme.

**Dr.-Ing. Marco Aurélio dos Santos Bernardes**
Researcher Energy & Environment and LCA,
CRP Henri Tudor, CRTE,
Luxembourg

# Part 1

## Biodiesel

# 1

# Abiotic Stress Diagnosis via Laser Induced Chlorophyll Fluorescence Analysis in Plants for Biofuel

Artur S. Gouveia-Neto et al.*
*Universidade Federal Rural de Pernambuco*
*Brazil*

## 1. Introduction

In the past few decades there has been a widespread scientific and technological interest in laser-induced remote techniques to monitor the status of terrestrial vegetation (Svanberg, 1995). The most employed nowadays are those which exploit the fluorescence emission from the plant leaves generated in the photosynthesis process. The fluorescence of terrestrial vegetation consists almost exclusively of the fluorescence of leaves, which account for the largest surface of plants above ground. A small part of the absorbed light energy in the photosynthesis process is lost during the migration from the pigment antenna to the reaction centers and are dissipated by a number of non-photochemical processes, including heat, and re-emission of a small but easily detectable amount (2-5% *in vivo*) of the absorbed radiation. This re-emission occurs at longer wavelengths in the red and far-red spectral regions and is termed as Chlorophyll Fluorescence (ChlF) (Shreiber, 1983; Backer & Bradbury, 1981). Chlorophyll fluorescence represents an intrinsic signal emitted by plants that can be employed to monitor their physiological state including changes of the photosynthetic apparatus, developmental processes of leaves, state of health, stress events, stress tolerance, and also to detect diseases or nutrient deficiency of plants. In particular, the application of laser induced chlorophyll fluorescence spectroscopy has drawn much attention recently owing to the non-invasive and nondestructive nature of the technique (Svanberg, 1995; Lang & Lichtenthaler, 1991; Chappelle et al., 1984). The technique can be applied for chlorophyll level monitoring in basic photosynthesis research, agriculture, horticulture, and forestry. Abiotic stress (water deficit, salinity, heat, heavy metals soil contamination, intense light, etc) affects significantly crop growth and yield in agricultural areas all over the world. Thus, it is imperative to study their effect upon the crops and discriminate among abiotic stresses using new noninvasive and nondestructive remote sensing precision diagnostic techniques. These procedures allow one to employ intervention measures that will prevent damage to the crop and will not provoke economical losses. Our aim here, is to exploit laser-induced fluorescence signatures from plants to evaluate the effect of abiotic stresses (water

* Elias Arcanjo da Silva-Jr, Patrícia C. Cunha, Ronaldo A. Oliveira-Filho, Luciana M. H. Silva, Ernande B. da Costa, Terezinha J. R. Câmara and Lilia G. Willadino
*Universidade Federal Rural de Pernambuco, Brazil*

deficit and soil salinity) upon the evolution and characteristics of *in vivo* chlorophyll emission spectra of leaves of *Saccharum officinarum* and *Jatropha curcas* L. plants. The main interest in these plants species resides in the fact that they are used for large scale biofuel production in Brazil. Brazil is the world's second largest producer of ethanol fuel and the world's largest exporter. Together, Brazil and the United States lead the industrial production of ethanol fuel, accounting together for 89% of the world's production in 2009. Besides, *Jatropha curcas* L. plant is considered the poster child among many proponents to biodiesel production.

## 2. Photosynthesis

Photosynthesis is a biophotonic mechanism by which green plants exploit solar energy to reduce $CO_2$ and oxidize $H_2O$, as indicated in the process pictured in Fig. 1. Within the plant tissue, visible and near-infrared (NIR) light is absorbed (>80%) by photosynthetic pigments (Chlorophyll *a*, *b*, and carotenoids) and used to drive photosynthetic light reactions and associated electron transport reactions to reduce carbon and oxidize water in the Calvin cycle (Allen, 1992). Photosynthesis occurs in the chloroplasts where the photosynthetic pigments occupy. Chlorophyll molecules organized into two groups of pigments called photosystem I (PSI) and photosystem II (PSII), each containing "antennae" chlorophyll molecules and a central chlorophyll molecule (P680 and P700). The numbers are associated with the wavelengths corresponding to the maxima of the absorption spectra of the two species of Chlorophyll *a*.

$$6CO_2 + 12H_2O + \text{photons} + \text{chlorophyll} \rightarrow C_6H_{12}O_6 + 6H_2O + 6O_2 + \text{photons}$$

Fig. 1. Photosynthesis process in green plants

The physical mechanism of light energy absorption and migration to the reaction center is better visualized using the energy level and block diagram of Fig. 2.
The pigments antenna absorb much of the visible portion of the electro-magnetic spectrum, mainly in the near UV-blue region, as can be seen in the absorption spectra shown in Fig. 3.
There exist a very strong energy transfer mechanism taking place among pigment antenna where the light energy absorbed by carotenoids and chlorophyll *b* pigments resonantly transfer their energy to neighbours chlorophyll *a* molecules and the total energy is conveyed to reaction centers in which the migration process will occur as pictured in Fig. 2 (top). The chlorophyll fluorescence re-emitted light occurs in the red around 680-690 nm and far-red

730-740 nm spectral regions (Lang & Lichtenthaler, 1991; Chappelle et al., 1984). When excited with either UV or blue radiation, plants exhibit a fluorescence emission spectrum in

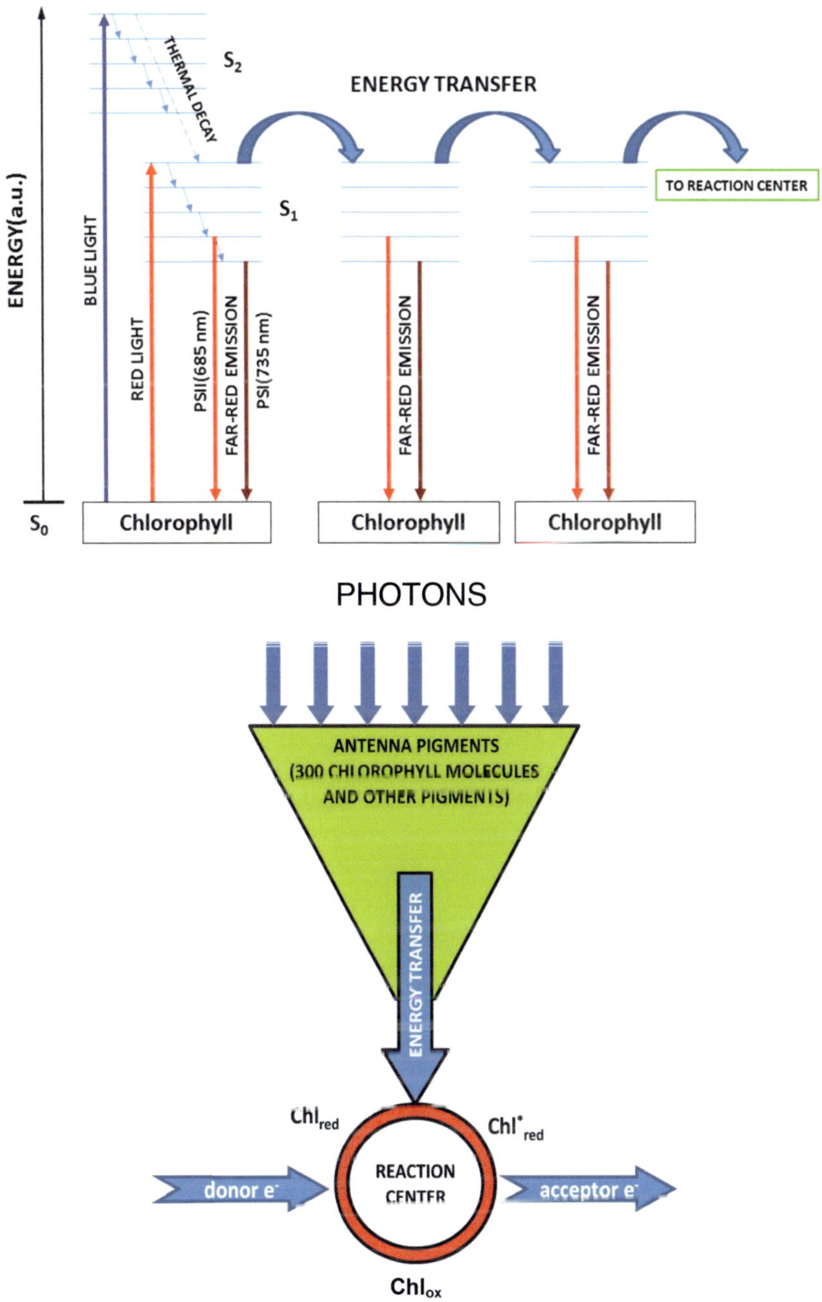


Fig. 2. Energy transfer (top) and light energy funneled to the reaction center (bottom)

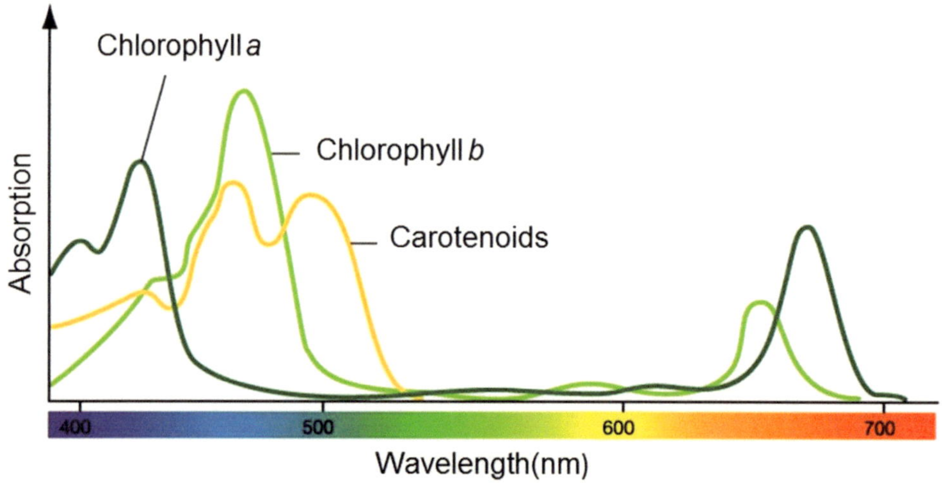


Fig. 3. Absorption spectrum of pigments antenna of green leaves

two distinct spectral regions blue-green (400-550 nm) and red-far-red (650-800 nm). However, in our case the fluorescence intensity of the blue-green spectral region was too small to be used as reliable chlorophyll fluorescence signatures, and was detected only in extreme cases, i.e., in plants under high degree of stress damage. On the other hand, the red fluorescence is characterized by a maximum in the red region (680-700 nm) which is attributted to the PSII antenna system and referred to as Fr, and one in the far-red (FFr) region (730-740 nm) owing the PSI photosystem.

## 3. Laser induced chlorophyll fluorescence

The substance emitting the red (Fr) and far-red (FFr) fluorescence of leaves, the red fluorophore, has been identified as Chlorophyll *a*. Although isolated Chlorophyll *b* dissolved in an organic solvent exhibits a red fluorescence, it does not do so *in vivo* because in a leaf the excitation energy is transferred completely to Chlorophyll *a*. At low Chl concentrations, the Fr and FFr increases with increasing Chl concentration (Bushmann, 1981; Stober & Lichtenthaler, 1992; Gitelson et al., 1998; Hák et al., 1990). At higher concentrations, the increase of Chl fluorescence with increasing Chl concentration is mainly detected in the FFr while Fr levels off with rising content. The re-absorption is caused by the overlapping of the short-wavelength range of Chl fluorescence emission spectra with the long-wavelength range of the Chl absorption spectrum. The Fr emission is much more affected by the re-absorption than the FFr, leading to the fluorescence ratio Fr/FFr decrease with increasing Chl content (Gitelson et al., 1998). The simultaneous measure of chlorophyll fluorescence in both red and far-red spectral region allows then the approximate determination of the Chl content of the leaves in a non-destructive way using the ChlF ratios (Hák et al., 1990; Lichtenthaler et al., 1990). In Fig. 4, one illustrates typical chlorophyll emission spectra for *Saccharum officinarum* excited with a blue 2.0 mW LED at 405 nm, for samples under different stages of salinity stress (NaCl concentration).

In a healthy plant (control) the spectrum presents the two distinct emission bands around 685 nm and 735 nm. For plants that experienced very intense stress (200 mM of NaCl), and presenting very low chlorophyll levels, one can clearly see the presence of two additional

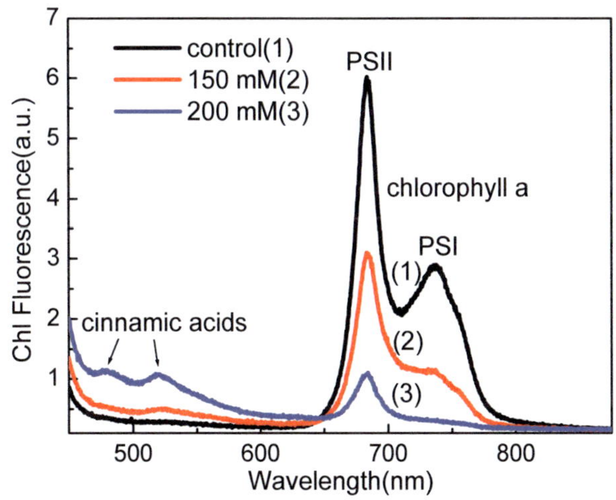


Fig. 4. Emission spectrum of *Saccharum officinarum* plants under intense salinity stress

fluorescence peaks around 440-450 nm and 520-530 nm. In green leaves the blue-green fluorescence is primarily emitted by cinnamic acids (Lichtenthaler & Schweiger, 1998) of the cells walls of the chlorophyll-free epidermis cells. The red and far-red fluorescences, in turn, are emitted by chlorophyll *a* in the chloroplasts of the leaves´ mesophyll cells. We have analyzed the dependence of the chlorophyll fluorescence upon the excitation wavelength and the results revealed that employing either UV (385 nm) or blue (405 nm) excitation light, the red fluorescence around 685 nm is higher when compared to the ones obtained employing blue-green (470 nm), orange (590 nm), and red (627 nm) excitation light, as indeed shown in the spectra of Fig. 5. The re-absorption of the chlorophyll fluorescence on its path towards the leaf surface, leads to different fluorescence spectral shapes for different excitation wavelengths as demonstrated by Agati (Agati, 1998), and Louis and co-workers

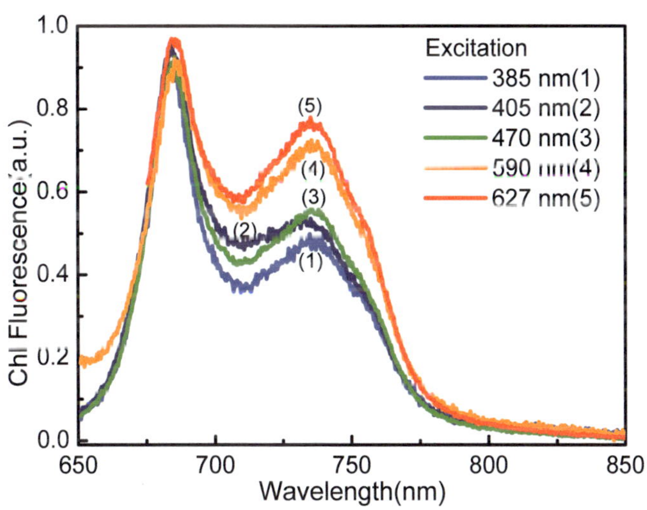


Fig. 5. Emission spectrum of *Saccharum officinarum* at different excitation wavelengths

(Louis et al, 2006) for bean leaves measurements. This is due to the fact that in green leaves, the chlorophylls and carotenoids have a broad absorption band in the 400-500 nm spectral region and blue light does not penetrate very deeply into the leaf tissue, and as a result the fluorescence associated to blue light excitation is mainly generated in the green mesophyll cells close to the leaf`s surface, therefore little absorption occurs. On the other hand, blue-green and orange excitations are not absorbed by carotenoids and penetrates more deeply into the green leaf mesophyll resulting in a chlorophyll fluorescence being generated deeper inside the leaf, from where on its way towards the leaf surface, resulting in a longer pathway and hence the re-absorption is stronger, leading to a less intense red emission compared to the far-red one.

## 4. Abiotic stress and plants

A number of abiotic stresses impose damage to crop and provokes reduction of yield productivity. Amongst several abiotic stresses, water deficit and soil salinity are the most commonly investigated owing to the extent of cultivated area affected by them. Water scarcity and increase competition for water resources involving several sectors of the production segment (agriculture, industry, hydroelectric energy, etc.) and also for human basic necessities, imposes the study of new concepts of irrigation, in order to adapt the crops to water shortage and maintain satisfactory levels of productivity. On the other hand, salinity affects 7-9 % of the world's land area (Szabolcs, 1994), and the area is increasing (Ghassemi et al., 1995). Nowadays, one of the major technological goals of the energy production is the replacement of the fossil-based fuel for biofuel, mainly due to environmental issues. Bearing that in mind, it is mandatory to investigate the effect of water deficit and salinity stress in plant species with high potential for application in large scale production of nonfossil based fuels.

### 4.1 Biofuel plants

Brazil is the world's second largest producer of ethanol fuel and the world's largest exporter (Renewable Fuel Association report, 2010). Together, Brazil and the United States lead the industrial production of ethanol fuel, accounting together for 89% of the world's production in 2009. In 2009 Brazil produced 37.7% of the world's total ethanol used as fuel. Brazil is considered to have the world's first sustainable biofuels economy and the biofuel industry leader, a policy model for other countries, and its sugarcane ethanol "*the most successful alternative fuel to date*". Concerning alternative proponents to renewable energy biofuels based on the use of plant oil as a fuel in stationary and mobile engines are the subject of much attention recently. One of the main crops currently being proposed as a diesel/kerosene substitute or extender, is *Jatropha curcas* (Linnaeus) (Openshaw, 2000; Francis et al., 2005) as will be discussed next.

#### 4.1.1 *Saccharum officinarum* (sugarcane)

Sugarcane has been cultivated in Brazil since 1532, and as sugar was one of the first commodities exported to Europe by the Portuguese settlers (Allah n.d.). The first use of sugarcane ethanol as a fuel in Brazil dates back to the late twenties and early thirties of the twentieth century, with the introduction of the automobile in the country. Sugarcane refers to any of 6 to 37 species (depending on which taxonomic system is used) of tall perennial grasses of the genus *Saccharum* (family Poaceae, tribe Andropogoneae). Native to warm temperate to tropical regions of Asia, they have stout, jointed, fibrous stalks that are rich in sugar, and measure two to six meters tall. All sugarcane species interbreed, and the major commercial cultivars are complex hybrids sugarcane products include table sugar, falernum, molasses, rum, *cachaça* (the national spirit of Brazil), bagasse and mainly ethanol.

#### 4.1.2 *Jatropha curcas* (physicnut)

*Jatropha curcas* (Linnaeus) is one of the most versatile plants with many attributes and notable potential. It is a small tree belonging to the family of *Euphorbiaceae*. *Jatropha* is easily settled, grows fast and is hardy, and in some way drought tolerant. Thus, it remedies soil degradation, desertification, and deforestation. *Jatropha* is native of tropical America, but now flourish in many parts of the tropics and sub-tropics in Africa/Asia. Various parts of the plant are of medicinal(both human and veterinary purposes) value for instance, and are under intensive scientific investigation. The oil is a strong purgative, widely employed as an antiseptic for cough, skin diseases, and a pain reliever from rheumatism. *Jatropha* latex can heal wounds and also has antimicrobial properties (Openshaw, 2000). Of particular scientific and/or technological interest is that, the fruit of jatropha contains viscous oil that can be used for soap making, in the cosmetic industry, and mainly as a diesel/kerosene substitute or extender (Francis et al., 2005).

## 5. Material and methods

### 5.1 Plant material and growing process

Both experiments(water and salinity stress) were conducted in a greenhouse at the Federal Rural University of Pernambuco (UFRPE), in Recife, Brazil, in the period October 2009 to January 2010 for the salinity stress study, and September 2010 to October 2010 for water deficit stress measurements. The seeds, provided by the Center for Technology and Natural Resources (CTRN), Federal University of Campina Grande (UFCG), Brazil, were sown in polyethylene tray containing washed sand as substrate, and samples were watered daily. Following germination, seedlings were irrigated daily in the morning, with nutrient solution containing 742.86 mg $L^{-1}$ soluble fertilizer (Brown Kristalon ®: 3% N, 11% P2O5, K2O 38%, 4% MgO, 11 % S, 0.025% B, 0.004% Mo, 0.01% Cu-EDTA, 0.025% Zn-EDTA, 0.07% Fe-EDTA and 0.04% Mn-EDTA) and 840 mg $L^{-1}$ nitrate Calcium (Viking Ship ® - 15.5% 19.0% N and Ca). This procedure was carried out throughout the whole investigation. After five days of germination, seedlings were selected based upon health and similarity in height and leaf number, and then were transferred to pots made of polyethylene with 10 kg maximum capacity, and containing washed sand substrate. The sand was covered with gravel to prevent soil water evaporation. After 27 days of acclimation, we have established seven treatments defined by the addition of NaCl to the nutrient solution: 0 (control), 25, 50, 75, 100, 150, and 200 mM. The treatment was carried out gradually in order to avoid osmotic shock in the plants. It was conducted by the addition of 25 mM of NaCl per day until the desired salt concentration was attained. The control of salt concentration in the substrate was performed every three days, by measuring the electrical conductivity of the solution drained from the pots. The daily drainage of the solution prevented the accumulation of salts in the substrate. The analysis of chlorophyll a and b content in the leaves was effectuated according to Arnon's methodology (Arnon, 1949). We have performed our experiment in a completely randomized design, with five replicates per treatment, producing a total of 35 experimental units during the period of 32 days. A randomized design experiment was also carried out for the water stress evaluation of the physicnut plants. We have examined the response through three levels of water stress. The pots were kept at field capacity during 21 days, after which irrigation treatments of drought (nonwatered), medium (50% of water capacity) and slightly below 100% of water field capacity. The 03 treatments were applied during 22 days for sugarcane and 10 days for physicnut plants, in 5 replicates, yielding a total of 15 samples for *Saccarum officinarum*, and 20 for *Jatropha curcas*. The pots were weighted before and after watering and in order to record their mass. The irrigation water contained a balanced nutrient mixture, as the one described in detail in the salinity stress experiment.

In order to evaluate the status of damage caused by the stresses on the plants growing process before visible damage is noticed, we have followed the evolution of the Chl content in the plant leaves using the Fr/FFr chlorophyll fluorescence ratio. The absolute emission signal of leaves can vary from sample to sample due to small differences such as excitation and sensing angles of the fluorescence, and the roughness and scattering properties of the leaf surface. Thus, the absolute fluorescence usually varies to a large extent than the fluorescence ratio. The fluorescence ratio turns out to produce much lower variations from leaf to leaf, resulting in a reliable and reproducible method for the quantification of changes in the fluorescence characteristics of leaves.

### 5.2 Experimental

In the experiments, the chlorophyll fluorescence was measured under steady-state conditions, in 20 min predarkened intact leaves, and we have employed as the excitation source, a blue LED at 405 nm with 10 nm of bandwidth and delivering a maximum power of 2.2 mW. The choice relies upon the fact that its wavelength resides within the main absorption band of Chl *a*, producing much higher fluorescence emission intensity. Red and far-red chlorophyll fluorescence emission around 685 nm and 735 nm, respectively, were observed and analyzed as a function of the stress intensity (NaCl concentration and amount of irrigation water). The LICF experiments were carried out within a time interval of 32 and 22 days (sugarcane) for the salinity and water stress evaluation, and 10 days for both sudies in physicnut. The measurements were performed every 4 days in order to monitor the evolution of the ChlF ratio during the NaCl treatment of plants. For the water stress a 2 days time intervals was utilized between measurements. The Fr/FFr ratio was evaluated using Gaussian shaped fluorescence fitting curves and analyzed as a function of time, and salinity intensity and water irrigation amount. Excitation and sensing were performed on the adaxial leaf surface. The ChlF experimental apparatus consisted of a fiber integrated LED source, spectrometer and light detector (Ocean Optics USB2000). The detection system had an overall operating spectral resolution of ~ 1.0 nm. The excitation source was directed to the leaf surface by means of a 200 μm diameter fiber cable which possessed a mechanical system at the fiber cable output extremity in order to prevent any ambient light of reaching the leaf surface during the measurements. Moreover, as the fiber itself was in contact with the leaf surface, it effectively shadowed away any leakage of ambient light. All spectra presented in this study were handled employing appropriate (Ocean Optics-SpectraSuite) software of the spectrometer. The data was stored and analyzed in a personal computer using a commercially available software (Origin 6.0).

## 6. Results and discussion

### 6.1 Salinity stress

Amongst abiotic stresses, salt stress is known to disturb the normal physiological processes and chloroplast ultrastructure at various levels (Allakhverdiev & Murata, 2008; Hasegawa et al., 2000; Munns, 2002; Sayed, 2003). The extent of the disturbance by NaCl ions depends upon the concentration and the plant tolerance. The decline in productivity observed in several plant species under salt distress is commonly associated with reduction in the photosynthetic capacity. Although the factors that limit photosynthesis in salt stressed plants are unclear, the effect of salinity stress on a number of species is quite evident and have been investigated in the past few years by several research groups worldwide (Yamane et al., 2003; Liu & Shen, 2006; Jimenez et al., 1997; Meloni et al., 2003; Lin et al., 2007; Zribi et al., 2009; Mehta et al., 2010).

#### 6.1.1 *Saccharum officinarum*

In this experiment we have followed the time evolution and the shape of the ChlF spectral profiles during a 35 days period of time, evaluating the effect of soil salinity in sugarcane plants. The dependence of the ChlF ratio upon the NaCl concentration was carried out for two varieties of *Saccharum officinarum* and the results are depicted in Fig. 6a (RB863129) and 6b (RB867515). The results clearly show an evident decrease in the chlorophyll content corresponding to a increase in the Fr/FFr ratio of the leaves experiencing an intense salinity stress, while the control sample undergo a steady increase in the chlorophyll concentration as time evolves. The results also indicates that, in the early stage (2-3 weeks) of the NaCl treatment, both plants follow the expected increase in the Chl content. After the 3rd week, however, a competition between the normal chlorophyll concentration evolution and the counter effect of the salinity distress takes place, and plants start to debilitate rapidly with time. The missing data at the fifth week in the graph of Fig. 6a (RB863129 variety) is because the samples did not resist to the salinity distress and samples remaining in the experiment would not provide reliable data for the Fr/FFr ratio.

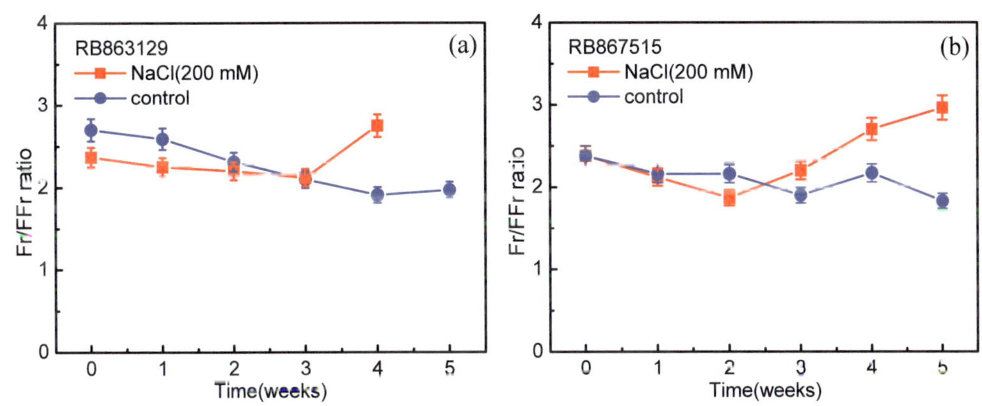


Fig. 6. Chlorophyll fluorescence ratio as a function of time

The chlorophyll fluorescence spectrum during the induction kinetic, the so called Kautsky effect, have been also investigated. When a 20 min pre-darkened plant leaf is submitted to excitation light, the onset of the photosynthetic process can be analyzed through the decrease of the Chl fluorescence from the initial maximum reached in the first second to the stead-state value after a few minutes of illumination. During the induction kinetics the wavelength position of the Chl fluorescence does not change, but the two emission bands decline at different time rates, leading to a decrease in the Fr/FFr ratio as can be observed in the results depicted in Fig. 7.

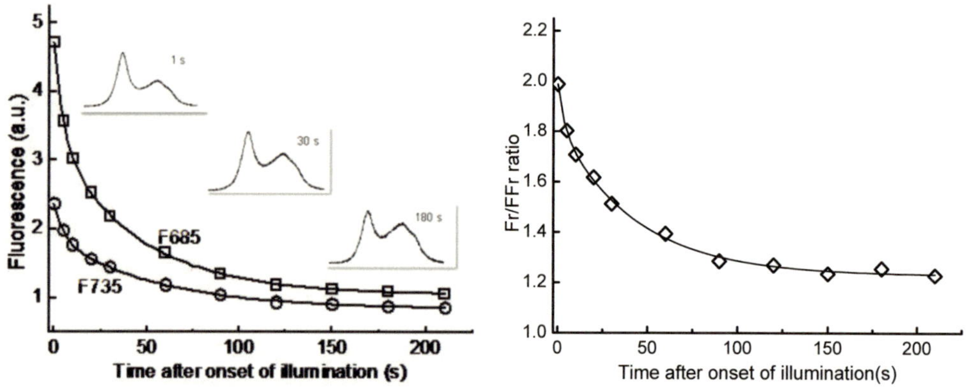


Fig. 7. Chlorophyll fluorescence ratio as a function time after onset of illumination

The dependence of the Chl fluorescence ratio upon the NaCl concentration was carried out and the results are depicted in Fig. 8. The results indicate that the salinity plays a very important role in the chlorophyll concentration of leaves tissues in both plants spieces, with a significant reduction in the Chl content for NaCl concentrations in the 70 - 100 mM range, where the fluorescence ratio curve exhibits a noticeable decrease for 100 mM NaCl concentration, which is a clear indication of Chl content decrease. This is corroborated by the spectrophotometric analysis presented in the same graph, which determines the chlorophyll content through *in vitro* absorption spectrum of leaves pigments in acetone extract.

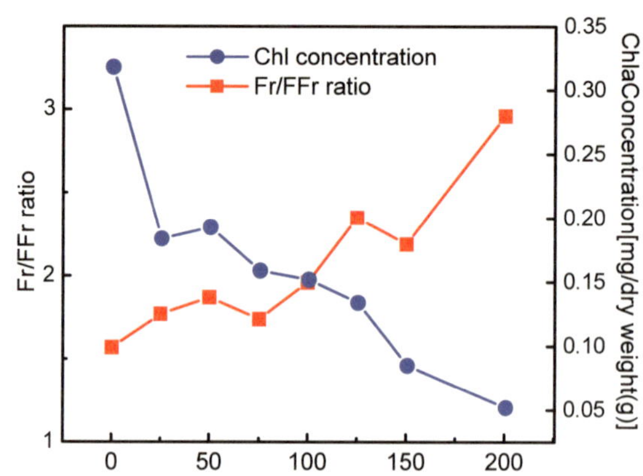


Fig. 8. ChlF ratio (left) and Chl *a* content (right) versus NaCl concentration

### 6.1.2 *Jatropha curcas*

The spectra shown in Fig. 9 are associated with *Jatropha curcas* plants treated with the maximum NaCl concentration of 200 mM at three different stages of the salinity stress time evolution. In the first day of experimentation, both the healthy plant (control) and the plants under high salt concentration presented spectra showing the two distinct emission bands around 685 nm and 735 nm. After 16 days of treatment(not shown in Fig. 9), on the other hand, the samples under intense stress exhibited a distinct reduction in the chlorophyll content as demonstrated by the noticeable increase in the Fr/FFr fluorescence ratio. The control sample, however, showed a significant Fr/FFr ratio reduction owing to the increase in the Chl content of the leaves.

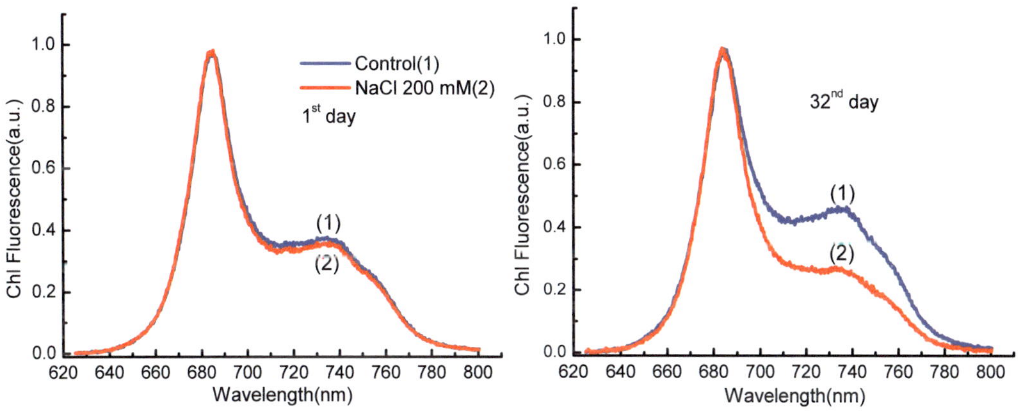


Fig. 9. Typical chlorophyll emission spectra of *Jatropha curcas* plants excited at 405 nm

The ChlF ratio time evolution for a 32 days period of time and several stress intensities (NaCl concentration) was studied and the results are depicted in Fig. 10.

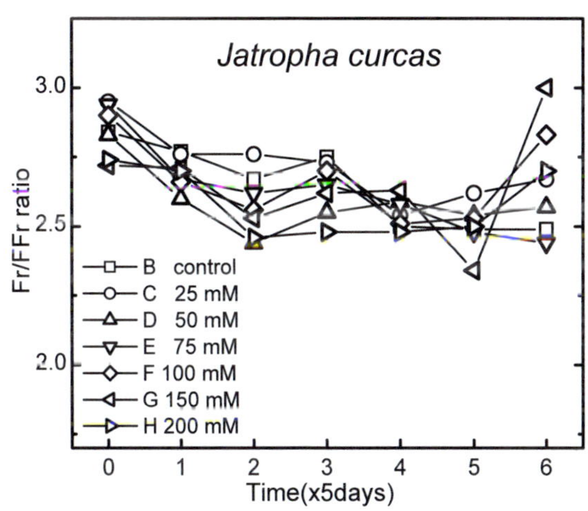


Fig. 10. Chlorophyll fluorescence ratio as a function of time

The results clearly show that there exists an inicial decrease in the ratio during the first few days of salt stress indicating increase in the chlorophyll content of the leaves submitted to salinity stress. This behavior observed for all stressed plants is attributed to the reaction of the plants to minimize the effect of distress caused by the high salinity of the soil. As time evolves, one observes that all samples under stress experience a steady behavior with the Fr/FFr ratio presenting basically the same value up to the 22nd day of stress. The control sample however, undergo a steady increase in the chlorophyll concentration as time evolves reaching the maximum value 24 days after the measurements had commenced. Following the 3rd week, however, a competition between the normal chlorophyll concentration evolution and the counter effect of the salinity distress takes place, and plants start to debilitate with time. It is important to point out that the salinity stress provokes a minor effect in the chlorophyll a content of *Jatropha curcas* leaves for NaCl concentrations up to 100 mM. This salinity distress resistance of *Jatropha curcas* indicates that this species can be considered as a main alternative crop for biofuel production in high salinity soil regions.

In order to demonstrate in detail the effect of the soil salinity on the *Jatropha* plants, it is presented in Fig. 11, the time evolution of the stress in the control and the plant under extreme distress (200 mM), and results clearly show that high salinity provokes detectable damage in the plants only after 20 days of stress exposure. In order to evaluate the effect of the salt stress on the chlorophyll content of leaves, we have carried out measurements at the end of the experimentation period (dismount), and the dependence of the Chl content, using the nondestructive fluorescence ratio and the conventional technique upon the NaCl concentration, was examined and the results are depicted in Figure 11. The results follow the trend presented in the time evolution of the salinity distress imposed to the plants and as such, the salinity plays a minor role in the chlorophyll concentration of leaves tissues. This is corroborated by the spectrophotometric analysis presented in the same graph. The chlorophyll content do not vary substantially for concentrations in the 25 to 200 mM, presenting a variation of less than 10 % of the initial value for the stressed plants.

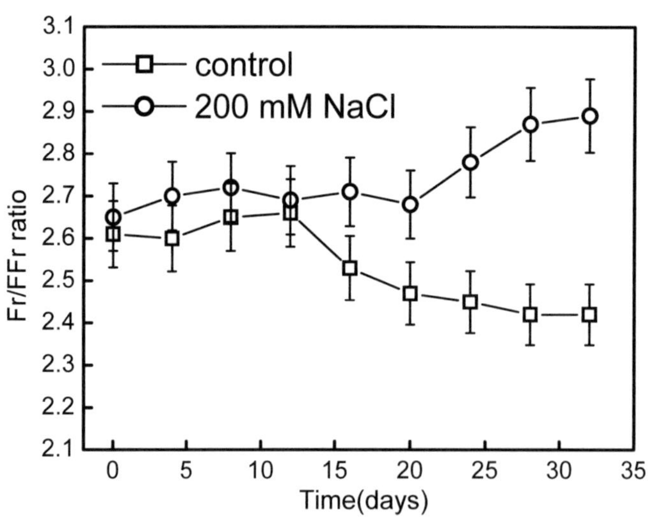


Fig. 11. Time evolution of Fr/FFr ratio for control and plant under extreme salt stress

The dependence of the Chl content, using the nondestructive fluorescence ratio and the conventional technique upon the NaCl concentration, was carried out and the results are depicted in Figure 12. The results follow the trend presented in the time evolution of the salinity distress imposed to the plants and as such, the salinity plays a minor role in the chlorophyll concentration of leaves tissues. This is corroborated by the spectrophotometric analysis presented in the same graph. The chlorophyll content do not vary substantially for concentrations in the 25 to 200 mM, presenting a variation of less than 10 % of the initial value for the stressed plants.

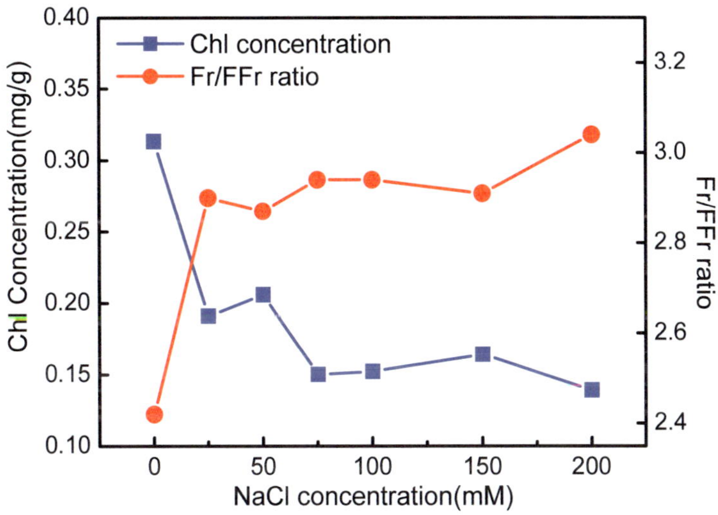


Fig. 12. ChlF ratio and Chl concentration as a function of NaCl concentration

### 6.2 Water stress

Nowadays, one of the major technological goals of the energy production, is the replacement of the fossil-based fuel for biofuel, mainly due to environmental issues. Bearing these concepts in mind, it is imperative to study the effects of water deficit in plant species with high potential for application in mass production of nonfossil based fuels. One of the main crops currently being proposed as a diesel/kerosene substitute or extender, is *Jatropha curcas* (Linnaeus) (Openshaw 2000, Francis et al., 2005). Water stress studies have been already carried out in several plant species, seeking for responses of different mechanisms in leaves under water distress (Theisen, 1988; Chappelle et al., 1984; Dahn et al., 1992; Broglia, 1993; Munns, 2002; Marcassa et al., 2006; Abou Kheira & Atta, 2009; Maes et al., 2009; Caires et al., 2010; Robredo et al., 2010; Patane & Cosentini, 2010; Tushar et al., 2010; Silva et al., 2010). In this section, the effect of water deficit in jatropha plants is investigated using chlorophyll fluorescence spectroscopy. To this end, we have investigated the response of *Jatropha* plants to water stress within three levels of water deficit. Fig. 13 shows the evolution of ChlF spectral profile of the samples under maximum water stress (nonwatered plants) within a 10 day time interval. As can be observed from data, in the very begining of the experiment both control and nonwatered samples present basically the same ChlF spectral profile. In the fifth day of investigation one observes a discrete change in the spectral profile with the control sample presenting a lower Fr/FFr ratio while the stressed ones maintain the initial profile. As time evolves, the stressed sample presents a more pronounced spectrum with the fluorescence ratio decreasing even further and the control sample exhibiting basically the same ChlF ratio and spectral profile.

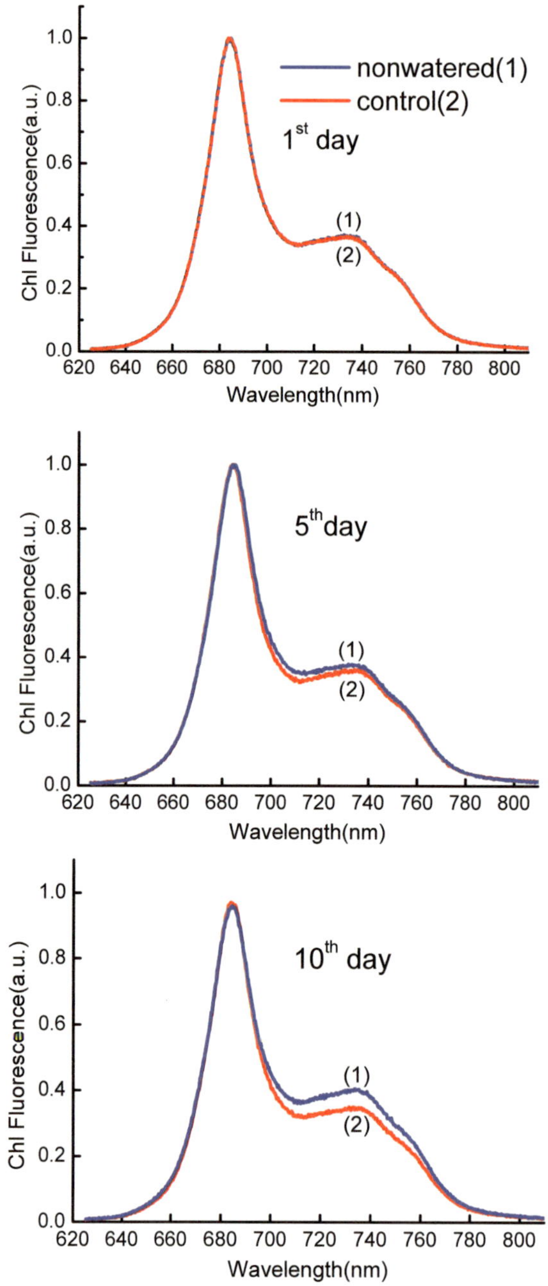


Fig. 13. Typical chlorophyll fluorescence spectrum of *Jatropha curcas* samples

This behavior of the *Jatropha curcas* under water stress is better visualized examining the time evolution of the ChlF ratio, as depicted in graph of Fig. 14. The data illustrated in Fig. 14, show a very unusual behavior with a decrease of the Fr/FFr ratio for the samples under maximum water stress as time evolves in the water distress case. This is to be compared with the behavior of the plants undergoing salinity stress, which exhibits an opposite tendency. The ChlF ratio decreases by approximately 18% within the first 10 days of the experiment for samples under highly intense water stress. It is also important to point out that the samples under mild stress (50% field capacity) did not undergo detectable changes either visual or in the Fr/FFr ratio along the 10 days period. These results would indicate, in principle, that the chlorophyll content of the highly stressed samples are increasing as the time evolves, while the control and mildly stressed samples maintained their initial concentrations. Nevertheless, the Chl concentrations obtained using conventional spectrophotometric techniques based upon Arnon's method (Arnon, 1949) showed no appreciable variation in the Chl concentration for all samples. The measured concentrations were 1.5 mg/g, 1.52 mg/g, and 1.53 mg/g for the control, 50% field capacity, and 0%(nonwaterd) field capacity, respectively. The decrease of the Fr/FFr ratio was observed previously by Chappelle and co-workers in soybeans (Chappelle et al., 1984), Dahn and co-workers in maize (Dahn et al., 1992), and by Marcassa and co-workers in orange trees (Marcassa et al., 2006). The most visible sign of water stress in the majority of plants is wilting. But, in our observations this process was not evident. The ChlF spectral profile, however, presented clearly detectable changes, particularly in the ChlF ratio. One possible reason for that is the efficiency of photosynthesis appears to be impaired. Another possible reason is that the quenching effect of water upon chlorophyll fluorescence is reduced due to the decrease in leaf water (Chappelle et al., 1984).

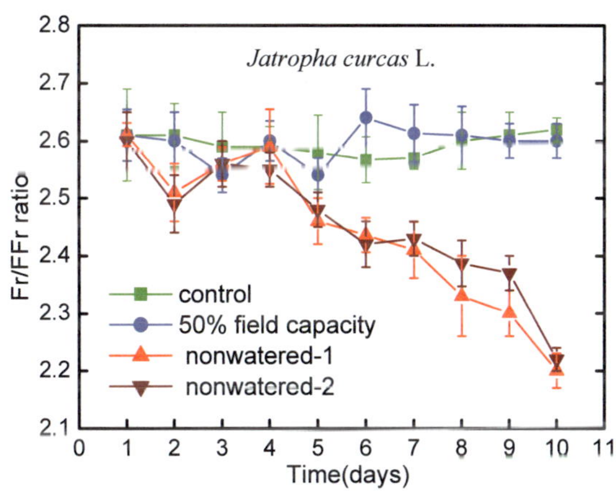


Fig. 14. Fluorescence ratio time evolution for *Jatropha curcas* under water stress

We have also examined the water deficit stress on *Saccharum officinarum* plants and the results exhibited a similar behavior as the *Jatropha curcas* plants, with the Fr/FFr ratio decreasing with stress intensity and time. It was also possible to detect the stress in the early stages and prior to visual inspection, as can be inferred from graph depicted in Fig. 15. The results for the sugarcane samples corroborates the behavior shown in our measurements with *Jatropha curcas* and the ones reported elsewhere (Chappelle et al., 1984; Dahn et al., 1992; Broglia, 1993).

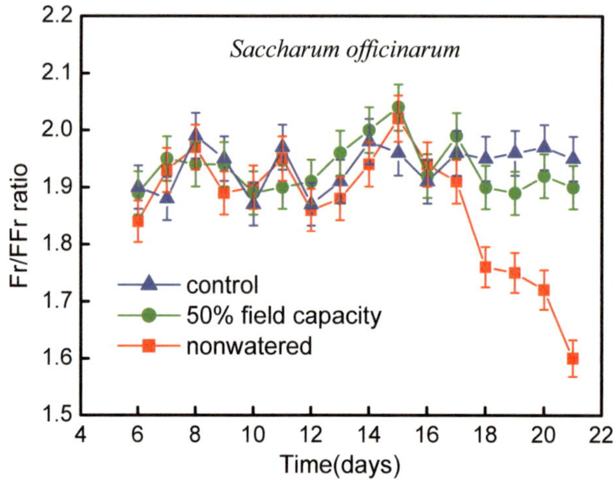


Fig. 15. Fluorescence ratio time evolution for *Saccharum officinarum* under water stress

## 7. Conclusion

Light-emitting-diode induced chlorophyll fluorescence analysis was employed to investigate the effect of water deficit and salt stress upon the growth process of *in vivo* leaves tissues of Brazilian biofuel plants species. The chlorophyll fluorescence emission spectra of 20 min predarkened intact leaves were studied employing several excitation wavelengths in the UV-VIS spectral region. We have chosen *Saccharum officinarum* and *Jatropha curcas* L. plants owing to their application in large scale industrial production of biofuel. Red and far-red chlorophyll fluorescence emission signals around 685 nm and 735 nm, respectively, were examined as a function of the stress intensity, and time. The chlorophyll fluorescence data indicated that the soil salinity plays a major hole in the chlorophyll concentration of *Saccharum officinarum* leaves, with a significant reduction in the Chl content for NaCl concentrations of a few tens of miliMolar. On the other hand, concerning *Jatropha curcas* plants, the soil salinity plays a minor role in the chlorophyll concentration of leaves tissues for NaCl concentrations in the 25 to 200 mM range, and in both cases, results agreed quite well with those obtained using conventional destructive spectrophotometric methods. The technique was also employed to investigate the effect of water deficit on the growing process of the biofuel plants species. The Chl fluorescence ratio analysis permitted detection of damage caused by water deficit in the early stages of the plants growing process with a significant variation of the Fr/FFr ratio as compared to the control sample in the first 10 days of the plant growing process. The results suggested that the technique can potentially be used as an early-warning indicator of stress caused by water deficit. It is also important to emphasize that salinity stress produced a minor effect in the chlorophyll content of *Jatropha curcas* leaves for NaCl concentrations up to 100 mM. The resistance of *Jatropha curcas* to salinity distress indicates that this species is a viable alternative crop for biofuel production in high salinity soil regions. The technique has also been applied to detect and monitor early stages of distress caused by heavy metal (Cd, Pb, Ni, Zn, etc) soil contamination (Gopal et al, 2002; Maurya et al, 2008; Ventrella et al, 2009)

## 8. Acknowledgment

The financial support for this research by CNPq, CAPES, UFRPE, and FACEPE, Brazilian research and development agencies is gratefully acknowledged. We would like to thank Mr. Diógenes Moura from UFRPE for his valuable contribution with the acquisition data apparatus for this research.

## 9. References

Abou Kheira, A. A., Atta, N. M. M. (2009). Response of *Jatropha curcas* L. to water deficits: yield, water use efficiency and oilseed characteristics. *Biomass and Bioenergy*, Vol. 33, No. 10, pp. 1343-1350, 0961-9534

Agati, G. (1998). Response of the *in vivo* chlorophyll fluorescence spectrum to environmental factors and laser excitation wavelength. *Pure and Applied Optics*, Vol. 7, No. 4, pp. 797-807, 0963-9659

Allah, D.K. (2010). In: Sugarcane history. www.npfs_minfa.gov.pk/PageFiles/226_Sugarcane-History.pdf

Allakhverdiev, S.I., Murata, N. (2008). Salt stress inhibits photosystem II and I in cynobacteria. *Photosynthesis Research*, Vol. 98, pp. 529-539, 1573-5079

Allen, J. F. (1992). How does phosphorilation regulate photosynthesis?. *Trends in Biochemical Sciences*, Vol. 17, No. 1, pp. 12-17, 0968-0004

Arnon, D. I. (1949). Copper enzymes in isolated chloroplasts polyphenoloxidase in *Beta vulgaris*. *Plant Physiology*, Vol. 24, No. 11, pp. 1-15, 0321-0889

Backer, N. R., and Bradbury, M. (1981). Possible applications of chlorophyll fluorescence technique for studying photosynthesis *in vivo*. In *Plants and Daylight spectrum*, Smith, H. (Eds.), pp. 355-373, Academic Press, London.

Broglia, M. (1993). Blue-green laser-induced fluorescence from intact leaves: actinic light sensitivity and subcellular origins. *Applied Optics*, Vol. 32, No. 3, pp. 334-338, 0003-6935

Bushmann, C. (1981). The characterization of the developing photosynthetic apparatus in greening barley leaves by means of(slow) fluorescence kinetic measurements. In *Photosynthesis*, Akoyunoglou, G. (Eds.), Balaban International Science Services, pp. 417-426, Philadelphia, 0-86689-012-2

Caires, A. R. L., Scherer, M. D., Santos, T. S. B., Pontim, B. C. A., Gavassoni, W. L., Oliveira, S. L. (2010). Water stress response of conventional and transgenic soybean plants monitored by chlorophyll fluorescence. *Journal of Fluorescence*, Vol. 20, No. 3, pp. 645-649, 1573-4994

Chappelle, E. W., Wood, F. M., McMurtrey, J. E., and Newcomb, W. W. (1984). Laser-induced fluorescence of green plants 1: A technique for remote detection of vegetation stress and specie differentiation. *Applied* Optics, Vol. 23, No. 1, pp. 134-138, 0003-6935

Dahn, H. G., Gunther, K. P., Ludeker, W. (1992). Characterisation of drought stress of maize and wheat canopies by means of spectral resolved laser induced fluorescence. *Advances in Remote Sensing*, Vol. 1, pp. 12-19, 1017-4613

Francis, G., Edinger, R., Becker, A. (2005). A concept for simultaneous wasteland reclamation, fuel production, and socioeconomic development in degraded areas in

India. Need, potential, and perspective of jatropha plantations. *Natural Resources Forum*, Vol. 29, No. 1, pp. 12-24, 0165-0203

Ghassemi, F. ; Jakeman, A. F. ; Nix, M. A. (1995) Salinization of land and water resources : Human causes, extent, management and case studies, England CAB International, Wallingford, England , p381, 085198990X

Gitelson, A. A., Bushmann, C., and Lichtenthaler, H. K. (1998). Leaf chlorophyll fluorescence corrected for re-absorption by means of absorption and reflectance measurements. *Plant Physiology*, Vol. 152, pp. 283-296, 0032-0889

Gopal, R., Mishra, K.B., Zeeshan, M., Prasad, S.M., Joshi, M.M. (2002). Laser induced chlorophyll fluorescence spectra of mung plants growing under nickel stress. *Current Science*, Vol. 83, No 7, pp.880–884, 0011-3905

Hák, R., Lichtenthaler, H. K., and Rinderle, U. (1990). Decrease of the chlorophyll fluorescence ratio F690/F730 during greening and development of leaves. *Radiation* and *Environmental Biophysics*, Vol. 29, pp. 329-336, 0301-634x

Hasegawa, P. M., Bressan, R. A., Zhu, J., Bohnert, H. J. (2000). Plant cellular and molecular responses to high salinity. *Annual Review of Plant Physiology and Plant Molecular Biology*, Vol. 51, No. 1, pp. 463-499, 1040-2519

Jimenez, M. S., Gonzalez-Rodriguez,A. M., Morales, D., Cid, M. C., Socorro, A. R., and Caballero, M. (1997). Evaluation of chlorophyll fluorescence as a tool for salt stress detection in roses. *Photosynthetica*, Vol. 33, No. 2, pp. 291-301, 0300-3604

Lang, M., and Lichtenthaler, H. K. (1991). Change in the blue-green and red fluorescence emission spectra of beaches leaves during autumnal chlorophyll breakdown. *Journal* of *Plant Physiology*, Vol. 138, pp. 550-553, 0176-1617

Lichtenthaler, H. K., Hák, R., Rinderle, U. (1990). The chlorophyll fluorescence ratio F690/F730 in leaves of different chlorophyll contents. *Photosynthesis Research*, Vol. 25, No. 3, pp. 295-298, 1573-5079

Lichtenthaler, H. K., and Schweiger, J. (1998). Cell wall bound ferulic acid, the major substance of the blue-green fluorescence emission of plants. *Journal* of *Plant Physiology*, Vol. 152, pp. 272-282, 0176-1617

Liu, X. D., and Shen, Y. D. (2006). Salt shock induces state II transition of the photosynthetic apparatus in dark-adapted *Dunaliella salina* cells. *Environmental and Experimental Botany*, Vol. 57, pp. 19-24, 0098-8472

Lin, Z-F., Liu, N., Lin, G-Z, Pan, X-P., and Peng, C-L. (2007). Stress-induced alteration of chlorophyll fluorescence polarization and spectrum in leaves of Al*ocasia macrorrhiza* L. Schott. *Journal of Fluorescence*, Vol. 17, No. 2-3, pp. 663-669, 1573-4994

Louis, J., Cerovic, Z., and Moya, I. (2006). Quantitative sudy of fluorescence excitation emission spectra of bean leaves. *Journal of Photochemistry and Photobiology B*, Vol. 84, No. 16, pp. 65-711011-1344

Maes, W. H., Achten, W. M. J., Reubens, B., Raes, D., Samson, R., Muys, B. (2009). Plant-water relationships and growth strategies of *Jatropha curcas* L. seedings under different levels of drought stress. *Journal of Arid Environments*, Vol. 73, No. 10, pp. 877-884, 0140-1963

Marcassa, L. G., Gasparoto, M. C. G., Belasque-Junior, J., Lins, E. C., Dias Nunes, F., Bagnato, W. S. (2006). Fluorescence spectroscopy applied to orange trees. *Laser Physics*, Vol. 16, No. 4, pp. 884-888, 1612-202x

Maurya, R., Prasad, S.M., Gopal, R. (2008). LIF technique offers the potential for the detection of cadmium-induced alteration in photosynthetic activities of Zea Mays L. *Journal of Photochemistry and Photobiology C*, Vol. 9, No. 1, pp. 29-35,

Mehta, P., Jajoo, A., Mathur, S., Bharti, S. (2010). Chlorophyll *a* fluorescence study revealing effects of high salt stress on Photosystem II in wheat leaves. *Plant Physiology and Biochemistry*, Vol. 48, No. 1, pp. 16-20, 0981-9428

Meloni, D. A., Oliva, M. A., Martinez, C. A., and Cambraia, J. (2003). Photosynthesis and activity of superoxide dismutase, peroxidase and glutathione reductase in cotton under salt stress. *Environmental and Experimental Botany*, Vol. 49, No. 1, pp. 69-76, 0098-8472

Munns, R. (2002). Comparative physiology of salt and water stress. *Plant, Cell and Environment*, Vol. 25, No. 2, pp. 239-250, 0140-7791

Openshaw, K. (2000). A review of *Jatropha curcas*: an oil plant of unfulfilled promise. *Biom. & Bioene*, Vol. 19, pp. 1-15, 0961-9537

Patane, C., Cosentini, S. L. (2010). Effects of soil water deficit on yield and quality of processing tomato under a Mediterranean climate. *Agricultural Water Management*, Vol. 97, pp. 131-138, 0378-3774

Renewable fuel association report (2010). 2010 Ethanol industry outlook. WWW.EthanolRFA.org

Robredo, A., Perez-Lopez, U., Lacuesta, M., Mena-Petite, A., Munoz-Rueda, A. (2010). Influence of water stress on photosynthetic characteristics in barley plants under ambient and elevated $CO_2$ concentrations. *Biologia Plantarum*, Vol. 54, No. 2, pp. 285-292, 1573-8264

Sayed, O. H. (2003). Chlrophyll fluorescence as a tool in cereal crop research. *Photosynthetica*, Vol. 41, No. 3, pp. 321-330, 0300-3604

Shreiber, U. (1983). Chlorophyll fluorescence as a tool in plant physiology, I: The measuring system. *Photosynthesis Research*, Vol. 4, pp. 361-375, 1573-5079

Silva, E. N., Ferreira-Silva, S. L., Fontenele, A. V., Ribeiro, R. V., Viégas, R. A., Silveira, J. A. G. (2010). Photosynthetic changes and protective mechanisms against oxidative damage subjected to isolated and combined drought and heat stresses in *Jatropha curcas* plants. *Journal* of *Plant Physiology*. Vol. 167, No. 14, pp. 1157-1164, 0176-1617

Stober, F., and Lichtenthaler, H. K. (1992). Changes of the laser-induced blue, green and red fluorescence signatures during greening of stiolated leaves of wheat. *Journal* of *Plant Physiology*, Vol. 140, pp. 673-680, 0176-1617

Svanberg, S. (1995). Fluorescence lidar monitoring of vegetation status. *Physica Scripta*, Vol. T58, pp. 79-85, 0031-8949

Szabolcs, I. (1994). The concept of soil resilience, In : *Greenland*, D. J. and Szabolcs, I., pp. 33-39, Soil Resilence and sustanable land use CAB International, Wallingford, UK.

Theisen, A. F. (1988). Fluorescence changes of a drying marple leaf observed in the visible and near-infrared. In *Applications of chlorophyll fluorescence in photosynthesis research, stress physiology, hydrobiology and remote sensing*, Lichtenthaler, H. K. (Eds.), pp. 197-201 (Dordrech: Kluwer).

Tushar, B., Manoj, K., Sushana, C. (.2010). Evaluation and genetic polymorphism studies of jatropha (*Jatropha curcas)* for water stress tolerance. *Journal of Stress Physiology & Biochemistry*, Vol. 6, pp. 11-18, 1997-0838

Ventrella, A., Catucci, L., Piletska, E., Piletsky, S., Agostiano, A. (2009). Interactions between heavy metals and photosynthetic materials studied by optical techniques. *Bioelectrochemistry*, Vol. 77, No. 1, pp. 19-25, 1567-5394

Yamane, K., Kawasaki, M., Taniguchi, M., and Miyake, H. (2003). Differential effect of NaCl and polyethylene glycol on the ultrastructure of chloroplasts in rice seedlings. *Journal of Plant Physiology*, Vol. 160, No. 5, pp. 573-575, 0176-1617

Zribi, L., Fatma, G., Fatma, R., Salwa, R., Hassan, N., Nejib, R. M. (2009) Application of chlrophyll fluorescence for the diagnosis of salt stress in tomato *Solanum lycopersicum*(variety Rio Grande). *Scientia Horticulturae*, Vol. 120, No. 3, pp. 267-372, 0304-4238

# 2

# Is It Possible to Use Biodiesel as a Reference Material?

Dalni Malta do E. Santo Filho et al.*
*Inmetro*
*Brazil*

## 1. Introduction

Beyond the complex international metrology structure and its conection to what has been applied in Brazil, using Biodiesel as a reference material constitutes an important factor to the growth of metrological services in the energy sector. Such fact is directly related to the national energy matrix. The fuels' market becomes more demanding each day forcing national companies and laboratories to act systematically, aiming towards reliability and traceability of measures, avoiding any potential barriers unrelated to taxes imposed at national products. It is worth mentioning that comercialization of petroleum based products influences many different sectors of the economy, especially transportation, guaranteeing excelence in the provided services or the complete insuficiency of economic growth.

The necessity of working with biofuels with the purpose of diminishing environmental impacts leads to an extremely relevant pondering for a National Metrology Institute. Any country must be aligned with international requirements and must also guarantee a products' quality through laboratory analysis whose major objective is the maintenance of its reliability and the metrological traceability of their measures. Furthermore, in Brazil, for example, biodiesel is one of the most interesting solutions to diminish, or at least stop, diesel oil importation. While motors that use electrical, wind and solar energy do not provide immediate results that prove to be as useful as petroleum, biodiesel has a certain advantage, since there is no need for great modification in diesel motors to be used with, still providing good efficiency (Lyra, 2008).

However, in order to use biodiesel in a motor and not endanger it, it is essential that previous studies need to be done in testing laboratories with the purpose of guaranteeing the products' specification following certain rules kept by a regulatory agency. For this, using certified reference materials in the measure process is of the utmost importance to guarantee metrological traceability to national laboratories. Because of this, apart from tests to analyse whether the fluid is in accordance with standards ISO EN 14214 and ASTM D6751, this chapter has the purpose of approaching the steps in ISO Guide 34 and ISO Guide 35, discussing the project of producing biodiesel as a reference material, such as

* José Renato R. Siqueira[1], Renata Martins H. Borges[1], Claudio Roberto da C. Rodrigues[1], Alex Pablo F. Barbosa[1], José Júlio P. dos Santos Júnior[1] and Roberto G. Pereira[2]
*1 Inmetro, Brazil*
*2 UFF/TEM/PGMEC/MSG, Brazil*

definition of the material candidate, planning the steps for the preparation of the material, the fulfillment of reference material characterization, homogeneity and stability tests, and, as consequence, the emission of a certificate for the reference material with the presentation of the obtained results. This chapter will also present studies of the physical-chemical properties that are able to keep their characteristics for a longer period if some recommendations, presented in this study, are followed.
Such studies can show the recommended time for stocking this kind of reference material and also for how long it can still be used.

## 2. Metrology as a tool to assure quality of measurement's results

Whether it is in an international or national scope, it can be pointed out that harmonizing norms and technical regulations is of extremely important to commercial relations, other than the necessity to invest in proceedings that aim towards quality assurance of commercialized products. Thus, when looking towards guaranteeing products' quality it is appropriate to fit two subjects of the utmost importance to international commerce: metrology and conformity assessment, building the foundation to any system applied to promoting commercial relations, that is, national products with reliability and metrological traceability meeting the requirements of international technical standards and regulations.
According to Hufbauer et al. (2000), the conformity assessment procedures of a certain product consists in any procedure used, direct or indirectly, to evaluate if certain demands regarding standards or technical regulations are taken into account. Apart from that, such procedures include sampling, testing, inspection, evaluation, verification and assessment assurance, registering, accreditation and approval, and combinations of these possibilities. The procedures for conformity assessment are considered essential in regulations related to health, security and environment contributing to the improvements on productivity and efficiency in commerce economy.
Reliability in the measurements' results is essential in the process of decision making regarding adjustment of different products or services. Measurements support quality controls in industrial processes and applied research, with the emphasis on conformity assurance being responsibility of national and international regulations. With this, metrology encloses, among other characteristics, measurement proficiency, result reliability and metrological traceability (Siqueira, 2006). Any laboratory may present proficiency in its measurements if it can present adequate procedures and have qualified technicians to execute calibration and measurement services, apart from installations regarding each service executed. In the same way, any laboratory that judges itself competent may calibrate equipment from other laboratories and issue calibration certificates and/or test reports. However, results from these laboratories, no matter how competent they may be, can be questioned if their equipment was not calibrated and, thus, the traceability chain may not be assured.
Fig. 1 present metrology as essential in assuring commercial relations and its direct relation with the scientific and technological development of a country, apart from its interaction with the normalization processes and accreditation bodies.
According to the International Vocabulary of Metrology (VIM, 2008; JCGM, 2008), the term metrological traceability consists in "property of a measurement result whereby the result can be related to a reference through a documented unbroken chain of calibrations, each contributing to the measurement uncertainty."

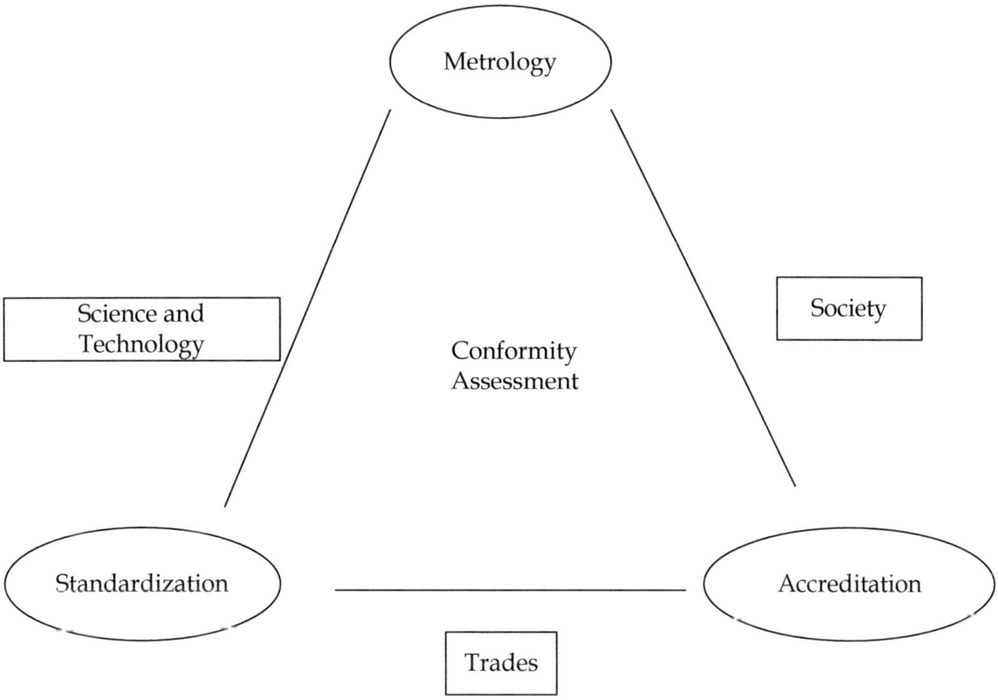


Fig. 1. Support of the conformity assessment infrastructure to economical development (Dennehy, 2006).

The measurement result consists in a "set of quantity values being attributed to a measurand together with any other available relevant information", being usually depicted by one sole measured value and its associated measurement uncertainty. Still according to VIM, the measurement uncertainty characterizes itself as a "non-negative parameter characterizing the dispersion of the quantity values being attributed to a measurand, based on the information used". In this way at each step of the traceability chain, the uncertainty associated to the standards is smaller, implying a crescent technological sophistication, which is the objective of scientific metrology. Fig. 2 shows an example of a established metrological traceability chain.

Based on the exposed facts, the establishment of metrological traceability through an uninterrupted calibration chain is crucial to the operation of equipment and measurement instruments in different sectors of economy. When dealing with test laboratories that analyze biodiesel, the lack of traceability in measurements can generate incorrect conclusions in determining parameters of extreme significance in this products' commercialization, such as, kinematic viscosity and density.

## 3. The methodology for preparing reference materials based on ISO guide 34:2009

Since this chapter analyzes the possibility of developing a reference material using as a matrix the biodiesel that is nationally commercialized, it is of worth to evaluate all parameters regarding production and the possibility of certification of the reference material

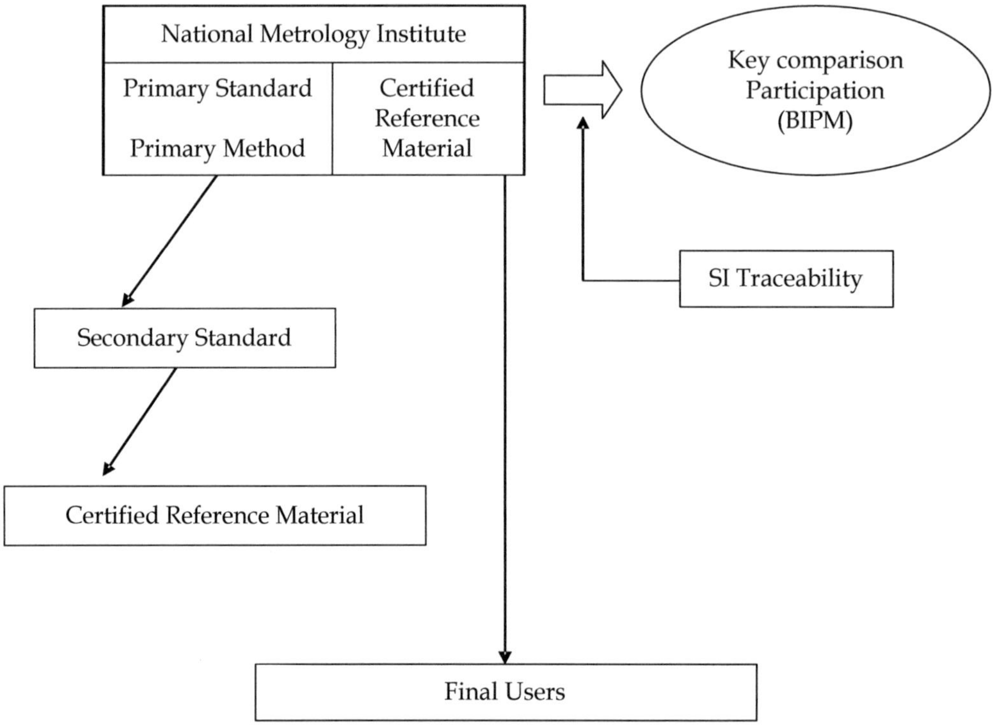


Fig. 2. Example of an established metrological traceability chain (Imai, 2007).
BIPM is the International Bureau of Weights and Measures,
SI is the International System of Units.

using ISO Guide 34. It is worth mentioning that the application of the requirements in ISO Guide 34 take the producer of the reference materials to an in-depth study of the concepts and guidelines of ISO Guide 35:2006, since in the aforementioned are presented the different approaches to be used by the producer in the production planning, attribution of the property value and its uncertainty, to the declaration of metrological traceability and the issue of pertinent information to the user of the reference material.

According to Emons (2006), a reference material consists on a "material sufficiently homogenous and stable regarding one or more specific properties, which have been defined adequate to the purpose of the measurement process." It is worth noting that this new definition is tied to four notes, presented as follows.

Note 1: Reference material is a generic term.

Note 2: Properties may be quantitative or qualitative, for example, identification of substances or species.

Note 3: Usages may include calibration of a measurement system, evaluation of a measurement process, assignment of values to other materials ("assigned values") and quality control.

Note 4: A reference material may be used only for one purpose in a given measurement.

That same publication presents the new definition of certified reference material, which consists of a "reference material, characterized by a metrologically validated method to one or more specified properties, accompanied of a certificate which provides a value of the certified property, its associated uncertainty and an established metrological traceability." The new definition also presents three notes, presented as follows.

Note 1: The concept of value includes qualitative attributes such as identity or sequence. Uncertainties for such attributes may be expressed as probabilities.

Note 2: Metrologically valid procedures for production and certification of reference materials are obtained from, among other, ISO Guide 34 and ISO Guide 35.

Note 3: ISO Guide 31 provides guidelines about the content of certificates.

Thus, for an institute to be able to produce a CRM it is necessary to use validated methods and calibrated equipment in order to guarantee the traceability of the measurements and also to thrust the calculation of the estimate of the uncertainty of measurement. Fig. 3 presents the "family" of Reference Materials which may include certified reference materials (CRM) usually represented in smaller quantity due to the difficulties in the process of characterization and certification, the materials known as calibrants (CAL) and, lastly, the quality control materials (QCM). Reference materials such as the ones used for quality control Consist on reference materials used, especially, for statistical control of a measurement process. Calibrants constitute measurement standards used for calibration of a measurement system, being characterized for possessing an established metrological traceability.

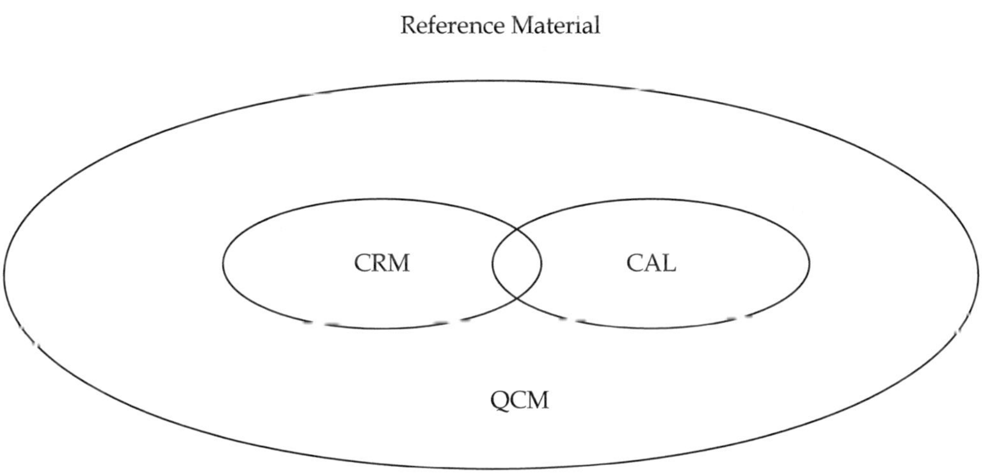


Fig. 3. The "Family" of Reference Materials (Emons, 2006).

It is worth noting that the new definitions were elaborated by the ISO committee that deals with Reference Materials (ISO/REMCO) on the meeting in Genebra, in 2005 (Emons et al., 2005). Such definitions were inserted in the ISO Guide 35 version published in 2006. It is also worth noting that, according to the last annual report made available by ISO, among the 18 members elected to the board responsible for coordinating all ISO, the Brazilian Association for Technical Standards (ABNT) is a part of.

### 3.1 Feasibility study for the production of a reference material

Following ISO Guides 34 and 35, it was possible to establish a viability study for production of a reference material for two physical properties: kinematic viscosity and density. In the aforementioned study samples of beef tallow transesterified with ethanol biodiesel were used, since, according with literature data, they would present higher stability since they are a more saturated matrix. As the chapter progresses, the parameter kinematic viscosity will be replaced by viscosity.

### 3.2 Accommodation and environmental conditions

With the purpose of demonstrating the importance of measurements to the energy sector, especially the conformity assessment of the biodiesel which is commercialized nationally, a case study done by Fluids Laboratory (Laflu) from Brazil's National Metrology Institute (Inmetro), which holds the national standards for density and kinematic viscosity, among other quantities.

Laflu is responsible for guaranteeing traceability to measurements in Brazil in density and viscosity. Its competence is confirmed through international interlaboratory comparisons [Maggi et al., 2009] and its services are registered in Appendix C of BIPM (Key Comparison Data Base [KCDB]).

Its installations are such that facilitate the correct accomplishment of tests and calibrations, assuring that environmental conditions do not invalidate results or adversely affect the needed quality for any measurement.

### 3.3 Measuring equipment

As a reference material producer, a National Metrology Institute should follow ISO Guide 34 in combination with ISO/IEC 17025. Considering the measuring equipments, all must be used in compliance with ISO/IEC 17025. Every measurement instrument was calibrated, verified and kept properly. This way, the reference materials produced by Laflu were characterized using a metrologically valid procedure, with an associated uncertainty, and a statement of metrological traceability, following the guidelines from ISO Guide 31, providing the final user of these reference materials a wide range of relevant information to ensure quality in the measurement results that will be obtained.

The following topics offer a brief description of the equipment used in this study and the reason for their choice.

#### 3.3.1 Capillary viscometers

Viscosity is a function of the viscometer constant and of the liquid flow time, in seconds, from the top line to the bottom line of the measurement bulb.

Capillary viscometers are used when measuring Newtonian fluids. The studied biodiesel, beef tallow biodiesel transesterified with ethanol, presented this behavior. The Ubbelohde viscometer is a suspended-level viscometer, with an uniform driving head of liquid, independent of the sample quantity inserted into the viscometer, making the viscometer constant independent of temperature. An Ubbelohde capillary viscometer type I (range of 2.0 $mm^2/s$ to 10.0 $mm^2/s$) (Fig. 4) was used for the studies.

This type of viscometer was chosen since international standards ISO EN 14214 and ASTM D 6751 indicate that biodiesels, to be considered in their specifications, must present viscosity between 3.0 $mm^2/s$ and 6.0 $mm^2/s$ at 40 °C.

It is important to emphasize that the metrological traceability of capillary viscosimeters must be obtained from the attributed value for the viscosimeter constant calibrated at the nth step in the stepping-up procedure based on the absolute value of 1.0034 $mm^2s^{-1}$ for the kinematic viscosity of distilled water at 20 °C (ISO 3104, ASTM D445-11 and ISO 3105) and the normal atmospheric pressure as stated in ISO TR 3666.

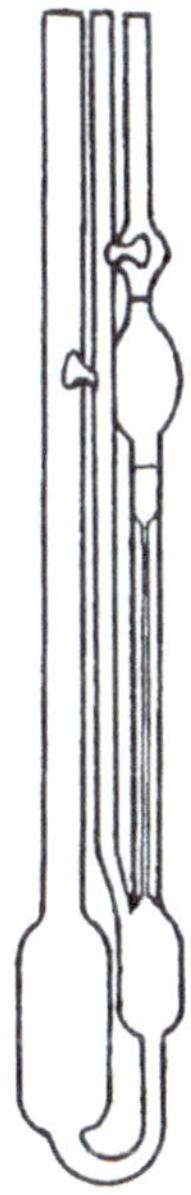

Fig. 4. Ubbelohde capillary viscometer.

### 3.3.2 Digital density meters

Density was measured with a digital density meter Anton Paar DMA 5000 (Fig. 5). The device is equipped with a vibration transducer for digital density meters which is able to indicate sample density when the fluid reaches its natural frequency.

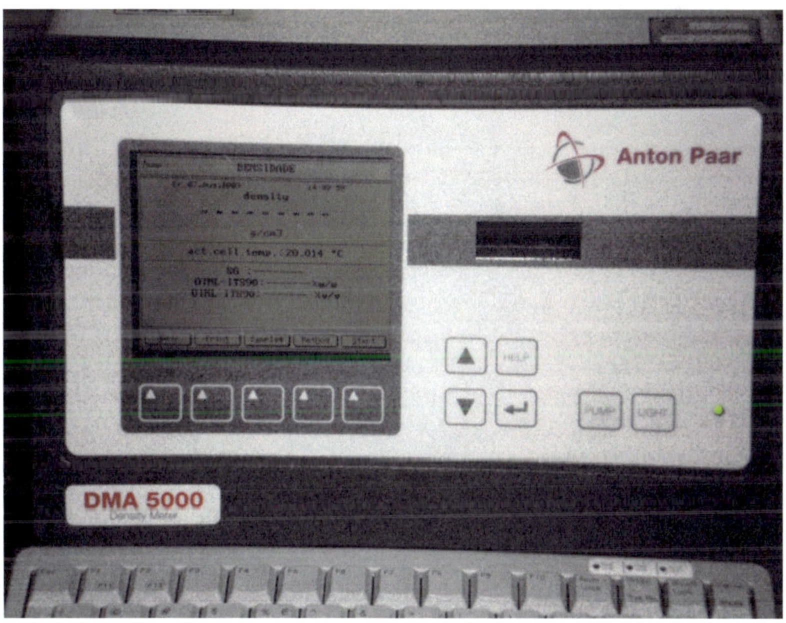


Fig. 5. Digital Density Meter.

The reasons why density meters were chosen in this work over other measuring devices are explained as follows. Density meters have been used more frequently nowadays due to their simple operational procedures. They allow measurements to be performed with very small sample amounts compared to hydrometers, and they also provide fast results. Furthermore, the temperature of the measurement tube in which the fluid is placed can be changed very quickly, minimizing, in this way, the amount of time spent in replacing solutions and stabilizing bath temperatures.

The metrological traceability for this digital density meter was done in the temperature range of 4 °C to 40 °C by the following solutions: n-nonane, mineral oil, and distilled water used as reference material which was measured by the hydrostatic weighing system, the primary method used in Brazil. Air density during the calibration was 0.00120 $g/cm^3 \pm$ 0.00002 $g/cm^3$. The internal adjust factor is determined using distilled water and dry air.

### 3.4 Production planning

For this study, 79 samples of biodiesel were bottled in 50 mL glass bottle, amber colored with a sealed nitrilic lid, with the purpose of avoiding modification of the studied properties due to light incidence and to allow the withdrawal of the sample with a hypodermic syringe, so it will not contaminate the environment. Such methodology is fundamental for biodiesel samples, since they are known for having little hygroscopy and big oxidation (Knothe et al., 2005] with the environment, especially at higher temperatures.

According to ISO Guide 35, the steps for the homogeneity, stability and characterization tests were planned for the desired properties. For the aforementioned studies, the following temperatures were used: 4 °C, 20 °C e 40 °C; with 20 °C being the reference temperature.

In order to establish the strategy for the homogeneity study, it is important to consider that a material is said to be homogeneous with respect to a property if this properties' value, when determined by testings with specific sized samples, is within uncertainty range, specified beforehand. Homogeneity is the condition of uniformity of a composition structure with respect to one or more specific properties. Homogeneity with respect to the determination of viscosity and density was evaluated employing variance analysis, and also evaluating the value of the measurement obtained within the specified uncertainty range. It is worth emphasizing that the chosen method to evaluate the reference material candidate's homogeneity presents proper repeatibility and sensitivity. Such parameters are fundamental so that the reference material producer is able to evaluate the variation of the production process, which, in this case, is related to producing a batch and choosing samples randomly in order to represent the whole batch in a correct homogeneity evaluation. It is noteworthy that what is known as a batch is a definite quantity of some commodity produced by one supplier at one time under conditions that are presumed uniform [ISO Guide 34 and 35].

One of the most important factors for a proper homogeneity evaluation of a produced batch of reference material is the number of samples to be selected. Considering ISO Guide 35, one may notice that the number of samples depends on the batch size, so the number of samples withdrew from such batch must be considered as a representation of the whole batch. The minimum number of randomly selected bottles must not be smaller than ten, usually between ten and thirty (ISO Guide 35, 2006). Twenty biodiesel bottles were randomly selected for the homogeneity evaluation. From the 20 analysed bottles, 10 were used for the kinematic viscosity evaluation at 40 °C and 10 were used for the density evaluation at 20 °C, so the requirements from standards ASTM D 6751 and ISO EN 14214 are met, apart from the criteria established in ANP's Resolution n° 007, publish in March 19, 2008.

In the same way as the homogeneity study, the stability evaluation was performed in order to evaluate the capacity of a product to maintain the value of a certain property within specified limits for a known and previously established period of time when stocked in specified conditions. It is important to emphasize that the degree of instability of the biodiesel was evaluated after being prepared. In this production planning two conditions were evaluated: storage conditions (long term: 20 °C for density and viscosity) and transport conditions of the materials (short term: 4 °C and 40 °C for density and viscosity).

It is worth emphasizing that the selected method to the stability evaluation of the reference material candidate presents adequate selectivity and reproducibility, since the samples were evaluated at different times. Upon finishing the studies, the reference material producer may not only guarantee the stability of the material, but is also capable of specifying its storage and transport conditions.

Considering ISO Guide 35, there are two long-standing approaches for the stability study: classic and isochronous (van der Veen 2001a, 2001b; ISO Guide 35]. To the short term stability evaluation of the material focusing on the viscosity as a parameter, the viability study planned for an evaluation using the classic approach. Such approach was chosen since the determination of the viscosity of the samples may not be made in repeatibility conditions, that is, according to the method established by the producer, only two samples may be studied each day. With this, 15 samples of the reference material candidate that were stored in the study temperatures were used, kept at different times (15, 30, 45, 60 and 90 days after the study had begun) and kept in the following temperatures: 6 samples kept at ((4.0 ± 1.5) °C and 9 kept at (40.0 ± 1.5) °C) (Santo Filho, 2010a).

With respect to the short term stability evaluation focusing on the density as a parameter, 21 samples of the reference material candidate were submitted to the isochronous approach. The isochronous approach emphasizes that measurements must be made, as best as possible, with the smallest time gap. For such study, the storage periods were of 10, 30, 45, 60 and 90 days after the study had begun and the study temperatures were ((4.0 ± 1.5) °C, (20 ± 1) °C and (40.0 ± 1.5) °C). In each of these periods, a reference material candidate sample was stored in the study temperatures. In the end of the period, all samples were measured at the reference temperature of ((20 ± 1) °C), simultaneously.

Following the stability evaluation of the reference material candidate, the long term stability studies were planned and executed, using the classic approach. The samples were stored and analyzed at the reference temperature of ((20 ± 1) °C) for 360 study days for both parameters in question, being 12 samples for the viscosity parameter (Santo Filho, 2010c) and 7 samples of density.

To characterize the reference material candidate, the density and viscosity parameters must be found, regarding their intended usage for biodiesel (ANP 7, 2008). Viscosity characterization was made using a viscometer calibrated in the 20 °C to 40 °C range, which was also the chosen range to make an analysis of density with kinematic viscosity possible and also enable, if needed, the dynamic viscosity for the studied biodiesel. To characterize density, a density meter was used, calibrated in the temperature range of 20 °C to 40 °C, with 5 °C increments for each measurement. To the viability study, 4 bottles were randomly selected, being 2 for the determination of viscosity and 2 for density.

In general, Laflu used a capillary viscometer which has a measurement bulb (Fig. 6). This viscometer must be immersed in a thermostatic bath and have the flow time measured using a calibrated chronometer.

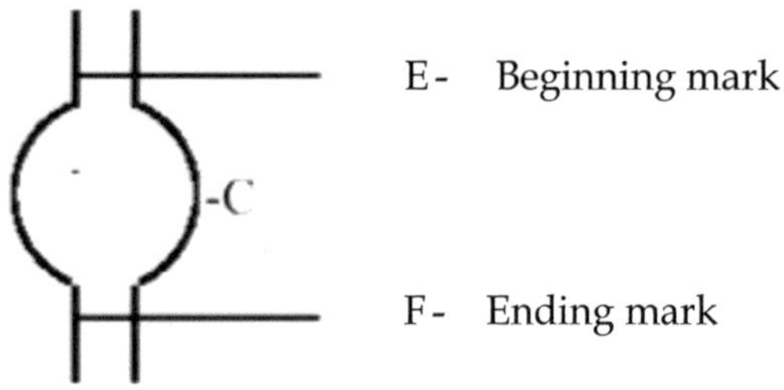


Fig. 6. Measurement bulb.

The viscometer with the reference material candidate for biodiesel had been immersed in a thermostatic bath until equilibrium temperature is reached. The thermostatic baths used were Lauda D-40 (as the main bath), with a Tamsom TLC-15 (as the fine tunning bath), controlled with 1/1000 °C precision thermometers. The testing temperatures were 20 °C, 25 °C, 30 °C, 35 °C and 40 °C. A time of at least 30 min is required for thermal equilibrium to be reached. During this period, the channels of the viscometer were protected from light to avoid any viscosity changes, such as oxidation.
Considering the density value, samples were placed into the digital density meter with the aid of a hypodermic syringe. The compartment that stores the samples is a U-shaped tube (Fig. 7). It is important to stress that care should be taken in avoiding bubbles when filling the tube since the presence of bubbles leads to incorrect readings. For each sample, at least four density readings were made for a given temperature value. The temperatures for the tests were 20 °C, 25 °C, 30 °C, 35 °C and 40 °C.

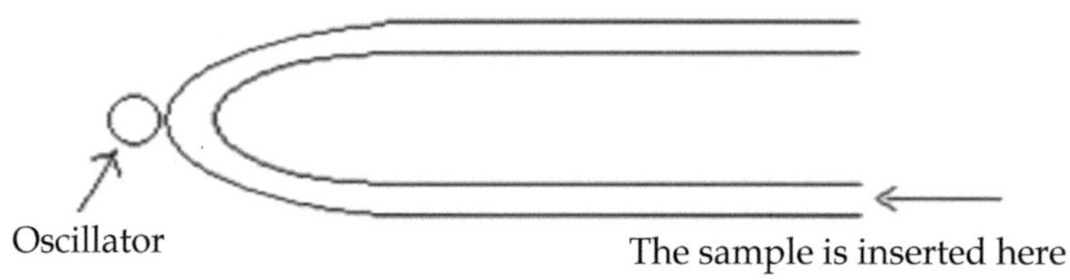


Fig. 7. U-shaped tube of digital density meter.

## 4. Results obtained from the case study: Candidate refrence material for beef tallow biodiesel transesterified with ethanol

### 4.1 Homogeneity evaluation of samples

Results from pure beef tallow biodiesel transesterified with ethanol are shown below and the guidelines of ISO Guide 35 were also applied. Table 1 and Fig. 8 shows the viscosity results from the randomly selected samples. Table 2 and Fig. 9 shows the density results from the randomly selected samples.
It can be observed that the results found through all the tests for each quantity are very close and are all placed within the range of the expanded uncertainty. With the variance analysis it was possible the find the uncertainty contribution regarding the homogeneity of the produced batch, considering the two analyzed parameters. The uncertainty regarding the homogeneity for density was 0.00004 $g/cm^3$ and for viscosity was 0.0006 $mm^2/s$.
The uncertainty values were found using equations 1 and 2.

$$S_a^{\,2} = \frac{MQ_{between} - MQ_{within}}{n_0} \tag{1}$$

And

$$S_{bs} = \sqrt{S_a^2} \tag{2}$$

where:

- $MQ_{between}$ is the squared average between samples
- $MQ_{within}$ is the squared average in each sample;
- $S_{bs}$ is the standard deviation between samples;
- $S_a^2$ is the variance between samples.

It is worth noting that the homogeneity uncertainty contribution for density was found considering the standard uncertainty of the measurement method.

| Sample | Viscosity (mm²/s) |
|---|---|
| 1 | 5.0261 |
| 2 | 5.0277 |
| 3 | 5.0266 |
| 4 | 5.0263 |
| 5 | 5.0273 |
| 6 | 5.0264 |
| 7 | 5.0277 |
| 8 | 5.0269 |
| 9 | 5.0264 |
| 10 | 5.0259 |

Table 1. Homogeneity results of beef tallow biodiesel transesterified with ethanol viscosity, at 40 ºC.

| Samples | Density (g/cm³) |
|---|---|
| 1 | 0.86475 |
| 2 | 0.86473 |
| 3 | 0.86481 |
| 4 | 0.86481 |
| 5 | 0.86482 |
| 6 | 0.86481 |
| 7 | 0.86481 |
| 8 | 0.86479 |
| 9 | 0.86478 |
| 10 | 0.86477 |

Table 2. Homogeneity results of beef tallow biodiesel transesterified with ethanol density, at 20 ºC.

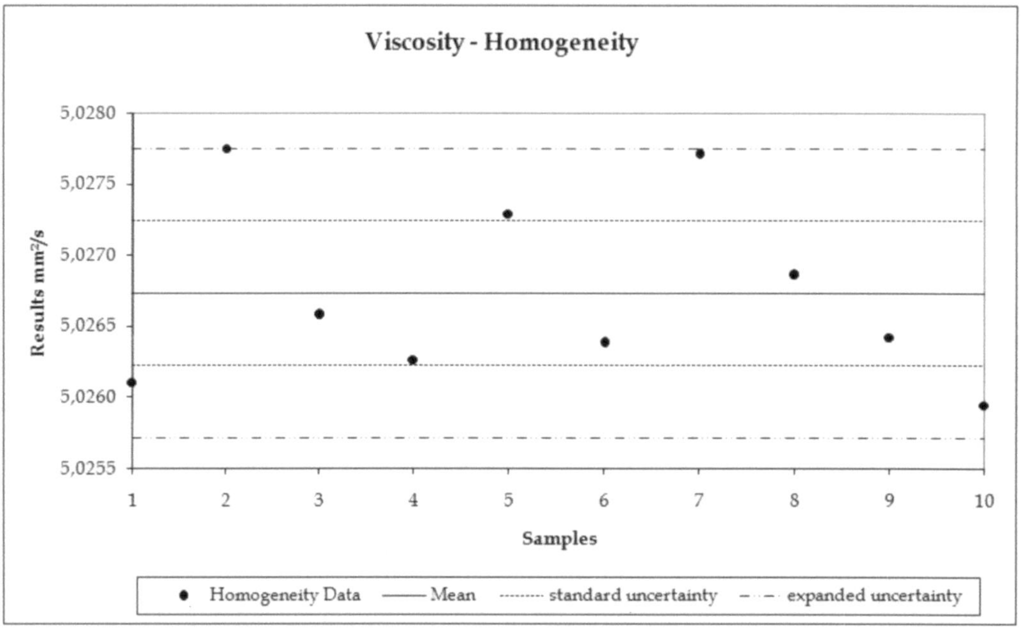


Fig. 8. Graph for homogeneity results of beef tallow biodiesel transesterified with ethanol viscosity, at 40 °C.

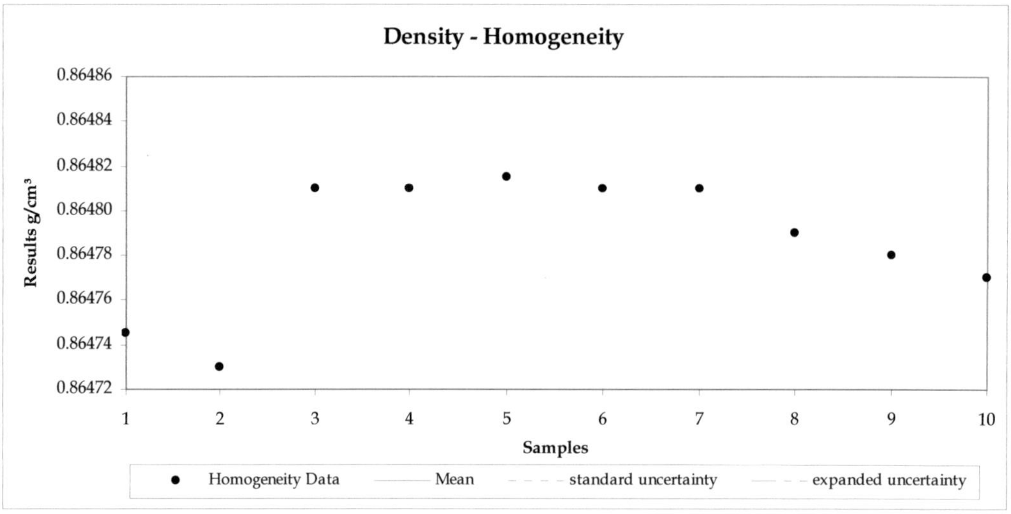


Fig. 9. Graph for homogeneity results of beef tallow biodiesel transesterified with ethanol density, at 20 °C.

### 4.2 Stability evaluation of samples

Results from pure beef tallow biodiesel transesterified with ethanol (Santo Filho 2010a, 2010b) are shown below and the guidelines of ISO Guide 35 were also applied. Fig. 10 shows the viscosity results from the randomly selected samples. Fig. 11 shows the density results from the randomly selected samples.

It can be observed that the results found through all the tests for each quantity were analyzed in a time period, applying different conditions of storage trying to evaluate the impact of temperature in transportation and storage of the reference material candidates.

Following the stability studies, the inclination of the curve, for all the measurements, was calculated using equation 3 (ISO Guide 35, 2006; Lisinger, 2001):

$$b_1 = \frac{\sum \left(X_i - \overline{X}\right)\left(Y_i - \overline{Y}\right)}{\sum_{i=1}^{n} \left(X_i - \overline{X}\right)^2} \tag{3}$$

Where:

- $X_i$ is the time when the measurement was made;
- $\overline{X}$ is the average time for all measurements, in months (it is possible to use days or years, but, later, it must be converted to months);
- $Y_i$ is the viscosity, density measurements;
- $\overline{Y}$ is the average of all viscosity, density measurements

This equation is found through the least mean square method (Taylor, 1997), and it is understood that the variation of viscosity and density is linear regarding time.

The inclination of the curve was calculated using equation (4):

$$b_0 = \overline{Y} - b_1 \overline{X} \tag{4}$$

The variance and the standard deviation were calculated using equations 5 and 6 (ISO Guide 35, 2006; Taylor, 1997, p.196-198.):

$$s^2 = \frac{\sum_{i=1}^{n} \left(Y_i - b_0 - b_1 X_i\right)^2}{n - 2} \tag{5}$$

$$s(b_1) = \frac{s}{\sqrt{\sum_{i=1}^{n} \left(X_i - \overline{X}\right)^2}} \tag{6}$$

Meaning of the inclination of the curve from equation (7)

$$|b_1| < t_{0,95,n-2} \cdot s(b_1) \tag{7}$$

Since the absolute value of $b_1$ was less than the product between the student's $t$ value (parameterized normal curve) with the standard deviation, it can be concluded that there was no relevant instability for the two studied properties.

The uncertainty regarding stability for density was 0.00009 g/cm³ and for viscosity was 0.0008 mm²/s.

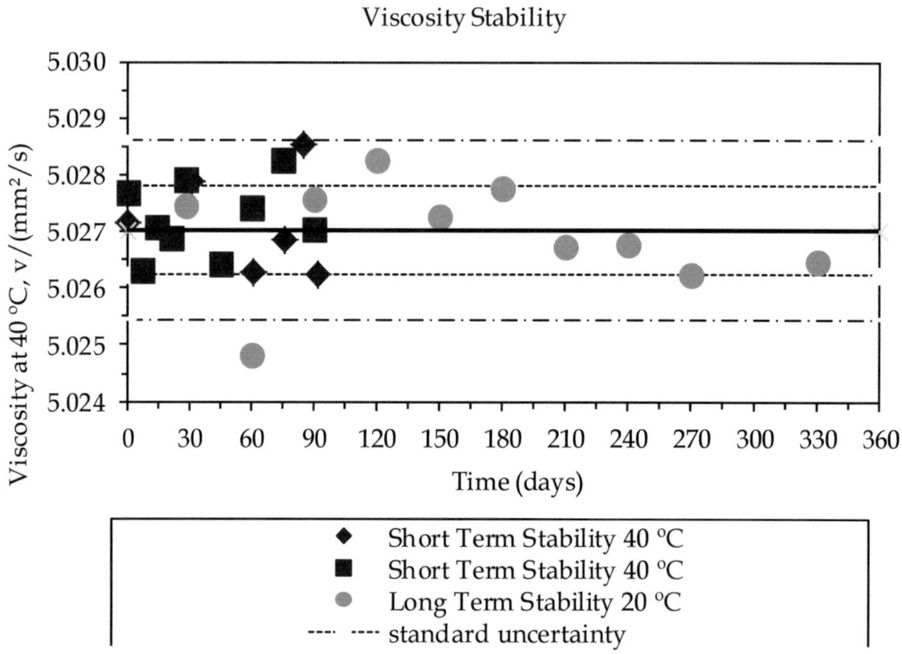


Fig. 10. Stability results of beef tallow biodiesel transesterified with ethanol viscosity, at 40 °C.

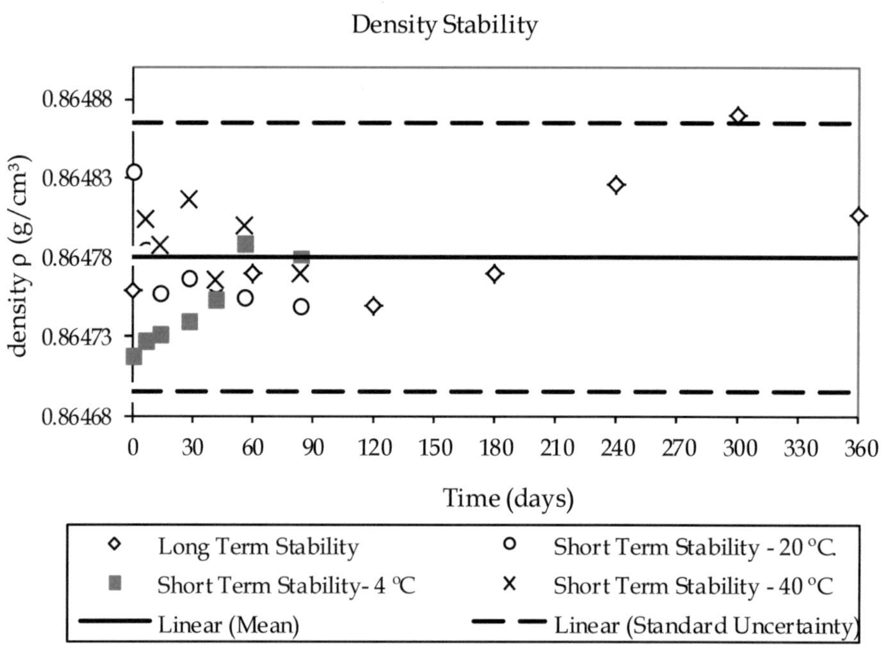


Fig. 11. Stability results of beef tallow biodiesel transesterified with ethanol density, at 20 °C.

### 4.3 Characterization of the produced reference material

Table 3 shows the characterization of the beef tallow biodiesel's viscosity at 20 °C, 25 °C, 30 °C, 35 °C and 40 °C. The temperatures were studied with an Ubbelohde type I viscometer, which is Brazil's reference national standard (Santo Filho, 2009).

| Temperature | Maximum Viscosity | Average Viscosity | Minimun Viscosity | Expanded Uncertainty | Ubbelohde Viscometer Type |
|---|---|---|---|---|---|
| $T$ (°C) | $\nu_{max}$ (mm²/s) | $\nu$ (mm²/s) | $\nu_{min}$ (mm²/s) | $U$ | |
| 20 | 8.2923 | 8.2830 | 8.2737 | 0.0093 | I |
| 25 | 7.2329 | 7.2248 | 7.2167 | 0.0081 | I |
| 30 | 6.3692 | 6.3620 | 6.3548 | 0.0072 | I |
| 35 | 5.6740 | 5.6675 | 5.6610 | 0.0065 | I |
| 40 | 5.0326 | 5.0269 | 5.0212 | 0.0057 | I |

Table 3. Viscosity results for beef tallow biodiesel transesterified with ethanol, in a temperature measurement range of 20 °C to 40 °C.

To determine the kinematic viscosity in each temperature it is necessary to calculate the measurement uncertainty of the kinematic viscosity measurement (Rodrigues et al., 2008 ; Barbosa et al., 2009). Since kinematic viscosity is measured indirectly, it must first be determined the mean flow time of the studied fluid, allowing then the calculation of the viscosity.

Regarding density, the viability study allowed the obtainment of the characterization data presented in tab. 4. The reference material candidate samples of biodiesel were measured in a digital density meter.

The combined uncertainty is the squared sum of all the uncertainties of the input quantities. This result will provide the uncertainty of the measurand (i.e. quantity intended to be measured, according to VIM), which in this case is density. The expanded uncertainty is the combined uncertainty multiplied by a coverage factor, $k$. In this case $k$ is considered 2, which means the results are in the range of two standard deviations.

| Measurement temperature | Liquid's calculated density |
|---|---|
| °C | g/cm³ |
| 20.00 | 0.86476 |
| 25.00 | 0.86105 |
| 30.00 | 0.85734 |
| 35.00 | 0.85369 |
| 40.00 | 0.85001 |

Table 4. Density results for beef tallow biodiesel transesterified with ethanol, in a temperature measurement range of 20 °C to 40 °C.

#### 4.3.1 Uncertainty of measurement estimates – approach used to assign uncertainties to the property values

Studies for uncertainty of measurement estimates will be shown for the two physical-chemical quantities used for the characterization.

**a) Viscosity equations and calculation of uncertainty of measurement**

The equations to determinate the viscosity of each measured are shown in equations 8 and 9. Viscosity is a function of the viscometer constant and of the liquid flow time, in seconds, from the top mark to the bottom mark of the measurement bulb.

$$\nu = C_1 \times \left( t - \frac{0.00166 \times \sqrt{V^3}}{C_2 \times L \times \sqrt{C_2 d}} \times \frac{1}{t^2} \right) + \beta \qquad (8)$$

$$C_1 = C_2 \times \left\{ \left[ 1 + \alpha \times (T_r - Temp) \times \left( \frac{\cos\phi_1}{\cos\phi_2} \right) \times \left( \frac{g_1}{g_2} \right) \times \left( 1 + \frac{2}{g_1 h} \right) \times \left( \frac{1}{r_u} - \frac{1}{r_l} \right) \times \left( \frac{\sigma_1}{\rho_1} - \frac{\sigma_2}{\rho_2} \right) \right] \right\} \qquad (9)$$

Where:

$\nu$ is the kinematic viscosity ($mm^2/s$);
$t$ is the flow time (s);
$C_1$ is the corrected constant of the calibrated viscometer ($mm^2/s^2$);
$C_2$ is the constant of the calibrated viscometer ($mm^2/s^2$);
$V$ is the volume of the flown liquid ($mm^3$);
$L$ is the capillary length (mm);
$d$ is the capillary diameter (mm);
$g_1$ is the acceleration of gravity at the measurement place ($m/s^2$);
$g_2$ is the acceleration of gravity at the calibration place ($m/s^2$);
OBS: In this study, $g_1$ and $g_2$ are considered equal because the place where the measurement took place is the same of the calibration.
$h$ is the hydrostatic pressure height (m);
$r_u$ is the inner radius of the upper tube (m);
$r_l$ is the inner radius of the lower tube (m);
OBS: In practice, the variation of both these quantities is negligible.
$\sigma_1$ is the surface tension of the measured oil (N/m), in this study measured with a Krüss K100MK2 tensiometer;
$\sigma_2$ is the surface tension of the oil used in the calibration (N/m), in this study measured with a Krüss K100MK2 (N/m) tensiometer;
$\rho_1$ is the density of the measured oil ($kg/m^3$), in this study measured with an Anton Paar DMA 4500 digital density meter;
$\rho_2$ is the density of the oil used in the calibration ($kg/m^3$), in this study measured with an Anton Paar DMA 4500 digital density meter;
$\phi_1$ is the verticality angle in the measurement (close to zero);
$\phi_2$ is the verticality angle in the calibration (close to zero);
$T_{emp}$ is the measurement temperature (°C);
$T_r$ is the reference temperature of the viscometer (°C);
$\alpha$ is the glass volumetric thermal expansion coefficient ($°C^{-1}$);
$\beta$ is the estimate of variation of the fluid's viscosity ($mm^2/s.°C$).

Note: The term $\beta$ is considered as 1 % of the value of the measured viscosity multiplied by the value of the gradient of temperature from the characterized thermostatic bath. This value is used because there is a temperature gradient between the thermostatic bath's

temperature and the studied oil's (in this case, biodiesel) temperature inside the viscometer. It is only considered for the calculation of the estimative of uncertainty of measurement.
Mensurand: Kinematic Viscosity

$$\nu\left(\frac{mm^2}{s}\right) \tag{10}$$

Used instrument: Ubbelohde viscometer, type I;
Nominal measuring range:

$$(2{,}0000 - 10{,}0000)\frac{mm^2}{s}$$

Therefore,

$$\nu = f(C_1, t) \tag{11}$$

Where:
$C_1$ is the viscometer constant ($mm^2/s^2$);
$t$ is the fluid flow time, in the measured temperature (s).

$$\nu = f(C_1, \alpha, t, \phi_1, \phi_2, g_1, g_2, h, r_u, r_l, \sigma_1, \sigma_2, \rho_1, \rho_2) \tag{12}$$

For each input quantity, the sensibility coefficients must be calculated.
After the calculation of the sensitivity coefficients the items 4.2 and 4.3 from the ISO Gum must be used in order to determine whether the evaluations are Type A or Type B, respectively.
The Type A evaluation is based in a set of observations of $Xi$, through a statistical treatment, and is usually obtained through the average between n measurements, shown in (13):

$$\bar{\nu} = \frac{1}{n}\sum_{k=1}^{n} \nu_k \tag{13}$$

Where:
$\bar{\nu}$ is the arithmetic mean or average;
$n$ is the number measurements;
$\nu_k$ is the number of independent observations.
The other components, which may be evaluated by Type B evaluation of measurement uncertainty, can also be characterized by standard deviations, evaluated from probability density functions based on experience or other information.
All of the Type B evaluations of this study were considered rectangular, as shown in table 4.
Then, item 5.1.3 from the ISO Gum is used to obtain the combined standard uncertainty.

$$u_c^2(\nu) = \sum_{i=1}^{N}\left[c_i \cdot u(x_i)\right]^2 = \sum_{i=1}^{N} u_i^2(\nu) \tag{14}$$

And

$$u_i(\nu) \equiv |c_i| \cdot u(x_i) \tag{15}$$

And the degrees of freedom, obtained from the Welch-Satterthwaite formula, as in item G.4 from ISO Gum.
Where:

$$\nu_{eff} = \frac{u_c^4(\nu)}{\sum_{i=1}^{N} \frac{u_i^4(\nu)}{\nu_i}} = \frac{u_c^4(\nu)}{\sum_{i=1}^{N} \frac{\left(u(x_i)\times c_i\right)^4}{\nu_i}} \tag{16}$$

$N$ is the number of input quantities;
$\nu_i$ is the number of degrees of freedom for each quantity $X_i$.
OBS: It was considered that the number of degrees of freedom is infinite if the calculated degrees of freedom is higher than 500.
A number n of measurements were made in the allotted time. With them, the standard deviation $st$ can be calculated. The chronometer uncertainty $I_t$ is obtained from its calibration certificate. In the case that the coverage factor $k$ has not been declared, a rectangular distribution is considered. The combined standard uncertainty will then be:

$$u_c(t) = \sqrt{\frac{s_t^2}{n} + \left(\frac{I_t}{k}\right)^2} \tag{17}$$

And the degrees of freedom

$$\nu_{eff}\,(t) = \frac{\left(u_c(t)\right)^4}{\frac{s_t^4}{n-1}} \tag{18}$$

**b) Density equations and calculation of uncertainty of measurement**
The equation to determine the density of the liquid indicated in the digital density meter (ISO 15212-1:1998; ISO 15212-2:1998; ASTM D4052-09) is:

$$\rho_{Li} = \rho_a - F_v \times \left(t_{va}^2 - t_{vL}^2\right) + \delta\rho_a + \delta\rho_L(T) + \delta R(\rho_L) + \delta D \tag{19}$$

Where:
$\rho_{Li}$ is the density of the liquid indicated in the digital density meter (g/cm³);
$\rho_a$ is the density of the air indicated in the digital density meter (g/cm³);
$F_v$ is the indicated value to determine the internal calibration factor (g/cm³);
$t_{va}$ is the indicated value directly proportional to the oscillation period of the air;
$t_{vl}$ is the indicated value directly proportional to the oscillation period of the liquid;
$\delta\rho_a$ is the variation of the air density during the solution's measurement (g/cm³);
$\delta\rho_L$ is the variation of the liquid density;
$\delta R(\rho_L)$ is the repeatability of measurements (random);
$\delta D$ is the Correction due to the fluid's damping (oscillation).

#### 4.4 Presentation of the final results
Tables 5 and 6 show the calculations of the measurement uncertainty estimative of each quantity. The values are rounded to two significative digits of the uncertainty.

The results were depicted assuming the associated uncertainty as a value of the viscosity and density properties as:

$$u = \sqrt{(u_{char})^2 + (u_{bs})^2 + (u_{lts})^2 + (u_{sts4})^2 + (u_{sts40})^2} \quad (20)$$

Where:
$u$ is the combined standard measurement uncertainty (or, the measured uncertainties in the usual metrological studies);
$u_{char}$ is the measurement uncertainty for the characterization;
$u_{bs}$ is the standard measurement uncertainty due to the (non) homogeneity between bottles;
$u_{lts}$ is the standard measurement uncertainty due to the long term (in)stability;
$u_{sts4}$ is the standard measurement uncertainty due to the short term (in)stability of the bottles that were stored in a freezer;
$u_{sts40}$ is the standard measurement uncertainty due to the short term (in)stability of the bottles that were stored in a thermostatic bath or hot air oven.

| | Average value mm²/s | Standard Uncertainty mm²/s | Uncertainty (%) |
|---|---|---|---|
| Homogeneity | 5.0267 | 0.0006 | 0.01 % |
| Stability | 5.0269 | 0.0008 | 0.02 % |
| Characterization | 5.0269 | 0.0029 | 0.06 % |
| | Average value mm²/s | Combined Uncertainty ($u$) mm²/s | Combined Uncertainty (%) |
| Declared value | 5.0269 | 0.0031 | 0.06 % |

Table 5. Summary of results for the viscosity of beef tallow biodiesel transesterified with ethanol, at a temperature of 40 °C.

| | Average value g/cm³ | Standard Uncertainty g/cm³ | Uncertainty (%) |
|---|---|---|---|
| Homogeneity | 0.86479 | 0.00004 | 0.004 % |
| Stability | 0.86478 | 0.00009 * | 0.010 % |
| Characterization | 0.86476 | 0.00004 | 0.005 % |
| | Average value g/cm³ | Combined Uncertainty ($u$) g/cm³ | Combined Uncertainty (%) |
| Declared value | 0.86478 | 0.00010 | 0.01 % |

*: Long term stability valued was used

Table 6. Summary of results for the density of beef tallow biodiesel transesterified with ethanol, at a temperature of 20 °C.

With this, the biodiesel reference material candidate presented the following values to the studied properties and the associated uncertainties:

$$\nu = (5.027 \pm 0.007)\, mm^2 / s^1 \quad (21)$$

$$\rho = (0.8648 \pm 0.0002)\, g / cm^3 \quad (22)$$

## 5. Conclusions

Considering the importance of the energy sector to the country, production and disponibilization of reference materials to this is of the utmost importance. Such reference materials have become extremely important for test laboratories that are responsible for issuing test reports with information regarding biofuels specification (Borges, 2007). As a National Metrology Institute, Inmetro aims towards production and certification of reference materials, assuring comparability and reliability of the measurement results.

The results presented in this chapter have demonstrated the possibility of producing a reference material aimed towards internal quality control of test laboratories that provide service to the Regulating Agency for Petroleum, Natural Gas and Biofuels (ANP - Agência Reguladora de Petróleo, Gás Natural e Biocombustíveis) from Brazil. The production of reference materials for quality control is preponderant when then reference materials present similarities to the samples that are usually handled in the laboratory and, mostly, when are no certified reference materials available. It is also possible to point out the application of the reference material as an homogenous and stable study subject to be used in proficiency studies carried out nationally or internationally. Such activities have great importance in assuring quality of the measurement results from Brazilian laboratories and are important tools to the Brazilian accreditation body, since the proficiency test reports allow relevant informations regarding the performance of accredited laboratories and those who wish to be accredited. Still as a conclusion for the viability study presented in this chapter, as an activity of the National Metrology Institute, the production of a certified reference material for biodiesel studies is possible following the used methodology. These certified reference materials could be used in the validation process of a method of a test laboratory, as it would allow the development of technical standards regarding biodiesel analysis. Following this approach, the certified reference materials could be used in a collaborative study to obtain parameters such as repeatibility and reproducibility limit to be used in the creating of Brazilian standards.

In conclusion, the studied biodiesel may be used as reference material for physical-chemical quantities (viscosity and density) in some cases, such as materials quality control (Dube, 2001) and to evaluate the performance of a laboratory that routinely analyses biofuels (Emons, 2006).

Biodiesels are not recommended for calibration of viscosity primary standards, such as capillary viscometers, since the variation of this property, in comparison to the same property in mineral oils (which are used as certified reference materials) is much greater. They are also not recommended to calibrate density devices such as digital density meters.

## 6. Acknowledgement

The authors would like to thank Eng. Felipe de Oliveira Baldner from Fluids Laboratory of Inmetro for the fruitful discussions.

The authors would also like to thank the Scientific Metrology Division of Inmetro (Dimci) for the financial support.

The author Roberto G. Pereira is grateful to the National Research Council of Brazil - CNPq for the financial support received.

## 7. References

ANP's *Resolution nº 007*, AGÊNCIA NACIONAL DO PETRÓLEO, GÁS NATURAL E BIOCOMBUSTÍVEIS (ANP), publish in March 19, 2008.

ASTM D445 - 11 *Standard Test Method for Kinematic Viscosity of Transparent and Opaque Liquids (and Calculation of Dynamic Viscosity)*.2011.

ASTM D4052 - 09 *Standard Test Method for Density, Relative Density, and API Gravity of Liquids by Digital Density Meter*.2009.

ASTM D6751 - 11 *Standard Specification for Biodiesel Fuel Blend Stock (B100) for Middle Distillate Fuels*.2011.

BARBOSA, A.P.F. et al., Metrological Approach in the Characterization of Viscosity of Corn Biodiesel Relative to Temperature, Using Capillary Viscometers. *XIX IMEKO World Congress*, Lisboa, Portugal, 2009.

Borges, R. M. H., Metrology as a laboratory instrument - The case of biofuels analysis laboratories "A METROLOGIA COMO UM INSTRUMENTO AOS LABORATÓRIOS - O CASO DE LABORATÓRIOS DE ANÁLISES DE COMBUSTÍVEIS", *UFRJ, Escola de Química, 2007.*

Dennehy, Peter. CONFORMITY ASSESSMENT and REGULATIONS & THE ISO/CASCO TOOLBOX. *WTO Workshop on CONFORMITY ASSESSMENT.* Geneva, 16th - 17th March 2006.

Dube, G. Metrology in Chemistry - A public Task. *Accreditation and Quality Assurance*, 2001, 6:3–7.5p.

Emons, H. et al., New Definitions on Reference Materials. *Accreditation and Quality Assurance*, 2006, 10:576-578, 3p.

Emons, H. The 'RM family'-Identification of all of its members. *Accreditation and Quality Assurance*, 2006, 10:690-691.2p.

Hufbauer, G., Kotschwar, B., Wilson, J., Trade, standards and development; perspectives from Latin America, *Washington, World bank*, 2000.

Imai, H., Overall view of ISO REMCO Guides, *RM producer accreditation training course*, Tsukuba, Japan, 2007,

ISO Guide 31:2000. *Reference materials - Contents of certificates and labels.*

ISO Guide 34:2009.*General Requirements for the Competence of Reference Material Producers*, 2009.32p.

ISO Guide 35:2006.*Reference Materials- General and Statistical Principles For Certification*, 2006.64p.

ISO 3104:1994 - *Petroleum products -- Transparent and opaque liquids -- Determination of kinematic viscosity and calculation of dynamic viscosity.*

ISO 3105:1994 - *Glass Capillary Viscometers Specification and Operating Instructions.*

ISO EN 14214:2008 – *Automotive Fuels – Fatty Acid Methyl Esthers (FAME) for diesel engines – Requirements and test methods.*

ISO 15212-1 *Oscillation-type density meters – Part 1: Laboratory Instruments*, First Edition, 1998-10-01.

ISO 15212-2 *Oscillation type density meters – Part 2. Process homogeneous liquids*, First Edition, 2002-03-01.

ISO 17025:2005 - *General requirements for the competence of testing and calibration laboratories.*

ISO/TR 3666:1998. *Viscosity of Water.*

JCGM 100:2008. *Evaluation of measurement data – Guide to the expression of uncertainty in measurement.*

KCDB - http://kcdb.bipm.org/AppendixC/M/BR/M_BR.pdf.

Knothe, G.; Van Gerpen, J. & Krahl, J. *The Biodiesel Handbook.* ISBN: 1893997790. AOCS Press, 2005, 303p.

Lisinger, T et al., Homogeneity and stability of reference materials. *Accreditation and Quality Assurance,* 2001. 6:20–25 Springer-Verlag.6p.

Lyra, P.S.F. Diesel Engine: Injection, Lubrification, Combustion and Emissions "Motor Diesel: Alimentação, Lubrificação, Combustão e Emissões". *Laboratório de Motores do Inmetro,* 2008.96p.

Maggi, Patrick C et al., Final report on CCM.V-K2 comparison. *Metrologia* 46 07003, 2009.

Rodrigues, C.R.R. et al., Metrological Approach in the Variation Study of Soy Biodiesel Viscosity related to the temperature using a capillary viscometer"Abordagem Metrológica no Estudo da Variação da Viscosidade de Biodiesel de Soja com Relação à Temperatura, Utilizando-Se um Viscosímetro Capilar", *I Mechanical Metrology Congress (Cimmec),* Rio de Janeiro, 2008.

Santo Filho, D.M.E. et al., Viscosity Measurement of Cappilary Viscometers. *SIM Metrology School.* 2009, 40p.

Santo Filho, D.M.E., Metrology Applied to Biodiesel Analysis "METROLOGIA APLICADA À ANÁLISE DE BIODIESEL", *UFF, Engenharia Mecânica,* PGMEC, 2010a.

Santo Filho, D.M.E. et al., Characterization of Density Of Biodiesel From Soybean, Sunflower, Canola and Beef Tallow in Relation to Temperature, Using a Digital Density Meter, With a Metrological Point of View. *Journal of ASTM International,* Vol. 7, No. 2 Paper ID JAI102430, 2010b.8p.

Santo Filho, D.M.E. et al., The Possibility of Using Beef Tallow Biodiesel as a Viscosity Reference Material. *Accreditation and Quality Assurance.* 10-3-2010, 2010c.4p.

Siqueira, J.R.R. - Accreditation of Reference Materials Producers - Profile Identification of Laboratories that Serve New Propostions "Acreditação de Produtores de Material de Referência - Identificação do Perfil de Laboratórios que Atendam às Novas Propostas", *UFF, Engenharia de Produção,* 2006.

Taylor, R.J. *An Introduction to Error Analysis – The Study of Uncertainties in Physical Measurements.* ISBN 978-0-935702-75-0 .Second Edition, University Science Books,1997.327p.

van der Veen A.; Lisinger T. & Pauwels J. Uncertainty calculations in the certification of reference materials. 2. Homogeneity study. *Accreditation and Quality Assurance* 6:26–30, 2001a.5p.

van der VEEN A et al. Uncertainty calculations in the certification of reference materials. 4. Characterisation and Certification. *Accreditation and Quality Assurance* 6:290–294, 2001b.5p.

VIM – *International Vocabulary of Metrology – Basic and general concepts and associated terms –* JCGM 200-2008.

# 3

# First Generation Biodiesel

Guomin Xiao and Lijing Gao
*Southeast University*
*P. R. China*

## 1. Introduction

Recent years, with the fossil oil crisis, biodiesel is no longer a new word for everyone. The internationally generally accepted definition is according to ASTM biodiesel standard D6751, which biodiesel is defined as a "fuel comprised of mono-alkyl esters of long-chain fatty acids derived from vegetable oils or animal fats, designated B100."

### 1.1 History

Using vegetable oil to make diesel is not a new idea, which can date back to 1853 by E. Duffy and J. Patrick (Feofilova et al., 2010). At the Paris International Exhibition in 1900, R. Diesel demonstrated a test engine sample working on peanuts oil. In 1912, Rudolf Diesel said, "The use of vegetable oils for engine fuels may seem insignificant today. But such oils may become in course of time as important as petroleum and the coal tar products of the present time"(Murugesan et al., 2009). During the following decades, because of fossil diesel's low price, the demands and research on biodiesel was not that much required. On 31 August 1937, G. Chavanne of the University of Brussels (Belgium) was granted a patent for a "Procedure for the transformation of vegetable oils for their uses as fuels" (Belgian Patent 422,877). This patent described the alcoholysis (often be called as transesterification) of vegetable oils using ethanol (and mentions methanol) in order to separate the fatty acids from the glycerol by replacing the glycerol with short linear alcohols. This appears to be the first account of the production of what is known as "biodiesel" today(Knothe, 2001).
The 1970s' energy crisis and people's more attention in environment re-sparked the study of biodiesel and boomed it rapidly. In the period 2001 - 2009, the production of biodiesel in the world increased by more than 16 times (from 9.57 billion tons to 157.6 billion tons). Except the traditional biodiesel production countries – the EU and US, a significant increase of market has been expected in developing countries as China, Brazil, Japan, Indonesia, and Malaysia.

### 1.2 Properties of biodiesel

Biodiesel is a mixture of monoalkyl (mostly methyl or ethyl) esters of fatty acids obtained from renewable resources, such as plant oils or animals fats.
Many industrial devices to produce biodiesel had been set up in Europe and USA, and certain standard of biodiesel had been formulated. EU and ASTM Standards on biodiesel listed in Table 1.

| Properties | EN14214 | ASTM D6751 |
|---|---|---|
| FAME content | ≥96.5% (m/m) | - |
| Density at 15°C | ≥ 860, ≤ 900 (kg/m)$^3$ | - |
| Viscosity at 40°C | ≥ 3.5, ≤ 5.0 ($mm^2/s$) | ≥ 1.9, ≤ 6.0 ($mm^2/s$) |
| Flash point | ≥ 101°C | ≥ 130°C |
| Sulfur content | ≤ 10 mg/kg | ≤ 50 mg/kg |
| Carbon residue remnant (at 10% distillation remnant) | ≤0.3% (m/m) | ≤0.05% (m/m) |
| Cetane number | ≥51.0 | ≥47 |
| Sulfated ash content | - ≤ 0.02% (m/m) | - ≤ 0.02% (m/m) |
| Water content | ≤ 500 mg/kg | ≤ 0.05%(v/v) |
| Total contamination | ≤ 24 mg/kg | - |
| Copper band corrosion (3 hours at 50 °C) | Class 1 max | No. 3 max |
| Oxidation stability, 110°C | ≥6 hours | ≥3 hours |
| Acid value | ≤ 0.5 | ≤ 0.8 |
| Iodine value | ≤ 120 | - |
| Linolenic Acid Methylester | ≤ 12% (m/m) | - |
| Polyunsaturated (≥4 Double bonds) Methylester | ≤1% (m/m) | - |
| Methanol content | ≤0.2% (m/m) | - |
| Monoglyceride content | ≤0.8% (m/m) | - |
| Diglyceride content | ≤0.2% (m/m) | - |
| Triglyceride content | ≤0.2% (m/m) | - |
| Free Glycerine | ≤0.02% (m/m) | ≤ 0.02 |
| Total Glycerine | ≤0.25% (m/m) | ≤ 0.25 |
| Group I metals (Na+K) | ≤ 5 mg/kg | ≤ 5 |
| Group II metals (Ca+Mg) | ≤ 5 mg/kg | - |
| Phosphorus content | ≤ 4 mg/kg | ≤ 0.001%(m/m) |

Table 1. Biodiesel standards of EU and US

### 1.3 Advantages & disadvantages of biodiesel

Compared with fossil diesel, biodiesel has the following advantages (Feofilova et al., 2010; Murugesan et al., 2009) :

1. Biodiesel is a renewable energy source as opposed to oil, the reserves of which are finite as the reserves of other fossil fuels.
2. Biodiesel can decompose easily under natural conditions, and over 90% pure biodiesel can be degrade in a few weeks.
3. Compared with common diesel and petrol, biodiesel has higher combustible value that makes it relatively safe to be stored and transport.
4. Biodiesel contains much less sulfur which not only provides lower share of toxic substances in the exhaust but also enables to provide the lubrication of movable parts during the work of the engine(Knothe & Steidley, 2005). The decrease of other harmful

compounds like PAHs and NOx occurs due to a big percentage of oxygen and more complete combustion of fuel. And pure or blend biodiesel also could suppress the net production of carbon dioxide.

Although biodiesel is "greener" than fossil fuels, it still has a number of disadvantages:

1. High viscosity and surface stress would lead to bigger drops which may cause problems with the system of fuel injection.
2. Vegetable oil contains much more unsaturated compounds than diesel, so biodiesel from it is much easier subjected to oxidation. This parameter correlates with the iodine number.
3. More expensive due to the raw material. Nowadays, the raw material of biodiesel usually soybean oil in US and peanuts oil in EU.

## 2. Transesterification

Generally, the main contents of vegetable oil and animal fats were triglycerides. The common and industrial method to produce biodiesel is chemically described as the transesterification of oil with short chain alcohol. The overall reaction equation is listed in Fig. 1. And this transesterification of triglycerides with alcohol is a three steps reversible reaction (Fig. 2.). This reaction proceeds essentially by mixing the reactants, however, it may accelerate with the presence of a catalyst.

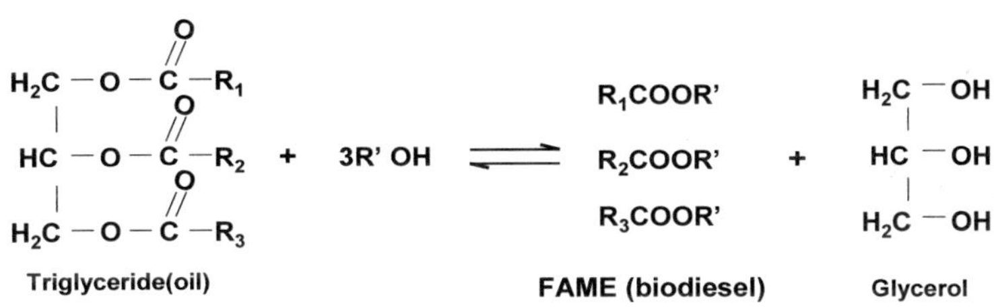


Fig. 1. The overall transesterification reaction of oil with alcohol (left)

Triglyceride + R'OH ⇌ Diglyceride + RCOOR'

Diglyceride + R'OH ⇌ Monoglyceride + RCOOR'

Monoglyceride + R'OH ⇌ Glycerol + RCOOR'

Fig. 2. Chemistry of transesterification process

Methanol is most frequently used, mainly because the reaction rate is higher and the price is cheaper than any other commercial alcohols. The fuel qualities of alkyl esters have received varying evaluations in terms of alcohol used. Methyl ester was better than ethyl ester from the standpoint of engine performance: higher power and torque were achieved from the engine when methyl ester was used as fuel (Knothe, 2005).

### 2.1 Mechanism

The mechanism of alkali-catalyzed transesterification is described in Fig.3 (Ma & Hanna, 1999). The first step is an attack on the carbonyl carbon atom of the triglyceride molecule by the anion of methoxide ion to form a tetrahedral intermediate. In the second step, the tetrahedral intermediate reacts with a methanol to regenerate the anion of the alcohol methoxide ion. In the last step, rearrangement of the tetrahedral intermediate results in the formation of a fatty acid ester and a diglyceride.

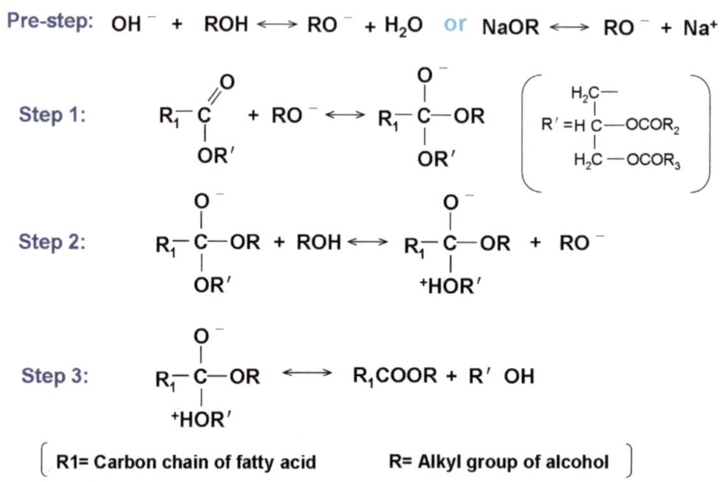


Fig. 3. The mechanism of alkali-catalyzed transesterification of triglycerides with alcohol (Ma & Hanna, 1999, as cited in Scridharan & Mathai, 1974)

Transesterification can also be catalyzed by Brønsted-Lowry acids. These catalysts give very high yields in alkyl esters but reactions are slow, requiring typically temperature above 100 ℃ and hours to complete the conversion (Schuchardt et al., 1998). The mechanism of acid catalyzed transesterification of vegetable oil (for a monoglyceride) is shown in Fig. 4 (Meher et al., 2006). However, it can be extended to di- and tri-glycerides. The protonation of carbonyl group of the ester leads to the carbocation, which after a nucleophilic attack of the alcohol produces a tetrahedral intermediate. This intermediate eliminates glycerol to form a new ester and to regenerate the catalyst.

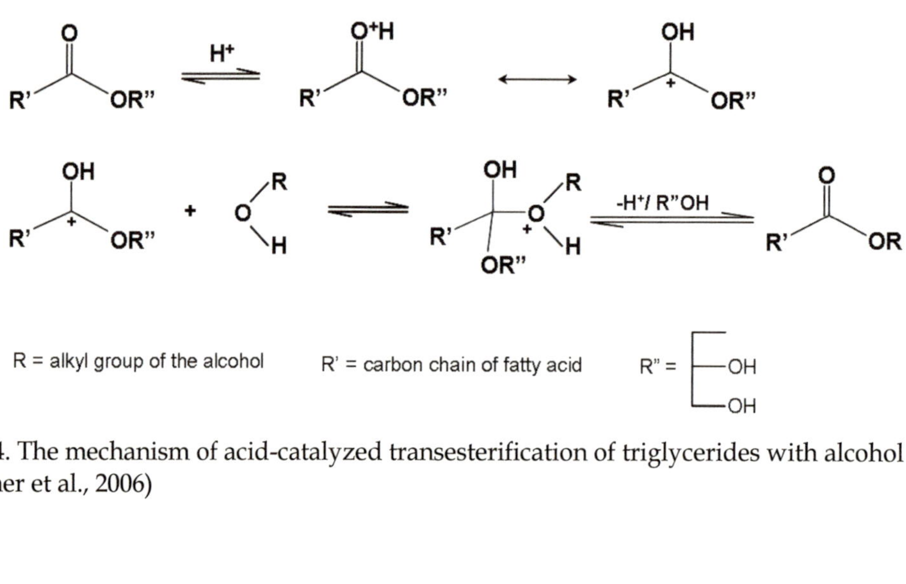


Fig. 4. The mechanism of acid-catalyzed transesterification of triglycerides with alcohol (Meher et al., 2006)

### 2.2 Main factors affecting the transesterification reaction

Except the effect of different catalysts, the process of transesterification is affected mainly by the following factors: temperature, molar ratio of methanol to oil, reaction time, mixing condition, amount of free fatty acid and moisture in the raw oil.

#### 2.2.1 Molar ratio of methanol to oil

The methanol/oil molar ratio is one of the most important variables affecting the ester yield. The stoichiometric ratio for transesterification is 3:1 (methanol/oil). Since this is an equilibrium reaction, an excess of methanol will increase the oil conversion by shifting this equilibrium to producing FAME. An acid catalyzed reaction usually needs much more alcohol than an alkali catalyzed one.

#### 2.2.2 Temperature & reaction time

Transesterification can occur in different temperatures depending on the type of oil employed (Ma & Hanna, 1999). A few works reported the reaction at room temperature (Encinar et al., 2002; Graboski & McCormick, 1998). With sodium and potassium hydroxides as catalysts, the transesterification reaction between *Cynara cardunculus L.* oils and ethanol could obtain a conversion of 91.6% at room temperature(Encinar et al., 2002). Being an equilibrium reaction, the equilibrium constant is influenced by temperature and pressure. High temperature and pressure is benefit for the conversion of oil, however, as usually this reaction occurs under atmosphere and the oil has a relatively higher boiling point, the boiling point of alcohol used in transesterification is considered as the best reaction temperature (Bo et al., 2007; Cui et al., 2007; Gao et al., 2008).
The conversion rate increases with reaction time. Ma et al. (Ma et al., 1999) studied the effect of reaction time on transesterification of beef tallow with methanol. The reaction was very slow during the first minute due to mixing and dispersion of methanol into beef tallow. From one to 5 min, the reaction proceeds very fast. The production of beef tallow methyl esters reached the maximum value at about 15 min.

#### 2.2.3 Mixing condition

The transesterification reaction employing methanol commences as two immiscible phases as a result of the very low solubility of TAG in methanol (Boocock et al., 1996a; Boocock et al., 1996b; Zhou & Boocock, 2006a, 2006b), which is about only 7.5 g of soybean oil soluble in 1 L of methanol at 30°C (Boocock et al., 1996b). Sufficient magnitude Stirring can make TAG transport into small drops which contact the methanol phase more effectively, and then convert into FAME and glycerin (Moser, 2009). The rate at which FAME are produced during the transesterification reaction is thus controlled by mass-transfer limitations, which results in a lag time before conversion to FAME begins (Boocock et al., 1998; Doell et al., 2008; Zhou & Boocock, 2006b). This condition is more obvious when the reaction is catalyzed by solid catalysts. In order to omit the mass transferring resistance between oil phase and alcohol phase, cosolvents like tetrahydrofuran (THF), 1, 4-dioxane, isopropyl ether and diethyl ether were added into the reaction system to obtain a one phase reaction (Meher et al., 2006).

#### 2.2.4 Free fatty acid and moisture

The free fatty acid and moisture content are key parameters for determining the viability of the vegetable oil transesterification process. The starting materials used for base catalyzed alcoholysis should meet certain specification which is that a free fatty acid (FFA) value lower than 3% is needed to carry the base catalyzed reaction to completion (Meher et al., 2006). If the reaction conditions do not meet the above requirements, ester yields are significantly reduced. When catalyzed by NaOH, without adding FFA and water, the apparent yield of beef tallow methyl esters (BTME) was highest, and when only 0.6% of FFA was added, the apparent yield of BTME got to the lowest, less than 5% (Ma et al., 1998). When 0.9% of water was added, without addition of FFA, the apparent yield was about 17% (Ma et al., 1998).

### 2.2.5 Catalyst concentration

Catalyst concentration can affect the yield of the biodiesel product(Leung et al., 2010). Usually, the conversion of triglycerides and the yield of biodiesel increase with the catalyst concentration increasing. This is because an insufficient amount of catalysts result in an incomplete conversion of the triglycerides into the fatty acid esters(Leung & Guo, 2006). However, if the catalyst amount over the optimal concentration, the biodiesel yield would decrease a little with a further increase, which is due to the excess alkali catalyst causing more triglycerides to react with the alkali catalyst and form more soap (Dorado et al., 2002).

### 2.2.6 Case study: Effect of reaction parameters in the transesterification of palm oil with methanol by KF/ hydrotalcite

The following part describe the main parameters effect of the transesterification reaction between palm oil and methanol, which catalyzed by a solid base KF/hydrotalcite. In this case, it is clearly shown the effect of methanol/oil molar ratio, temperature, catalyst amount, and reaction time.

Figure 5 a) showed the influence of methanol to oil molar ratio on FAME yields. The ester yields increased as the amount of methanol increased, and reach the maximum value of 85 % with a molar ratio of 12:1. The increased ester yield with increasing methanol amount below 12:1 can be explained by the pushing effect of excess methanol on the reaction balance. And the decreasing of yield should be due to the large amount of methanol diluting the oil and reducing the reaction rate.

Figure 5 b) showed the yields of FAME obtained over various amount of catalyst. As can be observed, the best results were reached with a concentration of 3 %. The excess catalyst over 3 % did not raise the FAME yield but caused a little decrease. This can be explained by the theory that the catalyst only changes the reaction rate and do not affect the balance of an equilibrium reaction. Therefore, in a fixed reaction time, sufficient amount catalyst is necessary to obtain a reaction rate fast enough. In this case, with 3 % catalyst, reacting 3h is enough to reach the balance, and the more catalyst could not shifting the balance to get higher FAME yield.

In Figure 5 c), as indicated, reaction temperature was varied between 318 ~ 348 K. For the same final reaction time, yield of FAME increased with the increasing of reaction temperature. The equilibrium constant of a reaction is influenced by temperature and pressure. And in this case, which carried out under atmospheric pressure, the former factor, reaction temperature, affected the equilibrium constant much. Therefore, as the temperature rose, the conversion of the oil went up. Moreover, mass-transfer effect was another factor which hinders the transesterification. High temperature is benefit to the mass transfer.

Based on the two reasons above, higher temperature could get higher yield. However, from the results in Figure 5 c), when the temperature exceeded 338 K, the FAME yield dropped obviously. That voluminous methanol gasified and reduced the amount of it in the liquid, when the temperature rose over 338 K (the boiling point of methanol), might be the reason of lower yield.

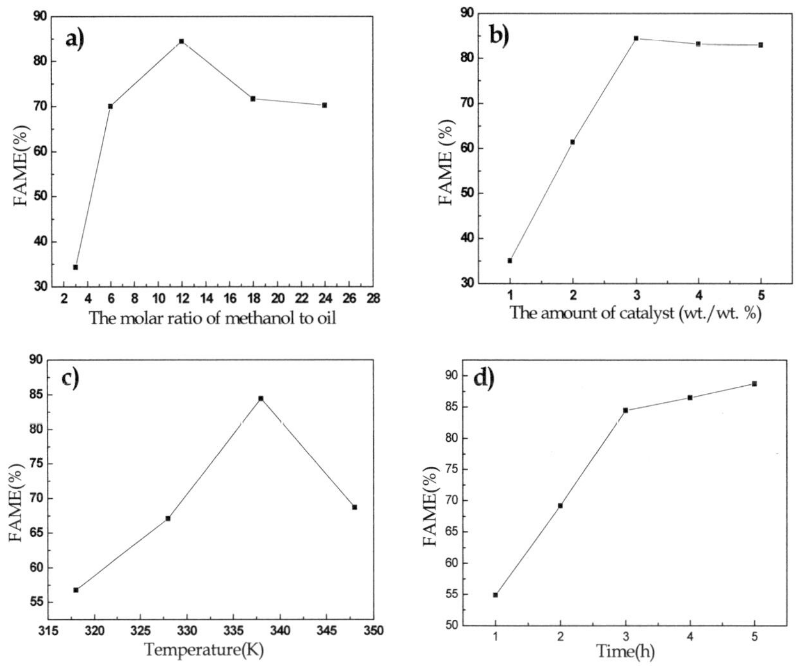


Fig. 5. a) Effect of the molar ratio of methanol/oil on the FAME yield; b) Effect of catalyst amount used on the FAME yield; c) Effect of reaction temperature on the FAME yield; d) Effect of reaction time on the FAME yield. (The mass ratio of KF/HT is 80 %).(Gao et al., 2008)

In Figure 5 d), the influence of reaction time has also been studied. The FAME yield increased with the prolonging of reaction time. From this figure, 3 h is the proper contact time, and longer time didn't enhance the yields obviously.

## 3. Catalysts in transesterification

Both base and acid can catalyze the transesterification reaction between oil and alcohol. Generally, the reaction catalyzed by base is faster than the one by acid. On the other hand, base catalysts have much more demands on the raw materials, especially the amount of free fatty acid and water, while acid catalyst don't need. Hundreds of researches have been done on each kind of catalyst.

### 3.1 Homogeneous catalyst

Industrially, Biodiesel is commonly produced using homogenous basic catalysts such as sodium (or potassium) hydroxide or methoxide because the transesterification reaction is

generally faster, less expensive, and more complete with these materials than with acid catalysts (Boocock et al., 1996a). The biodiesel industry currently uses sodium methoxide, because methoxide cannot form water upon reaction with alcohol such as with hydroxides, which influence the reaction and the quality of the production biodiesel (Zhou & Boocock, 2006a). Furthermore, base-catalyzed reactions are performed at generally lower temperatures, pressures, and reaction times and are less corrosive to industrial equipment than acid-catalyzed methods (Moser, 2009). Therefore, fewer capital and operating costs are incurred by biodiesel production facilities in the case of the base-catalyzed transesterification method (Demirbas & Dincer, 2008; Freedman et al., 1986) .
The liquid acid-catalyzed transesterification process is not much popular as the base-catalyzed process. Homogeneous acid catalyzed reaction is about 4000 times slower than the homogeneous base-catalyzed reaction(Srivastava & Prasad, 2000). However, the performance of the acid catalyst is not strongly affected by the presence of FFAs in the feedstock. Actually, acid catalysts simultaneously catalyze both esterification of FFAs with alcohol and transesterification of oil with alcohol. Thus, acid catalysts can directly produce bio-diesel from low-cost lipid feedstocks, generally associated with high FFA concentrations, which expands the raw materials to low-cost feedstocks, such as used cooking oil and greases, commonly have FFAs levels of >6% (Lotero et al., 2005). For acid-catalyzed systems, sulfuric acid(Al-Widyan & Al-Shyoukh, 2002; Wang et al., 2006), HCl, $BF_3$, $H_3PO_4$, and organic sulfonic acids, have been used by different researchers (Lotero et al., 2005).

### 3.2 Heterogeneous catalysts

Although homogeneous catalysts are cheap and showed great performance toward transesterification, the drawbacks of them are evident. The first is corrosion of the reactor and pipelines by dissolved acid/base species, which inevitably raises the material cost for process construction. The second is the impossibility of catalyst recovery from the reactant-product mixture, which also causes the problems of product separation. A third drawback of homogenously catalyzed transesterification is the limitation in establishing a continuous process. The heterogeneous catalysts was introduced and studied to solve the drawbacks that homogeneous catalysts caused (Lotero et al., 2005). Heterogeneous catalysts can be separated more easily from reaction products(Di Serio et al., 2008). Saponification reactions can also be avoided by using heterogeneous acid catalysts, which expand the raw materials to vegetable oils or animal fats with high contents of FFAs, such as deep-frying oils from restaurants and food processing (Garcia et al., 2008). Bio-diesel synthesis using solid catalysts could potentially lead to cheaper production costs because of reuse of the catalyst and the possibility for carrying out both transesterification and esterification simultaneously (Lopez et al., 2005).

#### 3.2.1 Solid base

i. Metal oxides

The early studies on heterogeneously catalyzed transesterification were focused on the catalysis by single metal oxides. The structure of metal oxides is made up of positive metal ions (cations) which possess Lewis acid and negative oxygen ions (anions) which possess Bronsted base. In methanolysis of oils, it provides sufficient adsorptive sites for methanol, in which the O-H bonds readily break into methoxide anions and hydrogen cations. And methoxide anions then react with triglyceride molecules to form corresponding FAMEs (Zabeti et al., 2009).Bancquart compared the activities of $La_2O_3$, MgO, CaO, and ZnO (Bancquart et al., 2001), for the transesterification of glycerol with fatty acid methyl esters (FAME) at 220 °C. The author concluded that the reaction rates by single metal oxides directly depend on the basicity of the oxide, especially of the strong basic sites. The order of activity followed that of the intrinsic basicity of oxides is $La_2O_3$ >MgO >> ZnO. Magnesium oxide which is produced by direct heating of magnesium carbonate or magnesium hydroxide has the weakest basic strength and solubility in methanol among group II oxides and has been rarely used for biodiesel production. This catalyst showed activity at high temperature and pressure. Under supercritical temperature of 523 °C and high pressure of 24 MPa, Nano magnesium oxide catalyzed transesterification of soybean oil and yields of 99% were obtained in 10 min (Wang & Yang, 2007).

Calcium oxide is the single metal oxide catalyst catches more attention for biodiesel synthesis, due to its cheap price, minor toxicity and relatively high availability. Gryglewicz (Gryglewicz, 1999) compared heterogeneous catalysis by CaO with typical homogeneous catalysts like alkaline-earth metal hydroxides and alkoxides for the transesterification of rapeseed oil by methanol at the boiling point of methanol. The reaction rate over the heterogeneous catalysts, however, was much lower than that of the homogeneous catalysts such as NaOH. The rate of CaO catalysis is accelerated in the presence of water, because methoxide ions, which are thought to be the true catalytic agent for transesterification, are increased through the hydrolysis of monoglyceride molecules (Liu et al., 2008). However, if too much water (more than 2.8% by weight of soybean oil) is added to methanol, the FAME will hydrolyze under basic conditions to generate fatty acid, which can react with CaO to form soap (Lee et al., 2009). Demirbas believes that the calcium oxide catalytic performance is quite weak at low temperatures since only 5% methyl ester yields were obtained at 60 °C after 3 h (Demirbas, 2007). However, the active surface sites of CaO were easily poisoned with $CO_2$ and covered with $H_2O$ (Granados et al., 2007). Therefore, some careful handling is required in order to use CaO as a base catalyst. Increasing the surface basicity of CaO by chemical treatment is a method to increase its catalytic activity (Zhu et al., 2006). Immersed CaO into ammonium carbonate solution and calcined the catalyst at high temperature of 900 °C. With this catalyst, a FAME yield of 94% was obtained for the transesterification of *jatropha curcas* oil with a relatively lower methanol/oil ratio (9:1) and catalyst amount (1.5 wt%) at a reaction temperature of 70 °C (Zhu et al., 2006).

ii. Layered Double Hydroxides (LDHs) & mixed Metal Oxides

Layered double hydroxides (LDHs) , which also be called as hydrotalcite or hydrotalcite-like compound, is a kind of based upon layered double hydroxides with brucite ($Mg(OH)_2$) like hydroxide layers containing octahedrally coordinated $M^{2+}$ and $M^{3+}$ cations (Roelofs et al., 2002). The general formula of LDHs is $[M^{2+}_{(1-x)}M^{3+}_x(OH)_2]^{x+}(A^{n-})_{x/n} \cdot yH_2O$, and its idealized layered structure is shown in Fig. 6. In this formula, $M^{2+}$ (M = Mg, Ca, Fe, Co, Cu, Ni, or Zn) and $M^{3+}$ (M = Al, Cr, Ga, Mn or Fe) are di- and trivalent cations, respectively, the value of x is equal to the molar ratio of $M^{2+}/(M^{2+} + M^{3+})$ and is generally in the range 0.2–0.33; $A^{n-}$ is an anion to balance the charge. As a result, a large class of isostructural materials with versatile physical and chemical properties can be obtained by changing the nature of the metal cations, the molar ratios of M2+/M3+, as well as the types of interlayer anions.

Conventionally, HT is synthesized by co-precipitation, wherein metal nitrates and precipitants are added slowly and simultaneously at a fixed pH under stirring, followed by a long (about 1 day) ageing time and/or hydrothermal treatment in order to improve the

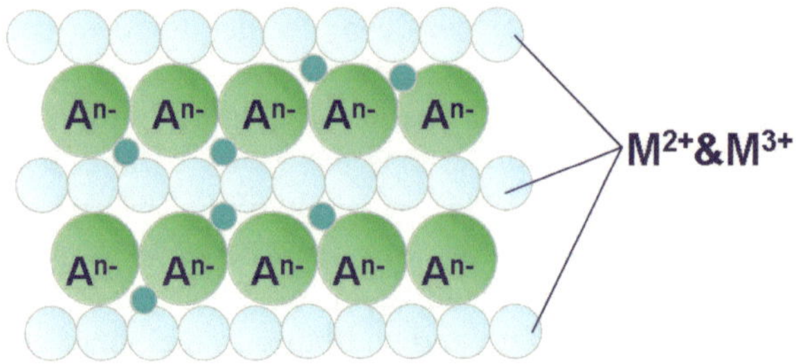


Fig. 6. The structure of LDHs

crystallinity. A particular chemical composition and the method of synthesis, i.e., temperature, solution pH, and ageing time of the gels, have a strong influence on the final basicity of the mixed oxides (Morato et al., 2001; Schulze et al., 2001). The change of the Mg/Al ratio in Mg-Al hydrotalcite leads to the variation of the basic property that is induced by the redistribution of acid-base sites, change of structure and transformation of the cation environment (Diez et al., 2003). Increasing Al content, the relative abundance of low and medium strength basic sites increased. For the Mg-Al hydrotalcites featuring Al contents of 1<Mg/Al<5, the basic site density increased because the $Al^{3+}$ cations within the MgO lattice created a defect in order to compensate for the positive charge generated, and the adjacent oxygen anions became coordinatively unsaturated (Di Cosimo et al., 1998). In the biodiesel synthesis process, the Mg/Al molar ratio of hydrotalcites is usually set from 2 to 4, with the ratio 3 being chosen as the best in terms of basic activity by many authors (Cantrell et al., 2005; Fishel & Davis, 1994; Xie et al., 2006a; Zeng et al., 2008). However, the transesterification catalyzed with Mg-Al HT catalysts prepared by coprecipitaion, the best ester conversions from soybean oil and triglycerides were below 80% (Barakos et al., 2008; Cantrell et al., 2005; Xie et al., 2006a). But, if the transesterification reaction occurs under high temperature and pressure, the uncalcined hydrotalcite could show higher activity (Barakos et al., 2008).

Calcining hydrotalcite materials at high temperature, the interlayer water is lost first, followed by dehydroxylation and decomposition of interlayer carbonate to $CO_2$, which generate an interactive, high surface (ranging from 150 to 300 $m^2/g$) area and well-dispersed mixed oxides that completely destroyed the layered structure (Corma et al., 2005; Mckenzie et al., 1992). These effects combine to make the catalyst quite competitive as a heterogeneous basic catalyst. The basic properties of these sites depend on the Mg-Al ratio in the precursor hydrotalcite (Di Cosimo et al., 1998). These mixed mental-oxides show higher activity in the transesterification reaction than LDHs. The Calcined Li-Al and Mg-Al LDHs (Corma et al., 1998)are able to catalyze the glycerolysis of fatty acid methyl esters to monoglycerides (the reverse of biodiesel synthesis). Shumaker(Shumaker et al., 2007; Shumaker et al., 2008) also used calcined Li-Al LDHs to catalyzed soybean oil with methanol, and Liu et al. (Liu et al., 2007) used calcined Mg-Al hydrotalcite to catalyzed poultry fat with methanol. All of these studies showed that LDHs, except the Li-Al LDHs, performed low activities at the lower temperature, and when the reaction was performed in autoclave at high temperature, the conversion of soybean oil or acid cotton oil could reach 90 %(Barakos et al., 2008; Di Serio et al., 2006).

If the mixed oxides obtained by calcined under certain temperature (generally below 550 °C) encounter water, the typical layered structure of hydrotalcite would represent. This is the well known "memory" property of hydrotalcite(Corma et al., 2005). The reconstruction of decomposed Mg–Al HT by rehydration at room temperature reportedly enhanced the catalytic activity(Rao et al., 1998).And during this reformed process, some other cations or anions could be introduced into the structure. Therefore, based on the hydrotalcite properties, loading some active substance can improve the activity of the HT. $KNO_3$/HT (Sun et al., 2006) showed stronger alkalescency and more active in methylation of cyclopentadiene than $KNO_3/Al_2O_3$. Catalyzed by the hydrotalcite loaded with $CH_3COOK$ (Trakarnpruk & Porntangjitlikit, 2008), at the condition of a 30:1 methanol to oil molar ratio at 373 K for 6 h and 7 wt.% catalyst, the FAMEs content could get to 96.9 %. Gao et al. studied KF loaded hydrotalcites KF/Mg-Al hydrotalcite(Gao et al., 2008), KF/Ca-Al hydrotalcite(Gao et al., 2010b) and KF/Ca-Mg-Al hydrotalcite(Gao et al., 2010a), found that different kinds and ratio of cations in LDHs leaded different catalytic activities. And they obtained a FAMEs highest yield over 99% (by KF/Ca-Mg-Al hydrotalcite) in 10 mins in the transesterification reaction between palm oil and methanol at 65 °C. $K_2CO_3$/Mg-Al hydrotalcite also showed very encouraged activity in biodiesel synthesis (Teng et al., 2010).

iii. Metal salt on porous support

Loading alkali metal or alkali-earth metal salt is the most familiar way to synthesis solid base catalysts. Na, K, Li, Ba, and Mg are frequently used in the metallic form or as various ionic forms of halide, carbonate, hydroxide and nitrate. The metal ion-supported catalysts are usually calcined at 400– 600 °C to obtain active sites. The catalysts generally exhibit the preferential dependence of activity on the surface basicity, rather than on the other properties such as specific surface area and pore volume (Lee et al., 2009). All researchers believe that the different activities of these catalysts were all mainly attributed to the difference in impregnated metal amount (Bo et al., 2007; Cui et al., 2007). Another key factor determining the surface basicity of alkali metal salt-supported catalysts is the calcination temperature. The support for alkali metal species could be diversified from alumina (Bo et al., 2007; Cui et al., 2007; Teng et al., 2009; Xie & Li, 2006; Xie et al., 2006b) to basic oxides such as ZnO (MacLeod et al., 2008; Xie & Huang, 2006), CaO (MacLeod et al., 2008; Watkins et al., 2004) and MgO(MacLeod et al., 2008). In every case, supported alkali metals on alkaline earth oxides are partly dissolved into the liquid phase and the catalysis section invariably becomes homogeneous, regardless of whether the calcination step was included in the preparation procedure of the catalysts or not.

### 3.2.2 Solid acid

Researches on the direct transesterification of lipid feedstocks into biodiesel by solid acid catalysts are not examined extensively. Among the catalysts reported, sulfuric acid prepared by impregnation method has shown the highest activity. Impregnation method prepared solid acid catalysts showed higher activities. However, leaching of sulfate species restricted the reusability of the catalyst, and on another hand, the use of solid acids still need high temperatures and high methanol-to-oil molar ratio(60:1) for a feasible process (Jothiramalingam & Wang, 2009).

Zirconium oxide, titanium oxide and zinc oxide can all be used as solid acid catalyst in biodiesel production. In transesterification reaction of palm kernel oil at supercritical methanol, zinc oxide and zirconium oxide both showed activity as solid acid catalysts. After 1 h of reaction time, using 3 wt.% catalyst and 6:1 molar ratio of alcohol/oil, 86.1% FAMEs yields were obtained for zinc oxide while only 64.5% for zirconium oxide(Jitputti et al., 2006). However, using sulfated zirconia ($SO_4{}^{2-}/ZrO_2$), the yields considerably increased to 90.3%.

$SO_4{}^{2-}/ZrO_2$ and $WO_3{}^{2-}$ / $ZrO_2$ was considered as super acid catalysts in the methanolysis of triacetin was compared (Lopez et al., 2005). The results indicate that $SO_4{}^{2-}/ZrO_2$ showed more activity(conversion of 57%) than $WO_3{}^{2-}$ / $ZrO_2$ (conversion of 10%) under the same conditions of 60 °C and 8 h of reaction time. The activity of $WO_3{}^{2-}$ / $ZrO_2$ was attributed to the formation of tetragonal phase of $ZrO_2$. Beyond 500 °C the tetragonal phase transferred to the monoclinic phase which caused a decrease in the activity(Ramu et al., 2004). However, the transesterification reaction catalyzed by $SO_4{}^{2-}/ZrO_2$ at low temperature is very slow. It was taken over 8h to obtain a FAMEs yield over 80% at 65 °C, while only 2h at 120 °C (Fu et al., 2009).

Fe-Zn double metal cyanide complex has been studied as a solid acid catalyst for methanolysis of sunflower oil (Sreeprasanth et al., 2006). The specific surface area of the catalyst was 51.6$m^2/g$.When the transesterification reaction was performed at 170 °C, with oil/alcohol molar ratio of 1:15 and 3 wt.% of catalyst, the oil conversion reached 97% after 8 h of reaction. The catalyst activity was attributed to the Lewis acid active sites of probably $Zn^{2+}$ on the surface of catalyst. Moreover, the catalyst converted the oil with up to 20% of water content which implies the surface hydrophobicity of the catalyst. The activity of catalyst was successfully tested for esterification of high amount of FFA in the oil. In addition, the catalyst was stable after many cycles since no significant loss of activity was detected.

## 4. Other transesterification methods

Except traditional chemical catalyzed method (base and acid), new methods has been introduced into biodiesel synthesis process. These new method mostly focus on the following goals. First, "greener", means causing less pollution as waste water. Second, "faster", indicate obtaining high FAMEs yield in relatively short time.

### 4.1 Enzyme

Enzymatic transesterification catches attentions for reasons of easy product separation, minimal wastewater treatment needs, easy glycerol recovery and the absence of side reactions (Jegannathan et al., 2008). The transesterification is typically catalyzed by lipases such as *Candida antarctica* (Watanabe et al., 2002), *Candida rugasa* (Linko et al., 1998), *Pseudomonas cepacia* (Shah & Gupta, 2007), *Pseudomonas spp.* (Lai et al., 1999) or *Rhizomucar miehei* (Lai et al., 1999). The yield of biodiesel from this process can vary depending on the type of enzyme used. The enzyme-catalyzed system normally requires a much longer reaction time than the base catalyzed systems.

While enzyme reactions are highly specific and chemically clean, the main problem of the lipase-catalyzed process is the high cost of the lipases. Du et al. reported that there are two ways to reduce the lipase cost. One is to reduce the production cost of the lipase, which can be realized through new lipase development, fermentation optimization, and downstream processing improvement. Another way is to improve/extend the operational life of the lipase, and this can be achieved through enzyme immobilization, alcoholysis reaction optimization, etc. (Du et al., 2008).

### 4.2 Supercritical & subcritical alcohol

Mass transfer between oil and alcohol phases inhibits the transesterification. As is known, when a fluid or gas is subjected to temperatures and pressures in excess of its critical point, a number of unusual properties are exhibited. Under the supercritical conditions, the mixture becomes a single homogeneous phase, which will accelerate the reaction because there is no interphase mass transfer to limit the reaction rate(Pinnarat & Savage, 2008). Another positive effect of using supercritical conditions is that the alcohol is not only a reactant but also an acid catalyst(Alenezi et al., 2010).

Supercritical transesterification is carried out in a high pressure reactor, with heat supplied from an external heater. Reaction occurs during the heating period. After the reaction is complete, the gas is vented and the product in the reactor is poured into a collecting vessel. The remaining contents are removed from the reactor by washing it with methanol (Bunyakiat et al., 2006). During the whole process, several variables (i.e. reaction pressure and temperature) affect the yield of the biodiesel product and the highest yield can be obtained under the optimal conditions.

Synthesis of bio-diesel by supercritical methanol has a drawback with the high cost of apparatus due to the high temperature and pressure, which are not viable in the large scale practice in industry. So, researches have focused on how to decrease the severity of the reaction conditions. Co-solvents and subcritical alcohol with small amount of catalyst, can decrease the operating temperature, pressure and the amount of alcohol (Vyas et al., 2010).

## 5. Biodiesel Industry: Opportunity & challenge

Sooner or later, petroleum will become the huge barrier to human development. Searching substitute is an extremely urgent thing. Although hydrogen energy or solar energy has the chance to use as main energy resource, the day of their wide application is still far away. Biodiesel is the most ideal substitute for fossil oil in a relatively short time. The primary market for biodiesel in the near to long term future is likely to be as a blend component in petrodiesel.

Despite its many advantages as a renewable alternative fuel, biodiesel presents a number of technical problems that must be resolved before it will be more attractive as an alternative to petrodiesel. These problems include improving the relatively poor low-temperature properties of biodiesel as well as monitoring and maintaining biodiesel quality against degradation during long-term storage. The raw material of biodiesel is also a restraining factor. The 70% cost of biodiesel is coming from its raw material oil. Therefore, development of alternative feedstocks is another important area research. Additionally, genetic modification oil plants may provide a solution of this problem.

## 6. References

Al-Widyan, M. I. & Al-Shyoukh, A. O. (2002). Experimental evaluation of the transesterification of waste palm oil into biodiesel, *Bioresource Technology*, Vol. 85, No. 3, pp. 253-56, ISSN 0960-8524

Alenezi, R., Leeke, G. A., Winterbottom, J. M., Santos, R. C. D. & Khan, A. R. (2010). Esterification kinetics of free fatty acids with supercritical methanol for biodiesel production, *Energy Conversion and Management*, Vol. 51, No. 5, pp. 1055-59, ISSN 0196-8904

Bancquart, S., Vanhove, C., Pouilloux, Y. & Barrault, J. (2001). Glycerol transesterification with methyl stearate over solid basic catalysts I. Relationship between activity and basicity, *Applied Catalysis a-General*, Vol. 218, No. 1-2, pp. 1-11, ISSN 0926-860X

Barakos, N., Pasias, S. & Papayannakos, N. (2008). Transesterification of triglycerides in high and low quality oil feeds over an HT2 hydrotalcite catalyst, *Bioresource Technology*, Vol. 99, No. 11, pp. 5037-42, ISSN 0960-8524

Bo, X., Xiao, G. M., Cui, L. F., Wei, R. P. & Gao, L. J. (2007). Transesterification of palm oil with methanol to biodiesel over a KF/Al2O3 heterogeneous base catalyst, *Energy & Fuels*, Vol. 21, No. 6, pp. 3109-12, ISSN 0887-0624

Boocock, D. G. B., Konar, S. K., Mao, V., Lee, C. & Buligan, S. (1998). Fast formation of high-purity methyl esters from vegetable oils, *Journal of the American Oil Chemists Society*, Vol. 75, No. 9, pp. 1167-72, ISSN 0003-021X

Boocock, D. G. B., Konar, S. K., Mao, V. & Sidi, H. (1996a). Fast one-phase oil-rich processes for the preparation of vegetable oil methyl esters, *Biomass & Bioenergy*, Vol. 11, No. 1, pp. 43-50, ISSN 0961-9534

Boocock, D. G. B., Konar, S. K. & Sidi, H. (1996b). Phase diagrams for oil/methanol/ether mixtures, *Journal of the American Oil Chemists Society*, Vol. 73, No. 10, pp. 1247-51, ISSN 0003-021X

Bunyakiat, K., Makmee, S., Sawangkeaw, R. & Ngamprasertsith, S. (2006). Continuous production of biodiesel via transesterification from vegetable oils in supercritical methanol, *Energy & Fuels*, Vol. 20, No. 2, pp. 812-17, ISSN 0887-0624

Cantrell, D. G., Gillie, L. J., Lee, A. F. & Wilson, K. (2005). Structure-reactivity correlations in MgAl hydrotalcite catalysts for biodiesel synthesis, *Applied Catalysis a-General*, Vol. 287, No. 2, pp. 183-90, ISSN 0926-860X

Corma, A., Abd Hamid, S. B., Iborra, S. & Velty, A. (2005). Lewis and Bronsted basic active sites on solid catalysts and their role in the synthesis of monoglycerides, *Journal of Catalysis*, Vol. 234, No. 2, pp. 340-47, ISSN 0021-9517

Corma, A., Iborra, S., Miquel, S. & Primo, J. (1998). Catalysts for the production of fine chemicals - Production of food emulsifiers, monoglycerides, by glycerolysis of fats with solid base catalysts, *Journal of Catalysis*, Vol. 173, No. 2, pp. 315-21, ISSN 0021-9517

Cui, L. F., Xiao, G. M., Xu, B. & Teng, G. Y. (2007). Transesterification of cottonseed oil to biodiesel by using heterogeneous solid basic catalysts, *Energy & Fuels*, Vol. 21, No. 6, pp. 3740-43, ISSN 0887-0624

Demirbas, A. (2007). Biodiesel from sunflower oil in supercritical methanol with calcium oxide, *Energy Conversion and Management*, Vol. 48, No. 3, pp. 937-41, ISSN 0196-8904

Demirbas, A. & Dincer, K. (2008). Sustainable green diesel: A futuristic view, *Energy Sources Part a-Recovery Utilization and Environmental Effects*, Vol. 30, No. 13, pp. 1233-41, ISSN 1556-7036

Di Cosimo, J. I., Diez, V. K., Xu, M., Iglesia, E. & Apesteguia, C. R. (1998). Structure and surface and catalytic properties of Mg-Al basic oxides, *Journal of Catalysis*, Vol. 178, No. 2, pp. 499-510, ISSN 0021-9517

Di Serio, M., Ledda, M., Cozzolino, M., Minutillo, G., Tesser, R. & Santacesaria, E. (2006). Transesterification of soybean oil to biodiesel by using heterogeneous basic catalysts, *Industrial & Engineering Chemistry Research*, Vol. 45, No. 9, pp. 3009-14, ISSN 0888-5885

Di Serio, M., Tesser, R., Pengmei, L. & Santacesaria, E. (2008). Heterogeneous catalysts for biodiesel production, *Energy & Fuels*, Vol. 22, No. 1, pp. 207-17, ISSN 0887-0624

Diez, V. K., Apesteguia, C. R. & Di Cosimo, J. I. (2003). Effect of the chemical composition on the catalytic performance of MgyAlOx catalysts for alcohol elimination reactions, *Journal of Catalysis*, Vol. 215, No. 2, pp. 220-33, ISSN 0021-9517

Doell, R., Konar, S. K. & Boocock, D. G. B. (2008). Kinetic parameters of a homogeneous transmethylation of soybean oil, *Journal of the American Oil Chemists Society*, Vol. 85, No. 3, pp. 271-76, ISSN 0003-021X

Dorado, M. P., Ballesteros, E., De Almeida, J. A., Schellert, C., Lohrlein, H. P. & Krause, R. (2002). An alkali-catalyzed transesterification process for high free fatty acid waste oils, *Transactions of the Asae*, Vol. 45, No. 3, pp. 525-29, ISSN 0001-2351

Du, W., Li, W., Sun, T., Chen, X. & Liu, D. H. (2008). Perspectives for biotechnological production of biodiesel and impacts, *Applied Microbiology and Biotechnology*, Vol. 79, No. 3, pp. 331-37, ISSN 0175-7598

Encinar, J. M., Gonzalez, J. F., Rodriguez, J. J. & Tejedor, A. (2002). Biodiesel fuels from vegetable oils: Transesterification of Cynara cardunculus L. oils with ethanol, *Energy & Fuels*, Vol. 16, No. 2, pp. 443-50, ISSN 0887-0624

Feofilova, E. P., Sergeeva, Y. E. & Ivashechkin, A. A. (2010). Biodiesel-fuel: Content, production, producers, contemporary biotechnology (Review), *Applied Biochemistry and Microbiology*, Vol. 46, No. 4, pp. 369-78, ISSN 0003-6838

Fishel, C. T. & Davis, R. J. (1994). Characterization of Mg-Al Mixed Oxides by Temperature-Programmed Reaction of 2-Propanol, *Langmuir*, Vol. 10, No. 1, pp. 159-65, ISSN 0743-7463

Freedman, B., Butterfield, R. O. & Pryde, E. H. (1986). Transesterification Kinetics of Soybean Oil, *Journal of the American Oil Chemists Society*, Vol. 63, No. 10, pp. 1375-80, ISSN 0003-021X

Fu, B. S., Gao, L. J., Niu, L., Wei, R. P. & Xiao, G. M. (2009). Biodiesel from Waste Cooking Oil via Heterogeneous Superacid Catalyst SO42-/ZrO2, *Energy & Fuels*, Vol. 23, No. 1, pp. 569-72, ISSN 0887-0624

Gao, L. J., Teng, G. Y., Lv, J. H. & Xiao, G. M. (2010a). Biodiesel Synthesis Catalyzed by the KF/Ca-Mg-Al Hydrotalcite Base Catalyst, *Energy & Fuels*, Vol. 24, No., pp. 646-51, ISSN 0887-0624

Gao, L. J., Teng, G. Y., Xiao, G. M. & Wei, R. P. (2010b). Biodiesel from palm oil via loading KF/Ca-Al hydrotalcite catalyst, *Biomass & Bioenergy*, Vol. 34, No. 9, pp. 1283-88, ISSN 0961-9534

Gao, L. J., Xu, B., Xiao, G. M. & Lv, J. H. (2008). Transesterification of palm oil with methanol to biodiesel over a KF/hydrotalcite solid catalyst, *Energy & Fuels*, Vol. 22, No. 5, pp. 3531-35, ISSN 0887-0624

Garcia, C. M., Teixeira, S., Marciniuk, L. L. & Schuchardt, U. (2008). Transesterfication of soybean oil catalyzed by sulfated ziconia, *Bioresource Technology*, Vol. 99, No. 14, pp. 6608-13, ISSN 0960-8524

Graboski, M. S. & Mccormick, R. L. (1998). Combustion of fat and vegetable oil derived fuels in diesel engines, *Progress in Energy and Combustion Science*, Vol. 24, No. 2, pp. 125-64, ISSN 0360-1285

Granados, M. L., Poves, M. D. Z., Alonso, D. M., Mariscal, R., Galisteo, F. C., Moreno-Tost, R., Santamaria, J. & Fierro, J. L. G. (2007). Biodiesel from sunflower oil by using activated calcium oxide, *Applied Catalysis B-Environmental*, Vol. 73, No. 3-4, pp. 317-26, ISSN 0926-3373

Gryglewicz, S. (1999). Rapeseed oil methyl esters preparation using heterogeneous catalysts, *Bioresource Technology*, Vol. 70, No. 3, pp. 249-53, ISSN 0960-8524

Jegannathan, K. R., Abang, S., Poncelet, D., Chan, E. S. & Ravindra, P. (2008). Production of Biodiesel Using Immobilized LipaseA Critical Review, *Critical Reviews in Biotechnology*, Vol. 28, No. 4, pp. 253-64, ISSN 0738-8551

Jitputti, J., Kitiyanan, B., Rangsunvigit, P., Bunyakiat, K., Attanatho, L. & Jenvanitpanjakul, P. (2006). Transesterification of crude palm kernel oil and crude coconut oil by different solid catalysts, *Chemical Engineering Journal*, Vol. 116, No. 1, pp. 61-66, ISSN 1385-8947

Jothiramalingam, R. & Wang, M. K. (2009). Review of Recent Developments in Solid Acid, Base, and Enzyme Catalysts (Heterogeneous) for Biodiesel Production via Transesterification, *Industrial & Engineering Chemistry Research*, Vol. 48, No. 13, pp. 6162-72, ISSN 0888-5885

Knothe, G. (2001). Historical perspectives on vegetable oil-based diesel fuels, *INFORM*, Vol. 12, No. 11, pp. 1103-07, ISSN

Knothe, G. (2005). Dependence of biodiesel fuel properties on the structure of fatty acid alkyl esters, *Fuel Processing Technology*, Vol. 86, No. 10, pp. 1059-70, ISSN 0378-3820

Knothe, G. & Steidley, K. R. (2005). Lubricity of components of biodiesel and petrodiesel. The origin of biodiesel lubricity, *Energy & Fuels*, Vol. 19, No. 3, pp. 1192-200, ISSN 0887-0624

Lai, O. M., Ghazali, H. M., Hashim, D. M., Cho, F. & Chong, C. L. (1999). Viscoelastic properties of table margarine prepared from lipase-catalyzed transesterified mixtures of palm stearin and palm kernel olein, *Journal of Food Lipids*, Vol. 6, No. 1, pp. 25-46, ISSN 1065-7258

Lee, D. W., Park, Y. M. & Lee, K. Y. (2009). Heterogeneous Base Catalysts for Transesterification in Biodiesel Synthesis, *Catalysis Surveys from Asia*, Vol. 13, No. 2, pp. 63-77, ISSN 1571-1013

Leung, D. Y. C. & Guo, Y. (2006). Transesterification of neat and used frying oil: Optimization for biodiesel production, *Fuel Processing Technology*, Vol. 87, No. 10, pp. 883-90, ISSN 0378-3820

Leung, D. Y. C., Wu, X. & Leung, M. K. H. (2010). A review on biodiesel production using catalyzed transesterification, *Applied Energy*, Vol. 87, No. 4, pp. 1083-95, ISSN 0306-2619

Linko, Y. Y., Lamsa, M., Wu, X. Y., Uosukainen, E., Seppala, J. & Linko, P. (1998). Biodegradable products by lipase biocatalysis, *Journal of Biotechnology*, Vol. 66, No. 1, pp. 41-50, ISSN 0168-1656

Liu, X. J., He, H. Y., Wang, Y. J., Zhu, S. L. & Piao, X. L. (2008). Transesterification of soybean oil to biodiesel using CaO as a solid base catalyst, *Fuel*, Vol. 87, No. 2, pp. 216-21, ISSN 0016-2361

Liu, Y., Lotero, E., Goodwin, J. G. & Mo, X. (2007). Transesterification of poultry fat with methanol using Mg-Al hydrotalcite derived catalysts, *Applied Catalysis a-General*, Vol. 331, No., pp. 138-48, ISSN 0926-860X

Lopez, D. E., Goodwin, J. G., Bruce, D. A. & Lotero, E. (2005). Transesterification of triacetin with methanol on solid acid and base catalysts, *Applied Catalysis a-General*, Vol. 295, No. 2, pp. 97-105, ISSN 0926-860X

Lotero, E., Liu, Y. J., Lopez, D. E., Suwannakarn, K., Bruce, D. A. & Goodwin, J. G. (2005). Synthesis of biodiesel via acid catalysis, *Industrial & Engineering Chemistry Research*, Vol. 44, No. 14, pp. 5353-63, ISSN 0888-5885

Ma, F., Clements, L. D. & Hanna, M. A. (1998). The effects of catalyst, free fatty acids, and water on transesterification of beef tallow, *Transactions of the Asae*, Vol. 41, No. 5, pp. 1261-64, ISSN 0001-2351

Ma, F. R., Clements, L. D. & Hanna, M. A. (1999). The effect of mixing on transesterification of beef tallow, *Bioresource Technology*, Vol. 69, No. 3, pp. 289-93, ISSN 0960-8524

Ma, F. R. & Hanna, M. A. (1999). Biodiesel production: a review, *Bioresource Technology*, Vol. 70, No. 1, pp. 1-15, ISSN 0960-8524

Macleod, C. S., Harvey, A. P., Lee, A. F. & Wilson, K. (2008). Evaluation of the activity and stability of alkali-doped metal oxide catalysts for application to an intensified method of biodiesel production, *Chemical Engineering Journal*, Vol. 135, No. 1-2, pp. 63-70, ISSN 1385-8947

Mckenzie, A. L., Fishel, C. T. & Davis, R. J. (1992). Investigation of the Surface-Structure and Basic Properties of Calcined Hydrotalcites, *Journal of Catalysis*, Vol. 138, No. 2, pp. 547-61, ISSN 0021-9517

Meher, L. C., Sagar, D. V. & Naik, S. N. (2006). Technical aspects of biodiesel production by transesterification - a review, *Renewable & Sustainable Energy Reviews*, Vol. 10, No. 3, pp. 248-68, ISSN 1364-0321

Morato, A., Alonso, C., Medina, F., Cesteros, Y., Salagre, P., Sueiras, J. E., Tichit, D. & Coq, B. (2001). Palladium hydrotalcites as precursors for the catalytic hydroconversion of CCl2F2 (CFC-12) and CHClF2 (HCFC-22), *Applied Catalysis B-Environmental*, Vol. 32, No. 3, pp. 167-79, ISSN 0926-3373

Moser, B. R. (2009). Biodiesel production, properties, and feedstocks, *In Vitro Cellular & Developmental Biology-Plant*, Vol. 45, No. 3, pp. 229-66, ISSN 1054-5476

Murugesan, A., Umarani, C., Subramanian, R. & Nedunchezhian, N. (2009). Bio-diesel as an alternative fuel for diesel engines-A review, *Renewable & Sustainable Energy Reviews*, Vol. 13, No. 3, pp. 653-62, ISSN 1364-0321

Pinnarat, T. & Savage, P. E. (2008). Assessment of noncatalytic biodiesel synthesis using supercritical reaction conditions, *Industrial & Engineering Chemistry Research*, Vol. 47, No. 18, pp. 6801-08, ISSN 0888-5885

Ramu, S., Lingaiah, N., Devi, B. L. A. P., Prasad, R. B. N., Suryanarayana, I. & Prasad, P. S. S. (2004). Esterification of palmitic acid with methanol over tungsten oxide supported on zirconia solid acid catalysts: effect of method of preparation of the catalyst on its structural stability and reactivity, *Applied Catalysis a-General*, Vol. 276, No. 1-2, pp. 163-68, ISSN 0926-860X

Rao, K. K., Gravelle, M., Valente, J. S. & Figueras, F. (1998). Activation of Mg-Al hydrotalcite catalysts for aldol condensation reactions, *Journal of Catalysis*, Vol. 173, No. 1, pp. 115-21, ISSN 0021-9517

Roelofs, J. C. A. A., Van Bokhoven, J. A., Van Dillen, A. J., Geus, J. W. & De Jong, K. P. (2002). The thermal decomposition of Mg-Al hydrotalcites: Effects of interlayer anions and characteristics of the final structure, *Chemistry-a European Journal*, Vol. 8, No. 24, pp. 5571-79, ISSN 0947-6539

Schuchardt, U., Sercheli, R. & Vargas, R. M. (1998). Transesterification of vegetable oils: a review, *Journal of the Brazilian Chemical Society*, Vol. 9, No. 3, pp. 199-210, ISSN 0103-5053

Schulze, K., Makowski, W., Chyzy, R., Dziembaj, R. & Geismar, G. (2001). Nickel doped hydrotalcites as catalyst precursors for the partial oxidation of light paraffins, *Applied Clay Science*, Vol. 18, No. 1-2, pp. 59-69, ISSN 0169-1317

Shah, S. & Gupta, M. N. (2007). Lipase catalyzed preparation of biodiesel from Jatropha oil in a solvent free system, *Process Biochemistry*, Vol. 42, No. 3, pp. 409-14, ISSN 1359-5113

Shumaker, J. L., Crofcheck, C., Tackett, S. A., Santillan-Jimenez, E. & Crocker, M. (2007). Biodiesel production from soybean oil using calcined Li-Al layered double hydroxide catalysts, *Catalysis Letters*, Vol. 115, No. 1-2, pp. 56-61, ISSN 1011-372X

Shumaker, J. L., Crofcheck, C., Tackett, S. A., Santillan-Jimenez, E., Morgan, T., Ji, Y., Crocker, M. & Toops, T. J. (2008). Biodiesel synthesis using calcined layered double hydroxide catalysts, *Applied Catalysis B-Environmental*, Vol. 82, No. 1-2, pp. 120-30, ISSN 0926-3373

Sreeprasanth, P. S., Srivastava, R., Srinivas, D. & Ratnasamy, P. (2006). Hydrophobic, solid acid catalysts for production of biofuels and lubricants, *Applied Catalysis a-General*, Vol. 314, No. 2, pp. 148-59, ISSN 0926-860X

Srivastava, A. & Prasad, R. (2000). Triglycerides-based diesel fuels, *Renewable & Sustainable Energy Reviews*, Vol. 4, No. 2, pp. 111-33, ISSN 1364-0321

Sun, L. B., Wu, Z. Y., Kou, J. H., Chun, Y., Wang, Y., Zhu, J. H. & Zou, Z. G. (2006). Catalytic performance of supported KNO3 solid bases for methylation of cyclopentadiene, *Chinese Journal of Catalysis*, Vol. 27, No. 8, pp. 725-31, ISSN 0253-9837

Teng, G. Y., Gao, L. J., Xiao, G. M. & Liu, H. (2009). Transesterification of Soybean Oil to Biodiesel over Heterogeneous Solid Base Catalyst, *Energy & Fuels*, Vol. 23, No., pp. 4630-34, ISSN 0887-0624

Teng, Guangyuan, Gao, Lijing, Xiao, Guomin, Liu, Hu & Lv, Jianhua (2010). Biodiesel preparation from Jatropha curcas oil catalyzed by hydrotalcite loaded with K2CO3, *Appl Biochem Biotechnol*, Vol. 162, No. 6, pp. 1725-36, ISSN 1559-0291

Trakarnpruk, W. & Porntangjitlikit, S. (2008). Palm oil biodiesel synthesized with potassium loaded calcined hydrotalcite and effect of biodiesel blend on elastomer properties, *Renewable Energy*, Vol. 33, No. 7, pp. 1558-63, ISSN 0960-1481

Vyas, A. P., Verma, J. L. & Subrahmanyam, N. (2010). A review on FAME production processes, *Fuel*, Vol. 89, No. 1, pp. 1-9, ISSN 0016-2361

Wang, L. Y. & Yang, J. C. (2007). Transesterification of soybean oil with nano-MgO or not in supercritical and subcritical methanol, *Fuel*, Vol. 86, No. 3, pp. 328-33, ISSN 0016-2361

Wang, Y., Ou, S. Y., Liu, P. Z., Xue, F. & Tang, S. Z. (2006). Comparison of two different processes to synthesize biodiesel by waste cooking oil, *Journal of Molecular Catalysis a-Chemical*, Vol. 252, No. 1-2, pp. 107-12, ISSN 1381-1169

Watanabe, Y., Shimada, Y., Sugihara, A. & Tominaga, Y. (2002). Conversion of degummed soybean oil to biodiesel fuel with immobilized Candida antarctica lipase, *Journal of Molecular Catalysis B-Enzymatic*, Vol. 17, No. 3-5, pp. 151-55, ISSN 1381-1177

Watkins, R. S., Lee, A. F. & Wilson, K. (2004). Li-CaO catalysed tri-glyceride transesterification for biodiesel applications, *Green Chemistry*, Vol. 6, No. 7, pp. 335-40, ISSN 1463-9262

Xie, W. L. & Huang, X. M. (2006). Synthesis of biodiesel from soybean oil using heterogeneous KF/ZnO catalyst, *Catalysis Letters*, Vol. 107, No. 1-2, pp. 53-59, ISSN 1011-372X

Xie, W. L. & Li, H. T. (2006). Alumina-supported potassium iodide as a heterogeneous catalyst for biodiesel production from soybean oil, *Journal of Molecular Catalysis a-Chemical*, Vol. 255, No. 1-2, pp. 1-9, ISSN 1381-1169

Xie, W. L., Peng, H. & Chen, L. G. (2006a). Calcined Mg-Al hydrotalcites as solid base catalysts for methanolysis of soybean oil, *Journal of Molecular Catalysis a-Chemical*, Vol. 246, No. 1-2, pp. 24-32, ISSN 1381-1169

Xie, W. L., Peng, H. & Chen, L. G. (2006b). Transesterification of soybean oil catalyzed by potassium loaded on alumina as a solid-base catalyst, *Applied Catalysis a-General*, Vol. 300, No. 1, pp. 67-74, ISSN 0926-860X

Zabeti, M., Daud, W. M. A. W. & Aroua, M. K. (2009). Activity of solid catalysts for biodiesel production: A review, *Fuel Processing Technology*, Vol. 90, No. 6, pp. 770-77, ISSN 0378-3820

Zeng, H. Y., Feng, Z., Deng, X. & Li, Y. Q. (2008). Activation of Mg-Al hydrotalcite catalysts for transesterification of rape oil, *Fuel*, Vol. 87, No. 13-14, pp. 3071-76, ISSN 0016-2361

Zhou, W. Y. & Boocock, D. G. B. (2006a). Phase behavior of the base-catalyzed transesterification of soybean oil, *Journal of the American Oil Chemists Society*, Vol. 83, No. 12, pp. 1041-45, ISSN 0003-021X

Zhou, W. Y. & Boocock, D. G. B. (2006b). Phase distributions of alcohol, glycerol, and catalyst in the transesterification of soybean oil, *Journal of the American Oil Chemists Society*, Vol. 83, No. 12, pp. 1047-52, ISSN 0003-021X

Zhu, H. P., Wu, Z. B., Chen, Y. X., Zhang, P., Duan, S. J., Liu, X. H. & Mao, Z. Q. (2006). Preparation of biodiesel catalyzed by solid super base of calcium oxide and its refining process, *Chinese Journal of Catalysis*, Vol. 27, No. 5, pp. 391-96, ISSN 0253-9837

# Part 2

## Bioethanol

4

# Bioethanol – What Has Brazil Learned About Yeasts Inhabiting the Ethanol Production Processes from Sugar Cane?

Maria da Graça Stupiello Andrietta[1], Sílvio Roberto Andrietta[2] and Érika Nogueira Andrade Stupiello[3]
[1]*Universidade de Campinas/CPQBA/DBP*
[2]*Biocontal Engenharia de Bioprocessos*
[3]*ENAS Consultoria Linguística*
*Brazil*

## 1. Introduction

According to Datagro (the largest Brazilian sugar and ethanol consulting firm), Brazil will produce nearly 25.3 billion liters of ethanol in the year 2011 (UDOP, 2011). In spite of the astonishing amount, the country does not figure as the world's largest producer, but certainly the bioethanol produced in Brazil stands out in the worldwide clean energy scenario. Ethanol is classified by the US Environmental Agency as an advanced biofuel, since it is capable of reducing greenhouse gas emissions up to 61% when compared to gasoline (UNICA, 2011). This significant production started in the 1970s, leveraged by the first oil crisis. The Brazilian government launched the Proálcool (National Alcohol Program). The required feedstock, sugar cane, has always been plentiful in a country with sugar production tradition. Many plants installed for sugarcane production already had fermentation units to treat molasses, a sugar manufacturing effluent. The government offered incentive to the creation of autonomous distilleries (using sugarcane juice as the only feedstock).

In 1986, out of the 758,965 passenger vehicles manufactured in Brazil, 697,049 were fueled by bioethanol. The end of government incentives to bioethanol production and Brazilian consumers' disbelief in face of bioethanol scarcity in fuel station pumps, associated with the reversion of the worldwide oil crisis, changed the Brazilian scenario. In 1998, out of the 1,389,958 passenger vehicles manufactured in Brazil, only 1,224 were fueled by bioethanol (UNICA, 2011). It seemed as the end of the bioethanol as an alternative fuel source. Many autonomous distilleries were converted into sugar factories, fermenting molasses only as an alternative for effluent treatment. Others were simply closed.

In the 1990s, the eminent global warming threat awakened the world to the need to promote the use of renewable fuels. Ethanol production was quickly resumed in Brazil. In 2001, 16 billion liters of ethanol were produced in the country, and in 2009 this figure escalated to 27 billion liters (UNICA, 2011). In the 1990s, the flex-fuel car (running on ethanol or gasoline) was launched for consumers of passenger vehicles. In 2009, Brazil produced 2,874,077 passenger vehicles, 92% of which were flex-fuel.

The benefits of the introduction of Proálcool were not limited to give highlight to Brazil as a producer and consumer of clean energy. Those benefits extended to research centers and public and private universities. Today, 35 years after the introduction of Proálcool, Brazil has excelled in different areas of knowledge comprising the sugar and ethanol sector. Average agroindustrial yield grows 3.7% a year in Brazil, and 60% of this figure is a result of research and development originated from science produced in the country (DATAGRO, 2009). This research comprises the study of yeasts inhabiting alcoholic fermentation processes.

Brazil has been producing ethanol since the 1930s, and at that time both this product and sugarcane liquor were produced from spontaneous fermentation. Spontaneous fermentation does not make use of an inoculum, and fermentation takes place by the action of naturally inhabiting yeasts in sugarcane fields that are introduced in the process by sugarcane juice. In 1935, Brazil started to use in its ethanol fermentation processes yeasts referred to as pure, that is, purchased by the manufacturer and with the required characteristics for the industrial alcoholic fermentation process (Amorim *et. al.*, 2005). The Sugar and Alcohol Institute (IAA) reports to ethanol producers "*Everyone is aware that yield losses faced by ethanol industries have reached appalling figures. These losses are mostly originated from spontaneous fermentation with impure wild yeasts which act on musts highly contaminated by competing microbes against which industry workers, unaware of the modern methods of rational fermentation, do not know what to do in order to weaken their virulence*". The author of this report could not have guessed that in the year 2011 yeasts referred to as "wild" and consequently harmful to the process are currently being qualified as indigenous and are largely propagated to be used in Brazilian fermentors.

Argueso and Pereira (2010) suggest the use of yeasts isolated from Brazilian ethanol fermentation as a biological platform for application of a wide range of new biorefining technologies. According to these authors, these yeast strains present heterogeneous genomic architectures and establish a strong connection between this genomic complexity and its adaptation to the industrial environment.

Regardless of the time or the fermentation method, the yeast responsible for the transformation of cane sugar or molasses belongs to the *Saccharomyces* genus. *Saccharomyces* is the widely used yeast genus in the industry of fermented products which uses alcohol as final product, whether for fuel or production of alcoholic beverages. This microorganism is mostly indicated for this purpose because it gathers all attributes required for conducting the alcohol production process. The ability to rapidly turn sugars into ethanol, high tolerance to the formed product, osmotolerance (tolerance to great temperature variations), and cell activity in acid environments are the main desirable traits for an industrial strain. All of these attributes are found in representatives of the *Saccharomyces* genus (Andrietta *et al*, 2007).

In Brazil, most industrial processes installed for biofuel production use yeast cell recycling, thus allowing an operation with high cell concentration in the process. This strategy makes fermentation fast and promotes competition between contaminating bacteria and yeast, which is favorable to yeast cells. High conversion speed of sugars into ethanol, though desirable, causes some problems, such as a great amount of heat generated per time unit. This heat should be removed, otherwise, temperature in fermentors may reach levels that are physiologically harmful to yeast cells. Heat is removed with the installation of plate heat exchangers. They are installed outside fermentors. Cooling water originates from a closed circuit and heat is cooled by cooling towers. Since Brazil is a tropical country, the temperature of wet bulbs in sugarcane producing areas range throughout the year from 24 to 27°C, which allows cooling water temperatures from 27 to 30°C, which in turn keep temperatures in fermentors from 31 to 34°C. This process characteristic becomes the first strong factor in yeast selection, leading to the installation of strains capable of developing at higher temperatures. Although these strains develop well at these temperatures, they become more sensitive to ethanol concentration. From 11°GL in temperatures practiced in Brazilian industries, yeasts already compromise their performance.

Even though cell recycling is a requirement in fermentation industry, it is important to highlight that along with yeast cell recycling, this type of operation also recycles contaminating bacteria.

These bacteria are usually acid producing Gram-positive rods, which develop well in pH near 5.0. Aiming to decrease the action of these bacteria, Brazilian distilleries treat recycled yeast with sulphuric acid in pH between 2.0 and 2.5 at times from 1 to 2 hours. Low pH in inoculum makes possible to keep fermentation pH between 3.8 and 4.2, creating an environment less favorable to the growth of bacteria found in the medium. This acidified medium also becomes a selection factor of yeast population in the process. Thus, only yeast strains which develop better in mediums with pH values within this range are capable of dominating the process.

## 2. Bioethanol fermentation processes

Alcoholic fermentation processes in Brazil are conceived similarly to those found in other countries. They are made up of three basic units: fermentation per se, cell separation unit and recycled yeast treatment unit. Fermentation per se is constituted of fermentors, where sugars are converted into ethanol. These fermentors are usually built of carbon steel, with lining properly painted to prevent corrosion and make walls less wrinkled, thus making cleaning easier. They are equipped with external heat exchangers and a shaking system that uses the kinetic energy from the fermentation medium, which is recycled by the heat exchangers. They are closed fermentors with a cleaning system. The conversion that takes place in these fermentors is fully anaerobic and gases produced in fermentation are sent to a washing column for recovery of ethanol found in them. The cell separation unit is made up of centrifuge separators. The fermentation medium containing between 10 and 13%(v/v) of yeast cells is sent to the centrifuges, which generate two product lines, the centrifuged wine, whose cell concentration ranges between 0.2 and 1%, and the yeast cream, whose cell concentration is between 60 and 75%. The centrifuged wine is sent to a lung tank of the distillation devices and the cream is sent to treatment tanks. The treatment unit consists of well-shaken tanks, equipped with an aeration system and built the same way as the fermentation tanks, that is, in carbon steel and with lining properly painted to avoid corrosion and make walls less wrinkled, so as to make cleaning easier. Treated water is added to the yeast cream in this tank with a dilution for cell concentration between 30 and 40% (v/v), then adding concentrated sulphuric acid for pH adjustment in the range from 2 to 2.5. Yeast cells remain in this treatment for 1 to 2 hours. The treated yeast cells return to the fermentors after treatment.

Although the fermentation processes have the same processing stages, they may differ in the operation method, with the fed-batch and continuous processes.

### 2.1 Fed-batch fermentation process

The fed-batch process presents basically the following characteristics: the treated ferment (inoculum) is transferred from the treatment tank to the fermentors through pumping. With transfer concluded, feeding of substratum to be fermented starts until the final fermentor volume is reached. After fermentor is filled, sugar draining is expected. After sugar is converted into alcohol and other products, the fermented medium is sent to the centrifuges where yeast cells are separated from the wine and sent to treatment tanks, where they are diluted and acidified in order to start a new cycle.

The fed-batch fermentation process was conceived to be used in cases where the agent microorganism is submitted to sharp inhibition by the substratum. In this case, in order to obtain higher final product concentrations, substratum is added to the fermentor in a controlled way so as to decrease this inhibition. In general, in plants using this type of process, substratum feed curves are exponentials and aim at keeping substratum in fermentation medium constant and lower than inhibiting concentration. In Brazilian bioethanol producing distilleries, feed curves differ from this standard, and are usually constant throughout the filling time. This happens because the yeast strains used in these processes present little inhibition by the substratum. These strains develop in concentrations from 65 g/L of glucose and 70 g/L of fructose, figures hardly achieved in ordinary operations. Low concentrations of these substrata significantly affect fermentation speed, characterizing the limitation effect by the substratum presented by these microorganisms. For that reason, higher fermentation speeds are obtained when the fermentor is fed quickly, due to the higher glucose and fructose concentration in the medium, without which ethanol yield would be affected (Ferreira, 2005). On the other hand, high fermentation speed leads to higher carbon dioxide production, which increases the foam formation and, consequently, the consumption of antifoam and the generation of heat, which to be removed, require heat exchangers with larger thermal exchange area. These facts make fermentor fill-up time in Brazilian industrial units take approximately 75% of the total fermentation time, which allows better distribution of the heat produced during the fermentation time.

### 2.2 Continuous fermentation process

As for the fed-batch evolution process, the continuous fermentation process is usually conceived with 4 or 5 fermentors connected in series, wherein a cultivation medium and treated yeast cells are added simultaneously and in a continuous and controlled manner in the first fermentor of the system. The purpose of the use of this configuration is to approach the system's behavior to a piston-flow fermentor, thus minimizing the inhibition effect by the product (ethanol) in the transformation speed of sugars into ethanol. The amount of TRS (Total Reducing Sugars) fed into the first fermentor is proportional to the amount of yeast cells, keeping an average rate of 3.5 g of TRS / g cell mass (dry base). This control is important to keep the conversion rate of each fermentor, providing a total fermentation time between 7 to 8 hours depending on the strain used and the feedstock employed. The fermentation medium flows from one fermentor to the other until it reaches the last one, from where it is taken to centrifuge separators, where ferment is separated from wine and sent to treatment tanks. The treatment unit of this type of process is made up of 3 tanks connected in series. Sulphuric acid, water and yeast cream are mixed in a small tank, from where the acidified ferment follows to the treatment tank. This material flows from one tank to the other until it reaches the third tank, from where it is taken to the first tank of the system. Ferment treatment is carried out in three tanks with the purpose of distributing its volume, making the size of the tanks more suitable to the plant layout.

### 2.3 Comparison between the two types of process

Since it is a process that works in stationary state, equipment use rate in continuous fermentation is 100%. This guarantees higher productivity in the process. The occupation rate of the batch system is compromised as a result of operations such as: inoculum load, centrifugation of the fermented medium and cleaning of fermentors.

With the increase in productivity, it is possible to produce the same amount of ethanol of a fed-batch plant, with significant reduction in equipment and physical area. As a result of this reduction, continuous fermentation plants have reduced installation cost when compared with the fed-batch system. The initial investment for installation of a continuous fermentation process ranges from 60 to 70% when compared with the total installation cost of a fed-batch fermentation unit with the same capacity.

As for automation of the continuous fermentation process, since it works in stationary state, it is simpler and more economical when compared with the fed-batch process.

On the other hand, fermentor cleaning is easier in fed-batch processes. In this process, the fermentor is completely emptied at the end of the cycle, which guarantees more efficient equipment cleaning. All yeast cell mass from fermentor centrifugation is sent to juice treatment. Since 100% of inoculum undergoes acid treatment, ferment sent to the next fermentation cycle is cleaner when compared with ferment from continuous fermentation. For this type of operation, only 15% of yeast cells in process are being treated, returning to the fermentor, which is never emptied. This fact contributes to an increase in bacteria contamination in continuous processes. However, with the new cleaning systems of line, fermentors, treatment tanks, heat exchangers, it is possible to keep this contamination under control and no significant difference in ethanol yield values has been observed between the two processes when properly operated.

## 3. The *Saccharomyces sensu strict* strains isolated from Brazilian industrial units

According to Basso *et al.*, (2008) in 2006 there were 329 bioethanol plants in Brazil, and 190 distilleries opted for starting their processes with selected yeasts. In this specific case, selected yeasts are understood as those isolated from industrial fermentation processes. These yeasts are indigenous of their environments, and they all belong to the *Saccharomyces sensu stricto* group. Four strains (CAT1 PE2, SA1 and BC1) are being produced in large scale and commercialized to be used as inoculum to start up alcoholic fermentation processes. The names of these strains correspond to their original units. Thus, strain CAT 1 was isolated at the Catanduva unit, belonging to the Virgolino de Oliveira S/A Açúcar e Álcool Group, whereas PE2 is originally from da Pedra Agroindustrial S/A unit, SA1 from Usina Santa Adélia S.A unit and BG1 from Usina Barra Grande de Lençóis S/A. All these units are located in the State of São Paulo. Some units chose to use a mix of these four strains to start up their processes. Figure 1 presents the karyotype profile of these four strains.

This information is not capable of revealing the fermentative performance of the strains, but it is an efficient tool to evaluate the permanence of these strains in industrial processes.

From this profile, it is possible to separate the different strains present in a sample of a fermentation process. Information concerning indigenous yeasts isolated from Brazilian alcoholic fermentation processes is still scarce, but Argueso et al. (2010) describes the complete genome of a haploid descendent of the PE2 strain and reports the following: "*Here we report the molecular genetic analysis of PE-2 derived diploid (JA Y270) and the complete genome sequence of a haploid derivative (JA Y291). The JA Y270 is highly heterozygous (2 SNPs/kb) and has several structural polymorphisms between homologous chromosomes. These chromosomes have breakpoints within repetitive DNA sequences. Despite its complex karyotype, this diploid, when sporulated, had a high frequency of viable spores. Hybrid diploids formed by outcrossing with the laboratory strain S288c also displayed good spore viability. Thus, the rearrangements that exist near the end of chromosomes do not impair meiosis, as the not span regions that contain essential genes. This observation is consistent with a model in which the peripheral regions of chromosomes represent plastic domains of the genome that are free to recombine ectopically and experiment with alternatives structures. We also explore feature of the JA Y270 and JA Y291 genomes that help explain their high adaptation to industrial environmental, exhibiting desirable phenotypes such as high ethanol and cell mass production and high temperature and oxidative stress tolerance*".

Fig. 1. Karyotyping profile of strains (from left to right): row 1 and 2, strain SA1; row 3 and 4, strain PE2; row 5 and 6, strain CAT1 and row 7 and 8 strain BG1

Stambuck *et al.*, (2011) evaluated, by using microarray-based comparative genome hybridization (aCGH), five different yeast strains (BG-1; CAT-1 PE-2, AS-1 and VR-1) isolated from Brazilian ethanol fermentation processes. Their results suggest that with regard to the reference laboratory *S. cerevisiae* strain, S288C, all five fuel strains showed significant amplification of the telomeric SNO and SNZ genes, which are involved in the biosynthesis of vitamins B6 (pyridoxine) and B1 (thiamin). These authors report that the increase in the number of copies of these genes confer on lab yeast strains the ability to grow more efficiently under the repressing effect of thiamin, especially in the absence of pyridoxine and in high sugar concentration, suggesting that the these gene amplifications provide an important adaptive advantage under the industrial fermentation conditions in which the yeast are propagated.

Although few works present the genetic and technological characteristics of yeasts from Brazilian processes, there is already a consensus that these yeasts are robust and, consequently, are deemed as suitable to be domesticated in order to make them easily handled. Galzerani (2010) domesticated an isolate of the PE-2 industrial strain. This author states that at the end of the domestication process, a strain presenting auxotrophy to uracil was obtained, since the gene URA3 was deleted from the genome of the diploid strain. According to this author, the auxotrophic yeast generation will facilitate the application of genetic engineering techniques.

As for the technological properties of the yeasts, some works have been carried out with yeasts isolated from fermentation processes. Tosetto (2008) studied the influence of organic substances known to be present in sugarcane molasses on the performance of two industrially used yeast strains. One of the strains tested was the Y904 (Mauri - Brazil) which, in spite of being used as inoculum by some industrial units, is a strain developed to be used in bread making. Another strain used was SA1. The study was carried out using eight fermentation cycles in sucrose-base synthetic medium with addition of the following acids: caffeic, lactic, syringic, vanillic, butyric, acetic, formic and HMF(hydroxymethylfurfural). The results suggest that there are behavioral differences between the two strains. The performance of the SA1 strain was compromised by the presence of formic and gallic acids whereas the performance of the Y904 strain was affected by the presence of acetic and butyric acids, as well as by HMF. The presence of lactic acid was a negative interference in the yield and productivity parameters for the two strains studied. Observing these data, it is possible to infer that the presence of a certain substance in the feedstock content of a unit may be a strong microbiota selection factor in fermentation tanks.

Alcarde (2001) studied the flocculation of yeasts cultivated in the presence of bacteria known to promote this phenomenon (*Lactobacillus* and *Bacillus*). The data presented show that the PE2 yeast strain is less sensitive to the flocculation phenomenon than the VR1 strain (yeast strain isolated from the process at Santelisa Vale Bioenergia S/A - Morro Agudo - São Paulo, Brazil) when in the presence of ethanol (7 to 9%).

Steckelberg (2001) characterized strains isolated from nineteen Brazilian alcoholic fermentation processes. This work was carried out under standardized conditions so as to guarantee the same cultivation conditions for all strains, which allows the comparison of yeasts between them. These data are presented in the following table, demonstrating the diversity between the strains isolated from different units. Table 1 presents the kinetic parameters, trehalose and protein percentages, respectively.

The data presented by the author elucidates the diversity of yeast strains inhabiting the Brazilian industrial processes. Even considering all these variations in the data obtained for these strains, it is important to highlight that none of these units presented yield and productivity problems with regard to the presence of these strains as dominant in their processes.

The values concerning all parameters compiled are significantly different and lead to the inference that, although they are yeasts belonging to the *Saccharomyces sensu strict* group, they differ in terms of strain. In theory, strains producing larger amounts of $Y_{x/s}$ cell shave higher chances of dominating fermentation processes. The analysis of the data in Table 1 shows that, under industrial conditions, not all isolated strains present high values for that parameter. The $Y_{x/s}$ of dominating yeasts in their original processes vary up to 25%. A similar variation is observed for the *μmax* value, a parameter which is also a strong influence on the dominion of a strain in the process, since the higher the parameter, the faster the strain develops. Thus, only strains presenting high values for this parameter were expected to be able to dominate the processes, but this fact has not taken place, and this value ranged from 0.35 to 0.60 $h^{-1}$. Considering trehalose, which is intimately related to external conditions, such as nutrient deprivation, temperature variation, high ethanol concentrations and osmotic pressure (Lilie, 1980), it is possible to assume that the processes where the strains were isolated presenting high concentrations of these sugars in their cell composition, were operating under extreme conditions, which may be temperature, osmotic pressure or high alcoholic contents. These yeasts were only capable of surviving these conditions because they were able to accumulate this sugar as a protection factor. Some strains present values near 6%, whereas others are lower than 0.02% in their cell composition. Following this theory, strains presenting lower trehalose levels in their cell compositions are probably originated from processes not operating under extreme conditions.

| ***Strain origin (industrial unit)*** | **Yx/s** | **VCS** | **NCO** | **μmax** | **∅** | **Yp/s** | % Trehalose | % Protein |
|---|---|---|---|---|---|---|---|---|
| *Alvorada* | 0.0436 | 5.8128 | 98.31 | 0.5136 | 2.6487 | 0.4712 | 4.07 | 42.02 |
| *Unialco* | 0.0401 | 5.4880 | 96.58 | 0.4867 | 2.3798 | 0.4512 | < 0.02 | 42.63 |
| *Diamante* | 0.0490 | 5.8885 | 99.59 | 0.3756 | 2.6528 | 0.4617 | 4.03 | 42.18 |
| *Diana* | 0.0409 | 6.5118 | 99.64 | 0.4172 | 2.8080 | 0.4502 | 6.33 | 42.84 |
| *Jalles Machado* | 0.0474 | 5.8823 | 99.48 | 0.4129 | 2.6102 | 0.4589 | 2.07 | 44.00 |
| *Junqueira* | 0.0401 | 5.8808 | 99.46 | 0.3809 | 2.6558 | 0.4719 | 1.27 | 44.74 |
| *Goiasa* | 0.0463 | 5.5473 | 99.64 | 0.4578 | 2.4686 | 0.4607 | 6.17 | 44.93 |
| *Vale Rosário* | 0.0391 | 5.5216 | 99.17 | 0.3537 | 2.3267 | 0.4399 | 0.45 | 48.79 |
| *Bonfim* | 0.0452 | 5.6389 | 99.24 | 0.5439 | 2.5122 | 0.4610 | 1.74 | 40.53 |
| *Costa Pinto* | 0.0469 | 5.5467 | 99.63 | 0.4994 | 2.3839 | 0.4466 | 1.52 | 45.40 |
| *Guarani* | 0.0342 | 5.1921 | 93.26 | 0.5695 | 2.2160 | 0.4472 | 1.94 | 48.24 |
| *Andrade* | 0.0404 | 5.5444 | 99.59 | 0.5694 | 2.4629 | 0.4608 | 5.61 | 44.41 |
| *Dacal* | 0.0368 | 5.5467 | 99.63 | 0.6045 | 2.4898 | 0.4660 | 1.76 | 45.07 |
| *Dacalda* | 0.0489 | 5.5467 | 99.63 | 0.5985 | 2.5167 | 0.4721 | < 0.02 | 42.14 |
| *Alcoeste* | 0.0424 | 5.5615 | 97.88 | 0.5857 | 2.3967 | 0.4458 | 0.71 | 44.41 |
| *Corol* | 0.0466 | 6.5118 | 99.64 | 0.5640 | 2.8419 | 0.4563 | 5.13 | 39.33 |
| *Barra Grande* | 0.0399 | 5.6500 | 99.58 | 0.5767 | 2.4332 | 0.4421 | 2.78 | 47.25 |
| *Santa Cruz* | 0.0450 | 6.5118 | 99.64 | 0.4316 | 2.8420 | 0.4563 | 1.94 | 44.84 |
| *Clealco* | 0.0440 | 5.8000 | 99.50 | 0.5000 | 2.5000 | 0.4600 | 4.23 | 45.20 |

Table 1. Kinetic parameters, yield and productivity of 19 isolated strains, wherein: Yx/s = g produced cell mass (dry mass)/g substratum; VCS = g substratum consumed/Lxh; NCO = % substratum conversion level; μmax = maximum specific growth speed ($h^{-1}$); ∅ = g ethanol produced/Lxh; Yp/s = g ethanol produced/ g substratum, % de trehalose (g/100g of dry mass) and protein (g/100 grams of dry mass)

As for protein concentration, it is possible to assume that yeasts presenting higher concentrations of this compound in their cell compositions were isolated from processes using a must with greater nutrient availability. In general, fermentations carried out with substrata at lower protein concentrations do not influence on yeast performance with regard to ethanol production, but the fact that most Brazilian units sell their surplus yeast, as a protein source for preparation of animal feed, must be taken into account. Thus, units presenting strains with high protein levels obtain a product with higher aggregate value.
The diversity of yeast strains found in processes theoretically equal between themselves - since they originate from the same feedstock, sugar cane, sugar and its byproducts and obtain the same product, ethanol, - suggests that there is not one sole factor responsible for the installation and permanence of a yeast strain in a certain process, but a series of variables which add up to determine the selection and dynamics of the yeast population in the process. This dynamics is unique for each one of the processes.
These data elucidate the diversity of strains inhabiting a restrictive process, such as fermentors, and suggest that this artificial ecosystem is an inexhaustible source of microorganisms, which present particular characteristics and may be used to obtain products with higher aggregate power.

## 4. Population dynamics of yeasts in fermentors

Even though the selected yeast strains are more suitable when compared with the bread-making strains, few industrial units end up the season with the same strain they started. Based on the practical data, PE2 seems to be the most persistent strain among the indigenous strains used. The reasons for this strain to be different from the others are still unknown, but high cell yields must be taken into account for this strain. Amorim *et al.*, (2008),studied the permanence of indigenous strains in industrial processes during 12 harvest seasons. The results presented show that the PE2 strain was capable of remaining in 58% of the distilleries where it was used as inoculum at the beginning of the season.
Another important point, which must be highlighted in the population dynamics in fermentors refers to the number of strains in process. One sole yeast strain is rarely observed in process samples. The presence of two or more strains is quite common. In most cases, the yeast population in fermentors is made up by a consortium of yeasts. These yeast strains belong to the *S. sensu stricto* group, since no other yeast group is capable of surviving under the extreme conditions found in fermentor environments. Some authors report the presence of non-*Saccharomyces* strains in fermentors. This fact is not surprising since the fermentation must, which is a non restrictive environment to microorganism growth, carries an indigenous yeast load at levels of about $10^5$ cells/ml. This fact explains the presence of strains that do not belong to the *Saccharomyces* group in process samples. However, these strains are naturally eliminated. On the other hand, the indigenous *Saccharomyces* present in the feedstock are those strains that will end up dominating the process, replacing the selected strains used at the start-up. This replacement does not harm the process, since the performance of indigenous strains capable of dominating the process is usually similar to that of strains now referred to as selected, but which used to be native.
A study carried out by Cabrini *et al.*, (1999) at a Brazilian distillery producing alcohol for fuel has reported the presence of a diversified yeast flora in sugarcane juice and mash samples. Isolates belonging to the *Saccharomyces, Candida, Torulopsis* e *Rhodotorula* genus have been found in these samples. When the same survey was carried out in yeast samples

undergoing acid treatment to eliminate part of the microorganisms, the *Saccharomyces* genus represented 88% of the isolates. The remaining 12% belong to the *Candida* genus. The presence of these microorganisms is not expected at this stage of the process; however, since this yeast inhabits the raw material, its presence in low concentrations may be explainable in the sample.

## 5. Why feedstock is a determinant factor in yeast population selection in fermentors

The main characteristic of yeast cells inhabiting bioethanol production fermentors is their ability to withstand great biotic and abiotic oscillations, since they are submitted to a process operating uninterruptedly (some Brazilian units operate up to 250 days per season) and in extreme conditions. In general, this process yeast develops well at high temperatures, low pH, presenting low inhibition by substratum and product and satisfactory ethanol yield values (Yp/s) and low nutritional requirement. In addition to these limitations, it must be capable of withstanding feedstock variations, which are quite a few in an agroindustrial unit such as the sugar and ethanol industry.

Special attention must be given to the type of feedstock used in ethanol production, since it is, in most units, a consequence of the requirements of the producing unit, which is controlled by the world's sugar market. Thus, Brazilian ethanol is produced from a substratum, which may not be standardized so as to meet the needs of the yeast.

The material to be fermented may vary from sugarcane juice itself to molasses and, most of the time, the combination of these two substrata. In terms of sugarcane juice, it may vary depending on the sugarcane variety, the soil where it was planted, the harvest period, crops pests, harvest method (manual or mechanical), microbial load contamination, juice extraction method and others. As for molasses, all factors mentioned as affecting sugarcane juice are added to the operation conditions of the sugar factory, since molasses is a byproduct of the latter. Molasses has different exhaust levels (rate between total reducing sugars concentration and total soluble solids concentration), which is determined by the sugar factory operation method. Molasses differs from sugarcane juice for presenting higher nutrient content. On the other hand, since it is a product resulting from a manufacturing process, where it is exposed to high temperatures, molasses presents a series of secondary products which may interfere in the fermentative process, such as: low-molecular-weight organic acids (lactic, acetic, formic and others), hydroxymethylfurfural (HMF), melanoidines and others. As a rule, fermentations using only molasses in must content present more problems than those using sugarcane juice, mainly for presenting more inhibiting substances and higher osmotic pressure, suggesting that the harmful effects of inhibitors are predominant over the beneficial effects of higher nutrient availability.

## 6. Why indigenous yeasts prevail in fermentors

In a system where the input of contaminating microorganism (yeasts and bacteria) is constant as in alcoholic fermentation processes in Brazilian distilleries, strains originated from feedstock and which dominate the process are usually the ones with the best set of required characteristics for survival in this environment at that moment. Since operation conditions vary from one industrial unit to the other, from one season to the other in the same industrial unit and even in different periods in the harvest season of the same industrial unit, differences in yeast cell populations between units, from season to season and also along the season are also expected. Considering the operation conditions of bioethanol producing plants, it is easy to understand why it is impossible to choose one sole yeast strain as the most suitable for ethanol production. As far as it is presently known, there is one yeast strain, isolated from a Brazilian alcoholic fermentation process and which has the ability to remain, not always as the dominant strain, in most part of distilleries where it is used as inoculum. This strain is known as PE2. Its ability to dominate the process seems to be connected to the high yield in cell (Yx/s) and high substratum consumption speed this strain presents. Argueso & Pereira (2010) accredit this permanence to the genomic complexity of this strain, which allows it to adapt to the industrial environment.

In order to illustrate the strong influence of the yield parameter in cell (Yx/s) in the process dominance, Figure 1 presents a simulation of the population dynamics of a fermentation process whose initial inoculum was a certain hypothetical strain with kinetic behavior similar to those observed in strains regularly isolated from industrial fermentors. The study has adopted the usual operation conditions of Brazilian industrial units for a fed-batch fermentation process: a fermentor with 100,000 l of useful volume, with yeast inoculum mass from a previous cycle of 2,700 kg (dry base) and 12-hour fermentation cycles. The study assumed the input of indigenous yeast with the same kinetic profile presented by the strain used as inoculum, differing only in the amount of cell yield, 20% higher to the one presented by the strain used as inoculum. It also considered that the feed must introduced in this fermentor carried a load of indigenous yeasts approximately 100 times lower than the yeast used as inoculum (27 kg dry base). In this profile, it was assumed that there was no introduction of new strains as inoculum and no input of a second indigenous strain.

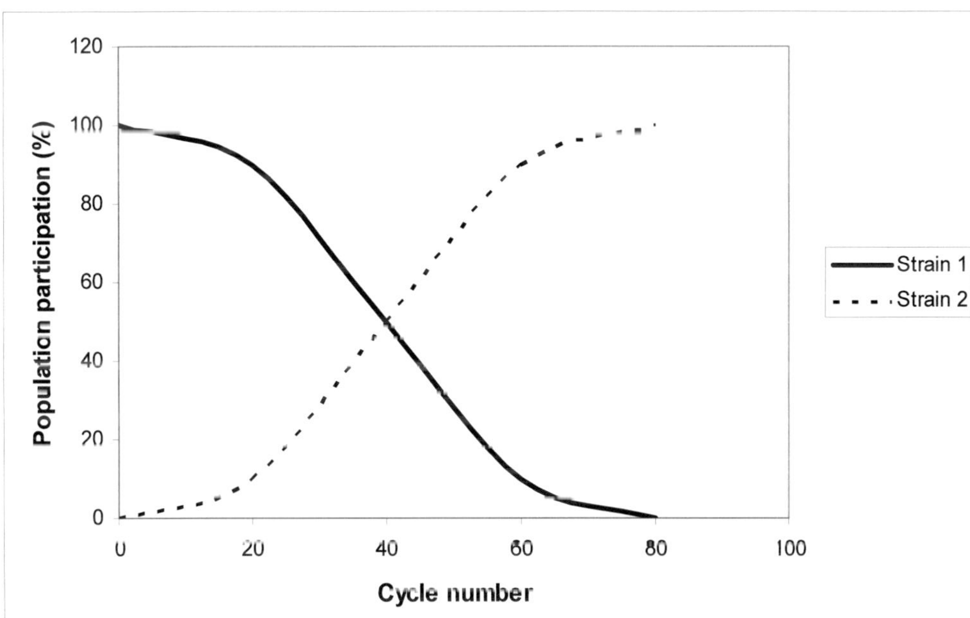


Fig. 2. Population participation profile of two strains with regard to the number of cycles.

The dominance of indigenous strain (yeast 2) in a fermentation process is illustrated in Figure 2. The time for yeast concentration to match the inoculated strain concentration is of 40 cycles, that is, 20 days. On the 70th day of harvest season, which corresponds to 35 days, the indigenous strain fully replaces the strain used as inoculum. These data illustrate the relevance of the cell yield parameter in the definition of the installation, dominance and permanence of a strain in the fermentation process. Thus, yeast strains selected to the used in fermentation units must combine satisfactory characteristics as for the fermentative performance associated to the capacity of presenting high cell yield (Yx/s). The installation and dominance of a certain yeast strain depends on a number of factors, which shall not be assessed apart, but strains presenting high cell yield will certainly be privileged in this dynamics.

The assertions made when using a mathematical model are reinforced when we assess the results obtained in the industrial units. A follow-up of the dynamics of this yeast population carried out in two different industrial units during the 2010 season, which started its processes with the PE2 strain. This strain presents, in lab conditions, $Y_{x/s}$ = 0.0479 grams of dry mass/g of reducing sugar consumed, which is considered high. For comparison purposes, Table 2 presents the values for the four selected commercial strains used in Brazilian distilleries.

| Strain | ($Y_{x/s}$)<br>Grams of cells/total reducing sugar consumed |
|---|---|
| PE2 | 0.0479 |
| CAT1 | 0.0409 |
| SA1 | 0. 0440 |
| BG1 | 0.0463 |

Table 2. Cell mass yield ($Y_{x/s}$) for four yeast strains isolated from Brazilian alcoholic fermentation processes.

The names of the two assessed units were kept confidential, but both units use the fed-batch processes and must with cane molasses base diluted in water as fermentation substratum. Collections were made during the months of April through October, with intervals of 30 ±2 days. Yeast colonies were selected based on their cell morphology in Wallerstein Laboratory Nutrient Agar cultivation medium (WL nutrient medium - DIFCO 0424-17-9) when grown for 7 days at 32°C. Yeast strains present at concentrations higher than $10^6$ CUF/ml of sample were selected. Strain differentiation was performed with the use of karyotyping (Andrietta et al., 2008).

Yeast population dynamics during the season, for a certain industrial plant, herein designated Unit A, is shown in Figure 3.

In the first collection, the PE2 strain is the only strain found in fermentors, since it was used as inoculum. After 30 days, this strain is still present in the process, but the installation of an indigenous strain (*Indigenous 1*) was observed. This strain started to dominate the process, representing 62.5% of the total yeast population. In the third collection (approximately 60 days into the season), the presence of the PE2 strain was not observed. At that point, the yeast population is made up of three indigenous strains, *Indigenous 1*, which still dominated the process (59.3%) and two more strains (*Indigenous 2 and Indigenous 3*), which appeared for the first time in the process. In the fourth collection (approximately 90 days into the season),

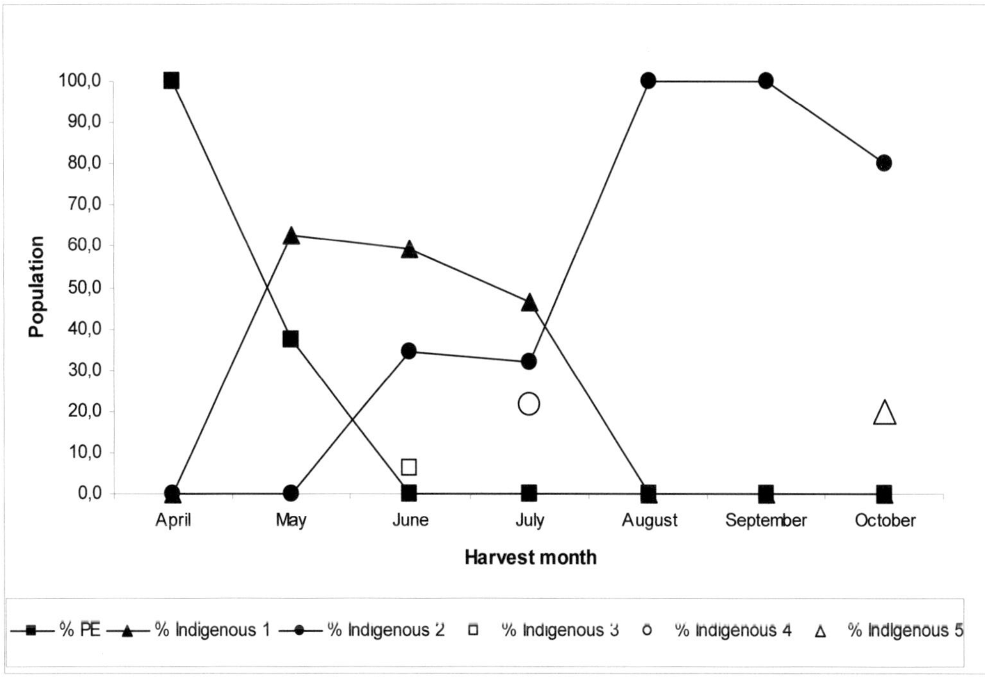


Fig. 3. Yeast population dynamics in industrial unit A

the process kept on operating with three yeast strains in the process, *Indigenous 1* continued to dominate (46.4%), *Indigenous 2* was still present in the sample, but another strain (*Indigenous 4*) came up. *Indigenous 3* strain was no longer detected in the process. In the fifth collection (approximately 120 days into the season), *Indigenous 2* became the only yeast strain in process and thus remained until the sixth collection (approximately 150 days into the season). The seventh and last collection (approximately 180 days into the season) presented yeast *Indigenous 2* as dominant, now inhabiting with another indigenous strain (*Indigenous 5*), which had not been detected in any of the previous collections. Yeast population dynamics during the season of unit A had the presence of six different yeast strains, five of them Indigenous and PE2. *Indigenous 2* was capable of installing, dominating and remaining in the process for the longest time and it was detected in five of the seven collections, predominantly in the three last ones. This strain was present in 120 days out of a total of 180 days. These statements are based on the results obtained from the data; however, some questions remain to be answered: Why hasn't *Indigenous 2* yeast come up right at the beginning of the season? ; Why did this yeast take 60 days after its installation to become dominant? How did this yeast manage to eliminate other process yeast strains and represent 100% of the populations for a 60-day period? Why does *Indigenous 5* yeast install into the process only in the last period of the season? Would it be eliminated should the season extend beyond October, or would it dominate the process in the stead of *Indigenous 2* ? Why do *Indigenous 3* and *4* come up in one collection and are quickly eliminated from the process?

The yeast population dynamics of another assessed unit, herein referred to as B, is presented in Figure 4.

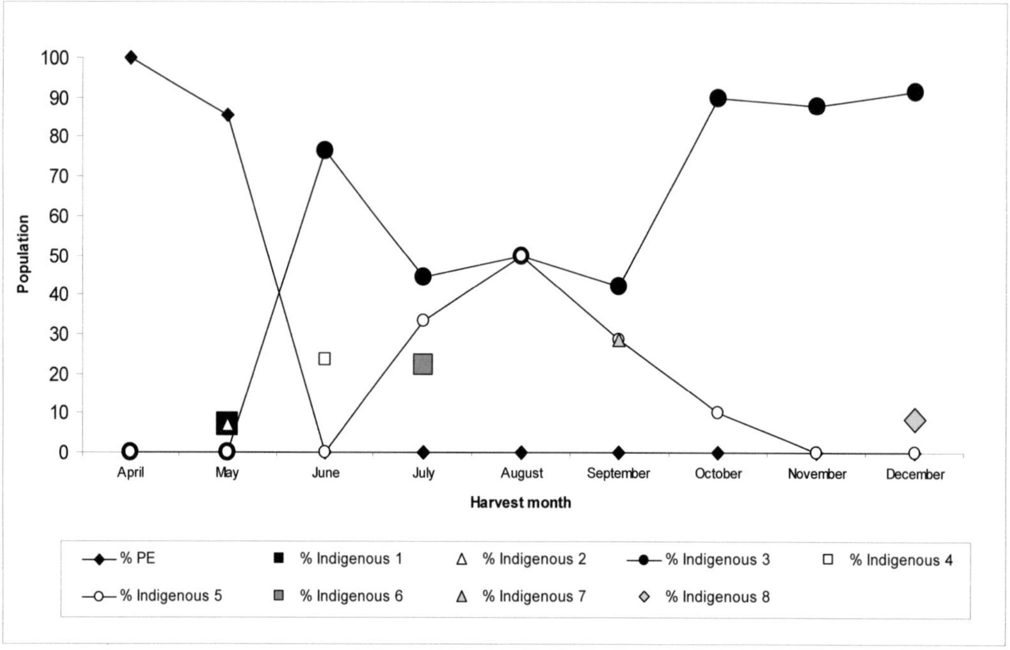


Fig. 4. Yeast population dynamics in industrial unit B

In the first collection, as expected, the yeast population in tanks was 100% of the PE2 strain. After 30 days into the season, PE2 dominated the process, representing 85.6% of the population, but, with participation of two more indigenous yeast strains (*Indigenous 1* and *Indigenous 2*). At 60 days into the season, a completely different scenario was found. PE2 strain was completely eliminated from the process and the two indigenous strains, present in the prior sample (30 days into the season) were also eliminated from the process. The population was now made up of two indigenous strains (*Indigenous 3* and *Indigenous 4*), which completely replaced the PE2 strain within 30 days. *Indigenous 3* presented an aggressive behavior, since it represented 76.5% of the yeast population in the tanks at the first time it was detected in the process. The yeast population present in the tanks in 90 days was made up of three different yeast strains. *Indigenous 3*, which presented as aggressive at the moment of its installation, still dominated, but it did no longer represent 76.5% of the population, but less than half (44%) of the total population. This strain shared the fermentation environment with two other strains (*Indigenous 5* and *Indigenous 6*), which were present for the first time in that process. *Indigenous 4* disappeared from the process from this period on. At 120 days into the season, *Indigenous 3* represented 50% of the population in the fermentors, together with *Indigenous 5*, which also contributed to 50% of the population. At 150 days into the season, *Indigenous 3*, even at a lower concentration, resumed dominance of the process and shared the population with two more strains, *Indigenous 5*, which remained in the process and one new strain observed, *Indigenous 7*. At 180 days into the season, *Indigenous 3* strain once again presented an aggressive behavior and was able to reach 90% of the total yeast strains in the process. *Indigenous 5* strain, even at a low 10% concentration, was able to remain in the process. This same scenario is observed at 210 days into the season, and *Indigenous 3* significantly dominated the process (88%), although *Indigenous 5* was still present. At the last collection (240 days into the season), the scenario for *Indigenous 3* strain persisted and it represented 91.6% of the population. However, at that time *Indigenous 5* disappeared from the process and a new indigenous strain (*Indigenous 8*) came up. The yeast population dynamics during the season for this unit included the presence of nine different yeast strains, eight of which of *Indigenous* origin and the PE2, which was used as inoculum. The yeast presenting the ability to install, dominate and remain for the longest time in the process was *Indigenous 3*, which was detected in seven out of nine collections. It seems that the presence of this strain was responsible for the elimination of PE2, since the latter came up for the first time in June, with a significant participation of 76.6% of the total population. Although it presented an aggressive behavior, in no period of the season did it represent 100% of the population and, in spite of entering the process at high concentration, it did not maintain this standard throughout the fermentation operation period, and in 90 days (July, August and September) it did not represent more than 50% of the total population. As discussed for unit A, some questions remain to be answered to elucidate the yeast population dynamics in fermentors: Why wasn't *Indigenous 3* strain already present at 60 days into the season? Was this strain really responsible for the elimination of the PE2 strain? Why has this strain lowered its percentage of participants during a period? Why was it in no time able to constitute 100% of the population in the tanks?

Many questions are left unanswered regarding the yeast population dynamics in fermentors used in ethanol production, but based on the results obtained thus far, it is possible to make some statements concerning this dynamics. It will be very difficult to isolate one strain which is able to persist in all, or at least in most, industrial processes. The units rarely work with one sole yeast strain in process. This population is usually made up of two or more yeast strains. This fluctuation is closely connected to the characteristics of the feedstock processed. Biodiversity, with regard to the yeast strain, found in industrial fermentors, is an inexhaustible source used to obtain a microorganism with specific characteristics which may be used in other segments of the bioprocess industry, generating products with higher aggregate value than ethanol. Some yeast strains with special characteristics, isolated from fermentation processes, have been used in unconventional alcoholic fermentation processes. One example of this type of application is the ethanol obtainment process, which makes used of autoimobilized strains and allows the operation without the separation unit. In this process, it is possible to keep yeast cells "stuck" in the fermentor for their ability to flocculate. This ability leads to the formation of high-density pellets, which are dragged by the must flow, thus originating a stable bed through which the must percolates. In this percolation, sugar is turned into ethanol. This fermented must leaving the fermentor and containing ethanol is sent directly to distillation. This process, in spite of some limitations, has proven to be a viable alternative for ethanol production in small-scale units, where the use of centrifuges may turn their installations unviable. This type of process will have great value to meet the demand in countries with no tradition in ethanol production and, therefore, with no infrastructure of equipment and maintenance. Yeast strains isolated from the Brazilian sugar and alcohol environment are used in a process of this kind, described in patent PCT/BR2009/000030. This process uses this strain for its ability of flocculate and forming pellets. This is just one of the applications for yeasts originated in the biodiversity of the

Brazilian sugar and alcohol industry. Although this text is about yeast strains, a reservation is required concerning the biodiversity of bacteria also inhabiting the sugar and alcohol environment. Many works have approached this topic, but always considering these microorganisms as unwanted in the process, since the presence of bacteria is always associated with some type of problem, whether regarding yield or operations. Although these bacteria are responsible for many problems in the industry, there is another approach which considers these bacteria as sources for the obtainment of new molecules with differentiated properties. Some gums, with unique properties were obtained by isolating the *Leuconostoc* strains, which naturally inhabit Brazilian sugarcane fields (Vieira, 2005, Vieira , 2009).

The study of this artificial ecosystem, the agroindustrial environment, not only leads to the understanding of the population dynamics of yeast strains inhabiting fermentation tanks, but also increases the possibility of obtaining microorganisms with specific characteristics that allow the achievement of bioproducts not yet known. The discovery of these biomolecules may benefit the most distinct industrial segments, including the pharmaceutical, food, petroleum and cosmetics industries, among others.

## 7. References

Alcarde V.E. (2001) Avaliação de parâmetros que afetam a floculação de leveduras e bactérias isoladas de processos industriais de fermentação alcoólica. [*Assessment of parameters affecting flocculation of yeast and bacteria isolated from alcoholic fermentation processes.*] Thesis, Universidade Estadual de Campinas, Campinas.

Amorim H.V.O., Leão R.M, Basso C., Oliveira A.J. & Lopes M.L (2005) *Fermentação Alcoólica: Ciência e Tecnologia*. Fermentec, Piracicaba.

Amorim Neto H.B., Yohannan B.K, Bringhurst T.A, Brosnan J.M, Pearson S.Y, Walker J.W & Walker G.M (2009) Evaluation of a Brazilian Fuel Alcohol Yeast Strain for Scotch Whisky Fermentations. *Journal of Institute of Brewing* 115: 198-207.

Andrietta M.G.S, Andrietta S.R, Steckelberg C & Stupiello E.N.A (2007) Bioethanol - Brazil, 30 years of Proálcool. *International Sugar Journal* 109: 195-200.

Andrietta S.R., Steckelberg, C., & Andrietta,M.G.S.(2008) Study of flocculent yeast performance in tower reactors for bioethanol production in a continuous fermentation process with no cell recycling. *Bioresour. Technol.*, 99(8) 3002-3008, 2008

Andrietta, Silvio Roberto (Autor); Andrietta, Graca (Autor); Process of fermentation for production of bioethanol, 01/2009. Patente de Invenção (No. Patente: PCT/BR2009/000030). Patente Unicamp. Data Pedido ou Depósito: 30/01/2009. Órgão de registro: WIPO - World Intellectual Property Organization, BRASIL

Argueso J.L & Pereira G.A.G. (2010) Perspective: Indigenous sugarcane yeast strains as ideal biological platforms for the delivery of next generationbiorefining technologies. *International Sugar Journal* 112: 86-89.

Argueso J.L.; Carazzolle, M.F. Mieczkowki, P.A.;Duarte, F.M. Netto,O.V.C.; Missawa,S.K.;Galzerani,F.;Costa,G.G.L.;Vidal,R.O.;Noronha,M.F.,Dominska,M.;Andrietta,M.G.S.;Andrietta,S.R.;Cunha,A.F.;Gomes,L.H.;Tavares, F.C. Alcarde,A.R.; Dietrich,F.S.;Mccusker,J.H.; Petes,T.D.; & Pereira,G.A.C. (2009) Genome structure of a *Saccharomyces cerevisiae* strain widely used in bioethanol production. *Genome Research* 19: 2258-2270.

Atala D. (2011) DATAGRO 2009 - THE FUTURE: "Targeting Sustainability: New Technologies in the Biofuels Industry".

Basso L.C., Amorim H.V., Oliveira A.J. & Lopes M.L (2008) Yeast selection for fuel ethanol production in Brazil. *FEMS Yeast Research* 8: 1155-1163.

Cabrini, K.T. & Gallo, C.R. Identificação de cepas de leveduras no processo de fermentação das destilaria Mandú, SP. In: 7º Congresso Nacional da STAB. 227-231, Londrina, PR, 1999.

Ferreira E. (2005) Contribuição para o estudo da otimização da fermentação alcoólica operando em batelada alimentada [*Optimization of fermentation operating in FED-BATH - a contribution]*.. Thesis, Universidade Estadual de Campinas - UNICAMP, Campinas.

Galzerani F. (2010) Desenvolvimento de plataformas industriais: domesticação de leveduras para o melhoramento da fermentação alcoólica. [*Development of industrial platforms: domestication of yeasts for improvement of alcoholic fermentation.*] Thesis, Universidade Estadual de Campinas, Campinas.

Godoy A., Amorim H.V., Lopes M.L. & Oliveira A.J. (2008) Continuous and batch fermentation processes: Advantages and disadvantages of these processes in the Brazilian ethanol production. *International Sugar Journal* 110: 175-181.

Lilie SH & Pringle J.R (1980) Reserve carbohydrate metabolism in Saccharomyces cerevisiae : responses to nutrient limitation. *Journal of Bacteriology* 143: 1384-1394.

Rainieri S., Zambonelli C. & Kaneko Y. (2003) *Saccharomyces sensu stricto* : Systematics, Genetic Diversity and Evolution. *Journal of Bioscience and Bioengineering* 96: 1-9.

Stambuck B.U., Dunn B, Alves Jr SL, Duval EH & Sherlock G (2009) Industrial fuel ethanol yeasts contain adaptive copy number changes in genes involved in vitamin B1 and B6 biosynthesis. *Genome Research* 19: 2271-2278.

Steckelberg C. (2001) Caracterização de leveduras de processos de fermentação alcoólica utilizando atributos de composição celular e características cinéticas [*Characterization of yeasts from alcoholic fermentation processes using cell composition attributes and kinetic characteristics.*] Thesis, Universidade Estadual de Campinas, Campinas.

Tosetto G.M (2008) Comportamento de linhagens industriais de *Saccharomyces* frente a compostos inibitórios presentes no melaço de cana-de-açúcar na produção de bioetanol. [*Behavior of Saccharomyces industrial strains before inhibiting compounds present in sugarcane molasses in bioethanol production.*] Thesis, Universidade Estadual de Campinas, Campinas.

UDOP (2011) http://www.udop.com.br/index.php?item=noticias&cod=1073044. Vol. 2011 ed.), p.^pp. Produção de álcool deve crescer 45% até 2020. UDOP - União dos Produtores de Bioenergia.

UNICA (2011) http://unica.com.br/noticias. Vol. 2011 ed. Brasil poderá quase dobrar exportações de etanol para o Japão.

Vieira E.D. (2003) Seleção de linhagens produtoras de goma isoladas da cana de açúcar [*Screening to gum producing strains from sugar cane industry]*. Thesis, Universidade Estadual de Campinas - UNICAMP, Campinas.

Vieira E.D. (2008) Produção de goma xantana em reator de coluna de bolha utilizando processo de batelada repetida. [*Xanthan gum production in a buble column reactor using a repeated bath process]*. Thesis, Universidade Estadual de Campinas - UNICAMP, Campinas.

5

# Ethanol Production in Brazil: The Industrial Process and Its Impact on Yeast Fermentation

Luiz Carlos Basso[1], Thiago Olitta Basso[2,3] and Saul Nitsche Rocha[4]
*[1]University of São Paulo, Escola Superior de Agricultura Luiz de Queiroz, Departamento de Ciências Biológicas, Piracicaba (SP)*
*[2]University of São Paulo, Escola Politécnica, Departamento de Engenharia Química, São Paulo (SP)*
*[3]University of São Paulo Programa de Pós-Graduação Inter-unidades em Biotecnologia, São Paulo (SP)*
*[4]Positivo University, Programa de Pós-Graduação em Biotecnologia Industrial Curitiba (PR)*
*Brazil*

## 1. Introduction

Ethanol is the most consumed biofuel in the world. Brazil is the country that first introduced this renewable fuel in its energy matrix. A large industry arose from this pioneering initiative, and nowadays detains the most economically feasible process for bioethanol production. For decades, Brazil was the main producer but was recently surpassed by the United States. Nevertheless, Brazil is currently the greatest ethanol exporter. Many factors contributed for the efficiency of this industry, as: feedstock, fermentation and process improvements. In this chapter, general aspects of the Brazilian ethanol fermentation process will be exhibited, many improvements will be highlighted, and the impacts of the peculiarities of this process on yeast fermentation will be discussed.

## 2. A brief history of the Brazilian ethanol industry

Ethanol is the main biofuel used for transportation and Brazil is the second largest bioethanol producer and the greatest exporter. The United States (the greatest producer) and Brazil are responsible for 70 % of the world ethanol production (RFA, 2011). The sugar and ethanol industry in Brazil make up 2.3 % of the Domestic Gross Product, generating 4.5 million jobs. Additionally, fuel ethanol represents almost 50 % of the total fuel volume consumed by cars (and light vehicles).

Sugar cane was introduced in Brazil by the Portuguese colonizers at the beginning of the 16th century. The first mills for cane sugar production (consumed in Europe) were established around the year 1530 (Amorim and Leão, 2005). Today, sugar cane is an important crop in Brazil occupying ca. 8 million hectares with a production of more than 600 millions tons per year, making the country the largest sugar cane producer worldwide.

The Brazilian knowledge in ethanol production from sugarcane began to be developed in colonial period, when farmers used to produce the Brazilian sugarcane distilled spirit, the "cachaça" (Basso and Rosa, 2010). Then, since the beginning of the 20th century, Brazil has been

using ethanol for energetic purposes. In 1905, the first tests of using ethanol as fuel for vehicles engines were performed. These tests provoked official attention and resulted in a law published in 1931, which determined that ethanol should be mixed to the gasoline at a rate of 5 % (v/v). As any new technology, ethanol needed economic and political investments to keep its viability in national market and also, to fight against a major competitor: petrol.
Winds had changed in early 1970's, when oil crisis tripled oil imports costs to Brazil in 1973, due to the Arab oil embargo. In addition, world sugar prices, which had been climbing upward since the mid-60's, declined sharply in 1974 (Sandalow, 2006). During these drastic changes in global fuel market, the country launched the Brazilian National Alcohol Program (PROALCOOL) in 1975, aiming at large scale ethanol generation by local distilleries and engines adaptation to consume the E20 mix (20 % and 80 % ethanol, gasoline, respectively) or even pure anhydrous ethanol (Amorim and Leão, 2005).
To improve ethanol competitiveness, government offered low-interest loans for construction of new refineries and gasoline prices were set to give ethanol a competitive advantage. Brazilian state-owned oil company (Petrobras), began making investments for distribution of ethanol throughout the country. After this intervention, ethanol production boomed more than 500 % (Sandalow, 2006).
In the 1980´s, the major car companies accorded to install assembly lines for 100 % ethanol cars. This resulted in a high demand, which reached the peak at the half of the decade, when ethanol supply as fuel for vehicles was the half of the total fuel consumed in Brazil.
Oil prices dropped sharply in 1985-86. In parallel, a bad economic period (illustrated by a high inflation) led the government to cut subsides to the ethanol industry, leading to the fuel shortage in the market. However, during the 1990's the Brazilian economy has lifted. In addition, the energy world market, especially the oil price and Asia crisis, were favourable for the ethanol production (Moreira, 2000).
At the first decade of 2000, flex-fuel engines were exclusively designed locally to Brazilian market. These engines let the consumers to choose between ethanol and gasoline, depending on its price on the market. Indirectly, this also facilitated the ethanol market regulation. Currently most of the produced ethanol is consumed internally. Of all automobiles used in Brazil, eight out of ten are flex fuel vehicles - more than in any other country in the world (Pilgrim, 2009).

## 3. The industrial process

### 3.1 Feedstock

Technically, ethanol can be produced from a wide variety of renewable feedstock, which can be roughly classified into three main groups: (1) those containing considerable amounts of readily fermentable sugars (sugar cane, sugar beets, sweet sorghum), (2) starches and fructosans (corn, potatoes, rice, wheat, agave) and (3) cellulosics (stover, grasses, corn cobs, wood, sugar cane bagasse). Sugar cane, beet and sweet sorghum provide the simple sugars, as sucrose, glucose and fructose that can be readily fermented by yeasts (Amorim et al. 2009). This differs from ethanol production processes based on starchy or lignocellulosic feedstock, where prior hydrolysis of polysaccharides is necessary with increased ethanol production costs (Dien and Bothast, 2009).
The feedstock has a great impact on ethanol production costs, which is also influenced by the region and by the processing. The production costs are affected by improvement in technologies and vary during time-to-time, making a realistic comparison a difficult task (Figure 1).

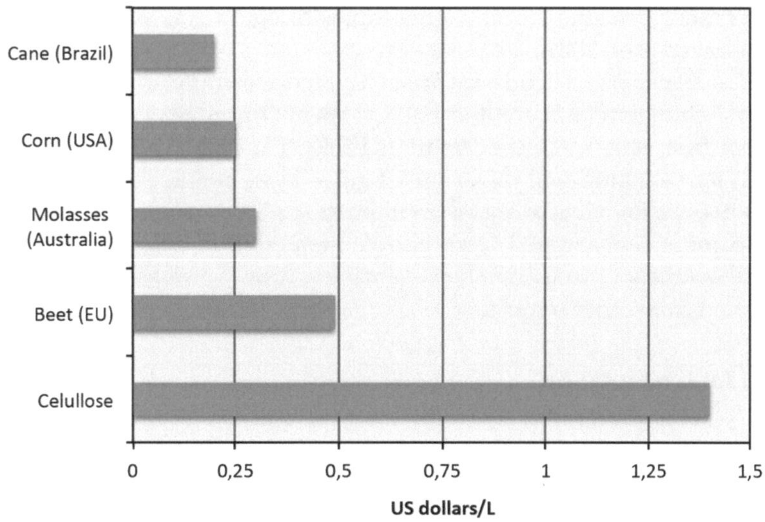


Fig. 1. Gross estimation of ethanol production costs in US dollars/litres from different feedstock and regions, from several sources available in 2007 (Burnquist, 2007).

As a C-4 photosynthetic species, sugar cane presents very high biomass productivity, amounting to 80-120 ton/ha.year with an industrial ethanol production of 8,000 litres/ha, higher when compared to 3,000 litres/ha from maize. Nitrogen-fixing endophytic bacteria (mainly *Acetobacter diazotrophicus*, *Azospirillum* spp. and *Herbaspirillum* spp., *Gluconacetobacter diazotropicus*) have been discovered in both sugar cane and in maize. It has been suggested that at least 60 % of the plant's nitrogen requirement, is supplied endogenously when sugar cane is grown in low fertility soil (James and Olivares, 1997; Boddley, 1995; Wheals et al., 1999; Bertalan et al., 2009). As nitrogen fertilizers are quite expensive and require huge amounts of fossil energy for its production, sugar cane presents economic and environmental benefits when compared to other crops. Its drought tolerance also contributes for some agricultural advantages.

Sugarcane contains 12-17 % total sugars (90 % sucrose and 10 % glucose plus fructose) on a wet-weight basis with 68-72 % moisture. The sugar average extraction efficiency by crushing (or diffusing) is approximately 95-97 % (Wheals et al., 1999). Bagasse plays an important role in the energy balance of sugar cane ethanol. It comprises 20 to 30 % (with 50 % moisture) of sugar cane on a wet-weight basis and it is used for steam generation for milling, heating, distilling and more recently for electricity co-generation, making an ethanol plant not only self-sufficient but an energy exporter.

After the beginning of the new millennium, several Brazilian distilleries began to invest in the acquisition of more efficient high-pressure boilers (steam operation pressure up to 80 bar), yielding 120 kWh/ton of bagasse. Electricity co-generation pulled up the already positive energy balance. According to Coelho et al. (2006), energy balance in ethanol industry can be higher than 10:1 (output:input). Leite et al. (2005) estimated 8:1. The main energy spent in sugarcane ethanol is due to agricultural features, mainly to the use of fertilizers and transportation. Regarding maize, Shapouri et al. (2008) concluded that energy balance for corn-based ethanol distilleries was 2.3:1 or even 2.8:1. These values are much higher than the balance shown by Pimentel (2003) who proposed that there were no energy gains to produce ethanol from corn, as the energy balance was 1:1. In addition, energy

balances for ethanol generated from lignocellulosics and sugar beet were 5.6:1 and 2:1, respectively (Elsayed et al, 2003).
Furthermore, as transport of loads in Brazil is predominantly done by road using low capacity trucks, this item pull up the energy spent. Improvements in this area would become ethanol from sugarcane even more feasible. It is also expected that the very high gravity (VHG) fermentation will make such figures even better, due to energy savings in distilling and mainly for vinasse transportation to the field, since this by-product will be generated in a more concentrated form and consequently in a lower volume. Therefore, energy balance in ethanol production by sugarcane is highly positive and much higher than ethanol generated from other sources.

### 3.2 Substrate for fermentation

Traditionally ethanol production was coupled with sugar industries in Brazil (Basso and Rosa, 2010). The cane sugar is pressed (some plants use diffusion), resulting in sugar cane juice and a solid fibrous residue, the cane bagasse. After clarification the juice is concentrated by evaporation till sucrose crystallization. The sucrose crystals are collected by centrifugation, generating a sucrose saturated viscous phase, called cane molasses with 45 to 60 % sucrose and 5 to 20 % glucose plus fructose. Initially, the production of ethanol was established as a way to process the resulting molasses from the sugar industry, but due to the increasing importance of ethanol in the 80's, many mills began as autonomous ethanol plants. Over time, additional investment enabled mills to direct the cane sugar to either row sugar or ethanol, and today only a few facilities were dedicated to produce exclusively ethanol and no sugar. Thereafter, the Brazilian industrial process for fuel ethanol production started to use sugar cane juice and molasses as substrates, mixed in different proportions. Indeed, some distilleries use only juice, while others only molasses, but the mixture is considered to be a better substrate, since the juice has some nutritional deficiencies, whereas molasses has inhibitory compounds for yeast fermentation.
The mineral composition of sugar cane substrates vary widely, depending on the molasses proportion used to formulate the media, sugar cane variety and maturity, soil, climate, and processing of cane juice (Table 1).

| Nutrient | Range concentration* (mg/L) | Optimum level* (mg/L) |
|---|---|---|
| Nitrogen ($NH_4^+$ and $R\text{-}NH_2$) | 70-350 | 100-300 |
| Phosphorus | 20-200 | 50-250 |
| Potassium | 300-12,000 | 700-1,300 |
| Magnesium | 80-3,900 | 100-200 |
| Sulphur | 80-3,900 | as low as possible |
| Calcium | 150-2,000 | as low as possible |
| Zinc | 0.45-9 | 1-5 |
| Copper | 0,20-8 | 1-5 |
| Manganese | 2-8 | 1-5 |
| Aluminium** | 2-500 | <10 (in juice substrate) |

Table 1. Mineral composition of sugar cane based substrates and the recommended levels for yeast fermentation (Amorim and Leão, 2005). *Concentration of the element; **Not a nutrient, but a toxicant.

### 3.3 The fermentation

The process operates in fed-batch (75 % of the distilleries), also mentioned as Melle-Boinot process, or in continuous mode, both utilizing yeast cell recycling. In both process, after the end of fermentation, yeast cells are collected by centrifuging and re-used in a next fermentation cycle. Up to 90-95 % of the yeast cells are recycled, resulting in high cell densities inside the fermentor [10 to 14 % (wet weight basis/v)]. Cell reuse reduces the need for intensive yeast propagation, and less sugar is deviated for biomass formation. It is estimated that yeast biomass increases 5 to 10 % (in relation to initial biomass) during a fermentation cycle, which is enough to replace the yeast cells lost during the centrifugation step. This high yeast biomass inside the fermentor is responsible for a very short fermentation time of 6 to 10 h, when compared to 40 to 50 h in corn fermentation process. Normally, temperature is kept around 32 to 35 °C, yet due to the short fermentation time, cooling is not always efficient enough in removing heat, and temperature may reach up to 40 °C, especially in the summer season (Lima et al., 2001; Laluce, 1991; Wheals et al., 1999; Lima et al., 2001; Amorim et al, 2004; Andrietta et al., 2002).

Generally, fermentation starts by adding cane must (prepared by mixing cane juice and molasses at any proportion), which contains 18 to 22 % (w/w) total reducing sugars (TRS), to a yeast cell suspension. This yeast suspension (with ca 30 % of yeast cell, on wet basis) represents 25 to 30 % of the total volume of fermentation, and is performed in tanks of 300 to 3,000 $m^3$. This large inoculum is normally prepared by mixing 2 to 12 ton of baker's yeast with 10 to 300 kg of selected strains in active dry yeast form. The feeding time normally lasts for 4 to 6 h and fermentation is finished within 6 to 10 h, resulting in ethanol titres of 8 to 12 % (v/v). When fermentation ceases, yeast cells are separated by centrifugation, resulting in a concentrated yeast cell suspension (the yeast "cream") with 60 to 70 % (wet weight basis/v) of cells. The yeast cream is diluted with water (1:1), and treated with sulphuric acid (pH 1.8-2.5) for 2 h, in order to reduce bacterial contamination and to be re-used as starter for a next fermentation cycle. This recycling trait makes the Brazilian process quite peculiar in that yeast cells are reused at least twice a day over the production season of 200-250 days. Tank stirring is also desirable at a low power, only to avoid cells compacting at the bottom of the tank and to keep a higher contact surface with substrate.

After centrifuging, the "beer" or "wine" is driven to distillation for ethanol recovery, normally using distillation tray technology. After distilling a liquid stream called stillage or "vinasse" is produced at the ratio of 10-15 litters per litter of produced ethanol and is delivered into the cane fields for use as irrigation water and fertiliser (adding to soil potassium, calcium, magnesium, others micronutrients and some organic matter).

When compared to the continuous mode, Melle-Boinot process (Figure 1) presented higher yields and productivity, lower levels of contamination and it was easier to clean (Zhang, 2009), but according to Andrietta et al. (2007) beneficial traits of a continuous version are masked by the improper engineering conception of low-cost adaptations of batch to continuous plants.

## 4. The impact of industrial conditions on yeast physiology and population dynamics

### 4.1 Industrial stresses

As seen before, industrial yeast strains encounter several simultaneous or sequential stress conditions imposed by the process itself, such as: high ethanol titters, high osmotic pressure, low pH, high temperature, and many others, all of them intensified by the practice of cell recycling.

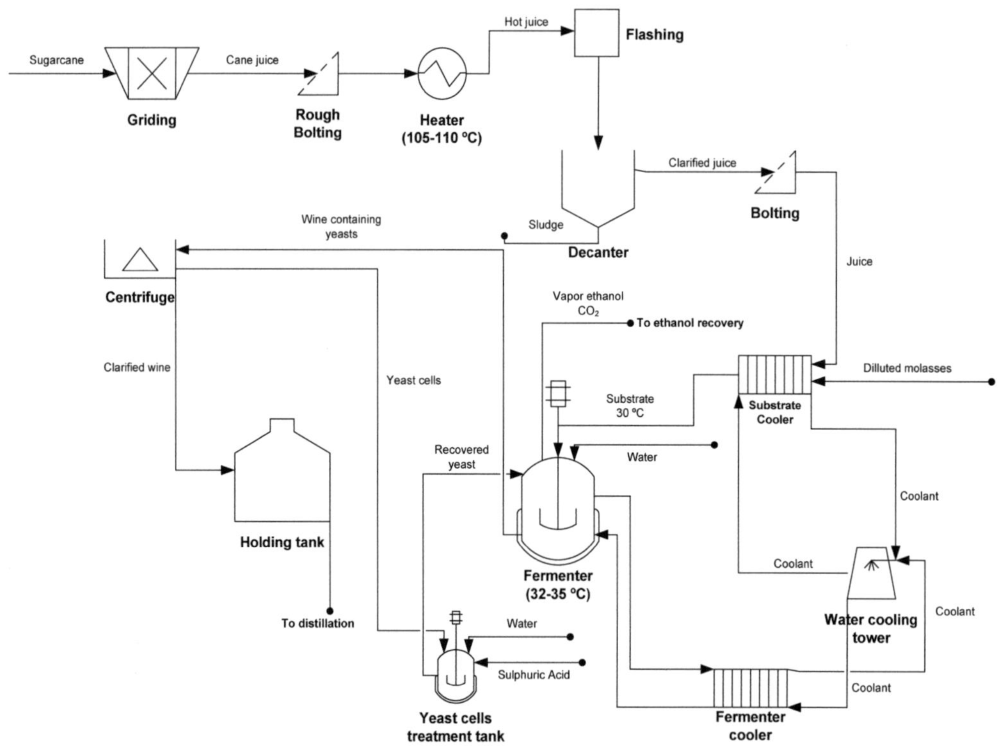


Fig. 1. Scheme of the fed-batch process with cell recycling (Melle-Boinot).

### 4.1.1 Osmotic stress

Osmotic stress is frequently mentioned in different industrial processes. In sugar cane-based fuel ethanol fermentation, yeasts are often exposed to high sugar concentrations, since a final ethanol content of 8-12 % (v/v) is very common to be achieved. In theory, the sugar concentration needed to result in ethanol titers of 12 % will be around 20 % (w/v) of TRS, considering an ethanol yield of 90 %. However, yeasts never face such sugar concentrations during industrial fermentations. This is because during the fed-batch process, with 4-6 h of feeding time, total sugar concentrations hardly exceeds 5-7 % (w/w). Indeed, in some occasions, sugar consumption rate by yeast cells equals the sugar-feeding rate, and very low levels of sugars [< 2 % (w/v)] are attained during fermentation. Therefore, one can deduce that the osmotic stress imposed by sugars can be practically neglected, especially in the case of fuel ethanol production performed in fed-batch mode.

On the other hand, the osmotic stress caused by salts, which are present in great amounts in sugar cane molasses, is a matter of concern. High levels of potassium, calcium and magnesium found in this substrate, all exceed by far the requirements for yeast nutrition. Average levels of potassium (4,000 mg/L) are high enough to induce stress responses in fermenting yeasts, increasing glycerol formation, reducing yeast storage carbohydrates and lowering ethanol yield (Alves, 2000).

It is often observed an increased formation of glycerol during osmotic stress conditions (Brumm & Hebeda, 1988; Myers et al, 1997), as well as during other stress situations, such as bacterial contamination (Alves, 1994; Gomes, 2009). This suggests that increased formation of this metabolite might be an indicative of general stress conditions during industrial fermentations. Glycerol is the major by-product during fermentation, mainly formed as a result of re-oxidation reactions to consume the excess NADH formed during biosynthesis under anaerobic conditions (van Dijken e Scheffers, 1986). As a result, during industrial fermentations 5 to 8 % of the sugars consumed by yeast cells are diverted into glycerol (Oura, 1977). Lowering the amount of this polyol is considered a promising route to increase ethanol yield. This could be achieved by adjusting the feeding rate during the fed-batch process, or by selecting strains that produce less glycerol. Both approaches are considered promising routes to increase ethanol yield, and these have been observed both in laboratory conditions as well as in the industrial practice.

### 4.1.2 Ethanol stress

In view of the high titers of ethanol towards the end of each fermentation cycle [8-12 % (v/v)], this alcohol is one of the major stress factors that act upon yeast. The inhibitory role of ethanol upon *S. cerevisiae* is not fully understood until now. Even so, the main target of ethanol is considered to be the cytoplasmatic membrane of yeast cells (Thomas et al., 1978; Alexandre et al., 2001). The membrane fluidity, which is related to its lipid composition, is profoundly altered in the presence of ethanol, and, as a result, membrane permeability to some ions (especially ions $H^+$) is significantly affected. As ions enters the cell, there is a dissipation of the electro-chemical gradient across the membrane, which in turns affects formation and maintenance of the proton driving force with subsequent decrease in intracellular pH. Apart from affecting yeast membrane composition, there are several other effects of ethanol upon yeast physiology during fermentation, including growth inhibition, and enzymatic inactivation, which leads to a decreased cell viability.

Cell re-use (yeast recycling) imposes a harsh condition towards industrial strains. Yeast cells must keep high viability at the end of each fermentation cycle in order to be able to cope with the following one. This is why a given strain can perform well in one fermentation cycle with a final ethanol content of 18 % (v/v), but can not be recycled for subsequent fermentation cycles. Indeed, if yeast physiological condition, namely the vitality of the cells at the end of the fermentation, is not a matter of concern, higher ethanol contents could easily be achieved. This is observed in some corn and cereal-based fermentation processes, where 17 to 23 % (v/v) of ethanol titers are obtained when wheat and formulated mashes are used (Jones et al., 1994; Bayrock and Ingledew, 2001).

Despite these problems, high ethanol fermentations are desirable in order to reduce water consumption and energy costs during the distillation step. It is also expected that this fermentation condition will favour the energy balance of the produced ethanol and improve the sustainability of the industrial process. However, in most distilleries the final ethanol content is limited by the ethanol tolerance of the fermenting strain. Additionally, ethanol stress can be intensified by high temperature and acidity, and all these stressing factors are imposed simultaneously or sequentially towards yeasts in industrial fermentations (Dorta et al., 2006).

A great research effort is now being conducted to search for ethanol tolerant yeast strains, in order to be used in high ethanol fermentations, specifically for the case of cell recycling.

The genetic basis for ethanol tolerance is polygenic and complex. More than 250 genes seem to be involved with this trait (Alexandre et al., 2001; Hu et al., 2007). The majority of these genes are related to energy metabolism, lipid synthesis, ionic homeostasis, trehalose synthesis, etc. This means that yeast improvement for ethanol tolerance through a rational approach is a difficult task, and a more "blind" approach (such as genome shuffling, mutagenesis, directed evolution) could be helpful in the search for such strains (Stephanopoulos, 2002; Çakar et al., 2005; Giudici et al., 2005).

#### 4.1.3 Acid stress

Although it is well known that yeasts generally can tolerate low pH, the sulphuric acid treatment of "yeast cream" (pH 1.8 to 2.5 during a couple of hours), in order to reduce bacterial contamination, causes physiological disturbances in yeast cells. This is observed by minerals (N, P, K, Mg) leakage and a decreasing level of yeast cellular trehalose in parallel with cell viability drop (Ferreira et al. 1999). Yeast strains that tolerate the stressing conditions of industrial fermentations normally present higher trehalose levels (Basso et al., 2008). Undoubtedly, undissociated organic weak acids present in the substrate or produced by contaminating bacteria can lower intracellular pH followed by ion translocation across the yeast plasma membrane. These events are intensified by the low pH of the fermentation medium. Residual levels of sulphite ($SO_2$), used for cane juice clarification, can be found particularly in molasses substrates. Although considered a toxicant towards yeast at a level of 200 $mg.L^{-1}$, sulphite can be considered beneficial to the fermentation, once it can reduce bacterial contamination (Alves, 1994).

#### 4.1.4 Other stresses

The presence of toxic levels of aluminium in cane based industrial substrates is also responsible for decreasing fermentation performance. Due to the acidic condition of fermentation, aluminium (absorbed by sugarcane in acid soils) is mainly present as its toxic form ($Al^{3+}$), leading to serious problems during fermentation. It negatively affects yeast viability, cellular trehalose levels, fermentation rate, and ethanol yield. The toxic effects of aluminium can be partially alleviated by magnesium ions and completely abolished in a molasses rich medium, suggesting the presence of chelating compounds in this substrate (Basso et al., 2004). Industrial yeast strains differ greatly regarding aluminium tolerance, in terms of cell viability, ethanol yield and aluminium cell accumulation. *S. cerevisiae* CAT-1 is less sensitive in comparison to PE-2 and commercial baker's yeast. In industrial practice, high levels of aluminium are associated with low cell viability and reduced ethanol yield.

Very low levels of cadmium were found in sugar cane from an area fertilized with treated municipal sewage sludge (Silva et al. 2010). Even so, the cell recycling and the yeast capacity of cadmium accumulation led to a toxic level of this metal to yeast fermentation. Yeast cells presented low viability, reduced sugar uptake, reduced trehalose levels and low ethanol yield. Vinasse (the effluent generated after ethanol distillation), when used as a substrate supplement, showed to be efficient in reducing the toxic action of aluminium (Oliveira et al., 2009) and cadmium (Mariano-da-Silva, 2001; Mariano-da-Silva and Basso, 2004) towards yeast. It is believed that chelating compounds present in vinasse were responsible to mitigate the toxic effects of heavy metals.

Several attempts were done in order to select thermo-tolerant strains for industrial fermentation (Laluce, 1991). On the other hand, high temperature is known to intensify ethanol and acidic stresses upon yeasts. As bacterial contamination is strongly stimulated by temperature above 32 °C, it is argued whether higher temperatures would be feasible. Nowadays, due to the advantages of the use of very high gravity (VHG) fermentations, lower temperature (27 °C) is appointed as a way to reduce the alcohol toxicity. All these suggestions are still to be evaluated in industrial scale.

Other stressors, such as herbicides used in sugar cane fields, phenols found in sugar cane juice, excessive amounts of iron in molasses, and others, are likely to be affecting yeasts during industrial fermentations, but their effects are still to be demonstrated. Additionally, the drastic changes in physiological conditions during fermentation as the rate of feeding, acid treatment and the frequent stoppage, all contribute to severely affect yeast fermentative performance.

### 4.1.5 Bacterial contamination

In view of the nature of the industrial process and its large volumes of processed substrates, aseptic conditions are very difficult to be achieved, and fermentation normally operates with bacterial contamination. This is often regarded as a major drawback during industrial ethanol fermentation. Besides deviating feedstock sugars from ethanol formation, there are also detrimental effects of bacterial metabolites (such as lactic and acetic acids) upon yeast fermentative performance, resulting in reduced ethanol yield, yeast cell flocculation, foam formation and low yeast viability (Yokoya et al., 1997; Narendranath et al., 1997; Bayrock & Ingledew, 2004; Eggleston et al. 2007). Induced yeast flocculation impairs the centrifuge efficiency and reduces the contact surface between yeast and the medium. Excessive foam formation increases costs due to the use of more antifoam chemicals in order to reduce headspace. The antibiotics, used to decrease contamination, increase costs and their residual levels make dried yeast (a by-product of ethanol industry) improper for commercialization (for human consumption or animal feed).

Most of the bacterial contaminants during the fermentative step of ethanol production are lactic acid bacteria (LAB), probably because of their higher ability to cope with the low pH values and high ethanol concentrations compared to others microorganisms (Kandler, 1983; Skinner and Leathers, 2004).

In a survey that investigated the identity of these contaminants during industrial fermentation in ethanol plants (distilleries) located in Brazil, Gallo et al. (1990) found that *Lactobacillus* was the most abundant isolated genera. Recently, Lucena et al. (2010) showed that LAB are the most common contaminants in distilleries located at the Northeast region of Brazil.

LAB are traditionally classified in two metabolic sub-groups according to the pathway used to metabolise hexose sugars: homo- and hetero-fermentative (Kandler, 1983), and bacterial isolates from industrial fermented sugarcane substrates, encompasses both homo- and hetero-fermentative *Lactobacilli* (Costa et al., 2008). It was also found that such contaminating bacteria produce both L(+)- and D(-)-lactic acid optical isomers. Some strains produce only the L(+) form, while others only the D(-) form, but the majority of the strains produce different proportions of both isomers.

As lactic acid titer is an indicator of bacterial contamination level, and the enzymatic kit used for this purpose only detects the L(+) isomer, one can expect that the negative impact of contamination is underestimated, since the D(-) isomer is not computed (Costa et al., 2008).

### 4.2 The importance of yeast glycogen and trehalose during stress conditions

As said before, yeast viability is of great importance in a Melle-Boinot-based process, since cells will be re-used in subsequent fermentation cycles.

Yeast cells, when fermenting in stressful conditions, usually show drops in cell viability, increased glycerol formation, lower biomass yield, and also diminished levels of storage carbohydrates, such as glycogen and trehalose. All these parameters are extremely helpful indicators that can be used to select new tolerant strains. Hence, glycogen and trehalose [the two major storage carbohydrates in *S. cerevisiae*, accounting to up 25 % in yeast biomass (w/w dry weight)] have been involved with tolerance towards several stresses (Attfield, 1997; D'Amore et al, 1991; Parrou et al., 1997; Singer & Lindquist, 1998).

During fed-batch fermentation, trehalose and glycogen exhibit great variations. These two compounds are initially degraded in response to increased sugar levels in the medium, and restored when sugar levels decline towards the end of the fermentation (Alves, 2000). The levels of these reserves are higher at the end when compared to the beginning of the fermentation. Therefore, a significant amount of sugars are stored as glycogen and trehaloseat the expense of ethanol production. Fortunately, high levels of carbohydrate reserves are of paramount importance for yeasts to withstand the stressful acid washing treatment, imposed by the industrial process. During treatment, part of the glycogen and trehalose are dissimilated through glycolysis, and ethanol is produced during the acid treatment with no sugar in the medium (Ferreira et al, 1999). At the end of the acid treatment, the levels of these reserves are adequate to guarantee high viability. It is believed that the high tolerance of industrial strains is partially explained by their higher levels of glycogen and trehalose when compared with less persistent strains.

### 4.3 The starter strain and the competition with indigenous *Saccharomyces* during fermentation

Molecular techniques, such as karyotyping and PCR-fingerprinting, have been used to monitor yeast population dynamics during industrial fermentations.

At the beginning of the 1990's, Basso et al. (1993) verified - for the first time - that some traditionally starter strains (baker's yeast and other two *S. cerevisiae* strains TA and NF), were all replaced by indigenous (contaminating) yeasts in a period of up to 40 days of recycling. This study was performed in 5 distilleries at Sao Paulo State (Southeast region of Brazil) during two crop seasons (1991-1993). In fact, a succession of different indigenous *S. cerevisiae* strains was detected all over the fermentation seasons in the majority of distilleries. Although starter strains were not able to persist during the cell recycling, it was shown that a wild strain (JA-1), formerly isolated from one of the distilleries, was capable to dominate when re-introduced in the process. This was an important indicative for selecting strains from the great biodiversity found in distilleries. Lately, it was verified in a universe of up to 70 distilleries, that baker's yeast were rapidly replaced by wild strains in a very short period (from 20 to 60 days) of cell recycling (Basso et al., 2008). During this study, it was also demonstrated that no strains (from brewing, wining, distilling) other than those isolated from the industrial process had the capability of being implanted in the distilleries investigated.

Da Silva et al. (2005b) employed PCR-fingerprinting as a method to illustrate yeast population dynamics during industrial fermentation. It was again showed that indigenous strains replaced commercial starter strains during the recycling process. Using this technique coupled with physiological assays it was possible to isolate an indigenous strain (JP1) to be used as starter culture for Northeast distilleries (Da Silva et al., 2005a).

A varying proportion of non-*Saccharomyces* strains contaminates the fermentation step. It was found that *Dekkera bruxellensis, Candida tropicalis, Pichia galeiformis, Schizosaccharomyces pombe* and *C. krusei* are major contaminants in acute contamination episodes, being responsible for decreased ethanol yields. Only few distilleries (< 5 %) located at Central and South-eastern regions of Brazil (responsible for the majority of the ethanol production) exhibits contamination by non-*Saccharomyces* yeasts (Basso et al., 2008). On the other hand, ca. 30 % of distilleries located at the Northeast region of the country suffer with this special contamination (Basillo et al., 2008).

### 4.4 Selection of suitable strains from the biodiversity found in distilleries

By monitoring yeast population dynamics during industrial fermentations, dominant and persistent indigenous strains have been selected and employed as starter strains.

One example (mentioned above) is the case of strain JP1. It is a highly dominant strain that displays stress tolerance towards acidic pH, high ethanol levels and high temperature. Moreover, it presents fermentative performance similar to other commercial industrial strains. This strain was re-introduced as starter in one of the distilleries it was isolated, and the population dynamics of this plant was followed on a monthly-basis for two consecutive production seasons. It was found that the strain was able to dominate the yeast population while conferring high ethanol yields (> 90%) during this period (Da Silva et al., 2005b).

The most extensive study on yeast population dynamics was performed by Basso et al. (2008), with a monthly yeast sampling scheme from a universe of 70 distilleries, responsible for most of the produced Brazilian ethanol and covering a period of 12 years. During this period, strains with prevalence (which dominate the yeast population in fermentation) and persistence (present along a 200-day season) were identified by karyotyping, isolated and screened in laboratory trails for desirable fermentation traits (no flocculation, low foam formation, high ethanol yield, low glycerol formation, high viability during recycling and high cellular glycogen and trehalose content). As soon a promising strain was identified, it was propagated in laboratory and re-introduced in many distilleries (up to 54, depending on the season) and followed by karyotyping. Along the study, up to 14 selected strains were re-introduced in industrial processes. Some of them (as PE-2) were followed during 10 years in different distilleries, comprising different regions, different processes (fed-batch or continuous), different substrates (molasses/juice), and thus encompassing great variations.

Most of the evaluated strains assessed were not able to permanently perpetuate among yeast population. On the other hand, some strains were able to dominate the fermentation population for a couple of seasons. Probably, variations in process conditions, climate, substrate, etc would be responsible by the lowered capacity of implantation of these strains.

Few strains were able to persist in many distilleries for many seasons. PE-2 and CAT-1 strains showed the highest implantation capabilities. These strains could be implanted in 51 to 58 % of the distilleries where they were introduced. They also presented higher competitiveness in relation to contaminating strains, representing an average of 45 to 54 % of yeast biomass during the fermentation season. In some distilleries, these strains represented the total biomass of the reactor during the whole season (more than 200 days of recycling). Due to these remarkable traits for an industrial fermentation, CAT-1 and PE-2 are the most used starter strains, representing today 80 % of the commercialized active dry yeast for fuel ethanol in Brazil. They are annually used in more than 200 distilleries, responsible for 60 % of the total country's ethanol production.

Even though selecting indigenous strains is considered an attractive strategy to guarantee high product yields and population homogeneity during industrial fermentations, this approach is laborious and not always successful. As appointed by Basso et al. (2008), after a 12 years yeast selection program, among 14 selected strains exhibiting high performance in laboratory trails, only few presented high implantation capability when re-introduced to the industrial process.

Probably, laboratory-screening procedures do not simulate all the stressing agents faced by the yeast in industrial-scale process. Other important factor could be the variations that occur in industrial fermentations from one season to another. It is worthwhile to point that each distillery has peculiar process features, imposing different stresses with different magnitudes to the fermenting yeast. This may explain that a given strain performs well only in a few distilleries (sometimes in only one), and just for one or a couple of seasons.

In conclusion, it can be said that any selected strain must be evaluated in industrial conditions in as many as possible distilleries for many seasons. The most important and rare attribute of superior yeast starters is the implantation capability, since good fermentation profile (at laboratory screening experiments) is more easily found among indigenous and even laboratory strains.

## 5. Future trends and concluding remarks

It can be foreseen that bioethanol production is expected increase continuously in Brazil, due to the growing investments in this field. New distilleries are being built, resulting in almost 400 operating plants all over the country (MAPA, 2011). They will be producing near 30 billion liters of ethanol in the next season. Although bioethanol production in Brazil is considered a mature process, there is plenty room for improvements.

The current broad interest for the use of very high gravity (VHG) fermentations in the industrial scenario is mainly focused in reducing production costs. It is also expected that this technology will bring benefits to the overall environmental sustainability of the process by decreasing water and energy consumption. Thus, it continues being possible to increase the efficiency of first generation fuel ethanol process by embracing this technology. Fermentations resulting in high ethanol titers would not only benefit the energy balance, but would also result in a significant reduction of vinasse volume. There will be a great economical and environmental impact due to reduction costs of vinasse transportation and application as fertilizer in sugar cane plantations. Besides that, higher ethanol levels would repress bacterial growth during fermentation and for that reason, decrease antibiotics usage for controlling such contaminations.

Despite all these advantages, the implementation of VHG technology in the Brazilian industrial scenario is limited by the availability of very high ethanol tolerant strains. It is expected that for VHG fermentations, substrates will be formulated with more molasses, where sugars are in a concentrated form. As a result, apart from enduring ethanol stress, yeasts will be subjected to others stresses from molasses.

The high international sugar price is leading sugar industry to prioritize sucrose production and more molasses will be generated. Molasses will also be more exhausted (*i.e.* possessing a lower sugar concentration in relation to total solids) and undoubtedly will exert a more pronounced toxic effect upon yeast fermentation. The new yeast strains must also be able to cope with high molasses content substrates, even operating with normal ethanol titers.

Sugar cane bagasse is considered a promising feedstock for second generation ethanol. Nevertheless, the produced ethanol from this lignocellulosic by-product must prove to be more advantageous than the use of bagasse as fuel for steam generation used for milling, heating, distilling, and electricity co-generation.

## 6. References

Alexandre H, Ansanay-Galeote V, Dequin S, Blondin B. (2001). Global gene expression during short-term ethanol stress in *Saccharomyces cerevisiae*. FEBS Letters 498(1):98-103.

Alves, D.M.G. (1994). Fatores que afetam a formação dos ácidos orgânicos, bem como outros parâmetros da fermentação alcoólica. Dissertação (mestrado), Escola Sup. Agric. Luiz de Queiroz/USP, Piracicaba, 251p.

Alves, D.M.G. (2000). Respostas fisiológicas de duas linhagens de *Saccharomyces cerevisiae* frente ao potássio durante a fermentação alcoólica. Tese (doutorado), Instituto de Biociências/UNESP, Rio Claro, 118p.

Amorim, H.V.; Basso, L.C. e Lopes, M.L. (2004). Evolution of ethanol production in Brazil, in Distiled Spirits - Tradition and Innovation, ed. Bryce, J.H. e Stewart, G;G., Nottingham Univ. Press (UK), v. 1, n. 1, p. 143-148.

Amorim, HV e Leão, RM. (2005) Fermentação Alcoólica - ciência e tecnologia. Fermentec, Piracicaba, 448p, ISBN 8599011-0104.

Amorim, HV, Basso, LC e Lopes, ML. (2009). Sugar cane juice and molasses, beet molasses and sweet sorghum: composition and usage, p 39-46. In The Alcohol Text Book (5th edition), Ingledew,WM, Kelsall, DR, Austin, GD and Kluhspies, C (ed.), Nottingham University Press, Nottingham, 541p.

Andrietta, MGS, Andrietta, SR, Steckelberg, C and Stupiello, ENA. (2007). Bioethanol - Brazil, 30 years of Proalcool. International Sugar Journal, v 109, p 195-200.

Andrietta, MGS, Andrietta, SR, Steckelberg, C, Stupiello, ENA. (2002). Bioethanol - 30 years after Proalcool. International Sugar Journal, 109 (1299), p. 195-200.

Attfield PV. (1997). Stress tolerance: The key to effective strains of industrial baker's yeast. Nature Biotechnology 15(13):1351-1357.

Basillo ACM, de Araujo PRL, de Morais JOF, de Silva EA, de Morais MA, Simoes DA. (2008). Detection and identification of wild yeast contaminants of the industrial fuel ethanol fermentation process. Current Microbiology 56(4):322-326.

Basso LC, de Amorim HV, de Oliveira AJ, Lopes ML. (2008). Yeast selection for fuel ethanol production in Brazil. FEMS Yeast Research 8(7):1155-1163.

Basso LC, Oliveira AJ, Orelli VFDM, Campos AA, Gallo CR, Amorim HV. (1993). Dominância das leveduras contaminantes sobre as linhagens industriais avaliada pela técnica da cariotipagem. Congresso Nacional da STAB, 5, Águas de São Pedro, 1993. Anais, vol. 1, p. 246-250.

Basso LC and Rosa CA. (2010). Sugar cane for potable and fuel ethanol. *In*. Worldwide Distilled Spirits Conference - New Horizons: energy, environment and enlightenment. Walker GM and Hughes PS (eds). Nottingham: Nottingham University Press, p.1-7.

Bayrock DP, Ingledew WM. (2001). Application of multistage continuous fermentation for production of fuel alcohol by very-high-gravity fermentation technology. Journal of Industrial Microbiology & Biotechnology 27(2):87-93.

Bayrock DP, Ingledew WM. (2004). Inhibition of yeast by lactic acid bacteria in continuous culture: nutrient depletion and/or acid toxicity? Journal of Industrial Microbiology & Biotechnology 31(8):362-368.

Bertalan M, et al. (2009). Complete genome sequence of the sugarcane nitrogen-fixing endophyte *Gluconacetobacter diazotrophicus* pal5. BMC Genomics. Vol. 10. pp.450-461. doi: 10.1186/1471-2164-10-450.

Boddey RM, Oliveira OC, Urquiaga S, Reis VM, Olivares FL, Baldani VLD, Döbereiner J. (1995). Biological nitrogen fixation associated with sugar cane and rice: Contributions and prospects for improvement. Plant and Soil 174 (1-2) 195-209.

Brumm PJ, Hebeda RE. (1988). Glycerol Production in Industrial Alcohol Fermentations. Biotechnology Letters 10(9):677-682.

Burnquist, H.L. (2007). Porque o Brasil deve apoiar a internacionalização do etanol? Revista Opiniões [sobre o setor sucroenergético], Jan-Mar 2007, p.22-23.

Cakar ZP, Seker UOS, Tamerler C, Sonderegger M, Sauer U. (2005). Evolutionary engineering of multiple-stress resistant Saccharomyces cerevisiae. FEMS Yeast Research 5(6-7):569-578.

Coelho ST, Goldemberg J, Lucon O, Guardabassi P (2006) Brazilian sugarcane ethanol: Lessons learned. Energy for Sustainable Development. Vol. 10, No. 2. pp. 26-39

Costa VM, Basso TO, Angeloni LHP, Oetterer M, Basso LC. (2008). Production of acetic acid, ethanol and optical isomers of lactic acid by *Lactobacillus* strains isolated from industrial ethanol fermentations. Ciencia E Agrotecnologia 32(2):503-509.

da Silva EA, de Melo WF, Antunes DF, dos Santos SKB, Resende AD, Simoes DA, de Morais MA. (2005a) Isolation by genetic and physiological characteristics of a fuel-ethanol fermentative *Saccharomyces cerevisiae* strain with potential for genetic manipulation. Journal of Industrial Microbiology & Biotechnology 32(10):481-486.

da Silva EA, dos Santos SKB, Resende AD, de Morais JOF, de Morais MA, Simoes DA. (2005b) Yeast population dynamics of industrial fuel-ethanol fermentation process assessed by PCR-fingerprinting. Antonie Van Leeuwenhoek International Journal of General and Molecular Microbiology 88(1):13-23.

D'Amore T, Crumplen R, Stewart GG. (1991). The Involvement of Trehalose in Yeast Stress Tolerance. Journal of Industrial Microbiology 7(3):191-195.

Dien, BS and Bothast, RJ. (2009). A primer for lignocellulose biochemical conversion to fuel ethanol. *In:* The Alcohol Text Book (5th edition), Ingledew, WM, Kelsall, DR, Austin, GD and Kluhspies, C (ed.), Nottingham University Press, Nottingham, 541p.

Dorta C, Oliva-Neto P, de-Abreu-Neto MS, Nicolau-Junior N, Nagashima AI. (2006). Synergism among lactic acid, sulfite, pH and ethanol in alcoholic fermentation of *Saccharomyces cerevisiae* (PE-2 and M-26). World Journal of Microbiology & Biotechnology 22(2):177-182.

Eggleston G, Basso LC, de Amorim HV, Paulillo SCD, Basso TO. (2007). Mannitol as a sensitive indicator of sugarcane deterioration and bacterial contamination in fuel alcohol production. Zuckerindustrie 132(1):33-39.

Elsayed MA, Matthews, R, Mortimer, ND. (2003) Carbon and energy balances for a range of biofuels options. Accessed: Feb 20th 2011. Available from: http://www.rms.lv/bionett/Files/File/ BioDE-2003-101%20DTI%20biodiesel%20LCA%20study.pdf

Ferreira, LV, Amorim, HV and Basso, LC. (1999). Fermentação de trealose e glicogênio endógenos em *Saccharomyces cerevisiae.* Ciência e Tecnoogia de Alimentos (Campinas), v. 19 (10: 29-32.

Gallo, C.R (1990) Determinação da microbiota bacteriana de mosto e de dornas de fermentação alcoólica. PhD Thesis, University of Campinas, Campinas.

Giudici P, Solieri L, Pulvirenti AM, Cassanelli S (2005) Strategies and perspectives for genetic improvement of wine yeasts. Applied Microbiology and Biotechnology 66(6):622-628.

Gomes, F.S. (2009) Antagonismo entre leveduras e bactérias láticas na fermentação alcoólica (Dissertação de Mestrado), Escola Sup. De Agric. Luiz de Queiroz/USP, Piracicaba, 171p.

Hu XH, Wang MH, Tan T, Li JR, Yang H, Leach L, Zhang RM, Luo ZW. (2007). Genetic dissection of ethanol tolerance in the budding yeast *Saccharomyces cerevisiae.* Genetics 175(3):1479-1487.

James EK, Olivares, FL (1997) Infection and colonization of sugar cane and other graminaceous plants by endophytic diazotrophs. Critical Reviews in Plant Sciences 17 (1): 77-119.

Jones AM, Thomas KC, Ingledew WM. (1994) Ethanolic Fermentation of Blackstrap Molasses and Sugarcane Juice Using Very High Gravity Technology. Journal of Agricultural and Food Chemistry 42(5):1242-1246.

Kandler O. (1983). Carbohydrate metabolism in lactic acid bacteria. Antonie Van Leeuwenhoek Journal of Microbiology 49(3):209-224.

Laluce, C. (1991). Current aspects of fuel ethanol production in Brazil. Critical Reviews in Biotechnology, 11(2), p. 149-161.

Leite RCC (2005) Biomassa, a esperança verde para poucos. Access: Feb 23rd 2011. Available: http://www.agrisustentavel.com/san/biomassa.htm.

Liberal ATD, Basilio ACM, Resende AD, Brasileiro BTV, da Silva EA, de Morais JOF, Simoes DA, de Morais MA. (2007). Identification of *Dekkera bruxellensis* as a major contaminant yeast in continuous fuel ethanol fermentation. Journal of Applied Microbiology 102(2):538-547.

Lima, U.A.; Basso, L.C. e Amorim, H.V. (2001). Produção de etanol, in Biotecnologia Industrial: processos fermentativos e enzimáticos. Ed. Lima, U.A.; Aquarone, E., Borzani, W. e Schmidell, W., ed. Edgard Blucher, São Paulo, v.3, p.1-43.

Lingle SE, Wiegand CL (1997). Soil salinity and sugarcane juice quality. Field Crops Research. Vol. 54, No. 2-3. pp. 259-268.

Lucena BTL, dos Santos BM, Moreira JLS, Moreira APB, Nunes AC, Azevedo V, Miyoshi A, Thompson FL, Morais MA. (2010). Diversity of lactic acid bacteria of the bioethanol process. BMC Microbiology 10:298.

Mariano-da-Silva, S. (2001). Acúmulo de cádmio por *Saccharomyces cerevisiae* fermentando mosto de caldo de cana. Dissertação (Mestrado em Microbiologia Agrícola) - Escola Superior de Agricultura "Luiz de Queiroz", Universidade de São Paulo. 52p.

Mariano-da-Silva, M and Basso LC. (2004). Efeitos do cádmio sobre o crescimento das leveduras *Saccharomyces cerevisiae* PE-2 e IZ-1904, e a capacidade da vinhaça em atenuar a toxicidade. Ciência e Tecnologia de Alimentos (Campinas), v 24 (1), p 16-22,

MAPA, 2011. Ministério da Agricultura, Pecuária e Meio Ambiente. Produção Brasileira de Etanol. Disponível: http://www.agricultura.gov.br/portal/page/portal/Internet-MAPA/pagina-inicial/desenvolvimento-sustentavel/agroenergia. Acessado em 07.apr.2011.

Moreira JR (2000). Sugarcane for energy - recent results and progress in Brazil. Energy for Sustainable Development. Vol. 4, No.3, pp. 43-54

Muchow RC, Robertson MJ, Wood AW, Keating BA (1996). Effect of nitrogen on the time-course of sucrose accumulation in sugarcane. Field Crops Research. Vol. 47. No. 2-3. pp.143-153.

Myers DK, Lawlor DTM, Attfield PV. (1997). Influence of invertase activity and glycerol synthesis and retention on fermentation of media with a high sugar concentration by *Saccharomyces cerevisiae*. Applied and Environmental Microbiology 63(1):145-150.

Narendranath NV, Hynes SH, Thomas KC, Ingledew WM. (1997). Effects of lactobacilli on yeast-catalyzed ethanol fermentations. Applied and Environmental Microbiology 63(11):4158-4163.

Oliveira, RPS, Torres, BR, Silli, M, Marques, DAV, Basso, LC and Converti, A. (2009). Use of sugar cane vinasse to mitigate aluminum toxicity to *Saccharomyces cerevisiae*. Archives of Environmental Contamination and Toxicology, v. 57(3): 488-494.

Oura E. (1977). Reaction-Products of Yeast Fermentations. Process Biochemistry 12(3):19.

Parrou, J.L., Teste, M.A. e Francois, J. (1997). Effects of various types of stress on the metabolism of reserve carbohydrates in *Saccharomyces cerevisiae*: genetic evidence for a stress-induced recycling of glycogen and trehalose. Microbiology, 143:1891-1900.

Peres JG, Souza CF, Lavorenti NA (2010). Avaliação dos efeitos da cobertura de palha de cana-de-açúcar na umidade e na perda de água do solo. Engenharia Agrícola. Vol. 30. pp. 875-886.

Pilgrim, C. Status of the worldwide fuel alcohol industry. (2009) In: The Alcohol Text Book (5th edition), Ingledew,WM, Kelsall, DR, Austin, GD and Kluhspies, C (ed.), Nottingham University Press, Nottingham, 541p.

Pimentel D (2003). Ethanol fuels: Energy balance, economics, and environmental impacts are negative. Natural Resources Research. No. 12, Vol. 2. pp. 127-134.

RFA (2011). Renewable Fuels Association. *In*: Statistics. Access Feb 17th 2011. Available from: http://www.ethanolrfa.org/pages/statistics

Sandalow, DB (2006). Ethanol lessons from Brazil. Access: Feb 11th 2011. Available from: http://www.brookings.edu/~/media/Files/rc/articles/2006/05energy_sandalow/sandalow_20060522.pdf

Shapouri H, Gallagher PW, Nefstead W, Schwartz R, Noe S, Conway R (2008). 2008 Energy Balance for the Corn-Ethanol Industry. Accessed Feb 27th 2011. Available from: http://www.usda.gov/oce/reports/energy/2008Ethanol_June_final.pdf.

Silva, FC, Boaretto, AE, Abreu-Jr, CH, Berton, RS, Basso, LC e Barbieri, V. (2010). Impactos da aplicação de lodo de esgoto na cultura da cana-de-açúcar e no ambiente. Holos Environment, v 10, n. 1 p 62-82.

Silva, SMG.(2001). Efeitos do cádmio sobre a fermentação alcoólica e o uso da vinhaça para atenuar a sua ação tóxica. Piracicaba. 142 p Tese (Doutorado), Escola Superior de Agricultura Luiz de Queiroz, Universidade de São Paulo.

Singer MA, Lindquist S. (1998). Thermotolerance in *Saccharomyces cerevisiae*: the Yin and Yang of trehalose. Trends in Biotechnology 16(11):460-468.

Skinner KA, Leathers TD. (2004). Bacterial contaminants of fuel ethanol production. Journal of Industrial Microbiology & Biotechnology 31(9):401-408.

Stephanopoulos G. (2002). Metabolic engineering by genome shuffling - Two reports on whole-genome shuffling demonstrate the application of combinatorial methods for phenotypic improvement in bacteria. Nature Biotechnology 20(7):666-668.

Thomas DS, Hossack JA, Rose AH. (1978). Plasma membrane lipid composition and ethanol tolerance in *Saccharomyces cerevisiae*. Archives of Microbiology 117(3):239-245.

Vandijken JP, Scheffers WA. (1986). Redox Balances in the Metabolism of Sugars by Yeasts. FEMS Microbiology Reviews 32(3-4):199-224.

Wheals AE, Basso LC, Alves DMG, Amorim HV. (1999). Fuel ethanol after 25 years. Trends in Biotechnology 17(12):482-487.

Yokoya F, Manfio GP & Variane SF (1997). Curso de treinamento: bactérias lácticas na fermentação alcoólica, Fundação Tropical de Pesquisas e Tecnologia "André Tosello", Campinas.

Zhang, Z. (2009) Batch fermentation and fermentor design. In The Alcohol Text Book (5th edition), Ingledew,WM, Kelsall, DR, Austin, GD and Kluhspies, C (ed.), Nottingham University Press, Nottingham, 541p.

# 6

# Ethanol Reforming in the Dynamic Plasma - Liquid Systems

Valeriy Ya. Chernyak[1] et al.*
*Taras Shevchenko National University of Kyiv*
*Ukraine*

## 1. Introduction

Today hydrogen ($H_2$) is considered as one of the most perspective energy sources for the future that can be renewable, ecologically clean and environmentally safe. The demand for hydrogen energy has increased tremendously in recent years essentially because of the increase in the world energy consumption as well as the recent developments in fuel cell technologies. The Energy Information Administration has projected that world energy consumption will increase by 59% over the next two decades, and the largest share will still be dominated by fossil fuels (EIA, 2011).

The interest to alternative fuels research in the last two decades is increased by the depletion of the traditional fossil fuels. Today, ethanol is considered the most perspective fuel for internal-combustion engines (Kakami, 2010).

Alcohols are especially appealing as primary fuels for fuel processors because they can be obtained from renewable biomass: methanol trough gasification and synthesis, and ethanol trough fermentation. Ethanol is easier and safer to store and transport due to its low toxicity and volatility, it is biodegradable, and since water is also consumed during its conversion into hydrogen, there is no need for absolute ethanol to be produced as it would be required if it were to be used in conventional engines, either alone or mixed with gasoline.

Among possible technologies for $H_2$ production, including steam reforming and partial oxidation of liquid hydrocarbons, the low-temperature plasma reforming of biomass-derived ethanol (ethyl alcohol $C_2H_5OH$) is believed to be a good alternative approach (Bromberg, 2006). There are various electric-discharge techniques of plasma conversion of ethanol into $H_2$ using thermal (equilibrium) and non-thermal (non-equilibrium) plasmas: arc, corona, spark, MW, RF, DBD, etc. (Matveev, 2007; Petitpas, 2007). Each plasma system has its merits and demerits, and even difficult to compare. Among them, one of the most efficient is the plasma processing in the dynamic plasma-liquid systems (PLS) using the DC and pulsed electric discharges in a gas channel with liquid wall (DGCLW) and the DC discharge in a reverse vortex gas flow of Tornado type with a "liquid" electrode (TORNADO-LE). Advantages of this technology are high chemical activity of plasma and

---

* Eugen V. Martysh[1], Sergei V. Olszewski[1], Vitalij V. Yukhymenko[1], Sergei M. Sidoruk[1], Oleg A. Nedybaliuk[1], Iryna V. Prysiazhnevych[1], Anatolij I. Shchedrin[2] and Dmitry S. Levko[2]
[1]*Taras Shevchenko National University of Kyiv*, [2]*Institute of Physics, National Academy of Sciences, Ukraine*

selectivity of plasma-chemical transformations, providing high-enough productivity and efficiency of conversion at a relatively low electric power consumption on the high-voltage discharging in a flow at atmospheric pressure.

The non-equilibrium plasma assists as an energetic catalyst containing charged particles and electronically excited atoms and radicals, which initiate fast chain-branching conversion of hydrocarbons that does not occur in usual conditions. The highly developed plasma-liquid interface with the large surface-to-volume ratio and the deep injection of active plasma particles into the liquid also favor to the intensification of conversion in the plasma-liquid system. At that, there is no problem with excess heat removal since such plasma system is thermally "cold". The main idea is that the discharge can burn directly in the liquid fuel without preliminary gasification. In this part we report new results of the experimental and theoretical studies of the process of plasma-assisted reforming of ethanol in the PLS with the DC and pulsed DGCLW and TORNADO-LE using available methods of diagnostics and numerical modeling.

In the present work the results of experimental and theoretical investigation of plasma kinetics in new plasma chemical reactor are discussed. This device uses the reverse vortex gas flow of "tornado" type similar to Fridman's group (Fridman, 2008). Discharge of such type operates in the transitional thermal to non-thermal plasma regime. It is characterized by a presence of electrons with average electron energy of few electron volts and a neutral gas temperature of ~ 1000-2000 K. High gas temperature increases hydrogen production as a result of the additional conversion of hydrocarbons generated in the discharge region. The presence of active O, OH and H in the mixture converts efficiently the hydrocarbons in the post-discharge region. The use of vortex air flow increases the stability of the discharge. An advantage of the proposed reactor over the one at (Prieto, 2001, Wang, 2004) is the use of a special working chamber. This chamber allows conversion of hydrocarbons either in the liquid or in the gas phase. In case of liquid, part of the input power goes into evaporation. The presence of liquid isolates the metallic electrode from the plasma region. It also prevents the electrode erosion and increases the working time of the reactor. The use of pyrolytic chamber allowed getting the energy efficiency value more than three times higher that such value for the similar plasma-liquid systems without pyrolytic chamber.

## 2. Equipment

Among possible types of electric discharges, which can produce non-thermal reacting plasma at high pressures, two specific cases are of research interest of scientists from Physic Electronics Department at the Taras Shevchenko National University of Kyiv. One source is the transverse arc in a blowing flow (BTA) that is an intermediate case of a high-voltage low-current self-sustained discharge with arc length adjusted by the transverse gas flow (Buchnev, 2000). It differs from the non-stationary gliding arc of Czernichowski type by the fixed arc length. It also has a convective cooling of the plasma by gas flow without conductive heat losses at the walls since it is a free arc jet. An intensive transverse gas ventilation of the BTA plasma increases its ionization nonequilibrium and non-isothermality (Chernyak, 2005). This factor has fundamental importance for plasma-chemical efficiency. Whereas the most of the discharge energy goes into the mean energy of electrons and not just to thermal heating, it gives desirable reactivity and selectivity of chemical transformations. BTA was tested successfully in the Kyiv University in different variants with the primary and secondary discharges for the plasma processing of various homo- and heterophase gas and liquid substances, including plasma-assisted fuel combustion (Yukhymenko, 2007). Another potential source of non-thermal plasma that can provide simultaneously a high level of non-equilibrium and high density of reacting species in the plasma-liquid system is the electric discharge in a flowing gas channel with liquid wall (DGCLW). Its main properties are (Chernyak, 2008):

i. large ratio of the surface of plasma-liquid contact to the plasma volume;
ii. possibility of controlling of plasma-created gas-phase and hetero-phase components, which specifies its potential opportunities during plasma-liquid processing;
iii. possibility of using non-classic liquids including colloidal solutions and mash (waste liquids with fine solid aggregates);
iv. possibility to realize both DC and AC modes of discharge (in contrast with quasi-stationary modes of diaphragm discharges and capillary discharges).

DGCLW has been initially realized and investigated in plasma treatment of organics in air-water, phenol and toluene systems]. For the purposes of fuel reforming, DGCLW has been proposed in the KNU for the first time at the end of 2006. The idea is that DGCLW can be burning directly within the liquid hydrocarbon fuels without preliminary gasification (Chernyak, 2007).

The following setups were prepared with the PLS reactors with the DGCLW working with one and two gas streams injected into the homogeneous work liquid as is shown in Fig.1a and Fig.1b. It consists of the cooper rod electrodes (*1*), plasma column (*2*), work liquid (*3*), electrode in liquid (*4*), and quarts tubes (*5*). The discharge channel in the work liquid was formed in two ways: with a constant gas flow and without it. As a work gas, the standard technical air of atmospheric pressure is used. Ethanol, water, and ethanol-water mixtures were used as work liquids. Various modes of the discharge operation were tested:

i. the mode where the voltage applied to the electrodes mounted into the lower and upper flanges and the discharge was initiated between them;
ii. the mode where "+" was applied to the electrode mounted into the lower flange, whereas "–" was applied to the liquid ("liquid" cathode);
iii. the mode where "–" was applied to the electrode mounted into the lower flange, while "+" was applied to the liquid ("liquid" anode).

Fig. 2 shows the dc DGCLW working in the mode I in water.

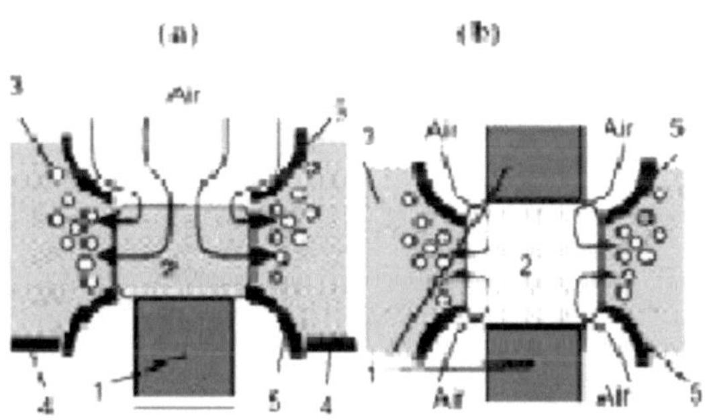


Fig. 1. Photo of discharge of air flow in water mode I with two solid electrodes.

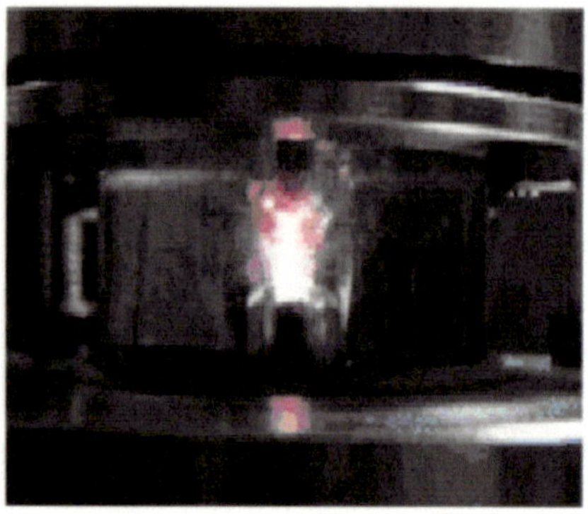

Fig. 2. Photo of discharge of air flow in water. Mode I with two solid electrodes.

Another PLS reactor was prepared with the DGCLW working with the air flow in the liquid under the induced microporous inhomogeneous conditions as is shown in Fig. 3. Here, the airflow was injected into the work liquid (*3*) through a copper tube (*1*) covered by a glass insulator (*2*) and it ran over a flat dielectric surface of the magnetostrictive transmitter (*5*) which produced ultrasonic cavitations, so the discharge channel (*4*) was formed by the air flow and water vapours (microbubbles). A high-voltage potential of ~4 kV was supplied between the gas input tube (*1*) and work liquid (*3*). The ultrasonic transmitter worked at the frequency of 18 kHz with the power ~ 20 W. Different modes of the discharge operation with the "liquid" cathode were tested:

i. with airflow and ultrasonic;
ii. with airflow and without ultrasonic;
iii. with ultrasonic and without airflow;
iv. and without airflow and ultrasonic.

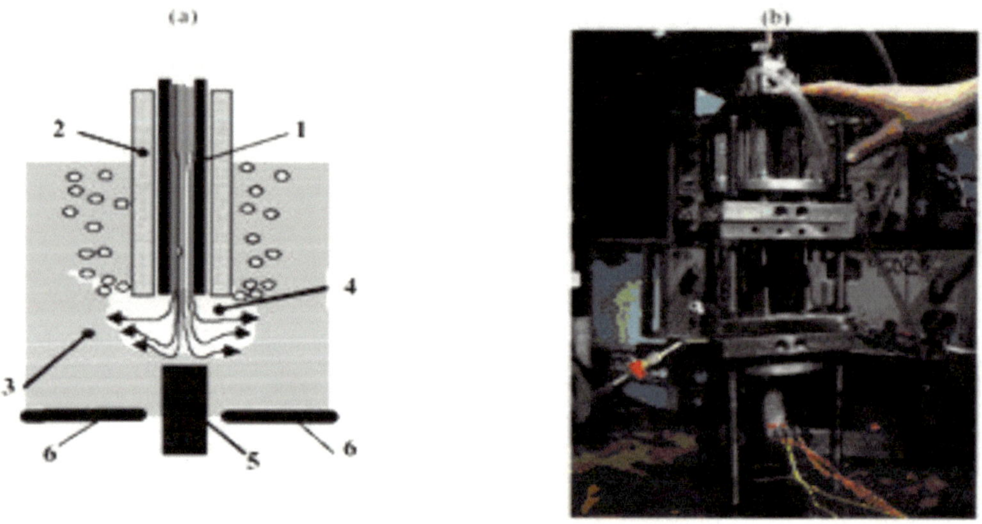


Fig. 3. DCGLW (dc) in the liquid with microbubbles induced by ultrasonic cavitation. (**a**) - Scheme of discharge with liquid electrode; (**b**) – photo of pulsed reformer.

In addition to direct plasma reforming, also pyrolysis of ethanol after initial plasma-assisted reforming was studied by using the unit shown in Fig.4. The installation consists of two main parts: 1) plasma reactor, which treats ethanol-water mixture in the pulsed DGCLW, and 2) pyrolytic reactor, which treats ethanol-air vapors mixed with products generated by plasma reactor, where (*1*) is the Teflon insulator around the steel pins, (*2*) are steel pins through which voltage is applied, (*3*) are copper electrodes, conical bottom and top cylinder, (*4*) is a discharge plasma zone between electrodes, (*5*) is a vortex zone in the discharge, (*6*) is a bubbling zone in the liquid, (*7*) is the work liquid (solution of 96% pure ethanol and distilled water), (*8*) are mixing inlet and outlet chambers, (*9*) is the steel pyrolytic chamber; (*10*) are electric heaters, (*11*) is the cylindrical casing; (*12*) are thermocouples for temperature control, (*13*) is the glass vessel (0.5 l) for the output syngas collection.

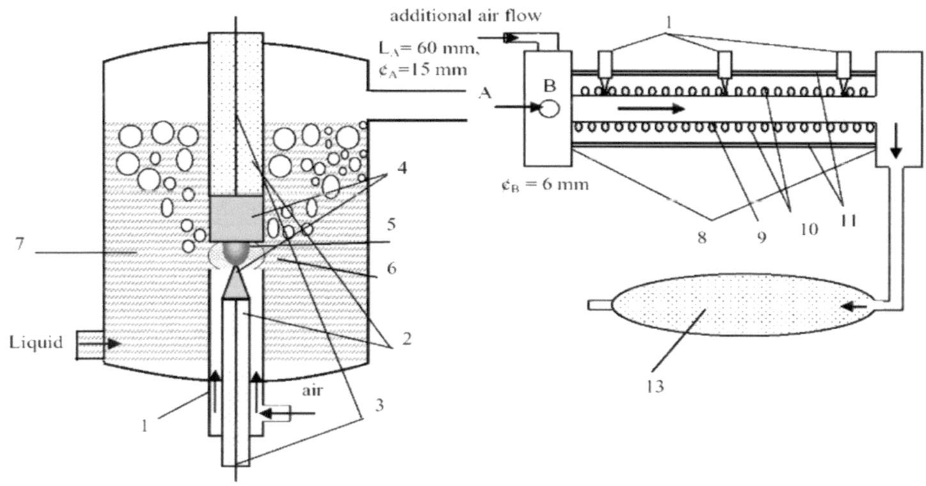


Fig. 4. Scheme of ethanol pyrolisis after plasma-assited reforming of ethanol water mixture.

## 3. Diagnostic methods

Diagnostics of the DGCLW plasma was prepared by means of optical emission and absorption spectroscopy. The scheme of optical measurements is shown in Fig. 5.
Here, (*1*) is a quartz cylinder of the reactor, (*2*) are duralumin flanges, (*3*) are cooper electrodes, (*4*) are glass tubes for air input; (*5*) is a plasma column, (*6*) is the work liquid, (*7*) is a water cooling system, (*8*) is an input tube for work liquid, (*9*) is an output tube for gas products, (*10*) is a tube for the maintenance of constant pressure inside the reactor and communicating vessels, (*11*) is a spectral lamp source. A high-speed CCD-based spectrometer "Plasmaspec" with a spectral resolution ~0.6 nm is used for the spectra registration in the range of wavelengths 200-1100 nm. According to spectral measurements, the emission spectra of the DGCLW plasma in the air-water system are multi-component (Fig. 6) and contain character atomic lines H (656.3; 486.1; 434.0 nm) and Cu (electrode material) (324.7; 327.4; 465.1; 510.5; 515.3; 521.8; 578.2 nm), molecular bands of the 2+-system of $N_2$ ($C^3\Pi_u$-$B^3\Pi_g$), UV system of OH ($A^2\Sigma$-$X^2\Pi$, (0-0): 306.4-308.9 nm), also NH band ($A^3\Pi^+$-$X^3\Sigma^-$) and NO $\gamma$-system ($A^2\Sigma^+$-$X^2\Pi$).

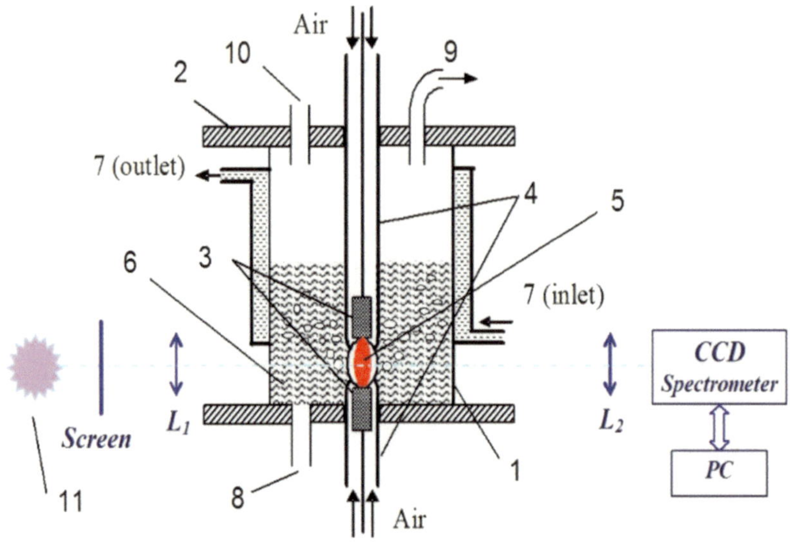


Fig. 5. Scheme of optical spectroscopy diagnostics of discharge plasma in the DGCLW.

The characteristic temperatures corresponding to excited states of atoms (electronic temperature $T_e^*$), and molecules (vibrational $T_v^*$ and rotational $T_r^*$ temperatures) in discharge plasma were determined by different methods. The electronic temperature $T_e^*$ was determined by relative intensities of hydrogen emission lines $H_\alpha$ (656.3 nm) and $H_\beta$ (486.1 nm) because these lines did no overlap with other spectral lines and bands. $H_\gamma$ peak (434.0 nm) was not used because of its low intensity. Emission spectra of H lines were simulated by using a developed code Spec-Elements.

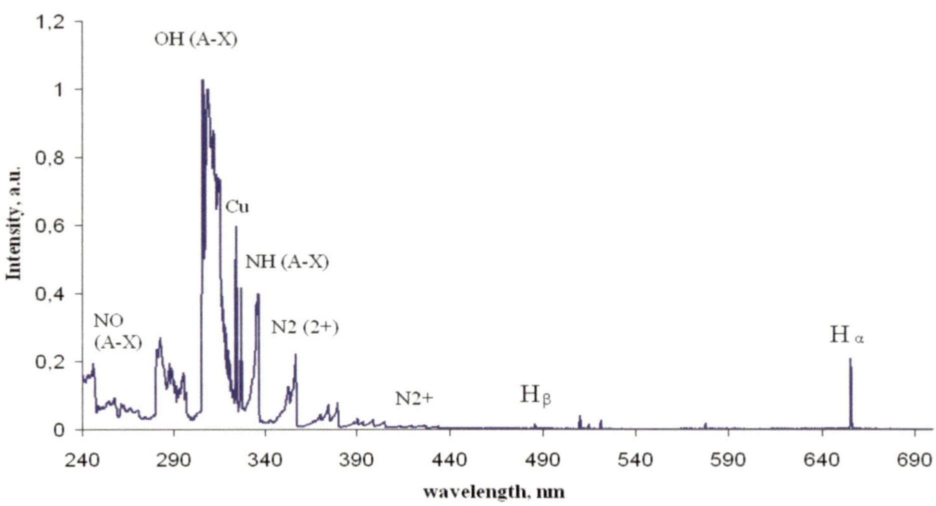


Fig. 6. Typical emission spectrum of air-water discharge plasma in the DCGLW.

This code allows calculations of absolute intensities of emission lines of elements (N, Ar, Cu, Fe, Ni, Co, Cr, etc) with using spectroscopic constants from the database of the Spectrum

Analyzer program (Navratil, 2006) and taking into account the instrumental function of the spectrometer. To determine $T_e^*$ both experimental and simulated spectra were normalized on the $H_\alpha$ peak. In this case the height of $H_\beta$ peak will directly depend on the $T_e^*$ value.
The electronic temperature $T_e^*$ was determined also by relative intensities of emission of oxygen multiplet lines (777.2; 844.6; 926.6 nm). To increase the accuracy of $T_e^*$ determining it is necessary to chose spectral lines corresponding to electronic transitions from the energy levels with the maximal energy discrepancies of upper excited states $E_2$. Among OI lines observed in spectra the best pair is OI 777.2 nm ($E_2$=10.74eV) and OI 926.6 nm ($E_2$=12.08eV). Emission spectra of OI lines were simulated by using the SPECAIR program (SPECAIR, 2011). The dependence of the ratio of relative intensities of OI multiplet lines on the $T_e^*$ values was plotted as a calibration curve. The spectral sensitivity of the used spectrometer was taken into account during the obtaining of corresponding intensities from experimental spectra. To determine vibrational $T_v^*$ and rotational $T_r^*$ temperatures, an original technique with using the SPECAIR program was developed. Since the emission of the OH ($A^2\Sigma$-$X^2\Pi$) bands is very intensive and its spectral structure is well characterized, it is a good monitor of $T_r^*$ in plasma (Levin, 1999). It is a well-known method to use the *P*- and *R*- branches intensity ratio in the OH (0-0) band at wavelengths $\lambda$=306.4-308.9 nm for determination of $T_r^*$ in plasma [16]. Because the peak of the *P*-branch is more self-absorbed than the peak of the *R*-branch, the *P/R* intensity ratio at given $T_r^*$ becomes smaller as the optical thickness of plasma increases. A smaller *P/R* ratio would suggest $T_r^*$ higher than its actual value. To avoid mistakes related with reabsorption it was proposed to use low intensive OH bands for determining $T_r^*$. A technique based on the SPECAIR simulation of the OH (1-0) and (2-1) bands, which are free from reabsorption, was suggested (Pryshiazhnevich, 2009).
Emission spectra were measured by the CCD spectrometer "Plasmaspec" as described before. The absolute intensities of calculated spectrum $I_{cal}(A_i)$ by using the SPECAIR at the fixed wavelengths (where the corresponding experimental signals $I_{exp}(A_i)$ were estimated) were determined. Then, the ratio of concentrations of radiating species A and B was evaluated by following formula:

$$\frac{[A_1]}{[A_2]} = \frac{I_{\exp}(A_1)\cdot I_{cal}(A_2)}{I_{\exp}(A_2)\cdot I_{cal}(A_1)}, \quad (1)$$

This makes possible to determine the relative concentration of each component in the investigated discharge plasma:

$$[A_1]^* = \frac{[A_1]}{\sum_i [A_i]} \quad (2)$$

Diagnostics by mass-spectrometry using a monopole mass-spectrometer MX 7301 and by gas-phase chromatography using a gas chromatograph 6890 N Agilent with the calibrated thermal conductivity detectors were prepared for the analysis of component content of the output gas products after the processing in the plasma-chemical reactor.
The typical mass spectrum of output gas products registered after the processing of ethanol-water mixture is shown in Fig. 7. The character components related to the mass ratios M/e = 2 ($H_2^+$), 12 ($C^+$), 14 ($N^+$), 16 ($O^+$, $CH4^+$), 18 ($H_2O^+$), 28 ($CO^+$, $N_2^+$), and 32 ($O_2^+$) have been recognized.

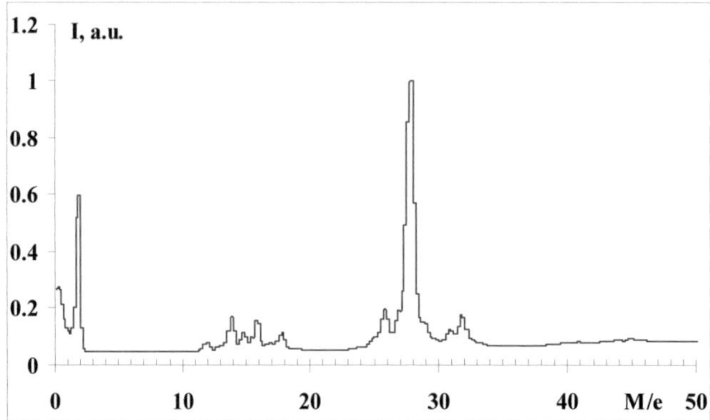


Fig. 7. Typical mass-spectrum of output gas products after the plasma reforming of ethanol - water mixture in the PLS with the DGCLW with the liquid cathode. $I_d$=300 mA, G= 55 $cm^3/s$.

## 4. Numerical models

The physical model of DGCLW-PLS was based on the next assumptions:

- electric power introduced in the discharge is immediately averaged in the discharge volume;
- internal electric field in the discharge does not vary in space and time;
- during the pass of air through the discharge into the reactor volume its content is totally refreshed and its flow rate in the reactor volume is the same as in the discharge gap.

The mathematical modelling of DGCLW-PLS was developed with the next features:

- calculation of the electron energy distribution function on the base of solution of the Boltzmann equation in two-term approximation (Raizer, 1997);
- hydrodynamic modeling in quasi-1D fluid volume averaged approximation (Shchedrin, 2008).
- kinetic modeling by solution of the system of chemical kinetic equations for kinetically valuable components of air-ethanol-water plasma system (Shchedrin, 2011).

The kinetic mechanism includes 59 components ($C_2H_5OH$, $N_2$, $O_2$, CO, etc), 76 electron-molecular processes and 364 chemical reactions with a set of corresponding cross sections and rate constants compiled according to update NIST kinetic databases (NIST, 2011).

As a case study, the comparative analysis of the developed method of simulation of kinetic processes in the plasma-liquid system in the electrical discharge in the gas channel with liquid wall and other known methods of calculations of plasma kinetics in micro-discharges was conducted. It was shown that the averaging of the energy that deposited into the discharge in the whole gas channel volume is the principal feature of the developed method.

### 4.1 Modeling of dusty plasmas with liquid microparticles

Complex (dusty) plasmas are gas plasmas consisting of electrons, ions, and neutral atoms that additionally contain microscopic particles with sizes ranging from 10 nm to 10 $\mu$m. These micro-particles may be solid or liquid. It is well-known (Frenkel, 1975) that many features of these states are very similar. From this point of view will be useful produce a review of generalities of both types of microparticles and give a special account of distinctive features of liquid substances. Essential differences of liquid particle from solid one mostly connected with special features of liquid surface. It is clear that addition of electric charge makes new peculiarities in this situation.

The main parameters of liquid drop are:

1. geometrical - what type of approximation is selected: ball-shaped (main factor is its radius) or spheroid (main factor is its eccentricity);
2. physical - mass/volume density; dielectric permittivity; surface tension coefficient; viscosity; specific conductivity; evaporation and coagulation;
3. chemical - composition; existence of surface-active substance (SAS).

The theoretical model should describe basic phenomena in dusty plasma. They are:

i. Elementary processes in dusty plasma;
ii. Dynamical processes in dusty plasma;
iii. Waves and instabilities in dusty plasmas.

Some quality conclusions can be made:

1. The droplet surface is unstable, because thermal capillary waves are disturbed.
2. The maximal surface charge ($Q_{max}$) exists.
3. This charge ($Q_{max}$) is the reason of drop decay. The method of this problem solving arises from Bohr - Frenkel theory of nuclear decay.
4. The decay path is a function of viscosity ($\gamma$) and dielectric permittivity ($\varepsilon$) = f($\varepsilon$, $\gamma$). If liquid in drop has a conductivity and its viscosity is small, the drop will be disintegrated on some hundreds little droplets with radius $a_i$ << R. If liquid in drop is dielectric, the drop has 2-particle decay. Their full charge is equal to initial $Q$ and their radius $a_i \leq$ R.
5. Changes in chemical composition or SAS additions make the surface charge variability.
6. Interaction of charged drop with laminar air flow is resulted decay criterion lower.
7. Linear vibration of charged drop surface can produce the electromagnetic radiation.
8. Surface of uncharged drop (incompressible) can produce the acoustic wave in definite frequency range. Surface charging will intensify this effect. It can be used for diagnostics of liquid dusty particles in plasma of the DGCLW.

### 4.2 Kinetics: Modeling and calculations

A global model is used in our calculations, with an assumption that the discharge is homogeneous over the entire volume. It is justified at zero approximation, because the time of the gas mixing in radial direction is less than the times of characteristic chemical reactions. Also neglected is the processes in the transitive zone between the discharge to post-discharge. The volume of the transitional zone is much smaller than the volume of discharge and post-discharge zones. Thus, the time of gas pumping through the transition region is too short for the chemical reactions to have a sufficient influence on the concentration of neutral components.

The total time of calculation is divided into two time intervals: the first one is the calculation of the kinetic processes of fast generation of active atoms and radicals in the discharge region. Those components accelerate the formation of molecular hydrogen, carbon oxides and other hydrocarbons production. The second time interval is the oxidation of the gas mixture in the post-discharge region as a result of the high gas temperature and the presence of O and OH. These components remain in the mixture after the dissociation of water and oxygen molecules by electrons impacts in plasma. The oxidation of generated hydrocarbons (mainly $C_xH_y$ and formaldehyde $CH_2O$) has noticeable influence on kinetics in the investigated mixture due to high gas temperature (1000 K). Under aforementioned conditions, the characteristic time of oxidation is approximately equal to the air pumping time through the discharge region ($10^{-3}$-$10^{-2}$ s). Therefore, the low temperature plasma model (Chernyak, 2008, Shchedrin, 2008), where the continuous discharge is divided into the sequence of quasi-constant discharges, is not applicable here. The following system of kinetic equations is used in order to account the constant air pumping through the system:

$$\frac{dN_i}{dt} = S_{ei} + \sum_j k_{ij} N_j + \sum_{j,l} k_{ijl} N_j N_l + \ldots + K_i - \frac{G}{V} N_i - kN_i \,. \tag{3}$$

It is calculated using a solver developed at the Institute of Physics National Academy of Science, Ukraine was verified many times on other systems and has demonstrated good results. $N_i$, $N_j$, $N_l$ in equation (3) are concentrations of molecules and radicals; $k_{ij}$, $k_{iml}$ are rate constants of the processes for $i$-th component. The rates of electron-molecule reactions are:

$$S_{ei} = \frac{W}{V} \frac{1}{\varepsilon_{ei}} \frac{W_{ei}}{\sum_i W_{ei} + \sum_i W_i} , \tag{4}$$

where $W$ is the discharge power and $V$ is the discharge volume. In this model, $W$ is the full power $W_0$ divided by factor of two. This division corresponds to the most stable regime of the discharge (Fridman, 2008). The value of $W_0$ is determined by the current-voltage characteristic. $W_{ei}$ is the specific power deposited into the inelastic electron-molecular process with threshold energy $\varepsilon_{ei}$:

$$W_{ei} = \sqrt{\frac{2q}{m}} n_e N_i \varepsilon_{ei} \int_0^\infty \varepsilon Q_{ei}(\varepsilon) f(\varepsilon) d\varepsilon \,. \tag{5}$$

Here $q = 1.602 \cdot 10^{-12}$ erg/eV, $m$ is the mass of electron and $n_e$ is the concentration of electrons. The variable $Q_{ei}$ is inelastic process cross section, $f(\varepsilon)$ is the electron energy distribution function (EEDF); $W_i$ is the specific power deposited into elastic processes:

$$W_i = \frac{2m}{M_i} \sqrt{\frac{2q}{m}} n_e N_i \int_0^\infty \varepsilon^2 Q_i(\varepsilon) f(\varepsilon) d\varepsilon \,. \tag{6}$$

Here $M_i$ are the molecules masses, $Q_i$ are the transport cross sections for nitrogen, oxygen, water and ethanol molecules.

The last three terms in equation (3) describe the constant inflow and outflow of gas from the discharge region. The term $K_i$ is the inflow of molecules of the primary components (nitrogen, oxygen, water and ethanol) into the plasma, $G/_V N_i$ and $kN_i$ are the gas outflow as the result of the air pumping and the pressure difference between the discharge region and the atmosphere.

In order to define the initial conditions, the ethanol/water solution is assumed to be an ideal solution. Therefore, the vapors concentrations are linear functions of the ethanol-to-water ratio in the liquid. The evaporation rates $K_i$ of $C_2H_5OH$ and $H_2O$ are calculated from the measured liquids consumption. The inflow rates $K_i$ of nitrogen and oxygen are calculated by the rate of air pumping through the discharge region:

$$K_i = \frac{G}{V} N_i^0 , \tag{7}$$

where $N_i^0$ correspond to $[N_2]$ and $[O_2]$ in the atmospheric pressure air flow.

In the non-equilibrium plasma almost the entire energy is deposited into the electron component. The active species, generated in the electron-molecular processes, lead to chain reactions with ethanol molecules. The released energy heats the gas. Calculations show that the fastest reactions of that type are:

$$C_2H_5OH + O \rightarrow C_2H_5O + OH \ (\Delta H = 6.2 \text{ kJ/mol}),$$

$$C_2H_5OH + OH \rightarrow C_2H_5O + H_2O \ (\Delta H = -65.4 \text{ kJ/mol}).$$

The first reaction is endothermic and the second is exothermic. By comparing their specific enthalpies, one can estimate the average gas temperature, which is near 1200 K. The gas temperature in the discharge region is taken to be constant in the model. In reality the gas temperature $T$ is dependent on the gas pumping rate and the heat exchange with the environment. Therefore, in order to account of those influences, $T$ is varied in the interval 800-1500 K (similarly to experimentally obtained temperature spread).

After ~$10^{-2}$ s the balance between the generation and the decomposition of the components leads to saturation of concentrations of all species. It allows to stop the calculations in the discharge region and to investigate the kinetics in the post-discharge region. The system (1) is solved without accounting of the last three terms on the time interval without plasma. The calculations are terminated when the molecular oxygen concentration reaches zero level. That time interval equals to a few milliseconds. The gas temperature in this region is found using the equation (Kosarev, 2008):

$$\frac{dT}{dt} = -\frac{1}{\rho C_p} \sum_i H_i(T) \cdot \mu_i \frac{dN_i}{dt} . \tag{8}$$

Here $\rho$ is the gas density, $C_p$ is the average specific gas heat capacity under constant pressure, $H_i$ and $\mu_i$ are molar enthalpy and molar mass of $i$-th component respectively.

It is seen from equations (5-6) the specific powers $W_{ei}$ and $W_i$ are influenced by the electron energy distribution function. EEDF is calculated from Boltzman kinetic equation in the two-term approximation for breakdown field in the air (Soloshenko, 2007). The last assumption was made due to the fact that nitrogen is a plasma-forming gas. Only the processes with the primary components (see table 1) are taken into account in EEDF calculations. The cross sections of processes 17-19 are absent in the recent literature. Therefore, in order to estimate these cross sections, an approximation (Soloshenko, 2007) was used. At this approximation the cross section of ethanol is equal to the cross section of oxygen biased on the doubled threshold energy.

Fig.8 shows the calculated EEDFs for different gas temperatures and 6.5% ethanol concentration in solution. It is shown that EEDF is defined by vibrations of water molecules (two levels ((100)+(010)) and (010)), excited by electrons, whose energy is less than 0.05 eV. In the 1.5-2 eV region EEDF is defined by the excitation of electron-vibrational levels of nitrogen and water by electron impact. When the electrons' energy is higher than 8 eV EEDF looks like Maxwell distribution function. At this region the function's behavior is defined by the dissociation and ionization of molecules by electrons' impacts. The average electrons' energy is 0.55 eV for 800 K, 0.46 eV for 1200 K and 0.43 eV for 1500 K.

| № | Reaction |
|---|---|
| 1.1 | $H_2O + e \rightarrow H_2O((100)+(010)) + e$ |
| 2.1 | $H_2O + e \rightarrow H_2O(010) + e$ |
| 3.1 | $H_2O + e \rightarrow OH + H + e$ |
| 4.1 | $H_2O + e \rightarrow H_2O^+ + 2e$ |
| 5.1 | $H_2O + e \rightarrow H_2O(J = 0\text{-}0) + e$ |
| 6.1 | $H_2O + e \rightarrow H_2O(J = 0\text{-}1) + e$ |
| 7.1 | $H_2O + e \rightarrow H_2O(J = 0\text{-}2) + e$ |
| 8.1 | $H_2O + e \rightarrow H_2O(J = 0\text{-}3) + e$ |
| 9.1 | $N_2 + e \rightarrow N_2(A^3{}_u{}^+) + e$ |
| 10.1 | $N_2 + e \rightarrow N_2(a^1{}_g) + e$ |
| 11.1 | $N_2 + e \rightarrow N_2(v) + e$ |
| 12.1 | $N_2 + e \rightarrow N + N + e$ |
| 13.1 | $N_2 + e \rightarrow N_2^+ + 2e$ |
| 14.1 | $O_2 + e \rightarrow O + O + e$ |
| 15.1 | $O_2 + e \rightarrow O_2(^1\Delta_g) + e$ |
| 16.1 | $O_2 + e \rightarrow O_2^+ + 2e$ |
| 17.1 | $C_2H_5OH + e \rightarrow CH_3 + CH_2OH + e$ |
| 18.1 | $C_2H_5OH + e \rightarrow C_2H_5 + OH + e$ |
| 19.1 | $C_2H_5OH + e \rightarrow CH_3CHOH + H + e$ |
| 20.1 | $C_2H_5OH + e \rightarrow C_2H_5OH^+ + 2e$ |

Table 1. Reactions which were taken into account in the EEDF calculations

There is no generally accepted kinetic mechanism for modeling plasma kinetics in air/water/ethanol mixture. In the previous works (Chernyak, 2008, Shchedrin, 2009), a mechanism for the low temperature region was proposed.

However, this mechanism does not describe kinetics in "tornado" type electrical discharge, where the gas temperature is much higher. Main mechanism (Marinov, 1999) and ethanol sub-mechanism (Dagaout, 2008) could be more appropriate for describing the chemical reactions in the investigated mixture. The preliminary calculations have shown that these schemes include many unneeded components which do not contribute to the end product in the investigated regimes. They may be ignored in the scheme of reactions. Dunphy proposed a kinetic mechanism for high-temperature ethanol oxidation in (Dunphy, 1991). The investigated temperature interval (1080-1660 K) in that article is close to the one at present work. Therefore, Dunphy's scheme was chosen as the basic mechanism in the research. It was expanded by Held's methanol sub-mechanism (Held, 1998) and by additional important processes in hydrogen oxidation (Konnov, 2008): $OH+OH+M \rightarrow H_2O_2+M$, $OH+OH \rightarrow H_2O+O$, $H+H+M \rightarrow H_2+M$, $H_2+O_2 \rightarrow OH+OH$, $H_2+O_2 \rightarrow H+HO_2$. The full mechanism developed in this work is presented in table 2. It is composed of 30 components and 130 chemical reactions between them. The charged particles (electrons and ions) were ignored in the mechanism, because of low degree of ionization of the gas ($\sim10^{-6}$-$10^{-5}$). Nitrogen acts as the third body in the recombination and thermal dissociation reactions. Additionally, the nitrogen-containing species were removed from the mechanism, since they are not the subject of this study. Also, the research (Levko, 2010) has shown that the processes between these components and hydrocarbons were third-order reactions.

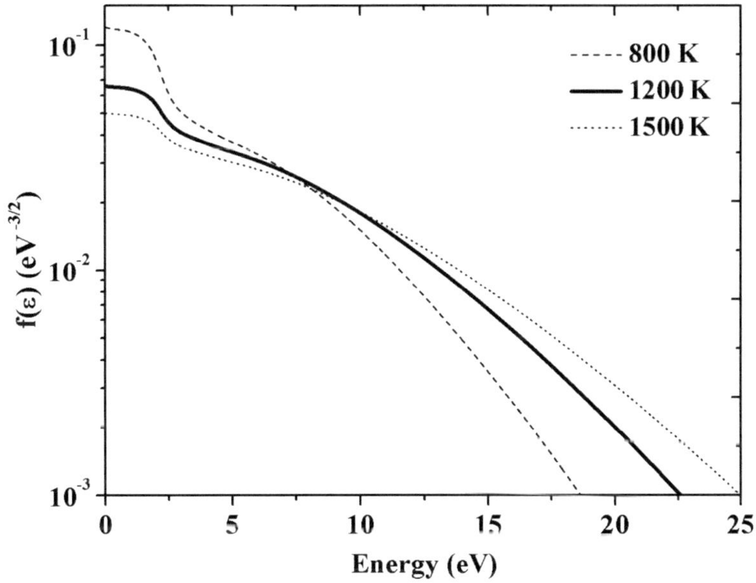


Fig. 8. Calculated electron energy distribution functions at 6.5% ethanol concentration in the liquid.

## 5. Results and discussion

The applicability of the developed methods for diagnostics of the discharge plasma parameters was tested in the DGCLW in different modes of operation: with two solid electrodes or with one liquid electrode of different polarity, i.e., liquid anode or liquid cathode.

Typical operating parameters in nominal regime were following: discharge current $I_d$= 200 mA, voltage $U$= 1.6 kV, airflow rate $G$= 55 cm$^3$/s. Emission spectra of investigated plasma were registered by the CCD spectrometer "Plasmaspec" in the range of wavelengths λ= 200-1100 nm.

For estimation of electronic excitation temperatures $T_e^*$ in plasma, the relative intensities of atomic emission lines of Cu (λ= 465.1; 510.5; 515.3; 521.8; 578.2 nm) and H (λ= 486.1, 656.6 nm) were analyzed. As a result, a set of characteristic electronic temperatures $T_e^*$(H) = 4300 K and $T_e^*$(Cu) = 7200 K was obtained for the investigated regime of the discharge operation.

For estimation of vibrational and rotational excitation temperatures $T_v^*$ and $T_r^*$ in plasma, the relative intensities of molecular emissions of the OH (A-X) bands (1-0) at 283.1 nm and (2-1) at 288.4 nm and the $N_2$ (C-B) bands (0-2) at 380.1 nm and (1-3) at 375.1 nm were analyzed on the base of the SPECAIR simulation. Results of the fitting of experimental and simulation spectra in the selected spectral intervals are shown in Fig. 9.

The best fit in Fig.9 was found for the following set of temperatures: $T_v^*$(OH) = 3800K, $T_r^*$(OH) = 3200K, $T_v^*$($N_2$) = 4000K, $T_r^*$($N_2$) = 2000K. The discrepancy between measured and calculated spectra in Fig. 9a in contrast to Fig. 9b can be explained by the fact that the simulation of the OH (A-X) emission was made without taking into account the $N_2$ (C-B) emission that occurs in the same spectral region.

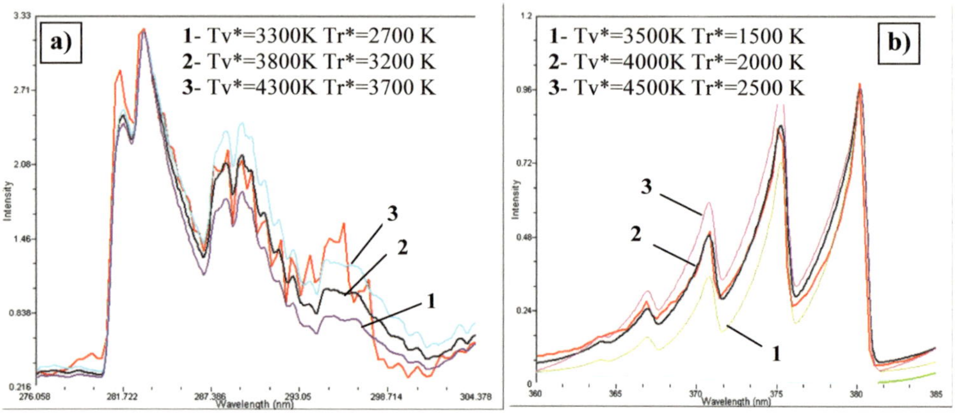


Fig. 9. SPECAIR simulation (curves 1, 2, 3) and experimental (red curve) spectra of the DGCLW in water: (**a**) OH (A-X) bands at 275-304 nm; (**b**) $N_2$ (C-B) bands at 360-385 nm.

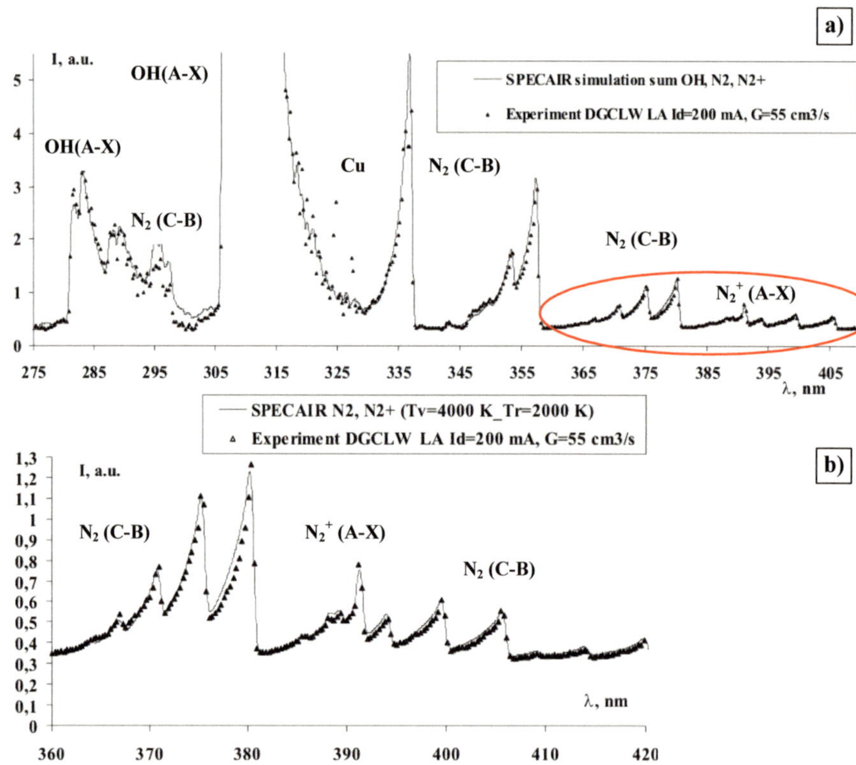


Fig. 10. SPECAIR simulation and experimental spectra of the DGCLW in water: (**a**) emission in the range 275-405 nm; (**b**) magnified spectral range 360-420 nm in the red oval. Characteristic temperatures: $T_v^*$(OH)=3800 K, $T_r^*$(OH)=3200 K, $T_v^*$($N_2$, $N_2^+$)=4000 K, $T_r^*$ ($N_2$, $N_2^+$)=2000 K.

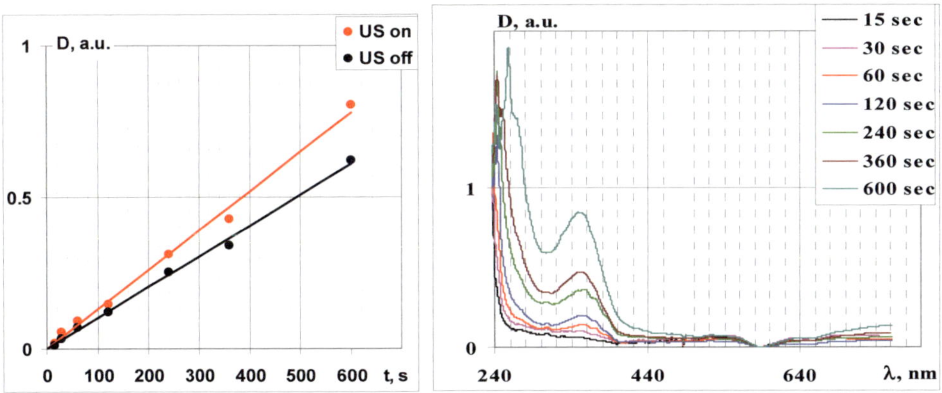


Fig. 11. Absorption spectra (left) of water solution after the plasma treatment in the DGCLW for different processing times (curves 1-7 are from 15 to 600 s) and dependences (right) of UV absorption at 380 nm on the exposure time in the discharge with (grey curve) and without ultrasonic cavitations (black curve).

Applying the absorption spectrophotometry, the absorption spectra of work liquids after the plasma treatment in the DGCLW-PLS reactor with and without ultrasonic cavitations were measured. According to data obtained (Fig. 11), the absorption spectra of the treated liquid consist of UV absorption bands of ions of nitrous acid $HNO_2$. Its spectral distribution does not change very much with and without US treatment, but the intensity of the absorption band at 350 nm permanently grows with the exposure time (Fig. 10.a). It means that the concentration of the $HNO_2$ products in the treated liquid grows proportionally to the time of the plasma treatment, and this process proceeds more intensively under the influence of the ultrasonic cavitations (Fig. 11b).

Analyzing the emission spectra in plasma of the pulsed DGCLW working in different modes with and without ultrasonic field, the dependences of relative intensities of molecular OH (A-X) and $N_2$ (C-B) bands and atomic $H_\alpha$, $H_\beta$ lines were obtained. Also, using the ratio of $H_\alpha$ and $H_\beta$ lines, the temperature of the population of H electronic states $T_e^*$(H) was estimated.

It was found that the ultrasonic field differently affects the behavior of various components of plasma emission in the DGCLW. The relative intensity of OH (A-X) emission non-monotonic increases with increasing exposure time, but it quickly reduces under the ultrasonic cavitations. Relative intensity of $N_2$ (C-B) emission is almost independent on exposure time with and without ultrasonic field.

Relative intensity of $H_\alpha$ emission linearly increases with increasing exposure time in both modes with and without ultrasonic; however, the ultrasonic field reduces the rate of this growth for half time. Despite the growth of the relative $H_\alpha$ emission intensity, the electronic temperatures $T_e^*$(H) are virtually the same, i.e. the ultrasonic field has no visual effect on the behavior of $T_e^*$.

It is important to note that all components of emission spectra observed have lower relative emission intensities in the presence of the ultrasonic field. This can be explained that the ultrasonic at 17 kHz may be easily transmitted in air. Moreover, the features of the plasma-liquid surface interaction at the boundary between the phases can favor to the penetration of the ultrasonic field into the gas phase. Then, the ultrasonic cavitations in plasma can increase the probability of the quenching of excited states of atoms due to enhanced collisions. As a result, total intensity of radiation in plasma may decrease with the time under the ultrasonic field.

The distributions of character temperatures $T_e^*, T_r^*, T_v^*$ in plasma in the DGCLW working in water at air flow rate $G$=55cm$^3$/s at currents $I_d$=50-400 A for three different modes: with solid electrodes, with solid cathode + liquid and with liquid cathode + solid anode were investigated. Electronic temperatures $T_e^*$(H) and $T_e^*$(Cu) were determined by using relative intensities of atomic lines $H_\alpha$, $H_\beta$ (656.5, 486.1 nm) and Cu I (465.1, 510.5, 515.5, 521.8, 578.2 nm). The vibrational and rotational temperatures, $T_v^*$ and $T_r^*$ were determined by using relative intensities of molecular bands $N_2$ (C-B) (0-2) at 380.1 nm and (1-3) at 375.1 nm and OH (A-X) (1-0) at 283.1 nm and (2-1) at 288.4 nm.

In common, the temperatures $T_e^*$(Cu) in the DGCLW with the liquid electrode gradually increase with increasing discharge current, however, for the mode of liquid anode the $T_e^*$(Cu) are considerably larger than for the mode of liquid cathode. This difference in the values of character temperatures in the discharge modes with the liquid electrodes is also observed for all other plasma components with the exception of $T_e^*$(H). With the liquid cathode the $T_e^*$(H) noticeable increases, whereas with the liquid anode the $T_e^*$(H) slightly decreases. Larger values for character temperatures in plasma of the DGCLW with the liquid cathode for almost all components (except Cu) may be explained by greater energy input in discharge plasma. For $T_e^*$(Cu), its larger value for the liquid anode follows by the decrease of the absolute intensity of Cu emission lines (if compared with other two modes); this may indicate a reduction in output of Cu from the electrodes in this regime.

Analyzing the obtained values and dependences, one can conclude that the investigated DGCLW generates non-equilibrium reacting plasma whose parameters are close to non-equilibrium plasma of the gliding arc, which is effectively used in non-thermal atmospheric-pressure plasma-assisted initiation and plasma-supported combustion of hydrocarbon fuels (Fridman, 2008). The results of OES diagnostics of plasma in the DGCLW working in pure ethanol and ethanol/water mixture (5/1) at the air flow rate $G$=55cm$^3$/s at currents $I_d$=50-400 A in the mode with two solid electrodes are shown in Fig. 11.

From the temperature dependencies in Fig. 12 one can reveal that the transition from the water to the pure ethanol in the DGCLW does not affect very much character temperatures $T_e^*$(Cu), $T_e^*$(H) and $T_r$ (OH), except CN, when noticeable difference between $T_r$ and $T_v$ takes place. The last fact demonstrates the effect of thermalization and reduction of non-equilibrium in plasma. Possible reason may be additional energy input from ethanol due to initiation of combustion processes in the plasma-fuel system.

The results of OES diagnostics of plasma in the DGCLW working in the mode with the liquid anode in ethanol/water mixture (5/1) at the air flow rate $G$=55cm$^3$/s at currents $I_d$=50-400 A are shown in Fig. 12. From the temperature dependences obtained one can reviled that in the investigated regimes in the DGCLW working with the liquid anode, the characteristic temperatures for radical CN are close $T_v$ (CN) $\approx T_r$ (CN) and for molecule $C_2$ are differ $T_v$ ($C_2$) >$T_r$ ($C_2$). At that, $T_v$ (CN) and $T_r$ (CN) exceed $T_v$ ($C_2$) a little bit (~20%). With increasing discharge current from 100 mA to 400 mA, all temperatures are decreased from 4000-5000 K to 3000-4000 K.

Fig. 14. shows the fragments of emission spectra related to the $C_2$ ($d^3\Pi_g$-$a^3\Pi_u$) Swan band emission in the spectral range 516-570 nm as obtained in plasma of the DGCLW with the liquid cathode working in ethanol/water mixture (5/1) with and without US field. One can see that fragments are almost identical, so the effect is practically absent. It can be explained by the possible depletion of ultrasonic cavitations in the work liquid caused by additional gas evolution in plasma-liquid system due to reforming.

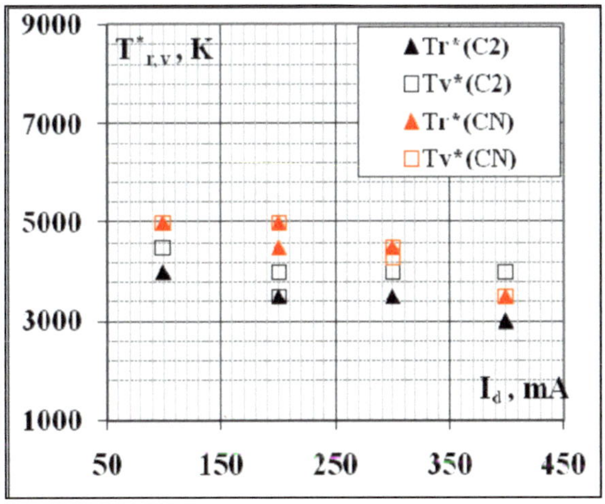


Fig. 12. Character temperatures $T_r$, $T_v$ in discharge plasma vs. current in the DGCLW with the liquid anode working in ethanol/water mixture (5/1). Air flow rate $G$=55 cm³/s.

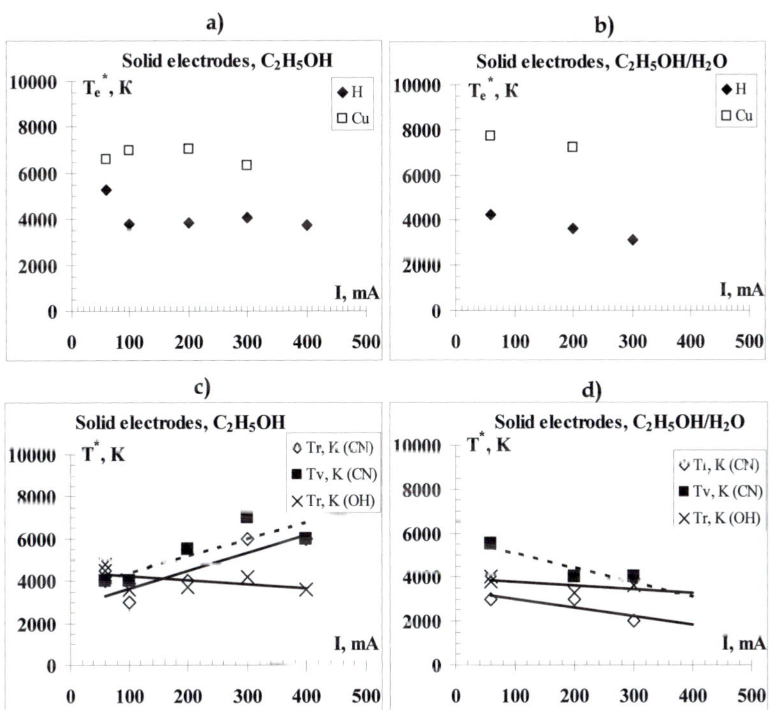


Fig. 13. Character temperatures $T_e^*$,$T_r^*$,$T_v^*$ in discharge plasma vs. current in the DGCLW with solid electrodes: (a, c) are pure ethanol, (b, d) are ethanol/water mixture (5/1). Air flow rate $G$=55 cm³/s.

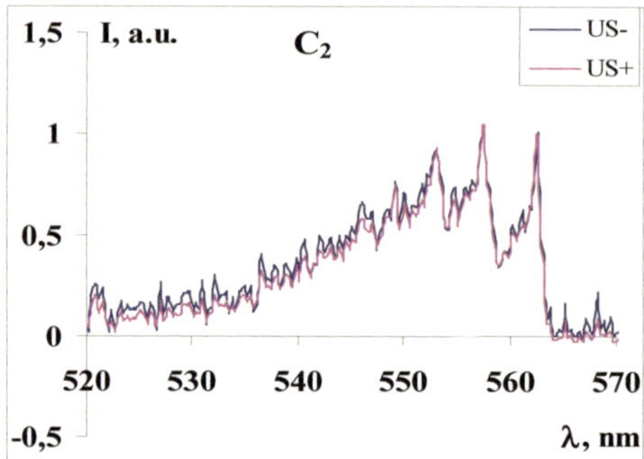


Fig. 14. Spectral distribution of the $C_2$ ($d^3\Pi_g$-$a^3\Pi_u$) emission in plasma of the DC DGCLW with liquid cathode working in ethanol/water mixture (5/1) with and without US field. Discharge current $I_d$ = 100 mA, air flow rate $G$=55 cm³/s.

The current-voltage and power characteristics of the DC DGCLW working in ethanol-water solution are shown in Fig. 15. A dropping character of *I-V* curves at discharge currents from 100 to 400 mA indicates the transition regime from the abnormal glow to the arc discharge.

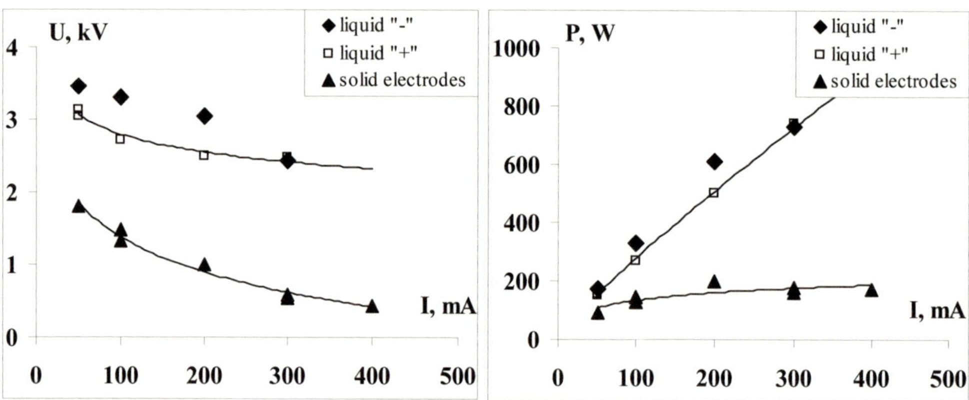


Fig. 15. Current-voltage (left) and power (right) characteristics of the DGCLW working in different modes in ethanol-water solution (5/1). Air flow rate $G$= 55 cm³/s.

The results of mass-spectroscopic and gas-chromatographic measurements of concentrations of basic components in output gas products after the ethanol processing in the DGCLW at different discharge currents are shown in Figs. 16-17. These data are given for the case of mixture $C_2H_5OH : H_2O$ = 5:1 and airflow rate $G$=55 cm³/s.

Fig.19 shows the results of numerical modeling and calculations of concentrations of $H_2$, $CO_2$ and other main stable components in output gas products after the ethanol processing in the PLS with the DGCLW. The qualitative and quantitative agreement between calculated and measured data is quite good, at least, for main components. One can see that the output concentration [$H_2$] grows linearly with the discharge current and it reduces exponentially with the gas flow rate.

In the discharge conditions, the kinetics of the $H_2$ formation is determined mainly by the reaction $C_2H_5OH + H \rightarrow CH_3CH_2O + H_2$. Since the ethanol concentration [$C_2H_5OH$] in solution changes slowly, the [$H_2$] production is determined entirely by the concentration of

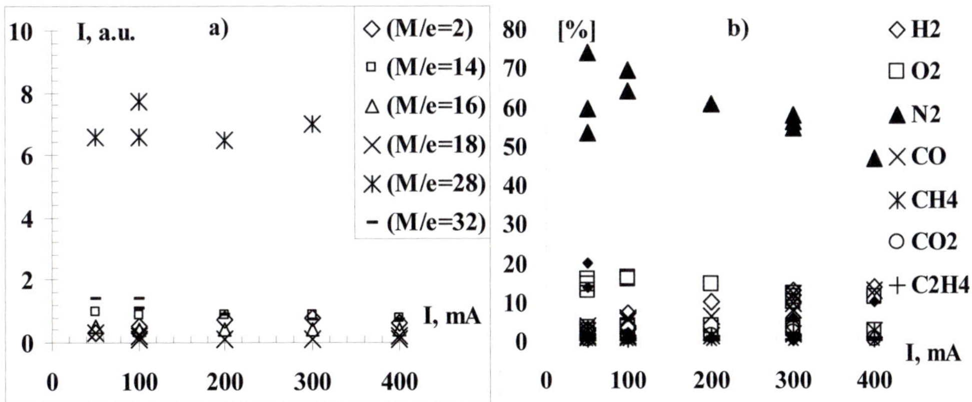


Fig. 16. Data of mass-spectrometry (a) and gas-chromatography (b) analysis of output gas products after the ethanol processing in the DGCLW. The mass ratio M/e = 2 is $H_2^+$, 12 is $C^+$, 14 is $N^+$, 16 is $O^+$, $CH4^+$, 18 is $H_2O^+$, 28 is $CO^+$, $N_2^+$, and 32 is $O_2^+$.

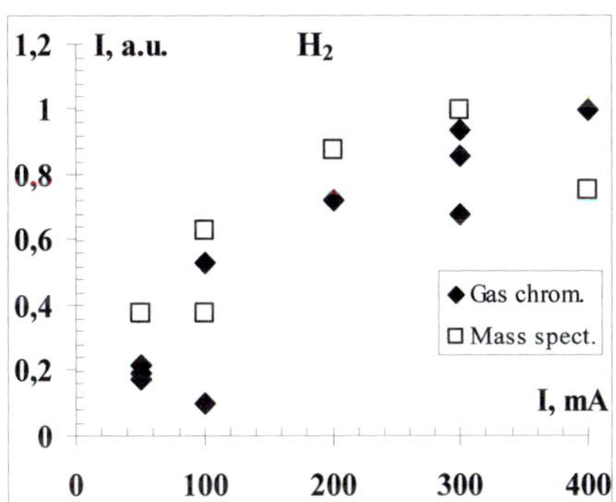


Fig. 17. Content of $H_2$ in output gas products after the ethanol processing in the DGCLW. Ethanol-water solution (5/1), airflow rate $G$=55 cm³/s.

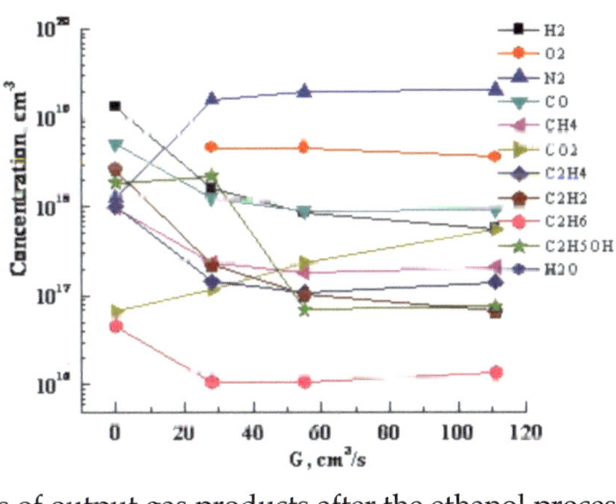


Fig. 18. Concentrations of output gas products after the ethanol processing in the DGCLW as function of air flow rate. Ethanol-water solution (5/1), $I_d$= 100 mA.

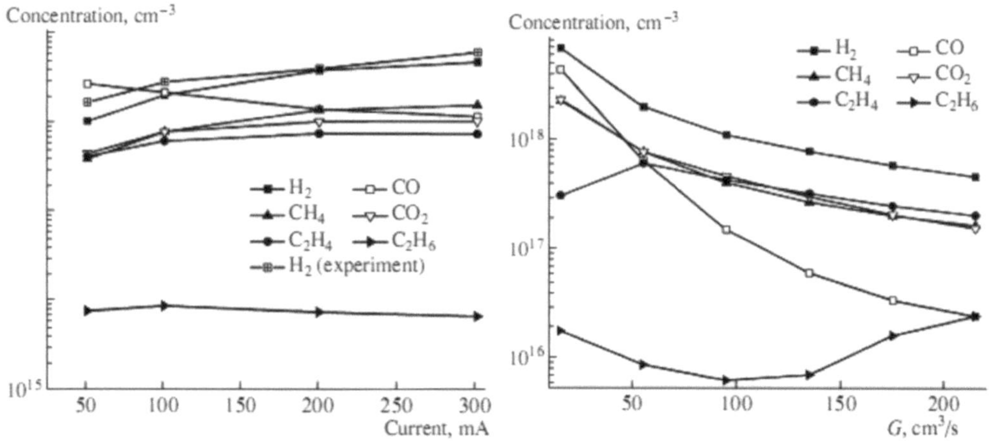


Fig. 19. Calculated concentrations of output gas products after the ethanol processing in the DGCLW as functions of discharge current (left, $G$= 55 cm³/s) and air flow rate (right, $I_d$= 100 mA). $C_2H_5OH : H_2O = 5:1$.

atomic hydrogen [H]. In the case under consideration, the main process responsible for the generation of H is the dissociation of water molecules $H_2O$ by the direct electron impact. The rate of this process is proportional to the specific electric power deposited to discharge (i.e., discharge current). Therefore, the [$H_2$] production is also a linear function of the discharge current in accordance with experimental data. Outside the discharge, the only process that influences the $H_2$ concentration is the water-gas shift reaction $CO + H_2O \rightarrow H_2 + CO_2$. Via this process, the system reaches the complete conversion of CO into $CO_2$ and $H_2$.

The estimation of efficiency of the proposed method of the plasma reforming of liquid ethanol into synthesis gas in the PLS-DGCLW reformer was performed on the basis of thermo-chemical calculations using criteria: (a) energy cost of 1 m³ syngas products; (b) productivity of conversion; (c) specific heat of 1 m³ syngas combustion, and (d) energy efficiency.

Calculations were made with taking into account standard thermo-chemical constants of hydrocarbons (NIST, 2011) using the coefficient of energy transformation (Chernyak, 2007):

$$\alpha = \frac{\sum_i Y_i \times LHV\,(Y_i)}{IPE}, \tag{9}$$

and also for the conversion efficiency by (Petitpas, 2007):

$$\eta = \frac{(Y_{H2} + Y_{CO}) \times LHV\,(H_2)}{IPE + Y_{HC} \times LHV\,(HC)} \tag{10}$$

Here, $IPE$ is the input plasma energy, $Y$ is the molar fraction, $LHV$ is the lower heating value of syngas components, $HC$ is the hydrocarbon fuel (ethanol). The formula (4) assumes that CO can be totally transformed into $H_2$ by the water-gas shift reaction with zero energy cost.

The results of estimations in the form of $\alpha$ ($I$) and $\eta$ ($I$) dependencies for the ethanol reforming in the DGCLW for different discharge modes against the discharge power are presented in Fig.20.

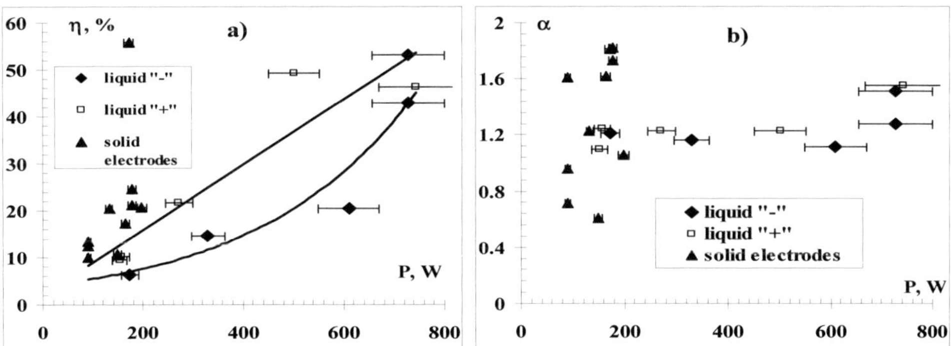


Fig. 20. Dependencies of (a) efficiency of conversion η and (b) coefficient of energy transformation α for the ethanol reforming in the DGCLW in different discharge modes as a function of the discharge power (currents vary from 50 to 400 mA). $C_2H_5OH : H_2O = 5:1$

Fig. 21 shows the values of coefficient *a* and parameter *η* together with data of mass-spectrometry of $H_2$ in the DGCLW for different modes at different discharge currents. One can see that that the coefficient α for the mode of solid electrodes has the same growth trend with increasing current as the $H_2$ yield. And for modes of liquid electrode the parameter of efficiency η has the same trend of growth as the $H_2$ yield. This difference in behavior of parameters may be caused due to the fact that the power of discharge using the liquid electrode is proportional to the current. At smaller currents the efficiency is of 20%, at large currents the efficiency is at 45-55%. For the modes of liquid electrode with increased current the level of efficiency of reforming is higher than for the mode of solid electrodes. At that, each mode demonstrates increased efficiency with increasing current.

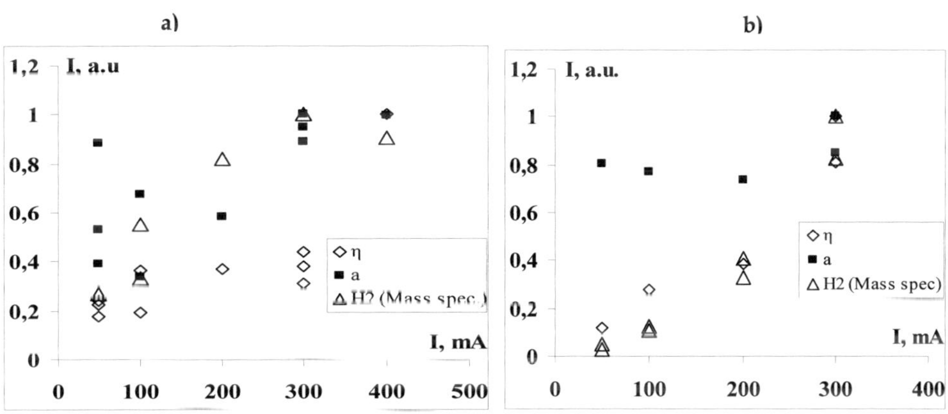


Fig. 21. Coefficient of energy transformation α, conversion efficiency η and $H_2$ yield as measured by mass-spectrometry for the ethanol reforming in the DGCLW at different currents: a) mode of solid electrodes; b) mode of liquid cathode.

One can see that the net $H_2$ yield in the discharge at $I$ = 300 mA is ~15% whereas the energy efficiency of the ethanol conversion into the syngas is up to 50%. (input energy is 4-10% of *LHV*= 26.8 MJ/kg).These numbers correlate with the earlier results (Chernyak, 2008) and are comparable with other known plasma-aided ethanol reforming methods (Petitpas, 2007).

The impact of the plasma-forming gas on the ethanol reforming in the DGCLW was also studied. For that, the composition of gas-phase products of conversion in the reactor and the coefficient of energy transformation were studied at different gas flow rates. Research was conducted for the mode of solid electrodes. The composition and mixture ratio under the ethanol reforming were taken the same as in previous research. The discharge current varied between 100 and 400 mA, the air flow rate varied from 0 to 110 $cm^3/s$.

Fig. 22. shows the results for $I$= 100 mA that demonstrates a good matching between gas chromatography and mass-spectrometry data. For other currents the same matching is observed. It should be noted that with increasing air supply in the discharge the concentration of $H_2$ in syngas products decreases. In fact, the highest yield of $H_2$ is observed in the discharge mode without air supply. But the time of $H_2$ production in this case increases considerably, and the power consumption also increases. All this reduces the coefficient of energy transformation (Fig. 22). Moreover, this decreases the lifetime of the system. Therefore, the total system performance without air supply seems to be not very good.

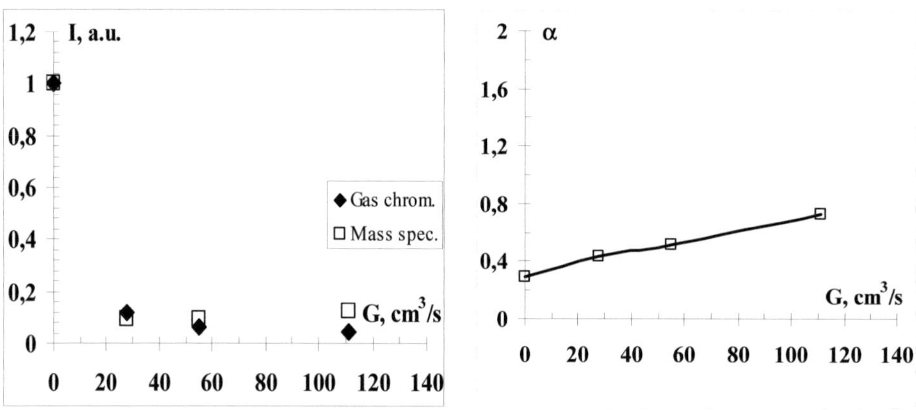


Fig. 22. Coefficient of energy transformation α for the ethanol reforming in the DGCLW with solid electrodes at different air flow rates. $I_d$ =100 mA.

The results of the studies of the post-discharge pyrolysis of ethanol after initial plasma-assisted ethanol reforming are presented in Figs.23-24. The experimental parameters were following: discharge frequency 420 Hz, air flow rate 17-28 $cm^3/s$, time of treatment up to 10 min (600 s); the temperature in the pyrolytic chamber varied from 0 to 870 K. The principal feature of the pulsed discharge power supply in comparison with dc mode is the ability to work with higher energy input in plasma at comparable power capacity. This gives more intensive plasma stimulation of pyrolysis due to deep injection of plasma in the pyrolytic chamber.

Fig.24. shows the $H_2$ as measured by mass-spectrometry and partial $H_2$ content in syngas as measured by gas-chromatography after the ethanol pyrolysis in a pyrolytic chamber. One can see a good correlation between GC and MS data. Also noticeable that the $H_2$ production in the pyrolytic chamber without discharge is very low (indicated by empty signs).

Fig. 25 shows the coefficient of energy transformation α for the reforming system as a function of temperature in the pyrolytic chamber. It is seen that values α increase with increasing temperature. Some modes with the change of air flow (modes 5+2 and 5+3 correspond to additional air supply into the pyrolytic chamber comparably with the air supply in the discharge) have lower energy efficiency than the mode with the constant air flow because of variation of partial output of isobutane $C_4H_{10}$. Fig. 45 shows the rate of syngas production in the reforming system. One can conclude that the investigated combination of electric discharge + post-discharge pyrolysis for the ethanol reforming demonstrates the smart efficiency of this approach.

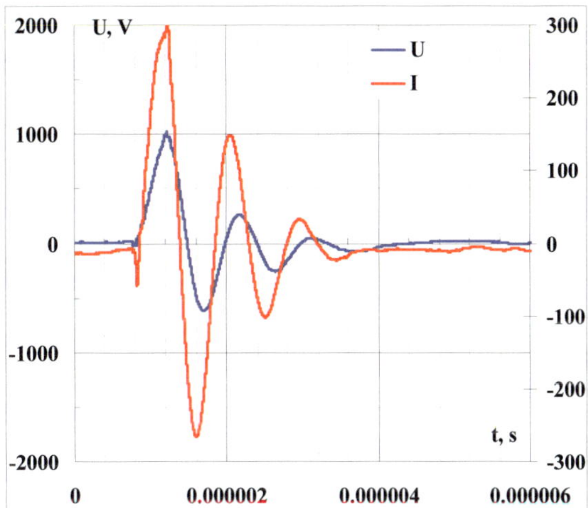


Fig. 23. Voltage and current oscillograms in the pulsed DGCLW.

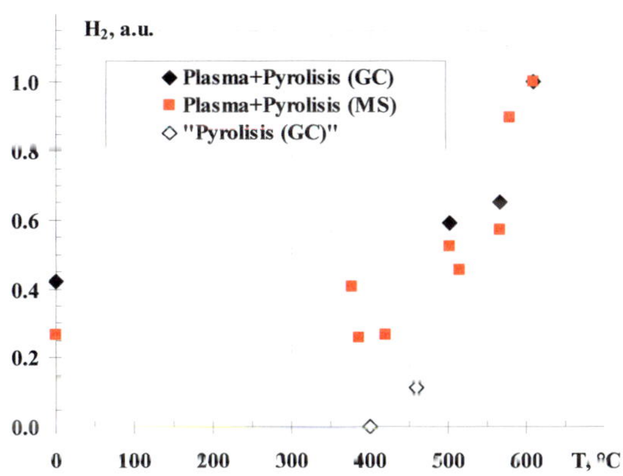


Fig. 24. $H_2$ as measured by mass spectrometry and partial $H_2$ content in syngas as measured by gas chromatography after the ethanol pyrolysis vs. temperature in the pyrolytic chamber.

### 6.1 TORNADO-LE: Experimental results

As it is mentioned above, various methods using thermal and non-thermal plasma for fuel reforming are known. Thermal plasma, which is thermodynamically equilibrium, has characteristics of high ionization by higher energetic density. This has merits of good rate of fuel decomposition but demerits of poor chemical selectivity and high specific energy

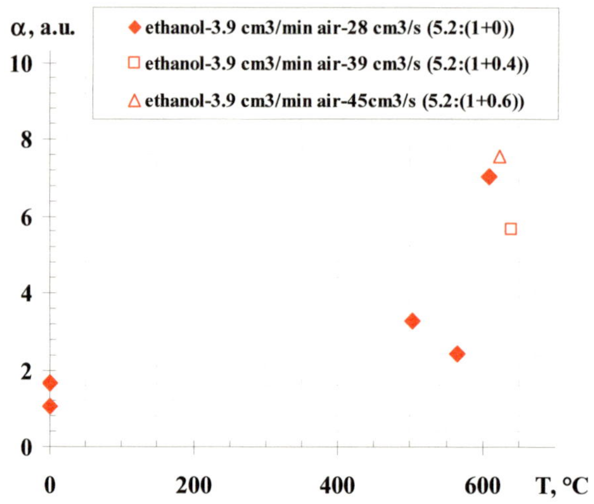


Fig. 25. Coefficient of energy transformation *a* vs. temperature in the pyrolytic chamber.

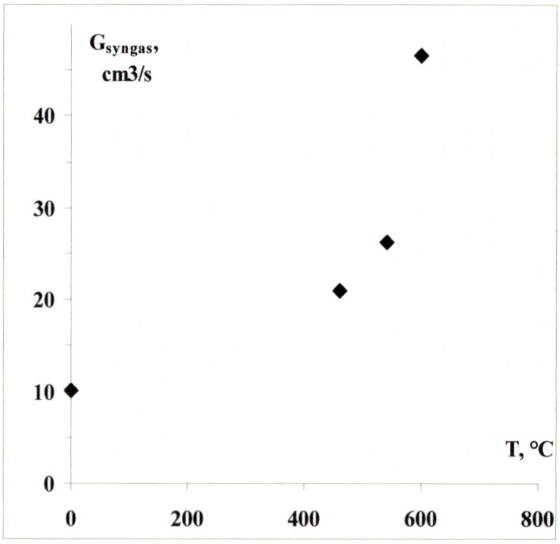


Fig. 26. The rate of syngas production vs. temperature in the pyrolytic chamber.

consumption. Non-thermal (low-temperature) plasma, which is kinetically non-equilibrium, has characteristics of low ionization but benefits of high reactivity and selectivity of chemical transformations providing high enough productivity at relatively low energy consumption; this can be obtained by high voltage discharge in a flow at low or high atmospheric pressures. or reforming with plasma support (pyrolysis, steam reforming, partial oxidization) it is preferable to utilize the high plasma flow rate generators: the pulsed systems and systems on the base of TORNADO discharge type (Chernyak, 2010).

PLS reactor with the DC discharge in a reverse vortex gas flow of Tornado type with the "liquid" electrode (TORNADO-LE) as is shown in Fig.27 has been prepared. It consists of a cylindrical quartz vessel (1) by diameter of 9 cm and height of 5 cm, sealed by the flanges at the top (2) and at the bottom (3). The vessel was filled by the work liquid (4) through the inlet pipe (5) and the level of liquid was controlled by the spray pump. The basic cylindrical T-shaped stainless steel water-cooled electrode (6) on the lower flange (3) made from stainless steel is fully immersed in the liquid. The electrode on the upper flange (2) made from duralumin had a special copper hub (11) with the axial nozzle (7) by diameter 2 mm and length of 6 mm. The gas was injected into the vessel through the orifice (8) in the upper flange (2) tangentially to the cylinder wall (l) and created a reverse vortex flow of tornado type, so the rotating gas (9) went down to the liquid surface and moved to the central axis where flowed out through the nozzle (7) in the form of jet (10) into the quartz chamber (12). Since the area of minimal static pressure above the liquid surface during the vortex gas flow is located near the central axis, it creates the column of liquid at the gas-liquid interface in the form of the cone with the height of ~1 cm above the liquid surface (without electric discharge).

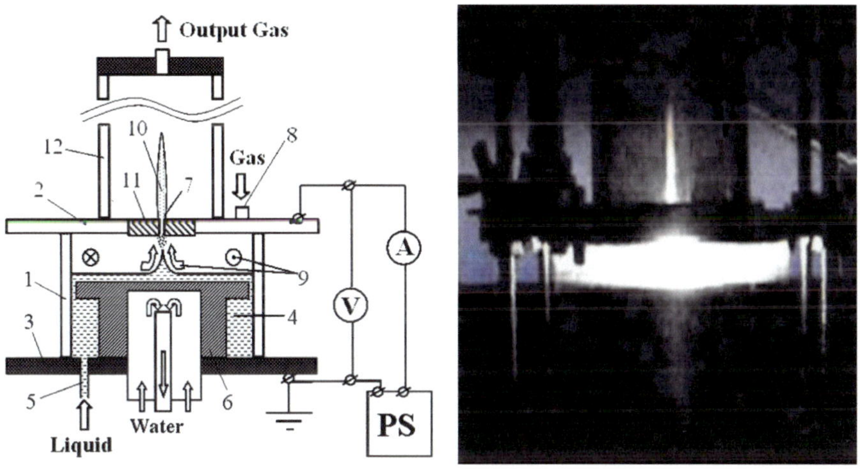


Fig. 27. Schematic (left) of the PLS reactor with the DC discharge in a reverse vortex gas flow of Tornado type with a "liquid" electrode and photo (right) of the TORNADO-LE working in ethanol-water solution.

The voltage was supplied between the upper electrode (2) and the lower electrode (6) in the liquid with the help of the DC power source powered up to 10 kV. Two modes of the discharge operation were studied: the mode with "liquid" cathode (LC) and the mode with "liquid" anode (LA): "+" is on the flange (2) in the LC mode, and "-" is on the flange (2) in the LA mode. The conditions of breakdown in the discharge chamber were regulated by three parameters: by the level of the work liquid; by the gas flow rate $G$; and by the value of voltage $U$. The ignition of discharge usually began from the appearance of the axial streamer; the time of establishment of the self-sustained mode of operation was ~1-2 s. The range of discharge currents varied within 100-400 mA. The pressure in the discharge chamber during the discharge operation was ~1.2 atm, the static pressure outside the reactor was ~1 atm. The elongated ~5 cm plasma torch (10) was formed during the discharge burning in the camera.

The typical current-voltage characteristics of the TORNADO-LE with the liquid anode working in water at different airflow rates are shown in Fig. 28.

Typical emission spectra of plasma in TORNADO-LE inside and outside of reactor are shown in Fig.29. All this spectra were normalize on maximum at wavelength $\lambda_n$=306,7 nm.

Hydroxyl OH and nitrogen $N_2$ bands, hydrogen $H_\alpha$ (656.3 nm), $H_\beta$ (486.1 nm), copper Cu and oxygen multiplet O (777.2; 844.6; 926.6 nm) lines are on emission spectra. Nitrogen band $N_2$(C-B) and copper lines Cu was presented only outside of system.

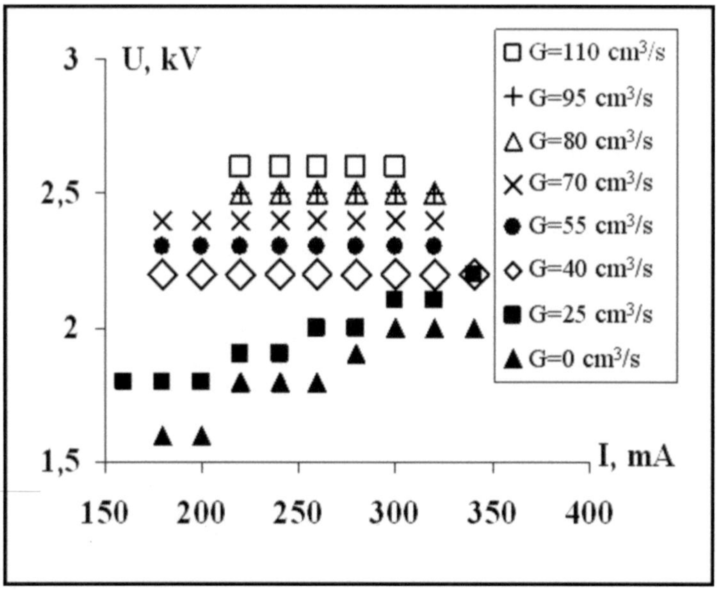


Fig. 28. Current-voltage characteristics of TORNADO-LE with the liquid anode working in ethanol-water solution at diffrent airflow rates.

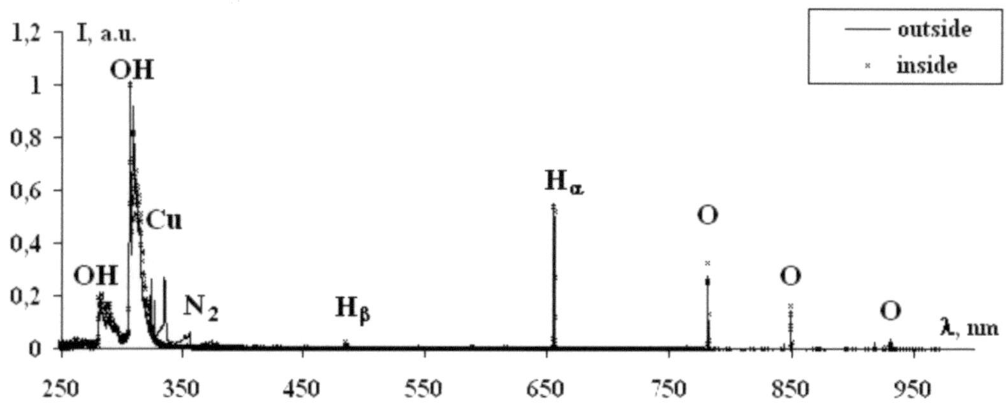


Fig. 29. Typical emission spectrum of discharge plasma inside and outside of reactor in the TORNADO-LE: I=300 mA; G=55 $cm^3/s$; U=2,3 kV; working liquid - distilled water; gas flow - air; mode with solid cathode.

Emission spectra in PLS TORNADO-LE working on distilled water measured at different discharge currents are shown in Fig.30. The electronic temperature $T_e^*$ was determined by relative intensities of hydrogen emission lines $H_\alpha$ (656.3 nm), $H_\beta$ (486.1 nm), relative intensities of emission of oxygen multiplet lines (777.2; 844.6; 926.6 nm)

Excitation temperatures for regime I = 300 mA, G=55 $cm^3/s$, U=2,3 kV in the mode SC were measured. Excitation temperatures inside of reactor was $T_r^*$=4000±500 K, $T_v^*$=4000±500 K and $T_e^*$ = 5000±500 K. Plasma temperatures outside of reactor was $T_r^*$=4500±500 K, $T_v^*$=3000±500 K and $T_e^*$=5000±500 K. Excitation temperatures vibrational $T_v^*$ and rotational $T_r^*$ was obtained by band of OH. Plasma in PLS TORNADO-LE inside of reactor was isothermal. But outside of reactor plasma was non-isothermal.

Experimental and calculated emission spectra by the SPECAIR are shown in Fig. 31. Results of mass-spectrometry of output syngas products after the ethanol reforming in the TORNADO-LE are shown in Fig.32. One can see that content of $H_2$ and CO in output syngas products is quite high.

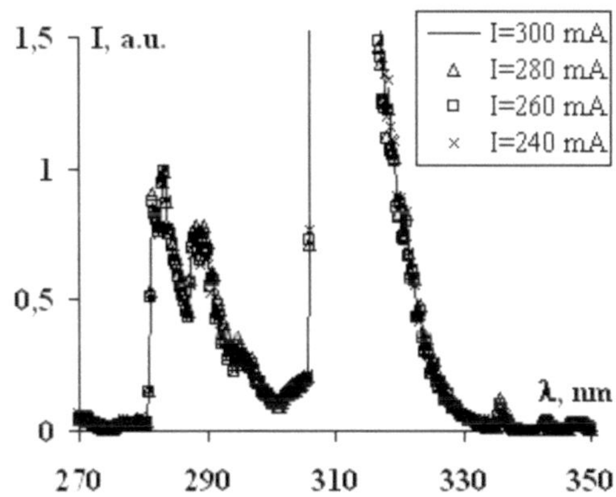


Fig. 30. Emission spectra of discharge plasma in the TORNADO-LE inside of reactor at different discharge currents: working liquid - distilled water; mode with SC; air flow G=55 $cm^3/s$.

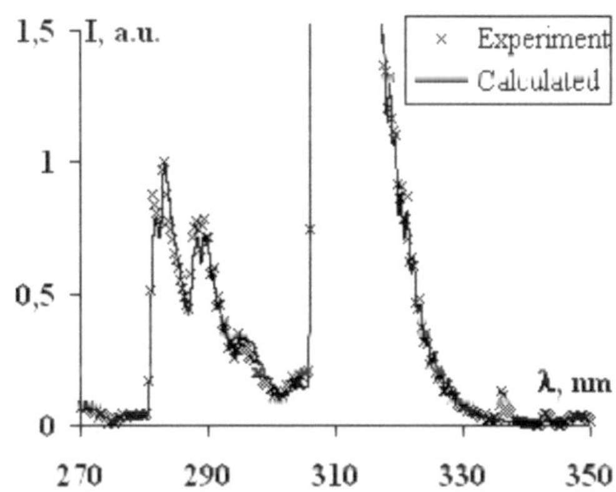


Fig. 31. Experimental emission spectrum of discharge plasma inside of system in the TORNADO-LE and calculated emission spectra by the SPECAIR.

The estimated coefficients of energy transformation (Petitpas, 2007) for the ethanol reforming in the PLS with the TORNADO-LE depending on the initial ethanol concentration in the ethanol-water mixture are presented in Fig.33. One can see that values α are quite high and reach ~0.8 at 25% ethanol-water solution.

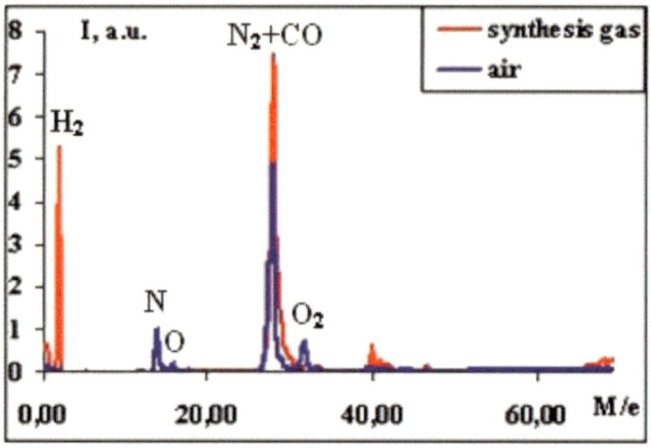


Fig. 32. Mass-spectrometry of gas products after the reforming in the TORNADO-LE with liquid cathode. Voltage 2 kV, current 320 mA, air flow rate 55 cm³/s, mixture $C_2H_5OH/H_2O = 1/4$.

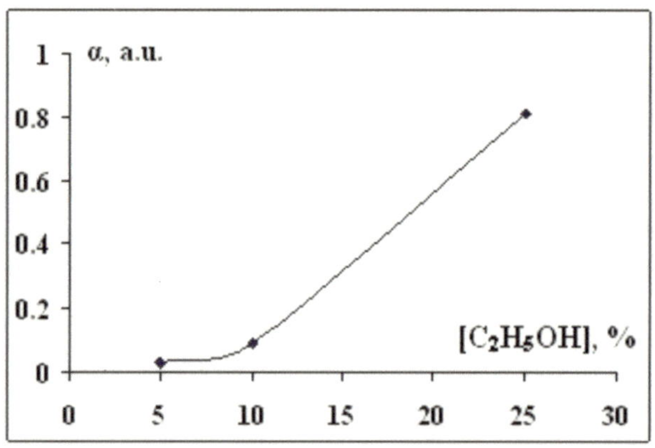


Fig. 33. Coefficient of energy transformation for the ethanol reforming in the TORNADO-LE as a function of the initial ethanol concentration.

### 5.1.1 TORNADO-LE: Experimental results

The composition of the gas at the output of the reactor is analyzed by the gas-phase chromatography using a gas chromatograph 6890 N Agilent with calibrated thermal conductivity detectors. The measurements show, that the main components of the mixture that leave the reactor are $H_2$, $CO$, $CO_2$, $CH_4$, $C_2H_4$, $C_2H_6$ and $C_2H_2$. This result is in good agreement with the results of numerical simulation. Fig.12 shows the comparison between experimental and computational data for different ethanol-to-water ratios in the solution. It is seen that the measured data are in a good agreement with computed ones. When ethanol concentration is 6.5% and 13% at $T$ = 1500 K, then hydrogen and carbon monoxide are the main components of the gas mixture. Fig.34 shows the sum of concentrations of $H_2$ and CO, since CO converts fully into $H_2$ by water gas shift reaction ($CO+H_2O \rightarrow H_2+CO_2$). This process was not taken into account in the kinetic mechanism, because its rate constant is too low to have sufficient influence on kinetics in discharge and post-discharge regions of the subject reactor. The best agreement between measured and calculated data at 26% ethanol concentration is reached at $T = 1000$ K.

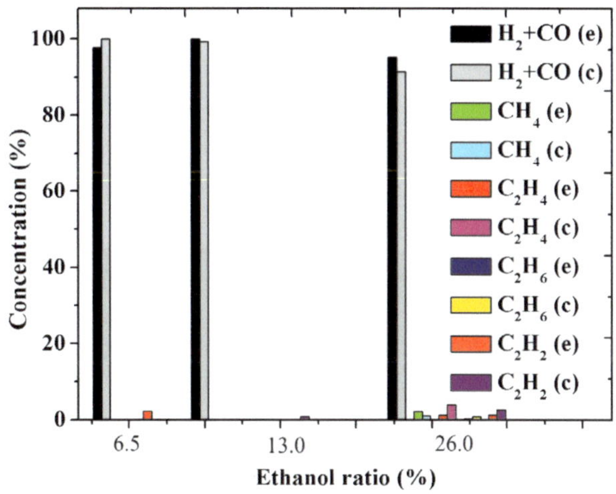


Fig. 34. Comparison between calculated (c) and experimentally (e) obtained concentrations of main components.

In order to define molecular hydrogen yield, the following expression is used (Petitpas, 2007):

$$Y(H_2) = \frac{[H_2]}{3 \cdot [C_2H_5OH] + [H_2O]}. \qquad (10)$$

Here $H_2$ concentration is taken on the outlet of the reactor, and the values of ethanol and water concentrations are taken at the beginning of the discharge phase. Fig.35 shows the calculated dependences of hydrogen yield from gas temperature for different ethanol concentrations. It is seen that in all three cases the yield $Y(H_2)$ grows when $T$ increases. Additionally, the highest yield is reached when ethanol concentration in the solution is 13% and $T = 1500$ K. Under these conditions $H_2$ generation process is efficient through the abstraction of hydrocarbons (7) and through the water H-abstraction (8).

Fig.36 presents the increase of $H_2$ yield at the post-discharge region as compared to the yield at the discharge region. It is seen that the use of the combustion zone is more beneficial in case of low ethanol concentration in liquid at high temperatures. On one hand, the increase of the ethanol-to-water ratio increases ethanol's flow rate (from 1 ml/min to 2 ml/min) and the initial ethanol vapor concentration. On the other hand, the growth of $[C_2H_5OH]$ leads to a decrease of both $[H_2O]$ and the rate of the channel (9). Fig. 36 shows, that the factor of water concentration decrease is much more dominant than the factor of ethanol concentration increase. Therefore, the highest molecular hydrogen yield is reached for 13% concentration of ethanol. These results are in good agreement with the results on Fig.35.

An important characteristic of plasma chemical reactor is its efficiency:

$$\eta = \frac{(Y_{H2} + Y_{CO}) \times LHV\ (H_2)}{IPE + Y_{HC} \times LHV\ (HC)} \qquad (11)$$

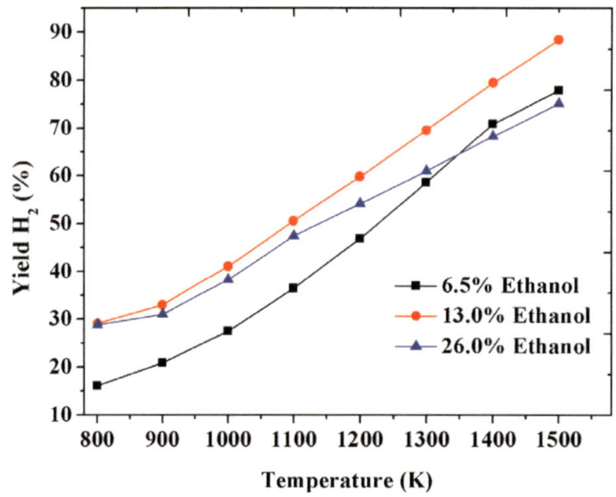


Fig. 35. The calculated dependence of the molecular hydrogen yield from the gas temperature.

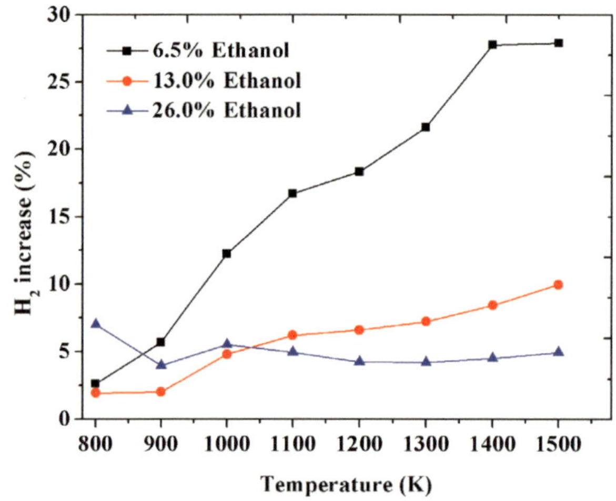


Fig. 36. The increase of yield of the molecular hydrogen as a result of the use post-discharge region

Here the sum of hydrogen and carbon oxide yields due to water gas shift reaction is taken. Also in (11) *LHV* is the lower heating value of the hydrogen and the fuel (ethanol), *IPE* is the introduced plasma energy. It is seen from Fig. 37 that the highest $\eta$ value is 6.5% at ethanol concentration of 26%. It is much lower than the values obtained by the Massachusetts Institute of Technology and the Waseda University groups for pure ethanol (Bromberg, 2006). However, this reactor has a higher value of conversion rate than the named groups' reactors. This characteristic is calculated from the expression (Petitpas, 2007):

$$\chi = \frac{[CO + CO_2 + CH_4 + ...]_{produced}}{2 \cdot [C_2H_5OH]_{injected}}. \tag{12}$$

The last formula describes the efficiency of breaking C-C and C-H bonds in ethanol molecule. Fig. 38 shows that the highest $\chi$ is reached at 6.5% of [$C_2H_5OH$] in the solution. Such behavior is associated with the post-discharge region, where ethanol is fully oxidized. The reactions between the stable hydrocarbons and the active species do not change the sum in the numerator of (12). They lead to the redistribution of carbon atoms between carbon oxides and $C_xH_y$. Let us note that $\chi$ value, which can be larger than 100%, is attributed to the constant ethanol pumping through the discharge zone.

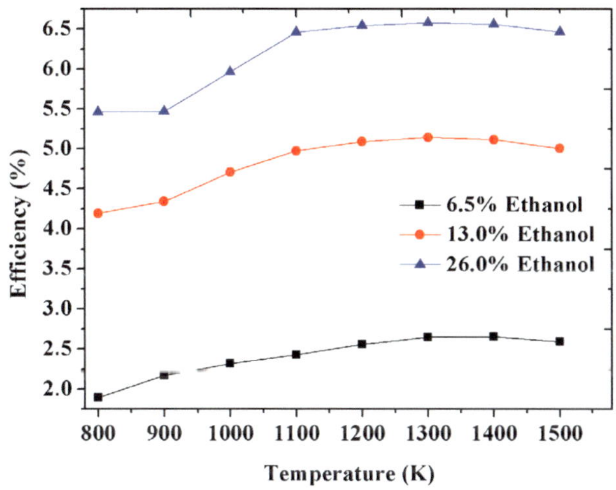


Fig. 37. The efficiency of researched plasma chemical reactor for different mixtures

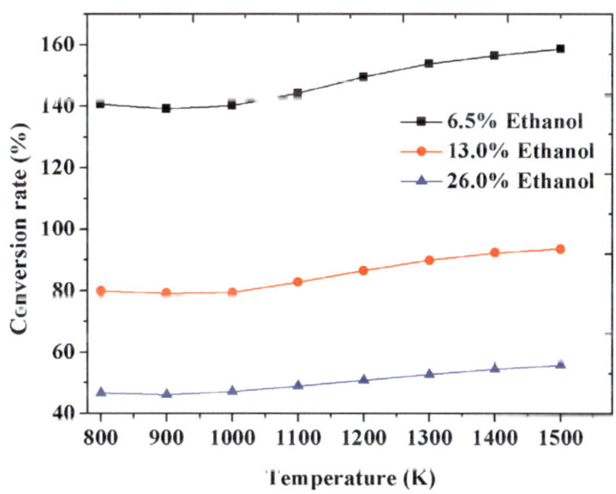


Fig. 38. Conversion rate of researched plasma chemical reactor for different mixtures

Thus, the results of the preliminary studies demonstrate that the glow discharge in the reverse vortex flow of Tornado type with the liquid electrode is very perspective for the plasma-assisted reforming of liquid hydrocarbon fuels.

## 6. Conclusions

It is shown that plasma in the investigated regimes is non-thermal: $T_e^*(Cu) > T_e^*(H) \geq T_v^*(N_2) \approx T_v^*(OH) \geq T_r^*(OH) > T_r^*(N_2)$. The observed discrepancy between electronic temperatures of Cu and H atoms is explained by the additional mechanism of population of the excited electronic states of Cu atoms due to the electron-ion recombination. The observed discrepancy between rotational temperatures determined by the OH (A-X) and $N_2$ (C-B) bands is not well understood and needs further careful examination.

The study of emission spectra of plasma in the blowing transverse arc discharge in a regime of the plasma-assisted fuel combustion in the system of utilization of syngas products after the plasma fuel reforming has shown the following:

- Significant differences in distributions of intensities of spectral atomic lines and molecular bands in the plasma plume along the flow depending on the fuel presence in the air flow. The addition of fuel in the mixture reduces the maxima in the distributions.
- Considerable influences on distributions of electronic temperatures $T_e^*$ of atomic species (H, O) and vibrational $T_v^*$ and rotational $T_r^*$ temperatures of molecular species ($N_2$, OH, CN, $C_2$) in the plasma plume along the flow from both the electrode materials (Cu) and the components of air/fuel mixture (O, N, H). All of them correlate within the measurement uncertainty approximating by linear dependences. Their relationships indicate the changes of the level of the plasma plume non-isothermality along the flow with the increase of the fuel fraction in the initial mixture.

The study of the plasma-liquid system with the discharge in a gas channel with liquid wall in the microporous media under the ultrasound cavitations has shown the following:

- The action of the ultrasound field in the liquid phase increases the efficiency of the nitrous acid production in the work liquid approximately in 1.5 times.
- The ultrasound field in the work liquid influences differently on the content of the emission spectra of discharge plasma. Thus, a part of the emission of OH radicals is reduced comparably to the emission of $N_2$ molecules. There was a linear increasing of relative intensities of H atoms emission with the time of plasma-liquid processing. But the presence of the ultrasound field reduces this rate approximately in 1.5 times.

The comparison of the developed method of numerical modeling of kinetics in air-water-plasma of atmospheric pressure in the plasma-liquid system in the electrical discharge in the gas channel with liquid wall using the assumption of the averaging of the energy that is deposited in the discharge volume without micro-details of the temporal-spatial structure of the discharge and other method of calculations of plasma kinetics in micro-discharges based on the assumption of the multi-channeling of the current in the plasma volume has shown that both approaches give good results in calculations of the component content and concentrations of main components.

The main components of syngas produced from ethanol in the PLS-DGCLW reactor are molecular hydrogen $H_2$ and carbon monoxide CO, which relative yield is many times higher than for hydrocarbons $CH_4$, $C_2H_2$, $C_2H4$, and $C_2H_6$.

The composition content of syngas and the power inputs on the ethanol conversion in the DGCLW discharge depends on the initial gas that forms the plasma and on the ethanol-water ratio in the solution. The output hydrogen concentration grows linearly with discharge current.

The kinetic plasma-chemical modeling is in a fairly good agreement with experimental data, at least, for the main syngas components, $H_2$ and CO, predicting a non-thermal

plasmachemical mechanism of the ethanol conversion in the investigated plasma-liquid system.
The combination of electric discharge plasma-assisted ethanol reforming and post-dischargepyrolysis of ethanol for hydrogen-rich syngas production is proposed and tested.
The synergetic effect of increasing of total energy efficiency of fuel reforming is demonstrated. PLS system with the glow discharge in the reverse vortex flow of Tornado type with the liquid electrode is proposed and tested. Preliminary results of plasma-assisted reforming of ethanol-water solutions have demonstrated rather high efficiency of fuel conversion and energy transformation.
Kinetics in "tornado" type electrical discharge in ethanol/water/air mixture in the discharge and post-discharge regions were investigated. It was shown, that the ethanol conversion was taking place at both regions. The discharge region acts as a catalyst for the generation of active species (O, OH, H, etc), gas heating and ethanol/water conversion into molecular hydrogen, carbon oxides and hydrocarbons $C_xH_y$. In the post-discharge region the conversion of ethanol continues, with the additional process of conversion of hydrocarbons $C_xH_y$.
Numerical simulation of kinetics showed, that the main channels of $H_2$ generation in plasma were ethanol abstraction for the first 10-100 μs, and hydrocarbons abstraction afterwards. Additionally, the conditions when the reaction between $H_2O$ and hydrogen atoms was the main channel of $H_2$ production were found.
A kinetic mechanism, which used to describe adequately the chemistry of main components, was proposed. The model did not account for nitrogen-containing species, and nitrogen was considered only as a third body in recombination and dissociation reactions. The comparison between experiments and calculations showed, that the mechanism can describe adequately the concentrations of the main components ($H_2$, CO, $CO_2$, $CH_4$, $C_2H_4$, $C_2H_6$ and $C_2H_2$). Sensitivity analysis for the two main components ($H_2$ and CO) revealed the most important chemical reactions.
The highest hydrogen yield was reached when concentration of ethanol in the solution was 13%. However, the use of post discharge region was more beneficial at 6.5% $C_2H_5OH$ concentration, because it increased [$H_2$] by 30% as compared to the discharge region. Additionally, this regime had the highest conversion rate among the investigated regimes. At the same time, the plasma reactor had the highest efficiency when the concentration of ethanol in the solution was 26%.

**Future Work**

We are planning the following research activities in the future:

- To study the regimes and parameters of non-thermal plasma reforming of liquid fuels in the pulsed discharge in a gas channel with liquid wall.
- To study the regimes and parameters of non-thermal plasma reforming of liquid fuels in the plasma-liquid system with reverse vortex flow using pulse and DC discharge of tornado type with "liquid" electrode.
- To study the reforming of liquid hydrocarbons in plasma-liquid systems with microporous media (liquid + microbubbles) and aerosols (gas + microdroplets) using low-frequency (~20 kHz) and high-frequency (~800 kHz) ultrasound.
- To study the regimes and parameters of plasma-supported combustion of paraffin fuels by using gas dynamic electric discharges.

## 7. References


*AEO2011* Early Release Overview, available on the sight: http:// www.eia.gov/forecast/aeo/pdf/0383er(2011).pdf

Bromberg L, Cohn D R, Rabinovich A, Alexeev N, Samokhin A, Hadidi K, Palaia J, Margarit-Bel N (2006). Onboard Plasmatron Hydrogen Production for Improved Vehicles. *MIT Plasma Science and Fusion Centre*, (February 6 2006), PSFC JA-06-3 (173 pp).

Buchnev V V, Koval S D, Chernyak V Ya (2000). Nonequilibrium arc plasma in transverse air flow. *Bull. Univ. of Kyiv: Series: Phys. & Math.*, No 1, pp. 315-319, ISSN: 1812-5409.

Chernyak V Ya, Naumov V V, Yukhymenko V V, Babich I L, Zrazhevskyy V A, Voevoda Yu V, Pashko T V (2005). Arc discharge in a cross flow of gas. *Probl. Atom. Sci. Technol., Ser. Plasma Phys.*, Vol. 11, No 2, pp. 164-166, ISSN: 1562-6016.

Chernyak V Ya, Yukhymenko V V, Slyusarenko Y I, Solomenko E, Olshevskii S V, Prisyazhnevich I V, Naumov V V, Lukyanchikov V, Demchina V P, Kudryavtsev V S (2007). Plasma conversion of ethanol. *Ind. Heat Eng.*, Vol. 29, No 7, pp. 165-169, ISSN: 0204-3602.

Chernyak V. Ya., Olszewski S. V., Yukhymenko V. V., Solomenko E. V., Prysiazhnevych I V, Naumov V V, Levko D S, Shchedrin A I, Ryabtsev A V, Demchina V P, Kudryavtsev V S, Martysh E V, and Verovchuck M.A. (2008). Plasma-Assisted Reforming of Ethanol in Dynamic Plasma-Liquid System: Experiments and Modeling. *IEEE Trans. Plasma Sci.* Vol.36, No 6, pp.2933-2939, ISSN: 0093-3813.

Dagaout P, Togbe C.2008. Experimental and Modeling Study of the Kinetics of Oxidation of Ethanol−Gasoline Surrogate Mixtures (E85 Surrogate) in a Jet-Stirred Reactor. *Energy & Fuels*, Vol.22, No 5, (July 2008), pp.3499-3505.

Dunphy M P, Patterson P M, Simmie J M 1991. High-temperature oxidation of ethanol. Part 2.—Kinetic modeling, *Journal of the Chemical Society, Faraday Transactions*, Vol.87, No 16, pp.2549-2559.

Fridman A (2008). *Plasma Chemistry* (Cambridge: Cambridge University Press), ISBN-10: 0521847354.

Frenkel Ya. I., 1975. *Kinetic theory of liquids* (Moscow, Nauka).

Held T J, Dryer F L 1998 A comprehensive mechanism for methanol oxidation. *International Journal of Chemical Kinetics*, Vol.30, No11, pp.805-830. ISSN: 1097-4601.

Kakami A., Egawa T., Yamamoto N., Tachibana T., 2010. Plasma-assisted combustion of $N_2O$/Ethanol propellant for Space Propulsion, *Proceedings of 46th AIAA/ASME/SAE/ASEE Joint Propulsion Conference*, (5 - 28 July 2010), Nashville.

Konnov A. 2008. Remaining uncertainties in the kinetic mechanism of hydrogen combustion. *Combustion and Flame.* Vol.152, No 4, (March 2008), pp.507-528. ISSN: 0010-2180.

Kosarev I N, Aleksandrov N L, Kindysheva S V, Starikovskaia S M and Starikovskii A Yu 2008 Kinetic mechanism of plasma-assisted ignition of hydrocarbons. *J. Phys. D: Appl. Phys.* 41 No 3 (7 February 2008) 032002 (6pp). ISSN 0022-3727.

Levin D A, Laux C O, Kruger C H 1999. A general model for the spectral calculation of OH radiation in the ultraviolet. *J. Quant. Spectrosc. Radiat. Transfer.* Vol. 61, No 3. pp.377-392, ISSN: 0022-4073.

Levko D S, Shchedrin A I, Chernyak V Ya, Olszewski S V 2010.The effectivity of ethanol conversion in thermal and non-thermal plasma, *Technical Physics* , Vol.80, No 11, pp.152-155, ISSN: 1063-7842.

Marinov N. M. (1999). A detailed chemical kinetic model for high temperature ethanol oxidation. *International Journal of Chemical Kinetics*, Vol.31, No 3, pp. 183-220, ISSN: 1097-4601.

Matveev I.B., Rosocha L.A. (eds.) (2007). Special issue on plasma-assisted combustion. *IEEE Trans. Plasma Sci.*, Vol. 35, No 6, pp. 1605-1606, ISSN: 0093-3813.

Navratil Z., Trunek D., Smid R., Lazar L. (2006). A software for optical emission spectroscopy - problem formulation and application to plasma diagnostics. *Czechoslovak J. Phys.* Vol. 56, No suppl. B, pp.944-951, ISSN: 0011-4626.

NIST Standard Reference Databases Online. Available: http://www.nist.gov/srd/

Petitpas G, Rollier J-D, Darmon A, Gonzales-Aguilar J, Metkemeijer R, Fulcheri L (2007). A comparative study of non-thermal plasma assisted reforming technologies. *Int. J. Hydrogen Energy*, Vol. 32, No 14, pp. 2848-2867, ISSN: 360-3199.

Prieto G, Okumoto M, Shimano K, Takashima K, Katsura S, Mizuno A (2001). Reforming of heavy oil using non-thermal plasma. *IEEE Trans. Ind. Appl.*, Vol. 37, No 5, pp. 1464-1467, ISSN: 0093-9994.

Prysiazhnevych I.V., Chernyak V.Ya., Yukhymenko V.V., Olszewski S.V., Lisitchenko T.E. (2009). Emission spectroscopy of transversal atmospheric pressure discharges. *Problems of Atomic Science and Technology.* Vol. 15 (Series: Plasma Physics), № 1, pp.168-170.

Raizer Yu.P. (1997). *Gas Discharge Physics* (Berlin, Springer-Verlag), ISBN-10: 3540194622.

Shchedrin A. I., Levko D. S., Chernyak V. Ya, Yukhymenko V. V, Naumov V. V., 2008. Effect of air on the concentration of molecular hydrogen in the conversion of ethanol by a non-equilibrium gas-discharge plasma. *JETP Lett.* Vol. 88, No 2,pp. 99-102, ISSN: 0021-3640.

Shchedrin A. group Web site. Available: http://www.iop.kiev.ua/~plasmachemgroup

SPECAIR. Available: http://www.specair-radiation.net/

Shchedrin A I, Levko D S, Chernyak V Ya, Yukhymenko V V, and Naumov V V 2009.Conversion of air mixture with ethanol and water vapors in nonequilibrium gas-discharge plasma.*Technical Physics Lett.* Vol. 35, No 5, (May 2009), pp. 449-451. ISSN: 1063-7850.

Soloshenko I. A., Tsiolko V. V., Pogulay S. S., Terent'eva A. G., Bazhenov V. Yu., Shchedrin A. I., Ryabtsev A. V. and Kuzmichev A. I. 2007. The component content of active particles in a plasma-chemical reactor based on volume barrier discharge. *Plasma Sources Sci. Technol.* Vol.16, No 1, (February 2007), pp.56-87, ISSN 0963-0252.

Wang F, Liu J B, Lee C L, Ronney O D, Gundersen M A (2004). Effect of fuel type on flame ignition by transient plasma discharges. *42nd AIAA Aerospace Sci. Meeting & Exhibit,* (5 - 8 January 2004), Reno, Nevada, AIAA Paper 2004-0837 (10pp), ISSN: 1533-385X.

Yukhymenko V.V., Chernyak V.Ya., Naumov V.V., Veremij Yu.P., Zrazhevskij V. A. (2007). Combustion of ethanol + air mixture supported by transverse arc plasma. *Probl. Atom. Sci. Technol., Ser. Plasma Phys.*, Vol. 13, No 1, pp. 142-144, ISSN: 1562-6016.

# 7

# Developing Organisms for Consolidated Bioprocessing of Biomass to Ethanol

Willem H. van Zyl, Riaan den Haan and Daniel C. la Grange
*Department of Microbiology, University of Stellenbosch, Stellenbosch*
*South Africa*

## 1. Introduction

Lignocellulosic biomass is an abundant, renewable feedstock for sustainable production of biofuels and chemicals. The main technological barrier that impedes widespread utilization of this resource for production of fuels and other commodity products is the lack of low-cost technologies to overcome the recalcitrance of lignocellulose. Organisms that hydrolyse the cellulose and hemicelluloses in biomass and produce a commodity product such as ethanol at a high rate and titre would significantly reduce the costs of biomass conversion. This would allow steps that are currently accomplished in different reactors, often by different organisms, to be combined in a consolidated bioprocess (CBP). While there is still no ideal organism to use in one-step biomass conversion, several candidates have been identified that are in various stages of development for establishment of a cellulolytic system and/or improvement of product-forming attributes. This chapter assesses the *status quo* for CBP organismal development either by enabling non-cellulolytic organisms to grow on cellulosic substrates or by improving product forming abilities of native cellulose utilizing organisms. The authors also discuss feedstocks that are available for the production of biofuels using CBP and assess how process integrations can make CBP economically feasible in the near future.

The increasing demand for oil coupled to the premium many governments place on greater energy security and environmental concerns have led to the development of an active biofuels industry (Van Zyl et al., 2011). First generation biofuels such as ethanol from starch or sugar already contribute considerable amounts of liquid fuels in several countries. However these technologies suffer from a shortage in the availability of feedstock in order to displace a more significant amount of petroleum based fuels. Lignocellulose represents the most widespread and abundant source of carbon in nature and is the only source that could provide a sufficient amount of feedstock to satisfy the world's energy and chemicals needs in a renewable manner (Hill et al., 2006; Van Zyl et al., 2011). Second generation biofuels such as ethanol form cellulosic biomass therefore seeks to overcome the problem of feedstock supply shortage by utilizing the energy contained in total plant biomass. Current technologies for conversion of biomass to ethanol commences with a pretreatment step during which physical and/or chemical processes are used to render the polymeric sugar fractions more accessible to conversion by enzymatic processes (Stephanopoulos, 2007). The type of feedstock will predetermine the optimal type of pretreatment which in turn defines the optimal enzyme mixture to be used in subsequent hydrolysis steps and the composition of the hydrolysis products. Four biologically mediated events occur during conversion of pretreated lignocellulose to ethanol via processes featuring enzymatic hydrolysis namely: (i) production of depolymerising enzymes (cellulases and hemicellulases), (ii) hydrolysis of the polysaccharide constituents of pretreated biomass, (iii) fermentation of the hexose sugars present, and (iv) fermentation of pentose sugars present (Lynd et al., 2002). Improvements of biomass conversion technology generally entail the consolidation of two or more of these steps. Hydrolysis and fermentation steps can be combined in simultaneous saccharification and fermentation (SSF) of hexoses or simultaneous saccharification and co-fermentation (SSCF) of both hexoses and pentoses. The ultimate objective would be a one-step "consolidated" bioprocessing (CBP) of lignocellulose to bioethanol, where all four of these steps occur in a single reactor and a single microorganism or microbial consortium converts pretreated biomass to a commodity product such as ethanol without added saccharolytic enzymes. CBP would represent a breakthrough for low-cost biomass processing, due to economic benefits of process integration (Galbe et al., 2005; Hahn-Hägerdal et al., 2007; Hamelinck et al., 2005; Robinson, 2006) and avoiding the high costs of enzymes that make the biochemical conversion route unattractive (Anex et al., 2010; Kazi et al., 2010).

Lignocellulosic plant biomass represents the largest source of renewable carbon on earth and consists of 40–55% cellulose, 25–50% hemicellulose and 10–40% lignin, depending on whether the source is hardwood, softwood, or grasses (Sun & Cheng, 2002). The main polysaccharide present is water-insoluble cellulose that represents the major fraction of fermentable sugars. Full enzymatic hydrolysis of crystalline cellulose requires synergistic action of three major types of enzymatic activity (i) endoglucanases, (ii) exoglucanases, including cellodextrinases and cellobiohydrolases, and (iii) β-glucosidases (Zhang & Lynd, 2004). Endoglucanases are active on the non-crystalline or amorphous regions of cellulose and their activities yield cellobiose and cellooligosaccharides as hydrolysis products (Figure 1). Cellobiohydrolases are processive enzymes that are active on the crystalline regions of cellulose and most yield almost exclusively cellobiose as their main hydrolysis product. In turn, β-glucosidases convert cellobiose and some cello-oligosaccharides to glucose. Hemicellulose refers to a number of heterogeneous structures, such as (arabino)xylan, galacto(gluco)mannan, and xyloglucan (Sun & Cheng, 2002). These chemically diverse polymers are linked together through covalent and hydrogen bonds, as well as being intertwined and can be chemically bound to the lignin fraction. Although many pretreatment protocols remove variable amounts of hemicelluloses, it remains imperative from an economic perspective that sugars contained in the hemicellulose fraction of lignocellulose are also converted to ethanol (Hahn-Hägerdal et al., 2001). The compositions of the major and minor types of hemicelluloses present in lignocellulosic feedstocks and the enzymes required to hydrolyze them are reviewed elsewhere (Girio et al., 2010; Van Zyl et al., 2007)

## 2. Feedstocks and conversion technologies

Many countries are embarking on ambitious biofuels policies resulting in a rapid global increase in the demand and supply of biofuels (Sastri et al., 2008). In recent years significant progress has been made towards the development of different technologies for the production of particularly bioethanol, but also butanol, alkanes and terpenes from lignocellulosic material (Fortman et al., 2008). Sugar-based ethanol is the least expensive biofuel and its production is mainly constrained by the availability of feedstock (Somerville et al., 2010). Grain-based ethanol is hampered by high feedstock prices and competition

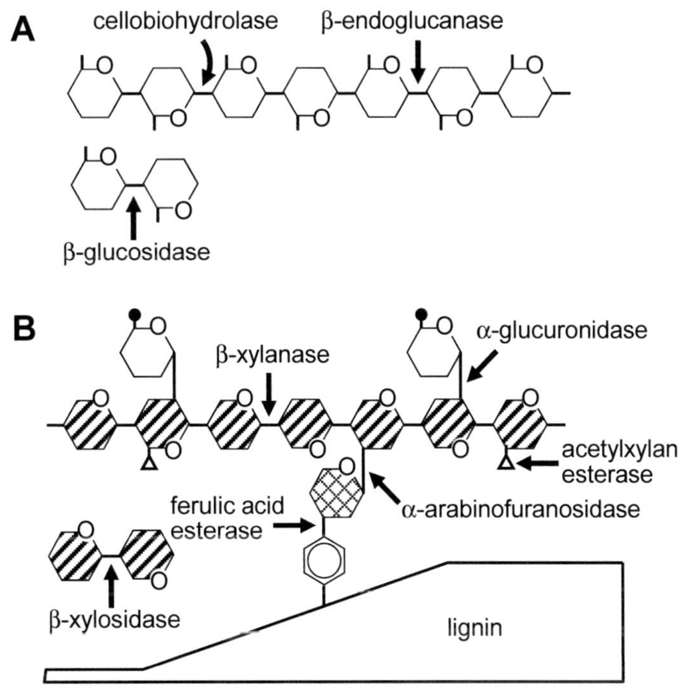


Fig. 1. Illustration of the complexity of cellulose (A) and arabinoxylan (B) as major polysaccharides in lignocellulose and the enzymes involved in their degradation. Hexoses are distinguished from pentoses by the presence of a protruding line from the cyclic hexagon (pyranose ring), depicting the $CH_2OH$ group. Hydrolase enzymes and the bonds targeted for cleavage in the polysaccharides structures are indicated by arrows (Van Zyl et al., 2007).

with food markets. Cellulosic biofuels hold great promise, but the necessary technology advances to overcome the recalcitrance of lignocellulose are needed to enable profitable production of biofuels like ethanol, (Fortman et al., 2008). Potential cellulosic feedstock are numerous and widespread and include woody biomass, perennial grasses, and agricultural and forest residues (Table 1).

Woody biomass can and has been harvested sustainably for lumber and paper for many years (Somerville et al., 2010). Furthermore, the electronic media and paper recycling help to reduce the demand for paper pulp. All this biomass is therefore potentially available for energy. The potential energy available in this biomass source is enormous. It is estimated that biomass harvested in the Northern Hemisphere from wood products has an energy content equal to 107% of the liquid fuel consumption of the United States (Goodale et al., 2002).

Perennial plants such as switchgrass, *Miscanthus* and Napier grass have high photosynthetic capacity, as well as water and nitrogen use efficiency (Somerville et al., 2010; Ansah et al., 2010). They are fast growing and have efficient root systems allowing them to reach deep into the soil for water. The root produces a network of stems and roots that holds onto soil to prevent erosion. These and other perennial grasses are capable of averaging around 30 metric tons of dry matter per hectare per year.

Maize is the largest crop in the world in terms of grain production at around 820 million metric tons per annum (Somerville et al., 2010). A more or less equal amount of stems and stripped cobs (stover) is potentially available for the production of fuel. If half of the stover can be converted to ethanol it would double the amount of ethanol produced from maize. However, the removal of this much stover would lead to significant losses of carbon from the soil and would aggravate erosion. It would also increase the amount of fertilizer needed to maintain good crop yields.

| Crop | Growth cycle (months) | Water needs (mm/season) | Average productivity (dry t/ha/year) | Ethanol yield (l/ha) |
|---|---|---|---|---|
| Perennial grass | | | | |
| Switchgrass | 12 | 700 | 15 | 5000 |
| Miscanthus | 12 | 750 | 25 | 7500 |
| Napier grass | 3 | 1500 | 40 | 12500 |
| Wood | | | | |
| Poplar | 36 | 900 | 8 | 2000 |
| Agricultural crop | | | | |
| Sugarcane | 15 | 2000 | 21 | 10000 |
| Sweet sorghum | 4 | 600 | 20 | 6000 |
| Corn | 4 | 750 | 10 | 3800 |
| Drought resistant crop | | | | |
| Agave | 60 | 400 | 20 | 7500 |

Ethanol yield refers to the total amount produced from a feedstock, including grain and stover or sugar and bagasse.

Table 1. Summary of biofuels feedstock (adapted from La Grange (2007) and Summerville et al., 2010)

Ethanol from sugarcane constitutes one of the largest sources of biofuels in the world (Somerville et al., 2010). Currently only 4.6 million hectares of sugar cane are used for bioethanol production in Brazil. In a recent announcement the Brazilian government stated that the area used for sugarcane cultivation would be increased substantially, but it would be limited to 63.5 million hectares (Decree No.6.961 2009). Approximately 60 million hectares of this allocated land would be available for biofuels production. Currently only the sugar component is used for the production of bioethanol. Once the technology becomes available, the cellulosic component in bagasse could also be used for the production of ethanol or other liquid fuels (Fortman et al., 2008). Estimates, based on the expected increase in sugarcane crops and cellulosic fuel from the 60 million hectares available land, are that Brazil could produce 14% of the current world transportation fuel demand of 4900 Gl by the year 2030. (Somerville et al., 2010). Currently, South Africa produces about 20 million tonnes of cane (about 50% of Africa's production) on 325 000 hectares of land. If the full potential of the estimated 6 million hectares of land suitable for sugarcane production in Angola, Malawi, Mozambique, Tanzania, Zambia and Zimbabwe are also realised, about 400 million tonnes of cane can be produced, which could yield 49 Gl ethanol, about 20% of Africa's current total petroleum consumption (Somerville et al., 2010; Watson, 2011).

Almost a fifth of the terrestrial surface on earth is semi-arid and prone to droughts with a rainfall of between 200 and 800 mm per year. If this is combined with agricultural land that has fallen out of production, the amount of land available for the production of biomass using drought resistant species such as *Agave* is vast. *Agave* spp. thrive under these conditions and produce between 1 and 34 dry tons of biomass per hectare per year (Davis et al., 2011). Obtaining biomass in sufficient quantities to merit the construction of commercial scale facilities will be a major concern in future, fortunately there are a number of different crops suitable for different environmental conditions that could enable sustainable production of sufficient quantities of biomass.

One of the major challenges of biomass harvesting and delivery to conversion facilities remains yield and density, which determines the volume of the biomass. The density of grassy feedstock to woody feedstocks can vary between ~70 $kg/m^3$ to ~300 $kg/m^3$. For a 200 - 1 000 million liter per annum cellulosic ethanol plant, 0.8 - 4.0 million tonnes of dry biomass are required, which would require 50 - 250 trucks per day to deliver the biomass (Richard, 2010). Innovative ways of harvesting and delivering biomass to conversion facilities have to be developed to ensure cost-effective production of cellulosic ethanol at significant quantities. These could include dedicated production of biomass (e.g. as found in the sugar and paper-and-pulp industries) or the development of biomass commodity markets, parallel to agricultural commodity markets, such as grain and livestock. Defining of strict specifications for biomass delivered will be crucial to ensure a uniform feedstock for take-off by biomass conversion industries.

## 3. CBP organismal development

While several microorganisms can be found in nature with the ability to produce the required enzymes to hydrolyse all the polysaccharides found in lignocellulose, there is no organism with the ability to directly hydrolyze these polysaccharides and ferment the liberated sugar to a desired product such as ethanol at rates and titers required for economic feasibility (Hahn-Hägerdal et al., 2006; Lynd et al., 2005). Strain development is therefore the most important technical obstacle towards the conversion of lignocellulose to commodity products in a CBP configuration (Bothast et al., 1999; Alfenore et al., 2002). Organisms with broad substrate ranges and cellulolytic and/or hemicellulolytic abilities generally suffer from poor growth characteristics or poor product producing characteristics. These include poor yield, titer and rate or producing mixtures of products where desirable products are produced along with undesirables. In comparison, organisms with desirable product producing qualities often suffer from limited substrate range including lack of cellulolytic ability, poor fermentation qualities, and sensitivity to the inhibitors present in pretreated lignocellulosic biomass. The ideal CBP organism should be robust with regards to inhibitor tolerance, able to degrade lignocellulose and utilize hexose and pentose sugars at high efficiency. Furthermore characteristics such as the ability to simultaneously utilize sugars, GRAS (Generally Regarded as Safe) status, minimal nutrient supplementation and tolerance of low pH and high temperature would also be desirable in a CBP organism (Zaldivar et al., 2001). Four different approaches have been followed to develop such an organism, these are summarized in Figure 2. Due to the variety of feedstocks likely to be used, the diversity in pretreatment methods and the difference in desired products produced; there is scope for development of organisms with a range of different traits (La Grange et al., 2010).

### 3.1 Eukaryotic cellulolytic organisms for CBP

Several species of cellulolytic fungi, such as *Trichoderma reesei*, naturally produce a large repertoire of saccharolytic enzymes to digest lignocellulose efficiently, assimilate all lignocellulosic sugars, and convert these sugars to ethanol, showing that they naturally

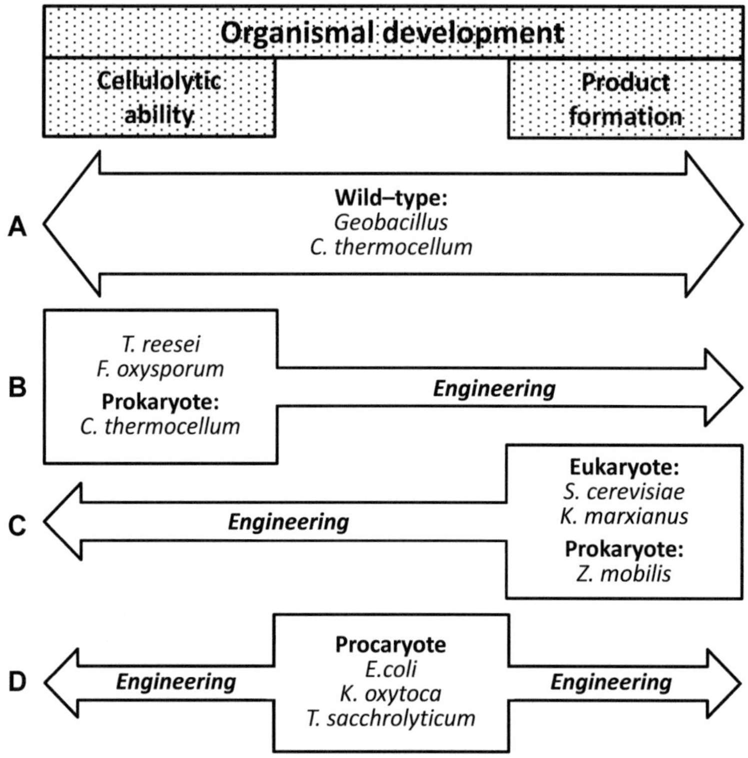


Fig. 2. Different recombinant strategies to engineer cellulolytic and product formation organisms for the CBP process. The four strategies are; (A) isolating microbes with both cellulolytic and product formation properties, (B) engineering superior cellulolytic microbes to produce desired products, (C) engineering cellulolytic activity into superior product forming organisms and (D) select organisms with special features for fermenting lignocellulosics and engineering both cellulolytic and product formation properties.

possess all pathways for conversion of lignocellulose to bioethanol (Chambergo et al., 2002; Lynd et al., 2002). Despite recent advances in engineering cellulases to be more efficient and less costly, the complete saccharification of pretreated lignocellulose still requires a long time for digestion and high loadings of enzyme (30–50 mg enzyme per g of crystalline cellulose) (Xu et al., 2009). Therefore, a biorefinery consuming thousands of tonnes of biomass per day will require many tonnes of cellulase preparation to operate. Currently, only fungi naturally produce the required amounts of cellulase to meet this need. Some strains of *T. reesei* are reportedly able to produce more than 100 g cellulase enzyme per liter of culture broth (Cherry & Fidantsef, 2003). The primary advantages of *T. reesei* as a CBP organism are: (i) the production of cellulases in sufficient quantities and at reasonable cost (ii) that it is already established commercially, and specific mutants are available that can be grown at a low cost and in quantities needed for the emerging biorefinery industry and (iii) that it has all the metabolic pathways necessary to utilize all lignocellulose sugars for production of ethanol (Xu et al., 2009). However, there are significant challenges to overcome before *T. reesei* can be considered as a CBP organism such as (i) ethanol yield and rate of production are low, (ii) ethanol tolerance is low, and (iii) mixing during fermentation may require more energy owing to its filamentous cell morphology.

*T. reesei* is an obligate aerobe making its survival for long periods without oxygen difficult (Rautio et al., 2006). The foremost reason for this is that the genes encoding glycolytic enzymes are strongly repressed in the absence of oxygen. However, preliminary studies indicated that *T. reesei*, could produce cellulases when grown aerobically on cellulose that continued to degrade cellulose to sugars and ferment these sugars to ethanol when cultures were rendered anaerobic, although acetic acid was produced as a major by-product (Xu et al., 2009). It was also shown that *T. reesei* could convert the five primary lignocellulosic sugars to ethanol but the ethanol yields and production rates were low. Therefore the major limitation for efficient ethanol production by *T. reesei* do not lie in the absence of the relevant genes and pathways but are more likely related to the low expression of these genes or the activity of the enzymes encoded. Approaches to solving these problems are to enhance the expression of the relevant genes at the transcriptional level and/or to introduce heterologous genes that encode enzymes with higher activities. It is expected that ethanol formation and tolerance could be improved using the following strategies: (i) identification and modification of genes involved in ethanol tolerance; (ii) introduction of heterologous genes, such as *S. cerevisiae* pyruvate decarboxylase and alcohol dehydrogenase to enhance the classical ethanol synthetic pathway and (iii) knockout of *T. reesei* genes responsible for the production of byproducts. Furthermore, *T. reesei* is multi-cellular with a growth mode that results in an extended cellular chain forming hyphae. This requires more energy input for mixing and handing in fermentation tanks, compared to unicellular yeast. Another challenge for the application of *T. reesei* as a CBP organism is the modification of its growth into a compact pelleted form rather than as extended hyphae.

Another filamentous fungus, *Fusarium oxysporum*, also produces the enzymes required to break down cellulose and hemicellulose while simultaneously fermenting the corresponding hexoses and pentoses to ethanol albeit at relatively low yields (Anasontzis et al., 2011; Panagiotou et al., 2005). In SSF of cellulose a *Fusarium oxysporum* wild type strain F3 was able to grow at a maximum specific growth rate of 0.023 $h^{-1}$ on cellulose in aerobic conditions and produced ethanol with a yield of 0.35 g/g cellulose under anaerobic conditions. The cellulase system in *F. oxysporum* is well balanced as no cellobiose accumulated during growth on cellulose. The strain was further shown to effectively produce a complete system of hydrolytic enzymes when grown on various agro-industrial lignocellulose by-products, such as dry citrus peels, corn cob and brewer's spent grain and simultaneously ferment the corresponding oligosaccharides to ethanol (Anasontzis et al., 2011; Xiros et al., 2008). In these studies, the hydrolysis of the lignocellulosic material was shown to be the major bottleneck on the productivity of the overall bioconversion process. The corresponding hydrolases, mainly cellulases and xylanases are inducible enzymes and their efficient production in the fermentation medium is a time consuming step. Homologous overexpression of these enzymes under constitutive control, could provide a higher breakdown rate of the (hemi-)cellulosic biomass and thus increase the supply of sugars to the ethanol production pathway. To this end the endo-xylanase 2 of *F. oxysporum*, was overproduced in the F3 strain under control to the constitutive *Aspergillus nidulans gpdA* promoter (Anasontzis et al., 2011). The fermentative performance of the transformants were evaluated and compared to that of the wild type in simple CBP systems using corn cob or wheat bran as sole carbon sources. Transformants produced approximately 60% more ethanol compared to the wild type on corn cob and wheat bran likely due to the high extracellular xylanase activities in the transformants' fermentation broths that were maintained 2–2.5-fold higher compared to the wild type.

### 3.2 Prokaryotic cellulolytic organisms for CBP

Thermophilic bacteria as a group show great potential as CBP organisms (Xu et al., 2010). These organisms are capable of cellulose hydrolysis and ethanol production under thermophilic conditions. High temperature hydrolysis and fermentation potentially provide a significant energy saving since reactors would not have to be cooled to 30 or 37$^{0}$C before inoculation and then heated again for distillation. Furthermore, it has been shown that a 10$^{0}$C increase in temperature approximately doubles the reaction rate, which in turn decreases the amount of enzyme needed (Ibrahim & El-diwany, 2007). Because thermostable enzymes are able to tolerate higher temperatures they generally have longer half-lives. The use of higher reaction and fermentation temperatures (above 60$^{0}$C) also minimizes the risk of bacterial contamination. Since cellulose hydrolysis and sugar release is in most cases the rate limiting step in a typical CBP process, high temperature hydrolysis will be advantageous.

The thermophilic gram-positive anaerobic bacterium *Clostridium thermocellum* is regarded as a potential CBP-organism (Lynd et al., 2002). It is very efficient at hydrolysing crystalline cellulose, however growth of wild type strains are inhibited in the presence of ethanol concentrations above 2% (v/v) (Xu et al., 2010). Laboratory strains have been evolved that remained viable at ethanol concentrations of up to 8% (v/v). At the heart of *C. thermocellum*'s cellulose hydrolyzing ability lays the cellulosome.

Cellulosomes are extracellular multienzyme systems produced by some cellulolytic bacteria to degrade crystalline cellulose. The cellulosome concept was originally defined in *C. thermocellum* in 1983 by Lamed et al. ( 1983a, 1983b). They were searching for the molecular component on the cell surface of *C. thermocellum* responsible for specific binding to cellulose. Cellulosomes complexes typically consist of a scaffoldin molecule with enzymatic units attached to it. Scaffoldins contain cohesin domains to which enzymatic-units can bind by means of their respective dockerin domains. The cohesin-dockerin interaction is Ca-dependent and species-specific. Cellulases from *C. thermocellum* failed to interact with the scaffoldin protein from *C. cellulolyticum* and vice versa (Fierobe et al., 1999). In cells growing on cellulose, cellulosomes are typically attached to the host cell surface and they also contain at least one cellulose binding domain enabling the whole complex to effectively bind to cellulose. This arrangement enables enzyme proximity synergy as well as enzyme-substrate-microbe synergy. This results in cellulosomes being much more efficient at breaking down cellulose than free enzymes.

Until recently it was not certain what the optimal reaction conditions for cellulosomes were since host cell growth temperatures do not necessarily match those of the enzymes they produce. Furthermore, during pretreatment of lignocellulosic materials side reactions lead to the formation of compounds, which are inhibitors of cell growth. Xu et al (2010) tested the effect of some of these inhibitors on cellulosome activity of *C. thermocellum*. They found that organic acids like formate, acetate and lactate actually promoted cellulolytic activity and that the *C. thermocellum* cellulosome could tolerate certain concentrations of furfural (up to 5 mM), *p*-hydroxybenzoic acid (up to 50 mM) and catecol (up to 1 mM). The *C. thermocellum* cellulosomes were also able to tolerate higher ethanol concentrations and temperatures than the *T. reesei* enzymes used commercially. Using conditions optimal for cellulosomal activity *C. thermocellum* produced 22.6 g/L ethanol, the highest ever reported for *C. thermocellum*.

Not all cellulolytic bacteria produce cellulosomes. *Clostridium phytofermentans* with a genome encoding the highest number of cellulosic enzymes of all sequenced Clostridia, secretes enzymes enabling it to hydrolyze lignocellulose to fermentable sugars (Jin et al.,

2011). However, unlike *C. thermocellum, C phytofermentans* can consume most of the sugars present in lignocellulosic biomass, including xylose and produce ethanol and acetate from it. Jin et al (2011) used AFEX treated corn stover to test *C. phytofermentans* as a potential CBP organism. Under optimal fermentation conditions *C. phytofermentans* hydrolyzed 76% of glucan and 88.6% of xylan in 10 days and yielded 2.8 g/L ethanol as well as 2.6 g/L acetate. Another group of organisms with great CBP potential is from the genus *Geobacillus*. These are thermophilic bacilli with certain species being able to ferment sugars like glucose, xylose and arabinose at temperatures of between 55 and 70$^0$C, producing a mixture of lactate, formate, acetate and ethanol (Barnard et al., 2010). Certain species like *Geobacillus* R7 also have the ability to produce lignocellulose-degrading enzymes including cellulases, xylanases and lignases. All the above mentioned attributes make *Geobacillus* a very good candidate for CBP, however the production of lactate and formate is not desirable. Therefore genetic engineering of these strains has been carried out at a British company, TMO Renewables Ltd, to improve ethanol production by *Geobacillus*.

### 3.3 Engineering cellulolytic ability into eukaryotic process organisms

The yeast *Saccharomyces cerevisiae* has long been employed for the industrial production of ethanol from hexose sugars (Kuyper et al., 2005; Nissen et al., 2000; Van Dijken et al., 2000). *S. cerevisiae* has many positive attributes which makes it suitable for industrial ethanol production such as a high rate of ethanol production from glucose (3.3 g/L/h) and its GRAS status. However this yeast species also has a number of shortcomings in terms of a CBP processing organism such as its inability to hydrolyze cellulose and hemicellulose or utilize xylose or arabinose. A number of research groups around the world have been working on improving the substrate range of *S. cerevisiae* to include the monomeric forms of sugars contained in plant biomass (Hahn-Hägerdal et al., 2001; Hahn-Hägerdal et al., 2007; Karhumaa et al., 2006; Kuyper et al., 2005). A *S. cerevisiae* strain that expressed the xylose isomerase gene from the fungus *Piromyces* sp. E2 was further metabolically engineered to allow anaerobic growth on xylose in synthetic media (Kuyper et al., 2004). Furthermore, laboratory and industrial *S. cerevisiae* strains were also engineered to co-ferment the pentose sugars xylose and arabinose (Karhumaa et al., 2006).

There have been many reports detailing the expression of one or more cellulase encoding gene(s) in *S. cerevisiae* (Van Zyl et al., 2007). Strains of *S. cerevisiae* were created that could grow on and ferment cellobiose, the main product of the action of cellobiohydrolases on cellulosic substrates, at approximately the same rate as on glucose in anaerobic conditions (van Rooyen et al., 2005). Recently the high affinity cellodextrin transport system of the model cellulolytic fungus *Neurospora crassa* was reconstituted into *S. cerevisiae* (Galazka et al., 2010). This led to the efficient growth of a recombinant strain also producing an intracellular β-glucosidase on cellodextrins up to cellotetraose. Furthermore, strains of *S. cerevisiae* were engineered to co-ferment mixtures of xylose and cellobiose, using a xylose fermenting strain that also produced a high affinity cellodextrin transporter and an intracellular β-glucosidase to hydrolyse cellobiose (Ha et al., 2011). It was shown that intracellular hydrolysis of cellobiose minimised glucose repression of xylose fermentation allowing co-consumption of cellobiose and xylose that improved ethanol yields. This was partly due to circumventing the competition between xylose and glucose for transport into the cell. Sadie et al. (2011) recently showed that expression of the gene encoding lactose permease of *Kluyveromyces lactis* (*lac12*) also facilitated transport of cellobiose into a recombinant *S. cerevisiae* strain. This report further showed the successful expression of a

*Clostridium stercorarium* cellobiose phosphorylase (*cepA*) that hydrolyses cellobiose and simultaneously phosphorylates one of the glucose molecules with an inorganic phosphate group yielding one glucose molecule and one glucose-1-phosphate molecule that are both further metabolised through glycolysis. Strains co-producing the heterologous CepA and Lac12 were able to grow on cellobiose as sole carbohydrate source.

There have also been reports showing co-production of cellulases specifically with the aim of enabling the organism to grow on a polymeric substrate. Cho et al. (1999) showed that for SSF experiments with a strain producing a β-glucosidase and an enzyme with exo- and endocellulase activity, loadings of externally added cellulase could be reduced. Fujita et al. (2002; 2004) reported co-expression and surface display of cellulases in *S. cerevisiae.* High cell density suspensions of a recombinant strain displaying the *Trichoderma reesei* endoglucanase II, cellobiohydrolase II, and the *Aspergillus aculeatus* β-glucosidase were able to directly convert 10 g/L phosphoric acid swollen cellulose (PASC) to approximately 3 g/L ethanol. However, growth of this strain on the cellulosic substrate was not demonstrated. An *S. cerevisiae* strain co-expressing the *T. reesei* endoglucanase 1 (*cel7B*) and the *S. fibuligera* β-glucosidase 1 (*bgl3A*) was able to grow on and convert PASC to ethanol up to 1.0 g/L (Den Haan et al., 2007b). Jeon et al. (2009) constructed a similar strain expressing the *S. fibuligera bgl3A* and the *Clostridium thermocellum cel5E* endoglucanase genes that produced significantly more endoglucanase activity than the strain reported by Den Haan et al. (2007b) and notably improved conversion of PASC to ethanol was achieved. When the processive endoglucanase Cel9A of the moderately thermophilic actinomycete *Thermobifida fusca* was functionally produced in *S. cerevisiae* growth of the strain expressing only this one cellulase encoding gene could be demonstrated on media containing PASC as sole carbohydrate source (van Wyk et al., 2010). Growth by the recombinant strain on amorphous cellulose was possible due to the sufficient amount of glucose cleaved from the cellulose chain as it was shown that the enzyme released cellobiose and glucose from cellulosic substrates in a ratio of approximately 2.5:1. In an effort to construct an engineered yeast with efficient cellulose degradation, Yamada et al. (2010) developed a method to optimize cellulase expression levels, named cocktail delta-integration. Several different cellulase expression cassettes were integrated into yeast chromosomes simultaneously in one step, and strains expressing an optimum ratio of cellulases were selected for by growth on media containing PASC as carbon source. Although the total integrated gene copy numbers of an efficient cocktail delta-integrant strain was about half that of a conventional delta-integrant strain, the PASC degradation activity (64.9 mU/g-wet cell) was higher than that of a conventional strain (57.6 mU/g-wet cell) suggesting that optimization of the cellulase expression ratio improved PASC degradation activity more than overexpression. As exoglucanase activity is required for the successful hydrolysis of crystalline cellulose, it is hypothesized that the addition of successful, high-level expression of a cellobiohydrolases to these strains will enable conversion of crystalline cellulose to ethanol. While there have been reports of successful expression of CBH encoding genes in *S. cerevisiae* the titres achieved were generally too low to allow CBP (Den Haan et al., 2007a). Recently the expression of relatively high levels of exoglucanases in *S. cerevisiae* was reported for the first time (Mcbride et al., 2010). Using these, the authors were able to construct a yeast strain that was able to convert most of the glucan available in paper sludge to ethanol. The strain was also able to displace 60% of the enzymes required to convert the sugars available in pretreated hardwood to ethanol in an SSF configuration.

Several other yeast strains have innate properties that make them attractive as possible CBP organisms (Lynd et al., 2005). The multistress tolerant yeast *Issatchenkia orientalis* was recently engineered to produce *Aspergillus aculeatus* β-glucosidase (Kitagawa et al., 2010). The transformant could convert cellobiose to ethanol under acidic conditions and at temperatures exceeding 40°C. Strains of the yeast *Kluyveromyces marxianus* can grow at temperatures as high as 52°C and have a short generation time and high growth rate at elevated temperatures (Rajoka et al., 2003). *K. marxianus* can convert a wide range of substrates, including xylose, to ethanol and successful SSF with a variety of feedstocks at elevated temperatures was demonstrated with *K. marxianus* (Fonseca et al., 2007; Fonseca et al., 2008). Thermotolerant cellobiohydrolase, endoglucanase and β-glucosidase encoding genes were expressed in combination in a strain of *K. marxianus* (Hong et al., 2007). The resulting strain was able to grow in synthetic media containing cellobiose or carboxymethylcellulose as sole carbon source but the hydrolysis of crystalline cellulose was not shown. Recently, a *K. marxianus* strain was engineered to display *T. reesei* endoglucanase II and *Aspergillus aculeatus* β-glucosidase on the cell surface (Yanase et al., 2010). This strain successfully converted 10 g/l of a cellulosic β-glucan to 4.24 g/l ethanol at 48°C within 12 h.

Some strains of the methylotrophic yeast *Hansenula polymorpha* are also able to grow at elevated temperatures up to 48°C and ferment glucose, cellobiose and xylose to ethanol (Ryabova et al., 2003). Additionally, attributes such as process hardiness and a high capacity for heterologous protein production make this yeast an attractive candidate for CBP. A recent report highlighted the promise of *H. polymorpha* in biomass conversion when strains were constructed that could ferment starch and xylan (Voronovsky et al., 2009). *Pichia stipitis* is one of the best studied xylose-fermenting yeasts and has a substrate range including all the monomeric sugars present in lignocellulose (Jeffries & Shi, 1999). Some *P. stipitis* strains produce low quantities of various cellulases and hemicellulases to break down wood into monomeric sugars although it cannot utilize polymeric cellulose as carbon source (Jeffries et al., 2007). Among the enzymes that are naturally produced are a β-glucosidase that allows the yeast to ferment cellobiose. Endoglucanases were successfully produced in *H. polymorpha* (Papendieck et al., 2002) and *P. stipitis* (Piotek et al., 1998). As these yeasts are capable of growth on cellobiose these recombinant strains should theoretically have the ability to hydrolyse amorphous cellulose although this aspect was not tested. The xylanolytic ability of *P. stipitis* was enhanced by the co-expression of xylanase and xylosidase encoding genes (Den Haan & Van Zyl, 2003). The resulting strains displayed improved biomass production on medium with birchwood glucuronoxylan as sole carbohydrate source. Even though mutant strains of *P. stipitis* with increased ethanol tolerance were recently isolated, *P. stipitis* remains a relatively poor fermentor (Watanabe et al., 2011). However, its ability to consume acetic acid and reduce the furan ring in furfural and hydroxymethylfurfural (HMF) creates an opportunity for this yeast to clean up some of the toxins in cellulosic biomass conversion (Agbogbo & Coward-Kelly, 2008). This could be very beneficial in waste water treatment.

### 3.4 Engineering prokaryotic organisms to hydrolyze polysaccharides

Although *Escherichia coli* cannot hydrolyze cellulose or produce ethanol at appreciable quantities it has been shown to metabolize all major sugars present in plant biomass, producing a mixture of organic acids and ethanol (Alterthum & Ingram, 1989). Bräu and Sahm (1986) successfully modified *E. coli* metabolism by expressing the *Zymomonas mobilis* pyruvate decarboxylase at high levels. The resulting strain produced ethanol at levels comparable with *Z. mobilis*. Subsequent work has focused on improving ethanol yields, growth rate, strain stability and ethanol tolerance (Ingram et al., 1987; Ohta et al., 1991a; Ingram et al., 1991; Chen et al., 2009; Da Silva et al., 2005a; Yamano et al., 1998). Wild type *E. coli* strains are incapable of rapid growth on cellobiose (Moniruzzaman et al., 1997). *Klebsiella oxytoca* contains a phosphoenol-dependent phosphotransferase system (PTS) enabling it to utilize cellobiose. The *K. oxytoca casAB* operon coding for an enzyme $II^{cellobiose}$ and a phospho-β-glucosidase was expressed in the ethanol producing strain of *E. coli*. While expression was initially poor, spontaneous mutants were isolated which exhibited over 15-times higher specific activities for cellobiose metabolism. The best mutant produced 45 g/L ethanol - a yield of 94% of the theoretical maximum. Several endoglucanases have been expressed in *E. coli* allowing it to hydrolyze amorphous and soluble cellulose to shorter cello-oligosaccahrides (Da Silva et al., 2005b; Seon et al., 2007; Srivastava et al., 1995; Wood et al., 1997; Yoo et al., 2004; Zhou et al., 2001). Among these are Cel5Z and Cel8Y from *Erwinia chrysanthemi*. Zhou et al. (2001) successfully reconstructed the type II secretion system, the predominant secretion system type in Gram negative bacteria, encoded by the *out* genes from *E. chrysanthemi*, in *E. coli*. This enabled *E. coli* to secrete more than 50% of the recombinant Cel5Z it produced. Recently, Shin et al. (2010) demonstrated a binary strategy for CBP of xylan. Two *E. coli* strains were designed to function cooperatively in the process of transforming xylan into ethanol. The first strain was engineered to co-express *axeA*, the acetylxylan esterase gene from *Streptomyces violaceoruber* and *xyl11A*, the xylanase gene from *Bacillus halodurans*. The recombinant enzymes were secreted into the growth medium by a method of *lpp* deletion with over 90% efficiency. Secreted enzymes hydrolyzed xylan into xylo-oligosaccharides, which were taken in by the second strain, designed to use the xylo-oligosaccharides for ethanol production. The second strain was based on the KO11 strain optimized for ethanol production. Into this strains the *KxynB* gene encoding β-xylosidase from *Klebsiella pneumonia* and *KxynT* encoding xyloside permease from *Klebsiella pneumoniae* were introduced. Co-cultivation of the two strains converted xylan to ethanol with a yield of about 55% of the theoretical value.

*Klebsiella oxytoca* is a hardy prototrophic bacterium with the ability to transport and metabolize cellobiose, cellotriose, xylobiose, xylotriose, sucrose, and all other monomeric sugars present in lignocellulosic biomass (Zhou & Ingram, 1999b). Four fermentation pathways are present in *K. oxytoca* producing formate, acetate, ethanol, lactic acid, succinate and butanediol (Ohta et al., 1991b). Through metabolic engineering and expression of the *Z. mobilis pdc* and *adhB* genes it was possible for a recombinant *K. oxytoca* strain to produce ethanol from soluble sugars at 95% of the maximum theoretical yield (Wood & Ingram, 1992). Unlike most other ethanol producing organisms *K. oxytoca* has the ability to ferment xylose and glucose at equivalent rates (Ohta et al., 1991b). This significantly shortens the time required to ferment the mixtures of glucose and xylose typically present in lignocellulosic hydrolysates. Zhou and Ingram (1999a) constructed a *K. oxytoca* strain expressing the *E. chrysanthemi cel8Y* and *cel5Z* endoglucanase genes. By also introducing the genes that encode the type II secretion system from *E. chrysanthemii*, both Cel8Y and Cel5Z were secreted effectively by *K. oxytoca*. This strain was capable of fermenting amorphous cellulose and producing a small amount of ethanol without the addition of cellulases.

*Z. mobilis* is a well known fermenting bacterium that produces ethanol at high rates (Zhang et al., 1997). However, *Z. mobilis* cannot ferment or utilize xylose as carbon source and it cannot hydrolyze polysaccharides. Zhang et al. (1997) engineered a *Z. mobilis* strain capable of fermenting both xylose and arabinose, the major pentose sugars present in plant material. Co-fermentation of 100 g/L sugar (glucose:xylose:arabinose - 40:40:20) yielded a final ethanol concentration of 42 g/L in 48 hours. Brestic-Goachet et al. (1989) expressed the *E. chrysanthemi cel5Z* in *Z. mobilis*. The maximum activity obtained was 1000 IU/L with 89% of the recombinant endo-glucanase secreted to the extracellular medium. Expression of the *Ruminococcus albus* β-glucosidase enabled *Z. mobilis* to ferment cellobiose to ethanol very efficiently in two days and most of the recombinant enzyme was transported across the cytoplasmic membrane (Yanase et al., 2005). Recently, numerous strains of *Z. mobilis* were shown to possess endogenous extracellular activities against carboxymethyl cellulose (Linger et al., 2010). Furthermore, two cellulolytic enzymes, E1 and GH12 from *Acidothermus cellulolyticus*, were heterologously produced as soluble, active enzymes in *Z. mobilis*. While the E1 enzyme was less abundantly expressed, the GH12 enzyme comprised as much as 4.6% of the total cell protein. Additionally, fusing predicted secretion signals native to *Z. mobilis* to the N-termini of E1 and GH12 was found to direct the extracellular secretion of significant levels of active E1 and GH12 enzymes though a significant portion of both resided in the periplasmic space.

The thermophilic anaerobic bacterium *Thermoanaerobacterium saccharolyticum* is also under development for biomass conversion. *T. saccharolyticum* grows in a temperature range of 45 - 65°C and a pH range of 4.0 – 6.5 and is able to ferment a wide range of sugars present in cellulosic biomass including cellobiose, glucose, xylose, mannose, galactose, and arabinose (Shaw et al., 2008a). Unlike most organisms *T. saccharolyticum* metabolizes xylose and glucose essentially at the same rate (Shaw et al., 2008a; Shaw et al., 2008b) but it produces organic acids in addition to ethanol. Knockout mutants were created that produced almost exclusively ethanol from xylose. Furthermore, a strain with *hfs* and *ldh* deletions exhibited an increased ethanol yield from consumed carbohydrates and represents a new strategy for engineering increased ethanol yields in *T. saccharolyticum* (Shaw et al., 2009). *T. saccharolyticum* naturally produces both a xylanase and a β-xylosidase (Lee et al., 1993a; Lee et al., 1993b) enabling it to ferment xylan directly to ethanol. Furthermore, *T. saccharolyticum* was able to produce as much ethanol from Avicel with 4 filter paper units (FPU) of externally added enzyme as *S. cerevisiae* was with 10 FPU in SSF, the result of improved enzyme efficiency at higher temperatures (Shaw et al., 2008b). This shows the potential of this thermophile as CBP organism if a cellulolytic system can be established.

## 4. Integrating consolidated bioprocessing with existing bio-based industries

Although major advances have been made, the cost of second generation biofuels still remains high. Integrating cellulosic ethanol technologies with first generation bio-based and thermochemical processes helps to minimize the capital investment, maximize energy efficiency and improve overall economics (Van Zyl et al., 2011). Various biological and thermochemical processes will be discussed and their integration in a few bio-based industries highlighted.

Three thermochemical options are available for the conversion of biomass: combustion, pyrolysis, and gasification. Combustion involves burning of biomass in the presence of air, which generates hot gases at temperatures of around 800-1000°C and energy that can be harvested as heat. Pyrolysis is the conversion of biomass to liquid (bio-oil), solid (char) and gaseous fractions by heating the biomass in the absence of air to about 500°C. Bio-oils can be upgraded to transport fuels, bio-oils and char can be gasified or used to improve soil quality. In contrast, gasification is the conversion of biomass by partial oxidation at higher temperatures (in the range of 800-900°C) to generate syngas that can be used for synthesis of different synthetic fuels (using the Fischer-Tropsch process) or burned for heat production (McKendry, 2002; Bridgwater, 2011).

In the biological process for lignocellulose hydrolysis-fermentation, large amounts of energy remain in the non-fermentable lignin-rich residues from the distillation process. Conversion of these residues through high-efficiency processes, such as a high pressure boiler coupled with a multi-stage steam turbine (Aden & Foust, 2009; Piccolo & Bezzo, 2009) can provide all the heat and electricity needed for cellulosic ethanol production, together with surplus electricity production for sale (Cardona & Sanchez, 2007; Leibbrant 2010; Reith et al., 2002). Energy consumption in the biochemical process can be reduced further by performing enzymatic hydrolysis and/or SSF processes at high substrate loadings (as typically used in high-gravity brewing), together with recycling of the process streams, both of which have substantial benefits in terms of process energy efficiency and economics (Martin et al., 2010; Wingren et al., 2003). Anaerobic digestion for wastewater treatment can be used to lower organic loadings while simultaneously producing methane-rich biogas that can be captured and used to generate electricity and/or process heating (Banerjee et al., 2009). Similarly, the integrated production of synthetic biofuels and electricity from lignocellulose in the gasification-synthesis process route will provide higher energy efficiencies than production of synfuels alone (Leibbrant 2010; Swanson et al., 2010). As an example, heat integration within biological (Aden & Foust, 2009; Kazi et al., 2010) and thermochemical routes for second generation biofuels production have the potential to increase overall energy efficiency by as much as 15% (Leibbrant 2010) and can reduce capital and operational costs substantially (Galbe et al., 2005).

### 4.1 Integration between lignocellulosic conversion processes and electricity production

Optimum use of cellulosic feedstocks can be achieved by integrating cellulosic ethanol processes with electricity production to achieve economies of scale and reduce capital investment per unit of electricity substantially (Easterly 2002; Hahn-Hägerdal et al., 2006; Laser et al., 2009a; Laser et al., 2009b; Sassner et al., 2008; Sims et al., 2008). Integration and scale-up of electricity and steam production can be achieved by combining feedstocks for electricity generation, such as lignin-rich residues from biological processing and using heat recovery/integration in both biofuel and electricity generation for steam production and distillation (Easterly 2002; Laser et al., 2009b; Sassner et al., 2008). By maximizing electricity production, increased revenue from second generation biofuels production and a reduction of GHG emissions from these processes can be achieved (Eriksson & Kjellström, 2010). Sharing of feedstock supply and handling infrastructure and logistics will bring about further savings.

### 4.2 Integration with biomass processing for pulp or sugar production

Both the sugar and pulp-and-paper industries only process biomass in part. Substantial quantities of residues from both these industries, not suitable or useful in the primary biomass processing, could be an attractive feedstock supply for the production of cellulosic ethanol. The cost of raw material and the capital costs of raw material handling adds substantially to the total production cost of cellulosic ethanol (Aden & Foust, 2009; Anex et al., 2010; Gnansounou et al., 2005; Hahn-Hägerdal et al., 2006; Kazi et al., 2010; Piccolo & Bezzo, 2009). Therefore, integrating cellulosic ethanol production from lignocellulose residues emanating from with these industries can improve efficient use of the residues and waste streams and savings in feedstock supply and/or energy integration (Goh et al., 2010; Hahn-Hägerdal et al., 2006; Soccol et al., 2010).

Highly efficient sugar mills can liberate up to 50% of the bagasse present in cane supply as surplus (Botha & von Blottnitz, 2006). However, the availability of bagasse can be highly variable and often limiting because many conventional sugar mills are designed to dispose of bagasse residues by inefficient burning. The cost of converting sugarcane bagasse to cellulosic ethanol is therefore inevitabilty coupled to the cost of capital investments required to improve the energy efficiency. However, optimum use of energy generated in both processes for the co-production of electricity and heat could result in economic benefits (Leibbrant 2010). Swedish researchers suggested that such an integration could reduce the cost of cellulosic ethanol production with up to 20 percent in Sweden (Hahn-Hägerdal et al., 2006; von Sivers & Zacchi, 1995).

### 4.3 Integration of biological first and second generation biofuels production

Integration of second generation cellulosic ethanol production with first generation production from sugars or starches can have multiple benefits, including reduction of capital costs and investor risk, increased economic attractiveness and environmental acceptance (Gnansounou et al., 2005). Such integration can provide joint feedstock supply, fermentation, water and nutrient recycle, distillation, and further opportunities for energy integration (Easterly 2002; Galbe et al., 2007). Sugar-rich crops for first generation ethanol production, such as sugarcane, sweet sorghum and sugarbeet, are particularly attractive for integration with cellulosic ethanol processes that ensured optimum use of the feedstock and its logistics (Gnansounou et al., 2005; Sims et al., 2008). These crops also allow flexibility of switching between the production of crystallized sugar and ethanol, as practiced in some Brazilian sugar mills (Gnansounou et al., 2005). Further benefits of the combined fermentation-distillation processes for ethanol production from lignocellulose and sugar streams could be (i) to replace exogenous nutrient supplements (Banerjee et al., 2009), (ii) mixing of sugars from juice and lignocellulose to increase sugar concentrations and resulting ethanol levels at the end of the cellulose fermentation, and (iii) scale-up of ethanol purification/distillation to achieve economies of scale and improve energy efficiency (Soccol et al., 2010). Similar integration possibilities also exist for starches (small grains, corn, etc.), where ethanol production could be supplemented with sugars from bran (starch fibre) and polysaccharide-rich waste streams such as thin stillage (Cardona & Sanchez, 2007; Linde et al., 2010).

## 5. Discussion

There are several types of feedstock that can be considered viable options for the production of cellulosic biofuels. The choice of feedstock will vary between geographical areas and depend on the availability of arable land and prevailing climatic conditions and will influence downstream processes such as pre-treatment and the CBP organism used for bioconversion. To date no ideal organism has been developed for CBP conversion of biomass. Bacteria generally have a high growth rate but lack process robustness. Yeasts are often sufficiently robust, but lack substrate range. Filamentous fungi often have a wide substrate range, but grow relatively slowly and do not produce enough of a desirable product. While the advantages of using the yeasts *P. stipitis*, *K. marxianus* and *H. polymorpha* are well appreciated, the engineered cellulolytic ability of these strains are currently rudimentary. None of the strains are as yet capable of utilizing crystalline cellulose and the high level production of an exocellulase remains a requirement. New information on secretion pathways, chaperones and metabolic engineering should help alleviate this problem in future. The *S. cerevisiae* strain developed by the Mascoma Corporation represents the best CBP organism engineered thus far as this strain could convert several cellulosic substrates to ethanol with addition of minimal exogenous enzymes in an SSF configuration (Mcbride et al., 2010). Compared to *S. cerevisiae,* all of the bacterial species discussed above are relatively sensitive to inhibitors associated with lignocellulosic hydrolysates (Bothast et al., 1999; Yamano et al., 1998; Ohta et al., 1991b). Engineering enhanced protein secretion allowed the successful secretion of endoglucanases in *E. coli* (Ji et al., 2009) and *K. oxytoca* (Zhou & Ingram, 2001). *E. coli* and *K. oxytoca* strains capable of breaking down cellulose could also be modified to produce other commodity products such as lactic acid, succinic acid, acetic acid or 2, 3-butanediol (Ji et al., 2009). The *Geobacillus* strain used by TMO Renewables Ltd. is capable of producing ethanol at appreciable titers from pretreated lignocellulosic feedstock and represents a very promising organism for CBP.

Candidate CBP organisms are in various stages of development for establishment of a cellulolytic system or improvement of product forming attributes. It is likely that more than one organism may eventually be used in various biomass conversion processes and the choice may depend on the sugar composition of the feedstock, the pretreatment method used and the end product required. The cost disadvantage of current second-generation biofuels configurations may be partially addressed through innovative methods of process integration, in order to minimize capital investment and maximize energy efficiency and improve overall economics. Integration of second generation biofuel production processes into existing first generation biofuel production or into other biomass based industries with integration strategies to ensure optimal energy usage and synergy may be the most effective way to bring second generation biofuels to market.

## 6. References

Aden, A. & Foust, T. (2009). Technoeconomic analysis of the dilute sulfuric acid and enzymatic hydrolysis process for the conversion of corn stover to ethanol. *Cellulose,* Vol.16, (2009), pp. 535-545

Agbogbo, F.K. & Coward-Kelly, G. (2008). Cellulosic ethanol production using the naturally occurring xylose-fermenting yeast, *Pichia stipitis. Biotechnology Letters,* Vol.30, No.9, (September 2008), pp. 1515-1524

Alfenore, S.; Molina-Jouve, C.; Guillouet, S. E.; Uribelarrea, J.-L.; Goma, G. & Benbadis, L. (2002). Improving ethanol production and viability of *Saccharomyces cerevisiae* by a vitamin feeding strategy during fed-batch process. *Applied Microbiology and Biotechnology,* Vol.60, (2002), pp. 67-72

Alterthum, F. & Ingram, L. O. (1989). Efficient ethanol production from glucose, lactose, and xylose by recombinant *Escherichia coli. Applied and Environmental Microbiology,* Vol.55, (1989), pp. 1943-1948

Anasontzis, G.E.; Zerva, A.; Stathopoulou, P. M.; Haralampidis, K.; Diallinas, G.; Karagouni, A. D. & Hatzinikolaou, D. G. (2011). Homologous overexpression of xylanase in *Fusarium oxysporum* increases ethanol productivity during consolidated bioprocessing (CBP) of lignocellulosics. *Journal of Biotechnology*, Vol.152, No.1-2, (March 2011), pp. 16-23

Anex, R.P.; Aden, A.; Kazi, F. K.; Fortman, J.; Swanson, R. M.; Wright, M. M.; Satrio, J. A.; Brown, R. C.; Daugaard, D. E.; Platon, A.; Kothandaraman, G.; Hsu, D. D. & Dutta, A. (2010). Techno-economic comparison of biomass-to-transportation fuels via pyrolysis, gasification, and biochemical pathways. *Fuel*, Vol.89, (2010), pp. S29-S35

Ansah, T.; Osafo, E. L. K. & Hasen, H. H. (2010). Herbage yield and chemical composition of four varieties of Napier (*Pennisetum purpureum*) grass harvested at three different days after planting. *Agriculture and Biology Journal of North America*, Vol.1, No.5, (2010), pp. 923-929

Banerjee, S.; Mudliar, S.; Sen, R. & Giri, B. (2009). Commercializing lignocellulosic bioethanol: technology bottlenecks and possible remedies. *Biofuels, Bioproducts and Biorefining*, Vol.4, (2009), pp. 77-93

Barnard, D.; Casanueva, A.; Tuffin, M. & Cowan, D. (2010). Extremophiles in biofuel synthesis. *Environmental Technology*, Vol.31, No.8, (2010), pp. 871-888

Botha, T. & von Blottnitz, H. (2006). A comparison of the environmental benefits of bagasse-derived electricity and fuel ethanol on a life-cycle basis. *Energy Policy*, Vol.34, No.17, (November 2006), pp. 2654-2661

Bothast, R.J.; Nichols, N. N. & Dien, B. S. (1999). Fermentations with New Recombinant Organisms. *Biotechnology Progress*, Vol.15, (1999), pp. 867-875

Bräu, B. & Sahm, H. (1986). Cloning and expression of the structural gene for pyruvate decarboxylase of *Zymomonas mobilis* in *Escherichia coli*. *Archives of Microbiology*, Vol.144, (1986), pp. 296-301

Brestic-Goachet, N.; Gunasekaran, P.; Cami, B. & Baratti, J. C. (1989). Transfer and expression of an *Erwinia chrysanthemi* cellulase gene in *Zymomonas mobilis*. *J.Gen.Microbiol.*, Vol.135, (January 1989), pp. 893-902

Bridgwater, A.V. (2011). Review of fast pyrolysis of biomass and product upgrading. *Biomass and Bioenergy*, [Corrected Proof]

Cardona, C.A. & Sanchez, O. J. (2007). Fuel ethanol production: Process design trends and integration opportunities. *Bioresource Technology*, Vol.98, (2007), pp. 2415-2457

Chambergo, F.S.; Bonaccorsi, E. D.; Ferreira, A. J.; Ramos, A. S.; Ferreira, J., Jr.; Abrahao-Neto, J.; Farah, J. P. & El Dorry, H. (2002). Elucidation of the metabolic fate of glucose in the filamentous fungus *Trichoderma reesei* using expressed sequence tag (EST) analysis and cDNA microarrays. *Journal of Biological Chemistry*, Vol.277, No.16, (April 2002), pp. 13983-13988

Chen, J.; Zhang, W.; Tan, L.; Wang, Y. & He, G. (2009). Optimization of metabolic pathways for bioconversion of lignocellulose to ethanol through genetic engineering. *Biotechnology Advances*, Vol.27, (2009), pp. 593-598

Cherry, J.R. & Fidantsef, A. L. (2003). Directed evolution of industrial enzymes: an update. *Current Opinion in Biotechnology*, Vol.14, No.4, (August 2003), pp. 438-443

Cho, K.M.; Yoo, Y. J. & Kang, H. S. (1999). δ-Integration of endo/exo-glucanase and β-glucosidase genes into the yeast chromosomes for direct conversion of cellulose to ethanol. *Enzyme and Microbial Technology*, Vol.25, No.1-2, (July 1999), pp. 23-30

Da Silva, G.P.; De Araujo, E. F.; Silva, D. & Guimaraes, W. V. (2005a). Ethanolic fermentation of sucrose, sugarcane juice and molasses by *Escherichia coli* strain KO11 and

*Klebsiella oxytoca* strain P2. *Brazilian Journal of Microbiology*, Vol.36, (2005a), pp. 395-404

Da Silva, G.P.; De Araujo, E. F.; Silva, D. & Guimaraes, W. V. 2005b). Ethanolic fermentation of sucrose, sugarcane juice and molasses by *Escherichia coli* strain KO11 and *Klebsiella oxytoca* strain P2. *Brazilian Journal of Microbiology*, Vol.36, (May 2005b), pp. 395-404

Davis, S.C.; Dohleman, F. G. & Long, S. P. (2011). The global potential of *Agave* as a biofuel feedstock. *GCB Bioenergy*, Vol.3, (2011), pp. 68-78

Decree No.6.961. (2009). Zoning agricultural land of sugarcane. Diário Oficial União de 18.9.10, Brazil. (17 September 2009)

Den Haan, R.; McBride, J. E.; La Grange, D. C.; Lynd, L. R. & Van Zyl, W. H. (2007a). Functional expression of cellobiohydrolases in *Saccharomyces cerevisiae* towards one-step conversion of cellulose to ethanol. *Enzyme and Microbial Technology*, Vol.40, (2007a), pp. 1291-1299

Den Haan, R.; Rose, S. H.; Lynd, L. R. & Van Zyl, W. H. (2007b). Hydrolysis and fermentation of amorphous cellulose by recombinant *Saccharomyces cerevisiae*. *Metabolic Engineering*, Vol.9, No.1, (January 2007b), pp. 87-94

Den Haan, R. & Van Zyl, W. H. (2003). Enhanced xylan degradation and utilisation by *Pichia stipitis* overproducing fungal xylanolytic enzymes. *Enzyme and Microbial Technology*, Vol.33, (October 2003), pp. 620-628

Easterly,J. (2002) *AES Greenidge Bioethanol Co-Location Assessment: Final Report*. National Renewable Energy Laboratory Report NREL/SR-510-33001,

Eriksson, G. & Kjellström, B. (2010). Assessment of combined heat and power (CHP) integrated with wood-based ethanol production. *Applied Energy*, Vol.87, (2010), pp. 3632-3641

Fierobe, H.P.; Pages, S.; Belaich, A.; Champ, S.; Lexa, D. & Belaich, J. P. (1999). Cellulosome from *Clostridium cellulolyticum*: Molecular study of the dockerin/cohesin interaction. *Biochemistry*, Vol.38, No.39, (1999), pp. 12822-12832

Fonseca, G.G.; Bombert, A. K.; Heinzle, E. & Wittmann, C. (2007). Physiology of the yeast *Kluyveromyces marxianus* during batch and chemostat cultures with glucose as the sole carbon source. *FEMS Yeast Research*, Vol.7, No.3, (2007), pp. 422-435

Fonseca, G.G.; Heinzle, E.; Wittmann, C. & Gombert, A. K. (2008). The yeast *Kluyveromyces marxianus* and its biotechnological potential. *Applied Microbiology and Biotechnology*, Vol.79, No.3, (June 2008), pp. 339-354

Fortman, J.L.; Chhabra, S.; Mukhopadhyay, A.; Chou, H.; Lee, T. S.; Steen, E. & Keasling, J. D. (2008). Biofuel alternative to ethanol: pumping the microbial well. *Trends in Biotechnology*, Vol.26, No.7, (2008), pp. 375-381

Fujita, Y.; Ito, J.; Ueda, M.; Fukuda, H. & Kondo, A. (2004). Synergistic saccharification, and direct fermentation to ethanol, of amorphous cellulose by use of an engineered yeast strain codisplaying three types of cellulolytic enzyme. *Applied and Environmetal Microbiology*, Vol.70, (February 2004), pp. 1207-1212

Fujita, Y.; Takahashi, S.; Ueda, M.; Tanaka, A.; Okada, H.; Morikawa, Y.; Kawaguchi, T.; Arai, M.; Fukuda, H. & Kondo, A. (2002). Direct and efficient production of ethanol from cellulosic material with a yeast strain displaying cellulolytic enzymes. *Applied and Environmetal Microbiology*, Vol.68, No.10, (October 2002), pp. 5136-5141

Galazka, J.M.; Tian, C.; Beeson, W. T.; Martinez, B.; Glass, N. L. & Cate, J. H. (2010). Cellodextrin transport in yeast for improved biofuel production. *Science*, Vol.330, No.6000, (October 2010), pp. 84-86

Galbe, M.; Liden, G. & Zacchi, G. (2005). Production of ethanol from biomass - research in Sweden. *Journal of Scientific and Industrial Research*, Vol.64, (2005), pp. 905-919

Galbe, M.; Sassner, P.; Wingren, A. & Zacchi, G. (2007). Process Engineering Economics of Bioethanol Production. *Advances in Biochemical Engineering and Biotechnology*, Vol.108, (2007), pp. 303-327

Girio, F.M.; Fonseca, C.; Carvalheiro, F.; Duarte, L. C.; Marques, S. & Bogel-Lukasik, R. (2010). Hemicelluloses for fuel ethanol: A review. *Bioresource Technology*, Vol.101, No.13, (July 2010), pp. 4775-4800

Gnansounou, E.; Dauriat, A. & Wyman, C. E. (2005). Refining sweet sorghum to ethanol and sugar: economic trade-offs in the context of North China. *Bioresource Technology*, Vol.96, (2005), pp. 985-1002

Goh, C.S.; Tan, K. T.; Lee, K. T. & Bhatia, S. (2010). Bio-ethanol from lignocellulose: Status, perspectives and challenges in Malaysia. *Bioresource Technology*, Vol.101, No.13, (July 2010), pp. 4834-4841

Goodale, C.L.; Apps, M. J.; Birdsey, R. A.; Field, C. B.; Heath, L. S.; Houghton, R. A.; Jenkins, J. C.; Kohlmaier, G. H.; Kurz, W.; Liu, S.; Nabuurs, G. J.; Nilsson, S. & hvidenko, A. Z. (2002). Forest carbon sinks in the northern hemisphere. *Ecological Applications*, Vol.12, No.3, (June 2002), pp. 891-899, ISSN 1051-0761

Ha, S.J.; Galazka, J. M.; Rin, K. S.; Choi, J. H.; Yang, X.; Seo, J. H.; Louise, G. N.; Cate, J. H. & Jin, Y. S. (2011). Engineered *Saccharomyces cerevisiae* capable of simultaneous cellobiose and xylose fermentation. *Proceeding of the National Academy of Sciences*, Vol.108, No.2, (January 2011), pp. 504-509

Hahn-Hägerdal, B.; Galbe, M.; Gorwa-Grauslund, M. F.; Lidén, G. & Zacchi, G. (2006). Bio-ethanol - the fuel of tomorrow from the residues of today. *Trends in Biotechnology*, Vol.24, No.12, (December 2006), pp. 549-556

Hahn-Hägerdal, B.; Karhumaa, K.; Fonseca, C.; Spencer-Martins, I. & Gorwa-Grauslund, M. F. (2007). Towards industrial pentose-fermenting yeast strains. *Applied Microbiology and Biotechnology*, Vol.74, No.5, (April 2007), pp. 937-953

Hahn-Hägerdal, B.; Wahlbom, C. F.; Gardonyi, M.; Van Zyl, W. H.; Cordero, O. R. & Jonsson, L. J. (2001). Metabolic engineering of *Saccharomyces cerevisiae* for xylose utilization. *Advances in Biochemical Engineering and Biotechnology*, Vol.73, (2001), pp. 53-84

Hamelinck, C.N.; van Hooijdonk, G. & Faaij, A. P. C. (2005). Ethanol from lignocellulosic biomass: techno-economic performance in short-, middle- and long-term. *Biomass and Bioenergy*, Vol.28, (2005), pp. 384 410

Hill, J.; Nelson, E.; Tilman, D.; Polasky, S. & Tiffany, D. (2006). Environmental, economic, and energetic costs and benefits of biodiesel and ethanol biofuels. *Proceedings of the National Academy of Sciences USA*, Vol.103, No.30, (July 2006), pp. 11206-11210

Hong, J.; Wang, Y.; Kumagai, H. & Tamaki, H. (2007). Construction of thermotolerant yeast expressing thermostable cellulase genes. *Journal of Biotechnology*, Vol.130, No.2, (June 2007), pp. 114-123

Ibrahim, A.S.S. & El-diwany, A. (2007). Isolation and identification of new cellulases producing Thermophilic Bacteria from an Egyptian hot spring and some properties of the crude enzyme. *Australian Journal of Basic and Applied Sciences*, Vol.1, No.4, (2007), pp. 473-478

Ingram, L. O., Conway, T., and Alterthum, F. Ethanol production by *Escherichia coli* strains co-expressing *Zymomonas* PDC and ADH genes. Filed: 15 May 1989, Published 19 March 1991 [WO 5000000]

Ingram, L.O.; Conway, T.; Clark, D. P.; Sewell, G. W. & Preston, J. F. (1987). Genetic engineering of ethanol production in *Escherichia coli*. *Applied and Environmental Microbiology*, Vol.53, No.10, (1987), pp. 2420-2425

Jeffries, T. & Shi, N. Q. (1999). Genetic Engineering for Improved Xylose Fermentation by Yeasts. *Advances in Biochemical Engineering and Biotechnology*, Vol.65, (1999), pp. 117-161

Jeffries, T.W.; Grigoriev, I. V.; Grimwood, J.; Laplaza, J. M.; Aerts, A.; Salamov, A.; Schmutz, J.; Lindquist, E.; Dehal, P.; Shapiro, H.; Jin, Y. S.; Passoth, V. & Richardson, P. M. (2007). Genome sequence of the lignocellulose-bioconverting and xylose-fermenting yeast *Pichia stipitis*. *Nature Biotechnology*, Vol.25, No.3, (March 2007), pp. 319-326

Jeon, E.; Hyeon, J. E.; Suh, D. J.; Suh, Y. W.; Kim, S. W.; Song, K. H. & Han, S. O. (2009). Production of cellulosic ethanol in *Saccharomyces cerevisiae* heterologous expressing *Clostridium thermocellum* endoglucanase and *Saccharomycopsis fibuligera* beta-glucosidase genes. *Molecular Cells*, Vol.28, No.4, (October 2009), pp. 369-373

Ji, X.J.; Huang, H.; Zhu, J. G.; Ren, L. J.; Nie, Z. K.; Du, J. & Li, S. (2009). Engineering *Klebsiella oxytoca* for efficient 2, 3-butanediol production through insertional inactivation of acetaldehyde dehydrogenase gene. *Applied Microbiology and Biotechnology*, Vol.85, No.6, (2009), pp. 1751-1758

Jin, M.; Balan, V.; Gunawan, C. & Dale, B. E. (2011). Consolidated bioprocessing (CBP) performance of *Clostridium phytofermentans* on AFEX-treated corn stover fro ethanol production. *Biotechnology and Bioengineering*, (2011

Karhumaa, K.; Wiedemann, B.; Hahn-Hägerdal, B.; Boles, E. & Gorwa-Grauslund, M.-F. (2006). Co-utilization of L-arabinose and D-xylose by laboratory and industrial *Saccharomyces cerevisiae* strains. *Microbial Cell Factories*, Vol.5, (2006a), pp. 1-11

Kazi, F.K.; Fortman, J. A.; Anex, R. P.; Hsu, D. D.; Aden, A.; Dutta, A. & Kothandaraman, G. (2010). Techno-economic comparison of process technologies for biochemical ethanol production from corn stover. *Fuel*, Vol.89, (2010), pp. S20-S28

Kitagawa, T.; Tokuhiro, K.; Sugiyama, H.; Kohda, K.; Isono, N.; Hisamatsu, M.; Takahashi, H. & Imaeda, T. (2010). Construction of a beta-glucosidase expression system using the multistress-tolerant yeast *Issatchenkia orientalis*. *Applied Microbiology and Biotechnology*, Vol.87, No.5, (August 2010), pp. 1841-1853

Kuyper, M.; Hartog, M. M.; Toirkens, M. J.; Almering, M. J.; Winkler, A. A.; Van Dijken, J. P. & Pronk, J. T. (2005). Metabolic engineering of a xylose-isomerase-expressing *Saccharomyces cerevisiae* strain for rapid anaerobic xylose fermentation. *FEMS Yeast Research*, Vol.5, No.4-5, (February 2005), pp. 399-409

Kuyper, M.; Winkler, A. A.; Van Dijken, J. P. & Pronk, J. T. (2004). Minimal metabolic engineering of *Saccharomyces cerevisiae* for efficient anaerobic xylose fermentation: a proof of principle. *FEMS Yeast Research*, Vol.4, (2004), pp. 655-664

La Grange, D.C. (2007). Bioethanol as renewable transportation fuel for the future. MBA thesis, University of Stellenbosch, Stellenbosch, Available from

http://hdl.handle.net/10019.1/837
La Grange, D.C.; Den Haan, R. & Van Zyl, W. H. (2010). Engineering cellulolytic ability into bioprocessing organisms. *Applied Microbiology and Biotechnology*, Vol.87, No.4, (May 2010), pp. 1195-1208
Lamed, R.; Setter, E. & Bayer, E. A. (1983a). Characterization of a cellulose-binding, cellulase-containing complex in *Clostridium thermocellum*. *Journal of Bacteriology*, Vol.156, (1983a), pp. 828-836
Lamed, R.; Setter, E.; Kenig, R. & Bayer, E. A. (1983b). The cellulosome – a discrete cell surface organelle of *Clostridium thermocellum* which exhibits separate antigenic, cellulose-binding and various cellulolytic activities. *Biotechnology and Bioengineering Symposia*, Vol.13, (1983b), pp. 163-181
Laser, M.; Jin, H.; Jayawardhana, K.; Dale, B. E. & Lynd, L. R. (2009a). Coproduction of ethanol and power from switchgrass. *Biofuels, Bioproducts and Biorefining*, Vol.3, (2009a), pp. 195-218
Laser, M.; Jin, H.; Jayawardhana, K.; Dale, B. E. & Lynd, L. R. (2009b). Projected mature technology scenarios for conversion of cellulosic biomass to ethanol with coproduction thermochemical fuels, power, and/or animal feed protein. *Biofuels, Bioproducts and Biorefining*, Vol.3, (2009b), pp. 231-246
Lee, Y.E.; Lowe, S. E.; Henrissat, B. & Zeikus, J. G. (1993a). Characterization of the active site and thermostability regions of endoxylanse from *Thermoanaerobacterium saccharolyticum* B6A-RI. *Journal of Bacteriology*, Vol.175, No.18, (1993a), pp. 5890-5898
Lee, Y.E.; Lowe, S. E. & Zeikus, J. G. (1993b). Regulation and characterization of xylanolytic enzymes of *Thermoanaerobacterium saccharolyticum* B6A-RI. *Applied and Environmental Microbiology*, Vol.59, (1993b), pp. 763-771
Leibbrant,N. (2010) *Techno-economic study for sugarcane bagasse to liquid biofuels in South Africa: A comparison between biological and thermochemical process routes*. PhD dissertation, Stellenbosch University, South Africa, April 2010
Linde, M.; Galbe, M. & Zacchi, G. (2010). Bioethanol production from non-starch carbohydrate residues in process streams from a dry-mill ethanol plant. *Bioresource Technology*, Vol.99, (2010), pp. 6505-6511
Linger, J.G.; Adney, W. S. & Darzins, A. (2010). Heterologous expression and extracellular secretion of cellulolytic enzymes by *Zymomonas mobilis*. *Applied and Environmetal Microbiology*, Vol.76, No.19, (October 2010), pp. 6360-6369
Lynd, L.R.; Van Zyl, W. H.; McBride, J. E. & Laser, M. (2005). Consolidated bioprocessing of cellulosic biomass: an update. *Current Opinion in Biotechnology*, Vol.16, (2005), pp. 577-583
Lynd, L.R.; Weimer, P. J.; Van Zyl, W. H. & Pretorius, I. S. (2002). Microbial cellulose utilization: fundamentals and biotechnology. *Microbiology and Molecular Biology Reviews*, Vol.66, (September 2002a), pp. 506-577
Martin, M., Ahmetovic, E., and Grossmann, I. E. Optimization of Water Consumption in Second Generation Bioethanol Plants. *Industrial & Engineering Chemistry Research* [In Press]
Mcbride, J., Brevnova, E., Ghandi, C., Mellon, M., Froehlich, A., Deleault, K., Rajgharia, V., Flatt, J, Van Zyl, E, Den Haan, R., La Grange, D., Rose, S., Penttila, M., Ilmen, M., Siika-aho, M., Uusitalo, J., Hau, H., Rice, C., Villari, J., Stonehouse, E., Gilbert, A., Keating, J., Xu, H., Willes, D., Shikhare, I., Thorngren, N., Warner, A., and Murphy,

D. Yeast expressing cellulases for simultaneous saccharification and fermentation using cellulose. Filed :23 November 2009, Published: 27 May 2010. [WO 2010/060056]

McKendry, P. (2002). Energy production from biomass (part 2): conversion technologies. *Bioresource Technology*, Vol.83, No.1, (May 2002), pp. 47-54

Moniruzzaman, M.; Lai, X.; York, S. W. & Ingram, L. O. (1997). Isolation and molecular characterization of high-performance cellobiose-fermenting spontaneous mutants of ethanologenic *Escherichia coli* KO11 containing the *Klebsiella oxytoca casAB* Operon. *Applied and Environmental Microbiology*, Vol.63, No.12, (1997), pp. 4633-4637

Nissen, T.L.; Kielland-Brandt, M. C.; Nielsen, J. & Villadsen, J. (2000). Optimization of ethanol production in *Saccharomyces cerevisiae* by metabolic engineering of the ammonium assimilation. *Metabolic Engineering*, Vol.2, (2000), pp. 69-77

Ohta, K.; Beall, D. S.; Mejia, J. P.; Shanmugam, K. T. & Ingram, L. O. (1991a). Genetic improvement of *Escherichia coli* for ethanol production: Chromosomal integration of *Zymomonas mobilis* genes encoding pyruvate decarboxylase and alcohol dehydrogenase. *Applied and Environmental Microbiology*, Vol.57, No.4, (1991a), pp. 893-900

Ohta, K.; Mejia, J. P.; Shanmugam, K. T. & Ingram, L. O. (1991b). Metabolic engineering of *Klebsiella oxytoca* M5A1 for ethanol production from xylose and glucose. *Applied and Environmental Microbiology*, Vol.57, No.10, (1991b), pp. 2810-2815

Panagiotou, G.; Christakopoulos, P. & Olsson, L. (2005). Simultaneous saccharification and fermentation of cellulose by *Fusarium oxysporum* F3--growth characteristics and metabolite profiling. *Enzyme and Microbial Technology*, Vol.36, No.5-6, (April 2005), pp. 693-699

Papendieck,A., Dahlems, U., & Gellissen, G. (2002) Technical enzyme production and whole-cell biocatalysis: application of *Hansenula polymorpha*. In: *Hansenula polymorpha: biology and applications,* Gellissen,G. (Ed), pp. 255-271. Wiley-VCH, Weinham

Piccolo, C. & Bezzo, F. (2009). A techno-economic comparison between two technologies for bioethanol production from lignocellulose. *Biomass and Bioenergy*, Vol.33, (2009), pp. 478-491

Piotek, M.; Hagedorn, J.; Hollenberg, C. P.; Gellissen, G. & Srasser, A. W. M. (1998). Two novel gene expression systems based on the yeasts *Schwanniomyces occidentalis* and *Pichia stipitis*. *Applied Microbiology and Biotechnology*, Vol.50, (1998), pp. 331-338

Rajoka, M.I.; Khan, S. & Shahid, R. (2003). Kinetics and regulation studies of the production of β-galactosidase from *Kluyveromyces marxianus* grown on different substrates. *Food Technology and Biotechnology*, Vol.41, No.4, (2003), pp. 315-320

Rautio, J.J.; Smit, B. A.; Wiebe, M.; Penttila, M. & Saloheimo, M. (2006). Transcriptional monitoring of steady state and effects of anaerobic phases in chemostat cultures of the filamentous fungus Trichoderma reesei. *BMC Genomics*, Vol.7, (2006), pp. 247

Reith,J.H., den Uil, H., de Laat, W. T. A. M., Niessen, J. J., de Jong, E., Elbersen, H. W., Weusthuis, R., Van Dijken, J. P., & Raamsdonk, L. (2002) *Co-production of bio-ethanol, electricity and heat from biomass residues*. Energy Research Centre, Netherlands, Report ECN-RX-02-030, July 2002.,

Richard, T.L. (2010). Challenges in scaling up biofuels infrastructure. *Science*, Vol.329, No.5993, (August 2010), pp. 793-796

Robinson,J. (2006) *Bio-ethanol as a household cooking fuel: A mini pilot study of the SuperBlu stove in peri-urban Malawi.* Loughborough University, UK

Ryabova, O.B.; Chmil, O. M. & Sibirny, A. A. (2003). Xylose and cellobiose fermentation to ethanol by the thermotolerant methylotrophic yeast *Hansenula polymorpha. FEMS Yeast Researh*, Vol.4, No.2, (November 2003), pp. 157-164

Sadie, C.J.; Rose, S. H.; Den Haan, R. & Van Zyl, W. H. (2011). Co-expression of a cellobiose phosphorylase and lactose permease enables intracellular cellobiose utilisation by Saccharomyces cerevisiae. *Applied Microbiology and Biotechnology*, [In press]

Sassner, P.; Galbe, M. & Zacchi, G. (2008). Techno-economic evaluation of bioethanol production from three different lignocellulosic materials. *Biomass Bioenergy*, Vol.32, (2008), pp. 422-430

Sastri, B., Lee, A., Kaarsberg, T., Alfstad, T., Curtis, M., and MacNeil, D. (2008). World biofuels production potential, 02.03.2011,Available from http://www.pi.energy.gov/documents/20080915_WorldBiofuelsProductionPotential.pdf

Seon, P.J.; Russell, J. B. & Wilson, D. B. (2007). Characterization of a family 45 glycosyl hydrolase from *Fibrobacter succinogenes* S85. *Anaerobe*, Vol.13, No.2, (April 2007), pp. 83-88

Shaw, A.J.; Hogsett, D. A. & Lynd, L. R. (2009). Identification of the [FeFe]-hydrogenase responsible for hydrogen generation in *Thermoanaerobacterium saccharolyticum* and demonstration of increased ethanol yield via hydrogenase knockout. *Journal of Bacteriology*, Vol.191, No.20, (October 2009), pp. 6457-6464

Shaw, A.J.; Jenney, F. E.; Adams, M. W. W. & Lynd, L. R. (2008a). End-product pathways in the xylose fermenting bacterium, *Thermoanaerobacterium saccharolyticum. Enzyme and Microbial Technology*, Vol.42, (2008a), pp. 453-458

Shaw, A.J.; Podkaminer, K. K.; Desai, S. G.; Bardsley, J. S.; Rogers, S. R.; Thorne, P. G.; Hogsett, D. A. & Lynd, L. R. (2008b). Metabolic engineering of a thermophilic bacterium to produce ethanol at high yield. *Proceedings of the National Academy of Sciences USA*, Vol.105, No.37, (2008b), pp. 13769-13774

Shin, H.D.; McClendon, S.; Vo, T. & Chen, R. R. (2010). *Escherichia coli* binary culture engineered for direct fermentation of hemicellulose to a biofuel. *Applied and Environmetal Microbiology*, Vol.76, No.24, (December 2010), pp. 8150-8159

Sims, R., Taylor, M., Saddler, J., and Mabee, W. (2008). From 1st- to 2nd-generation biofuel technologies - An overview of current industry and RD&D activities. International Energy Agency (IEA Bioenergy).

Soccol, C.R.; Vandenberghe, L. P. d. S.; Medeiros, A. B. P.; Karp, S. G.; Buckeridge, M.; Ramos, L. P.; Pitarelo, A. P.; Ferreira-Leitão, V.; Gottschalk, L. M. F.; Ferrara, M. A.; Bon, E. P. d. S.; de Moraes, L. M. P.; Araújo, J. d. A. & Torres, F. A. G. (2010). Bioethanol from lignocelluloses: Status and perspectives in Brazil. *Bioresource Technology*, Vol.101, (2010), pp. 4820-4825

Somerville, C.; Youngs, H.; Taylor, C.; Davis, S. C. & Long, S. P. (2010). Feedstocks for lignocellulosic biofuels. *Science*, Vol.329, (2010), pp. 790-792

Srivastava, R.; Kumar, G. P. & Srivastava, K. K. (1995). Construction of a recombinant cellulolytic *Escherichia coli. Gene*, Vol.164, (1995), pp. 185-186

Stephanopoulos, G. (2007). Challenges in engineering microbes for biofuels production. *Science*, Vol.315, (2007), pp. 801-804

Sun, Y. & Cheng, J. (2002). Hydrolysis of lignocellulosic material for ethanol production: a review. *Bioresource Technology*, Vol.83, (2002), pp. 1-11

Swanson, R.M.; Platon, A.; Satrio, J. A. & Brown, R. C. (2010). Techno-economic analysis of biomass-to-liquids production based on gasification. *Fuel*, Vol.89, (2010), pp. S11-S19

Van Dijken, J.P.; Bauer, J.; Brambilla, L.; Duboc, P.; Francois, J. M.; Gancedo, C.; Giuseppin, M. L.; Heijnen, J. J.; Hoare, M.; Lange, H. C.; Madden, E. A.; Niederberger, P.; Nielsen, J.; Parrou, J. L.; Petit, T.; Porro, D.; Reuss, M.; van Riel, N.; Rizzi, M.; Steensma, H. Y.; Verrips, C. T.; Vindelov, J. & Pronk, J. T. (2000). An interlaboratory comparison of physiological and genetic properties of four *Saccharomyces cerevisiae* strains. *Enzyme and Microbial Technology*, Vol.26, (June 2000), pp. 706-714

van Rooyen, R.; Hahn-Hagerdal, B.; La Grange, D. C. & Van Zyl, W. H. (2005). Construction of cellobiose-growing and fermenting *Saccharomyces cerevisiae* strains. *Journal of Biotechnology*, Vol.120, No.3, (November 2005), pp. 284-295

van Wyk, N.; Den Haan, R. & Van Zyl, W. H. (2010). Heterologous co-production of *Thermobifida fusca* Cel9A with other cellulases in *Saccharomyces cerevisiae. Applied Microbiology and Biotechnology*, Vol.87, No.5, (May 2010), pp. 1813-1820

Van Zyl, W.H.; Chimphango, A. F. A.; Den Haan, R.; Görgens, J. F. & Chirwa, P. W. C. (2011). Next-generation cellulosic ethanol technologies and their contribution to a sustainable Africa. *Interface Focus*, Vol.1, No.2, (April 2011), pp. 196-211

Van Zyl, W.H.; Lynd, L. R.; Den Haan, R. & McBride, J. E. (2007). Consolidated Bioprocessing for Bioethanol Production Using *Saccharomyces cerevisiae. Advances in Biochemical Engineering Biotechnology*, Vol.108, (2007), pp. 205-235

von Sivers, M. & Zacchi, G. (1995). A techno-economical comparison of three processes for the production of ethanol from pine. *Bioresource Technology*, Vol.51, No.1, (1995), pp. 43-52

Voronovsky, A.Y.; Rohulya, O. V.; Abbas, C. A. & Sibirny, A. A. (2009). Development of strains of the thermotolerant yeast *Hansenula polymorpha* capable of alcoholic fermentation of starch and xylan. *Metabolic Engineering*, Vol.11, (2009), pp. 234-242

Watanabe, T.; Watanabe, I.; Yamamoto, M.; Ando, A. & Nakamura, T. (2011). A UV-induced mutant of Pichia stipitis with increased ethanol production from xylose and selection of a spontaneous mutant with increased ethanol tolerance. *Bioresource Technology*, Vol.102, No.2, (January 2011), pp. 1844-1848

Watson, H.K. (2011). Potential to expand sustainable bioenergy from sugarcane in southern Africa. *Energy Policy*, [Corrected Proof]

Wingren, A.; Galbe, M. & Zacchi, G. (2003). Techno-Economic Evaluation of Producing Ethanol from Softwood: Comparison of SSF and SHF and Identification of Bottlenecks. *Biotechnology Progress*, Vol.19, (2003), pp. 1109-1117

Wood, B.E.; Beall, D. S. & Ingram, L. O. (1997). Production of recombinant bacterial endoglucanase as a co-product with ethanol during fermentation using derivitives of Escherichia coli KO11. *Biotechology and Bioengineering*, Vol.55, No.3, (1997), pp. 547-555

Wood, B.E. & Ingram, L. O. (1992). Ethanol production from cellobiose, amorphous cellulose, and crystalline cellulose by recombinant *Klebsiella oxytoca* containing chromosomally integrated *Zymomonas mobilis* genes for ethanol production and plasmids expressing thermostable cellulase genes from *Clostridium thermocellum*. *Applied and Environmental Microbiology*, Vol.58, No.7, (1992), pp. 2103-2110

Xiros, C.; Topakas, E.; Katapodis, P. & Christakopoulos, P. (2008). Evaluation of *Fusarium oxysporum* as an enzyme factory for the hydrolysis of brewer's spent grain with improved biodegradability for ethanol production. *Industiral Crop Production*, Vol.28, No.2, (September 2008), pp. 213-224

Xu, C.; Qin, Y.; Li, Y.; Ji, Y.; Huang, J. & Song, H. (2010). Factors influencing cellulosome activity in Consolidated Bioprocessing of cellulosic ethanol. *Bioresource Technology*, Vol.101, No.9560, (2010), pp. 9569

Xu, Q.; Singh, A. & Himmel, M. E. (2009). Perspectives and new directions for the production of bioethanol using consolidated bioprocessing of lignocellulose. *Current Opinion in Biotechnology*, Vol.20, No.3, (June 2009), pp. 364-371

Yamada, R.; Taniguchi, N.; Tanaka, T.; Ogino, C.; Fukuda, H. & Kondo, A. (2010). Cocktail δ-integration: A novel method to construct cellulolytic enzyme expression ratio-optimized yeast strains. *Microbial Cell Factories*, Vol.9, (May 2010), pp. 32

Yamano, L.P.; York, S. W. & Ingram, L. O. (1998). Isolation and characterization of ethanol-tolerant mutants of *Escherichia coli* KO11 for fuel ethanol production. *Journal of Industrial Microbiology & Biotechnology*, Vol.20, (1998), pp. 132-138

Yanase, H.; Nozaki, K. & Okamoto, K. (2005). Ethanol production from cellulosic materials by genetically engineered *Zymomonas mobilis*. *Biotechnology Letters*, Vol.27, (2005), pp. 259-263

Yanase, S.; Hasunuma, T.; Yamada, R.; Tanaka, T.; Ogino, C.; Fukuda, H. & Kondo, A. (2010). Direct ethanol production from cellulosic materials at high temperature using the thermotolerant yeast *Kluyveromyces marxianus* displaying cellulolytic enzymes. *Applied Microbiology and Biotechnology*, Vol.88, No.1, (September 2010), pp. 381-388

Yoo, J.S.; Jung, Y. J.; Chung, S. Y.; Lee, Y. C. & Choi, Y. L. (2004). Molecular cloning and characterization of CMCase gene (*celC*) from *Salmonella typhimurium* UR. *Journal Microbiology*, Vol.42, No.3, (September 2004), pp. 205-210

Zaldivar, J., Nielsen, J. & Olsson, L. (2001). Fuel ethanol production from lignocellulose: a challenge for metabolic engineering and process integration. *Applied Microbiology and Biotechnology*, Vol.56, (2001), pp. 17-34

Zhang, M., Chou, Y. C., Picataggio, S. K., and Finkelstein, M. (1997) Single *Zymomonas mobilis* strain for xylose and arabinose fermentation. Filed:6 May 1997, Published: 1 December 1998 [WO 5843760]

Zhang, Y.H. & Lynd, L. R. (2004). Toward an aggregated understanding of enzymatic hydrolysis of cellulose: noncomplexed cellulase systems. *Biotechnology and Bioengineering*, Vol.88, No.7, (December 2004), pp. 797-824

Zhou, S.; Davis, F. C. & Ingram, L. O. (2001). Gene integration and expression and extracellular secretion of *Erwinia chrysanthemi* endoglucanase CelY (*celY*) and CelZ (*celZ*) in ethanologenic *Klebsiella oxytoca* P2. *Applied and Environmetal Microbiology*, Vol.67, No.1, (January 2001), pp. 6-14

Zhou, S. & Ingram, L. (1999a). Engineering endoglucanase-secreting strains of ethanologenic *Klebsiella oxytoca* P2. *Journal of Industrial Microbiology and Biotechnology*, Vol.22, No.6, (June 1999a), pp. 600-607

Zhou, S. & Ingram, L. O. (1999b). Engineering endoglucanase-secreting strains of ethanologenic *Klebsiella oxytoca* P2. *Journal of Industrial Microbiology and Biotechnology*, Vol.22, (1999b), pp. 600-607

Zhou, S. & Ingram, L. O. (2001). Simultaneous saccharification and fermentation of amorphous cellulose to ethanol by recombinant *Klebsiella oxytoca* SZ21 without supplemental cellulase. *Biotechnology Letters*, Vol.23, (January 2001), pp. 1455-1462

# 8

# Latest Frontiers in the Biotechnologies for Ethanol Production from Lignocellulosic Biomass

De Canio Paola[1], De Bari Isabella[2] and Romano Patrizia[1]
[1]*Department of Biology*
*Agricultural and Environmental Biotechnology Basilicata University*
[2]*ENEA Italian National Agency for New Technologies*
*Energy and Sustanaible Economic Development*
*Italy*

## 1. Introduction

The development of the industrial and technological society together with the economic and environmental implications, such as global warming and decreasing oil reserves, have been driving worldwide interest in searching for renewable energies to replace fossil fuels. With respect to fossil fuels, biomass-based fuels have the advantage of decreasing greenhouse gas (GHG) emissions. In this context, ethanol produced from biomass, the so called "bioethanol", has become a major energy carrier for a sustainable transportation sector. Bioethanol is an oxygenate fuel with an high octane number (Moon *et al.*, 2009) and it can be used as biofuel either in its pure state (E100) or blended with petrol in various proportions, such as E85, E95, E10 containing 85%, 95% and 10% of ethanol respectively. Among these, E10 not requires any change in engine (Balat, 2009a). In addition, bioethanol has low toxicity and reduces urban air pollution because the carbon dioxide released during its combustion is virtually reused by plants during the chlorophyll photosynthesis. Currently, United States and Brazil are the largest bioethanol producers in the world from corn and sugarcane respectively. However, in some countries with low availability of agricultural lands, the production of biofuels from dedicated crops could lead to direct conflict with food productions. Lignocellulosic materials and, among them, agro-forest residues, could, offer a great potential as biomass source for bioethanol production. In fact, they are virtually abundant and low cost (Perlack *et al.*, 2005). Lignocellulosics materials can be classified in four groups: forest residues (chips and sawdust from lumber mills, dead trees, and tree branches), municipal solid wastes (household garbage and paper products), waste paper and energy crops (Balat, 2010). Lignocellulosic feedstocks are composed primarily of carbohydrate polymers (cellulose and hemicellulose) and phenolic polymers (lignin). Cellulose $(C_6H_{10}O_5)_x$ is a linear polysaccharide polymer of glucose made of cellobiose units that are packed by hydrogen bonds. The structure of this polymer is rigid and compact, so that in order to obtain glucose, the biomass needs pre-treatment that breaks its structure to facilitate the action of the enzymes. The individual cellulose chains are packed and organized into crystalline microfibrils. Within these microfibrils, cellulose is found in two forms, namely amorphous and crystalline. The crystalline form of cellulose is very difficult to degrade. Hemycellulose such as xylan $(C_5H_8O_4)_m$ is a short polymer of pentoses and hexoses sugars. The dominants sugars in hemicelluloses are mannose (six-carbon sugar) in softwoods and xylose (five carbon sugar) in hardwoods and agriculture residues (Persson *et al.*, 2006). Hemicellulose contains also, galactose, glucose and arabinose. This polymer is amorphous and easier to hydrolyse than cellulose. Lignin $[(C_9H_{10}O_3)(OCH_3)0.9\text{-}1.7]_n$ is a phenyl propane polymer that contains many functional groups such as hydroxyl, methoxyl and carbonyl. Unlike cellulose and hemicellulose, lignin cannot be utilized in the fermentation process. In fact, it has high resistance to chemical and enzymatic degradation. Low concentration of various other compounds, such as extractive and ash are also present. Ash consists of minerals such as silicon, aluminum, calcium, magnesium, potassium, and sodium. Extractives include resins, fats and fatty acids, phenolics, phytosterols, salts, minerals and other compounds. The proportions of these constituents vary between different species. Hardwood has a content of cellulose and hemicelluloses around 80% of total feedstock dry matter while softwood contains around 70% of total dry matter (Balat, 2010). On the other side, lignin is more in softwood than hardwood (Balat, 2009b).Table 1 shows the composition of several lignocellulosic materials and their potential ethanol output obtainable from 1 Kg dry biomass of each type. Cellulose generally accounts for 30-60% of the biomass dry weight while the hemicellulose content varies from 10% to 40%, and the lignin content from 10% to 25% except for olive husks in which the lignin content is higher (48.4%, Table 1). Actually, the world's largest ethanol producers are Brazil and USA, which together account for more than 65% of global ethanol production. In Europe (EU), the high oil prices and the ratification of the Kyoto Protocol in 2005 have provided additional incentives to promote the use of alternative fuels. Today, EU is the third producer of bioethanol in the world with a production that in 2009 amounted to 3.7 billion liters (www.plateforme-biocarburants.ch).

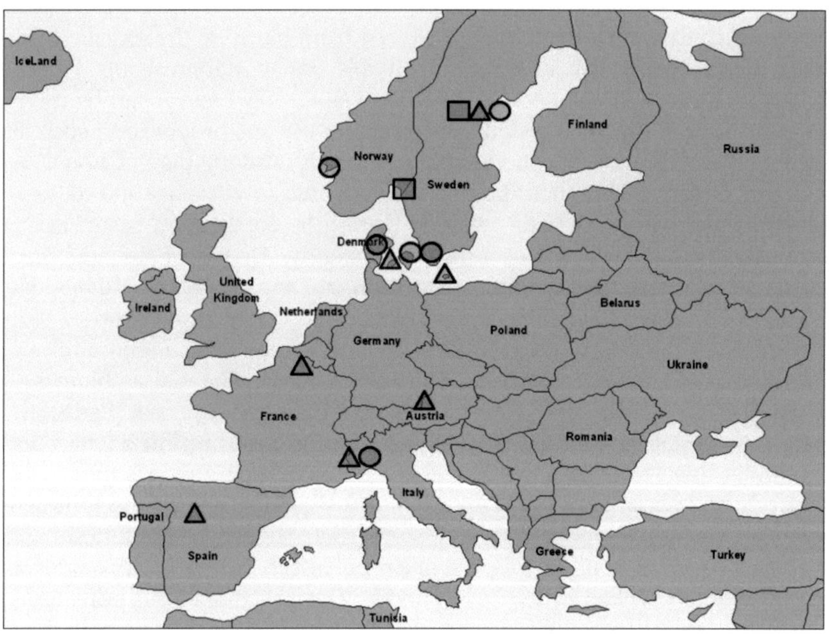


Fig. 1. European biochemical plants for bioethanol production. Demonstrative plants are marked with a triangle. Pilot plants are indicated with a circle and commercial plants are marked with a square. Information were taken from: http://biofuels.abc-energy.at/demoplants/projects/mapindex (TASK IEA 39).

Table 2 shows the detailed bioethanol production in EU for the year 2009 in the major countries.

| *Biomass* | *Ash (%)* | *Hemicellulose (%)* | *Cellulose (%)* | *Lignin (%)* | *Ethanol potential kg/kg** |
|---|---|---|---|---|---|
| Poplar | | 17 | 49 | 18 | 0.37 |
| Eucalyptus | | 31.8 | 43.3 | 24.7 | 0.42 |
| Maize stalk straw | 3 | 26 | 38 | 11 | 0.36 |
| Wheat straw | 1.3 | 27.6 | 34 | 18 | 0.35 |
| Rice straw | 18.9 | 22.7 | 37 | 13.6 | 0.34 |
| Oat straw | 2.6 | 24.9 | 37.1 | 15.4 | 0.35 |
| Rye straw | 1.2 | 25.7 | 37.1 | 17.6 | 0.35 |
| Barley straw | 7.1 | 44 | 37 | 11 | 0.46 |
| Potato rests | 5 | 11.8 | 26 | | 0.21 |
| Miscanthus straw | 2.7 | 29.6 | 44.7 | 21 | 0.42 |
| Kenaf | | | 41.9 | 12.3 | 0.24 |
| Hemp (wood fiber) | | 27.5 | 37.5 | 22 | 0.37 |
| Beet tail and beet green | 5 | 10 | 10 | 5 | 0.11 |
| Tobacco stalk | 2.4 | 28.2 | 42.4 | 27 | 0.40 |
| Wood, ailanthus | 0.5 | 26.6 | 46.7 | 26.2 | 0.41 |
| Soybean stalks and leaves | | 18.5 | 32.1 | | 0.29 |
| Bagasse | | 24.6 | 39.7 | 25.2 | 0.36 |
| Tomato plant waste | 20.2 | 6 | 25.7 | 19.5 | 0.18 |
| Garlic waste | 17.1 | 6.9 | 24.2 | 8.5 | 0.17 |
| Vines # | | 29.42 | 19.80 | | 0.28 |
| Olive husk | 4 | 23.6 | 24 | 48.4 | 0.27 |
| Agrarian residues | | 17 | 32 | | 0.27 |

Table 1. Composition of some lignocellulosic materials and theoretical ethanol yields. (*source: Phyllis database for biomass and waste*);* calculated as: 1) cellulose:glucan->glucose->ethanol, 2) hemicelluloses: xylane->xylose->_ ethanol; # data from ENEA

The EU's biggest producer is France with 1250 million liters mainly from beet and molasses. Germany comes second (750 million liters) followed by Spain with 465 million liters. In this country, the goal was reached also thanks the Abengoa's demonstration plant in

Babilafuente (Salamanca). In particular, Abengoa Bioenergy New Technologies has been developing the biorefinery concept to convert a wide range of biomass feedstocks into ethanol, chemicals and energy.
The feedstock includes agricultural residues, wood residues, and energy crops such as switchgrass and poplar (www.abengoabioenergy.com). Table 3 lists some bioethanol plants in the EU using lignocellulosic feedstocks while figure 1 displays the overall distribution of plants, including demonstrative pilot and commercial scale (figure 1), using biochemical conversions to obtain ethanol.

## 2. Bioethanol production from lignocellulosic raw material

The conversion of lignocellulosics materials to bioethanol via enzymatic hydrolysis can be simplified in four major steps: pretreatment, hydrolysis, fermentation and product separation (figure 2). In the next section the main pretreatment strategies will be overviewed.

| COUNTRY | Ethanol production (million liters) |
|---|---|
| Germany | *750* |
| Spain | *465* |
| France | *1250* |
| Poland | *166* |
| Sweden | *175* |
| Italy | *72* |
| Hungary | *150* |
| Lithuania | *30* |
| Austria | *180* |
| Belgium | *143* |
| Czech Republic | *113* |
| Slovakia | *118* |

Table 2. Bioethanol production in Europe for the year 2009 (source: www:plateform biocarburants.ch)

### 2.1 Pretreatments

The conversion of lignocellulosic biomass into ethanol requires a pretreatment step to change the physical and chemical structure of biomass and to enhance the hydrolysis rate. There are several pretreatment strategies, all aimed at opening the structure of the cell biomass and allow the enzymes to access the internal polysaccharides. The available pretreatments can be grouped in chemical, biological, physical and physicochemical processes.
*Chemical pretreatments* employ different chemical agents like ozone, acids and alkalis. The ozonolysis can degrade lignin and part of hemicellulose but this technology appears quite expensive.
Sulfuric acid is the most applied acid, but other acids such as HCl and $HNO_3$ were also reported (Taherzadeh *et al.*, 2008). Dilute-acid hydrolysis can be used either as a pretreatment of lignocellulose for enzymatic hydrolysis, or as the actual method of hydrolyzing to fermentable sugars (Taherzadeh *et al.*, 2007, 2008). In general, it has the

disadvantage of the toxicity due to the unspecific and, sometime, harsh degradation of the biomass matrix. Furthermore it is could be corrosive for employed facilities (Abril D. & Abril A., 2009).
Alkali pretreatment is based on the use of alkaline solutions such as NaOH, $Ca(OH)_2$ or ammonia to remove lignin and part of the hemicellulose, and increase the enzymes accessibility to the biopolymers.
Most promising is also the wet oxidation in which, the material are treated with water and air or oxygen at temperatures above 120°C for a period of e.g. 30 min. The process represents an effective method in separating the cellulosic fraction from lignin and hemicellulose (Taherzadeh *et al.*, 2008).
*Biological pretreatment* uses microorganisms such as brown, white and soft-rot fungi which degrade lignin and solubilize hemicelluloses (Sun & Cheng, 2002). In recent years, progresses in bioengineering have led to the development of microorganisms which can attack lignin in the biomass. The biological process is interesting for its low energy requirement. However, the rate of hydrolysis in the biological process is very low (Sun & Cheng, 2002).
Among the investigated pretreatment, the *steam explosion (SE)* appears one of the most interesting since it limits the use of chemicals mostly to the use of saturated steam (Ballesteros *et al.*, 1998; De Bari *et al.*, 2002; Ogier *et al.*, 1999).

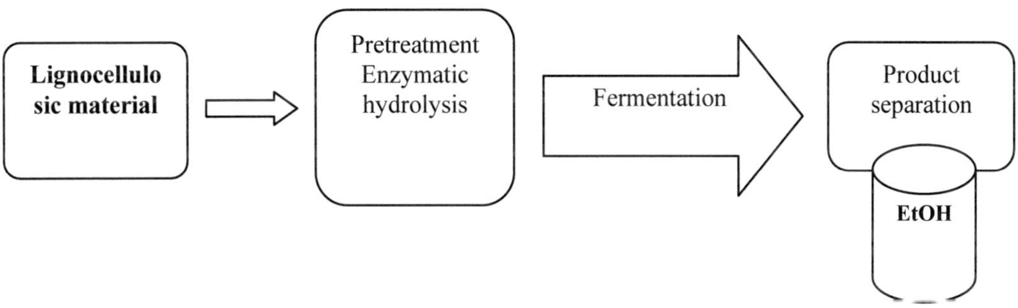


Fig. 2. Main steps of bioethanol production from lignocellulosic materials

Through the saturated water steam at high temperature, SE causes autohydrolysis reactions in which part of hemicellulose and lignin are converted into soluble oligomers. Thus, the lignocellulosic matrix is opened up, and the cellulose surface becomes more accessible to enzymes. The process employs high pressure steam with temperature typically ranging from 160 to 260 °C for few minutes. This is followed by explosive decompression of biomass (Banerjee *et al.*, 2010; Boussaid *et al.*, 1999; Sun *et al.*, 2004; Varga *et al.* 2004).
A number of studies have been already reported in literature describing positive effects in terms of enhancing the enzymatic hydrolizability of several materials (hardwood and softwood, corn stover, straws etc.) (Cara *et al* ,2008; Galbe *et al.* 2002; Kobayashi *et al.*,2004; Ohgren *et al.*, 2006; Sun *et al.* 2002; Viola *et al.*,2008) The steam explosion technology, investigated for several years in Italy at the ENEA research Center of Trisaia is now going to be developed at industrial scale thanks to investments from the Italian Mossi & Ghisolfi group. Another physicochemical pretreatment is the *ammonia fiber explosion (AFEX)* in which the biomass is exposed to liquid ammonia at temperature around 90-100 °C followed by instantaneous pressure release. The AFEX process at reduces the lignin fraction but has less effect on the hemicellulose and cellulose fractions. In order to develop improved

lignocelluloses pretreatment strategies the use of $CO_2$ explosion was also reported (Kumar *et al.*, 2009).

| **Location** | **Coordinating organization/Company** | **Input** | **Output** | **Technology** | **Start - up** |
|---|---|---|---|---|---|
| Örnsköldsvik (Sweeden) | SEKAB | | 4500 t/a | Enzymes with pretreatment of diluted acid in one step. | 2011 |
| Blomsterdalen (Norway) | Weyland AS | 0.075 t/h | 158 t/a | Strong Acid Process | 2010 |
| Ballerup (Denmark) | BioGasol | 0.5t/h | 10 t/a | Enzymatic hydrolysis and fermentation | 2008 |
| Fredericia (Denmark) | Inbicon (DONG Energy) | 1t/h | | hydrothermal pre-treatment, high gravity hydrolysis, yeast fermentation | 2005 |
| Tortona (Italy) | Chemtex-Ghisolfi (Italia) | 160.000 t/a | 40.000 t/a | Enzymatic conversion. Pretreatment in equipment specifically designed. | 2011 |
| Babilafuente, Salamanca (Spain) | Abengoa Bioenergy | 35.000 t/a | 3950 t/a | Steam-explosion biochemical conversion | 2009 |
| POMACLE (France) | PROCETHOL 2G | | 2700 t/a | Enzymatic hydrolysis followed by yeast fermentation | 2011 |

Table 3. Some bioethanol plants in the EU. (source: http://biofuels.abc-energy.at)

On the whole , however, there isn't one general method of pretreatment because different types of raw material require different approaches. For instance, methods such as AFEX and wet oxidation seem to be more successful for agricultural residues whereas steam pretreatment has resulted in high sugar yields for both forestry and agricultural residues (Hahn-Hagerdal *et al.*, 2006). Table 4 summarizes advantages and disadvantages of some pretreatment processes.

## 2.2 Hydrolysis step

After the pretreatment, biomass is hydrolyzed to syrups containing monomeric sugars that can be fermented. The most applied methods for hydrolysis can be grouped in two classes: chemical hydrolysis and enzymatic hydrolysis. The latter process is particularly interesting because it is selective in the biomass degradation and can be operated at mild temperature and pH conditions. For several years, the enzymatic hydrolysis of cellulose has been the major target of an international research activity. The main obstacles to the achievement of high process yields have been the existence of crystalline domains within the cellulose and the low efficacy of the enzymes used for the transformation. Considering the specificity of the enzymes action, several components with complementary functions are necessary to attack the different regions in the biopolymers chains. As result, the enzymatic preparations used for the hydrolysis process are complex mixtures of proteins with synergistic actions termed cellulases (Bayer *et al.*, 1998).

| Pretreatment | Advantages | Disadvantages |
|---|---|---|
| Steam-explosion | Chemical free | Generation of degradation products |
| AFEX | Low degradation products | Low hydrolysis yields with woody crops |
| Ozonolysis | Reduction of lignin content, doesn't produce toxic residue | Expensive |
| Wet-oxidation | Low degradation products | Use of oxygen |
| Alkalis | Removal lignin, increase accessible surface area | Use of chemicals, long residence time |
| Acids | Alteration of lignin structure | Equipment corrosion, toxicity |
| $CO_2$-explosion | Contamination free, increase of accessible surface area | Use of $CO_2$ |
| Biological | Low energy requirement | Low hydrolysis rate |

Table 4. Advantages and disadvantages of some pretreatment strategies

These are proteins with a molecular weight from 30000 to 60000 AMU with a typical size from min. 30 to max. 200 Å (Fan *et al.* 1987). The surface area of lignocellulosic material is unaccessible to enzymes molecules and this fact implies the need of an initial pretreatment. In fact, the rate of the cellulose enzymatic hydrolysis depends by the structure of cellulose (Balat, 2010) and its crystallinity. In effect, the rate of hydrolysis of amorphous cellulose is 3-30 times faster than that of high crystalline cellulose (Lynd *et al.*, 2002).

Cellulase production is common in a large variety of fungi like *Trichoderma, Aspergillus, Penicillum* (Galbe *et al.*, 2002). The most frequently reported sources of cellulose is the fungus *Trichoderma reesei* which produces an extracellular and efficient cellulase enzyme system (Jana *et al.* 1994)

In particular, the cellulases mix is constituted of *endo1,4β-D-glucanase, exo1,4β-glucanase* and *β-glucosidase*. The hydrolysis of hemicellulose is carried out by hemicellulolytic enzymes that include mostly endoxylanase, exoxylanase and β-xylosidase (Saha, 2004).
Most of these cellulolitic cocktails are present in commercial preparations supplied by several biotechnological companies such as Novozymes (Denmark), Genecor (Palo Alto, CA), Iogen (Canada). Recently Genecor launched a new class of enzyme called "*Accelerase 1500*", which have an enhanced β-glucosidase activity. Similarly, Novozymes has recently produced the Cellin CTEC mixtures having improved activities with respect to the traditional Celluclast.
The commercial preparations are often compared on the base of their activities assayed by standard protocols (e.g. FPU, filter paper units). However the complexity of the lignocellulosic substrates does not make easy the prediction of the enzymes dosage on the base of the standard activities (Kabel *et al.*, 2006). As consequence, the process must be tailored to the specific biomass used.
The enzymes activity mainly depends on the process temperature. An increase of temperature of 20-30°C can introduce a significant improvement of the hydrolysis rate. However, the enzymes are proteins and high temperatures cause their denaturation. In this regard, thermostable enzymes offer potential benefits in the hydrolysis of lignocellulosics. In particular, thermostable enzymes have several advantages like higher stability and higher activity that decrease the optimal dosage needed for the process. Some thermostable enzymes have been isolated from bacteria thermophilic including the *Rhodothermus* strains (Hreggvidsson *et al.*, 1996) and *Thermotoga* (Bok *et al.*, 1998; Bronnenmeier *et al.*, 1995; Evans *et al.*, 2000)
Recently, a new mix of three thermostable enzymes (*cellobiohydrolase, endoglucanase* and *β-glucosidase*) were cloned and produced in *Trichoderma reesei* (Viikari *et al.*, 2007). The obtained cellulases mixture was then added with thermostable xylanase and tested at high temperature for the hydrolysis of steam pretreated spruce and corn stover. The results showed that the new enzymatic formulation had an activity at 65°C, 25% higher than the maximum activity of commercial reference enzymes.

## 3. Fermentation of lignocellulosic hydrolyzates: Conversion of biomass to ethanol by microorganisms

Fermentation of enzymatic hydrolyzates can be carried out by various microorganisms such as several species of bacteria, yeasts and filamentous fungi. Depending on the overall process scheme, mixed or separate C5 and C6 sugars streams can be obtained. While the ethanolic fermentation of glucose, mannose and galactose is well established on large scale (Berg, 2002), the conversion of the pentose sugars, namely xylose and arabinose, is much difficult. However, it was estimated that the complete conversion of pentose sugars to ethanol would reduce the bioethanol production cost by as much as 22% (Sassner *et al.*, 2008). Other essential characteristics required in fermenting microorganisms are high ethanol yields and productivities, minimum formation of secondary metabolites and high tolerance to inhibitors produced during the pretreatment and hydrolysis steps.
The common yeast used for alcoholic fermentation is *Saccharomyces cerevisiae,* which has most of these characteristics. In particular, this specie of yeast catabolizes glucose to ethanol very efficiently by means of the Embden-Meyerhof and Parnas pathway (EMP) followed by alcoholic fermentation under anaerobic conditions (figure 3).

The stoichiometric reaction for glucose conversion into ethanol is described by equation 1:

$$C_6H_{12}O_6 \xrightarrow{alcholic\ fermentation} 2C_5H_5OH + 2CO_2 \quad (1)$$

Considering that the molecular weight of ethanol is 46 g/mole and that of glucose is 180 g/mole and that one mole of glucose produce 2 moles of ethanol, the theoretical yield for ethanol production from glucose is 0.51.

Another microorganism capable to convert glucose into ethanol is *Zymomonas mobilis*, a Gram-negative bacterium which produces ethanol at high yield. Choi *et al.* (2008) reported an ethanol yield of 90.4% from naked barley.

Nevertheless, both *Saccharomyces cerevisiae* and *Zymomonas mobilis* cannot ferment pentoses such as xylose present in the hydrolysates of several abundant lignocellulosic biomass such as residual straws (Keshwani *et al.*, 2009). This inability represents the major obstacle to use these microorganisms for the fermentation of mixed syrups from lignocellulosics.

However, in nature, there are some microorganisms (bacteria and yeasts) which have demonstrated a good capacity of using xylose (table 5). Figure 4 shows the xylose utilization pathways in bacteria and yeasts. Among yeasts, *Pichia stipitis, Candida shehatae* and *Pachisolen tannophilus* resulted very interesting for their capacity to ferment xylose. As shown in figure 4, yeasts metabolize xylose by means of the xylose reductase (XR) that converts xylose to xylitol and xylitol dehydrogenase (XDH) that convert xylitol to xylulose After phosphorylation, xylulose is methabolized through the pentose phosphate pathway (PPP) (Zaldivar *et al.*, 2001 ).

Generally, XR is an enzyme NADPH cofactor dependent while XDH is $NAD^+$ cofactor dependent (Agbogbo & Coward-Kelly, 2008). When the process is carried out under anaerobic conditions, the production of xylitol is favoured and this reduces the final ethanol yield. Among the wild type yeasts fermenting xylose, *Pichia stipitis* was considered the most promising (Agbogbo & Coward-Kelly, 2008) because it has a XR capable to use as cofactor both NADPH and NADH. For this reason, under anaerobic conditions, xylose fermentation in *Pichia stipitis* is carried out by using NADH.

Accordingly, *P. stipitis* produces less xylitol compared to others xylose fermenting yeasts (Agbogbo & Coward-Kelly, 2008). Nevertheless, the use of *P. stipitis* and others wild yeasts for the xylose fermentation is limited by the reduced capacity of using xylose when also glucose is present in the hydrolyzates. In fact, many of these microorganisms have a diauxic growth: when they are in a medium containing mixed sugars, glucose is consumed as first and the others sugars are metabolizes after its depletion resulting in a low productivity.

Furthermore, the fermentation capacity of natural *P. stipitis* depends in a critical way on the preservation, through the process of the microaerophilic conditions. Several investigation on the effects of aeration rate on the fermentation of glucose and xylose by *P.stipitis* have established that a low aeration rate is necessary for an optimal conversion of these sugars to ethanol. In detail, an ethanol production rates of 0.35 and 0.13 g $g^{-1}$ $h^{-1}$ were reached respectively on glucose and xylose by using oxygen uptake rates below 0.005 mol $l^{-1}$ $h^{-1}$ ; however because the substrate uptake rate is the rate-limiting step, a high cell concentration is needed to obtain high volumetric productivities (Grootjen *et al.*, 1990).

Unlike yeasts, bacteria fermenting xylose directly convert xylose to xylulose (Zaldivar *et al.*, 2001) through the xylose isomerase (XI) (figure 4).

There are some bacteria that have a natural capacity to use pentoses (table 5). *Escherichia coli*, for instance, is a bacterium gram-negative with a facultative anaerobic behavior which metabolizes pentoses via the PPP. However, the wild strain of this bacterium produces a small amount of ethanol. Thanks to the recombinant DNA technology, it has been possible to transform this microorganism into a bacterium, *E coli B KO11*, capable to produce ethanol with high yields (Ohta *et al.*, 1991).

Some *lactic acid bacteria* (LAB) were also investigated for their ability to produce ethanol. Among them, *L. buchneri* strain NRRL B-30929 can metabolize glucose and xylose simultaneously (Liu *et al.*, 2008) but produce undesirable bio-products such as acetate and lactate.

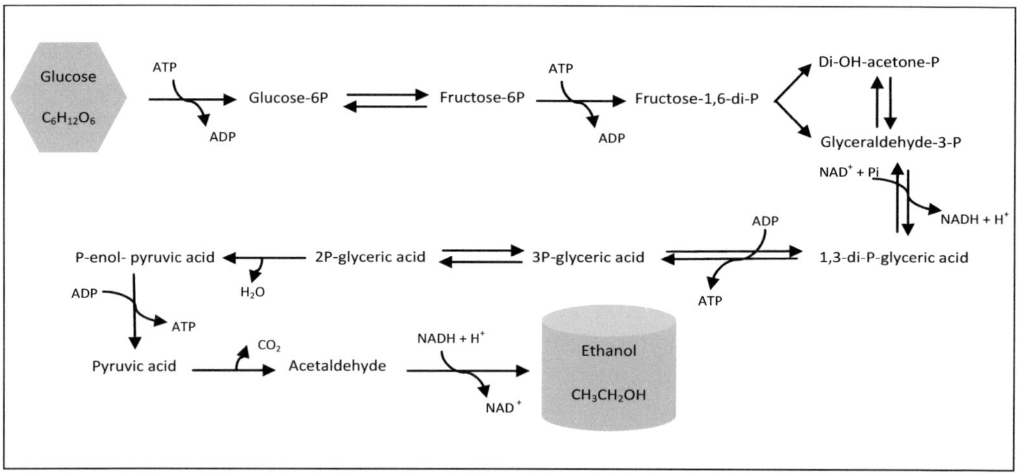


Fig. 3. Metabolic pattern from glucose to ethanol in *S. cerevisiae*. Under anaerobic conditions, piruvic acid is converted into ethanol by alcohol dehydrogenase.

*Thermophilic anaerobic bacteria* such as *Clostridium termohydrosulfuricum* and *Thermoanaerobacter ethanolicus* (table 5) have also been considered for their ethanol production (Balat, 2010). Using thermophilic microbes have several advantage such as the possibility to perform simultaneous hydrolysis and fermentation at high temperature (Knutson *et al.*, 1999). However the low ethanol tolerance of thermophilic anaerobic bacteria represents an obstacle for their industrial application.

Table 5 summarizes the performances of the most common wild type microorganisms. On the whole, the major part of these microorganisms has low productivities. Therefore the scientific community is trying new approaches to achieve the goal of using all the biomass carbohydrates with high efficiency. The following paragraphs describe some breakthroughs obtained in the fermentation of pentoses.

### 3.1 Cofermentation of mixed hydrolyzates

The simultaneous fermentation of glucose and xylose in the hydrolysates is one of the most ambitious challenges in the field of bioethanol production because this would simplify some process steps and, as consequence, could reduce capital and management costs.

Certainly, the use of wild yeast co-cultures is a mature approach for the fermentation of mixed syrups. In co-cultures experiments, various combinations of yeasts were tested: the most commonly used co-cultures were constituted by cells of *P.stipitis* and *S.cerevisiae* thanks to the ability of *P.stipitis* to metabolize xylose and the efficient consumption of glucose by *S.cerevisiae.*

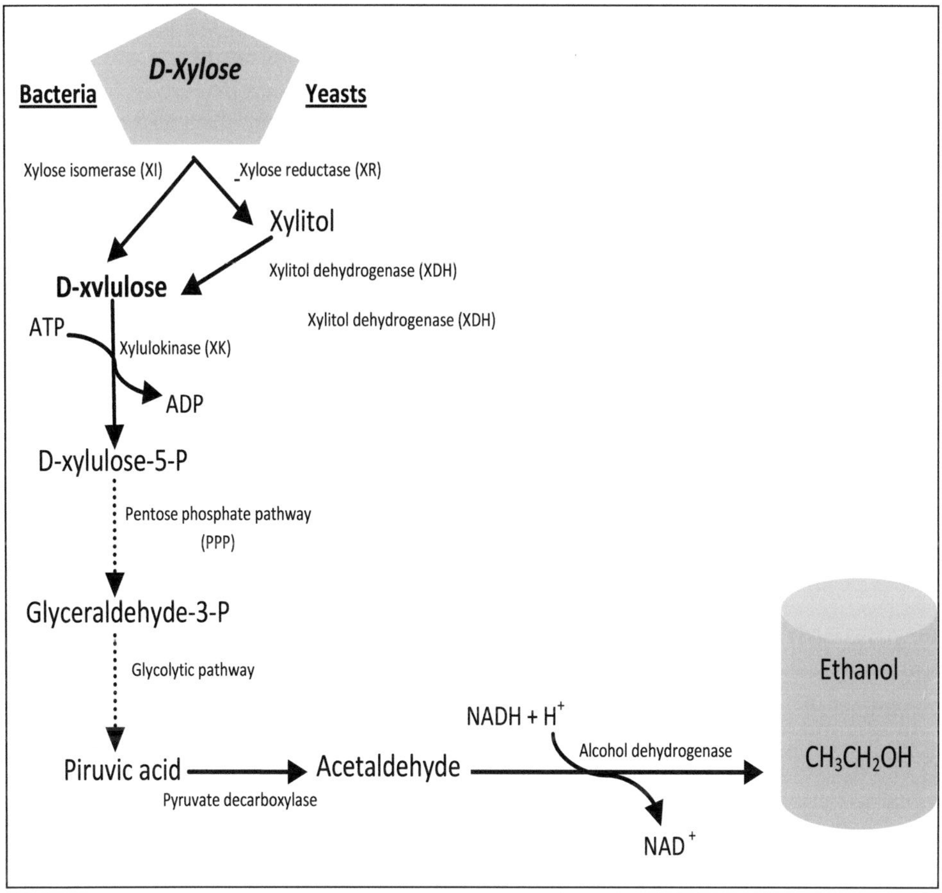


Fig. 4. Metabolism of xylose in yeasts and bacteria.

However, co-cultures of these yeasts do not always ensure the complete conversion of xylose because of the diauxic behavior of *P.stipitis* (Nakamura *et al.*, 2001) and, in batch co-cultures, the production of ethanol from *S. cerevisiae* could worsen the performance of *P. stipitis* whose ethanol toxicity threshold is around 3% (De Bari *et al.*, 2004; Delgenes *et al.*, 1996). Moreover, in the cofermentation process a compromise between the oxygen requirement of the two microorganisms must be used. Often, an efficient ethanol production by co-coltures depends on the competition for oxygen between the two species of yeasts in the medium (Laplace *et al.*, 1991). To favor the xylose consumption by yeast like *P.stipitis* in co-coltures, some researchers proposed the use of a respiratory-deficient strain of *S.cerevisiae* that cocultivated with *P.stipitis* in continuous cultures enabled a substrate conversion rate of 100% (Delgenes *et al.*, 1996). Other researchers proposed the fermentation in immobilized cells bioreactors. (Cuna *et al.* 2008 , De Bari *et al.*, 2004, Lebeau *et al.*, 1998). In particular, enzymatic hydrolyzates from steam treated aspen chips were fermented with *P. stipitis* and *S.cerevisiae* immobilized in Ca-alginate beads. In the best conditions, the process produced 77% of the theoretical yield (De Bari *et al.*, 2004). Moreover, when *P. stipitis* and *S. cereviasiae* are coimmobilized in calcium alginate gel beads, all the cells in the beads external shells metabolize glucose more rapidly than xylose. As consequence, the nutrients flux entering

the internal shells of the beads mainly contain xylose. Furthermore the oxygen level inside the bead is lower than that at the beads surface. These conditions could favor the conversion of xylose to ethanol thus by-passing the *P.stipitis* diaxuc behavior.
Recently, others combinations of yeasts were examined to improve the yield of fermentation and the use of xylose (Hamidimotlagh *et al.*, 2007). In detail, co-cultures of two xylose fermenting yeasts, *Kluyveromyces marxianus* and *P.stipitis*, showed process yields of 80% thanks the higher ethanol tolerance of *K.marxianus* than *P.stipitis* (Hamidimotlagh *et al.*, 2007).

### 3.2 Recombinant yeasts

To overcome the problems related to the inability of wild-type microorganisms to ferment all the sugars in the hydrolysates, several researches were devoted to the development of recombinant organisms which can use both glucose and xylose. In this regard, different metabolic engineering strategies have been explored. The major part of the engineering strategies were based on the construction of recombinants *S.cerevisiae* strains due to its intrinsic robustness and high stress tolerance (Almeida *et al.*, 2007).
The observation that *S.cerevisiae* can ferment xylulose to ethanol (Chiang *et al.*, 1981) led different research groups to develop recombinant strains, cloning the bacterial xylose isomerase (XI) gene in *S.cerevisiae* (table 6).
XI gene from the thermophilic bacterium *Thermus thermophilus* was expressed in *S.cerevisiae.* (Walfridsson *et al.*, 1996) However, the bacterial enzyme XI showed a low activity in the yeast (0.04 U/mg protein$^{-1}$ Walfridsson *et al.*, 1996) due to an improper folding of the protein in *S. cerevisiae* and to its intracellular precipitation (Gárdonyi & Hahn-Hägerdal, 2003). More recently, the gene XylA encoding the xylose isomerase was isolated from *Pyromyces* sp E2, an anaerobic cellullolytic fungus, and after expressed in *S.cerevisiae.* (Kuyper *et al.*, 2003), The obtained engineered strain, RWB 202, exhibited a xylose isomerase activity of about 1 U/mg protein$^{-1,}$ to say higher than that of the bacterium *Thermus thermophilus* . This finding could be due to the fact that the mechanism of protein folding in *S. cerevisiae* is similar to that of *Pyromyces* (Kuyper *et al.*, 2003). Additional improvements in the RWB 202 were achieved by further genetic modifications. In particular, the strain RWB 218 showed high fermentation rates in mixed syrups, even during anaerobic growth at high sugar concentrations with an ethanol yield of 0.40 g ethanol/g sugars and a low xylitol production (table 6). This engineered strain was obtained from a recombinant strain RWB 217, by prolonging its anaerobic cultivation in automated sequencing-batch reactors on glucose and xylose mixtures. In the recombinant RWB 217 strain the expression of the *Piromyces* XylA gene was combined with the overexpression of the native *S.cerevisiae* xylulokinase gene and the genes for the conversion of xylulose to glycolytic intermediates. In addition, the endogenous GRE3 gene encoding for a xylose aldolase, was deleted with the effect of reducing the flux of xylose to xylitol, a bio-product that inhibits the activity of XI and decreases the ethanol yields (Kuyper *et al.*, 2005, table 6).
Others strains of *S.cerevisiae* capable to use xylose were generated by expressing the *P.stipitis* genes XIL1 and XIL2 encoding XR and XDH respectively (Jeffries, 2006). The only insertion of these genes enabled *S.cerevisiae* to grow on xylose. However, in most cases, low levels of ethanol were achieved (Kötter & Ciriacy 1993, table 6). In fact, in order to improve the ethanol yields further modifications were necessary.
To obtain this goal, the gene XKS1, encoding xylulokinase XK, from *S.cerevisiae* and the genes XIL1 and XIL2 from *P.stipitis* were inserted into a hybrid host, obtained by breeding of *S.uvarum* and *S.diastaticus.*

The engineered strain obtained in this way, the so called 1400 pLNH32, showed higher yields with respect to recombinant strains containing only XYL1 and XYL2 genes (Ho *et al.*, 1998; Moniruzzaman *et al.* 1997, table 6). Over the years, several recombinant strains of *S.cerevisiae* were obtained by adopting the same approach (table 6, Eliasson *et al.*, 2000; Jeppsson *et al.*, 2002; Karhumaa *et al.*, 2007; Roca *et al.*, 2003; Wahlbom *et al.*, 2003; Zaldivar *et al.*, 2002).

More recently, further improvements of the engineered yeasts performances were obtained by improving the xylose uptake in *S.cerevisiae* through the insertion of genes for xylose transport. In this way, an interesting strain was obtained by the overexpression of the *Opinomyces* xylose isomerase, the *S.cerevisiae* xylulokinase and the *P.stipitis* gene SUT1 encoding for a sugar permease (Madhavan *et al.*, 2009). A more efficient xylose-utilizing strain was isolated from the recombinant strain so obtained, by serial cultivations in minimal media containing only xylose as carbon source.

The xylose adapted strain, ADAP28, showed good performances in the fermentation tests (table 6, Madhavan *et al.*, 2009). Recombinant strains obtained in laboratory were not always applicable at industrial scales because of their instability (Hann-Hägerdal *et al.*, 2007a). In fact only a limited number of engineered strains used at industrial scale have been described in literature. The major part of these strains are genetically modified to express the *P.stipitis* genes XIL1 and XIL2 in the *S.cerevisiae* host and overexpressing the endogenous XK (Hann-Hägerdal *et al.*, 2007b).

Some of the industrial recombinant *S.cerevisiae* strains used in the fermentation of lignocellulosic hydrolysates are summarized in the table 7. With the exception of F12, all of the strains reported in table 7, showed an ethanol yields of more 0.4 $g_{ethanol}/g_{sugars\ consumed}$ (Hann-Hägerdal *et al.*, 2007b). Finally, given the restriction on GM organisms in many countries, some researchers investigated non-GM strains of *S. cerevisiae* capable to use xylose efficiently (Attfield & Bell, 2006). Attfield and Bell developed a native strain of *S.cerevisiae* capable of using xylose as a sole carbon source by means of natural selection and breeding. The authors claimed that this innovative approach could open new attractive ways to develop yeasts for lignocellulosic substrates.

### 3.3 Fermentation schemes and technologies

Industrial fermentation processes are traditionally classified in batch, fed-batch and continuous process (figure 5). The choice of the suitable process depends on the type of lignocellulosic hydrolysate and on the properties of the microorganisms employed. Currently, most of the bioethanol process schemes follow the same process employed for centuries in the beverage industry. This strategy is based on the **batch** technology in which substrate and cells are introduced simultaneously into the bioreactor (figure 5). At the end of fermentation, the bioreactor is washed, sterilized and then new medium is introduced. The batch technology is low cost and provides easy operations with reduced risks of contamination given that nothing is added into reactor after the initial inoculation. However, when lignocellulosic biomass is processed, the presence of inhibitor compounds could make the batch process unsuitable (see section 1). The inhibitors effect in the batch reactor can be reduced by increasing the initial cell density in order to exploit the intrinsic capacity of many microorganisms to detoxify the lignocellulosic broths. In **fed-batch** fermentation the substrate is added progressively while fermentation proceeds (figure 5). This process is widely used in industrial applications (Balat, 2010).

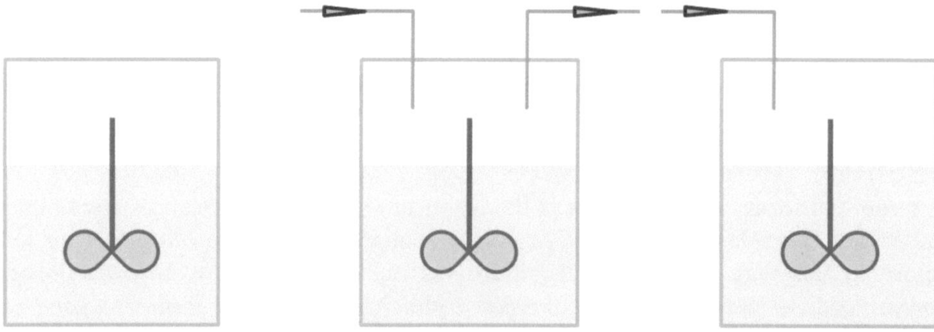


Fig. 5. Schematic representation of fermentation processes.

Fed-batch cultures provide better yields and productivity than batch cultures (Chandel *et al.*, 2007b) thanks to the high cells concentrations during the initial phase of the process. When applied to the fermentation of lignocellulosic hydrolyzates, this approach has the advantage of favoring an "in situ" detoxification through the action of the fermenting microorganisms. In the fed-batch fermentation, the process productivity is influenced by the feed rate of the substrate so that two low feed rates could yield low productivities. (Taherzadeh *et al.*, 1999). **Continuous fermentation** is an open system. Sterile medium is continuously added to the bioreactor and an equivalent amount of the converted nutrient solution with microorganisms is simultaneously subtracted from the system (figure 5). Continuous fermentation operations often give higher productivities than batch fermentation (Chandel *et al.*, 2007b), eliminate much of the downtime associated with cleaning and sterilization, and are easier to automate than batch and fed-batch processes. However, the continuous approach is often limited by difficulty of maintaining high cell concentrations in the bioreactor. The use of **immobilized cells** could overcome this problem (Chandel *et al.*, 2007b). Higher ethanol yields compared to free cells were reported in continuous fermentation processes with *S.cerevisiae* immobilized in calcium alginate (Taherzadeh *et al.*, 2001). The next subparagraph contains a survey of the most promising immobilizing matrices and immobilization techniques. In order to make an efficient conversion of biomass to ethanol, several process strategies have been explored, namely Separate Hydrolysis and Fermentation (SHF), Simultaneous Saccharification and Fermentation (SSF) and more recently the Consolidated BioProcessing (CBP). **SHF** consists of two steps: the first involves the enzymatic hydrolysis while the second converts the monomeric sugars into ethanol (Wingren *et al.*, 2003). It offers various advantages such as the possibility to carry out both hydrolysis and fermentation at optimal conditions. In detail, the enzymes can operate at high temperature increasing their performances while microorganisms can work at their optimal temperature and pH. The disadvantages of this method are the risk of contaminations during the process and the inhibition of cellulase and β-glucosidase enzymes by glucose (Xiao *et al.*, 2004). One way to solve the problem of inhibition by glucose is to carry out the hydrolysis and fermentation simultaneously. This process, called **SSF,** combines the hydrolysis step and fermentation in one vessel. As soon as hydrolysis starts, a fermenting microorganism is added into reactor. SSF represents a good strategy with several advantages such as high ethanol yield, lower required amounts of enzymes (Lin & Tanaka,

| **Microorganisms** | **Medium** | **Xilose (g/L)** | **Glucose (g/L)** | **Ethanol yield [gp/gs]** | **Productivity [g/Lh-1]** | **References** |
|---|---|---|---|---|---|---|
| Yeasts | | | | | | |
| *Candida shehatae* NRRLY12856 | Synthetic | 50 | n.r | 0.45 | 0.29 | Slininger *et al.*, 1985 |
| *Candida shehatae ATCC 22484* | Hydrolyzate of hardwood | 43.5 | 9.0 | 0.14 | 0.10 | Perego *et al.*, 1990 |
| *Pachysolen tannophilus* (NRRL Y2460) | Hydrolyzate of hardwood | 43.5 | 9.0 | 0.21 | n.r | Perego *et al.*, 1990 |
| *Pachysolen tannophilus* DSM70352 | Wheat straw | 10.38 | 16.62 | 0.44 | 0.25 | Zayed *et al.*, 1996 |
| *Pichia stipitis (NRLL-Y7124)* | Hydrolyzate of Eucaliptos | 30.5 | 1.5 | 0.35 | 0.16 | Ogier *et al.*, 1999 |
| *Pichia stipitis NRRLY-7124* | Synthetic | 150 | n.r | 0.39 | 0.28 | Slininger *et al.*,1985 |
| *Candida shehatae NCIM3501* | Hydrolyzate of Sugarcane bagasse treated with ion-exchange resin | 21.5 | 5.84 | 0.48 | 0.36 | Chandel *et al.*,2007a |
| *Pichia stipitis NRRL Y-7124 adapted* | Hydrolyzate of Wheat straw overlimed | 45 | 6.40 | 0.36 | 0.30 | Nigam, 2001a |
| *Pichia stipitis NRRL Y-7124* | Hydrolyzate of Eicchornia crassipies treated | 54 | 3.5 | 0.35 | 0.18 | Nigam, 2002 |
| *Pichia stipitis CBS 6054* | Synthetic | 120.3 | n.r | 0.381 | 0.214 | Agbogbo *et al.*, 2007 |
| *Candida shehatae FPL-Y-049* | Wood hydrolyzate | 121.7 (total fermentable sugars) | 121.7 (total fermentable sugars) | 0.32 | 0.45* | Sreenath *et al.*, 2000 |
| Bacteria | | | | | | |
| *Thermoanaerobacter ethanolicus* | Synthetic | 10 | n.r | 0.5 | 0.12 | Carreira *et al.*, 1983 |
| *Clostridium saccharolyticum* ATCC 35040 | Synthetic | 25 | n.r | 0.21 | 0.05 | Asther & Khan, 1985 |
| *Clostridium termohydrosulfuricum* 39E | Synthetic | 5 | n.r | 0.39 | n.r | Ng *et al.*, 1981 |

Table 5. Yeasts and bacteria capable to metabolize xylose; (*calculated from reference, n.r.: not reported)

| Recombinant *S.cerevisiae* Strain | Genotipe | Sugar composition | Fermentation conditions | Ethanol yield (g ethanol/ g sugars) | Xylitol yield (g xylitol/ g xylose) | References |
|---|---|---|---|---|---|---|
| *S.cerevisiae* | XIL1, XIL2 | 21.7 g/L xyl | | 0.07 | 0.07 | Kötter & Ciriacy, 1993 |
| **RWB217** | XI, XK, del GRE3, overexpressed PPP | 20 g/L glu + 20 g/L xyl | Anaerobic batch colture | 0.43 | 0.006 | Kuyper *et al.*, 2004 |
| **RWB218** | XI, XK, del GRE3, overexpressed PPP, selected for increase glucose uptake | 20 g/L glu + 20 g/L xyl | Anaerobic batch colture | 0.40 | 0.003 | Kuyper *et al.*, 2005 |
| **RWB202** | XI | 20 g/L glu + 10 g/L xyl | Anaerobic chemostat coltures | 0.39 | 0.07 | Kuyper *et al.*, 2003 |
| **1400 (pLNH32)** | XYL1, XYL2, XKS1 | 50 g/L xylosein YPD | Oxygen-limited batch colture | 0.33 | 0.10 | Ho *et al.*, 1998 |
| **TMB3001** | XYL1, XYL2, XKS1 | 50 g/L glu + 50 g/L xyl | Aerobic batch fermentation | 0.23 | 0.08 | Zaldivar *et al.*, 2002 |
| **TMB3001** | XYL1, XYL2, XKS1 | 50 g/L xyl | Oxygen-limited batch colture 70 h | 0.31 | 0.29 | Eliasson *et al.*, 2000, Jeppsson *et al.*, 2002 |
| **TMB3001** | XYL1, XYL2, XKS1 | 20 g/L glu + 50 g/L xyl | Anaerobic batch colture | 0.33 | 0.48 | Eliasson *et al.*, 2000, Roca et al., 2003 |
| **TMB3400** | XYL1, XYL2, XKS1 | 20 g/L xyl | Anaerobic batch colture | 0.18 | 0.25 | Wahlbom *et al.*, 2003 |
| **TMB3066** | XI, XSK1, PPP, del GRE | 50 g/L xyl | Anaerobic batch | 0.43 | 0.04 | Karhumaa *et al.*, 2007 |
| **ADAP28** | XI, XKS1, SUT1, xylose adapted | 50 g/L glu + 20 g/L xyl + borate | Fermentation in bottle with a bubbling CO2 outlet, 35°C, 40 h | 0.48 | 0.04 | Madhavan *et al.*, 2009 |

Table 6. Engineered *S.cerevisiae* strain for xylose conversion.

2006; Sun & Cheng, 2002) and contamination reduction during hydrolysis also thanks to the action of ethanol simultaneously produced. However SSF has the disadvantage to operate at temperature and pH conditions that represent a compromise between the optimal conditions for hydrolysis and fermentation. In particular, it is fundamental to consider temperature as the key parameter in the process. In fact, while the cellulase enzymes are more active at 50°C, the yeasts usually work at temperatures lower than 35°C. Several

| Strain | Hydrolysate | Fermentation strategy | References |
|---|---|---|---|
| **TMB3400** | Corn stover steam pretreated | Batch and fed-batch SSF | Ohgreen *et al.*, 2006 |
| **F12** | Still bottoms fermentation residue | Batch | Olsson *et al.*, 2006 |
| **TMB3400** | Spruce | Fed-batch | Hann- Hägerdal & Pamment 2004 |
| **TMB 3006** | Spruce | Fed-batch | Hann- Hägerdal & Pamment 2004 |
| **424ALNH-ST** | Corn stover | Batch | Sedlak & Ho, 2004 |

Table 7. Industrial *S.cerevisiae* strains fermenting xylose in lignocellulosic hydrolysates.

research efforts were concentrated on the isolation of strains able to work at high temperatures. Good performances have been recently obtained with the thermotolerant strain *Kluyveromyces marxianus* 6556 that showed promising results in the SSF of lignocellulosic agricultural wastes at 37°C (Zhang *et al.*, 2010). In fact various strains of the *K. marxianus* species have the ability to grow at temperature around 40°C and ferment mixed sugars such as glucose, xylose, mannose and galactose (Fonseca *et al.*, 2008). In this regard, Ballesteros *et al.*(2001) carried out several fed-batch SSF tests using *K.marxianus* at 42°C and obtaining ethanol yield of 76% for olive pulp. Rudolf *et al.* (2008), also demonstrated that undetoxified steam-pretreated bagasse could be successfully fermented to ethanol in a SSF process using both natural yeasts (*P.stipitis* CBS6054) that recombinant yeast (*S.cerevisiae* TMB3400). Interesting results were obtained using SSF with other materials such as industrial wastes (Kàdàr *et al.*, 2004), wheat straw, and sweet sorghum bagasse (Ballesteros *et al.*, 2004). To improve the ethanol yield through the overall consumption of sugars, a variant of SSF has been developed known as Simultaneous Saccharification and Co-Fermentation (**SSCF**) (Chandel *et al.*, 2007b; Pejo *et al.* 2008) that includes the cofermentation of multiple sugar substrates in the hydrolysates using pentose-fermenting yeast. In conclusion, either SSF or SSCF are preferred to SHF, because both can be performed in the same tank resulting , in lower capital costs, higher ethanol yield and shorter processing time (Chandel *et al.*, 2007b). Recently, a new integrated approach, so called Consolidate BioProcessing (**CBP**), has been developed. It combines the cellulase production, the cellulose hydrolysis and the sugar fermentation into a single unit operation (Lynd *et al.*, 2005). In other words, CBP combines all the biological steps required for the conversion of lignocellulosic materials to ethanol into one reactor. The process can be drive by a single microorganism or through a microbial consortium capable to ferment pretreated biomass directly (van Zyl *et al.*, 2007). Unfortunately, no natural microorganism exhibits all the features desired for CBP. There are two main strategies to make feasible the CBP process. The first approach consists of cloning and expressing the genes for ethanol production into cellulolytic microorganisms such as *Clostridium cellulolyticum* and *Clostridium thermocellum* (Lynd *et al.*, 2005). Conversely, the other approach constitutes of cloning the genes for cellulolytic activity in the efficient ethanol producing microorganisms such as *S.cerevisiae*. This latter strategy is more viable also thanks to develop of *S.cerevisiae* recombinant strains capable to express cellulases. Most of the cellulolytyc enzymes expressed in *S.cerevisiae* are of fungal origin, mainly from *Trichoderma* spp. and *Aspergillus* spp. (van Zyl *et al.*, 2007). Recently, a promising yeast have been constructed capable to grow on 10 g/L PASC (acid-swollen cellulose) with a subsequent production of 1 g/L ethanol. The recombinant *S.cerevisiae* strain co-expresses a *T.reesei* endoglucanase and a *Saccharomycopsis fibuligera* β-glucosidase (Den Haan *et al.*, 2007). Certainly, further improvements are necessary to optimize the heterologous enzyme expression in order to increase the ethanol yield. For instance, for the conversion of hemicellulose in the CBP process, the microorganism should have also hemicellulase activities. Katahira *et al.* (2004) introduced the genes encoding for xylose utilization from *P.stipitis* into a recombinant *S.cerevisiae* expressing xilanase II from *T.reesei* and β-xilosidase from *A. oryzae*. Despite the noticeable breakthroughs, the CBP approach seems still far from the industrial use and more studies are required to obtain microorganisms capable of producing ethanol from lignocellulosic materials in a single step.

### 3.4 Innovative bioreactors configurations: Fermentation in immobilized cells bioreactors

Fermentation can be carried in free or immobilized cells bioreactors. The use of the immobilized cells technology (**ICT**) in the bio-industry has recently received much attention thanks to several advantages: high cell load enhancing the fermentation productivity; feasibility of continuous processing without any interruption. Generally, four categories of immobilization techniques can be distinguished, based on the cell localization and on the interaction mechanisms between cells and supports : "attachment to a surface", "containment behind a barrier", "self-aggregation" and "entrapment within a porous matrix" (Karel *et al.* 1985). In the immobilization by **surface attachment**, yeast cells are allowed to attach to a solid support (Verbelen *et al.*, 2006). Cellular attachment to the carrier can be induced through linking agents such as metal oxides, glutaraldehyde or aminosilanes. **Containment of yeast cells behind a barrier** can be obtained through the use of microporous membrane filters or by entrapment into microcapsules. Several polymers can be used as microporus membranes: nylon, polystyrene and polyester. The drawback of this strategy is the membrane fouling caused by the cells growth (Lebeau *et al.*, 1998). The immobilization by **self-aggregation**, known as "flocculation" is based on the natural ability of yeast strains, such as *S.cerevisiae*, to adhere at inert surfaces (Oliveira, 1997). In this process, yeasts form a reversible flocs of thousands of cells (Bony *et al.*, 1997). In particular, adhesion is conferred by a class of special cell wall proteins called "adhesins" or "flocculins" that bind some amino-acid or sugar residues on the surface of the other cells or promote binding to abiotic surfaces (Verstrepen & Klis, 2006). Flocculation is dependent on several parameters, namely the calcium level, the pH and the fermentation temperature (Sampermans *et al.*, 2005). Furthermore, different yeast species present different families of adhesins. The brewer's yeast *Saccharomyces cerevisiae*, for example has five flocculation genes FLO (Teunissen & Steensma, 1995). For the industrial application flocculation profile was improved through recombinant DNA strategies (Pretorius & Bauer, 2002). Immobilization by **entrapment within porous matrix** is the most widely used method. The matrix is usually composed of agar, agarose, kappa-carrageenan, collagene, alginate, polyurethane, chitosan, plyacrylamide and cellulose. Among these, the most reported in literature is Ca-alginate that is commonly synthesized as spherical polymeric beads with diameter ranging from 0.3 to 3 mm around the cells (Verbelen *et al.*,2006). Besides the in situ synthesis around the cells, a second entrapment strategy was also reported in which cells are allowed to diffuse into a preformed porous matrix (Verbalen *et al.*, 2006). It was demonstrated that the immobilization of *S.cerevisiae* in the ICR (Immobilized Cell Reactor) column packed with Ca-alginate beads enables the conversion of concentrated syrups (150 g/L of glucose) with an ethanol yield of 38% in seven hours (Najafpour *et al.*, 2004). More recently, it was found that the use of the *S.cerevisiae* immobilized in the Ca-alginate beads coupled with a perm-selective separation of ethanol allow to convert diluted hydrolyzates into hydro-alcoholic solutions containing 9 wt% ethanol (De Bari *et al.*, 2009). This process could make to subsequent recovery of ethanol from the fermentation broths more sustainable. In fact, several energetic balances demonstrated that the lowest threshold to make feasible the bioethanol distillation from fermentation broths is 4 wt% (Zacchi & Sassner, 2008). However, the use at industrial scale of Ca-alginate beads is limited by the lack of stability through continuous processes. Other entrapping carriers such as mixed calcium alginate and silica beads, silica film, polyvinyl alcohol (PVA) and tetramethyl orthosilicate (TMOS) were investigated for bioethanol production from mixed sugars syrups by *P.stipitis* (Cuna *et al.*, 2008; De Bari *et al.*, 2007b). In particular, the use of this bioreactor configuration also help to overcome the diauxic behavior of this yeast (Cuna *et al.*, 2008; De Bari *et al.*, 2007b). The obtained results demonstrated that the silica films offer the advantages of immobilizing higher cell concentrations with respect to alginate beads (De Bari *et al.*, 2007b). It was also shown that the ethanol yields obtained by using the TMOS films were higher than those of PVA beads (70% for PVA against 80-82% for TMOS). Another material tested to immobilize cells for ethanol production was the ϒ-alumina that is a good promoter of ethanol fermentation because of its high porosity and high stability (Kanellaki et al., 1989). The immobilization was carried out by using the spray drier technology. In particular, it was demonstrated that the pre-soaking of ϒ-alumina particles in a resin solution before the cells immobilization improved the cells uptake and increased the sucrose conversion to ethanol (Isono *et al.*, 1994). Despite the interesting achievements some improvements are still necessary: carriers stability, reduced diffusion coefficients of nutrients and metabolites between the immobilization carrier and the fermentation broth. Furthermore industrial techniques must be optimized for the production of the immobilized biocatalytis at industrial scale.

## 4. Inhibitory compounds derived from biomass pretreatment: Effect on fermentation step

Fermentation of hydrolyzates represents a critical step in the lignocellulosics-to-bioethanol process not only for the efficient conversion of all the sugars but also for the microbial inhibition due to the pretreatment by-products. In fact, during the pretreatment and the hydrolysis step (chemical hydrolysis) many microbial inhibitors compounds are commonly generated. As reported in literature, these compounds can be classified in three major groups: furan derivates, weak acids and phenolic compounds (Almeida *et al.*, 2007; Palmqvist & Hann-Hägerdal , 2000). **Furan derivates** are mainly constituted by 5-hydroxymethyl-2-furaldehyde (5-HMF) and 2-furaldehyde generated by dehydration of hexoses and pentoses, respectively. These chemical compounds inhibit both the cell growth and ethanol production (Palmqvist & Hann-Hägerdal , 2000).

Several mechanisms were proposed to explain the effect of furfural and 5-HMF on the ethanol fermentation. It was found that in *S.cerevisiae* they inhibit alcohol dehydrogenase (ADH), pyruvate dehydrogenase (PDH) and aldehyde dehydrogenase (ALDH). Furthermore they cause the DNA breakdown resulting in a inhibition of the RNA and protein synthesys (Modig *et al.*, 2002). Additionally, furan derivates damage cell walls and membranes (Almeida *et al.*, 2007). However, the inhibition effect of these compounds is dose-dependent (Liu *et al.*, 2004). It was demonstrated that *S.cerevisiae* and *P. stipitis* strains were more sensitive to the inhibition by furfural than 5-HMF at the same concentration (table 8), while combined treatment with furfural and HMF suppressed cell growth (Liu *et al.*, 2004). Nigam (2001a) also found that a furfural concentration of 1.5 g/L interfered in respiration and growth of *P. stipitis*. Delgenes *et al.* (1996) showed that *P.stipitis* growth was reduced by 43%, 70% and 100% when the concentration of HMF was 0.5, 0.75 and 1.5 g/L respectively.

Some microorganisms such as *S.cerevisiae* (Liu, 2006) have the capacity to transform furfural and 5-HMF into less toxic compounds of furfuryl alcohol and 2,5-bishydroxymethylfuran respectively. This process is also known as "in situ-detoxification ".

The **weak acids** such as acetic, formic and levulinic are the most frequent acids present in the hydrolysate from lignocellulosic materials. Acetic acid is produced by de-acetylation of hemicellulose while levulinic and formic acid are formed through the 5-HMF breakdown (Palmqvist & Hann-Hägerdal, 2000). Undissociated acids are liposoluble and therefore can diffuse across the plasma membrane. Once inside the cell, because of neutral pH, dissociation of acids occurs resulting in the cytosolic pH decrease and the cell growth-inhibition. The decrease in the intracellular pH is compensated by the activity of the plasma membrane ATP-ase that pumps proton out of the cell and increases the ATP hydrolysis. This led to a reduction of ATP available for the yeast biomass formation (Russel, 1992, Verduyn *et al.*, 1992). In addition, weak acids reduce the uptake of aromatic aminoacids from the medium (Bauer *et al.*, 2003). The concentration of undissociated acids in the hydrolysate is pH-dependent and, as a result, pH is a crucial parameter during the fermentation step.

**Phenolic compounds** are produced following the lignin degradation and depend on the biomass source (Almeida *et al.*, 2007). The most common are phenol aldehydes (4-hydroxybenzaldehyde, syringaldehydes and vanillin), phenol ketones and alcohols (hidroquinone, cathecol, eugenol, guaiacol). The inhibitor mechanisms of phenolic compounds in the fermenting microorganisms have not yet been completely elucidated (Almeida *et al.*, 2007). Some researches indicated that these compounds partition into cells membranes cause loss of integrity (Palmqvist & Hann-Hägerdal, 2000). It was established that low molecular weight phenolic compounds are more inhibitory than those with high molecular weight (Klinke *et al.*, 2004) and furthermore, the substituent position influenced the compounds toxicity (Larsson *et al.*, 2000). Among the phenolic compounds, vanillin was shown to be a strong inhibitor of growth and ethanol production in *P.stipitis*, *C. shehatae* and *S.cerevisiae* at the concentration of 1 g/L (Delgenes *et al.*, 1996).

On the whole, inhibitor compounds in the hydrolysate from lignocellulosic biomass have a synergistic inhibitory effect (Mussatto & Roberto, 2004). Therefore, removal of inhibitors from hydrolysates is necessary for an efficient fermentation step. In this regard, several methods have been proposed to reduce the inhibitors concentrations (Larsson *et al.* 1999, Mussatto & Roberto, 2004). In general, the inhibitors content in the hydrolyzates can be reduced by using mild pretreatment/hydrolysis conditions; detoxifying the hydrolyzate before fermentation; developing inhibitor tolerant strains, and converting toxic compounds into harmlesses products (Taherzadeh *et al.*, 2000a). The major part of the detoxification methods are physical, chemical or biological (Mussatto & Roberto, 2004). Among the

physical methods, evaporation removes volatile compounds such as acetic acid, furfural and vanillin (Converti *et al.*, 2000). Chemical methods includes different strategies like overliming treatment and use of ion exchange resins and activated charcoal (De Bari *et al.*, 2004; Lee *et al.*, 1999; Martinez *et al.*, 2001; Nilvebrant *et al.*, 2001,). Biological detoxification is substantially based on the enzymatic treatment using peroxidase and laccase obtained from the lignolytic fungus *Trametes versicolor* (Palmqvist & Hann- Hägerdal, 2000). However, although the detoxification treatments are well established on large-scale processes, they could increase the process cost. According to some economical evaluations, the detoxification costs can constitute 22% of ethanol production cost (Von Sivers *et al.*, 1994). Thus, the use of resistant microorganisms, such as engineered or adapted strains, would be preferable. In particular, laccase gene from *Trametes versicolor* was expressed into *S.cerevisiae* resulting in higher ethanol productivity in spruce hydrolysates and in a media supplemented with coniferyl aldehyde (Larsson *et al.*, 2001). On the other hand, adaptation to toxins could make microorganisms more tolerant (Cuna *et al.*, 2004; Nigam, 2001b). Generally, adaptation is carried out by the sequential transfer of cells in media containing increasing concentrations of inhibitors (Cuna *et al.*, 2004; De Bari., 2005; Liu *et al.*, 2005; Nigam, 2001b). Following the adaptation of *P. stipitis,* De Bari *et al.* (2005) increased the xylose consumption and improved the ethanol yield by 17%. The adaptation strategy also increased the tolerance of *S. cerevisiae* to 5-HMF and furfural (Liu *et al.*, 2005).
The ability to adapt S. *cerevisiae* to lignocellulosic hydrolysates is strain dependent (Olsson & Hann-Hägerdal, 1996). For instance, a strain of *S. cerevisiae* isolated from a sulphite-spent liquor (SSL) was shown to be able to use glucose and galactose simultaneously in the presence of acetic acid in contrast to the behaviour of Bakers' yeast (Linden *et al.*, 1992). On the whole, recombinant technology together with the strains adaptation appears as the most promising approach to develop efficient processes to convert lignocelluloses biomass into ethanol. More insights in the inhibitors mechanisms and in the genomic characteristics of some resistant microorganisms could help the definition of protocols to enhance the yeast robustness.

## 5. Microbial inhibition by ethanol

The performances of the fermentation microorganism is also affected by the ethanol tolerance. In particular, at low concentrations, ethanol retards the growth rate of yeasts and inhibits cell division, while high ethanol concentrations reduce cells viability and increase their death (Stanley *et al.*, 2010). Furthermore, ethanol stress alters the metabolism (Hu *et al.*, 2007) mostly acting on the plasma membrane and on the cytosolic enzymes (Ansanay-Galeote *et al.*, 2001, Lopes & Sola-Penna, 2001). Exposure to ethanol causes also a disruption of the membrane structure resulting in a loss of electrochemical gradients and transport associated to the membrane (D'Amore *et al.*, 1990).
The xylose fermenting yeasts *P.stipitis* and *C.shehatae* are low tolerant to ethanol and are completely inhibited by ethanol concentration of 30 g/L (Laplace *et al.*, 1991). Conversely, *S.cerevisiae* reasonably tolerates higher levels of up to 70- 110 g/L (Casey *et al.*, 1992). This "ethanol tolerance" property is one of the reasons why *S.cerevisiae* is considered the alcoholic fermentation-organism for excellence. Generally, the yeasts such as *S.cerevisiae* have evolved some protective/adaptive responses to ethanol. One of the ethanol stress response, is the increase of unsaturated fatty acid and sterols in the cell membranes (Beaven *et al.*, 1982). Furthermore when *S.cerevisiae* is exposed to ethanol stress, an increase of the

heat shock proteins (HSPs) expression (i.e. HSP12 and HSP104) was observed (Glover & Lindquist, 1998; Sales *et al.*, 2000). These proteins protects the liposomal membrane integrity and act as remodeling agent in the disaggregation of denaturated proteins. Furthermore, following the exposure to ethanol, yeasts accumulate trehalose. However, this sugar is produced in response to many stresses (Attfield, 1997) and its function in stress tolerance is still not clear. Although this compound has long been considered as a protectant agent in the stress conditions (Ogawa *et al.*, 2000) it was found that a mutant which accumulate high levels of trehalose showed a reduced expression of other adaptive mechanisms (Singer & Lindquist, 1998). According to Lopes & Sola-Penna (2001), the pyrophosphatase inactivation promoted by alcohols is not prevented by the presence of trehalose while 1.5 M urea attenuated this effect.

| Inhibitor | Inhibitor concentration (g/L) | *Yeasts* Strain | Inhibition of ethanol yield (%) | Inhibition of ethanol productivity (%) | References |
|---|---|---|---|---|---|
| 5-HMF | 4 | *S.cerevisiae* Tembec T1 | 12 | 45 | Keating *et al.*, 2006 |
| 5-HMF | 4 | *S.cerevisiae* CBS 8066 | | 40 | Taherzadeh *et al.*, 2000b |
| 5-HMF | 4 | *S.cerevisiae* Y-1528 | 11 | 40 | Keating *et al.*, 2006 |
| Furfural | 4 | *S.cerevisiae* CBS 8066 | | 69 | Taherzadeh *et al.*, 1999 |
| Furfural | 1.6 | *S.cerevisiae* Tembec T1 | | 27 | Keating *et al.*, 2006 |
| Furfural | 1.6 | *S.cerevisiae* Y-1528 | | 25 | Keating *et al.*, 2006 |
| Acetic acid | 4.3 | *S.cerevisiae* | 50 | | Olsson *et al.*, 1996 |
| Acetic acid | 8 | *P.stipitis* | 98 | | Olsson *et al.*, 1996 |

Table 8. Effect of inhibitors compounds on fermentation by yeasts

It was established that during acclimatisation to ethanol stress, hundreds of genes are down-regulated and about 100 genes are up-regulated (Alexandre *et al.*, 2001; Chandler *et al.*, 2004).

Ethanol tolerance is not only genetically determined and can be influenced by many factors, such as plasma membrane composition (Mishra & Prasad, 1989), sugars concentration (Meyrial *et al.*, 1995), temperature, osmotic pressure, intracellular ethanol accumulation, byproduct formations. As consequence, it is very difficult to develop more tolerant yeast strains and still few studies are available on the construction of recombinant strains more tolerant to ethanol. In this regard Alper *et al.* (2006) combined mutagenesis and selection to isolate ethanol tolerant strains that showed an increased ethanol yield under a number of conditions and glucose concentrations. Kajiwara *et al.* (2000), also created a recombinant strain of *S.cerevisiae* with a higher unsaturated fatty acid content. This strain showed a higher survival rate than the wild-type strain in broths containing 15% (v/v) ethanol.

Interesting is also the *Saccharomyces diastaticus* (LORRE 316), an ethanol tolerant yeast capable of producing ethanol from corn starch, yielding a final concentration as high as 17.5% (v/v) (Wang & Sheu, 2000).

## 6. Conclusions

Lignocellulosic materials can be considered an important feedstock for the production of second generation bioethanol. Several breakthroughs have been achieved in the last years in all the process steps thus making this opportunity close to the industrial development. This is confirmed by several demo plants built around the world aiming at exploring the integrated process at significant scale. Depending on the specific biomass composition, some feedstocks, such as softwoods, can be more easily processed at demo scale. In fact, due to the hemicellulose composition of this biomass, the fermentable sugar streams do not contain pentoses and this reduces the process difficulties. The research in this field has given several microorganisms capable of fermenting diverse carbon sources and several process schemes. On the whole, the future development of bioethanol from lignocellulosics can be favored not only by the further optimization of some crucial process steps but also by the full implementation of the biorefinery concept. In this regard, further conversion options might be available for the various biomass streams, including the C5 fraction, and this could make more convenient the entire conversion.

## 7. Acknowledgment

We gratefully acknowledge Valentino Mannarino for his support in the preparation of the imagines.

## 8. References

Abril, D. & Abril, A. (2009). Ethanol from lignocellulosic biomass. *Cien.Inv.Agr.*, Vol.36 No 2, pp. 177-190

Agbogbo, F.K. & Coward-Kelly, G. (2008). Cellulosic ethanol production using the naturally occurring xylose-fermenting yeast, *Pichia stipitis*. *Biotechnol. Lett.* , Vol.30, pp. 1515-1524

Agbogbo, F.K., Coward-Kelly, G., Torry-Smith, M., Wenger, K. & Jeffries, T. W. (2007). The effect of initial cell concentration on xylose fermentation by *Pichia stipitis*. *Appl. Biochem. and Biotechnol.* Vol. 136-140

Alexandre, H., Ansanay Galeote, V., Dequin, S. & Blondin, B. (2001). Global gene expression during short term ethanol stress in *Saccharomyces cerevisiae*. *FEBS Letters* 498, pp. 98-103

Almeida, J.R.M., Modig, T., Petersson, A., Hahn-Hagerdal, B., Liden, G. & Gorwa-Grauslund, M.F. (2007). Mini-Review Increased tolerance and conversion of inhibitors in lignocellulosic hydrolysates by *Saccharomyces cerevisiae*. J *Chem. Technol. Biotechnol.*, Vol. 82, pp. 340–349

Alper, H., Moxley, J., Nevoigt, E., Fink, G.R. & Stephanopoulos, G. (2006). Engineering yeast transcription machinery for improved ethanol tolerance and production. *Science.*, Vol. 314, pp. 1565–1568

Ansanay-Galeote, V., Blondin, B., Dequin, S. & Sablayrolles, J-M. (2001). Stress effect of ethanol on fermentation kinetics by stationary-phase cells of . *Biotechnology Letters.*, Vol. 23, pp. 677-681

Asther, M. & Khan, A.W. (1985). Improved fermentation of cellobiose, glucose, and xylose to ethanol by *Zymomonas anaerobia* and a high ethanol tolerant strain *of Clostridium saccharolyticum. Appl. Microbiol. Biotechnol.*, Vol. 21, pp. 234-237

Attfield, P. (1997). Stress tolerance: the key to effective strains of industrial baker.s yeast. *Nature Biotechnology* ., Vol. 15, pp. 1351.1357

Attfield, P.V. & Bell, P. J. L. (2006). Use of population genetics to derive nonrecombinant *Saccharomyces cerevisiae* strains that grow using xylose as a sole carbon source. *FEMS Yeast Research.*, Vol. 6, pp. 862-868

Balat, M. (2009a). Global status of biomass energy use. *Energy source A*. Vol. 31, pp.1160-1173

Balat, M. (2009b). Gasification of biomass to produce gaseous products. *Energy Source A*. Vol. 31, pp. 516-526

Balat, M. (2010). Production of bioethanol from lignocellulosic materials via the biochemical pathway: a review. Energy Convers Manage. DOI:10.1016/j.enconman.2010.08.013

Ballerini, D., Desmarquest, J.P. & Pourquiè, J. (1994). Ethanol production from lignocellulosics: large scale experimentation and economics. *Bioresource Technology.*, 50, pp. 17-23

Ballesteros, M., Oliva, J.M., Negro, M.J., Manzanares, P. & Ballesteros, I. (2004). Ethanol from lignocellulosic materials by a simoultaneous saccharification and fermentation process (SFS) with *Kluyveromyces marxianus* CECT 10875. *Proc. Biochem.*, Vol. 39, pp. 1843-1848

Ballesteros, Martínez, Oliva, J.M., Navarro, A.A., Carrasco, J. & Ballesteros, M. (1998). Feasibility of steam explosion pretreatment to enhance enzymatic hydrolysis of municipal solid organic waste In Biomass for Energy and Industry. Proceeding of the International Conference of Würzburg Kopetz H., Weber T., Palz W., Chartier P., Ferrero G. L. Eds., C.A.R.M.E.N.: Germany, 1696-1699.

Ballesteros, Oliva, J.M., Sàez, F. & Ballesteros, M. (2001). Ethanol production from lignocellulosic byproducts of olive oil extraction. *Appl. Biochem. and Biotechnol.*, 91-93, 237–252

Banerjee, S., Mudliar, S., Sen, R., Giri, B., Satpute, D., Chakrabarti, T. & Pandey. R.A. (2010). *Biofuels, Bioprod. Bioref.*, Vol. 4, pp. 77–93

Bauer, B.E., Rossington, D., Mollapour, M., Mamnun, Y., Kuchler, K. & Piper, P.W. (2003). Weak organic acid stress inhibits aromatic amino acid uptake by yeast, causing a strong influence of amino aacid auxotrophies on the phenotypes of membrane transporter mutants. *Eur. J. Biochem.* ,270, pp. 3189-3195

Bayer, E.A., Chanzy, H., Lamed, R. & Shoham, Y. (1998) Cellulose, cellulases and cellulosomes. *Current Opinion in Structural Biology* , Vol. 8, pp. 548–557.

Beaven, M.J., Charpentier, C. & Rose, A.H. (1982). Production and tolerance of ethanol in relation to phospholipid fatty-acyl composition in *Saccharomyces cerevisiae* NCYC 431. *J. Gen. Microbiol.*, Vol. 128, pp. 1447–1455

Berg, C. (2002). World ethanol production 2001. Available at http://www.distill.com/world ethanolproduction.htm

Bok, J.D., Yernool, D.A. & Eveleigh, D.E. (1998). Purification, characterization, and molecular analysis of thermostable cellulases CelA and CelB from *Thermotoga neapolitana*. *Appl.Environ. Microbiol.*, Vol. 64, pp. 4774–4781

Bony, M., Thines-Sempoux, D., Barre, P. & Blondin, B. (1997). Localization and cell surface anchoring of the *Saccharomyces cerevisiae* flocculation protein Flo1p. *J. Bacteriol.*, Vol. 179, pp. 4929–4936

Boussaid, J., Robinson, Y., Cai, J., Gregg, D.J. & Saddler, J.R. (1999). Fermentability of the hemicellulose-derived sugars from steam-exploded softwood (Douglas fir). *Biotechnol. Bioeng.*, 64, pp. 284-289.

Bronnenmeier, K., Kern, A., Liebl, W. & Staudenbauer, W. (1995). Purification of *Thermotoga maritima* enzymes for the degradation of cellulosic materials. *Appl. Environ. Microbiol.*, Vol.61, pp. 1339–1407

Bura, R., Mansfield, S.D., Saddler, J.N. & Bothast, R.J. (2002). SO2-catalyzed steam explosion of corn fiber for ethanol production. *Appl. Biochem. Biotechnol.*, Vol. 98-100, pp. 59-71

Cara, C.; Ruiz, C.; Ballesteros, M.; Manzanares, P.; Negro, M. J.; Castro, E. (2008). Production of fuel ethanol from steam-explosion pretreated olive tree pruning. *Fuel, 87,* 692–700.

Carreira, L.H., Wiegel, J. & Ljungdahl (1983). Production of ethanol from biopolymers by anaerobic, thermophilic, and extreme thermophilic bacteria: Regulation of carbohydrate utilization in mutants of *Thermoanaerobacter ethanolicus*. *Biotech. And Bioeng. Symp.*, Vol.13, pp. 183-191

Casey, J.P. & Ingledew, W.M. (1992). Ethanol tolerance in yeasts. CRC Crit. *Rev. Microbiol.*, Vol.13, pp.219-277

Chandel, A.K., ES, C., Rudravaram, R., Narasu, M.L., Rao, L.V. & Ravindra, P. (2007b). Economics and environmental impact of bioethanol production technologies: an appraisal. *Biotechnol. Mol. Biol.*, Vol. 2, No 1, pp. 014-032

Chandel, A.K., Kapoor, R.K., Kuhad, A. S. R. C. (2007)a. Detoxification of sugarcane bagasse hydrolysate improves ethanol production by *Candida shehatae* NCIM 3501. *Bioresource Technology*, 98, pp. 1947–1950.

Chandler, M., Stanley, G.A., Rogers, P. & Chambers. P. (2004). A genomic approach to defining the ethanol stress response in the yeast *Saccharomyces cerevisiae*. *Annals of Microbiology.*, Vol. 54, pp. 427.454

Chiang, L.C., Gong, C.S., Chen, L.F. & Tsao, G.T. (1981). D-Xylulose Fermentation to Ethanol by *Saccharomyces cerevisiae*. *Appl. Environ. Microbiol.*, Vol. 42, No. 2, pp. 284-289

Choi, G.W., Kim, H.W.K.Y-R. & Chung, B.W. (2008). Ethanol Production by *Zymomonas mobilis* CHZ2501 from Industrial Starch Feedstocks. *Biotechnology and Bioprocess Engineering.*, Vol. 13, pp. 765-771

Converti, A., Dominguez, J.M., Perego, P., Silva, S.S. & Zilli, M. (2000). Wood hydrolysis and hydrolyzates detoxification for subsequent xylitol production. *Chem. Eng. Technol.* ,23, pp. 1013–1020

Cuna, D., De Bari, I., Liuzzi, F. & Braccio, G. (2008). Exploitation of PVA and Silica hydrogels for the production of Bioethanol by immobilized cells of *Pichia stipitis*. Proceedings of the 16th European Biomass Conference & Exhibition From Research to Industry and Markets Valencia, Spain, 2-6 June 2008

Cuna, D., De Bari, I., Nanna, F. & F. Liuzzi (2004). Bioethanol production from mixed sugar syrup of steam-exploded biomass: investigation of adapted strain of *Pichia stipitis* in immobilized-cell bioreactors. Proceeding 2nd World Conference on Biomass for Energy, Industry and Climate Protection, Rome, Italy, 10-14 May, 2004

D'Amore, T., Panchal, C.J., Russell, I. & Stewart, G.G. (1990). A study of ethanol tolerance in yeast. *Crit. Rev. Biotechnol.*, Vol. 9, pp. 287-304

De Bari, I., Cuna, D., Di Matteo, V. & Villone, A. (2005). Proceeding of the 14th European Biomass Conference & Exhibition From Research to Industry and Markets Paris, France, 17-21 October, 2005.

De Bari, I., Cuna, D., Nanna, F. & Braccio, G. (2004). Ethanol Production in Immobilized-Cell Bioreactors from Mixed Sugar Syrups and Enzymatic Hydrolysates of Steam-Exploded Biomass. *Appl. Biochem. Biotechnol.*, Vol.113-116.

De Bari, I., Nanna, F. & Braccio, G. (2007a). SO2-catalyzed steam fractionation of aspen chips for bioethahnol production: Optimization of the catalyst impregnation. *Ind. Eng. Chem. Res.*, Vol.46, pp. 7711-7720

De Bari, I., Cuna, D., Calabrò, V., Stillo, V., Liuzzi, F., Nanna, F. & Braccio, G. (2009). Determination of kinetic parameters of alcoholic fermentations in immobilized cells bioreactors coupled with a pervaporation system. Proceeding of the 17th European Biomass Conference & Exhibition From Research to Industry and Markets Hamburg, Germany, 29 June - 3 July, 2009.

De Bari, I., Cuna, D., Liuzzi, F. & De Tommaso, B. (2007b). Bioethanol production by *S.cerevisiae* and *P.stipitis* immobilized in hydrogels and si-gels. Proceeding of the 15th European Biomass Conference From Research to Market Deployment Berlin, Germany, 7-11 May, 2007.

De Bari, I., Viola, E., Barisano, D., Cardinale, M., Nanna, F., Zimbardi, F., Cardinale, G. & Braccio, G. (2002). Ethanol Production at Flask and Pilot Scale from Concentrated Slurries of Steam-Exploded Aspen. *Ind. Eng. Chem. Res.*, Vol. 41, pp. 1745-1753

Delgenes, J.P., Moletta, R. & Navarro, J.M. (1996). Effects of lignocellulose degradation products on ethanol fermentations of glucose and xylose by *Saccharomyces cerevisiae, Zymomonas mobilis, Pichia stipitis*, and *Candida shehatae. Enzyme Microb. Technol.*, Vol.19, pp. 220–225

Den Haan, R., McBride, J.E., La Grange, D.C., Lynd, L.R. & van Zyl, W.H. (2007). Functional expression of cellobiohydrolases in *Saccharomyces cerevisiae* towards one-step conversion of cellulose to ethanol. *Enzyme Microb. Technol.*, Vol.40, pp. 1291–1299

Eliasson, A., Christensson, C., Wahlbom, C.F. & Hahn-Hägerdal, B. (2000). Anaerobic Xylose Fermentation by Recombinant *Saccharomyces cerevisiae* Carrying XYL1, XYL2, and XKS1 in Mineral Medium Chemostat Cultures *Appl. Environ. Microbiol.*, 66, pp. 3381-3386

Evans, B., Gilman, A., Cordray, K. & Woodward, J. (2000). Mechanism of substrate hydrolysis by a thermophilic endoglucanase from *Thermotoga maritima. Biotechnol. Lett.* , 22, 735–740

Excoffier, G., Toussaint, B. & Vignon, M.R. (1991) Saccharification of steam-exploded poplar wood. *Biotechnol.Bioeng.* ,Vol.38, pp. 1308-1317

Fan, L.T., Gharpuray, M.M. & Lee, Y.H. (1987). Enzymatic hydrolysis in Cellulose Hydrolysis, Ed. by *Springer-Verlag* Berlin Heidelberg.

Fonseca, G.G., Heinzle, E., Wittmann, C. & Gombert, A.K. (2008). The yeast *Kluyveromyces marxianus* and its biotechnological potential. *Appl. Microbiol. Biotechnol.*, Vol. 79, pp. 339–354

Galbe, M. & Zacchi, G. (2002). A review of the production of ethanol from softwood. *Appl. Microbiol. Biotechnol.*, Vol. 59, pp. 618-628

Gárdonyi, M. & Hahn-Hägerdal, B. (2003). The *Streptomyces rubiginosus* xylose isomerase is misfolded when expressed in *Saccharomyces cerevisiae*. *Enzyme Microb Technol.*, Vol. 32, pp. 252-259

Glover, J.R. & Lindquist, S. (1998). Hsp104, Hsp70, and Hsp40: A novel chaperone system that rescues previously aggregated proteins. *Cell.*, Vol. 94, pp. 73–82

Grootjen, D.R.J., van der Lans R.G.J.M. & Luyben, K.Ch.A.M. (1990). Effects on the aeration rate on the fermentation of glucose and xylose by *Pichia stipitis* CBS 5773. *Enzyme Microb. Technol.* Vol. 12, pp. 20-23

Hahn-Hägerdal, B. & Pamment, N.B. (2004). Microbial pentose metabolism. *Appl. Biochem. Biotechnol.*, 113–116, pp. 1207–1209

Hahn-Hagerdal, B., Galbe, M., Gorwa-Grauslund, M.F., Lidén, G. & Zacchi, G. (2006). Bio-ethanol - the fuel of tomorrow from the residues of today. *TRENDS in Biotechnology.*, Vol. 24, No. 12, pp. 549-556

Hahn-Hägerdal, B., Karhumaa, K., Jeppsson, M. & Gorwa-Grauslund, M.F. (2007a). Metabolic Engineering for Pentose Utilization in *Saccharomyces cerevisiae*. *Adv.Biochem. Engin/Biotechnol.*, 108, pp. 147-177

Hahn-Hägerdal, B., Karhumaa, K., Fonseca, C., Spencer-Martins I. & Gorwa-Grauslund, M. F. (2007b). Towards industrial pentose-fermenting yeast strains. *Appl. Microbiol. Biotechnol.* ,74, pp. 937-953

Hamidimotlagh, R., Nahvi, I., Emitiazi, G. & Abedinifar, S. (2007). Mixed sugar fermentation by *Pichia stipitis, Saccharomyces cerevisiae,* and an isolated xylose-fermenting *Kluyveromyces marxianus* and their cocultures. *African Journal of Biotechnology*, Vol. 6, No 9, pp. 1110-1114

Ho, N.W., Chen, Z. & Brainard, A.P. (1998). Genetically Engineered *Saccharomyces* Yeast Capable of Effective Cofermentation of Glucose and Xylose. *Appl Environ Microbiol* , Vol. 64, pp. 1852

Hreggvidsson, G., Kaiste, E., Holst, O., Eggertsson, G., Palsdottir, A. & Kristjansson, A. (1996). An extremely thermostable cellulase from the thermophilic eubacterium *Rhodothermus marinus*. *Appl. Environ. Microbiol.*, Vol. 62, pp. 3047–3049

Hu, X., Wang, M., Tan, T., Li, J., Yang, H., Leach, L., Zhang, R.M. & Luo, Z.W. (2007). Genetic dissection of ethanol tolerance in budding yeast *S. cerevisiae*. *Genetics.*, Vol. 175, pp. 1479–1487

Isono, Y., Araya, G. & Hoshino, A. (1994). Immobilization of *Saccharomyces cerevisiae* for ethanol fermentation on γ-Alumina Particles using a spray-dryer. *Process Biochemestry.*, Vol. 30, No 8, pp. 743-746

Jana, S.K., Ghosh, V.K. & Singh, A. (1994). Production and hydrolytic potential of cellulose enzymes from a mutant strain of *Trichoderma reesei*. *Appl. Biochem. and Biotechnol.*, Vol. 20, pp. 233-239

Jeffries, T.W. (2006). Engineering yeasts for xylose metabolism. *Current Opinion in Biotechnology.*, Vol. 17, pp. 320-326

Jeppsson, M., Johansson, B., Hahn-Hägerdal, B. & Gorwa-Grauslund, M.F. (2002). Reduced oxidative pentose phosphate pathway flux in recombinant xylose-utilizing *Saccharomyces cerevisiae* strains improves the ethanol yield from xylose. *Appl.Environ. Microbiol.*, Vol. 68, pp. 1604-1609

Kabel, M.A., van der Maarel, M.J.E.C., Klip, G., Voragen, A.G.J. & Schols, H.A. (2006). Standard assays do not predict the efficiency of commercial cellulase preparations towards plant materials. *Biotechnol. Bioeng.*, Vol. 93, pp. 57–63

Kàdàr, Z., Szengyel, Z. & Réczey, K. (2004). Simoulateneous saccharification and fermentation (SSF) of industrial wastes for the production of ethanol. *Ind. Crop. Prod.*, Vol. 20, pp. 103-110

Kajiwara, S., Suga, K., Sone, H. & Nakamura, K. (2000). Improved ethanol tolerance of *Saccharomyces cerevisiae* strains by increases in fatty acid unsaturation via metabolic engineering. *Biotechnol. Lett.*, Vol. 22, pp. 1839–1843

Kanellaki, M., Koutinas, A.A., Kana, K., Nicolopoulou, M., Papadimitriou, A. & Lycourghiotis, A. (1989). Ethanol production by *Saccharomyces cerevisiae* promoted by γ-Alumina. *Biotechnol. Bioeng.*, Vol. 34, pp. 121-125

Karel, S.F., Libicki, S.B. & Robertson, C.R. (1985). The immobilization of whole cells: Engineering principles. *Chem. Eng. Sci.*, Vol. 40, pp. 1321–1354

Karhumaa, K., Sanchez, R.G., Hahn-Hägerdal, B. & Gorwa-Grauslund, M.F. (2007). Comparison of the xylose reductase-xylitol dehydrogenase and the xylose isomerase pathways for xylose fermentation by recombinant *Saccharomyces cerevisiae. Microbial. Cell. Factories*. 6:5.

Katahira, S., Fujita, Y., Mizuike, A., Fukuda, H. & Kondo, A. (2004). Construction of a Xylan Fermenting Yeast Strain through Codisplay of Xylanolytic Enzymes on the Surface of Xylose Utilizing *Saccharomyces cerevisiae* Cells. *Appl.Environ.Microbiol.* Vol.70 No 9, pp. 5407-5414

Keating J.D., Panganiban, C. and Mansfield, S.D. (2006). Tolerance and adaptation of ethanologenic yeasts to lignocellulosic inhibitory compounds. *Biotechnol. Bioeng.*, Vol.93, pp. 1196-1206

Keshwani, D.R. & Cheng, J.J. (2009). Switchgrass for bioethanol and other value-added applications: a review. *Bioresour Technol.* ,100, pp. 1515–23

Klinke, H.B., Thomsen, A.B. & Ahring, B.K. (2004). Inhibition of ethanol-producing yeast and bacteria by degradation products produced during pre-treatment of biomass. *Appl. Microbiol. Biotechnol.*, Vol.66, pp. 10-26

Knutson, B.L., Strobel, H.J., Nokes, S.E., Dawson, K.A., Berberich, J.A. & Jones, C.R. (1999). Effect of pressurized solvents on ethanol production by the thermophilic bacterium *Clostridium thermocellum. J. Supercritical Fluids.* ,16, pp. 149–56

Kobayashi, F.; Take, H.; Asada, C.; Nakamura, Y. (2004) Methane Production from Steam-Exploded Bamboo. *J. Biosci. Bioeng., 97* (6), 426–428

Kötter, P. & Ciriacy, M. (1993). Xylose fermentation by *Saccharomyces cerevisiae. Appl. Microbiol. Biotechnol.*, Vol. 38, pp. 776–783

Kumar, P., Barret, D.M:, Delwiche, M.J. & Stroeve, P. (2009). Methods for pretreatment of lignocellulosic biomass for efficient hydrolysis and biofuel production. *Ind.Eng. Chem. Res.*, Vol 48, No. 8, pp. 3713-3729

Kuyper, M., Harhangi, H.R., Stave, A.K., Winkler, A.A., Jetten, M.S.M., de Laat, W.T.A.M., den Ridder, J.J.J., Op den Camp, H.J.M., van Dijken, J.P. & Pronk, J.T (2003). High level functional expression of a fungal xylose isomerase: the key to efficient ethanolic fermentation of xylose by *Saccharomyces cerevisiae*? *FEMS Yeast Research.*, Vol. 4, pp. 69-78

Kuyper, M., Toirkens, M.J., Diderich, J.A., Winkler, A.A., van Dijken, J.P. & Pronk, J.T. (2005). Evolutionary engineering of mixed-sugar utilization by a xylose-fermenting *Saccharomyces cerevisiae* strain. *FEMS Yeast Research.*, Vol. 5, pp. 925-934

Kuyper, M., Winkler, A.A., van Dijken, J.P. & Pronk, J.T. (2004). Minimal metabolic engineering of *Saccharomyces cerevisiae* for efficient anaerobic xylose fermentation: a proof of principle. *FEMS Yeast Research*, Vol. 4, pp. 655–664

Laplace, J.M., Delgenes, J.P., Moletta, R. & Navarro, J.M. (1991). Simultaneous alcoholic fermentation of D-xylose and D-glucose by four selected microbial strains: process considerations in relation with ethanol tolerance. *Biotechnol. Lett.* Vol.13, pp. 445-450

Larsson, S., Cassland, P. & Jonsson, L.J. (2001). Development of a *Saccharomyces cerevisiae* strain with enhanced resistance to phenolic fermentation inhibitors in lignocellulosic hydrolysates by heterologous expression of laccase. *Appl. Environ. Microbiol.*, Vol. 67, pp. 1163-1170

Larsson, S., Quintana-Sainz, A., Reimann, A., Nilvebrant, N.O. & Jonsson, L.J. (2000). Influence of lignocelluloses-derived aromatic compounds on oxygen-limited growth and ethanolic fermentation by *Saccharomyces cerevisiae*. *Appl. Biochem. Biotechnol.*, Vol. 84-86, pp. 617-632

Larsson, S., Reimann, A., Nilvebrant, N.O. & Jonsson, L.J. (1999). Comparison of different methods for detoxification of lignocelluloses hydrolyzates of spruce. *Appl. Biochem. Biotechnol.*, Vol. 77-79

Lebeau, T., Jouenne, T. & Junter, G.A. (1998). Continuous alcoholic fermentation of glucose/xylose mixtures by co-immobilized *Saccharomyces cere*visiae and *Candida shehatae*. *Appl. Microbiol. Biotechnol.*, Vol. 50, pp. 309-313

Lee, W.G., Lee, J.S., Shin, C.S., Park, S.C., Chang, H.N. & Chang, Y.K. (1999). Ethanol production using concentrated oak wood hydrolysates and methods to detoxify. *Appl. Biochem. Biotechnol.*, Vol. 77-79, pp. 547-559

Lin, Y. and Tanaka, S. (2006). Ethanol fermentation from biomass resources: current state and prospects. A*ppl. Microbiol. Biotechnol.*, Vol. 69, pp. 629-642

Linden, T., Peetre, J. & Hahn- Hägerdal, B. (1992). Isolation and characterization of acetic acid-tolerant galactose fermenting strains of *Saccharomyces cerevisiae* from a spent sulfite liquor fermentation plant. *Appl. Environ. Microbiol.*, Vol. 58, pp. 1661-1669

Liu, S., Skinner-Nemec, K.A. & Leathers, T.D. (2008). Lactobacillus buchneri strain NRRL B-30929 converts a concentrated mixture of xylose and glucose into ethanol and other products. *J Ind Microbiol. Biotechnol.*, Vol. 35, pp. 75–81

Liu, Z.L. (2006). Genomic adaptation of ethanologenic yeast to biomass conversion inhibitors. A*ppl. Microbiol. Biotechnol.*, Vol. 73, pp. 27-36

Liu, Z.L., Slininger, P.J., Dien, B.S., Berhow, M.A., Kurtzman, C.P. & Gorsich, S.W. (2004). Adaptive response of yeasts to furfural and 5-hydroxymethylfurfural and new chemical evidence for HMF conversion to 2,5-bis-hydroxymethylfuran. *J. Ind. Microbiol. Biotechnol.*, Vol. 31:, pp. 345–352

Liu, Z.L., Slininger, P.J. & Gorsich, S.W. (2005). Enhanced biotransformation of furfural and hydroxymethylfurfural by newly developed ethanologenic yeast strains. *Appl. Biochem. Biotechnol.*, Vol. 121-124, pp. 451-460

Lopes, D.H.J. & Sola-Penna, M. (2001). Urea increases tolerance of yeast inorganic pyrophosphatase activity to ethanol: the other side of urea interaction with proteins. *Archives of Biochemestry and Biophysics*. Vol. 394, No1, pp. 61-66

Lynd, L.R., Van Zyl, W.H., McBride, J.E. & Laser, M. (2005). Consolidated bioprocessing of cellulosic biomass: an update. *Curr. Opin. Biotechnol.*, Vol. 16, pp. 577–583

Lynd, L.R., Weimer, P.J., van Zyl, W.H. & Pretorius, I.S. (2002). Microbial cellulose utilization: fundamentals and biotechnology. *Microbiology and Molecular Biology Reviews.*, Vol. 66, pp. 506-577

Madhavan, A., Tamalampudi, S., Srivastava, A., Fukada, H., Bisaria, V.S. & Kondo, A. (2009). Alcoholic fermentation of xylose and mixed sugars using recombinant *Saccharomyces cerevisiae* engineered for xylose utilization. *Appl. Microbiol. Biotechnol.*, 82, pp. 1037-1047

Martinez, Rodriguez, M.E., Wells, M.L., York, S.W., Preston, J.F. & Ingram, L.O. (2001). Detoxification of dilute acid hydrolysates of lignocellulose with lime. *Biotechnol. Prog.* , Vol. 17, pp. 287–293

Meyrial, V., Delgenes, J.P., Romieu, C., Moletta, R. & Gounot, A.M. (1995). Ethanol tolerance and activity of plasma membrane ATPase in *Pichia stipitis* grown on D-xylose or on D-glucose. *Enzymes Microb. Technol.*, Vol. 17, pp. 535-540

Mishra, P. & Prasad, R. (1989). Relationship between ethanol tolerance and fatty acyl composition of *Saccharomyces cerevisiae*. *Appl. Microbiol. Biotechnol.* , Vol. 30, pp. 294-298

Modig, T., Liden, G. & Taherzadeh, M. (2002). Inhibition effects of furfural on alcohol dehydrogenase, aldehyde dehydrogenase and pyruvate dehydrogenase. *Biochem. J.* ,Vol. 363, pp. 769-776

Moniruzzaman, M., Dien, B.S., Skory, C.D., Chen, Z.D., Hespell, R.B., Ho, N.W.Y., Dale, B.E. & Bothast, R.J. (1997). Fermentation of corn fibre sugars by an engineered xylose utilizing *Saccharomyces* yeast strain. *World Journal of Microbiology & Biotechnology*, Vol. 13, pp. 341-346

Moon, S.K.H.M., Kim, Y., Chung, B. & Choi, G.W. (2009). Bioethanol production from ammonia percolated wheat straw. *Biotechnol Bioprocess Eng.*, Vol. 14, pp. 606-611

Mussatto, S.I. & Roberto, I.C. (2004). Alternatives for detoxification of diluited-acid lignocellulosic hydrolyzates for use in fermentative processes: a review. *Bioresource Technology.*, Vol. 93, pp. 1-10

Najafpour, G., Younesi, H. & Ismail, K.S.K. (2004) Ethanol fermentation in an immobilized cell reactor using *Saccharomyces cerevisiae*. *Bioresource Technology.*,Vol.92, pp. 251-260

Nakamura, Y., Sawada, T. & Inoue, E. (2001). Mathematical model for ethanol production from mixed sugars by *Pichia stipitis*. *J Chem. Technol. Biotechnol.*, Vol. 76, pp. 586-592

Ng, T.K., Ben-Bassat, A. & Zeikus, J.G. (1981). Ethanol production by thermophilic bacteria: Fermentation of cellulosic substrates by co-cultures of *Clostridium thermocellum* and *Clostridium thermohydrosulfuricum*. *Appl. Environ. Microbiol.*, Vol. 41, pp. 1337-1343

Nigam ,J.N. (2001a). Ethanol production from wheat straw hemicelluloses hydrolysate by *Pichia stipitis*. *Journal of Biotechnology.*, Vol. 87, pp. 17-27

Nigam, J.N. (2002). Bioconversion of water-hyacinth (Eichhornia crassipes) hemicelluloses acid hydrolysate to motor fuel ethanol by xylose-fermenting yeast. *Journal of Biotechnology.*, Vol.97, pp. 107-116

Nigam, J.N. (2001b). Development of xylose-fermenting yeast *Pichia stipitis* for ethanol production through adaptation on hardwood hemicellulose acid prehidrolysate. *J. Appl. Microbiol.* , Vol.90, pp. 208-215

Nilvebrant, N.O., Reimann, A., Larsson, S. & Jonsson, L.J. (2001). Detoxification of lignocelluloses hydrolysates with ion exchange resins. *Appl. Biochem. Biotechnol.* Vol. 91–93, pp. 35–49

Ogawa, Y., Nitta, A., Uchiyama, H., Imamura, T., Shimoi, H. & Ito, K. (2000). Tolerance mechanism of the ethanol-tolerant mutant of sake yeast. *J. Biosci. Bioeng.* Vol.90, No 3, pp. 313-320

Ogier, J.C., Ballerini, D., Leygue, J.P., Rigal, L. & Pourquiè, J. (1999). Production d'ethanol à partir de biomasse lignocellulosique. *Oil & Gas Science and Technology-Revue de l'IFP.* , Vol.54, pp. 67-94

Ohgreen, K., Bengtsson, O., Gorwa-Grauslund, M.F., Galbe, M., Hahn-Hägerdal, B. & Zacchi, G. (2006). Simultaneous saccharification and co-fermentation of glucoseand xylose in steam-pretreated corn stover at high fiber content with *Saccharomyces cerevisiae* TMB3400. *J. Biotechnol.*, Vol.126, No4, pp. 488-498

Ohta, K., Beall, D.S., Mejia, J.P., Shanmugam, K.T. & Ingram, L.O. (1991). Genetic improvement of Escherichia coli for ethanol production: chromosomal integration of *Zymomonas mobilis* genes encoding pyruvate decarboxylase and alcohol dehydrogenase II. *Appl. Environ. Microbiol.*, Vol. 57, pp. 893–900

Oliveira, R. (1997). Understanding adhesion: a means for preventing fouling. Exp. Therm. Fluid. Sci., Vol. 14, pp. 699-704

Olsson, L. & Hahn- Hägerdal, B. (1996). Fermentation of lignocellulosic hydrolysates for ethanol production. *Enzyme Microb. Technol.*, Vol. 18, pp. 312-331

Olsson, L., Soerensen, H.R., Dam, B.P., Christensen, H., Krogh, K.M. & Meyer, A.S. (2006). Separate and simoultaneous enzymatic hydrolysis and fermentation of wheat emicellulose with recombinant xylose utilizing *Saccharomyces cerevisiae. Appl. Biochem. Biotechnol.* 129-132:117-129.

Palmqvist, E. & Hahn-Hägerdal, B. (2000). Fermentation of lignocellulosic hydroliydates II: inhibitors and mechanisms of inhibition. *Biores. Technol.*, Vol. 74, pp. 25-33

Pejo, E.T., Oliva, J.M & Ballesteros, M. (2008). Realistic approach for full-scale bioethanol production from lignocellulose: a review. J.*Sci.Ind.Res.*, Vol. 67, pp. 874-884

Perego, P., Converti, A., Palazzi, E., Del Borghi, M. & Ferraiolo, G. (1990). Fermentation of hardwood hemicellulose hydrolysate by *Pachysolen tannophilus, Candida sheshatae,* and *Pichia stipitis. J. Ind. Microbiol.*, Vol.6, pp. 157-164

Perlack, R.D., Wright, L.L., Turhollow, A.F., Graham, R.L., Sotcks, B.J. & Erbach, D.C. (2005). Biomass as feedstock for a bioenergy and bioproducts industry: the technical feasibility of a billion-ton annual supply. DOE/GO-10 2005-2135, April. U.S. Department of Energy and U.S. Department of Agriculture.

Persson, T., Matusiak, M., Zacchi, G. & Jonsson, A.S. (2006). Extraction of hemicelluloses from process water from the production of masonite. *Desalination.*, Vol. 199, pp. 411-412

Pretorius, I.S. & Bauer ,F.F. (2002). Meeting the consumer challenge through genetically customized wine-yeast strains. *Trends Biotechnol.*, Vol. 20, pp. 426-432

Roca, C., Nielsen, J. & Olsson, L. (2003). Metabolic engineering of ammonium assimilation in xylose-fermenting *Saccharomyces cerevisiae* improves ethanol production. *Appl. Environ Microbiol.* , 69, pp. 4732-4736

Rudolf, Baudel, H., Zacchi, G., Hahn-Hägerdal, B. & Lidén, G. (2008). Simultaneous Saccharification and Fermentation of Steam-Pretreated Bagasse Using *Saccharomyces cerevisiae* TMB3400 and *Pichia stipitis* CBS6054 *Biotechnology and Bioengineering*, Vol. 99, No. 4, March 1.

Russel, J.B. (1992) Another explanation for the toxicity of fermentation acids at low pH: anion accumulation versus uncoupling. *J. Appl. Bacteriol.*, Vol. 73, pp. 363-370

Saha, B.C. (2004). Lignocellulose biodegradation and application in biotechnology. ACS Symp Ser., 889, pp. 2-34

Sales, K., Brandt, W., Rumbak, E. & Lindsey, G. (2000). The LEA-like protein HSP12 in *Saccharomyces cerevisiae* has a plasma membrane location and protects membranes against desiccation and ethanol-induced stress. B*iochim. Biophys. Acta*, Vol. 1463, pp. 267–278

Sampermans, S., Mortier, J. & Soares, E.V. (2005) Flocculation onset in *Saccharomyces cere*visiae: the role of nutrients. *J. Appl. Microbiol.* Vol.98, pp. 525–531

Sassner, P., Martensson, C.G., Galbe, M. & Zacchi, G. (2008). Steam pretreatment of H2SO4 impregnated Salix for the production of bioethanol. *Biores. Technol.*, Vol. 99, pp. 137–145

Sedlak, M. & Ho, N.W. (2004). Production of ethanol from cellulosic biomass hydrolysates using genetically engineered *Saccharomyces cerevisiae* cells. *Appl. Biochem. Biotechnol.*,Vol. 113-116, pp. 403-416

Shevchenko, S.M., Chang, K., Robinson, J. & Saddler, J.N. (2000). Optimization of monosaccharide recovery by post-hydrolysis of the water-soluble hemicellulose component after steam explosion of softwood chips. *Bioresource Technology.*, Vol. 72, pp. 207-211

Singer, M.A. & Lindquist, S. (1998). Thermotolerance in *Saccharomyces cerevisiae*: the Yin and Yang of trehalose. *Trends Biotechnol.*, Vol. 16, pp. 460–468

Slininger, S.P., Bothast, R.J., Okos, M.R. & Ladish, M.R. (1985). Comparative evaluation of ethanol production by xylose-fermenting yeasts presented high xylose concentration. *Biotech. Lett.*, Vol. 7, pp. 431-436

Sreenath, H.K. & Jeffries, T.W. (2000). Production of ethanol from wood hydrolyzate by yeasts. *Bioresource Technology* , Vol.72, pp. 253-260

Stanley, D., Bandara, A., Fraser, S., Chambers, P.J. & Stanley, G.A. (2010). The ethanol stress response and ethanol tolerance of *Saccharomyces cerevisiae*. *J. Appl. Microbiol.* ,109, pp. 13-24

Stenberg, K., Tengborg, C., Galbe, M. & Zacchi, G. (1998). Optimization of steam pretreatment of SO2-impregnated mixed softwoods for ethanol production. *J. Chem. Technol. Biotechnol.*, Vol. 71, pp. 299-308

Sun, X.F., Xu, F., Sun, R.C., Wang, Y.X., Fowler, P. & Baird, M.S. (2004). Characteristics of degraded lignins obtained from steam exploded wheat straw. *Polym. Degrad. Stabil.*, Vol. 86, pp. 245-256

Sun, Y. & Cheng, J. (2002). Hydrolysis of lignocellulosic materials for ethanol production: A review. *Bioresource Technol.* Vol. 83, pp. 1-11

Taherzadeh, M.J. & Karimi, K. (2007). Acid-based hydrolysis processes for ethanol from lignocellulosic materials: A review. *BioResources* , Vol. 2, pp. 472-499

Taherzadeh, M.J. & Karimi, K. (2008). Pretreatment of Lignocellulosic Wastes to Improve Ethanol and Biogas Production: A Review. *Int. J. Mol. Sci.*, Vol. 9, pp. 1621-1651

Taherzadeh, M.J., Gustafsson, L., Niklasson, C. & Liden, G. (1999). Conversion of furfural in aerobic and anaerobic batch fermentation of glucose by *Saccharomyces cerevisiae. J. Biosci. Bioeng.* , Vol. 87, pp. 169–174

Taherzadeh, M.J., Niklasson, C. & Liden, G. (2000a). On-line control of fed-batch fermentation of diluite-acid hydrolyzates. *Biotechnol. Bioeng.* , Vol. 69, pp. 330-338

Taherzadeh, M.J., Gustafsson, L., Niklasson, C. & Liden, G. (2000b). Inhibition effects of furfural on aerobic batch cultivation of *Saccharomyces cerevisiae* growing on ethanol and or acetic acid. *J. Biosci. Bioeng.* , Vol. 90, pp. 374-380

Taherzadeh, M.J., Millati, R. & Niklasson, C. (2001). Continuous cultivation of diluite-acid hydrolysates to ethanol by immobilized *Saccharomyces cerevisiae. Appl. Biochem. Biotechnol.*, Vol.95

Teunissen, A.W. & Steensma, R.H. (1995). Review: the dominant flocculation genes of *Saccharomyces cerevisiae* constitute a new subtelomeric gene family. *Yeast.*, Vol. 11, pp. 1001-1013.

van Zyl, W.H., Lynd, L.R., den Haan, R. & McBride, J.E. (2007). Consolidated bioprocessing for bioethanol production using *Saccharomyces cerevisiae. Adv. Biochem. Eng. Biotechnol.*, Vol. 108, pp. 205–235

Varga, E., Reczey, K. & Zacchi, G. (2004). Optimization of steam pretreatment of corn stover to enhance enzymatic digestibility. *Appl. Biochem. Biotechnol.* , Vol. 113, pp. 509-523

Verbelen, P.J., De Schutter, D.P., Delveaux, F., Verstrepen, K.J. & Delveaux, F.R. (2006). Immobilized yeast cell systems for continuous fermentation applications. *Biotechnol. Lett.*, Vol. 28, pp. 1515-1525

Verduyn, C., Postma, E., Scheffers, W.A. & Van Dijken, J.P. (1992). Effect of benzoic acid on metabolic fluxes in yeasts: a continuous-culture study on the regulation of respiration and alcoholic fermentation. *Yeast.* ,Vol. 8, pp. 501-517

Verstrepen, K.J. & Klis, F.M. (2006). Flocculation, adhesion and biofilm formation in yeasts. *Molecular Microbiology*. Vol.60, No1, pp. 5-15

Viikari, L., Alapuranen, M., Puranen, T., Vehmaanperä ,J. & Matti Siika-aho. (2007). Thermostable Enzymes in Lignocellulose Hydrolysis. *Adv. Biochem. Engin/Biotechnol.*, Vol. 108, pp. 121–145

Viola, E.; Zimabardi, F.; Cardinale, M.; Cardinale, G.; Braccio, G.; Gamabacorta, E. (2008) Processing cereal straws by steam explosion in a pilot plant to enhance digestibility in ruminants. *Bioresour. Technol.*, 99, 681–689

Von Sivers, M., Olsson, L., Hahn- Hägerdal, B. & Zacchi, G. (1994). Cost analysis of ethanol production from willow using recombinant *Escherichia coli. Biotechnol. Prog.* , Vol. 10, pp. 555-560

Wahlbom, C.F., van Zyl, W.H., Jonsson, L.J., Hahn-Hägerdal, B. & Otero, R.C. (2003). Generation of the improved recombinant xylose-utilizing *Saccharomyces cerevisiae*

TMB 3400 by random mutagenesis and physiological comparison with Pichia stipitis CBS 6054. *FEMS Yeast Research*, 3, pp. 319-326

Walfridsson, M., Bao, X., Anderlund, M., Lilius, G., Bulow, L. & Hahn-Hagerdal, B. (1996). Ethanolic fermentation of xylose with *Saccharomyces cerevisiae* harboring the *Thermus thermophilus* xylA gene, which expresses an active xylose (glucose) isomerase, *Appl. Environ. Microbiol.*, Vol. 62, pp. 4648–4651

Wang, F.S. & Sheu, J.W. (2000). Multiobjective parameter estimation problems of fermentation processes using a high ethanol tolerance yeast. *Chemical Engineering Science.*, Vol. 55, pp. 3685-3695

Wingren, A., Galbe, M. & Zacchi, G. (2003). Techno-Economic evaluation of producing ethanol from softwood: comparison of SSF and SHF and identification of bottlenecks. *Biotechnol.Prog.*, 19, pp. 1109-1117

Wu, M.M., Chang, K., Gregg, D.J., Boussaid, A., Beatson, R.P. & Saddler, J.N. (1999). Optimization of steam explosion to enhance hemicellulose recovery and enzymatic hydrolysis of cellulose in softwoods. *Appl. Biochem. and Biotechnol.* Vol. 77-79, pp. 47-54

Xiao, Z., Zhang, X., Gregg, D.J. & Saddler, J.N. (2004). Effects of sugar inhibition on cellulases and beta-glucosidase during enzymatic hydrolysis of softwood substrates. *Appl. Biochem. Biotechnol.* Vol. 113-116, pp. 1115–1126

Zacchi, G. & Sassner, P. (2008). Biotechnology for biofuels.1176- 1186

Zaldivar, J., Borges, A., Johansson, B., Smits, H.P., Villas-Boas, S.G., Nielsen, J. & Olsson, L. (2002). Fermentation performance and intracellular metabolite patterns in laboratory and industrial xylose fermenting *Saccharomyces cerevisiae*. *Appl. Microbiol. Biotechnol.*, Vol. 59, pp. 436-442

Zaldivar, J., Nielsen, J. & Olsson, L. (2001). Fuel ethanol production from lignocellulose: a challenge for metabolic engineering and process integration. *Appl. Microbiol. Biotechnol.* , Vol. 56, pp. 17–34

Zayed, G. & Meyer, O. (1996). The single-batch bioconversion of wheat straw to ethanol employing the fungus Trichoderma viride and the yeast *Pachysolen tannophylus*, *Appl. Microbiol. Biotechnol.*, Vol.45, No4, pp. 551-555

Zhang, M., Shukla, P., Ayyachamy, M., Permaul, K. & Singh, S. (2010). Improved bioethanol production through simoultaneous saccharification and fermentation of lignocellulosic agricultural wastes by *Kluyveromyces marxian*us 6556. *World J. Microbiol. Biotechnol.*, Vol. 26, pp. 1041-1046

# Part 3

## Cellulosic Biofuel

# 9

# Novel Approaches to Improve Cellulase Biosynthesis for Biofuel Production – Adjusting Signal Transduction Pathways in the Biotechnological Workhorse *Trichoderma reesei*

Doris Tisch and Monika Schmoll
*Vienna University of Technology, Institute of Chemical Engineering*
*Austria*

## 1. Introduction

Every living organism is destined to adjust its life style to the ecological niche of its habitat. Thereby the rotation of the earth, bringing about daily changes in light conditions, humidity and temperature, as well as nutrient requirements and the need for reproduction dominate the existence of creatures from bacteria to man. Adjustment of literally every biochemical pathway to the environmental conditions of the very spot an organism sees the light of day is of crucial importance for survival and successful competition against a plethora of rivals.

However, not only adaptation as such but more importantly fast perception, correct interpretation, rating and an optimal response to environmental cues with efficient use of available resources for this task will ensure success in nature (Figure 1).

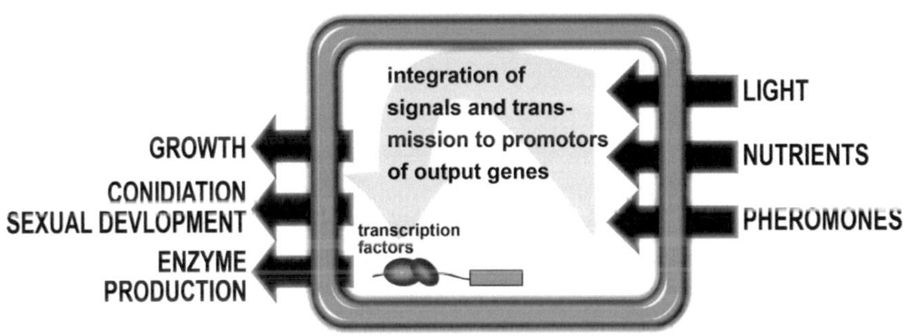


Fig. 1. Schematic representation of signal transduction and its consequences in fungi. Examples for environmental signals and possible adaptive reactions are shown.

In the kingdom of fungi, many species can be found that are masters of this game. One of them is *Trichoderma reesei* (anamorph of *Hypocrea jecorina*), a filamentous ascomycete possessing a very effective enzyme machinery for degradation of cellulose (Martinez et al., 2008; Schuster & Schmoll, 2010). If it would not have been for the so called cellulases of this fungus, its name would probably only be known to a few specialists. But during World War II, *T. reesei* chose the cellulosic material of military tents and uniforms in the south pacific as its habitat. The efficient degradation of these materials - initially considered chemical warfare - was soon found to be accomplished by a fungus living in the warm and humid tropical rain forest of the Solomon Islands (Reese, 1976). After initial research efforts aimed at preventing degradation, the focus soon changed to promotion of decomposition for industrial use of cellulases (Reese, 1976). Notably, this fungus not only efficiently degrades cellulose in its natural habitat, the tropics, but also in the spartan environment of a shake flask containing minimal medium or a solid steel fermenter, which reflects a high potential for adaptation. Consequently, decades of research and industrial application made *T. reesei* one of the most prolific biotechnological workhorses in industry nowadays (Kubicek et al., 2009; Schmoll & Kubicek, 2003). However, *Trichoderma* spp. were also investigated for light dependent phenomena (Schmoll et al., 2010).

The cellulolytic and hemicellulolytic enzymes of *T. reesei* have been used for countless applications thereafter (Buchert et al., 1998; Galante et al., 1998a; Galante et al., 1998b), but only recently the need for alternative and sustainable fuel again boosted research with this fungus. Recent concerns about crop based-"first generation" biofuels enhanced the research on economically feasible "second generation" biofuels (Solomon, 2010; Somerville, 2007). The advantage of second generation biofuels is the usage of cheap, renewable and non-food biomass like cellulose from plants (Wilson, 2009). In the respective industrial process cellulases are applied for initial degradation of cellulosic plant material to fermentable sugars, which can then be further converted into bioethanol (Naik et al., 2010). The conversion process can be based on thermochemical processes or biochemical processes (Naik et al., 2010). The advantages of biochemical production of bioethanol are lower energy costs and that the production at smaller scales is possible, albeit the hydrolysis of cellulose to fermentable sugars is still the most expensive step in this process (Gomez et al., 2008).

In the following we explain how signal transduction pathways impact regulation of cellulase gene expression. Both known mechanisms and promising candidate factors/processes are discussed in order to provide a basis for evaluation of the relevance of a given signalling pathway or environmental cue for cellulase gene expression.

## 2. Cellulose, cellulases and *Trichoderma reesei*

Cellulose is a polysaccharide made up from D-glucose subunits in a polymerization degree of 10 000 or even higher. Therefore, this substrate cannot be transported into the cell for degradation. Additionally, the presence of the substrate outside the cell in its original form cannot be sensed. Hence *T. reesei* has developed an intriguing system for initial degradation of cellulose immediately upon landing of a spore on degradable material: Spores of *T. reesei* are covered with the cellulolytic enzymes, predominantly CBHII, which allows for an initial attack on cellulose (Kubicek et al., 1988; Messner et al., 1991) and also to detection of substrate by uptake of the characteristic degradation products. Once the cellulose signal is received, the powerful enzymatic machinery for plant cell wall degradation is launched and enables the spore to germinate and grow. Yet, the precise nature of this signal as well as the mechanism of its perception are still unknown, although the transglycosylation product of cellobiose, sophorose, is considered the natural inducer (Sternberg & Mandels, 1979). Additionally, the mode of perception of the signal remains obscure. While a G protein coupled receptor for sensing carbon has been identified in *N. crassa* (Li & Borkovich, 2006) and also homologues in many other filamentous fungi, analysis of the *T. reesei* genome revealed a deletion in the respective genomic region (Schmoll, 2008). So far no GPCR was found in *T. reesei*, which could have assumed the task of the obviously lost receptor. Consequently, it is reasonable to believe that detection of the presence of cellulose outside of the cell occurs upon import of degradation products and the respective signal is likely to be transmitted starting from a carbohydrate transporter or degradative enzyme inside the cell.

In contrast, the enzymatic machinery, which is launched upon detection of cellulose outside the cell, is well studied (Schmoll & Kubicek, 2003). Interestingly, compared to other ascomycetes, *T. reesei* has a very limited number of genes encoding cellulases and hemicellulases (Martinez et al., 2008), which is a real surprise considering its great potential of cellulose degradation.

However, not only the insoluble substrate cellulose induces cellulase gene expression. Already in the late fifties of the last decade also soluble carbon sources were found to induce cellulase gene expression (Mandels & Reese, 1957). In the following for cellobiose (Mandels & Reese, 1960), sophorose, which is a transglycosylation product of cellobiose (Mandels et al., 1962; Sternberg & Mandels, 1979; Vaheri et al., 1979), lactose (Mandels & Reese, 1957; Seiboth et al., 2007), L-sorbose (Kawamori et al., 1985), L-arabitol and several additional substrates (Margolles-Clark et al., 1997), a cellulase inducing potential in *T. reesei* was detected. Interestingly, the set of cellulolytic enzymes produced on these carbon sources is not necessarily the same as on cellulose as has been shown for sophorose and lactose (Messner et al., 1988; Schmoll & Kubicek, 2005; Sternberg & Mandels, 1980). These findings have several implications when considering signal transduction processes: Detection of these carbon sources in the environment causes *T. reesei* to produce cellulolytic enzymes and hence these different signals must be combined to cause a common output. Nevertheless, since the array of cellulases produced is incomplete on lactose and sophorose, certain fine tuning mechanisms, which adjust the cellulase mixture to slightly different substrates, are likely to be at work.

The industrial application of cellulases necessitated not only the optimization of the cultivation media but also the fungus itself was treated for higher performance. In research, not only QM6a, but also an early mutant strain enhanced for cellulase production, QM9414 as well as the hyperproducer strain RUTC30 (Seidl & Seiboth, 2010) are frequently used.

## 3. Regulation of cellulase gene expression

Besides evaluation of the influence of different carbon sources on expression of cellulase genes, the mechanism which triggers this process was also investigated at a molecular level. Cloning of the major cellulase gene, *cbh1* (encoding cellobiohydrolase 1, today known as CEL7A) along with antibody-based analysis of this and further cellulolytic enzymes (Mischak et al., 1989; Shoemaker et al., 1983a; Shoemaker et al., 1983b; Teeri et al., 1987) established a crucial basis for elucidation of the mechanism of cellulase gene expression in *T. reesei*. The following studies concluded from the correlation of protein expression with mRNA abundance, that cellulase gene expression is predominantly regulated on the pre-translational level (El-Gogary et al., 1989; R. Messner & Kubicek, 1991; Morawetz et al., 1992). Nevertheless, recent results indicate that this assumption must be treated with caution and that posttranscriptional regulation in addition has to be considered (Gyalai-Korpos et al., 2010).

The main cellobiohydrolases *cbh1* and *cbh2*, as well as *egl1*, *egl2* and *egl5* are coordinately transcribed, with *cbh1* being the most highly expressed cellulase gene (Ilmén et al., 1997). Although the real inducer of cellulase production still remains to be determined, the transcription factors regulating the expression are well studied. Five transcription factors are implicated in this process: ACE1 and CRE1 are negative regulators and ACE2, XYR1 and the HAP2/3/5 complex are positive ones (Kubicek et al., 2009). These transcription factors not only bind to their target promotors under inducing conditions (for example cellulose or sophorose), but mostly also under repressing conditions (glucose) (Rauscher et al., 2006; Zeilinger et al., 1998).

Consequently, it is reasonable to assume that the activity of these transcription factors is (also) regulated by modification in response to changing conditions. These modifications, such as phosphorylation or ubiquitinylation can be the outcome of a signal transduction cascade. Phosphorylation in general is one of the central processes, by which environmental signals are transmitted or in other words by which switches on multiple proteins are moved into place to trigger a certain reaction to the environment. Kinases and phosphatases are thereby of crucial importance for signalling also in fungi (Dickman & Yarden, 1999; Kosti et al., 2010) and have even been termed the currency of signalling. In *T. reesei* DNA binding of ACE2 was suggested to require phosphorylation (Stricker et al., 2008b). However, neither the responsible kinase nor the signalling event causing its activation, are known. Another possible mechanism which predominantly acts via regulation of protein stability would be ubiquitinylation (Hochstrasser, 2009). Interestingly, first indications for an involvement of ubiquitinylation in regulation of cellulase gene expression have been found (Gremel et al., 2008). For both mechanisms, support comes from analysis of the amino acid sequences of the transcription factors known to be involved in regulation of cellulase gene expression: all of them comprise phosphorylation sites for cAMP dependent protein kinase, casein kinase II and protein kinase C (Table 1). XYR1 and HAP3 additionally contain probable PEST sequences (Table 1) that have been shown to act as ubiquitinylation and degradation signals (Rechsteiner & Rogers, 1996).

| | cAMP and cGMP dependent phosphorylation sites | casein kinase II phosphorylation sites | protein kinase C phorphorylation sites | PEST sequence |
|---|---|---|---|---|
| **XYR1** | 6 | 12 | 14 | 1 |
| **ACE1** | 1 | 7 | 2 | none |
| **ACE2** | 2 | 17 | 18 | none |
| **HAP2** | 2 | 2 | 7 | none |
| **HAP3** | 1 | 5 | 2 | 2 |
| **HAP5** | 1 | 2 | 1 | none |
| **CRE1** | 1 | 7 | 8 | none |

Table 1. Phosphorylation sites and PEST sequences of *T. reesei* transcription factors involved in regulation of cellulase gene expression. Phosphorylation sites were analyzed using GeneRunner v3.0 (Hastings). PEST-sequences were identified using the online tool epestfind (http://emboss.bioinformatics.nl/cgi-bin/emboss/epestfind).

A further important regulatory process acting on the level of the promotor is chromatin remodelling. Also this mechanism was found to be involved in regulation of cellulase gene expression in *T. reesei* (Zeilinger et al., 2003).

In summary, there are numerous hints indicating direct action of signalling pathways on the transcription factors regulating cellulases, but it remains to be determined how they precisely work together to cause signal specific adjustment of cellulase gene expression.

## 4. Evaluating the influence of signal transduction pathways on cellulase gene expression

Despite thorough investigation of cellulase induction, transcription, secretion, enzyme function and elucidation of important transcript factors, the role of classical signal transduction pathways in these processes only received little attention for quite some time. Although *T. reesei* protein kinase C (PKC) was among the first PKCs to be characterized in detail in filamentous fungi (Lendenfeld & Kubicek, 1998; Morawetz et al., 1996), this topic did not become a research focus in this fungus and information on characteristics and regulatory targets of protein kinases and phosphatases of *T. reesei* is still scarce. Nevertheless, the availability of the sequenced genome (Martinez et al., 2008) enabled an initial evaluation of the signalling inventory of this fungus (Schmoll, 2008), which for example revealed interesting differences to other ascomycetes in the number of available two component histidine kinases and casein kinases.

Additionally, early studies also revealed hints as to an involvement of $Ca^{2+}$-calmodulin signalling in regulation of plant cell wall degradation (Mach et al., 1998). Although in this study only transcription of a xylanase has been studied, coregulation of cellulases and xylanases under many conditions (Stricker et al., 2008a) suggests that also cellulases might be a target of $Ca^{2+}$-calmodulin signalling.

Detailed investigations aimed at an evaluation of the role of signal transduction pathways in cellulase regulation started less than 10 years ago. Basis for the first study were two *T. reesei* strains derived from QM6a. QM9414 represents an early improved cellulase producer strain, which produces higher amounts of cellulases than the wild-type, while QM9978 - resulting from the same mutation cycle - does not. Since the defect of QM9978 was neither found to be a mutation in the protein coding region of the major cellulase genes nor in their promotors (Torigoi et al., 1996; Zeilinger et al., 2000) the assumption was that a crucial signal transduction cascade must be perturbed in this strain. Consequently, a cDNA subtraction method (RaSH, rapid subtraction hybridization) was applied to compare transcription of QM9414 and QM9978 under conditions inducing cellulase expression (Schmoll et al., 2004). Although this study did not reveal the actual defect of QM9978, intriguing new insights into the signals influencing cellulase gene expression were gained, which are still subject to investigation.

### 4.1 Light positively influences cellulase transcription

The first and most surprising finding of the transcriptional comparison of QM9414 and QM9978 was the differential regulation of a putative light regulatory gene, encoding the orthologue of the *Neurospora crassa* photoreceptor VIVID (Schmoll et al., 2004). Detailed investigation of this gene, named *env1* (encoding ENVOY for "messenger") indeed confirmed a function in regulation of cellulase gene expression and moreover, also light itself was found to influence transcription of cellulase genes (Schmoll et al., 2005). However, this is not the only function of ENVOY. A follow up study showed that light regulated genes found in *T. reesei* can be ENVOY-mediated or independent and the effect can be both positive and negative. Intriguingly, also a function of ENVOY in darkness was detected (Schuster et al., 2007). From these findings, the question arises, which role light might play in the life of *T. reesei* and especially in cellulase expression.

In terms of signal transduction, this finding indicates that despite decades of laboratory cultivation and propagation, *T. reesei* still has not lost its evolutionary heritage. Although completely detached from its natural habitat, adjustment of organismal functions to light and darkness is still operational in *T. reesei* (Schmoll et al., 2010). The presence of this reaction to light in this industrial workhorse also suggests that a circadian rhythm is at work, which however remains to be proven. If so, the daily cycles of expression of certain genes, meant to anticipate dusk and dawn, would be an important feature to be considered and potentially exploited for strain improvement.

In general, light is used predominantly for two purposes. It serves as a source of energy by photosynthetic processes in plants and algae or as a source of information. Fungi are not able to exert photosynthesis and therefore they only use the information connected to the light signal. The presence of light indicates growth on an exposed surface in contrast to growing inside a rotting tree - one of the natural substrates of *T. reesei*. The fungus has to adapt to the different conditions on the surface, like altered humidity, temperature and different nutritional conditions. Furthermore, the chance to encounter a potential mating partner is higher outside a substrate and the entrainment of components of the circadian rhythm, which allows for adaptation to day and night is also induced by light (Brunner & Kaldi, 2008).

Fungi can sense different wavelengths of visible light – many of them developed perception machineries for blue, green and red light and therefore photoreceptor proteins are required. These are proteins that generate a signal in response to light and transfer this to appropriate targets to react to the changed light conditions (Herrera-Estrella & Horwitz, 2007). A photoreceptor consists of an apoprotein and a cofactor (chromophore). Dependent on the kind of chromophore, the photoreceptors have different absorption maxima for particular wavelength ranges (Herrera-Estrella & Horwitz, 2007). The importance of light for fungi becomes manifest in the widespread reaction of fungi to illumination. Thereby fungi not only react to the harmful effects of light, especially UV-light, by taking protective measures, they even adjust their metabolism - including virtually all metabolic pathways - in the presence of light (Tisch & Schmoll, 2010).

The immediate response to light at the molecular level involves early light responsive genes (ELRGs), transcription of which peaks around 45 minutes after illumination and late light response genes (LLRGs), which are transiently upregulated mainly around 90 minutes after the light pulse (Chen et al., 2009). Thereby, the early light response rather involves protective measures and thereafter also metabolic pathways seem to become adjusted to light conditions. In the light of these data, the fact that also cellulases are influenced by light perfectly fits and investigation of more factors of the light signalling machinery proved to be highly interesting.

In *N. crassa* three blue light photoreceptors were identified: WC-1 (white collar 1) and WC-2 (white collar-2), which are GATA family zinc finger transcription factors (Linden & Macino, 1997; Linden et al., 1997) and VIVID (Heintzen et al., 2001; Schwerdtfeger & Linden, 2003), the orthologue of *T. reesei* ENVOY. Upon blue light exposure, WC-1 and WC-2 form the white collar complex (WCC) and bind to light response elements in the promoter regions of target genes (He & Liu, 2005; Smith et al., 2010; Talora et al., 1999). VIVID interacts with the WCC and modulates its transcriptional activity (Chen et al., 2010a; Hunt et al., 2010).

The genome of *T. reesei* also comprises homologues of these two blue light receptors (encoded by *blr1* and *blr2*) (Castellanos et al., 2010; Schmoll, 2008). BLR1 possesses a PAS/LOV domain like ENVOY and WC-1 for blue light sensing and protein-protein interaction. Both BLR proteins are necessary for full induction of *env1* transcription in light (Castellanos et al., 2010). The effect of BLR1 and BLR2 on transcription of *env1* strongly suggests also a function in cellulase gene regulation for these photoreceptor genes. Indeed, BLR1 and BLR2 positively impact the transcription of the major cellobiohydrolase of *T. reesei*, *cbh1* not only in light, but also in darkness. Interestingly, this effect on *cbh1* does not seem to be accomplished via regulation of *env1* (Castellanos et al., 2010). Homologues of the white collar proteins were identified in *Aspergillus*, *Fusarium* and other *Trichoderma* spp. (Casas-Flores et al., 2004; Castellanos et al., 2010; Purschwitz et al., 2008; Ruiz-Roldan et al., 2008; Schmoll, 2008). In *Trichoderma atroviride* the homologues of the WC proteins, the BLR proteins (blue light receptor 1 and 2), were shown to be necessary for blue light induced conidiation (Casas-Flores et al., 2004) and for carbon source dependent conidiation (Casas-Flores et al., 2006; Friedl et al., 2008a), which confirms a connection between light response and nutrient assimilation in fungi. Interestingly, also the photoreceptors of *T. atroviride* are involved in cellulase gene expression: Deletion of BLR-1 or BLR-2 causes strongly enhanced cellulase activities secreted into the culture medium (M. Friedl and M. Schmoll, unpublished).

### 4.2 How does *Trichoderma reesei* react to light?

The unexpected importance of light in metabolic processes of *T. reesei* necessitated a more detailed investigation of this effect in order to evaluate the possible relevance of this environmental cue for further studies (Schmoll et al., 2010). Analysis of light response of *T. reesei* upon growth on glycerol already revealed an adjustment of processes involved not only in protection from the harmful effects of light, but also of signalling pathways and metabolic processes (Schuster et al., 2007) In this study the central position of the light regulatory protein ENVOY became obvious.

New results under conditions more relevant for cellulase production, i. e. from cultivations on cellulose in light and darkness by microarray analysis further support significant adjustments of *T. reesei* to light (Tisch et al., 2011b, ms submitted). This genome wide analysis revealed 2.7 % of the genes to be differentially regulated in light and darkness in the *T. reesei* wild-type strain. Among the genes upregulated in light, a significant enrichment in the functions of DNA photolyase activity, carbohydrate metabolic activity (including cellulase activity and cellulose binding), regulation of oxidoreductase activity and sulphur metabolism was detected. Additionally, 6 genes involved in sexual development and 16 glycoside hydrolase genes are upregulated in light. Interestingly, glycoside hydrolase genes have also been detected among the genes downregulated in light and consequently, light cannot be considered a strictly positive or negative factor for expression of enzymes, but rather an environmental signal which triggers production of an altered enzyme mixture.

One of the most obvious effects of light on many fungi is the induction or enhancement of sporulation, or more generally speaking: Light is a crucial signal for the decision to undergo sexual development or reproduce asexually (Bayram et al., 2008; Blumenstein et al., 2005; Mooney & Yager, 1990; Seidl et al., 2009). Nevertheless, the genes regulated by light in *T. reesei* and other organisms (Chen et al., 2010a, Smith et al., 2010), which also involve adjustments of metabolic pathways, clearly indicate that sporulation is only one of many outputs of the signalling cascade initiating the adaptation to light.

### 4.3 Light-dependent modulation of cellulase gene expression is a conserved phenomenon in fungi

As described above, the light regulatory protein ENVOY is a homolog of VIVID, a photoreceptor that is well characterized in *N. crassa,* the model organism for studies on

photobiology and the circadian clock (Davis & Perkins, 2002). One of the crucial functions of VIVID is to act as a universal brake and repress both early and late light responses after prolonged illumination (Chen et al., 2009). Despite the similarity of ENVOY and VIVID, they are not fully functionally homologous, because *env1* could not complement a mutation in *vvd*. Additionally, the phenotypes of non functional mutants in these genes are strikingly different: While deletion of *env1* in *T. reesei* causes strongly decreased growth and obvious problems with adaptation to light, *N. crassa* strains lacking function *vvd* do not show such a growth defect in light, but enhanced carotenoid production (Schmoll et al., 2005; Shrode et al., 2001).
Nevertheless, since also *N. crassa* possesses an efficient machinery for degradation of cellulose (Eberhart et al., 1964; Eberhart et al., 1977), which was recently studied in detail (Tian et al., 2009), a comparison of the influence of light and the machinery transmitting the light signal on cellulase gene expression proved topical.
For *N. crassa*, as a model organism, there are a lot of genetic tools available, including a sequenced genome (Galagan et al., 2003) and an almost complete library of gene knock out strains (Colot et al., 2006). So far *N. crassa* was predominantly used for investigations of photobiology and the circadian clock (Chen et al., 2010b). Therefore, the combination of *N. crassa* as a model for light response and *T. reesei* as the model fungus for studies on plant cell wall degradation is an ideal fit for investigation of light modulated cellulase gene expression. Genome wide analysis of the effect of WC-1, WC-2 and VVD on cellulase gene expression indeed revealed that the light response machinery of *N. crassa* influences cellulase gene expression in the same way as in *T. reesei*. Surprisingly, while also in *N. crassa* WC-1, WC-2 and VVD exert a positive effect on transcription of the major cellulase gene, the cellulase mixture secreted into the culture medium by mutants in the respective genes was even more efficient than that in the wild-type (Schmoll et al., 2011, ms submitted). A similar effect was recently also observed in *T. reesei* (Gyalai-Korpos et al., 2010). In both cases, the light response machinery was suggested to accomplish this enhancement of cellulase activity despite decreased transcription of *cbh1* (or its orthologue) by alteration of the composition of the secreted enzyme mixture for higher efficiency dependent on light.
Consequently, the phenomenon of light modulated cellulase gene expression is likely to be conserved in fungi. However, the different phenotypes of mutants in *T. reesei env1* and *N. crassa vvd*, as well as in the photoreceptor genes (Castellanos et al., 2010) indicate that despite similar outcome, the signalling cascade achieving this aim is not identical in these two fungi.
Obviously, the environmental signal "light" starts a signal cascade which crosstalks with other factors like nutrition conditions leading to elevated levels of cellulase transcripts. Up to now, three of the crucial components of the light signalling pathway could be shown to be involved in transduction of the positive light signal to the cellulase promoter. The question arises, how the light signals are transferred to the cellulase promoters.

### 4.4 The role of heterotrimeric G protein signalling in the light transferring pathway

For a signal to take effect, it first has to be perceived and thereafter its transmission is necessary to reach its target. Considering a complex environment like a tropical rain forest clearly countless signals, like light intensity, humidity, temperature, carbon source, nitrogen source, pH or the potential presence of a mating partner are to be noticed by a filamentous fungus growing in this habitat. However, not every signal has the same significance under every condition. The complex network of signal transduction pathways in fungi not only has to recognize these various signals. Also a rating and appropriate distribution of these signals is necessary in order to affect output pathways such as growth, development and enzyme production for optimal adaptation.

Results showing that nutritional and other environmental signals interfere with light signalling (Friedl et al., 2008a; Friedl et al., 2008b; Tisch & Schmoll, 2010), suggest a role for heterotrimeric G proteins in transduction of the light signal.

The G protein (guanine nucleotide binding protein) pathway is well studied and known as a connection pathway from the cell surface to intracellular targets (Figure 2) (Hamm, 1998; Li et al., 2007).

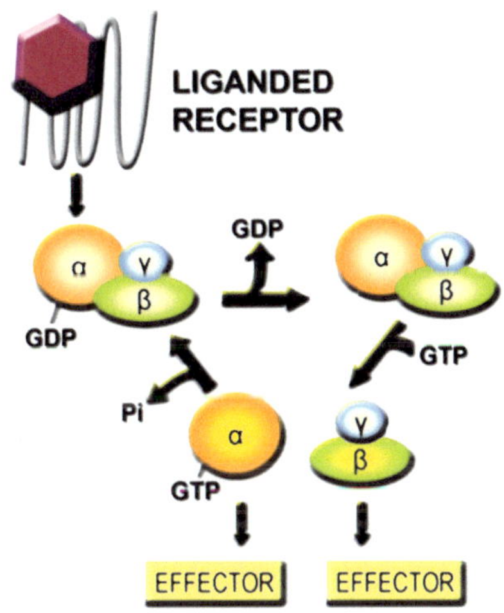


Fig. 2. Binding of a ligand to a G protein coupled receptor (GPCR) leads to a conformational change of the G protein alpha subunit and thereby exchanging GDP (guanosine diphosphate) to GTP (guanosine triphosphate). The GTP bound G alpha subunit is activated and releases the G protein beta and gamma subunits. G beta and gamma work as a dimer to affect further targets. The intrinsic GTPase activity of G alpha hydrolyzes GTP to GDP to terminate the signalling cycle and enables the G beta gamma dimer to reassociate with the G alpha subunit. As a heterotrimeric G protein it binds to the GPCR the circle of G protein signalling is closed (Hamm, 1998; Sprang, 1997).

Fungi have a versatile array of G protein coupled receptors for perception of environmental signals. They belong to six classes: Ste2-like pheromone receptors, Ste3 like pheromone receptors, carbon/amino acid receptors, putative nutrient receptors, cAMP receptor-like and microbial opsins (Xue et al., 2008), not including the PTH11-like receptors first identified in *Magnaporthe grisea* and involved in pathogenicity (Kulkarni et al., 2005). So far, only nutrients, pheromones and light could be confirmed as activating ligands of GPCRs (Li et al., 2007; Xue et al., 2008). The transmitting factors, the G protein alpha subunits, are classified in three groups (I, II, III) with group I and III being similar to mammalian G alpha-i and G alpha-s subunits. In contrast to the filamentous fungi *Neurospora crassa* and *Aspergillus nidulans*, the genome of which comprises three G alpha subunits each, *S. cerevisiae* only contains two different G alpha subunits (Bölker, 1998; Li et al., 2007).

The genome of *T. reesei* comprises three G alpha subunits (GNA1, GNA2 and GNA3), one G beta subunit (GNB1) and one G gamma subunit (GNG1) (Schmoll, 2008). In 1995 it was already shown that a G alpha subunit CPG-1 is required for cellulase mediated induction of *cbh1* transcription in *Cryphonectria parasitica* (Wang & Nuss, 1995). Additionally, an effect of cAMP on cellulase gene expression had been discovered (Sestak & Farkas, 1993). Since intracellular cAMP levels are modulated by class III G alpha subunits, also a participation of this subgroup in regulation of cellulase gene expression was likely.

Consequently, an involvement of the heterotrimeric G proteins in regulation of cellulase transcription was investigated in *T. reesei*. Indeed, GNA1 was found to exert an influence on cellulase gene expression (Seibel et al., 2009). However, the most interesting outcome of this study was that this influence is different in light and darkness. Deletion of *gna1* caused strongly enhanced *cbh1* transcript levels in darkness, but abolished *cbh1* transcription in light. Interestingly, also carbon source dependent feedback regulation of *gna1* upon activation was observed. These results reflect the need for integration of different environmental signals to adjust the output - for example cellulase gene expression. Obviously, the light signal is important for cellulase gene expression and for the signal transmitted by GNA1. Since despite constitutive activation of GNA1 (i. e. constant transmission of the signal), still an inducer is needed, GNA1 cannot be the transmitter of the cellulose signal and the actual nature of the signal transmitted by GNA1 remains to be determined (Seibel et al., 2009). Also, whatever this signal is, the carbon source on which the fungus grows is important for its relevance: Feedback regulation of the transcript of *gna1* is one step in setting the sensitivity of the GNA1-related signal(s) by increasing transcription upon activation and hence leads to stronger amplification of the signal.

The second G protein alpha subunit (belonging to subgroup III), GNA3, was also found to be involved in regulation of cellulase gene expression (Schmoll et al., 2009): Constitutive activation of GNA3 revealed a strong positive effect of this G protein and hence its cognate signal on *cbh1* transcript levels. Intriguingly, this positive effect was only observed in light. In darkness, neither constitutive activation nor knock-down nor overexpression had any effect.

This light-dependent effect of both GNA1 and GNA3 indicated that a light regulatory factor must be involved in their regulatory output - most likely acting as a checkpoint for signal transmission. First indications as to such a function came from the study on GNA3, which showed an influence of ENVOY (Schmoll et al., 2009). In the following, ENVOY indeed turned out to interfere with transcriptional regulation of both *gna1* and *gna3*, albeit in a different way. It was required for efficient feedback regulation of *gna1*, but not *gna3*, which nevertheless showed enhanced light response in the absence of *env1* (Tisch et al., 2011a). This central position of ENVOY in this regulatory mechanism is reflected by the phenotype of double mutant strains expressing constitutively active G alpha subunits but lacking ENVOY: In darkness, they exhibit the phenotype of constitutive activation of GNA1 or GNA3. In light ENVOY clearly is of major importance, since these strains show the typical poor growth and sporulation as strains lacking *env1*. Consequently, ENVOY represents a major checkpoint of signal transduction via the heterotrimeric G protein pathway in *T. reesei*. The mechanism by which this crucial impact is likely to be mediated involves the regulation of cAMP levels that are adjusted also by ENVOY which seems to contribute to this mechanism by exerting a negative effect on the function of phosphodiesterases.

Knowing that nutrient signalling and light response are connected, a further factor, known to be involved in both, transmission of light signals and the heterotrimeric G protein pathway, came into focus: phosducin like proteins. Phosducins were initially isolated from photoreceptor cells from the retina of mammals (Lee et al., 1984; Lee et al., 1987) and assume the function of a co-chaperone to regulate the correct folding of G beta and gamma (Lukov et al., 2006; Lukov et al., 2005), which makes them the ideal candidate for transduction of the light signal to the G proteins. Interestingly, these proteins had previously not been investigated for a function in light response in fungi.

In *T. reesei*, the class I phosducin like protein PhLP1 and the G protein beta and gamma subunits GNB1 and GNG1 act in the same pathway and additionally perform their function in a light dependent way (Tisch et al., 2011b, ms submitted). Indeed, cellulase gene transcription was found to be positively influenced by PhLP1 and hence G beta and gamma folding. Interestingly, this analysis showed that regulation of glycoside hydrolase transcript levels is among the major targets of PhLP1-GNB1-GNG1. Consequently, not only the major cellulases *cbh1* and *cbh2* (Schmoll et al., 2009; Seibel et al., 2009) are subject to regulation in response to nutrient signals transmitted by heterotrimeric G protein signalling, this important pathway targets carbon source utilization at a broad scale.

Additionally, the operation and output of this pathway is light dependent with the major function in light, but also some impact in darkness. Hence, these results emphasize that for research on carbon source degradation aiming at an improvement of cellulase production, consideration of light dependent processes is of crucial importance. Otherwise, beneficial mechanisms might remain undiscovered or misinterpreted due to an interference of random light pulses with transcript profiles.

### 4.5 The light signalling pathway and the cAMP pathway are interconnected

While initial results clearly confirmed the early steps of signal transmission (the heterotrimeric G protein pathway and the photoreceptors) to be important for adjustment of cellulase levels to be produced, the question remained how the respective signals might reach their target. Hence the gap between the signal being relayed in the upper part of the cascade and regulation of transcription of cellulase genes by transcription factors deserves closer attention.

One of the downstream targets of G protein signalling is the 3′-5′-cyclic adenosine monophosphate (cAMP) pathway (Bölker, 1998; Li et al., 2007). cAMP represents an important secondary messenger used for signalling processes from bacteria to man. In fungi, cAMP plays diverse roles in the cell - it is involved in development, virulence, pathogenicity, sporulation, and light responses (D'Souza & Heitman, 2001). With respect to plant cell wall degradation it is interesting that it was shown already decades ago, that cAMP affects xylanase production in *Cryptococcus albidus* - addition of exogenous cAMP results in a 1.5 to 2 fold increase (Morosoli et al., 1989).

Also for *T. reesei* cAMP is crucial in regulation of plant cell wall degrading enzymes: an early report showed an involvement of cAMP in regulation of cellulase gene expression (Sestak & Farkas, 1993) and is in accordance with the function of GNA3 in regulating cAMP levels and its function in this process (Schmoll et al., 2009). For *T. atroviride* it was shown that the light response pathway crosstalks with the cAMP pathway (Friedl et al., 2008a; Casas-Flores et al., 2006) and in *T. reesei*, GNA3 as well as ENVOY have an effect on cAMP levels (Schmoll et al., 2009; Tisch et al., 2011a). Therefore the cAMP pathway represents an important downstream regulatory pathway to be investigated for improvement of cellulase production.

The most important task of cAMP is to activate cAMP dependent protein kinase A (PKA), which in turn initiates a phosphorylation cascade and activates/inactivates further target genes. cAMP is produced by adenylate cyclase from ATP and degraded by phosphodiesterase (Houslay & Adams, 2003; Pall, 1981). Intriguingly, the steady state levels of intracellular cAMP are subject to a sophisticated fine tuning mechanism: PDE and PKA establish a negative feedback loop (Hicks et al., 2005; Wang et al., 2005), which seems to be targeted by ENV1 via negative impact on PDE in light (Tisch et al., 2011a). Adenylate cyclase is either stimulated by G alpha, beta and gamma subunits or inhibited by G alpha subunits (Ivey et al., 1999; Neves et al., 2002; Schmoll et al., 2009).

In *T. reesei*, the function of the two major components of the cAMP pathway has recently been studied with an emphasis on the involvement in regulation of cellulase gene expression (Schuster et al., 2011, ms submitted). Both adenylate cyclase (ACY1) and protein kinase A (catalytic subunit, PKAC1) were found to be important factors in the cellulase regulon. Intriguingly, both factors enhance the light responsiveness of cellulase gene transcription, i. e. the difference between cellulase transcription in light and darkness. Moreover, while ACY1 exerts a consistently positive effect, PKAC1 acts positively in light and negatively in darkness. Their effect on cellulase transcription is likely mediated by regulation of the transcription factor XYR1 via phosphorylation an upstream factor modulating the abundance of the *xyr1* transcript. Consequently, PKAC1 is a crucial light dependent checkpoint downstream of the heterotrimeric G protein pathway.

This differential effect of components of the cAMP pathway in dependence of light highlights its importance and a connection of this pathway to light dependent processes has been shown previously: Interestingly, cAMP levels and regulatory targets of cAMP react to different light conditions. The operation of adenylate cyclase of *T. reesei*, which is a membrane bound enzyme, is stimulated by light (Kolarova et al., 1992). PKA activity of *T. atroviride* is increased in light and overexpression of the regulatory subunit of PKA results in a blocking of light response genes (Casas-Flores et al., 2006). The protein kinase A of *N. crassa* is a part of the circadian clock by stabilizing the white collar complex (WCC) and the negative regulating element FREQUENCY (Huang et al., 2007; Mehra et al., 2009).

Due to the fact that PKA is involved in the circadian clock in *N. crassa* (Huang et al., 2007) and also for *Trichoderma* a connection between PKA activity and light response was shown (Casas-Flores et al., 2006), a feedback loop is likely (Figure 3). First, light is sensed and transduced by the BLR proteins and ENVOY. In parallel, the G proteins, which regulate

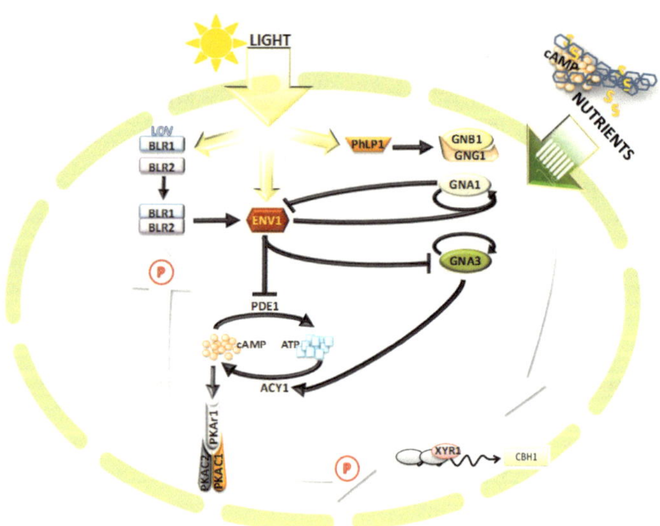


Fig. 3. Model for integration of nutrient signals and light signals regulating cellulase gene expression.

## 5. Nutrient availability and the output of signalling cascades

Early studies on *T. reesei,* which are the basis for the high levels of cellulase production possible today, provide detailed investigations on the nutrients and trace elements beneficial for cellulase induction and secretion. However, the molecular basis of the need and function of these nutrients is largely unknown, as is the interdependence with different environmental conditions.

With respect to the carbon source, it becomes obvious now, that the effect of one environmental cue cannot be considered isolated from other important conditions: Regulation of cellulase gene transcription is stimulated by light on cellulose (Schmoll et al., 2005), but decreased in light on lactose (Schuster et al., 2011, ms submitted).

The cost efficiency for cellulase production prompted investigations about cellulase production on different carbon sources. A very cheap renewable carbon source for cellulase fermentations is lactose, because it is a by-product from cheese production (Kubicek et al., 2009; Seiboth et al., 2007). Most interestingly, although the products of cellulose and lactose conversion are the same (glucose), the influence of light on cellulase transcription is in *T. reesei* is not (Schmoll et al., 2005; Schuster et al., 2011, ms submitted). On cellulose, which resembles the natural substrate, light positively influences cellulase transcription, whereas on lactose, which cannot be considered a natural carbon source, the cellulase transcription is enhanced in darkness.

Although this phenomenon is in agreement with studies already reported decades ago (Carlile, 1965), showing that the influence of light on growth is dependent on the carbon source, the mechanisms triggering this differential response will be intriguing targets of further research.

Only recently a few further studies provide a first glimpse on the relevance of chemicals other than the major carbon source for cellulase gene expression at the molecular level.

The screening of new regulators of cellulase gene transcription identified a protein involved in regulation of sulphur metabolism as a potential signalling factor (Gremel et al., 2008). The respective protein is called LIMPET (encoded by *lim1*) and encodes a WD-repeat/F-box protein, a putative E3 ubiquitin ligase. LIMPET shares similarity with *N. crassa* SCON-2, *A. nidulans* SconB and *S. cerevisiae* Met30p, which are members of a sequential control mechanism for regulation of catabolite repression of sulphur metabolism (Marzluf, 2001). Consequently, analysis of the role of the sulphur source for cellulase gene expression proved reasonable.

Interestingly, the uptake of sulphate was found to be dependent on the light status and even essential for growth of *T. reesei* in light. Moreover, the concentration of methionine, which can be used as alternative sulphur source, is important for regulation of cellulase gene transcription. But most strikingly, addition of 5 mM methionine to the medium abolished transcription of *cbh1* in light and even has a slightly positive effect on *cbh1* in darkness. Consequently, methionine or possibly the sulphur source in general is not only perceived as a nutritional signal. The presence of methionine is of different relevance for cellulase gene expression in light and darkness (Gremel et al., 2008). This surprising result further indicates that similar differences between the effect of a nutrient or even trace element under illumination and in constant darkness may occur. Consequently, routine evaluation of a given medium in open shakers can show effects, which are deleterious for the yield or maybe worse in a dark large scale fermentor or even remain undetected despite highly beneficial.

This study also provides further hints as to the regulatory mechanism of cellulase gene expression. LIMPET represents a component of the ubiquitinylation pathway, which can either activate a protein or destine it for degradation. The reaction of the *lim1* transcript to both cellulase inducing conditions and sulphur limitation indicates an interconnection between the respective regulatory pathways via a mechanism that acts on the cellulase transcription factors in response to changing availability of sulphur sources in the environment. However, the physiological significance of the differential and strong response of cellulase transcription to high levels of methionine in light and darkness remain to be determined.

## 6. Nutrition, reproduction and the rotation of earth – the footprints of evolution in *T. reesei*

It would be reasonable to assume that only nutritional signals would be relevant for the levels of plant cell wall degrading enzymes, needed to sustain growth and development of a fungus. However, an early screening for signal transduction genes involved in regulation of cellulase gene expression, identified a pheromone precursor gene in *T. reesei* (Schmoll et al., 2004). This signal - indicating the presence of a potential mating partner - is transmitted by the heterotrimeric G protein pathway (Bölker & Kahmann, 1993; Li et al., 2007; Xue et al., 2008), which was already shown to be important for cellulase regulation (Schmoll et al., 2009; Seibel et al., 2009; Tisch et al., 2011a).

This is in agreement with the assumption that a shared signalling pathway may also lead to shared output effects. In the following, this pheromone precursor gene (*hpp1*) turned out to represent a so far undescribed class of pheromones (h-type pheromones, in addition to a-type and alpha-type), which - at least in *T. reesei* - assume a-type function (Schmoll et al., 2010). With respect to relevance for cellulase gene expression, analyses are still in progress, and first results indicate that indeed, the availability of the pheromone precursor genes has an influence on regulation of cellulase gene expression (A. Schuster and M. Schmoll, unpublished results). Although the reaction to an extracellular substrate and the detection of a mating partner may seem to be unrelated events, one has to keep in mind that in many cases, deterioration of environmental conditions such as lack of nutrients initiates sexual development in an attempt to improve the genetic equipment for coping with the harsh conditions (Aanen & Hoekstra, 2007). Additionally, the encounter of a mating partner can lead to pheromone induced growth arrest (Skulachev, 2002), which cannot be considered isolated from metabolic regulation. In the special case of *T. reesei* the decision between asexual development (sporulation) or sexual development also has direct implications for cellulase gene expression, since the spores of this fungus become covered with cellulases (Kubicek et al., 1988; Messner & Kubicek, 1991). Therefore the sexual cycle and its relevance for production of plant cell wall degrading enzymes in *T. reesei* warrant further investigations.

The presence of a peptide pheromone precursor gene in *T. reesei*, indicated that sexual development is likely to be possible, although all previous attempts had failed and *T. reesei* had been considered asexual (Kuhls et al., 1996). Indeed, crossing of the original isolate from the Solomon Islands, QM6a, which is also the parental strain of all strains used in research and industry, with a wild-type isolate was achieved. Interestingly, light is required for mating of *T. reesei* (Seidl et al., 2009). Accordingly, also the transcription of the pheromone precursor gene *hpp1* is light responsive. Consequently, it is obvious that *T. reesei* - despite decades of application in research and industry and growth under most unnatural conditions - is still bound to its evolutionary obligations to adjust its life to the availability of nutrient, the need for reproduction and the necessary resources to ensure its success as well as to the rotation of earth by responding to light and darkness.

## 7. Perspectives and challenges

The relevance of signal transduction pathways for regulation of cellulase gene expression as reviewed here only uncovered a small part of the complex network, which determines the adjustment of *T. reesei* to its environment - be it a tropical rainforest, a shake flask or a steel fermentor (Table 2). Nevertheless, these studies also provide guidelines and directions for exploiting the natural signal transduction pathways to cheat *T. reesei* by mimicking conditions requiring huge amounts of cellulases to be produced.

| gene | function | delivered signal | light dependent | effect on cellulase gene transcription | | publication(s) |
|---|---|---|---|---|---|---|
| | | | | light | darkness | |
| *gna3* | G protein alpha subunit 3 | nutrients | yes | positive | no effect | Schmoll et al., 2009 |
| *gna1* | G protein alpha subunit 1 | nutrients | yes | positive | positive | Seibel et al., 2009 |
| *env1* | ENVOY, light regulatory protein | light | yes | positive | positive | Schmoll et al., 2005 |
| *blr1* | Blue light receptor 1 | light | yes | positive | positive | Castellanos et al., 2010 |
| *blr2* | Blue light receptor 2 | light | yes | positive | positive | Castellanos et al., 2010 |
| *phlp1* | Phosducin like protein | nutrients | yes | positive | positive | Tisch et al., 2011b |
| *gnb1* | G protein beta subunit | nutrients / pheromones | yes | positive | positive | Tisch et al., 2011b |
| *gng1* | G protein gamma subunit | nutrients / pheromones | yes | positive | positive | Tisch et al., 2011b |
| *acy1* | adenylate cyclase 1 | nutrients | yes | positive | positive | Schuster et al., 2011 |
| *pkac1* | protein kinase A, catalytic subunit | nutrients | yes | positive | negative | Schuster et al., 2011 |
| *lim1* | LIMPET, putative E3 ubiquitin ligase | sulphur | yes | unknown | unknown | Gremel et al., 2008 |

Table 2. Genes involved in cellulase signalling. Data obtained from indicated publications.

On the one hand there are still numerous signal transduction pathways (Bahn et al., 2007; Schmoll, 2008) and related regulatory mechanisms, which remain to be studied with respect to their impact on plant cell wall degradation. The reports summarized here can only be considered a beginning in unravelling the interdependence of signal transduction pathways, enzyme production and growth. Several more signalling pathways, such as MAP-kinase cascades, two component phosphorelays or calcium signalling still await evaluation with respect to regulation of plant cell wall degrading enzymes. Additionally, mechanisms of posttranslational modification of proteins involved in signal transduction and transcriptional regulation deserves increased attention. It would only be logical if different signal transduction pathways would influence each other by activation, deactivation or degradation of intermittent steps in their cascade.

Reports on the function of two component phosphorelay systems and MAPkinase pathways and their involvement with the circadian clock (Catlett et al., 2003; de Paula et al., 2008; Jones et al., 2007) indicate functions in regulation of metabolic pathways of these signalling cascades. Also recent findings on the functions of the COP9 signalosome in *A. nidulans*, which is a crucial regulator of ubiquitin ligases (Braus et al., 2010; Nahlik et al., 2010) open up excitingly new topics for investigation of cell signalling with the involvement of protein modification. Additionally, the recently described mechanisms of regulation of mRNA stability by decapping (Morozov et al., 2010) may also play a role in posttranscriptional regulation of cellulase gene expression in *T. reesei* (Gyalai-Korpos et al., 2010).

On the other hand, studies on signal transduction pathways and especially light response in recent years, clearly showed that light is an important environmental cue, which cannot be neglected when analyzing *T. reesei*. Even though light response is not a primary field of research when intending to improve industrial production of cellulases, *T. reesei* still obeys to its evolutionary heritage and reacts to light with unexpectedly broad adjustments of metabolic processes, which also involve alterations in transcript levels of plant cell wall degrading enzymes. Consequently, when investigating regulatory processes in cellulase production, establishment of controlled light conditions is equally important as careful preparation of growth media with respect to precise amounts of carbon source, nitrogen source or trace elements. Otherwise, random light effects - comparable to significant impurities in nutrient sources - may cause unreliable interpretations or unpredictable and costly problems when upscaling laboratory cultivations (in transparent shake flasks or fermentors) to the production scale in large, constantly dark steel fermentors.

Despite the intriguing insights into signal transduction pathways of *T. reesei* and their influence on cellulase gene expression, the sometimes significantly increased *cbh1* transcript levels in several mutant strains do not result in equally enhanced efficiency in plant cell wall degradation. So far only few hints as to the reason of this phenomenon are available. Interestingly, several reports indicate that pretranslational regulation of cellulase gene expression may not be the general standard. Therefore, it will be of crucial importance to elucidate the molecular mechanism behind the discrepancy between cellulase regulation on the transcriptional level and efficiency of the secreted plant cell wall degrading enzymes. At the same time, knowledge on this mechanism will uncover important bottlenecks in cellulase production and provide the means to harvest the real potential of *T. reesei* and likely also of other industrially important fungi.

The application of light-dependent results in an industrial setting can be a problem. Large scale steel fermenters can hardly be equipped with sufficiently effective light sources to reproduce laboratory conditions. Therefore one future challenge will be to uncover the interconnections between light response and enzyme expression in more detail to allow for modification of the respective pathways in a way that a light-dependent increase in cellulase production can be realized also in a dark steel fermentor.

It will be interesting to see, whether genes known to impact cellulase gene expression by acting as signal transmitters can still enhance the already efficient industrial production strains derived from the wild-type QM6a. Considering the multiple mutation cycles used for strain improvement, it is very likely that also signalling pathways contributed to the high production of cellulase mixtures, which can be achieved nowadays. Beneficial mutations detected in the course of studies on signal transduction may already be present in producer strains. Nevertheless, more detailed understanding of the mechanism of the complex signalling network in *T. reesei*, can still be used to optimize pathways, which may be perturbed by mutation, identify bottlenecks, improve the use of alternative substrates by targeted adjustment of signalling, and achieve optimal use of resources for most cost efficient cellulase biosynthesis in *T. reesei*.

## 8. Acknowledgements

This work was supported by the Austrian Science Fund FWF to MS (V152-B20 & P21072).

## 9. References


Aanen, D. K. & Hoekstra, R. F. (2007). Why sex is good: on fungi and beyond. In: *Sex in fungi: molecular determination and evolutionary implications.* J. Heitman, J. Kronstad, J. W. Taylor & L. Casselton (eds), pp. ASM Press, ISBN 9781555814212, Washington, D.C.

Bahn, Y. S., Xue, C., Idnurm, A., Rutherford, J. C., Heitman, J. & Cardenas, M. E. (2007). Sensing the environment: lessons from fungi. *Nat Rev Microbiol,* Vol. 5, No. 1, pp. 57-69, ISSN 1740-1534

Bayram, O., Krappmann, S., Ni, M., Bok, J. W., Helmstaedt, K., Valerius, O., Braus-Stromeyer, S., Kwon, N. J., Keller, N. P., Yu, J. H. & Braus, G. H. (2008). VelB/VeA/LaeA complex coordinates light signal with fungal development and secondary metabolism. *Science,* Vol. 320, No. 5882, pp. 1504-1506, ISSN 1095-9203

Blumenstein, A., Vienken, K., Tasler, R., Purschwitz, J., Veith, D., Frankenberg-Dinkel, N. & Fischer, R. (2005). The *Aspergillus nidulans* phytochrome FphA represses sexual development in red light. *Curr Biol,* Vol. 15, No. 20, pp. 1833-1838, ISSN 0960-9822

Bölker, M. (1998). Sex and Crime: Heterotrimeric G Proteins in Fungal Mating and Pathogenesis. *Fungal Genet Biol,* Vol. 25, No. 3, pp. 143-156, ISSN 1087-1845

Bölker, M. & Kahmann, R. (1993). Sexual pheromones and mating responses in fungi. *Plant Cell,* Vol. 5, No. 10, pp. 1461-1469, ISSN 1040-4651

Braus, G. H., Irniger, S. & Bayram, O. (2010). Fungal development and the COP9 signalosome. *Curr Opin Microbiol,* Vol. 13, No. 6, pp. 672-676, ISSN 1879-0364

Brunner, M. & Kaldi, K. (2008). Interlocked feedback loops of the circadian clock of *Neurospora crassa. Mol Microbiol,* Vol. 68, No. 2, pp. 255-262, ISSN 1365-2958

Buchert, J., Oksanen, T., Pere, J., Siika-Aho, M., Suurnäkki, A. & Viikari, L. (1998). Applications of *Trichoderma reesei* enzymes in the pulp and paper industry. In: *Trichoderma and Gliocladium, vol. 2.* G. E. Harman & C. P. Kubicek (eds), pp. 343-363, Taylor & Francis, London, United Kingdom

Carlile, M. J. (1965). The Photobiology of Fungi. *Annu Rev Plant Physiol,* Vol. 16, No. 1, pp. 175 202

Casas-Flores, S., Rios-Momberg, M., Bibbins, M., Ponce-Noyola, P. & Herrera-Estrella, A. (2004). BLR-1 and BLR-2, key regulatory elements of photoconidiation and mycelial growth in *Trichoderma atroviride. Microbiol,* Vol. 150, No. Pt 11, pp. 3561-3569, ISSN 1350-0872

Casas-Flores, S., Rios-Momberg, M., Rosales-Saavedra, T., Martinez-Hernandez, P., Olmedo-Monfil, V. & Herrera-Estrella, A. (2006). Cross talk between a fungal blue-light perception system and the cyclic AMP signaling pathway. *Eukaryot Cell,* Vol. 5, No. 3, pp. 499-506, ISSN 1535-9778

Castellanos, F., Schmoll, M., Martinez, P., Tisch, D., Kubicek, C. P., Herrera-Estrella, A. & Esquivel-Naranjo, E. U. (2010). Crucial factors of the light perception machinery and their impact on growth and cellulase gene transcription in *Trichoderma reesei. Fungal Genet Biol,* Vol. 47, No. 5, pp. 468-476, ISSN 1096-0937

Catlett, N. L., Yoder, O. C. & Turgeon, B. G. (2003). Whole-genome analysis of two-component signal transduction genes in fungal pathogens. *Eukaryot Cell,* Vol. 2, No. 6, pp. 1151-1161, ISSN 1535-9778

Chen, C. H., DeMay, B. S., Gladfelter, A. S., Dunlap, J. C. & Loros, J. J. (2010a). Physical interaction between VIVID and white collar complex regulates photoadaptation in

*Neurospora*. *Proc Natl Acad Sci U S A*, Vol. 107, No. 38, pp. 16715-16720, ISSN 1091-6490

Chen, C. H., Dunlap, J. C. & Loros, J. J. (2010b). *Neurospora* illuminates fungal photoreception. *Fungal Genet Biol*, Vol. 47, No. 11, pp. 922-929, ISSN 1096-0937

Chen, C. H., Ringelberg, C. S., Gross, R. H., Dunlap, J. C. & Loros, J. J. (2009). Genome-wide analysis of light-inducible responses reveals hierarchical light signalling in *Neurospora*. *EMBO J*, Vol. 28, No. 8, pp. 1029-1042, ISSN 1460-2075

Colot, H. V., Park, G., Turner, G. E., Ringelberg, C., Crew, C. M., Litvinkova, L., Weiss, R. L., Borkovich, K. A. & Dunlap, J. C. (2006). A high-throughput gene knockout procedure for *Neurospora* reveals functions for multiple transcription factors. *Proc Natl Acad Sci U S A*, Vol. 103, No. 27, pp. 10352-10357, ISSN 0027-8424

D'Souza, C. A. & Heitman, J. (2001). Conserved cAMP signaling cascades regulate fungal development and virulence. *FEMS Microbiol Rev*, Vol. 25, No. 3, pp. 349-364, ISSN 1574-6976

Davis, R. H. & Perkins, D. D. (2002). Timeline: *Neurospora*: a model of model microbes. *Nat Rev Genet*, Vol. 3, No. 5, pp. 397-403, ISSN 1471-0056

de Paula, R. M., Lamb, T. M., Bennett, L. & Bell-Pedersen, D. (2008). A connection between MAPK pathways and circadian clocks. *Cell Cycle*, Vol. 7, No. 17, pp. 2630-2634, ISSN 1551-4005

Dickman, M. B. & Yarden, O. (1999). Serine/threonine protein kinases and phosphatases in filamentious fungi. *Fungal Genet Biol*, Vol. 26, No. 2, pp. 99-117, ISSN 1087-1845

Eberhart, B., Cross, D. F. & Chase, L. R. (1964). Beta-Glucosidase System of *Neurospora crassa*. I. Beta-Glucosidase and Cellulase Activities of Mutant and Wild-Type Strains. *J Bacteriol*, Vol. 87, No. pp. 761-770, ISSN 0021-9193

Eberhart, B. M., Beck, R. S. & Goolsby, K. M. (1977). Cellulase of *Neurospora crassa*. *J Bacteriol*, Vol. 130, No. 1, pp. 181-186, ISSN 0021-9193

El-Gogary, S., Leite, A., Crivellaro, O., Eveleigh, D. E. & El-Dorry, H. (1989). Mechanism by which cellulose triggers cellobiohydrolase I gene expression in *Trichoderma reesei*. *Proc Natl Acad Sci U S A*, Vol. 86, No. 16, pp. 6138-6141, ISSN 0027-8424

Friedl, M. A., Kubicek, C. P. & Druzhinina, I. S. (2008a). Carbon source dependence and photostimulation of conidiation in *Hypocrea atroviridis*. *Appl Environ Microbiol*, Vol. 74, No. 1, pp. 245-250, ISSN 1098-5336

Friedl, M. A., Schmoll, M., Kubicek, C. P. & Druzhinina, I. S. (2008b). Photostimulation of *Hypocrea atroviridis* growth occurs due to a cross-talk of carbon metabolism, blue light receptors and response to oxidative stress. *Microbiology*, Vol. 154, No. Pt 4, pp. 1229-1241, ISSN 1350-0872

Galagan, J. E., Calvo, S. E., Borkovich, K. A., Selker, E. U., Read, N. D., Jaffe, D., FitzHugh, W., Ma, L.-J., Smirnov, S., Purcell, S., Rehman, B., Elkins, T., Engels, R., Wang, S., Nielsen, C. B., Butler, J., Endrizzi, M., Qui, D., Ianakiev, P., Bell-Pedersen, D., Nelson, M. A., Werner-Washburne, M., Selitrennikoff, C. P., Kinsey, J. A., Braun, E. L., Zelter, A., Schulte, U., Kothe, G. O., Jedd, G., Mewes, W., Staben, C., Marcotte, E., Greenberg, D., Roy, A., Foley, K., Naylor, J., Stange-Thomann, N., Barrett, R., Gnerre, S., Kamal, M., Kamvysselis, M., Mauceli, E., Bielke, C., Rudd, S., Frishman, D., Krystofova, S., Rasmussen, C., Metzenberg, R. L., Perkins, D. D., Kroken, S., Cogoni, C., Macino, G., Catcheside, D., Li, W., Pratt, R. J., Osmani, S. A., DeSouza, C. P. C., Glass, L., Orbach, M. J., Berglund, J. A., Voelker, R., Yarden, O., Plamann,

M., Seiler, S., Dunlap, J., Radford, A., Aramayo, R., Natvig, D. O., Alex, L. A., Mannhaupt, G., Ebbole, D. J., Freitag, M., Paulsen, I., Sachs, M. S., Lander, E. S., Nusbaum, C. & Birren, B. (2003). The genome sequence of the filamentous fungus *Neurospora crassa*. *Nature*, Vol. 422, No. 6934, pp. 859-868, ISSN 0028-0836

Galante, Y., Conti, A. D. & Monteverdi, R. (1998a). Application of *Trichoderma* enzymes in the food and feed industries. In: *Trichoderma and Gliocladium, vol.2*. G. E. Harman & C. P. Kubicek (eds), pp. 327-342, Taylor & Francis, London, United Kingdom

Galante, Y., Conti, A. D. & Monteverdi, R. (1998b). Application of *Trichoderma* enzymes in the textile industry. In: *Trichoderma and Gliocladium, vol. 2*. G. E. Harman & C. P. Kubicek (eds), pp. 311-325, Taylor & Francis, London, United Kingdom

Gomez, L. D., Steele-King, C. G. & McQueen-Mason, S. J. (2008). Sustainable liquid biofuels from biomass: the writing's on the walls. *New Phytol*, Vol. 178, No. 3, pp. 473-485, ISSN 1469-8137

Gremel, G., Dorrer, M. & Schmoll, M. (2008). Sulphur metabolism and cellulase gene expression are connected processes in the filamentous fungus *Hypocrea jecorina* (anamorph *Trichoderma reesei*). *BMC Microbiol*, Vol. 8, No. 1, pp. 174, ISSN 1471-2180

Gyalai-Korpos, M., Nagy, G., Mareczky, Z., Schuster, A., Reczey, K. & Schmoll, M. (2010). Relevance of the light signaling machinery for cellulase expression in *Trichoderma reesei* (*Hypocrea jecorina*). *BMC Res Notes*, Vol. 3, No. pp. 330, ISSN 1756-0500

Hamm, H. E. (1998). The many faces of G protein signaling. *J Biol Chem*, Vol. 273, No. 2, pp. 669-672, ISSN 0021-9258

He, Q. & Liu, Y. (2005). Molecular mechanism of light responses in Neurospora: from light-induced transcription to photoadaptation. *Genes Dev*, Vol. 19, No. 23, pp. 2888-2899, ISSN 0890-9369

Heintzen, C., Loros, J. J. & Dunlap, J. C. (2001). The PAS Protein VIVID Defines a Clock-Associated Feedback Loop that Represses Light Input, Modulates Gating, and Regulates Clock Resetting. *Cell*, Vol. 104, No. 3, pp. 453-464, ISSN 0092-8674

Herrera-Estrella, A. & Horwitz, B. A. (2007). Looking through the eyes of fungi: molecular genetics of photoreception. *Mol Microbiol*, Vol. 64, No. 1, pp. 5-15, ISSN 0950-382X

Hicks, J. K., Bahn, Y. S. & Heitman, J. (2005). Pde1 phosphodiesterase modulates cyclic AMP levels through a protein kinase A-mediated negative feedback loop in *Cryptococcus neoformans*. *Eukaryot Cell*, Vol. 4, No. 12, pp. 1971-1981, ISSN 1535-9778

Hochstrasser, M. (2009). Origin and function of ubiquitin-like proteins. *Nature*, Vol. 458, No. 7237, pp. 422-429, ISSN 1476-4687

Houslay, M. D. & Adams, D. R. (2003). PDE4 cAMP phosphodiesterases: modular enzymes that orchestrate signalling cross-talk, desensitization and compartmentalization. *Biochem J*, Vol. 370, No. Pt 1, pp. 1-18, ISSN 0264-6021

Huang, G., Chen, S., Li, S., Cha, J., Long, C., Li, L., He, Q. & Liu, Y. (2007). Protein kinase A and casein kinases mediate sequential phosphorylation events in the circadian negative feedback loop. *Genes Dev*, Vol. 21, No. 24, pp. 3283-3295, ISSN 0890-9369

Hunt, S. M., Thompson, S., Elvin, M. & Heintzen, C. (2010). VIVID interacts with the WHITE COLLAR complex and FREQUENCY-interacting RNA helicase to alter light and clock responses in *Neurospora*. *Proc Natl Acad Sci U S A*, Vol. 107, No. 38, pp. 16709-16714, ISSN 1091-6490

Ilmén, M., Saloheimo, A., Onnela, M. L. & Penttila, M. E. (1997). Regulation of cellulase gene expression in the filamentous fungus *Trichoderma reesei*. *Appl Environ Microbiol*, Vol. 63, No. 4, pp. 1298-1306, ISSN 0099-2240

Ivey, F. D., Yang, Q. & Borkovich, K. A. (1999). Positive regulation of adenylyl cyclase activity by a G alpha i homolog in *Neurospora crassa*. *Fungal Genet Biol*, Vol. 26, No. 1, pp. 48-61, ISSN 1087-1845

Jones, C. A., Greer-Phillips, S. E. & Borkovich, K. A. (2007). The response regulator RRG-1 functions upstream of a mitogen-activated protein kinase pathway impacting asexual development, female fertility, osmotic stress, and fungicide resistance in *Neurospora crassa*. *Mol Biol Cell*, Vol. 18, No. 6, pp. 2123-2136, ISSN 1059-1524

Kawamori, M., Morikawa, Y. & Takasawa, S. (1985). Inductive formation of celluases by L-sorbose in *Trichoderma reesei*. *Appl Microbiol Biotechnol*, Vol. 22, No. 3, pp. 235-236, ISSN 0175-7598

Kolarova, N., Haplova, J. & Gresik, M. (1992). Light-activated adenyl cyclase from *Trichoderma viride*. *FEMS Microbiol Lett*, Vol. 72, No. 3, pp. 275-278, ISSN 0378-1097

Kosti, I., Mandel-Gutfreund, Y., Glaser, F. & Horwitz, B. A. (2010). Comparative analysis of fungal protein kinases and associated domains. *BMC Genomics*, Vol. 11, No. pp. 133, ISSN 1471-2164

Kubicek, C. P., Mikus, M., Schuster, A., Schmoll, M. & Seiboth, B. (2009). Metabolic engineering strategies for the improvement of cellulase production by *Hypocrea jecorina*. *Biotechnol Biofuels*, Vol. 2, No. pp. 19, ISSN 1754-6834

Kubicek, C. P., Mühlbauer, G., Klotz, M., John, E. & Kubicek-Pranz, E. M. (1988). Properties of a Conidial-bound Cellulase Enzyme System from *Trichoderma reesei*. *J Gen Microbiol*, Vol. 134, No. 5, pp. 1215-1222

Kuhls, K., Lieckfeldt, E., Samuels, G. J., Kovacs, W., Meyer, W., Petrini, O., Gams, W., Borner, T. & Kubicek, C. P. (1996). Molecular evidence that the asexual industrial fungus *Trichoderma reesei* is a clonal derivative of the ascomycete *Hypocrea jecorina*. *Proc Natl Acad Sci U S A*, Vol. 93, No. 15, pp. 7755-7760, ISSN 0027-8424

Kulkarni, R. D., Thon, M. R., Pan, H. & Dean, R. A. (2005). Novel G-protein-coupled receptor-like proteins in the plant pathogenic fungus *Magnaporthe grisea*. *Genome Biol*, Vol. 6, No. 3, pp. R24, ISSN 1465-6914

Lee, R. H., Brown, B. M. & Lolley, R. N. (1984). Light-induced dephosphorylation of a 33K protein in rod outer segments of rat retina. *Biochemistry*, Vol. 23, No. 9, pp. 1972-1977, ISSN 0006-2960

Lee, R. H., Lieberman, B. S. & Lolley, R. N. (1987). A novel complex from bovine visual cells of a 33,000-dalton phosphoprotein with beta- and gamma-transducin: purification and subunit structure. *Biochemistry*, Vol. 26, No. 13, pp. 3983-3990, ISSN 0006-2960

Lendenfeld, T. & Kubicek, C. P. (1998). Characterization and properties of protein kinase C from the filamentous fungus *Trichoderma reesei*. *Biochem J*, Vol. 330 ( Pt 2), No. pp. 689-694, ISSN 0264-6021

Li, L. & Borkovich, K. A. (2006). GPR-4 is a predicted G-protein-coupled receptor required for carbon source-dependent asexual growth and development in *Neurospora crassa*. *Eukaryot Cell*, Vol. 5, No. 8, pp. 1287-1300, ISSN 1535-9778

Li, L., Wright, S. J., Krystofova, S., Park, G. & Borkovich, K. A. (2007). Heterotrimeric G protein signaling in filamentous fungi. *Annu Rev Microbiol*, Vol. 61, No. pp. 423-452, ISSN 0066-4227

Linden, H. & Macino, G. (1997). White collar 2, a partner in blue-light signal transduction, controlling expression of light-regulated genes in *Neurospora crassa*. *EMBO J*, Vol. 16, No. 1, pp. 98-109, ISSN 0261-4189

Linden, H., Rodriguez-Franco, M. & Macino, G. (1997). Mutants of *Neurospora crassa* defective in regulation of blue light perception. *Mol Gen Genet*, Vol. 254, No. 2, pp. 111-118, ISSN 0026-8925

Lukov, G. L., Baker, C. M., Ludtke, P. J., Hu, T., Carter, M. D., Hackett, R. A., Thulin, C. D. & Willardson, B. M. (2006). Mechanism of assembly of G protein betagamma subunits by protein kinase CK2-phosphorylated phosducin-like protein and the cytosolic chaperonin complex. *J Biol Chem*, Vol. 281, No. 31, pp. 22261-22274, ISSN 0021-9258

Lukov, G. L., Hu, T., McLaughlin, J. N., Hamm, H. E. & Willardson, B. M. (2005). Phosducin-like protein acts as a molecular chaperone for G protein betagamma dimer assembly. *EMBO J*, Vol. 24, No. 11, pp. 1965-1975, ISSN 0261-4189

Mach, R. L., Zeilinger, S., Kristufek, D. & Kubicek, C. P. (1998). Ca2+-calmodulin antagonists interfere with xylanase formation and secretion in *Trichoderma reesei*. *Biochim Biophys Acta*, Vol. 1403, No. 3, pp. 281-289, ISSN 0006-3002

Mandels, M., Parrish, F. W. & Reese, E. T. (1962). Sophorose as an inducer of cellulase in *Trichoderma viride*. *J Bacteriol*, Vol. 83, No. pp. 400-408, ISSN 0021-9193

Mandels, M. & Reese, E. T. (1957). Induction of cellulase in *Trichoderma viride* as influenced by carbon sources and metals. *J Bacteriol*, Vol. 73, No. 2, pp. 269-278, ISSN 0021-9193

Mandels, M. & Reese, E. T. (1960). Induction of cellulase in fungi by cellobiose. *J Bacteriol*, Vol. 79, No. pp. 816-826, ISSN 0021-9193

Margolles-Clark, E., Ihnen, M. & Penttilä, M. (1997). Expression patterns of ten hemicellulase genes of the filamentous fungus *Trichoderma reesei* on various carbon sources. *J Biotechnol*, Vol. 57, No. 1-3, pp. 167-179, ISSN 0168-1656

Martinez, D., Berka, R. M., Henrissat, B., Saloheimo, M., Arvas, M., Baker, S. E., Chapman, J., Chertkov, O., Coutinho, P. M., Cullen, D., Danchin, E. G., Grigoriev, I. V., Harris, P., Jackson, M., Kubicek, C. P., Han, C. S., Ho, I., Larrondo, L. F., de Leon, A. L., Magnuson, J. K., Merino, S., Misra, M., Nelson, B., Putnam, N., Robbertse, B., Salamov, A. A., Schmoll, M., Terry, A., Thayer, N., Westerholm-Parvinen, A., Schoch, C. L., Yao, J., Barabote, R., Nelson, M. A., Detter, C., Bruce, D., Kuske, C. R., Xie, G., Richardson, P., Rokhsar, D. S., Lucas, S. M., Rubin, E. M., Dunn-Coleman, N., Ward, M. & Brettin, T. S. (2008). Genome sequencing and analysis of the biomass-degrading fungus *Trichoderma reesei* (syn. *Hypocrea jecorina*). *Nat Biotechnol*, Vol. 26, No. 5, pp. 553-560, ISSN 1546-1696

Marzluf, G. A. (2001). Metabolic Regulation in Fungi. In: *Applied Mycology and Biotechnology*. G. K. George & K. A. Dilip (eds), pp. 55-72, Elsevier, ISBN 1874-5334

Mehra, A., Baker, C. L., Loros, J. J. & Dunlap, J. C. (2009). Post-translational modifications in circadian rhythms. *Trends Biochem Sci*, Vol. 34, No. 10, pp. 483-490, ISSN 0968-0004

Messner, R., Gruber, F. & Kubicek, C. P. (1988). Differential regulation of synthesis of multiple forms of specific endoglucanases by *Trichoderma reesei* QM9414. *J Bacteriol*, Vol. 170, No. 8, pp. 3689-3693, ISSN 0021-9193

Messner, R., Kubicek-Pranz, E. M., Gsur, A. & Kubicek, C. P. (1991). Cellobiohydrolase II is the main conidial-bound cellulase in *Trichoderma reesei* and other *Trichoderma* strains. *Arch Microbiol*, Vol. 155, No. 6, pp. 601-606, ISSN 0302-8933

Messner, R. & Kubicek, C. P. (1991). Carbon Source Control of Cellobiohydrolase I and II Formation by *Trichoderma reesei. Appl Environ Microbiol*, Vol. 57, No. 3, pp. 630-635, ISSN 0099-2240

Mischak, H., Hofer, F., Messner, R., Weissinger, E., Hayn, M., Tomme, P., Esterbauer, H., Kuchler, E., Claeyssens, M. & Kubicek, C. P. (1989). Monoclonal antibodies against different domains of cellobiohydrolase I and II from *Trichoderma reesei. Biochim Biophys Acta*, Vol. 990, No. 1, pp. 1-7, ISSN 0006-3002

Mooney, J. L. & Yager, L. N. (1990). Light is required for conidiation in *Aspergillus nidulans. Genes Dev*, Vol. 4, No. 9, pp. 1473-1482, ISSN 0890-9369

Morawetz, R., Gruber, F., Messner, R. & Kubicek, C. P. (1992). Presence, transcription and translation of cellobiohydrolase genes in several *Trichoderma* species. *Current Genetics*, Vol. 21, No. 1, pp. 31-36, ISSN 0172-8083

Morawetz, R., Lendenfeld, T., Mischak, H., Muhlbauer, M., Gruber, F., Goodnight, J., de Graaff, L. H., Visser, J., Mushinski, J. F. & Kubicek, C. P. (1996). Cloning and characterisation of genes (*pkc1* and *pkcA*) encoding protein kinase C homologues from *Trichoderma reesei* and *Aspergillus niger. Mol Gen Genet*, Vol. 250, No. 1, pp. 17-28, ISSN 0026-8925

Morosoli, R., Durand, S. & Boucher, F. (1989). Stimulation of xylanase synthesis in *Cryptococcus albidus* by cyclic AMP. *FEMS Microbiol Lett*, Vol. 57, No. 1, pp. 57-60, ISSN 0378-1097

Morozov, I. Y., Jones, M. G., Razak, A. A., Rigden, D. J. & Caddick, M. X. (2010). CUCU modification of mRNA promotes decapping and transcript degradation in *Aspergillus nidulans. Mol Cell Biol*, Vol. 30, No. 2, pp. 460-469, ISSN 1098-5549

Nahlik, K., Dumkow, M., Bayram, O., Helmstaedt, K., Busch, S., Valerius, O., Gerke, J., Hoppert, M., Schwier, E., Opitz, L., Westermann, M., Grond, S., Feussner, K., Goebel, C., Kaever, A., Meinicke, P., Feussner, I. & Braus, G. H. (2010). The COP9 signalosome mediates transcriptional and metabolic response to hormones, oxidative stress protection and cell wall rearrangement during fungal development. *Mol Microbiol*, Vol. 78, No. 4, pp. 964-979, ISSN 1365-2958

Naik, S. N., Goud, V. V., Rout, P. K. & Dalai, A. K. (2010). Production of first and second generation biofuels: A comprehensive review. *Renewable Sustain. Energy Rev*, Vol. 14, No. 2, pp. 578-597, ISSN 1364-0321

Neves, S. R., Ram, P. T. & Iyengar, R. (2002). G protein pathways. *Science*, Vol. 296, No. 5573, pp. 1636-1639, ISSN 1095-9203

Pall, M. L. (1981). Adenosine 3',5'-phosphate in fungi. *Microbiol Rev*, Vol. 45, No. 3, pp. 462-480, ISSN 0146-0749

Purschwitz, J., Müller, S., Kastner, C., Schöser, M., Haas, H., Espeso, E. A., Atoui, A., Calvo, A. M. & Fischer, R. (2008). Functional and Physical Interaction of Blue- and Red-Light Sensors in *Aspergillus nidulans. Current Biol*, Vol. 18, No. 4, pp. 255-259, ISSN 0960-9822

Rauscher, R., Wurleitner, E., Wacenovsky, C., Aro, N., Stricker, A. R., Zeilinger, S., Kubicek, C. P., Penttila, M. & Mach, R. L. (2006). Transcriptional regulation of *xyn1*, encoding xylanase I, in *Hypocrea jecorina. Eukaryot Cell*, Vol. 5, No. 3, pp. 447-456, ISSN 1535-9778

Rechsteiner, M. & Rogers, S. W. (1996). PEST sequences and regulation by proteolysis. *Trends Biochem Sci*, Vol. 21, No. 7, pp. 267-271, ISSN 0968-0004

Reese, E. T. (1976). History of the cellulase program at the U.S. army Natick Development Center. *Biotechnol Bioeng Symp*, Vol. No. 6, pp. 9-20, ISSN 0572-6565

Ruiz-Roldan, M. C., Garre, V., Guarro, J., Marine, M. & Roncero, M. I. (2008). Role of the white collar 1 photoreceptor in carotenogenesis, UV resistance, hydrophobicity, and virulence of *Fusarium oxysporum*. *Eukaryot Cell*, Vol. 7, No. 7, pp. 1227-1230, ISSN 1535-9786

Schmoll, M. (2008). The information highways of a biotechnological workhorse--signal transduction in *Hypocrea jecorina*. *BMC Genomics*, Vol. 9, No. pp. 430, ISSN 1471-2164

Schmoll, M., Franchi, L. & Kubicek, C. P. (2005). Envoy, a PAS/LOV domain protein of *Hypocrea jecorina* (anamorph *Trichoderma reesei*), modulates cellulase gene transcription in response to light. *Eukaryot Cell*, Vol. 4, No. 12, pp. 1998-2007, ISSN 1535-9778

Schmoll, M. & Kubicek, C. P. (2003). Regulation of *Trichoderma* cellulase formation: lessons in molecular biology from an industrial fungus. A review. *Acta Microbiol Immunol Hung*, Vol. 50, No. 2-3, pp. 125-145, ISSN 1217-8950

Schmoll, M. & Kubicek, C. P. (2005). *ooc1*, a unique gene expressed only during growth of *Hypocrea jecorina* (anamorph: *Trichoderma reesei*) on cellulose. *Curr Genet*, Vol. 48, No. 2, pp. 126-133, ISSN 0172-8083

Schmoll, M., Schuster, A., Silva Rdo, N. & Kubicek, C. P. (2009). The G-alpha protein GNA3 of *Hypocrea jecorina* (anamorph *Trichoderma reesei*) regulates cellulase gene expression in the presence of light. *Eukaryot Cell*, Vol. 8, No. 3, pp. 410-420, ISSN 1535-9786

Schmoll, M., Seibel, C., Tisch, D., Dorrer, M. & Kubicek, C. P. (2010). A novel class of peptide pheromone precursors in ascomycetous fungi. *Mol Microbiol*, Vol. 77, No. 6, pp. 1483-1501, ISSN 1365-2958

Schmoll, M., Esquivel-Naranjo, E. U. & Herrera-Estrella, A. (2010). *Trichoderma* in the light of day - physiology and development. Fungal Genet Biol, Vol. 47, No. 11, pp. 909-916

Schmoll, M., Tian, C., Sun, J., Tisch, D. & Glass, L. (2011). Unravelling the molecular basis for light modulated cellulase gene expression - the role of photoreceptors in *Neurospora crassa*. *Manuscript submitted*

Schmoll, M., Zeilinger, S., Mach, R. L. & Kubicek, C. P. (2004). Cloning of genes expressed early during cellulase induction in *Hypocrea jecorina* by a rapid subtraction hybridization approach. *Fungal Genet Biol*, Vol. 41, No. 9, pp. 877-887, ISSN 1087-1845

Schuster, A., Kubicek, C. P., Friedl, M. A., Druzhinina, I. S. & Schmoll, M. (2007). Impact of light on *Hypocrea jecorina* and the multiple cellular roles of ENVOY in this process. *BMC Genomics*, Vol. 8, No. pp. 449, ISSN 1471-2164

Schuster, A. & Schmoll, M. (2010). Biology and biotechnology of *Trichoderma*. *Appl Microbiol Biotechnol*, Vol. 87, No. 3, pp. 787-799, ISSN 1432-0614

Schuster, A., Tisch, D., Seidl-Seiboth, V., Kubicek, C. P. & Schmoll, M. (2011). cAMP signaling is involved in light-modulated cellulase regulation in *Trichoderma reesei*. *Manuscript submitted*

Schwerdtfeger, C. & Linden, H. (2003). VIVID is a flavoprotein and serves as a fungal blue light photoreceptor for photoadaptation. *EMBO J*, Vol. 22, No. 18, pp. 4846-4855, ISSN 0261-4189

Seibel, C., Gremel, G., do Nascimento Silva, R., Schuster, A., Kubicek, C. P. & Schmoll, M. (2009). Light-dependent roles of the G-protein alpha subunit GNA1 of *Hypocrea jecorina* (anamorph *Trichoderma reesei*). *BMC Biol*, Vol. 7, No. pp. 58, ISSN 1741-7007

Seiboth, B., Pakdaman, B. S., Hartl, L. & Kubicek, C. P. (2007). Lactose metabolism in filamentous fungi: how to deal with an unknown substrate. *Fungal Biol. Rev*, Vol. 21, No. 1, pp. 42-48, ISSN 1749-4613

Seidl, V., Seibel, C., Kubicek, C. P. & Schmoll, M. (2009). Sexual development in the industrial workhorse *Trichoderma reesei*. *Proc Natl Acad Sci U S A*, Vol. 106, No. 33, pp. 13909-13914, ISSN 1091-6490

Seidl, V. & Seiboth, B. (2010). *Trichoderma reesei*: genetic approaches to improving strain efficiency. *Biofuels*, Vol. 1, No. 2, pp. 343-354

Sestak, S. & Farkas, V. (1993). Metabolic regulation of endoglucanase synthesis in *Trichoderma reesei*: participation of cyclic AMP and glucose-6-phosphate. *Can J Microbiol*, Vol. 39, No. 3, pp. 342-347, ISSN 0008-4166

Shoemaker, S., Schweickart, V., Ladner, M., Gelfand, D., Kwok, S., Myambo, K. & Innis, M. (1983a). Molecular Cloning of Exo-Cellobiohydrolase I Derived from *Trichoderma Reesei* Strain L27. *Nat Biotech*, Vol. 1, No. 8, pp. 691-696

Shoemaker, S., Watt, K., Tsitovsky, G. & Cox, R. (1983b). Characterization and Properties of Cellulases Purified from *Trichoderma Reesei* Strain L27. *Nat Biotech*, Vol. 1, No. 8, pp. 687-690

Shrode, L. B., Lewis, Z. A., White, L. D., Bell-Pedersen, D. & Ebbole, D. J. (2001). *vvd* is required for light adaptation of conidiation-specific genes of *Neurospora crassa*, but not circadian conidiation. *Fungal Genet Biol*, Vol. 32, No. 3, pp. 169-181, ISSN 1087-1845

Skulachev, V. P. (2002). Programmed death in yeast as adaptation? *FEBS Lett*, Vol. 528, No. 1-3, pp. 23-26, ISSN 0014-5793

Smith, K. M., Sancar, G., Dekhang, R., Sullivan, C. M., Li, S., Tag, A. G., Sancar, C., Bredeweg, E. L., Priest, H. D., McCormick, R. F., Thomas, T. L., Carrington, J. C., Stajich, J. E., Bell-Pedersen, D., Brunner, M. & Freitag, M. (2010). Transcription factors in light and circadian clock signaling networks revealed by genomewide mapping of direct targets for *Neurospora* white collar complex. *Eukaryot Cell*, Vol. 9, No. 10, pp. 1549-1556, ISSN 1535-9786

Solomon, B. D. (2010). Biofuels and sustainability. *Ann N Y Acad Sci*, Vol. 1185, No. pp. 119-134, ISSN 1749-6632

Somerville, C. (2007). Biofuels. *Curr Biol*, Vol. 17, No. 4, pp. R115-119, ISSN 0960-9822

Sprang, S. R. (1997). G protein mechanisms: insights from structural analysis. *Annu Rev Biochem*, Vol. 66, No. pp. 639-678, ISSN 0066-4154

Sternberg, D. & Mandels, G. R. (1979). Induction of cellulolytic enzymes in *Trichoderma reesei* by sophorose. *J Bacteriol*, Vol. 139, No. 3, pp. 761-769, ISSN 0021-9193

Sternberg, D. & Mandels, G. R. (1980). Regulation of the cellulolytic system in *Trichoderma reesei* by sophorose: induction of cellulase and repression of beta-glucosidase. *J Bacteriol*, Vol. 144, No. 3, pp. 1197-1199, ISSN 0021-9193

Stricker, A. R., Mach, R. L. & de Graaff, L. H. (2008a). Regulation of transcription of cellulases- and hemicellulases-encoding genes in *Aspergillus niger* and *Hypocrea jecorina* (*Trichoderma reesei*). *Appl Microbiol Biotechnol*, Vol. 78, No. 2, pp. 211-220, ISSN 0175-7598

Stricker, A. R., Trefflinger, P., Aro, N., Penttila, M. & Mach, R. L. (2008b). Role of Ace2 (Activator of Cellulases 2) within the xyn2 transcriptosome of *Hypocrea jecorina*. *Fungal Genet Biol*, Vol. 45, No. 4, pp. 436-445, ISSN 1096-0937

Talora, C., Franchi, L., Linden, H., Ballario, P. & Macino, G. (1999). Role of a white collar-1-white collar-2 complex in blue-light signal transduction. *EMBO J*, Vol. 18, No. 18, pp. 4961-4968, ISSN 0261-4189

Teeri, T. T., Lehtovaara, P., Kauppinen, S., Salovuori, I. & Knowles, J. (1987). Homologous domains in *Trichoderma reesei* cellulolytic enzymes: gene sequence and expression of cellobiohydrolase II. *Gene*, Vol. 51, No. 1, pp. 43-52, ISSN 0378-1119

Tian, C., Beeson, W. T., Iavarone, A. T., Sun, J., Marletta, M. A., Cate, J. H. & Glass, N. L. (2009). Systems analysis of plant cell wall degradation by the model filamentous fungus *Neurospora crassa*. *Proc Natl Acad Sci U S A*, Vol. 106, No. 52, pp. 22157-22162, ISSN 1091-6490

Tisch, D., Kubicek, C. P. & Schmoll, M. (2011a). New insights into the mechanism of light modulated signaling by heterotrimeric G-proteins: ENVOY acts on *gna1* and *gna3* and adjusts cAMP levels in *Trichoderma reesei* (*Hypocrea jecorina*). *Fungal Genet Biol*, Vol. No. pp. ISSN 1096-0937

Tisch, D., Kubicek, C. P. & Schmoll, M. (2011b). The phosducin like protein PhLP1 is an important factor in light response and regulation of glycoside hydrolases in *Trichoderma reesei* (*Hypocrea jecorina*). *Manuscript submitted*

Tisch, D. & Schmoll, M. (2010). Light regulation of metabolic pathways in fungi. *Appl Microbiol Biotechnol*, Vol. 85, No. 5, pp. 1259-1277, ISSN 1432-0614

Torigoi, E., Henrique-Silva, F., Escobar-Vera, J., Carle-Urioste, J. C., Crivellaro, O., El-Dorry, H. & El-Gogary, S. (1996). Mutants of *Trichoderma reesei* are defective in cellulose induction, but not basal expression of cellulase-encoding genes. *Gene*, Vol. 173, No. 2, pp. 199-203, ISSN 0378-1119

Vaheri, M., Leisola, M. & Kauppinen, V. (1979). Transglycosylation products of cellulase system of *Trichoderma reesei*. *Biotechnol Lett*, Vol. 1, No. 1, pp. 41-46, ISSN 0141-5492

Wang, L., Griffiths, K., Jr., Zhang, Y. H., Ivey, F. D. & Hoffman, C. S. (2005). *Schizosaccharomyces pombe* adenylate cyclase suppressor mutations suggest a role for cAMP phosphodiesterase regulation in feedback control of glucose/cAMP signaling. *Genetics*, Vol. 171, No. 4, pp. 1523-1533, ISSN 0016-6731

Wang, P. & Nuss, D. L. (1995). Induction of a *Cryphonectria parasitica* cellobiohydrolase I gene is suppressed by hypovirus infection and regulated by a GTP-binding-protein-linked signaling pathway involved in fungal pathogenesis. *Proc Natl Acad Sci U S A*, Vol. 92, No. 25, pp. 11529-11533, ISSN 0027-8424

Wilson, D. B. (2009). Cellulases and biofuels. *Curr Opin Biotechnol*, Vol. 20, No. 3, pp. 295-299, ISSN 1879-0429

Xue, C., Hsueh, Y.-P. & Heitman, J. (2008). Magnificent seven: roles of G protein-coupled receptors in extracellular sensing in fungi. *FEMS Microbiol Rev*, Vol. 32, No. 6, pp. 1010-1032, ISSN 1574-6976

Zeilinger, S., Haller, M., Mach, R. & Kubicek, C. P. (2000). Molecular characterization of a cellulase-negative mutant of *Hypocrea jecorina*. *Biochem Biophys Res Commun*, Vol. 277, No. 3, pp. 581-588, ISSN 0006-291X

Zeilinger, S., Mach, R. L. & Kubicek, C. P. (1998). Two adjacent protein binding motifs in the *cbh2* (cellobiohydrolase II-encoding) promoter of the fungus *Hypocrea jecorina*

(*Trichoderma reesei*) cooperate in the induction by cellulose. *J Biol Chem*, Vol. 273, No. 51, pp. 34463-34471, ISSN 0021-9258

Zeilinger, S., Schmoll, M., Pail, M., Mach, R. L. & Kubicek, C. P. (2003). Nucleosome transactions on the *Hypocrea jecorina* (*Trichoderma reesei*) cellulase promoter *cbh2* associated with cellulase induction. *Mol Genet Genomics*, Vol. 270, No. 1, pp. 46-55, ISSN 1617-4615

10

# Detoxification of Lignocellulosic Hydrolysates for Improved Bioethanol Production

Anuj K. Chandel[1], Silvio Silvério da Silva[1] and Om V. Singh[2]
[1]*Department of Biotechnology, School of Engineering of Lorena, University of Sao Paulo*
[2]*Division of Biological and Health Sciences, University of Pittsburgh, Bradford*
[1]*Brazil*
[2]*USA*

## 1. Introduction

Lignocelluloses are the most abundant raw materials on Earth comprised of cellulose, hemicelluloses and lignin. After cellulose, hemicellulose is the principal fraction of the plant cell wall that could serve as a potential substrate for the production of value-added products under optimized conditions [Chandel & Singh, 2011; Chandel et al., 2010a; Hahn-Hagerdal et al., 2007; Saha, 2003]. Largely, the secondary cell wall of plants contains cellulose (40–80%), hemicellulose (10–40%), and lignin (5–25%). The carbohydrate fraction of the plant cell wall can be converted into fermentable monomeric sugars through acidic and/or enzymatic (hemicellulase/cellulase) reactions, which have been exploited to produce ethanol, xylitol, n-butanol and 2, 3-butanediol via microbial fermentation processes [Sun, 2009.; Chandel et al., 2010a; Carvalheiro et al., 2005; Saha, 2003]. Until now the pretreatment is unavoidable necessity, which has been examined and employed extensively in the past [Moiser et al., 2005, Taherzadeh & Karimi 2007; Kumar et al., 2009; Chandel et al., 2010b]. The acidic pretreatment of lignocellulosics hydrolyzes the hemicellulose fraction, enabling subsequent enzymatic digestion of the cellulose in fermentation reaction [Kumar et al., 2009; Chandel et al., 2007a; Chandel et al., 2007b; Chandel et al., 2007c]. However, the non-specificity of acidic treatment led to the formation of complex sugars and compounds inhibitory to the microorganisms for ethanol production [Parawira & Tekere, 2011].

The depolymerization of hemicellulose by chemical process yields xylose as the major fraction and arabinose, mannose, galactose, and glucose in smaller fractions in addition to potential microbial inhibitors [Chandel et al., 2010a; Círio et al., 2010; Chandel et al., 2009; Chandel et al., 2007a]. These inhibitors can be divided into three major groups (Fig. 1), i.e. organic acids (acetic, formic and levulinic acids), ii. furan derivatives [furfural and 5-hydroxymethylfurfural (5-HMF)], iii. phenolic compounds [Chandel et al., 2010a; Chandel et al, 2007b; Mussatto and Roberto, 2004; Palmqvist and Hahn-Hagerdal, 2000a], affecting overall cell physiology and often result in decreased viability, ethanol yield, and productivity [Chandel et al, 2007a; Palmqvist & Hahn-Hagerdal, 2000a]. The ethanologenic microorganisms have ability to degrade some of the inhibitors; however, the toxicity of hydrolysate was determined by the aggregate effect of compounds [Mussatto and Roberto, 2004; Zaldivar et al., 2001]. Progress has been made to achieve higher levels of sugars by diminishing the overall impact of fermentative inhibitors which in-turn improves the fermentability of lignocellulosic hydrolysates [Alriksson et al., 2011; Sun & Liu, 2011;

Parawira & Tekere, 2011]. The ion exchange resins, active charcoal, enzymatic detoxification using laccase, alkali treatments and overliming with calcium hydroxide are among selective detoxification strategies which have been investigated in the past [Jurado et al., 2009; Chandel et al., 2007a, b, c; Villarreal. et al., 2006]. Other strategies include changes in fermentation methodologies and metabolic engineering (incorporation of ligninase or laccase genes) have been introduced to overcome from the cell wall degrading inhibitors [Larsson et al., 2001]. Treatment with the soft-rot fungus Trichoderma reesei and other microorganisms to degrade inhibitors in a hemicellulase hydrolysate has also been proposed [Yu et al., 2011; Fonesca et al., 2011; Okuda et al., 2008; Tian et al., 2009; López et al., 2004]. This chapter aims to discuss the detoxification strategies which may assist to overcome the fermentation inhibitors of lignocellulosics. The biological pretreatment of lignocellulosic raw materials prior to their hydrolysis and the approaches of biotechnology routes making them resistant towards the inhibitors have also discussed.

## 2. Inhibitors profile and lignocellulose hydrolysates

Thermochemical degradation of hemicellulose liberates majority of sugar monomers i.e. xylose, mannose, acetic acid, galactose, and glucose, in conjunction with number of inhibitors toxic to the fermenting microorganism (Fig. 1, 2). These inhibitors include furans (furfurals and 5-Hydroxy methyl furfural (5-HMF)), phenolics, weak acids (acetic acid, levulinic acid, formic acid etc.), raw material extractives (acidic resins, tannic and terpene acids), and heavy metal ions (iron, chromium, nickel and copper) (Fig. 1). Among inhibitors Hibbert's ketones have also been noticed in the acid hydrolysates of pine and spruce [Klinke et al., 2004; Clarck & Mackie, 1984]. Table 1 summarizes the inhibitors profile derived from variety of lignocellulosic materials.

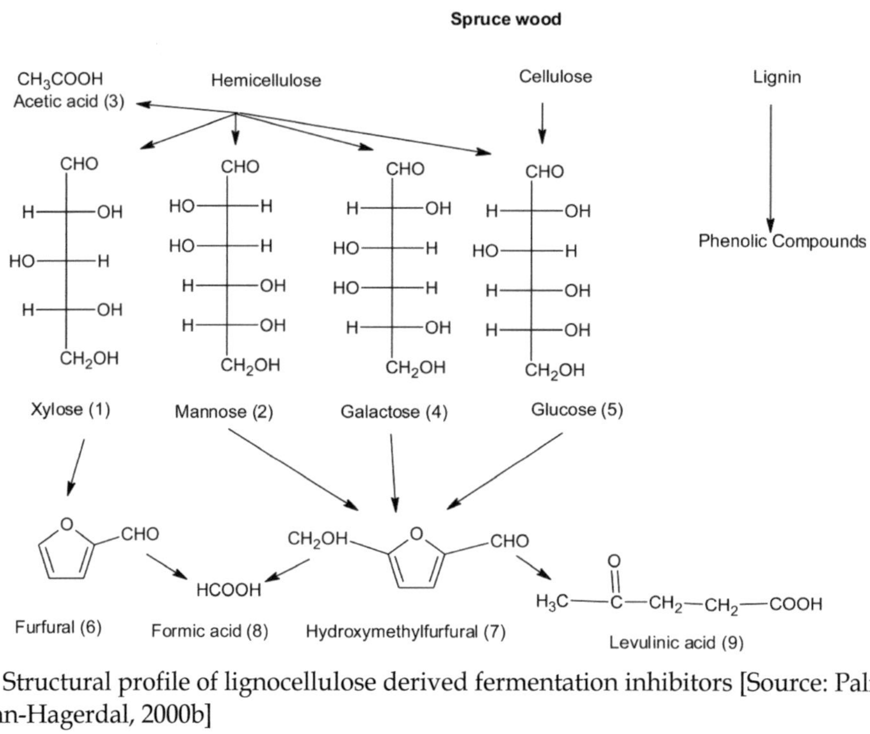


Fig. 1. Structural profile of lignocellulose derived fermentation inhibitors [Source: Palmqvist & Hahn-Hagerdal, 2000b]

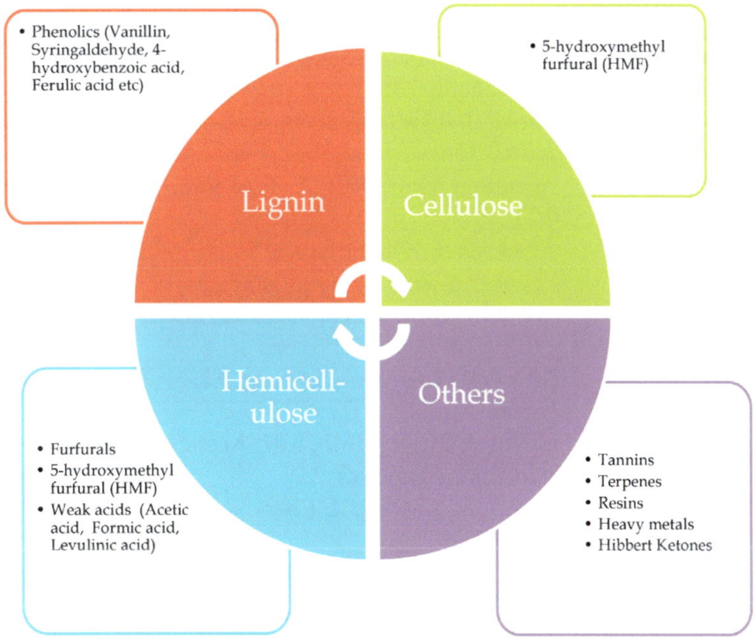


Fig. 2. Plant cell wall derived inhibitors. Cell wall component yields a variety of inhibitors in addition to sugar monomers after acid hydrolysis of lignocellulosic biomass at high temperature. Hemicellulose and cellulose breakdown releases weak acids, furans (5-HMF, furfurals) and 5-HMF respectively. Specialized plant cell wall components yield terpenes, tannins, Hibbert Ketones etc. A variety of phenolic monomers (syringaldehyde, vanillin, ferulic acid etc.) are generated from lignin in conjunction with hemicelluloses and cellulose-derived inhibitors.

Degradation of xylose at higher temperature and pressure reveal furfurals inhibitor. 5-(Hydroxymethyl) fufural (5-HMF) forms during hexose degradation, however the concentration of HMF in acid hydrolysates tends to low due to limited degradation of hexose in acidic saccharification. Among inhibitors most phenolics are being generated from lignin breakdown. These inhibitors include three forms of organic compounds such as acid, ketone and aldehyde (i.e. example, syringaldehyde, syringic acid and syringone, vanillic acid, vanillone, and vanildehyde) [Parawira & Tekere, 2011; Mussatto & Roberto, 2004; Palmqvist & Hahn-Hagerdal, 2000a, b]. 4-Hydroxy benzoic acid, ferulic acid and guiacol are among most common lignin derived inhibitors observed in lignocellulose acid hydrolysates [Klinke et al., 2004]. Raw material extractives are generated due to the presence of tannic acid, terpenes and other kind of ploymers present in plant cell walls [Mussatto & Roberto, 2004]. The heavy metal ions could form due to the corrosion of reaction vessel used for hydrolysis reactions or other chemical moieties [Mussatto & Roberto, 2004; Parajo et al., 1996; Watson et al., 1984]. The monomeric sugars and cell wall derived components of sugarcane bagasse, a model lignocellulosic substrate, has been summarized in Table 2. The formation of cell wall degrading components in the acid hydrolysate are depend upon multiple factors including nature of lignocellulosic material and its cell wall composition, thermochemical conditions, and reaction time of the hydrolysis [Palmqvist & Hahn-Hagerdal, 2000a, b]. These components are required to remove from the lignocellulose hydrolysates prior to the fermentation.

| Lignocellulosic material | Inhibitors profile (g/l) | References |
|---|---|---|
| Sugarcane bagasse | Furans, 1.89; Phenolics, 2.75; Acetic acid, 5.45 | Chandel et al., 2007[a] |
| Wheat straw | Furfural, 0.15±0.02; acetic acid, 2.70±0.33 | Nigam, 2001 |
| Rice straw | Acetate, 1.43; HMF, 0.15; Furfural, 0.25 | Baek & Kwon, 2007 |
| Corn stover | Acetic acid, 1.48; Furans, 0.56; Phenolics, 0.08 | Cao et al., 2009 |
| Spruce | Phenolics, 0.44 ± 0.05; Furfural, 1.0 ± 0.1; HMF,3.3 ± 0.2; Acetic acid, 5.0 ± 0.4; Levulinic acid, 0.2 ± 0.1; Formic acid, 0.7 ± 0.1 | Alriksson et al., 2010 |
| *Eucalyptus globulus* | Furfural, 0.26; 5-HMF, 0.07; Acetic acid, 3.41; Phenolics, 2.23 | Villarreal et al., 2006 |
| *Saccharum spontaneum* | Furfurals, 1.54 ± 0.04; Phenolics, 2.01 ± 0.08 | Chandel et al., 2011a |
| Poplar | 2-furoic acid, 0.3 microgram/g; 3,4-HBA, 2.5; Salicylic acid, 56; Syringaldehyde, 6.0; Ferulic acid, 4.7 | Balan et al., 2009 |
| Soft wood | Acetic acid, 5.3; Furfural, 2.2 | Qian et al., 2006 |

Table 1. Plant cell wall derived inhibitors profile from different lignocellulosic substrates

| HCl (%) | Xylose (g/l) | Arabinose (g/l) | Glucose (g/l) | Total sugars (g/l) | Total furans (g/l) | Total phenolics (g/l) | Acetic acid (g/l) |
|---|---|---|---|---|---|---|---|
| 0.5 | 16.5 | 1.98 | 1.85 | 20.33 | 0.94 | 0.65 | 3.50 |
| 1.5 | 17.2 | 2.56 | 3.82 | 23.58 | 1.36 | 1.58 | 4.19 |
| 2.5 | 21.5 | 2.95 | 5.84 | 30.29 | 1.89 | 2.75 | 5.45 |
| 3.5 | 19.5 | 1.82 | 2.09 | 23.41 | 3.41 | 3.01 | 6.69 |

Table 2. Acid hydrolysis of sugarcane bagasse at 140 °C, 30 min and initial solid: liquid ratio (1:10) [Source: Chandel et al., 2007a] (The values are mean of three replicates. Standard deviation was within 10%)

## 3. Plant cell-wall derived inhibitors and microorganisms

The toxicity of inhibitors depends upon the concentration, type of fermentative organism, the mode of cultivation, and cultivation conditions (i.e. pH, inoculum, dissolve oxygen and temperature). The toxic component may lead to stop the growth of microorganism by affecting the rate of sugar uptake with simultaneous decay in product formation. These inhibitors affect the cellular physiology by disturbing the function of biological membranes, causes poor microbial growth lingering towards extended incubation time with poor metabolite production. However, the yield may remain unaltered. The mechanism of inhibition of some compounds such as phenolics and plant cell wall derived extractives are yet to be known.

Among sugar derived inhibitors, furfurals have been found to inactivate the cell replication that reduces the growth rate and the cell mass yield on ATP, volumetric growth rate and specific productivities [Taherzadeh et al., 1999; Palmqvist et al., 1999a]. Furfurals have been found toxic to *Pichia stipitis* under aerobic condition, whereas the growth of *Saccharomyces cerevisiae* was less affected under anaerobic condition by converting into furoic acid [Palmqvist et al., 1999a; Taherzadeh et al., 1999]. Adaptation of microorganisms on high furfural concentration has been found a successful option to decrease the furfural effect on growth. It may be due to the synthesis of new enzymes or co-enzymes for furfural reduction [Boyer et al., 1992; Villa et al., 1992]. Furans (furfurals and 5-HMF) in conjunction with acetic acid have been reported highly affective to the growth of *P. stipitis*, *Pachysolen tannophilus* and *Escherichia coli* [Martinez et al., 2000; Lohmeier-vogel et al., 1998].

The presence of lignin derived inhibitors in the acid hydrolysate is highly effective for the growth of fermenting microorganisms. Lignin derived inhibitors include polyaromatic, phenolics and aldehydes. Their toxicity is thought to be proportional to the molecular weights. The lower molecular weight of phenolic compounds is generally lethal to the microorganisms than higher molecular weight compounds [Clarck & Mackie, 1984]. They are toxic to fermenting microorganism even more than furans and weak acids inhibiting the microbial growth. They cause a partition and disturb the membranes in turn affecting their ability to serve as selective barriers and enzyme matrices affecting the cell growth and sugar assimilation [Palmqvist & Hahn-Hagerdaal, 2000b]. Among lignin derived inhibitors vanillin, syringaldehyde, 4-hydroxybenzoic acid, catechol, acetosyringone, and 1-hydroxybenzotrizole (HBT) causes a partition and loss of integrity of biological membranes in microorganisms diminishing cell growth and sugar assimilation [Palmqvist et al., 1999]. Delgenes et al. [1996] reported that the inhibitory effect of lignin derived compounds on the sugar utilization efficiency of *C. shehatae, P. stipitis* and *S. cerevisiae and Z. mobilis*. Vanillin was found to be the strongest inhibitor of growth and ethanol production in both xylose and hexose fermenting yeasts. Hu et al., [2009] observed the p-hydroxy benzaldehyde and vanillin at 10 mM concentration inhibits the growth of *Rhodosporidium toruloides*.

Apart from the furans and phenolics, there are other compounds present in acid hydrolysates. These include acetic acid and raw material extractives (tannic and terpene acids). However their degree of severity on fermenting microorganisms is low compared to furans and phenolics [McMillan, 1994]. Among acidic inhibitors, the hemicellulose derived organic acid inhibitors, i.e. acetic acid, formic acid levulinic acid, acidic resins, tannic, and terpene acids mostly inhibit the growth and metabolism allowing the higher permeation of cell membrane in microorganisms [Zaldivar and Ingram, 1999; Takahashi et al., 1999; Imai and Ohno, 1995]. The toxicity of acetic acid also depends upon the culture conditions are being employed during the fermentative process. Felipe et al., [1995] reported limits of acetic acid concentrations (1g/l) in the fermentation medium that can improve the xylose-to-xylitol bioconversion, whereas the ethanol production was reported to be stimulated at higher concentration (10 g/l) when medium was free of other toxic compounds [Palmqvist et al., 1999]. Other week acids at low concentration have been found to exert a stimulating effect on the ethanol production by *S. cerevisiae* [Pampulha & Loureiro-Dias, 1989].

## 4. Strategies to overcome the fermentation inhibitors

There are several detoxification methods such as physical (evaporation, membrane mediated detoxification), chemical (neutralization, calcium hydroxide overliming, activated charcoal treatment, ion exchange resins, and extraction with ethyl acetate) and biological detoxification (enzymatic mediated using laccase, lignin peroxidase), in-situ detoxification, in-situ microbial detoxification etc. Table 3 summarizes the hydrolysate detoxification using various non-biological methods employed to the variety of lignocellulosic hydrolysates. Each method represents its specificity to eliminate particular inhibitor from the hydrolysate.

| Lignocellulose Hydrolysae | Detoxification methods | Changes in hydrolysate composition | References |
|---|---|---|---|
| Sugarcane bagasse | Neutralization | NA | Chandel et al., 2007a |
| *Saccharum spontaneum* | Over-liming | Removal of furfurals (41.75%), total phenolics (33.21%), no effect on acetic acid content. Reduction of reducing sugars (7.61%) | Chandel et al., 2011a |
| Oak wood | Activated charcoal | Removal of phenolics (95.40%) | Converti et al., 1999 |
| Wheat straw | Ion exchange-D 311 + over-liming | Removal of furfurals (90.36%), phenolics (77.44%) and acetic acid (96.29%) | Zhuang et al., 2009 |
| Wheat straw | Ethyl acetate + Over-liming | Removal of furfurals (59.76%), phenolics (48.23%) and acetic acid (92.19%) | Zhuang et al., 2009 |
| Aspen | Roto-evaporation | Removal of acetic acid (54%), furfural (100%) and vanillin (29%) | Wilson et al., 1989 |
| Spruce wood | Dithionite and sulfite | No major change in composition of hydrolysates | Alriksson et al., 2010 |
| Corn stover | Membrane based organic phases alamine 336 | 60% acetic acid removal | Grzenia et al., 2008 |

Table 3. Different detoxification strategies (Non-biological) applied to lignocellulose hydrolysates for the removal of fermentation inhibitors

It is difficult to compare detoxification methodologies based on the selection of lignocellulosic hydrolysates and types of the microorganisms been used in the fermentation media. Also, the lignocellulosic hydrolysates vary in their degree of inhibition, and microorganism reveals different inhibitor tolerances [Mussatto & Roberto, 2004; Palmqvist & Hahn-Hagerdal, 2000a, b].

### 4.1. Physical methods

#### 4.1.1 Evaporation

The evaporation under vacuum can eliminate volatile compounds such as acetic acid, furfural and vanillin from lignocellulosic hydrolysate. However, this method retains the concentration of non-volatile toxic compounds (extractives and lignin derivatives) in the hydrolysates. Palmqvist et al., [1996] observed the removal of most volatile fraction (10% v/v) from willow hemicellulose hydrolysate by roto-evaporation. Wilson et al., [1989] found a decrease in the concentration of acetic acid, furfural and vanillin by 54%, 100% and 29%, respectively, compared with the concentrations in the hydrolysate. Larsson et al., [1999] observed the removal of furfural (90%) and HMF (4%) using vacuum evaporation from wood hemicellulosic hydrolysate. The improved production of xylitol was reported from hemicelluloses hydrolysate after removal of acetic acid, furfural and other volatile compounds [Converti et al., 2000].

Another potential substrate sugarcane bagasse was hydrolyzed and vacuum evaporated followed by activated charcoal treatment, revealed 89% removal of furfural [Rodrigues et al., 2001] with partial elimination of acetic acid. Zhu et al., (2011) applied the complex extraction to detoxify the prehydrolysate corn stover using mixed extractant (30% trialkylamine-50% n-octanol−20% kerosene). The detoxification resulted into removal of 73.3% acetic acid, 45.7% 5-HMF and 100% furfural. The effect of evaporation on the removal of fermentation inhibitors has been summarized in Table 3.

#### 4.1.2 Membrane separations

Adsorptive micro porous membranes have surface functional groups attached to their internal pores, which may eliminate the cell wall derived inhibitors from the lignocellulose acid hydrolysates. During clarification of inhibitors, the feed is being pumped through the membrane pores that bind to the solute predominantly by convection. This phenomenon can greatly reduce the required processing time. Also, the drop in the pressure for flow through adsorptive membranes changes significantly compared to the typical packed beds. Wickramasinghe & Grzenia [2008] observed better performance of membrane assisted system for acetic acid removal from the biomass hydrolysates than ion-exchange resins. The effect of various diluted organic phases (alamine 336, aliquat 336) for the removal of acetic acid (60%) from corn stover hydrolysates from alamine 336 was further investigated [Grzenia et al., 2008]. Later, Grzenia et al. [2010] used the membrane extraction for removal of inhibitors from sulfuric acid derived hemicellulosic hydrolysate of corn stover. Extraction of sulphuric, acetic, formic and levulinic acid as well as 5 hydroxymethylfurfural and furfural was removed when alamine 336, octanol and oelyl alcohol used in the organic phase. Thus, the adsorptive membranes may offer significant improvements over traditional ion-exchange resins. Effect of membrane separation on the removal of fermentation inhibitors has been summarized in Table 3.

### 4.2 Chemical methods

#### 4.2.1 Neutralization

Due to highly acidic nature of hemicellulosic hydrolysates, the neutralisation of is unavoidable step before using the hydrolysate for fermentation. Alkali most preferably calcium hydroxide or sodium hydroxide are used for neutralization of hydrolysates (pH-6.0-7.0). During the process, furfurals and phenolics may be removed by precipitation to the some extent. Table 3 summarizes the neutralization effect on the removal of fermentation inhibitors from lignocellulose hydrolysates.

#### 4.2.2 Calcium hydroxide over-liming

Over-liming with a combination of high pH and temperature has for a long time been considered as a promising detoxification method for dilute sulfuric acid-pretreated hydrolysate of lignocellulosic biomass [Chandel et al., 2007a; Martinez et al., 2001]. This process has been demonstrated to help with the removal of volatile inhibitory compounds such as furfural and hydroxymethyl furfural (HMF) from the hydrolysate additionally causing a sugar loss (~10%) by adsorption [Chandel et al., 2011a, b; 2009; Martinez et al., 2000; Ranatunga et al., 2000]. The dried calcium hydroxide is added in acidic hydrolysates converting into gypsum which can be used as plaster of paris having many commercial values. The effect of overliming on the removal of fermentation inhibitors from the variety of lignocellulose hydrolysates has been summarized in Table 3.

#### 4.2.3 Activated charcoal treatment

The detoxification of hemicellulose hydrolysates, by activated charcoal is known to be a cost effective with high capacity to absorb compounds without affecting levels of sugar in hydrolysate [Canilha et al., 2008; Chandel et al., 2007a; Mussatto & Roberto, 2001]. The effectiveness of activated charcoal treatment depends on different process variables such as pH, contact time, temperature and the ratio of activated charcoal taken versus the liquid hydrolysate volume [Prakasham et al., 2009]. A summarized description of charcoal treatment on detoxification of lignocellulose hydrolysate has been summarized in Table 3.

#### 4.2.4 Ion exchange resins

Treatment with ion exchange resins has been known to remove lignin-derived inhibitors, acetic acid and furfurals respectively, leading to hydrolysate that show a fermentation similar to that of an inhibitor-free model substrate. The ion-exchange resins based separation of fermentative inhibitors may not be cost effective [Lee et al., 1999], however, it provides most effective means of inhibitor separation when the hydrolysate being adjusted to a pH of 10 which requires significant quantities of base chemicals [Wilson and Tekere, 2009; Ranjan et al., 2009]. Further, the anion treatment also helps to remove most inhibitors (i.e. levulinic, acetic, formic acids, and furfural and 5-HMF). Villarreal et al. (2006) investigated the effect of four different ion exchange resins (cation and anion) for the detoxification of *Eucalyptus* hemicellulosic hydrolysates for the improved xylitol production by *Candida guilliermondii*. The ion exchange detoxification drastically enhanced the fermentability of the hydrolysate. Total 32.7 g/l of xylitol was achieved after 48 h fermentation, which correspond to 0.68 g/l/ h volumetric productivity and 0.57 g/g xylitol yield factor [Villarreal et al. 2006]. The ion exchange resins also led to a considerable loss of fermentable sugars from the hydrolysate. Chandel et al., [2007a] observed that ion exchange resins diminish furans (63.4%) and total phenolics (75.8%) from sugarcane bagasse acid hydrolysates. Although the ion exchanges resins is effective, however is not cost effective that reflects its limited feasibility in commercial industrial purpose in lignocellulosics derived products synthesis. Table 3 summarizes the effect of different ion exchange resins treatment on detoxification of lignocellulose hydrolysate.

#### 4.2.5 Extraction with ethyl acetate

The extraction of fermentation inhibitors using ethyl acetate has been found to increase the ethanol yield in fermentation by *P. stipitis* from 0 to 93% of that obtained in the reference fermentation [Wilson et al., 1989]. The extraction procedure could eliminate acetic acid (56%), and total furfural, vanillin, and 4-hydroxybenzoic acid. 84% phenolics was removed from the *Eucalyptus* wood hemicelluloses hydrolysate with the extraction using ethyl acetate or diethyl ether [Cruz et al. 1999]. Ethyl acetate extraction has been shown to increase the rate of glucose consumption in hydrolysate of pine wood [Clark & Mackie, 1984]. The low molecular weight phenolic compounds were suggested to be the most inhibiting compounds in the ethyl acetate extract [Zhuang et al., 2009]. Pasha et al., [2007] detoxified the *Prosopis juliflora* hydrolysate with calcium hydroxide overliming in conjunction with ethyl acetate. In these studies, the ethanol yield of 0.459 ± 0.012 g /g, productivity of 0.67 ± 0.015 g/l/h and fermentation efficiency of 90% after fermentation of this detoxified sugar syrup with fusant *S. cerevisiae* $VS_3$. Table 3 shows a summarized effect of extraction using ethyl acetate on detoxification of lignocellulose hydrolysate.

### 4.3 Biotechnological routes for detoxification

Due to concerns of feasibility and affordability of physico-chemical treatments, the biotechnological methods encompass the application of living microorganisms and/or the microbial enzymatic applicability for the detoxification of lignocellulose hydrolysates. These microorganisms and/or the enzymes have potential to alter the chemical nature of inhibitors in hydrolysates. The biological methods of detoxification are more feasible, environmental friendly, with fewer side reactions and less energy requirements [Parawira and Tekere, 2011.]. The slow reaction time of microbial/ enzymatic detoxification and the loss of fermentable sugars make them unattractive [Yang & Wyman, 2008]. However, it is unavoidable to explore the biotechnological routes of detoxification towards process economization.

#### 4.3.1 Microbial pretreatment of lignocellulosics

The microorganisms are able to degrade lignin during the incubation by leaving cellulose and hemicelluloses behind with loose bonding of both in the raw material. The lignocellulosic substrate after pretreatment with microorganism could easily be hydrolyse into fermentable sugars requiring less acid load, lower temperature, and less hydrolysis time [Kuhar et al., 2008]. The microbial mediated pretreatment resulted into the maximum de-polymerisation of carbohydrate polymers into mixture of simple sugars with less fermentation inhibitors. The pretreated lignocellulose substrate when hydrolysed leads to maximum sugar recovery with minimum inhibitors in short period of time by eliminating the requirement of detoxification step [Liang et al. 2010; Kuhar et al., 2008; Keller et al. 2003].

### 4.3.2 Microbial acclimatization

Often, microorganisms can adapt to a variety of fermentation media. However, multiple essential nutrients or compounds along with suitable carbon and nitrogen sources are required for any fermentation reaction. The adaptation of microorganisms to lignocellulosic hydrolysates is another parameter to get the improved product yields [Silva and Roberto, 2001; Sene et al., 2001]. Acclimatization of fermentative microorganisms to the hydrolysates containing inhibitors prior to fermentation of lignocellulose hydrolysates provide improved productivities [Parawira & Tekere, 2011]. Microorganisms, *P. stipitis, C. shehatae* and *P. tanophillus* are being explored to overcome inhibition and improve fermentation ability of xylose-containing hydrolysates by adaptation [Tomas-Pejo et al., 2010; Zhu et al., 2009; Martin et al., 2007]. Parekh et al. [1987] employed *P. stipitis* on a steam-stripped hardwood hydrolysate that improved ethanol production (46%), with >90% xylose utilization compared with steam-unstripped hardwood hydrolysate. Microorganism *P. stipitis* NRRL Y-7124 adapted on wheat straw hemicellulosic hydrolysates, showed improved ethanol production (yield, 0.41 ± 0.01 g/g) equivalent to 80.4 ± 0.55% theoretical conversion efficiency [Nigam, 2001]. The adaptation of *C. guilliermondii* to rice straw hemicellulose hydrolysate for xylitol production was found to be an effective and inexpensive method to alleviate the inhibitory effect of toxic compounds on the xylose- to-xylitol bioconversion [Silva & Roberto, 2001].

### 4.3.3 In-situ microbial detoxification

Under in-situ microbial detoxification, the microorganisms are being grown in lignocellulose hydrolysate to detoxify the inhibitory substances by transforming their chemical nature [López et al., 2004]. The in-situ detoxification of impurities could be achieved at higher rate that may economize the overall ethanol production in the same vessel. Attempts were made to detoxify the lignocellulose hydrolysates directly with the employing wild (yeasts, fungi, bacteria) and/or recombinant microorganisms expressing the laccase or peroxidases. Palmqvist et al. [1997] used *Trichoderma reesei* to degrade the inhibitors from willow hemicellulosic hydrolysate that directly assisted improvements in ethanol productivity by three fold and yield four fold. Later, Larsson et al. [1999] detoxified the dilute-acid hydrolysate of spruce by *T. reesei* and found to be most efficient compared to anion exchange, over-liming, and treatment with laccase enzyme, however, with higher consumption of fermentable sugars (35%). A fungal isolate, *Coniochaeta ligniaria,* (NRRL30616), was also reported to metabolize furfural, HMF, aromatic and aliphatic acids, and aldehydes present in corn stover hydrolysate [Nichols et al., 2008].

The microbial mediated detoxifications are been considered effective than the soft rot fungi resulted into less sugar consumption with shorter incubation time. In a related study, López et al. [2004] isolated five bacteria related to *Methylobacterium extorquens, Pseudomonas sp., Flavobacterium indologenes, Acinetobacter sp., Arthrobacter aurescens,* and fungus *C. ligniaria* C8 (NRRL30616), capable of depleting toxic compounds from defined mineral medium containing a mixture of ferulic acid, HMF, and furfural as carbon and energy sources. Organism *C. ligniaria* C8 (NRRL30616) was effective in removing furfural and HMF from corn stover hydrolysate.

Okuda et al. [2008] investigated the biological detoxification of a waste house wood (WHW) hydrolysate by thermophilic bacterium *Ureibacillus thermosphaercus*. Chromatographic analysis confirmed that *U. thermosphaercus* degraded the furfural or HMF present in the synthetic hydrolysates, and the phenolic compounds present in the WHW hydrolysates. The bacterium grows rapidly and consumes less than 5% fermentable sugars. In another example of in-situ detoxification, Tian et al. [2009] isolated yeast strains namely Y1, Y4 and Y7 and evaluated their efficiency for ethanol production after in-situ detoxification of hydrolysates. Strains Y1 and Y4 yielded 0.49 g and 0.45 g ethanol/g glucose, equivalent to maximum theoretical values of 96% and 88.2%, respectively. Further, attempts are underway to detoxify the lignocellulose hydrolysates through direct application of microorganisms in the hydrolysate. Table 4 summarizes the microorganism used for in-situ detoxification of lignocellulose hydrolysates.

| Lignocellulose hydrolysate | Enzyme/ Microorganisms | Effect of the method | References |
|---|---|---|---|
| Sugarcane bagasse | Laccase | 80% removal of phenolics | Martin et al., 2002 |
| Lignocellulose hydrolysate | Peroxidase from *C. cinereus* IFO 8371 | 100% removal of p-coumaric acid, ferulic acid, vanillic acid and vanillin | Cho et al., 2009 |
| Spruce | Residue lignin | 53% removal of phenolics and 68% removal of furans | Bjorklund et al., 2002 |
| Willow | *T. reesei* | Considerable removal of phenolics, furans and weak acids | Palmqvist et al., 1997 |
| Corn stover | *Coniochaeta ligniaria* | 80% Removal of furfural and 5-HMF | López et al., 2004 |
| Sugarcane bagasse | *Issatchenkia occidentalis* CCTCC M 206097 | Reduction of syringaldehyde (66.67%), ferulic acid (73.33%), furfural (62%), and 5-HMF (85%) | Fonesca et al., 2011 |
| Spruce | Continuous fermentation | Elimination of detoxification step; improved ethanol yield, 0.42-0.46 g/g | Purwadi et al., 2007 |
| Willow | High cell density fermentation (10 g/l dry weight) | High ethanol productivity even in un-detoxified hydrolysate | Palmqvist et al., 1996 |

Table 4. Different Biotechnological strategies applied to lignocellulose hydrolysates for the removal of fermentation inhibitors

#### 4.3.4 Alterations in fermentation conditions

The presence of lignocellulose inhibitors in fermenting medium affects the ethanol and biomass productivities as microorganism take more incubation times to convert into products [Chandel et al., 2007a; Nilvebrant et al., 2001; Zaldivar et al., 2001]. Usually the ethanol productivity is determined by cell-specific productivity and cell mass concentration, cumbersome by lignocellulose-derived inhibitors. To overcome by inhibitors, high cell-mass inocula are effective to tolerate the stress of inhibitory substances [Purwadi et al., 2007]. The ethanol productivity has been increased by maintaining the initial cell-mass at higher density [Brandberg et al., 2007]. By altering the initial cell density, the increased production of ethanol (0.44 g/g) was reported at initial cell density (10 g/l dry weight) [Palmqvist et al., 1996]. The ethanol productivity in fed-batch fermentation was limited by the feed rate that in turn, was limited by the cell-mass concentration (Taherzadeh et al. 1999).

In continuous fermentation, the ethanol productivity also depends upon the rate of dilution. Since the microbial growth rate is known to decrease by the inhibitors, the productivity in continuous fermentation of lignocellulosic hydrolysates remains low [Lee et al., 1996; Palmqvist et al., 1998]. Purwadi et al. [2007] has achieved the ethanol yield of 0.42-0.46 g/g sugar utilized from the crude hydrolysates of spruce wood as carbon source under continuous fermentations using the flocculating *S. cerevisiae* CCUG 53310. Cellular re-circulation strategy was employed in the fermentation of an enzymatic hydrolysate of spruce [Palmqvist et al., 1998], and in fermentation of bagasse hydrolysate [Ghose and Tyagi, 1979].

#### 4.3.5 Enzymatic clarification

Enzymatic detoxification is the most preferred biotechnological route of inhibitors' clarification. Laccase and peroxidases enzymes derived from white rot fungi (*Trametes versicolor, Phenorochete chrysosporium, Cythus bulleri, C. stercoreous,* and *Pycnoporous cinnabarinus*) have been found effective for the removal of phenolics from the lignocellulose hydrolysates. Jonsson et al. [1998] explored laccase and peroxidase enzymes of the white-rot fungus *T. versicolor* to detoxify the willow hemicellulosic hydrolysate. The detoxification mechanism of these enzymes probably involves oxidative polymerization of low-molecular-weight phenolic compounds in turn increasing the ethanol productivity. Martin et al. [2002] compared the effect of laccase treatment and over-liming, on the composition and fermentability of enzymatic hydrolysates of sugarcane bagasse by the genetically engineered xylose-utilizing *S. cerevisiae* strains. About 80% of the phenolic compounds were specifically removed by the laccase treatment.

The effect of laccase enzyme derived from *C. bulleri* for the detoxification of sugarcane bagasse hemicellulosic hydrolysate showed 77% phenolics were eliminated from the acid hydrolysate [Chandel et al. 2007a]. The ethanol production efficiency was superior in the laccase treated hysrolysate (6.50 g/l, 0.37 g/g) compared to the untreated (3.46 g/l, 0.22 g/g) [Chandel et al. 2007a]. Cho et al., [2009] studied the effect of peroxidase enzyme derived from *Coprinus cinereus* IFO 8371 on the detoxification of phenolics enriched fermentation medium for the butanol production from *Clostridium beijerinckii* NCIMB 8052 with 100% removal of phenolic compounds (e.g. p-coumaric acid, ferulic acid, vanillic acid, and vanillin) using 0.01 μM of peroxidase enzyme. This peroxidase treated fermentation solution along with acidification and precipitation (A/P) revealed amelioration in butanol production from 0.6±0.15 to 8.9±0.43 g/l. Enzymatic detoxification could be cost effective, if immobilized laccases and peroxidases on robust matrices (Diaion-HP-87H, Sepa beads, Eupergit-C etc) are employed for detoxification of lignocellulose hydrolysates. The immobilized enzymes could be promising solution towards the development of a reliable process for detoxification at industrial scale in the biorefineries.

## 5. Technological integration of detoxification process

A single method may not be sufficient to remove variety of inhibitors from lignocellulosic hydrolysate. Based on the multitask process, a common detoxification strategy could provide a better performance and effective ethanol yield at industrial scale. The integration of multi-step processes of detoxification in one experiment will provide a better solution to overcome the inhibitors. However, increased cost is the inevitable if integrated steps are followed. A comparative account of the different detoxification steps (i.e. over liming, activated charcoal, ion-exchange and laccase treatment) in conjunction with neutralization for the detoxification of sugarcane bagasse hemicellulosic hydrolysate was explored and summarized in Table 5 that shows the effect of detoxification treatments on the ethanol production by *Candida shehatae* NCIM3501 [Chandel et al. 2007a]. Fermentation of these hydrolysates with *Candida shehatae* NCIM 3501 showed maximum ethanol yield (0.48 g/g) from ion exchange treated hydrolysate, followed by activated charcoal (0.42 g/g), laccase (0.37 g/g), over-liming (0.30 g/g), and neutralized hydrolysate (0.22 g/g).

| **Treatment** | **Sugar fermented (%)** | **Ethanol (gp/l)** | **Biomass (gx/l)** | **Ethanol yield (gp/gs)** | **Volumetric ethanol productivity (gp/l)/h** | **Specific ethanol production (gp/gx)** |
|---|---|---|---|---|---|---|
| Neutralization | 78.8 | 3.46 | 9.8 | 0.22 | 0.144 | 0.353 |
| Overliming | 85.9 | 5.19 | 12.3 | 0.302 | 0.216 | 0.421 |
| Laccase | 86.89 | 6.50 | 14.8 | 0.374 | 0.270 | 0.439 |
| Activated charcoal | 87.41 | 7.43 | 15.5 | 0.425 | 0.309 | 0.476 |
| Ion exchange | 89.93 | 8.67 | 16.0 | 0.482 | 0.361 | 0.535 |

Table 5. Fermentation profile of sugarcane bagasse hydrolysate detoxifed with different methods [Source: Chandel et al., 2007a] (The values are mean of three replicates. Standard deviation was within 10%. Initial total sugar concentration-20.0 g/l. Ethanol productivities were calculated after 24 h of fermentation. Biomass production rate was continuously increased till the completion of the fermentation batch.)

Over-liming in combination with sulphite addition was also tried by Olsson & colleagues [1995] that showed four times higher fermentation rate with recombinant *E. coli* than detoxification by overliming only, whereas the ethanol production from untreated willow hemicellulosic hydrolysate was uncomparable. The simultaneous detoxification and enzyme production has been reported to occur when the inhibitor-containing hemicellulose hydrolysate from the pretreatment stage was used as substrate for *T. reesei* [Palmqvist et al., 1997].

A similar concept, simultaneous detoxification and fermentation (SDF) was designed to improve production of ethanol from lignocellulose hydrolysates by appropriate ethanologen in conjunction with detoxification of sugar stream using detoxifying biocatalyst simultaneously together. This approach could have potential impact in bio-refineries, merging both individual steps detoxification and fermentation in one vessel. The enzyme-containing inhibitor-free liquid can then be used to hydrolyse the cellulose fraction. This detoxification method could further improve the process economy since all the wood-derived sugars were utilized. Gyalai-Korpos et al. [2010] detoxified the steam pretreated rice straw hydrolysate through dual detoxification strategy (vacuum evaporation and calcium hydroxide overliming) for the cellulase production by *Trichoderma reesei* RUT C30. These studies revealed maximum activity of filter paperase (1.87 ±0.05 FPU/ml) and Beta-glucosidase (1.74 ±0.03 IU/ml) after 11 days of incubation.

## 6. Detoxification and future perspectives

Beyond lignocellulosic hydrolysis, constituents releases different monomeric sugars and widen the ethanol fermentability in bio-refinery. However, a wide range of compounds that are inhibitory to microorganisms are formed or released during the fermentation reaction. Based on the origin, inhibitors are usually divided in to three major groups' i.e. weak acids, furan derivatives, and phenolic compounds. It is unavoidable to economize overall process, as the inhibitors directly affect cellular growth and kinetics of biocatalysts used in the fermentation reaction. These compounds have a significant role individually and more toxic if they are used synergistically.

A number of methods could effectively be optimized at industrial scale including physical (evaporation, membrane based filtrations), physico-chemical (Rota evaporation with organic solvents), chemical (Calcium hydroxide over-liming, application of other alkalis such as sodium hydroxide, sodium di thionite, sodium di sulphite, adsorption on activated charcoal, ion-exchange) and biological (changes in fermentation strategies, laccase, peroxidases, using of microorganisms such as *T. reesei, C. ligniaria, I. occidentalis* in the hydrolysates) for detoxification of lignocellulose hydrolysates. Apart from conventional methods, recent genetic engineering approaches and directed evolution methods to make the competitive strains combating the inhibitors are also in vague. Larsson et al. [2001] have successfully attempted for the heterologous expression of laccase in *S. cerevisiae*. All these methods have been quite successful in terms of removal of inhibitors and simultaneously improving the product titers from the fermentation reaction.

## 7. Conclusion

Presence of inhibitors in lignocellulosic hydrolysate is an industrial malaise. Efficient detoxification can be prescribed as medicine to eliminate fermentation inhibitors present in lignocellulosic hydrolysate. Among all the detoxification methods, the biological strategies to eliminate inhibitors are most promising. A search for novel microbial strains for detoxification of lignocellulose hydrolysates without affecting the sugar and other nutrient fraction from the fermented media is yet to come. *T. reesei* has been found affective; however, has limitations with inefficient sugar utilization proven to be an expensive affair for industrial bioprocess. An optimal design of the fermentation process, rate of bioconversion, and the adaptive response of the microorganism to the toxic compounds in the hydrolysate could be established. Technologies are required to utilize cheaper and highly affective chemicals, which have more affinity towards inhibitors without affecting the original sugar content in lignocellulosic hydrolysates.

## 8. Acknowledgements

Funding support from FAPESP and CNPq to AKC and SSS is gratefully acknowledged.

## 9. References

Baek, S.-C. & Kwon, Y-J. (2007) Optimization of the pretreatment of rice straw hemicellulosic hydrolyzates for microbial production of xylitol. *Biotechnol. Bioprocess Eng.,* Vol. 12, pp. 404-409.

Balan, V., Sousa, L.C., Chundawat, S.P.S., & Marshall, D. (2009) Enzymatic Digestibility and Pretreatment Degradation Products of AFEX-Treated Hardwoods (*Populus nigra*). *Biotechnol. Prog.*, Vol. 25, pp. 365-375.

Björklund, L., Larsson, S., Jönsson, L.J., Reimann, A., & Nivebrant, N.O. (2002) Treatment with lignin residue A novel method for detoxification of lignocellulose hydrolysates. *Appl. Biochem. Biotechnol.*, Vol. 98-100, No. 1-9, pp. 563-575.

Alriksson, B., Cavka, A., & Jönsson, L. J. (2011) Improving the fermentability of enzymatic hydrolysates of lignocellulose through chemical in-situ detoxification with reducing agents. *Bioresource Technol.,* Vol. 102, pp. 1254-1263.

Boyer, L.J., Vega, K., Klasson, K.T., Clausen, E.C., & Gaddy, J.L. (1992) The effects of furfural on ethanol production by *Saccharomyces cerevisiae. Biomass Bioeng.,* Vol. 3, No. 1, pp. 41-48.

Brandberg, T., Karimi, K., Taherzadeh, M.J., Franzén, C.J., & Gustafsson, L. (2007) Continuous fermentation of wheat-supplemented lignocellulose hydrolysate with different types of cell retention. *Biotechnol. & Bioeng.,* Vo. 98, pp. 80–90.

Canilha, L., Carvalho, W., Felipe, M.G.A., & Silva, J.B.A. (2008) Xylitol production from wheat straw hemicellulosic hydrolysate: hydrolysate detoxification and carbon source used for inoculum preparation. *Brazilian J Microbiol.,* Vol. 39, pp. 333-336.

Cao, G., Ren, N., Wang, A., Lee, D.J., Guo, W., Liu, B., Feng, Y., & Zhao, Q. (2009) Acid hydrolysis of corn stover for biohydrogen production using *Thermoanaerobacterium thermosaccharolyticum* W16. *Int. J. Hyd. Ener.*, Vol. 34, pp. 7182-7188.

Carvalheiro, F., Duarte, L.C., Lopes, S., Parajó, J.C., Pereira, H., & Gírio, FM. (2005) Evaluation of the detoxification of brewery's spent grain hydrolysate for xylitol production by *Debaryomyces hansenii* CCMI 941. *Proc. Biochem.,* Vol. 40, pp. 1215–1223.

Chandel, A.K. & Singh, O.V. (2011) Weedy lignocellulosic feedstock and microbial metabolic engineering: Advancing the generation of 'Biofuel'. *Appl. Microbiol. Biotechnol.*, Vol. 89, pp. 1289-1303.

Chandel, A.K., Chan, E.C., Rudravaram, R., Narasu, M.L., Rao, L.V., & Ravindra, P. (2007b) Economics and environmental impact of bioethanol production technologies: an appraisal. *Biotechnol. Mol. Biol. Rev.*, Vol. 2, pp. 14–32.

Chandel, A.K., Kapoor, R.K., Singh, A. & Kuhad, R.C. (2007a) Detoxification of sugarcane bagasse hydrolysate improves ethanol production by *Candida shehatae* NCIM 3501. *Bioresource Technol.*, Vol. 98, pp. 1947–1950.

Chandel, A.K., Narasu, M.L., Rudravaram, R., Ravindra, P., Narasu, M.L. & Rao, L.V. (2009) Bioconversion of de-oiled rice bran (DORB) hemicellulosic hydrolysate into ethanol by *Pichia stipitis* NCIM3499 under optimized conditions. *Int. J. Food Eng.*, Vol. 2, pp. 1-12.

Chandel, A.K., Singh, O.V., Chandrasekhar, G., Rao, L.V. & Narasu, M.L. (2011a) Bioconversion of novel substrate, *Saccharum spontaneum*, a weedy material into ethanol by *Pichia stipitis* NCIM3498. *Bioresource Technol.*, Vol. 102, pp. 1709-1714.

Chandel, A.K., Singh, O.V., Chandrasekhar, G., Rao, L.V. & Narasu, M.L. (2010b) Key-drivers influencing the commercialization of ethanol based biorefineries. *J Comm. Biotechnol.*, Vol. 16, pp. 239–257.

Chandel, A.K., Singh, O.V., Narasu, M.L. & Rao, L.V. (2011b) Bioconversion of *Saccharum spontaneum* (wild sugarcane) hemicellulosic hydrolysate into ethanol by mono and co-cultures of *Pichia stipitis* NCIM3498 and thermotolerant *Saccharomyces cerevisiae*-VS3. *N. Biotechnol.*, DOI 10.1016/j.nbt.2010.12.002.

Chandel, A.K., Singh, O.V. & Rao, L.V. (2010a) Biotechnological applications of hemicellulosic derived sugars: state-of-the-art. In: *Sustainable biotechnology: renewable resources and new perspectives*, Om V. Singh, & Steven P. Harvey (Eds.), pp. 63–81. Springer, ISBN 978-90-481-3294-2, Netherland.

Chandel, A.K., Kapoor, R.K., Narasu, M.L., Viswadevan, V., Kumaran, S.S.G., Ravinder, R., Rao, L.V., Tripathi, K.K., Lal, B. & Kuhad, R.C. (2007c) Economic evaluation and environmental benefits of biofuel: an Indian perspective. *Int. J. Global Energy Issues*, Vol. 28, pp. 357-381.

Cho, D.H., Lee, Y.J., Um, Y., Sang, B.I. & Kim, Y.H. (2009) Detoxification of model phenolic compounds in lignocellulosic hydrolysates with peroxidase for butanol production from *Clostridium beijerinckii. Appl. Microbiol. Biotechnol.*, Vol. 3, pp. 1035–1043.

Clark, T. & Mackie, K.L. (1984) Fermentation inhibitors in wood hydrolysates derived from the softwood *Pinus radiata*. *J. Chem. Biotechnol.*, Vol. B 34, pp. 101-110.

Converti, A., Perego, P. & Dominguez, J.M. (1999) Xylitol production from hardwood hemicellulose hydrolysates by *P. tannophilus, D. hansenii* and *C. guilliermondii. Appl. Biotechnol. Biochem.*, Vol. 82, pp. 141–151.

Converti, A., Dom_ınguez, J.M., Perego, P., Silva, S.S. & Zilli, M. (2000) Wood hydrolysis and hydrolyzate detoxification for subsequent xylitol production. *Chem. Eng. Technol.*, Vol. 23, pp. 1013–1020.

Cruz, J.M., Dominguez, J.M., Dominguez, H. & Parajo, J.C. (1999) Solvent extraction of hemicellulose wood hydrolysates: a procedure useful for obtaining both detoxified fermentation media and poylphenols with antioxidant activity. *Food Chem*, Vol. 67, pp. 147–153.

Delgenes, J.P., Moletta, R. & Navarro, J.M. (1996) Effects of lignocellulose degradation products on ethanol fermentations of glucose and xylose by *Saccharomyces cerevisiae, Zymomonas mobilis, Pichia stipitis,* and *Candida shehatae. Enzyme Microb. Technol.*, Vol. 19, pp. 220–225.

Felipe, M.G.A., Veira, M.V., Vitolo, M., Mancilha, I.M., Roberto, I.C. & Silva, S.S. (1995) Effect of acetic acid on xylose fermentation to xylitol by *Candida guilliermondii*. *J. Basic Microbiol.*, Vol. 35, pp. 171- 177.

Fonseca, B.G., Moutta, R.O., Ferraz, F.O., Vieira, E.R., Nogueira, A.S., Baratella, B.F., Rodrigues, L.C., Hou-Rui, Z. and Silva, S.S. (2011) Biological detoxification of different hemicellulosic hydrolysates using *Issatchenkia occidentalis* CCTCC M 206097 yeast. *J. Ind. Microbiol. Biotechnol.* Vol. 38, pp. 199–207.

Ghose, T.K. & Tyagi, R.D. (1979) Rapid ethanol fermentation of cellulose hydrolysate I Batch versus continuous systems, *Biotechnol. Bioeng.*, Vol. 21, pp. 1387-1400.

Girio, F.M., Fonseca, C., Carvalheiro, F., Duarte, L.C., Marques, S. & Bogel-Łukasik, R. (2010) Hemicellulose for fuel ethanol: A review, *Bioresource Technol.*, Vol. 10, pp. 4775-4800.

Grzenia, D.L., Schell, D.J. & Wickramasinghe, S.R. (2008) Membrane extraction for removal of acetic acid from biomass hydrolysates. *J. Membr. Sci.*, Vol. 322, pp. 189-195.

Grzenia, D.L., Schell, D.J. & Wickramsinghe, S.R. (2010) Detoxification of biomass hydrolysates by reactive membrane extraction, *J. Membr. Sci.*, Vol. 348, pp. 6-12.

Gyalai-Korpos, M., Mangel, R., Alvira, P., Dienes, D., Ballesteros, M. & Re´czey, K. (2010) Cellulase production using different streams of wheat grain- and wheat straw-based ethanol processes. *J Ind Microbiol Biotechnol.*, DOI: 10.1007/s10295-010-0811-9

Hahn-Hägerdal, B., Karhumaa, K., Fonseca, C., Spencer-Martins, I., Gorwa-Grauslund, M.F. (2007) Towards industrial pentose fermenting yeast strains. *Appl. Microbiol. Biotechnol.*, Vol. 74, pp. 937–953.

Hu, C., Zhao, X., Zhao, J., Wu, S. & Zhao, Z.K. (2009) Effect of biomass hydrolysis by-products on oleaginous yeast *Rhodosporidium toruloides*. *Bioresource Technol.*, Vol. 100: pp. 4843-4847.

Imai, T. & , Ohono, T. (1995) The relationship between viability and intracellular pH in the yeast *Saccharomyces cerevisiae*, *Appl. Environ. Microbiol.*, Vol. 61, pp. 3604-3608.

Jonsson, L.J., Palmqvist, E., Nilvebrant, N-O. & Hahn-Hagerdal, B. (1998) Detoxification of wood hydrolysate with laccase and peroxidase from the white-rot fungus *T. versicolor*. *Appl. Microbiol. Biotechnol.*, Vol. 49, pp. 691–697.

Keller, F.A., Hamilton J.E., & Nguyen, Q.A. (2003) Microbial pretreatment of biomass: potential for reducing severity of thermo-chemical biomass pretreatment. *Appl. Biochem. Biotechnol.*, Vol. 105, pp. 27-41.

Klinke, H.B., Thomsen, A.B. & Ahring, B.K. (2004) Inhibition of ethanol-producing yeast and bacteria by degradation products produced during pretreatment of biomass. *Appl. Microbiol. Biotechnol.*, Vol. 66, pp. 10–26.

Kuhar, S., Nair, L.M. & Kuhad, R.C. (2008) Pretreatment of lignocellulosic material with fungi capable of higher lignin degradation and lower carbohydrate degradation improves substrate acid hydrolysis and the eventual conversion, *Can. J. Microbiol.*, Vol. 54, pp. 305-313.

Kumar, P., Barrett, D.M., Delwiche, M.J. & Stroeve, P. (2009) Methods for pretreatment of lignocellulosic biomass for efficient hydrolysis and biofuel production. *Ind. Eng. Chem. Res.*, Vol. 48, pp. 3713–3729.

Larsson, S., Cassland, P. & Jönsson, L.J. (2001) Development of a *Saccharomyces cerevisiae* strain with enhanced resistance to phenolic fermentation inhibitors in lignocellulose hydrolysates by heterologous expression of laccase. *Appl. Environ. Microbiol.,*Vol. 67, pp. 1163–1170.

Larsson, S., Reimann, A., Nilvebrant, N. & Jonsson, L.J. (1999a) Comparison of different methods for the detoxification of lignocellulose hydrolysates of spruce. *Appl. Biochem. Biotechnol.*, Vol. 77–79, pp. 91–103.

Lee, W.G., Lee, J.S., Park, B.G. & Chang, H.N. (1996) Continuous ethanol production from wood hydrolysate by chemostat and total cell retention culture. *Korean J. Chem. Eng.*, Vol. 13, pp. 453-456.

Lee, W.G., Lee, J.S., Shin, C.S., Park, S.C., Chang, H.N. & Chang, Y.K. (1999) Ethanol production using concentrated oak wood hydrolysates and methods to detoxify. *Appl. Biochem. Biotechnol.*, Vol. 77–79, pp. 547–559.

Liang, Y.S., Yuan, X.Z., Zeng, G.M., Hu, C.L., Zhong, H., Huang, D.L., Tang, L. & Zhao, J.J. (2010) Biodelignification of rice straw by *Phanerochaete chrysosporium* in the presence of dirhamnolipid. *Biodegradation,* Vol.21, pp. 615-624.

Lohmeier-vogel, E.M., Sopher, C.R. & Lee, H. (1998) Intracellular acidification as a mechanism for the inhibition by acid hydrolysis-derived inhibitors of xylose fermentation by yeasts. *J. Ind. Microbiol. Biotechnol.*, Vol. 20, pp. 75–81.

López, M.J., Nichols, N.N., Dien, B.S., Moreno, J. & Bothast, R.J. (2004) Isolation of microorganisms for biological detoxification of lignocellulosic hydrolysates. *Appl. Microbiol. Biotechnol.*, Vol. 64, pp.125–131.

Martin, C., Galbe, M., Wahlbom, C.F., Hahn-Hagerdal, B. & Johnsson, L.J. (2002) Ethanol production from enzymatic hydrolysates of sugarcane bagasse using recombinant xylose-utilising *Saccharomyces cerevisiae. Enzyme Microbial Technol.*, Vol. 31, pp. 274–282.

Martín, C., Marcet, M., Almazán, O. & Jönsson, L.J. (2007) Adaptation of a recombinant xylose-utilizing *Saccharomyces cerevisiae* strain to a sugarcane bagasse hydrolysate with high content of fermentation inhibitors. *Bioresource Technol.*, Vol. 98, pp 1767–1773.

Martinez, A., Rodriguez, M.E., York, S.W., Preston, J.F. & Ingram, L.O. (2000) Effects of $Ca(OH)_2$ treatments ("overliming") on the composition and toxicity of bagasse hemicellulose hydrolysates. *Biotechnol. Bioeng.*, Vol. 69, pp. 526–536.

McMillan, J.D. (1994) Pretreatment of lignocellulosic biomass. In: *Enzymatic Conversion of Biomass for Fuels Production,* M.E. Himmel, J.O. Baker, R.P. Overend, (Eds.), American Chemical Society, Washington, DC, pp. 292–324.

Jurado, M., Prieto, A ., Martínez-Alcalá, Á., Martínez, Á.T. & Martínez, M.J. (2009) Laccase detoxification of steam-exploded wheat straw for second generation bioethanol. *Bioresource Technol.*, Vol. 100, pp. 6378–6384.

Moiser, N., Wyman, C., Dale, B., Elander, R., Lee, Y.Y., Holtzapple, M. & Ladisch, M. (2005) Features of promising technologies for pretreatment of lignocellulosic biomass. *Bioresource Technol.*, Vol. 96, pp. 673–686.

Mussatto, S.I. & Roberto, I.C. (2004) Alternatives for detoxification of diluted-acid lignocellulosic hydrolyzates for use in fermentative processes: a review. *Bioresource Technol.*, Vol. 93, pp. 1–10.

Nichols, N.N., Sharma, L.N., Mowery, R.A., Chambliss, C.K., van Walsum, G.P., Dien, B.S. & Iten, L.B. (2008) Fungal metabolism of fermentation inhibitors present in corn stover dilute acid hydrolysate. *Enzyme Microbial Technol.*, Vol. 42, pp. 624–630.

Nigam, J.N. (2001) Ethanol production from wheat straw hemicelluloses hydrolysate by *Pichia stipitis*. *J. Biotechnol.*, Vol. 87, pp. 17–27.

Nilvebrant, N-O., Reimann, A., Larsson, S. & Jönsson L.J. (2001) Detoxification oflignocellulose hydrolysates with ion-exchange resins. *Appl. Biochem. Biotechnol.*, Vol. 91/93, pp. 35–49.

Okuda, N., Sonuera, M., Ninomiya, K., Katakura, Y. & Shioya, S. (2008) Biological detoxification of waste house wood hydrolysate using Ureibacillus thermosphaericus for bioethanol production. *J Biosci. Bioeng.*, Vol. 106, pp. 128–133.

Olsson, L., Hahn-Hagerdal, B. & Zacchi, G. (1995) Kinetics of ethanol production by recombinant *Escherichia coli* K011. *Biotechnol. Bioeng.*, Vol. 45, pp. 356–365.

Palmqvist, E., Hahn-Hagerdal, B., Szengyel, Z., Zacchi, G. & Reczey, K. (1997) Simultaneous detoxification and enzyme production of hemicelluloses hydrolysates obtained after steam pretreatment. *Enzyme Microbial Technol.* Vol. 20, pp. 286–293.

Palmqvist, E. & Hahn-Hagerdal, B. (2000a) Fermentation of lignocellulosic hydrolysates.I: inhibition and detoxification: review. *Bioresource Technol.*, Vol. 74, pp. 17–24.

Palmqvist, E. & Hahn-Hagerdal, B. (2000b) Fermentation of lignocellulosic hydrolysates.II: inhibitors and mechanism of inhibition: review. *Bioresource Technol.*, Vol. 74, pp. 25–33.

Palmqvist, E., Almeida, J. & Hahn-Hagerdal, B. (1999a) Infuence of furfural on anaerobic glycolytic kinetics of *Saccharomyces cerevisiae* in batch culture. *Biotechnol. Bioeng.*, Vol. 62, pp. 447-454.

Palmqvist, E., Galbe, M. & Hahn-Hagerdal, B. (1998) Evaluation of cell recycling in continuous fermentation of enzymatic hydrolysates of spruce with *Saccharomyces cerevisiae* and on-line monitoring of glucose and ethanol. *Appl. Biochem. Biotechnol.*, Vol. 50, pp. 545-551.

Palmqvist, E. & Hahn-Hagerdal, B. (2000a) Fermentation of lignocellulosic hydrolysates. I: Inhibition and detoxification. *Bioresource Technol.*, Vol. 74, pp. 17–24.

Palmqvist, E., Hahn-Hagerdal, B., Galbe, M. & Zacchi, G. (1996b) The effect of water-soluble inhibitors from steam-pretreated willow on enzymatic hydrolysis and ethanol fermentation. *Enzyme Microb. Technol.*, Vol. 19, pp. 470-476.

Pampulha, M.E. & Loureiro-Dias, M.C. (1989) Combined effect of acetic acid, pH and ethanol on intracellular pH of fermenting yeast. *Appl. Microbiol. Biotechnol.*, Vol. 31, pp. 547-550.

Parajo, J.C., Dominguez, H. & Dominguez, J.M. (1996) Charcoal adsorption of wood hydrolysates for improving their fermentability: influence of the operational conditions. *Bioresource Technol.*, Vol. 157, pp. 179–185.

Parawira, W. & Tekere, M. (2011) Biotechnological strategies to overcome inhibitors in lignocellulose hydrolysates for ethanol production: review. *Crit. Rev. Biotechnol.*, Vol. 31, pp. 20-31.

Parekh, S.R., Parekh, R.S. & Wayman, M. (1987) Fermentation of wood derived acid hydrolysates in a batch bioreactor and in a continuous dynamic immobilized cell bioreactor by *Pichia stipitis* R. *Proc. Biochem.*, Vol. 22, pp. 85-91.

Pasha, C., Kuhad, R.C. & Rao, L.V. (2007) Strain improvement of thermotolerant *Saccharomyces cerevisiae* $VS_3$ strain for better utilization of lignocellulosic substrates. *J Appl. Microbiol.*, Vol. 103, pp. 1480–1489.

Prakasham, R.S., Rao, R.S. & Hobbs, P.J. (2009) Current trends in biotechnological production of xylitol and future prospects. *Curr. Trends Biotechnol. Phar.*, Vol. 3, pp. 8-36.

Purwadi, R., Brandberg, T., & Taherzadeh, M.J. (2007) A Possible Industrial Solution to Ferment Lignocellulosic Hydrolyzate to Ethanol: Continuous Cultivation with Flocculating Yeast. *Int. J. Mol. Sci.*,Vol. 8, pp. 920-932.

Qian, M., Tian S., Li, X., Zhang, J., Pan, Y. & Yang X. (2006) Ethanol production from dilute-acid softwood hydrolysate by co-culture. *Appl. Biochem. Biotechnol.*, Vol. 134, pp. 273–283.

Ranatunga, T.D., Jervis, J., Helm, R.F., McMillan, J.D. & Wooley, R.J. (2000) The effect of overliming on the toxicity of dilute acid pretreated lignocellulosics: the role of inorganics, uronic acids and ether-soluble organics. *Enzyme Microbial. Technol.*, Vol. 27, pp.240-247.

Ranjan, R., Thust, S., Gounaris, C.E., Woo, M., Floudas, C.A., Keitz, M., Valentas, K.J., Wei, J. & Tsapatsis, M. (2009) Adsorption of fermentation inhibitors from lignocellulosic biomass hydrolysates for improved ethanol yield and value added product recovery. *Microporous Mesoporous Mat.*, Vol. 122, pp. 143–148.

Rodrigues, R.C.L.B., Felipe, M.G.A., Almeida e Silva, J.B., Vitolo, M. & Villa, P.V. (2001) The influence of pH, temperature and hydrolysate concentration on the removal of volatile and non-volatile compounds from sugarcane bagasse hemicellulosic hydrolysate treated with activated charcoal before or after vacuum evaporation. *Brazilian J Chem. Engin.*, Vol. 18, pp. 299–311.

Saha, B.C. (2003) Hemicellulose bioconversion. *J. Ind. Microbiol. Biotechnol.*, Vol. 30, pp. 279–291.

Sene, L., Converti, A., Zilli, M., Felipe, M.G.A. & Silva, S.S. (2001) Metabolic study of the adaptation of the yeast *Candida guilliermondii* to sugar cane bagasse hydrolysate. *Appl. Microbiol. Biotechnol.*, Vol. 57, pp. 738–743.

Silva, C.J.S.M. & Roberto, I.C. (2001) Improvement of xylitol production by *Candida guilliermondii* FTI 20037 previously adapted to rice straw hemicellulosic hydrolysate. *Lett. Appl. Microbiol.*, Vol. 32, pp. 248–252.

Sun, Z. & Liu, S. (2010) Production of n-butanol from concentrated sugar maple hemicellulosic hydrolysate by *Clostridia acetobutylicum* ATCC824. *Biomass & Bioenergy*, DOI:10.1016/j.biombioe.2010.07.026

Taherzadeh, M.J., Gustafsson, L., Niklasson, C. & Lidén, G. (1999) Conversion of furfural in aerobic and anaerobic batch fermentation of glucose by *Saccharomyces cerevisiae*. *J Biosci. Bioeng.*, Vol. 87, pp. 169–174.

Taherzadeh, M.J. & Karimi, K. (2007) Acid-based hydrolysis processes for ethanol from lignocellulosic materials: a review. *Bioresource Technol.*, Vol. 2, pp. 472–499.

Takahashi, C.M., Takahashi, D.F., Carvalhal, M.L., Alterthum, F. (1999) Effects of acetate on the growth and fermentation performance of *Escherichia coli* KO11. *Appl. Biochem. Biotechnol.*, Vol. 81, pp. 193-203.

Tian, S., Zhou, G., Yan, F., Yu, Y. & Yang, X. (2009) Yeast strains for ethanol production from lignocellulosic hydrolysates during in situ detoxification. *Biotechnol. Adv.*, Vol. 27, pp. 656–660.

Tomas-Pejo, E., Ballesteros, M., Oliva, J.M. & Olsson, L. (2010) Adaptation of the xylose fermenting yeast Saccharomyces cerevisiae F12 for improving ethanol production in different fed-batch SSF processes. *J. Ind. Microbiol. Biotechnol.*, Vol. 37, pp. 1211-1220.

Villa, G.P. (1992) Microbial transformation of furfural to furfuryl alcohol by *Saccharomyces cerevisiae*. *Acta Biotechnol.*, Vol. 12, pp. 509-512.

Villarreal, M.L.M., Prata, A.M.R., Felipe, M.G.A., Almeida, E. & Silva, J.B. (2006) Detoxification procedures of eucalyptus hemicellulose hydrolysate for xylitol production by *Candida guilliermondii*. *Enzyme Microbial Technol.*, Vol. 40, pp. 17–24.

Watson, N.E., Prior, B.A., Lategan, P.M., Lussi, M. (1984) Factors an acid treated bagasse inhibiting ethanol production from D- xylose by *Pachysolen tannophilus*. *Enzyme Microbial Technol.*, Vol. 6, pp. 451–456.

Wickramasinghe, S.R. & Grzenia, D.L. (2008) Adsorptive membranes and resins for acetic acid removal from biomass hydrolysates, *Desalination*, Vol. 234, pp. 144-151.

Wilson, J.J., Deschatelets, L. & Nishikawa, N.K. (1989) Comparative fermentability of enzymatic and acid hydrolysates of steam pretreated aspen wood hemicellulose by *Pichia stipitis* CBS 5776. *Appl. Microbiol. Biotechnol.*, Vol. 31, pp. 592-596.

Yang, B. & Wyman, C.E. (2008) Pretreatment: the key to unlocking low-cost cellulosic ethanol. *Biofuels Bioproduct Bioref.*, Vol. 2, pp. 26–40.

Yu, Y., Feng, Y., Xu, C., Liu, J. & Li, D. (2011) Onsite biodetoxification of steam exploded corn stover for cellulosic ethanol production. *Bioresource Technol.*, DOI: 10.1016/j.biortech.2011.01.067.

Zaldivar, J. and Ingram, L.O. (1999) Effect of organic acids on the growth and fermentation of ethanologenic *Escherichia coli* LY01. *Biotechnol. Bioeng.*, Vol. 66, pp. 203-210.

Zaldivar, J., Nielsen, J. & Olsson, L. (2001) Effect of selected aldehydes on the growth and fermentation of ethanologenic *Escherichia coli* LY01. *Biotechnol. Bioeng.*, Vol. 65, pp. 24–33.

Zhu, J., Yong, Q., Xu, Y. & Yu, S. (2011) Detoxification of corn stover prehydrolyzate by trialkylamine extraction to improve the ethanol production with *Pichia stipitis* CBS 5776. *Bioresource Technol.*, Vol. 102, pp. 1663-1668.

Zhu, J.J., Yong, Q., Xu, Y, Chen, S.X. & Yu, S.Y. (2009) Adaptation fermentation of *Pichia stipitis* and combination detoxification on steam exploded lignocellulosic prehydrolyzate. *Natural Sci.*, Vol. 1, pp. 47-54.

Zhuang, J., Liu, Y, Wu, Z., Sun, Y. and Lin, L. (2009) Hydrolysis of wheat straw hemicellulose and detoxification of the hydrolysate for xylitol production. *Bio-Resources*, Vol. 4, pp. 674-686.

# 11

# Biofuel From Cellulosic Mass with Incentive for Feed Industry Employing Thermophilic Microbes

Javed Iqbal Qazi, Naureen Chaudhary and Saima Shahzad Mirza
*Department of Zoology University of the Punjab, Lahore Pakistan*

## 1. Introduction

Drastically depleting fossil fuels' supplies and the associated environmental concern dictate for immediate, renewable and environmental friendly alternatives. Cellulosic biomass has a great potential for bioethanol production. Many problems of the way have been solved by isolating and employing thermophilic, cellulolytic and ethanologenic microorganisms. Many workers have established simultaneous saccharification and ethanol fermentation from agro-industrial wastes rich in cellulosic material and some soluble sugars. The latter substances provide quick carbon and energy sources to the bacteria and yeasts inoculants. In Pakistan sugarcane bagasse is a very appealing potential agro-industrial waste in this regard. Owing to the great sun shine in majority of the country area throughout the year, solar energy has been dreamt for disinfecting the fermentation facility as well as providing steam for pretreatment of the substrate.

Another source of clean fuel is $H_2$. We have also been able to isolate and cultivate purple-non sulfur bacteria for the production of $H_2$ employing certain agro-industrial waste including the one just referred as major media ingredients. Biohydrogen can be obtained economically, by employing bacteria capable of fixing sun energy and utilizing agroindustrial wastes including cellulosic material heterotrophically.

A very appealing notion about the potential of microorganisms isolated from industrially contaminated aquatic and soil habitats form developing countries is their pollutants' resistance. This ability renders such microorganisms capable of biofuel generation from industrial effluents containing biomass as well as different chemical pollutants. It is very right time to conserve such pollutants' resistant microbial diversity before the developing countries progress for in situ treatments plants for their industries and the polluted areas recover back to their uncontaminated nature, alongwith losing the pollutants selective pressures mediated and thus evolved microbial communities. This journey is expected to be completed earlier than the time frame the developed countries had passed through. As the developing countries are benefitting from the experiences of the developed nations and thus are striving to escalate the process of progress.

This chapter outlines the possibilities of ethanol and hydrogen fermentations for the application of agro/food industrial wastes. The related issues have been dealt in depth and the chapter comprises two major sections i.e.bioethanol and biohydrogen.

## 2. Ethanol a renewable biofule

Energy needs of most nations of the world have increased over the time. Following industrial revolution in the late 18th century, societies that had been based largely on agriculture turned to industry to meet the needs of their growing populations. Energy plays an essential role in modern society. Fuel consumption has not only increased by factories, rather more fuel is required to distribute the market products. Ever increasing human population density and the desire for higher life standards, demanding more and more comforts, had necessitated large scale exploitation of fossil fuel energy resources, in the recent centuries. Between 1900 and 2000, word energy consumption increased by a factor of fourteen while the population increased threefold. Owing to the facts of ever increasing consumption and rapidly depleting resources of fossil fuels, scientists have rightly sensed that to feed and provide other requirements to the human population at a reduced environmental cost is a real target for future biotechnological improvements. Fueling both the humans and the required mechanical engines necessitates various developments in the agricultural and energy sectors, respectively. (Enger & Smith, 2002; Gray *et al.*, 2006; Smith, 1996).

Besides environmental deterioration, one of consequences of fossil fuels usage, their supply is being exhausted rapidly. Priorities are being shifted from building power stations, oil fields and coal mines, to active pursuit of energy and efficiency improvement and identifying renewable energy sources. One such resource is the bioconversion of plant biomass to ethanol. Motor cars in some countries are being driven by gasoline-alcohol mixture (4:1) called gasohol (Bernstein *et al.*, 1996; Preuss *et al.*, 1998; Van Haandel, 2005).

Biofuels represented by biologically produced alcohols, gasses, and oil represent renewable energy resources, unlike petroleum, coal and nuclear fuels. Rising energy and environmental problems have led to increased interest in the production from diverse routes and resources and utilization of alcohols as fuel (Atiyeh & Duvnjak, 2002; Lawford *et al.*, 2001; Von *et al.*, 1994).

The subject mater is reviewed here under the following headlines:

1. Ethanol as fuel
2. Ethanologenic fermentations
3. Ethanol from lignocellulosic biomass
4. Consolidated bioprocess: Simultaneous Saccharification and Fermentation (SSF).
5. Thermophilic ethanologenic microbes.
6. Sugarcane bagasse a resource rather than a waste.
7. Single Cell Protein (SCP) from agro industrial wastes.

Some of the highlights regarding the above referred topics are described in the forthcoming pages.

## 3. Ethanol as fuel

Ethanol has been used as biofuel in the United States, Europe and Brazil. In Brazil industrial scale ethanol is produced from sugarcane for blending with gasoline.While in the U.S. corn is used for ethanol production and is then blended with gasoline to produce gasohol (Enger & Smith, 2002; Lynd, 1995; Wheals *et al.*, 1999). Apart from being a renewable fuel made from plants, with high octane at low cost, ethanol is a much cleaner fuel than petrol. Ethanol blends dramatically reduce emissions of hydrocarbons, major sources of ground level ozone formation, cancer-causing benzene and butadiene, sulphur dioxide and particulate matter. Moreover, ethanol blends can be used in all petrol engines without modifications (Miller, 2003).

Lynd (1995) has condensed valuable information in his essay on biological fuel production.Accordingly, ethanol is the most widely used biologically produced transportation fuel. Major ethanol industries arose during the 1980s in Brazil and the United States. Ethanol has a higher economic value in low level (e.g., 10%) gasoline blends than in neat (unblended) form. However, the fuel properties of neat ethanol are in general excellent and decreased emissions of ozone precursors are expected for neat ethanol.

Brazil is the largest producer of bioethanol, and sugarcane is the main raw material. In this country ethanol has been used as an octane enhancer in gasoline in the form of 22% anhydrous ethanol at 99.6 Gay-Lussac (GL) and 0.4% water or in neat ethnaol engines in the form of hydrated ethanol at 95.5 GL. In other countries gasohol blends typically contain only 10% ethanol. Ethanol makes an excellent motor fuel: it has a research octane number of 109 and a motor octane number of 90, both of which exceed those of gasoline. Ethanol has a lower vapour pressure than gasoline, which results in lower evaporative emission. Ethanol's flammability in air is also much lower than that of gasoline, which reduces the number and severity of vehicle fires. These properties of ethanol have led to the development of dedicated (E-100) and modified (E-22) engines for the ethanol-gasoline mixture in Brazil (Goldemberg & Macedo, 1994; Zanin *et al.*, 2000).

Sixty eight percent of the ethanol produced in the world is used as fuel. Production of ethanol is not evenly distributed throughout the world. North America contributes for 66%, Asian and Pacific Ocean countries for 18%, Europe for 14% and Africa for 2%. Brazil and United States contribute a great share of global production with 53% and 19%, respectively. Brazilian sugarcane ethanol is now a global energy commodity that is fully competitive with motor gasoline and appropriate for replication in many countries (Goldemberg, 2007; Zanin *et al.*, 2000).

## 4. Ethanologenic fermentations

Ethanologenic fermentation is the microbial conversion of sugars into carbon dioxide and ethyl alcohol. Regarding the provision of sugars for ethanol fermentation, it is pertinent to note that development of several novel sweeteners, many times sweeter than sucrose could ultimately lead to a reduction in the traditional sugar market for sugarcane and sugar beet. In this way, these economics predominately in developing countries could experience severe financial and employment discretion with alternatives difficult to find (Smith, 1996). The ethanol fermentations meant to generate biofuel would then be amongst the considered alternatives. Sugars may also be derived from starches and cellulosic materials in addition to black strap molasses, a by-product of cane sugar manufacture. Once simple sugars, the monomeric units are formed, enzymes from yeasts and bacteria can readily ferment them into ethanol.

Moat *et al.* (2004) have summarized the fermentative pathways occurring in some of the major groups of microorganisms (Fig.1). They have described that a thorough evaluation of the pathways of carbohydrate fermentation requires qualitative identification of and quantitative accounting for the amount of products recovered. To assess the accuracy of the analytical determinations, a carbon balance or carbon recovery is calculated. Oxidation-reduction (O-R) reactions play a major role in the fermentative metabolism of carbohydrates.

The O-R balance provides an indication as to whether the formed products balance with regard to their oxidized or reduced states. It may not be possible to balance the hydrogen and oxygen of the substrate directly because hydrations or dehydrations may occur as intermediary steps in the fermentation pathways.

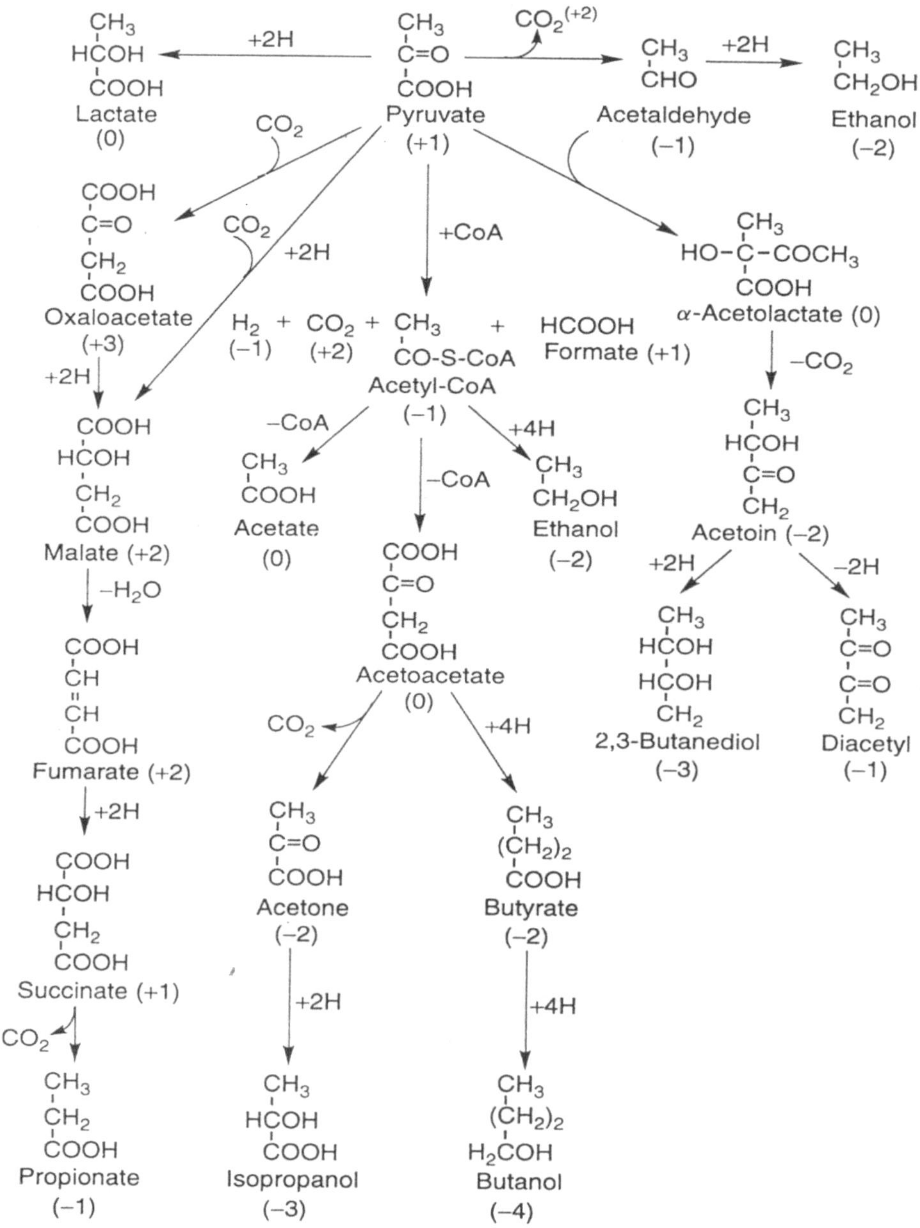


Fig. 1. Major pathways of fermentation products formation from pyruvate. Number in parentheses are the oxidation values calculated on the basis of the number of oxygen atoms less one-half the number of hydrogen atoms (Moat *et al.,* 2004).

If the ratio of oxidized products to reduced products is close to the theoretical value of 1.0, this provides further indication that the products are in balance. The oxidation value of a compound is determined from the number of oxygen atoms less one-half number of hydrogen atoms. For glucose, which has 6 oxygen atoms and 12 hydrogen atoms the oxidation value is 0. Thus, glucose is referred to as a neutral compound, and to be equally balanced the fermentation products should contain equivalent amounts of oxidized and reduced products. Another aspect of fermentation balances is the $C_1$ balance. The amount of expected $C_1$ product is calculated from the amounts of those products for which $CO_2$ or format is expected as an accompanying product. For example, if a $C_2$ compound such as ethanol or acetate is among the final products, an equal amount of $CO_2$ will be expected, since ethanol is derived from pyruvate by decarboxylation.

## 5. Ethanol from lignocellulosic biomass

Having technologically difficult boundaries but abundantly available cellulosic material has been conceived by contemporary biotechnologists for bioethanol production. For instance, Wheals *et al.* (1999) while discussing commercial viability of fuel ethanol had described that there will only be sufficient, low-cost ethanol if lignocellulose feddstock is also used. Similarly, Farrell *et al.* (2006) have explained that large-scale use of ethanol for fuel will almost certainly require cellulosic technology.

Amongst the plant biomass, cellulose is the primary substance that is produced following chemical transformation of the solar energy. It is one of most abundant organic compound in the biosphere. Some $10^{15}$ Kg of cellulose is synthesized by converting more than 100 billions metric tons of $CO_2$ and water into this and other plant products. A comparable amount of cellulose is also degraded on earth every year. It is an unbranched polymer of glucose residues joined by β-1,4 linkages consisting of 10,000-15,000 D-glucose units (Stryer, 1995). The β-1, 4-linked glucose polymer occurs in crystalline or amorphous forms and is usually found along with other oligosaccharides in the walls of plants and fungi. Cellulose-digesting microorganisms in the rumen of herbivorous animals are responsible for the ability of ruminants to use cellulose as a source of energy and building blocks for biosynthesis. The ubiquitous distribution of cellulose in municipal, agricultural and forestry wastes emphasizes its potential use for conversion to useful products such as single-cell protein or fermentation products such as methane or alcohol. As a consequence, the degradation of cellulose has been a continuing subject of intense study. Cellulose constitutes much of mass of wood and cotton is almost pure cellulose. Many manufactured products; including paper, cardboard, rayon and insulating tiles are derived from cellulose. (Moat *et al.*, 2004; Nelson & Cox, 2000; Stryer,1995). When first discovered, it was believed that polysaccharide bound and trapped in the cellulose structure of plant extractable with alkali, comprised of smaller molecules, which would be eventually converted to cellulose by plant. For this reason they were termed hemicelluloses. It is now known that this was an erroneous belief. Upon hydrolysis hemicelluloses may yield pentoses and hexoses or both, together with uronic acids. The most abundant polysaccharides in this group are the xylans, which occur particularly in all land and some marine plants. They constitute some 15 30% of corncobs, grains and nuts etc. They are composed of almost D-xylose (Oser, 1965). Five-carbon sugar xylose is stereo-chemically similar to glucose but one carbon shorter, bind to hexokinase, but in a position where it cannot be phosphorylated. Xylose is sufficient to induce a change in a hexokinase (Nelson & Cox, 2000).

Hemicelluloses are closely associated with cellulose and occur in matrix of plant cell wall. They include, for example, glactomonnas, glucomonnas, mixed beta glucans, xylans and xyloglucans etc. In plant cell wall the cellulose fibers are embedded in and cross-linked by a matrix containing other polysaccride and lignin, a plastic like phenolic polymer (Singleton & Sainbury, 2001; Voet *et al.*, 1999). Lignin is perhaps second to cellulose in term of biomass. It protects cellulose and hemicellulose from enzymatic attack. Lignin is also important structural component of plants. It provides structural rigidity and resistance to the compression, bending and resistance to pathogens. They are richest sources of aromatic compounds in nature (Coyne, 1999).

While considering production of ethanol from cellulosic biomass complete conversion of its suitable constituents is critical for the development of an efficient and economically feasible fermentation process. Since pentose sugar can comprise up to 30% of the biomass substrate, therefore there is considerable economic incentive to develop strains of yeast that will efficiently ferment this biomass component too. If xylose were converted, in addition to glucose to ethanol, the final ethanol yield would be expected to increase several folds (Ho *et al.*, 1999; Wilke *et al.*, 1981). Moat *et al.* (2004) have described that cellulose degradation requires the combined activities of three basic types of enzymes (Fig.2). Initially, an *endo-β-*1,4-glucanase cleaves cellulose to smaller oligosaccharides with free-chain ends. Then *exo-β-1*, 4-glucanases remove disaccharide cellobiose units from either the reducing or nonreducing ends of the oligosaccharide chains. Cellobiose is then hydrolyzed to glucose by *β*-glucosidases.

The cellulolytic enzymes may be produced as extracellular proteins by organisms such as *Trichoderma, Phanaerochete* (filamentous fungi), *Cellulomonas, Microbispora,* and *Thermomonaspora* (Actinomycetes). Rumen bacteria such as *Ruminococcus flavofaciens* and *Fibrobacter succinogenes,* or gram-positive anaerobes such as *Clostridium thermocellum, C. cellulovorans,* or *C. cellulolyticum,* produce a cell-bound multienzyme complex called the cellulosome. With the aid of the electron microscope, cellulosomes can be seen as protuberances on the cell surface. The cellulosome of *C. cellulovorans* contains three major subunits: a scaffolding protein, CipA; an exoglucanase, ExgS; and an endoglucanase, EngE. Also present are endoglucanases EngB, EngL, and EngY, and a mannanase, ManA. The scaffolding protein serves as a cellulose-binding factor. Another component, present in duplicate and referred to as dockerin, mediates the association of cellulose fibers with the scaffolding protein. Various models have been proposed to conceptualize the complete interaction of the cellulosome with cellulose fibers during the digestion process. A wide diversity of actively cellulolytic organisms is important in industrial applications, in the rumen of animals, and in the digestive systems of arthropods that degrade wood. Termites and other arthropods that degrade wood owe their ability to digest cellulose in the presence of specific microbial symbionts in their digestive tract (Moat *et al.*, 2004).

Future processes will increasingly make use of organic materials that are renewable in nature and/or occur as low value wastes, valueless or adding negative value to the produce, that may presently cause environmental pollution. Currently more than ten times more energy is generated annually by photosynthesis than is consumed by mankind. On a worldwide basis land plants produce 24 tons of cellulose per person per year. Definitely, lignocellulose is the most abundant and renewable natural resource available to humanity throughout the word. It has been documented that massive technological difficulties such as expensive energy-demanding pretreatment processes have to be overcome before economic use may be made of the plentiful renewable resource. Following its chemical and/or enzymatic hydrolysis soluble sugar products of cellulose can then be converted to form ethanol, butanol, acetone, single cell protein and methane, etc. (Anderson *et al.*, 2005; Nelson & Cox, 2000; Smith, 1996). Hill *et al.* (2006) have described that negative environmental consequences of fossil fuels and concerns about petroleum supplies have spurred the search for renewable transportation biofuels, but to be a viable alternative, a biofuel should provide a net energy gain, have environmental benefits, be economically competitive and be producible in large quantities without reducing food supplies. These authors have reported that dedicatedly even if all the U.S. corn and soybean productions were dedicated to produce the biofuels, it would only cover 12% of gasoline and 6% of diesel demands. Thus, fuels such as celllulosic ethanol produced from low-input biomass grown on agriculturally marginal land or from waste biomass could provide much greater supplies and environmental benefits than food-based biofuels. Likewise, Taherzadeh & Karimi (2007) have recently indicated that lignocelluloses can be expected to be major feedstocks for ethanol production in future. Evans (2005) has earlier explained that being 50% of the total dry matter of plants, the cellulose is potentially a huge renewable energy store, and vast amounts of this material are routinely thrown away. However, until recently, the prospect of realizing this potential fuel source was viewed as difficult and expensive; the combination of cellulase-resistant links and its close association with lignin discouraged its large-scale hydrolysis to sugars. Energy involved in rendering various cellulosic materials into acceptable from had been considered a major limiting factor. Nevertheless contemporary technologies employing whole organisms and isolated enzyme technique appear to be promising to make the commercial processing of cellulose to alcohol a reality.

Xylose is represented by 20 to 40% of the contents of different cellulosic materials (Bicho *et al.*, 1988). Economic ethanol fermentations of cellulose are required to use this pentose sugar along with glucose following the saccharification of the fibrous matter. As majority of the well known microbial diversity in this regard has been reported capable of utilizing only glucose. Previously reported scarcity of xylose fermenting pathways in the microorganisms has been discouraging for cellulosic materials to be employed for economic ethanol generation. However, recently naturally occurring as well as genetically engineered xylose fermenting microorganisms have also been well documented (Chaudhary & Qazi, 2006a; Sonderegger *et al.*, 2004; Toivari *et al.*, 2001). It has been, however, reported variously that glucose is preferred by fermenting microorganisms capable of fermenting the both categories of the monosaccharides i.e., the glucose and xylose. In such cases glucose depletion within a fermenting substrate may allow for xylose utilization (Govindaswamy & Vane, 2006). Many strategies can be attempted to overcome co-substrate inhibition of xylose consumption by glucose considering the nature and diversity of fermenting microorganism(s). For instance, in case of microbial consortia first the glucose be utilized and then the residual material be attempted with xylose utilizers. Regarding other sugars, microbes could be found capable of co-utilization. Karhumaa *et al.* (2006) have reported simultaneous co-utilization of xylose and arabinose in recombinant strains of *S. cerevisiae*. This is well clear that economic ethanol generation from lignocelluloses requires the maximum utilization of all the diverse sugars monomers derived through any feasible saccharification process.

Responding above referred situations requires the isolation and construction of microorganisms capable of fermenting glucose and xylose at appropriate levels. Fermenting microbes that would prefer xylose and/or be incapable of glucose utilizations may find increasing utilization in one or two-chambered fermentative processes. In the latter case, a fermented matter in which glucose has been used maximally would serve feed stock for xylose-fermenting microorganisms. Mutants or genetically modified organisms with derived characteristics would be required to develop processes for obtaining ethanol from cellulose.

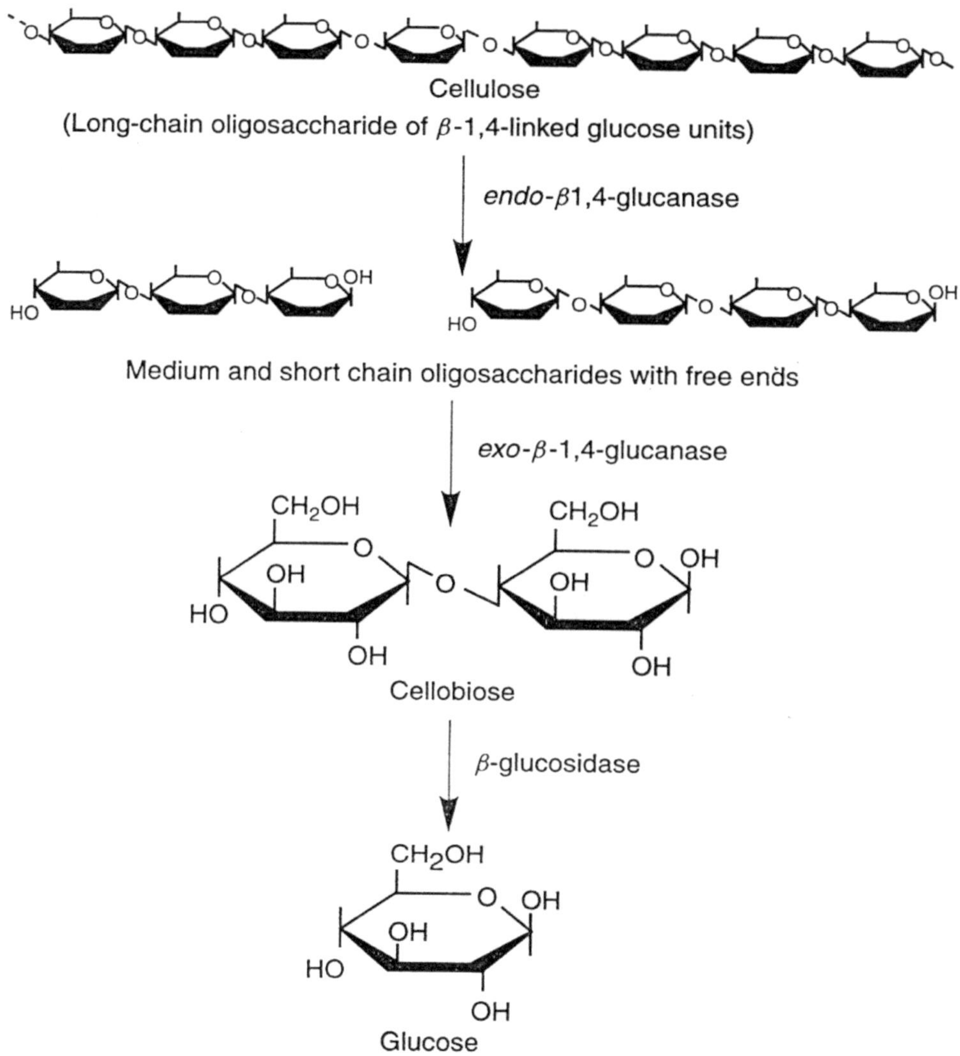


Fig. 2. Enzymatic degradation of cellulose (Moat *et al.*,2004).

Infact, the afore-mentioned approach could play a major part in addressing the largest environmental issue of our time, energy and waste. Energy extraction accompanied by environmentally safe disposable of a cellulosic waste may render the process economically feasible. A large number of point and non-point/plant biomass loads, to the natural water systems result in higher BOD values. Processing such wastes for ethanol fermentation is highly appealing to reduce the pollutants (Evans, 2005). Converting cellulose to ethanol is accomplished in essentially four stages, discussed below:

**Acid/microbial hydrolysis:** This process breakdowns the cellulose into a slurry of sugars in acid water and solid lignin particles. While reviewing acid-based hydrolysis processes for ethanol from lignocellulosic material, Taherzadeh & Karimi (2007) have summarized that concentrated-acid processes operated at low temperatures gave high sugar yield. On the other hand, dilute-acid processes operated at high temperature gave low sugar yield. Both categories of the hydrolyses result, however, into equipment corrosion.

**Acid recovery:** Following acid hydrolysis sugary liquid is separated from the previous stage and the process acid is recovered partly and reused. Van Groenestijin *et al.* (2006) have claimed that recovery of sulphuric acid in the form of $H_2S$ from anaerobic waste water treatment has a low overall cost for ethanol production.

**Fermentation:** The derived sugars are fermented by yeast and/or bacteria into alcohol. As already indicated efficient fermentation of lignocellulosic hydrolyzates requires employing microorganisms capable of fermenting the maximum variety of monomeric sugars and to resist inhibitory products of the hydrolyses process as well.

**Distillation:** It is required to collect the market grade ethanol. Economizing these four phases necessitates understanding and optimizing the diversity of the processes involved.

Fortunately, a large number of efforts in this regards have been continuously reported in the literature (Anderson *et al.*, 2005; Ballesteros *et al.*, 2001; Negro *et al.*, 2003). For instance, the latter authors have described that crops such as switch grass, bermudagrass, or napiegrass have the capacity to produce large quantities of lignocellulose for biofuel. To facilitate use of lignocellulosic material for the production of ethanol, it will be necessary to determine cost efficient pretreatments to enhance of the substrate conversion to fermentable sugars.

## 6. Consolidated bioprocess: Simultaneous Saccharification and Fermentation (S.S.F.)

Apart from the distillation step, which is required to separate the produce from the fermentation chamber, actual process of obtaining ethanol from lignocellulosic material essentially comprises of two phases i.e., saccharification of the substrate and the efficient utilization of all types of monosaccharides by suitable microbe(s) for producing the ethanol. Usually the two processes are accomplished in separate facilities. Science of process economics has dictated for developing some consolidated processes to allow the individual phases of the two processes to be completed simultaneously. This has been elaboratively described as Simultaneous Saccharification and Fermentation (SSF). Lynd (1995) described that biological conversion of cellulosic biomass typically involves four stages: production of cellulose enzymes, enzymatic hydrolysis of cellulase, hexose fermentation and pentose fermentation. And if a single organism or system of organisms were to carry out all four elements of ethanol production with high rates and yields, process economics would benefit profoundly.

The SSF has been conceptualized by many workers as a promising economical strategy for converting plant biomass to ethanol. In case of cellulosic ethanol production, developing genetically engineered microbes with the traits necessary for one-step processing of cellulosic biomass to ethanol appear to achieve the goal. Genetic improvements of microorganisms have been made either to enlarge the range of substrate utilization or to channel metabolic intermediates specifically toward ethanol for simultaneous saccharification and fermentation of lignocellulosic biomass from various sources (Chandrakant & Bisaria, 1998; Greer, 2005; Teixeira *et al.*, 2000). In their findings, Stenberg and co-workers (2000) pointed out that economic optimization of the production of ethanol by SSF requires knowledge about the influence of substrate and enzyme concentration on yield and productivity. These investigators obtained highest ethanol yield 68% of the theoretical based on the glucose and mannose present in the original wood at 5% substrate concentration. Compared with separate hydrolysis and fermentation, SSF gave a higher yield and doubled the productivity.

Lynd *et al.* (2002) have described that developing microorganisms capable of substrate utilization and product formation required for consolidated bioprocess can be pursued by two strategies. The native cellulolytic strategy involves engineering naturally occurring cellulolytic microorganisms to improve product related properties, such as yield and titer. While, the recombinant cellulolytic strategy involves engineering non-cellulolytic organisms that exhibit high product yields and titers so that they express a heterologous cellulose system that enables cellulose utilization.

Fujita *et al.* (2004) achieved efficient direct fermentation of amorphous cellulose to ethanol by developing a yeast strain and have reported the role of whole-cell biocatalysts for reducing the cost of ethanol production from cellulosic biomass. They described the advantages of the engineered yeast strain displaying three types of cellulolytic enzymes. The advantages included: conversion of cellobiose and glucose, which inhibit cellulase and β-glucosidase activities; lower sterilization requirements, as glucose was immediately taken up by the cells for ethanol production in a single cell reactor.

Considering process kinetics is very important. Microorganisms tend to disturb the optimum conditions provided to them due to their own growth and metabolic activities. Further feed back inhibition is an important limiting factor for both saccharification and ethanologenic levels. Thus microorganisms with a wide range of activities are to be worked out. Different bacteria and yeast have varying levels of ethanol tolerance. Here thermophilic ethanologenic microorganisms become important as the produce, recovery can be achieved under elevated temperatures, while not stopping or influencing negatively the fermentation process. Last but no the least, is the requirement for large scale microbial decontamination of the process material for controlled microbial hydrolysis and subsequent fermentation. This is surely energy demanding activity. A plant that has recently been proposed will work by intensifying solar radiations to provide heat for decontaminating the substrate and enabling cellulose substrate to be hydrolyzed as well as fermented to ethanol by thermophilic microorganisms (Chaudhary & Qazi, 2007). The proposed plant has technically been designed; its outcome is likely to be reported soon. The designed plant derives maximum benefits from sun heat and in this regard pivotal role of thermophilic ethanologenic microorgansisms has been discussed.

## 7. Potential of thermophilic ethanologenic microbes

Production of ethanol from low cost plant biomass is influenced by a number of phenomena. Majors of which are saccharification and fermentation efficiencies of the microbial culture(s) involved. The fermentation efficiency is in part influenced negatively with the raising levels of product accumulation and its inhibitory effect for further production. This can be overcome by employing thermophilic fermenting microorganisms as con-comittant removal of product, the bioethanol, at elevated temperature may delay or practically keep away accumulation of inhibitory level of the produce. Moreover, employing thermophilic microorganisms can bring support to the process economics by reducing the efforts required to keep the process facilities decontaminated by mesophilic bacteria and to reduce the cost of cooling that must be provided to maintain the correct temperature range required for optimal functioning of the mesophilic bacteria. These considerations have well earlier been taken into account by various workers (Budden & Spencer, 1993; Lamed & Zeikus, 1980; Thomas *et al.*, 1981).

While, commenting on yeast in their book "Thermophilic Microbes in Ethanol Production" Slapack *et al.* (1987) concluded that it is evident that thermotolerant yeasts would offer many advantages to the fermentation industry. Energy costs (cooling, distilling, and mixing) would be minimized and theoretically, productivity and growth rates should increase. Thermotolerant yeasts would be especially attractive in tropical countries where cooling costs are very expensive, and they are paramount for efficient simultaneous saccharification fermentation processes currently being investigated. In view of their many advantages, it is surprising that so few attempts have been made to select for thermotolerant yeasts and in particular, yeasts that can produce ethanol efficiently at high temperatures.

Sree *et al.* (1999) reported a novel solid substrate fermentation system to produce fuel ethanol from sweet sorghum and sweet potato using a thermotolerant *Saccharomyces cerevisiae* strain (VS#) and an isolate of amylolytic *Bacillus* sp. (VB9). They recorded maximum amount of ethanol production in co-culture with a mixed substrate as to be 5g/100 g of substrate at 37°C and 3.5 g/100g of substrate at 42°C. Likewise, Ueno *et al.* (2002) have evaluated a thermotolerant, fermentative yeast strain (RND 13) from a hot spring drainage for ethanol producing ability at elevated temperatures at 15% concentration of glucose. The RND 13 utilized glucose almost completely at 40°C with increasing inoculum size producing ethanol upto 6.6% (w/v). These workers found maximum rate of ethanol production of 9.00 g/L at 40°C with 5% inoculum size in batch fermentation.

It appears that further research on thermophilic microbes both prokaryotes and the eukaryotes with higher ethanologenic potential would continue. And the present, incipient successes are promising to dig more in this field for isolating, optimizing and developing thermophilic ethanologenic microbes for designing economically and environmentally friendly strategies. For obtaining the bioethanol form agro-indistrial low cost residues, various lignocellulosic materials are under trials. One of the such cellulosic materials under consideration by various workers for conversion to biofuel is sugarcane bagasse, a waste by product of sugar industry.

## 8. Sugarcane bagasse-a resource rather than a waste

Sugarcane bagasse is an important, renewable, abundant and cheap or even having negative value agricultural waste in many countries (Bustos *et al.*, 2003; Molina Junior *et al.*, 1995; Rodrigues *et al.*, 2001; Van Haandel, 2005). Composition of the fibrous residue may vary based on its different verities, age of cane at the time of harvesting and efficiency of milling operation for extracting the juice

Besides the compositional analysis, different fractions of bagasse can be separated employing suitable techniques. For example, Bustos *et al.* (2003), while describing sugarcane bagasse hydrolysis with HCl have mentioned 128°C, 2% HCl and 51.1 minutes as optimal conditions. At these conditions they obtained 22.6 g xylose, 3.31 g arabinose, 3.77g glucose, 3.59 g acetic acid and 1.54 g furfural/L.

Concerning the bioconversion of the substrate, products such as alcohol, alkaloids, mushrooms, protein enriched animal feed, enzymes L-glutamic acid, fruity aroma, and xylitol have been reported to be obtained from the waste sugarcane bagasse (Alonso *et al.*, 2007; Christen *et al.*, 1994; Liu *et al.*, 2006, 2007; Martinez *et al.*, 2000; Pereira *et al.*, 2007; Sasaki *et al.*, 2003; Van Haandel, 2005). Besides the above mentioned diverse and usually bench scale utilities of the substrate Meunchang *et al.* (2005) have rightly commented that one of the under utilized sources of organic materials, is the sugarcane industry. Global sugar production from sugarcane releases large amounts of sugar mill by-products as filter cake and bagasse. It had been reported that in Brazil at the end of last century during the ethanol production season more than $60x10^6$ tons of sugarcane bagasse containing 50% moisture were produced annually.

Like any other lignocellulosic material the sugarcane bagasse is a complex and stable substrate. Any significant and efficient utilization would require first its hydrolysis. Following acid, enzymatic or microbial hydrolysis of the sugarcane bagasse the monosaccharides yield can find many applications in different bioconversion processes. One consideration is their conversion into biofuel, the ethanol. Sugarcane bagasse has relatively earlier been considered a source of fermentable carbohydrates (Du Toit *et al.*, 1984). However, pretreatment of the bagasse has been found useful for the microbial attack, which may results into its saccharification, fermentation or the both processes simultaneously. Chemical as well as microbial enzymatic pre-treatments have been described by various workers (Chaudhary & Qazi, 2006a; Dominguez *et al.*, 1996; Laser *et al.*, 2002; Lavarack *et al.*, 2002; Martin *et al.*, 2002; Zheng *et al.*, 2002). Martin *et al.* (2002) have described that sugarcane bagasse is a potential lignocellulosic feedstock for ethanol production, since it is cheap, readily available and has a high carbohydrate content. These workers performed different pretratments of the substrate at 205°C for 10 minutes followed by its hydrolysis using cellulolytic enzymes. They found highest yield of xylose (16.2 g/100g dry bagasse), arabinose (1.5 g/100g) and total sugar (59.9 g/100g) in the hydrolysis of the $SO_2$-impregnated bagasse. The $H_2SO_4$ impregnated bagasse gave highest glucose yield (35.0 g/100g) but the lowest total sugar yield (42.3 g/100g). Sulfuric acid impregnation led to a three-fold increase in the concentration of the fermentation inhibitors, the furfural and 5-hydroxymethyl furfural and a two fold increase in the concentration of inhibitory aliphatic acids (formic, acetic and levulinic acids) compared to the without any impregnation and sulfur dioxide impregnation yields. They found no major differences in the content of inhibitors in the hydrozylates obtained from $SO_2$-impregnated and non-impregnated bagasse. When Martin and colleagues studied fermentability of the three hydrolyzates with a xylose utilizing *Saccharomyces cerevisiae* with and without nutrient supplementation they found that the $H_2SO_4$ impregnated bagasse fermented considerably poorer than the situations found in the other two categories of the bagasse. Cheng *et al.* (2007) have reported the ethanologenic fermentation of sugarcane bagasse hemicellulose hydrolyzates, pretreated by over-liming as well as electrodialysis and supplemented with nutrient materials employing *Pachysolen tannophilus* DW 06. These workers found that compared with detoxification by over-liming, detoxification by electrodialysis decreased the loss of sugars and increased the acetic acid removal. This lead to better fermentability and the Cheng's team found that a batch culture employing electrodialytically pretreated hydrolyzate substrate gave 21g ethanol $L^{-1}$ with a yield of 0.35 g $L^{-1}$ sugar and productivity of 0.59 g $L^{-1}$ $h^{-1}$. For better yield of the produce ethanol from sugarcane hydrolyzates, the above described studies highlight two important notions. That is detoxification of inhibitory substances, that may emerge within the hydrolyzates, of different nature and varying levels depending upon the specific pre-treatment employed. Secondly bagasse hydrolyzate would mainly consists of carbohydrates content, its supplementation with a suitable nutritive material is likely to enhance the growth and/or fermentative potential of the microorganism(s).

Fermentation inhibitors can be tackled at two levels i.e., their removal/detoxification or employing the inhibitors' resistant fermentative microorganisms. Martinez *et al.* (2000) reported that hemicellulose syrups from dilute sulfuric acid hydrolyzates of hemicellulose contain inhibitors that prevent efficient fermentation by yeast and bacteria. These workers have optimized overliming treatments for sugarcane bagasse hydrolyzates and found a substantial reduction in furfural, hydroxymethyl furfural and three un-identified high-performance liquid chromatography peaks. They further demonstrated that the extent of furan reduction correlated with increasing fermentability, although furan reduction was not found to be the sole cause of reduced toxicity. Rodrigues *et al.* (2001) studied the influence of pH, temperature and drgree of hydrolyzate concentration on the removal of volatile and non-volatile compounds from sugarcane bagasse hemicellulosic hydrolyzate treated with activated charcoal before and after the vacuum evaporation process. They found that furfural and 5-hydroxymethyl furfural were almost totally removed irrespective of pH, temperature and whether the charcoal was added before or after the vacuum evaporation process. Adding activated charcoal before the vacuum evaporation process favoured the removal of phenolic compounds for all values of pH. Acetic acid was most effectively removed when the activated charcoal was added after the vacuum evaporation process at an acid pH (0.92).

Regarding the use of fermentation inhibitory products' resistant microorganisms, Morita & Silva (2000) reported the fermentation of precipitated sugarcane bagasse hemicellulosic hydrolyzate containing acetic acid, employing *Candida guilliermondii* FT 120037 under different operational conditions for the production of xylose. At pH 7.0 and Kla of 35/h (4.5 vvm), the acetic acid was rapidly consumed and that the acetic inhibition was not important. They concluded that the acetic acid assimilation by the yeast inidicates the ability of this strain to ferment a partially detoxified medium and makes possible the utilization of the sugarcane bagasse hydrolyzate in this manner.

For simultaneous bioconversion of cellulose and hemicellulose to ethanol, need of xylose fermenting microorganisms has been established (Chandrakant & Bisari, 1998; Sedlak & Ho, 2004; Toivari *et al.*, 2001; Yang *et al.*, 1997). De-Castro *et al.* (2003) have described a new approach for the utilization of hemicellulosic hydrolyzates from sugarcane bagasse. They diluted the conventional feedstock, sugarcane juice; by the bagasse hydrozylate to the usual sugar concentration of 150 gm per liter that is employed for industrial production of ethanol. These workers used a pentose fermenting yeast strain, and achieved ethanol productivity of about 11.0 gm per liter per h and overall sugar conversion of more than 95%. Katzen and Fowler *et al.* (1994) reported first commercial application of unique fermenting organism capable of converting five carbon sugars and oligmers of cellulose directly to ethanol. These worker described conversion of hemicellulose content of sugarcane bagasse to the five-carbon sugar by mild acid prehydrolysis, followed by fermentation of the 5-carbon sugar extract with recombinant *Escherichia coli*. The process also recovered the majority of sucrose normally lost with the bagasse fibers to ethanol. Sun & Cheng (2002) have described the benefits of simultaneous saccharification and fermentation that it effectively removed glucose, which is an inhibitor to cellulase activity thus increasing the yield and rate of cellulose hydrolysis.

Various workers have reported different protocols and models for fermenting cellulosic biomass to ethanol and considered it the cleanest liquid fuel alternatives to fossil fuels (Gray *et al.*, 2006; Lawford & Rousseau, 2003; Lin & Tanaka, 2006; Sun & Cheng, 2002). From above cited literature it appears that relatively recently considered renewable resource, the sugarcane bagasse process much potential for the bioethanol production. In Brazil sugarcane cultivtion and its dependent sugar industry is well developed. Consequently, a huge amount of bagasse is generated. It consists mainly of 37% cellulose, 28% hemicellulose and 21% lignin (Bon, 1996). A reasonable number of cellulose saccharifying and/or ethanologenic bacteria as well yeast have been isolated and characterized in our laboratory (Chaudhary & Qazi, 2006a; Saeed, 2005).

Above referred studies suffice to highlight different achievements and areas that require more research concerning the developments of bioprocesses to utilize sugarcane bagasse lignocellulosic material for obtaining bioethanol at economically feasible levels. As in all such bioprocess developments subsidiary supports are very important. Benefits derived from appropriate utilization of auxiliary products/often process wastes, have an influential bearings on the main process economics. In this regard Pandey *et al.* (2000) pointed out an important aspect. Accordingly, developing associated or complimentary technologies, during the fuel ethanol production from sugarcane bagasse which could produce other value-added by-products would improve the overall economy of ethanol production. It is pertinent here to mention that the non-fermentable residues of variously processed sugarcane bagasse would contain the microorganisms employed for the saccharification and/or fermentation of the substrate. Thus the residue may attain the levels of protein (due to single cell protein) that may render them to the status of animal feed / supplement. This may bring additional support to the process economics. Following is a brief review of single cell protein in connection with microbiological utilization of lignocellulosic materials including sugarcane bagasse.

## 9. Single Cell Protein (SCP) from agro-industrial wastes; Sugarcane bagasse

Growth of microbial cells both bacterial and yeast on any material means that the substrate ingredients are being transferred or altered to proteins along with synthesis and accumulation of other contents of protoplasm. Upgradation of a large number of agro-industrial wastes, which after being fortified with S.C.P. may find their useful application in preparing or supplementing animal feed. The S.C.P. from various agro-industrial wastes has been well documented from several laboratories (Chaudhary & Sharma, 2005; Dimmling & Seipenbusch, 1978; Hongpattarakere & Kittikum, 1995; Kamel, 1979). Stabnikova *et al.* (2005) used extracts of cabbage, watermelon, a mixture of residual biomass of green salads and tropical fruits for yeast cultivation and concluded that the yield was comparable with the yield of yeast biomass grown in potato dextrose broth. These workers commented that the yeast biomass can be considered as protein source. Single cell protein production from sugarcane bagasse has relatively earlier been reported by various workers (Molina *et al.*, 1984; Sindhu & Sandhu, 1980). Molina and colleagues treated sugarcane bagasse pith with 1% NaOH solution at room temperature, at a NaOH/pith ratio of 10%. They used different contact times and found that the shortest period required for maximum protein production was 24h at 25°C. These workers used mixed culture of *Cellulomonas* sp. and *Bacillus subtilis*. Rodriguez *et al.* (1993) reported optimal production of *Cellulomonas* with 1% (w/v) bagasse pith pre-treated with either 0.2M NaOH for 1h at 80°C or 0.4M NaOH for 40h at 28°C to

30°C. With these milder pretreatments they obtained growth comparable to the one found for the substrate prepared with a more severe treatment. Growth was also comparable with other reports for cellulolytic bacteria cultivated on pre-treated bagasse pith. Rodriguez & Gallardo (1993) studied association of *Cellulomonas* sp. with an isolate of *Pseudomons* sp. for S.C.P. production from bagasse pith. They found a mutualistic symbiotic relationship during their mixed growth on bagasse pith, the *Cellulomonas* supplying carbon source (glucose produced from bagasse) to the *Pseudomonas* and the latter producing the vitamin supplements necessary for *Cellulomonas* growth. The metabolic symbiosis allowed the growth of the mixed culture in a minimal medium, without any growth factor supplement. Fed-batch cultivation of the mixed culture yielded high biomass production (19.4 g/L). Perez *et al.* (2002) while reporting use of sugarcane bagasse complemented with a mineral medium and inoculated with *Candida utilis* as bio-filter for ethanol concluded that 57% of the carbon from ethanol was converted to $CO_2$ and 8.7% into biomass. They found final yeast population of $7x10^9$ cells/g of dry matter corresponding to 56 mg protein/g dry matter. Perez *et al.* (2002) concluded that this much protein offers potential for using the protein enriched bagasse as feed too. The above described studies clearly indicate that the sugarcane bagasse or its pith can be upgraded with the generation of S.C.P. by employing the suitable microorganisms on untreated as well as pretreated substrates.
As has been introduced earlier, that being an agriculture country, sugarcane is cultivated at large commercial scale in Pakistan. The produce is largely used for obtaining sucrose. The bagasse is a waste of the sugar industries. Instead of other lignocellulosic material, its usage as substrate for biofuel ethanol production has two advantages. Tackling of a waste and presence of some amounts of soluble sugars that may be assimilated quickly by the inoculated microorganisms meant for saccharification and/or ethanol fermentation of the substrate. Moreover, the fermented residual material enriched with microbial cells may find its application as animal feed or its supplement there of. The latter notion is likely to bring support to the economic constrains regarding the process developments for obtaining ethanol from lignocellulosic materials in general and specifically from sugarcane bagasse. In our lab. Ahlam (2005) and Chaudhry (2008) conducted studies on the same lines and reported isolation, characterization and optimization of microorganisms both prokaryotic and eukaryotic, which are useful for saccharifying and fermenting fruits and vegetables' wastes and the sugarcane bagasse,respectively. Following maximum yield extraction, the fermented residue is likely to find its application to supplement animal feed with S.C.P.

## 10. Biohydrogen; Another potential for biofuel provision

Regarding the provision of the clean and sustainably available fuel, hydrogen gas ($H_2$) has been claimed as an alternative source of energy due to non-emission of pollutants (Das & veziroglu, 2001; Gest *et al.*, 1950; Prince & Kheshgi, 2005; Valdez-vazquez *et al.*, 2005). It is plentiful element in universe (Bockris, 1981; Levin *et al.*, 2004; Suzuki, 1982) and has a wide range of uses (Czuppon *et al.*, 1996; Kalia *et al.*, 2003; Ramachandran & Menon, 1998). Das & Veziroglu (2001) have summarized the uses of $H_2$ as reactant in hydrogenation processes, $O_2$ scavenger, fuel in rocket engines and coolant in electrical generators etc. Thus it is expected that commercial and domestic uses of hydrogen gas will increase in the coming next years. And there are signs that hydrogen may finally become an important component of the energy balance of a global economy (Benemann, 1996; Gregoire-Padro, 1998; Kalia *et al.*,

2003). Despite the green nature of hydrogen as a fuel, it is still primarily produced from nonrenewable sources such as natural gas, naphtha, and coal.

Different methods used for hydrogen production from fossil fuels include steam methane reforming of natural gas. Nearly 90% of hydrogen is produced by the reactions of natural gas or light oil fractions with steam at high temperature by the process of steam reforming (Armor, 1999; Casper, 1978; Cox & Williamson, 1979; Lodhi, 1987; Rosen & Scott, 1998; Sastri, 1989), coal gasification, thermal cracking of natural gas, and partial oxidation of heavier than naphtha hydrocarbons (Claassen *et al.*, 2006; Hawkes *et al.*, 2002). Pyrolysis or gassification is a method for the production of hydrogen from biomass (Claassen *et al.*, 1999; Hofbauer, 2007). Electrolysis, photolysis, thermochemical process, direct thermal decomposition and thermolysis are some of the conventional methods of hydrogen production from biomass. However, these are highly energy-intensive and not environmentally benign. For instance, electrochemical hydrogen production via solar battery-based water splitting requires the use of solar batteries with high-energy requirements (Basak & Das, 2007).

In order for hydrogen to become a more sustainable source of energy, it must be produced either through photosynthetic or fermentative routes using wastes or renewable substrates (Benemann, 1996; Czernik *et al.*, 2002; Dunn, 2001; Hawkes *et al.*, 2007).

The biological production of hydrogen is less energy intensive than chemical and electrochemical methods, because it is carried out largely at ambient temperatures and pressures (Benemann, 1997; Greenbaum, 1990; Miyamoto *et al.*, 1979; Sasikala *et al.*, 1993; Tanisho *et al.*, 1983).

Favoring process economics and rendering environmental improvement, the use of domestic/ agroindustrial wastes as substrates for cultivation of energy yielding microorganisms appears imperative conventional anaerobic treatment of organic pollutants results into the generation of methane (Fang & Liu, 2000; Hulshoff Pol & Lettinga, 1986) and solid wastes (Iglesias *et al.*, 1998). However, methane and its combustion products are themselves greenhouse gases (Cecchi *et al.*, 1989; Dickinson & Cicerone, 1986; Oleszkiewicz & Poggi-varaldo, 1997; Poggi-varaldo *et al.*, 1997, 1999, 2002). By contrast hydrogen gas is clean and produces no green house gases. It has high-energy yield of 122kJ/g, which is 2.75 fold greater than that of hydrocarbon fuels (Hart, 1997; Kirk *et al.*, 1985; Mizuno *et al.*, 2000; Onodera *et al.*, 1999).

Light dependent production of molecular hydrogen by photosynthetic bacteria was first observed in cultures of *Rhodospirillum rubum* growing photoheterotrophically (anaerobically) in media containing dicarboxylic acids of the citric acid cycle and either glutamate or aspartate as nitrogen source (Gest & Kamen, 1949a, 1949b). Efficiency of light energy conversion to hydrogen, and proper supply of an appropriate carbon source, are the key factors for hydrogen production by biological systems (Basak & Das, 2007; Hillmer & Gest, 1976; Miyake *et al.*, 2001, 2004; Rocha *et al.*, 2001; Zaborsky, 1998). The phototrophic purple non-sulfur bacteria (PNSB or PPNS) produce a high ratio of molecular hydrogen to carbon dioxide ranging from 85:15 to 98:2 (v/v). They can also utilize organic substrates as electron donors for hydrogen production (Hillmer & Gest, 1976; Ormerod *et al.*, 1961; Segers & Verstraete, 1983).

The purple non-sulfur bacteria (PNSB or PPNS) have been grown at temperature ranges from 28-32 °C (Jung *et al.*, 1999, Mehrabi *et al.*, 2001). Hydrogen production yield is much higher at extreme thermophilic conditions than mesophilic and thermophilic conditions ( Fang *et al.*, 2002a; Hussy *et al.*, 2003; Shin *et al.*, 2005; Wu *et al.*, 2006; Yokoyama *et al.*, 2007).

However, molecular hydrogen production by thermotolerant (purple non sulfur bacteria) *Rubrivivax gelatinosus* using raw cassava starch as an electron donor have been reported which may be suitable for out door cultivation using solar energy (Buranakarl *et al.*, 1985, 1988; Watanabe *et al.*, 1979). Mostly glucose, sucrose, molasses, lactate and cellulose have been used as substrates for hydrogen production (Kotsopoulos *et al.*, 2006; Van Niel *et al.*, 2003; Wu *et al.*, 2007; Zhang *et al.*, 2003, 2008).

Logan (2002) has reported that similar hydrogen conversion efficiencies for glucose and sucrose and lower for molasses and lowest for lactate and cellulose. It has also been demonstrated that very low pH's and high substrates concentrations can reduce biohydrogen production (Eroglu *et al.*, 2009). Increasing the substrate loading increases the relative production of volatile acids and decreases the pH, which can shift the reaction to solvent production (Jones & Woods, 1986).

Metabolic diversity of purple non-sulfur bacteria allows them to occupy a broad range of environments (Hiraishi & Ueda, 1994; Imhoff & Truper, 1992; Imhoff *et al.*, 2005; Madigan, 2003). It is one of the most diverse groups of the photoorganotrophic bacteria, as they utilize organic compounds as electron donors and carbon sources (Das & Veziroglu, 2001; Hiraishi *et al.*, 1984; Montgomery, 2004; Nandi & Sengupta, 1998). Facultatively microaerophilic to aerobic nature of these bacteria renders them versatile (Pfenning, 1977; Pfenning & Truper, 1974). However, their presence in nature, is evaluated from results obtained by enrichment techniques (Kaiser, 1966) or by membrane filtration (Biebl & Drews, 1969; Swoagar & Linderstrom, 1971). Extent of organic pollutants, mainly controls the presence of purple non sulfur bacteria in a given water body. Wide variety of organic compounds in a water body can be photoassimilated by purple non-sulfur bacteria (Cooper *et al.*, 1975; Holm & Vennes, 1970; Sunita & Mitra, 1993). Thermophilic nature, however, is limited to only among few genera of anoxygenic phototrophic bacteria (Fardeau *et al.*, 2004; Hanada, 2003). However, thermotolerant or mildly thermophilic PPNS bacteria, having optimum growth temperature, are well documented around 40°C (Ahn *et al.*, 2005; Fang *et al.*, 2002b; Hisada *et al.*, 2007; Madigan, 2003; Shin *et al.*, 2005; Ueno *et al.*, 2001b; Wu *et al.*, 2006). Low dissolved oxygen (DO) tension and availability of light and simple organic nutrients, as is the case in nutrient rich stagnant water bodies, are important factors promoting proliferation of PPNS bacteria. They occasionally occur in high numbers in wastewater treatment plants operating under highly aerated and illuminated conditions (Hiraishi *et al.*, 1989, 1991; Okubo *et al.*, 2006).

PPNS bacteria also have the capacity to grow rapidly in simple synthetic media under either anaerobic or aerobic photosynthetic conditions (Sojka & Gest, 1968). Hydrogenase, which catalyzes irreversible interconversion of hydrogen to protons and electrons is central metabolic feature of some microorganisms including most prokaryotic genera and some lower eukaryotes (Adams, 1990; Kleihues *et al.*, 2000; Noda *et al.*, 1998). Purple bacteria are also able to produce molecular $H_2$ catalyzed by nitrogenase under nitrogen limiting conditions (Tsygankov *et al.*, 1998).

A biological wastewater treatment process using purple non sulfur bacteria (PNSB), has been used for purifying various organic wastewater, especially food industrial wastes of high BOD strength, extreme thermophilic conditions offer better destruction for digested residues (Kobayashi & Tchan, 1973; Sahlstrom, 2003) and full scale wastewater treatment plants for food industry had been reported by Sasikala & Ramana, (1995). PNSB will grow selectively in a reactor under illumination and heterotrophically remove organic carbon. Wastewater treatment by PNSB is considered to be effective because they are metabolically the most versatile among all prokaryotes. Anaerobically photoautotrophic and photoheterotrophic in presence of light and aerobically chemoheterotrophic microbes in the dark can accomplish consumption of various types of organic matter (Levin *et al.*, 2004; Nakadomi *et al.*, 1999). Concentrated latex wastewater (Choorit *et al.*, 2002), aquarium wastewater (Nakadomi *et al.*, 1999) and agricultural waste (Arooj *et al.*, 2007; Fang & Liu, 2002; Hiraishi *et al.*, 1989; Liu & Shen, 2004; Wang *et al.*, 2007; Yang & Shen, 2006; Yokoi *et al.*, 2001) and sewage wastewater treatment (Li & Fang, 2007; Nagadomi *et al.*, 2000) are main types of wastes that are treated with purple non sulfur bacteria. These bacteria directly convert organic carbon into biomass that is suitable for direct reuse e.g., single cell protein (Kobayashi & Tchan, 1973). Cell mass of purple non sulfur bacteria has been reported to be a good alternative of manure, fish feed or agriculture supplement because of its richness in proteins and vitamins (Getha, *et al.*, 1998; Kobayashi & Tchan, 1973).

Single cell protein (SCP) is one of the very attractive facets of agricultural and industrial waste upgradation/ treatment. The microbial biomass surely attains higher levels of proteins and other nutritionally important parameters than the non-cultivated substrate/ waste. Many workers have advocated SCP as potential animal feed or supplement thereof. This aspect renders the waste treatment process economically feasible or at least cost-compatible. Following is a comprehensive review of some useful attributes of hydrogen producing bacteria.

### 10.1 $H_2$ gas production and phototrophic bacteria

Hydrogen gas has been perceived one of future renewable energy resources as well as an environmentally compatible one as it does not evolves the "greenhouse gas" $CO_2$ at combustion (Abraham, 2002; Hansel & Lindblad, 1998; Matsunaga *et al.*, 2000; Woodward *et al.*, 1996).Further combustion of $H_2$ liberates large amounts of energy per unit weight and is easily converted to electricity by fuel cells (Miyamoto, 1993; Ueno *et al.*, 1995; Rupprecht *et al.*,2006). Hydrogen may be produced by electrolysis of water, thermocatalytic reformation of hydrogen-rich organic compounds and biological processes. At present it is almost exclusively produced by electrolysis of water or by steam reformation of methane. Whilst, biological hydrogen production has several advantages over hydrogen production by photochemical or thermochemical processes. Various routes of biological hydrogen production such as direct biophotolysis, indirect biophotolysis, photo and dark fermentations has been described (Kapdan & Kargi, 2006; Levin *et al.*, 2004).

Numerous prokaryotes produce hydrogen gas, while some photosynthetic microorganisms also produce light dependent hydrogen from organic substrates. Among the photosynthetic microorganisms, photosynthetic bacteria exhibit a high level of hydrogen production. Nonetheless, low conversion efficiencies of other biological systems can be compensated for, by low energy requirements and reduced initial investment costs (Macler *et al.*, 1979; Oh *et al.*, 2002).

Moreover, in laboratory experiments, high light energy conversion efficiency upto 7% has been reported for photoheterotrophic processes (Nakada *et al.*, 1995; Sunita & Mitra, 1993; Ueno *et al.*, 1995)It is reported by Rupprecht *et al.* (2006) that continuously depleting oil reserves had necessitated for search of alternative energy sources. of clean fuels and less $CO_2$ emission to reduce the impact of global warming is meet by employing potoheterotrophic and photoautotrophic organisms in bio-$H_2$ production process.

In the mid-1920s, Van Niel, was attracted by the color of these purple and green sulfur bacteria which ranges from purple-red to yellowish green. These bacteria impart their coloration to their habitats too. Ecophysiology of these anoxygenic phototrophic bacteria became a topic of scientific interest in the late 1960s. Later on, Van Niel (1929), morphologically and physiologically characterized pure cultures of purple and green sulfur bacteria. These anoxygenic phototrophs play crucial roles in the biogeochemical cycling of sulfur. In addition to being photosynthetic purple sulfur bacteria also have a chemosynthetic metabolism (Bryant & Frigaard, 2006; Madigan, 2003; Overmann, 2001; Pfenning, 1987). Photosynthetic process of purple sulfur bacteria differs from that of green plants in respect of photochemical reduction of $CO_2$ with hydrogen which is ultimately derived, not from water as in green plants, but from $H_2S$. But purple non sulfur bacteria utilizing (needed for phototrophic growth by these organisms) no hydrogen sulfide, in addition to light source and anaerobicity. Hence the use of "non-sulfur" in purple non sulfur group name derived from no use of $H_2S$ under photoautotrophic conditions (Van Niel, 1929, 1931).

Van Niel (1944), reported that purple bacteria are characterized with a complex pigment system, made up of a green pigment, "bacteriochlorin" and one or more red pigments, "bacteriopurpurin." Diagnostically, these were distinguished by the occurrence of sulfur droplets in the cells by their accumulation in the external environment and fundamental difference of autotrophic and heterotrophic modes of life (Holt *et al.*, 1994; Imhoff & Truper, 1992; Jansen & Harfoot, 1991; Pfenning, 1977; Pfenning & Truper, 1974).

Hiraishi *et al.* (1984), examined fifty five strains of 13 species of purple bacteria *Rhodospirillaceae* and one strain of *Chromatium vinosum* for presence and composition of isoprenoid quinine. The author also discussed significance of quinine system in *Rhodospirillaceae* taxonomy and divided the bacteria into five categories based on their predominant quinine pattern.

## 10.2 Nature of the bacterial pigments

Presence of bacteriochlorophyll **a** and carotenoids of spirilloxanthin series have been also detected by spectroscopic analyses. Ubiquinone-10 was found to be major quinine. A clone library also showed that all of the clones derived from the members of the genera *Rhodobacter* and *Rhodopseudomonas*. The dominant phototrophic bacteria identified by studying 16S rRNA and *puf*M gene sequence information were *Rhodopseudomonas* and *Rhodobacter* (Brock & Madigan, 1991; Imhoff & Truper, 1998; Madigan, 2003). The investigation concluded that *Rhodopreudomonas* isolates had been reported to home a wider spectrum of carbon utilization and higher affinity for acetate than *Rhodobacter* isolates. This was demonstrated by oxygen uptake with lower fatty acids. A laminated mitochondria-like structure, resembling the chloroplast of higher plants has been found the pigment bearing portion of the *Rhodospirillum rubrum* (Nianzhi *et al.*, 2003; Niklowitz & Drews, 1955; Okubo *et al*, 2006). Disrupted cells of *Rhodospirillum rubrum* yield discrete subcellular particles with a sedimentation constant of 190S and a diameter, when dried on a membrane, of about 1100Å. Since these units were found to contain all of the photosynthetic pigments of the cell, they were accordingly named chromatophores (Padree *et al.*, 1952; Schachman *et al.*, 1952; Yasa *et al.*, 2006). Due to intracellular deposition of sulfur and presence of pigments the bacteria are called as purple sulfur bacteria with purple or violet growth

## 10.3 Purple non-sulfur bacteria and their natural occurrence

Laboratory culture of purple non-sulfur bacteria appear as pink, violet, to deep-red (Bryant & Frigaard, 2006; Melis, 2005; Proctor, 1997).

Lake Fryxell of Antarctica, show extensive diversity and highly stratified distribution of purple non-sulfur bacteria, as determined by analysis of a photosynthesis-specific gene, *puf*M. Enrichment cultures for purple bacteria from the Lake have been found to yield two morphotypes possessing gas vesicles and buoyancy structures previously not reported in purple non-sulfur bacteria. These structures have been explained necessary for the organisms to position themselves at specific depths within the nearly freezing water column (Jones *et al.*, 1998; Yasa *et al.*,2006). Wide variety of photosynthetic purple and green bacteria has been isolated from cool drink refilling stations' wastewater. A fastest growing isolate of a *Rhodopseudomonas* sp. has been reported to produce hydrogen gas in the presence of light and its Immobilized cells yielded significant amounts of hydrogen from both sewage and wastewater. A number of microbes can metabolize conversion of water and carbon monoxide into hydrogen and carbon dioxide. Rate of CO conversion and $H_2$ production performed on trickle-bed reactor (TBR). Overall performance of reactor controlled by liquid recirculation rate and reactor support material both affects the mass transfer coefficient. A good agreement was obtained between two reactor scales after comparing simple reactor and TBR (Karr *et al.*, 2003; Sunita, & Mitra, 1993; Wolfrum & Watt, 2002).

Purple non sulfur bacteria have been described to be present in relatively high numbers ($10^5$-$10^7$ viablecells/ ml) in surface water of ditches, paddy fields and tide pools under oxygen-limited conditions. Whereas in eutrophic pond and river waters where sufficient amount of oxygen is present they have been demonstrated to be absent or found in small numbers. However, activated sludge and other aquatic environments mainly have facultative aerobic phototrophic purple non sulfur bacterial flora richness and diversity of which depends on organic nutrients composition of the waste water (Hiraishi & Kitamura, 1984; Oda *et al.*, 2003; Van Ginkel & Sung, 2001).

Okubo *et al.* (2006), while studying microbial mat of swine waste ditch reported that wastewater of the ditch contained acetate and propionate as major carbon nutrients based upon independent biomarker and molecular methods as well conventional cultivation methods, these authors revealed that the microbial mats were dominated by rod shaped cells containing intracytoplasmic membranes and small number of oval cells with vesicular internal membrane.

### 10.4 Appealing characteristics of phototrophic bacteria

Photosynthetic bacteria evolve hydrogen at much higher rates than do other classes of photosynthetic microorganisms. In addition, they can metabolize common biomass waste material, genetically may be easily manipulated to enhance rate of hydrogen production, grow rapidly, use both visible and near infrared light rays and tolerate harsh environments. Photosynthetic bacteria show vigorous evolution of $H_2$ gas when DL-malate or DL-lactate is used as an electron donor (Fascetti & Todini, 1995; Federov *et al.*, 1998; He *et al.*, 2005). Various agricultural products and wastes have been used as substrates for photosynthetic bacteria and the non-sulfur purple photosynthetic bacteria producing hydrogen from raw starch, corn, potato, or cassava as well as soluble starch under phototrophic conditions have been described (Buranakarl *et al.*, 1985; Ike *et al.*, 1996,1998; Weaver *et al.*, 1979).

Hydrogen production by *Rhodospirillum rubrum* with lactate containing wastes, including yogurt waste and whey has been reported. While soluble starch and cooked cassava starch have been described for the growth of *Rhodopseudomonas gelatinosa* to produce vitamin $B_{12}$, ubiquinone and single cell protein respectively (Noparatnaraporn *et al.*, 1983; Sasaki & Nagar, 1979; Zurrer & Bachofen, 1979).

Besides the $H_2$ production, successful cultivation of the fastidious anaerobes on natural and preferably agroindustrial wastes has opened new gate for the production of single-cell protein (Honda, *et al.*, 2006; Robert & Wolfe, 1970; Weaver, *et al.*, 1975). Further purple non-sulfur bacteria like *Rhodobacter sphaeroides* O.U. 001 produce useful by product in addition to hydrogen gas when grown anaerobically. Poly-β-hyroxy (PHB) has been reported to be produced in maximum concentration when bacteria have anaerobically grown in standard growth media containing L-malic acid, sodium glutamate and 30% wastewater from sugar refinery (Yigit *et al.*, 1999). Wastewater from sugar refinery has been partially replaced carbon source of bacterial growth medium (Arik *et al.*, 1996; Yetis *et al.*, 1998). It is pertinent here to stress another applicable offshoot of the purple non-sulfur bacterial metabolism that PHB is a biodegradable thermoplastic that can be synthesized during unfavorable growth conditions by a wide range of bacteria (Byrom, 1987; Page, 1989; Kim *et al.*, 1994; Yamagishi, 1995).

Anderson and Dawes (1990) reported that in addition to some important medical applications PHB can be used to construct biodegradable carriers for long term dosage of herbicides and insecticides, packaging containers, bottles and bags.

### 10.5 Factors influencing production of $H_2$ by purple non-sulfur bacteria

Like any microbiological process performance parameters including pH, temperature, hydraulic retention time, seed sludge, nutrients, inhibitors, reactor design, and the means used for lowering hydrogen partial pressure are very important (Chen *et al.*, 2001; Fang & Liu, 2002; Khanal *et al.*, 2004; Lay *et al.*, 2000; Lay, 2001). As for example, for the optimizing the cultivation of the non-sulfur purple bacteria over accumulation of dissolved hydrogen in the liquid and high hydrogen partial pressures are thought to inhibit the process of hydrogen production (Chenlin & Herbert, 2007; Doremus *et al.*, 1985; Fennell & Gossett, 1998; Pauss *et al.*, 1990).

Infact, various factors have been studied in relation to bacterial hydrogen production including enzymes, of the process and inhibitory effects of $NH_4$ on gas production (Kim *et al.*, 1980; Koku *et al.*, 2003; Yoch & Gotto, 1982; Zhu *et al.*, 2001).

Besides the identification of proper cultivation strategies for obtaining higher and/or continuous yield potential/nature of the organism is of prime importance. For instance mutants of *Rhodopseudomonas palustris* capable of producing hydrogen constitutively, even in the presence of ammonium has been reported. It is pertinent to note here that wild-type cells do not accumulate detectable amounts of hydrogen in the presence of ammonium. Two purple non sulfur bacteria *Rhodopseudomonas palustris* and *Rhodobacter sphaeroides* grow with taurine as a sole electron donor, sulfur and nitrogen sources (Oda *et al.*, 2003; Qian *et al.*, 2002; Rey *et al.*, 2007).

Bao (2002) reported that *Rhodopseudomonas palustris* Z strain produced maximum $H_2$ by using acetate as carbon source and electron donor. The author also concluded that temperature, light intensity and acetate and glutamate concentrations significantly affected hydrogen photo evolution and cell growth. Rate of hydrogen production was found to be inversely related to substrate concentration, while directly to cell growth.

Efficiency of conversion of light energy to hydrogen is the key factor for the realization of hydrogen production from biological systems (Barbosa *et al.*, 2001; Kondo *et al.*, 2002; Shi & Yu, 2005). Batch tests using mixed cultures have demonstrated that very low pH's and high substrates, such as starch concentrations can reduce biohydrogen production (Liu & Shen, 2004; Van Ginkel & Sung, 2001; Zhang & Shen, 2006).

Regarding effect of light intensity on hydrogen production by photosynthetic bacteria *Rhodopseudomonas sp.* has been demonstrated to produce highest volume of hydrogen at a production rate of 25ml $H_2\,l^{-1}\,h^{-1}$, under a light intensity of 680 µmol photons $m^{-2}s^{-1}$. Decrease in the electron donor concentration organic acid resulted in decreased hydrogen evolution (Barbosa *et al.*, 2001; Fang *et al.*, 2006; Miyake *et al.*, 1999; Najafpour *et al.*, 2004; Oh *et al.*, 2004). Light dependent hydrogen production was first observed by *Rhodospirillum rubrum* in a media containing dicarboxylic acids of the citric acid cycle and either glutamate or aspartate as nitrogen sources. Hydrogen is not evolved, however, when nitrogen is provided in the form of ammonium salts, which repress synthesis of the hydrogen- evolving system (Barbosa *et al.*, 2001; Gest & Kamen, 1949a, 1949b; Hillmer & Gest, 1976; Ormerod *et al.*, 1961).

Concerning the illumination patterns for phototrophic bacterial growth it has been claimed that single-step illumination method provides an appropriate simulation of sunlight and its dependent hydrogen production. While indoor hydrogen production rate found to be independent of the mode of illumination. However, actual outdoor research has been considered difficult owing to the fluctuation of sunlight due to the weather time of the day solar spectrum etc. (Katsuda *et al.*, 2006; Kim *et al.*, 1987; Noparatnaraporn *et al.*, 1982; Novak *et al.*, 2004).

## 10.6 Mechanism of biological $H_2$ production

Amongst microbial diversity strict anaerobes and facultative anaerobic chemoheterotrophes are efficient producers of hydrogen. However, lower rate of hydrogen evolution by fermentative processes necessitates research for improving yield (Kim *et al.*, 2004; Nath & Das, 2004). Accumulation of hydrogen and other degradation byproducts during fermentation however can make the hydrogen-acetate reaction unfavorable leading to solvent production (Bahl *et al.*, 1982; Fond *et al.*, 1985; Grupe & Gottchalk, 1992; Jones & Wood, 1986).

Such investigations advocate the necessity of simultaneous removal of $H_2$ from a fermentation system. Thus for higher yield of hydrogen, a system is capable of removing $H_2$ before its accumulation to the level where repression of its production comes and to prevent interspecies hydrogen transfer leading to methanogenesis, would have to be developed.

Various attempts have been made to enhance hydrogen production by using proper cultivation conditions, such as provision of nitrogen atmosphere, and various organic acids as electron donors (Buranakarl *et al.*, 1988; Kim *et al.*, 1981; Mizuno *et al.*, 2000; Segers & Verstraete, 1983; Watanabe *et al.*, 1979).

Anaerobic photosynthetic bacterium, *Rhodospirillum* has been evaluated for bioconversion of syngas (synthesis gas) to hydrogen in continuous stirred tank bioreactor utilizing acetate as a carbon source upto period of two months. Such continuous process have been considered as alternative method of conventional Fischer-Tropsch synthetic reaction for the conversion of syngas into hydrogen (Miyake *et al.*, 1982; Najafpour *et al.*, 1995; Younesi *et al.*, 2008). Predominant commercial technology for syngas has been steam methane reforming, in which methane and steam were catalytically and endothermically converted to hydrogen and carbon monoxide. *Rhodospirillum rubrum* the purple non-sulfur anaerobic bacterium, is capable of catalyzing the Fisher-Tropsch type reaction and water gas shift reaction (Barbosa *et al.*, 2001; Czerink *et al.*, 2000).

*Rhodospirillum rubrum* has been reported for CO uptake with faster growth and higher cell dry weight as compared to other hydrogen producing microorganisms (Koku *et al.*, 2002,

2003; Lee *et al.*, 2002). It is interesting to note three times faster rate of carbon monoxide (CO) consumption than that of acetate. Two species of purple non sulfur bacteria, *Rhodopseudomonas gelatinosa* and *Rhodospirillum rubrum* are known to perform the water- gas shift reaction to produce hydrogen (Klasson *et al.*, 1990; Najafpour *et al.*, 2003, 2004; Oh *et al.*, 2002, 2004).

In certain bacterial strains, like massive non-sulfur photosynthetic bacterium (*Rhodovulum* sp.) rate of hydrogen production depends on light and addition of a small amount of oxygen (Maeda *et al.*, 2003; Matsunaga *et al.*, 2000). Four times higher rate of hydrogen production under microaerobic conditions as compared to anaerobic conditions, emphasizes the need of but little oxygen in such processes. Bacterium *gelatinosus* CBS has been described for CO oxidation and equal rates of $H_2$ production rate of $H_2$ in light and dark conditions in the presence of electron transport uncoupler carbonyl-cyanide m-Chlorophenylhydrazone (CCCP) (Fascetti & Todini, 1995; Federov *et al.*, 1998; He *et al.*, 2005; Maness *et al.*, 2004; Matsunaga *et al.*, 2000).

A modified pour plate technique with an overlay of wax has been worked out for isolation and enumeration of purple non-sulfur bacteria (PNSB) with equal efficiency as that of agar shake culture. Owen and respirometric methods can be choice for evaluating biological production of hydrogen from fermentation in batch tests by using different substrate. However, 43% more hydrogen gas has been reported to be produced by respirometric method. Comparable hydrogen conversion efficiencies for sucrose and molasses and much lower for lactate and cellulose have been documented (Archana *et al.*, 2004; Logan, 2002).

Regarding the provision of photons, Hai *et al.* (2000) have described use of a low-cost closed tubular glass photobioreactor for cultivations of axenic/ anoxygenic photrophic bacteria, oxygenic photrophic cyanobacterium and microalgae.

## 10.7 Treatment of wastewater by purple non-sulfur bacteria

Various agricultural/industrial wastes containing organic acids and sugar residues have been identified and used as feedstocks for microbial fermentations designed to yield value added products (Yetis *et al.*, 2000; Yigit *et al.*, 1999, Zhu *et al.*, 1999, 2002). Besides the availability of rich diversity of other microbes for such fermentative processes phototrophic, purple non-sulfur bacteria, have to been exploited for example they have been explored to treat odorous swine based water (Kim *et al.*, 2004; Nath & Das, 2004).

Likewise purple non-sulfur bacteria have been described for treatment of domestic and industrial effluents of varying chemical nature (Lay, 2001; Najafpour *et al.*, 2006; Nakashimada *et al.*, 2002; Nath & Shukla, 1997; Zhu *et al.*, 1995).

## 10.8 Role of different enzymes in hydrogen production

Direct and efficient photoproduction of $H_2$ gas by biological means have encountered several challenges. In this regard sensitivity of hydrogenases (the $H_2$ evolving enzyme) to $O_2$ and different approaches to overcome the enzyme's limitation has been emphasized. Simultaneous production of hydrogen and oxygen gas have been demonstrated for a nitrogen fixing cynobacterium, *Anabaena cylindrical* in an argon atmosphere for several hours (Asada *et al.*, 1985; Miyake *et al.*, 1989; Miyamoto *et al.*, 1979; Weissman & Benemann, 1977; Wolk *et al.*, 1994).

Biological nitrogen fixation is a major route for hydrogen production by purple photosynthetic bacteria in which nitrogenase act as reducer of atmospheric nitrogen to ammonia with the concomitant production of molecular hydrogen. However, it is rather inefficient process because about 75% of the reductant consumed by the nitrogenase is used to generate ammonia (Takabatake *et al.*, 2004; Zhu *et al.*, 2001). Albeit some of the strains have been isolated of purple photosynthetic bacteria in which hydrogen production is necessary for growth and independent of nitrogen fixation (Benemann & Weare, 1974; Ghirardi *et al.*, 2005; Miyake *et al.*, 1989; Proctor, 1997).

Many bacteria contain enzymes (hydrogenases) that can produce hydrogen during the fermentation of a variety of substrates. ATP is produced by substrate level or electron transport phosphorylation, but the ATP yields of fermentation are quite low as compared to those of aerobic oxidation reactions. Fermentation reactions can produce many different end products such as hydrogen, acetate, ethanol, and others. The hydrogen-acetate couple produces more ATP per mole of substrate than alcohols, such as ethanol and butanol, and represents energetically a "preferred" bacterial fermentation product for a sugar.The green alga, *Scenedesmus*, also produces molecular hydrogen under light after being kept under anaerobic conditions (Asada & Miyake, 1999; Florin *et al.*, 2001; Gaffron and Rubin, 1942; Gottschalk, 1986).

Unique type of hydrogenase activity found in some photosynthetic bacteria that function in darkness to shift CO and $H_2O$ into $H_2$ and $CO_2$ has been exploited in hollow-fiber and bubble-train bioreactors employing immobilized and free-living bacteria. These efforts proven effective for enhancing the mass transfer of CO (Mao *et al.*, 1986; Miyake & Kawamura, 1987). Occurrences of these biological processes at ambient temperatures and pressures requires minimum energy inputs and thus appear promising for low cost are need for the process developments (Klasson *et al.*, 1992; Markov & Weaver, 2008; Najafpour *et al.*, 2004).

## 10.9 Implication of phototrophs in environmental clean up and upgradation

Constructed microbial mats including photo-organotrophic purple non-sulfur bacteria have been used for bioremediation of heavy metals and organic chemical pollutants. A gram-negative rod shaped bacterium, characterized with production of dark red culture under phototrophic conditions, budding, bacteriochlorophyll a and carotenoids has been found to possess multiple metal resistances and to be effective in the reductive removal of Cr (VI) and the degradation of 2,4,6-trichlorophenol (Mehrabi *et al.*, 2001; Miyoshi, 1997; Moucha *et al.*, 2003; Okubo *et al.*, 2006).

Production of hydrogen gas has been reported for anaerobes, facultative anaerobes, aerobes, methylotrophs and photosynthetic bacteria. Anaerobic *Clostridia* are potential producers of $H_2$ (Cohen *et al.*, 1985; Taguchi *et al.*, 1995; Yokoi *et al.*, 1998). Spontaneous production of $H_2$ from formate and glucose by immobilized *Escherichia coli* showed 100% and 60% efficiencies, respectively. *Enterobactericiae* produce $H_2$ at similar efficiency range from different mono-saccharides (Fabiano & Perego, 2002; Nakashimada *et al.*, 2002; Tanisho & Ishiwata, 1995; Yokoi, *et al.*, 1997, 1998, 2001). Among methylotrophs, methanogenes, rumen bacteria and thermophilic archae, *Ruminococcus albus*, are promising. Photosynthetic *Rhodospirillum rubrum* produces 4, 7, and 6 mol of $H_2$ from acetate, succinate and malate, respectively. Excellent productivity (6.2 mol $H_2$/mol glucose) by co-cultures has been achieved. *A. cylindrica* produced $H_2$ (20 ml/g dry wt/h) continually for 1 year. *Synechococcus* sp. has a high potential for $H_2$ production in fermentors and outdoor cultures (Aoyama *et al.*, 1996; Miyake & Asada, 1996; Miyake *et al.*, 1992, 1996). Simultaneous productions of oxychemicals and $H_2$ by *Klebseilla* sp. and by enzymatic methods have also been attempted. The fate of $H_2$

biotechnology is presumed to be dictated by the stock of fossil fuel and state of pollution in future (Gorman, 2002; Koku *et al.*, 2002; Nandi & Sengupta, 1998). Mirza (unpublished data) has successfully cultivated PNSB employing sugarcane bagasse and food industrial waste waters for the production of biohydrogen, essentially following the scheme depicted in Fig.3.

Of the above described nutrient diversity of the phototrophic bacteria (Photoheterotrophes) are highly relevant for upgradation of food industrial wastes. These heterotrophes from nature can use organic remnants in the effluents that can yield the photohydrogen as well as cell biomass to single cell protein.This process supporting aspect of biofules fermentations has been described in detail at the end of bioethanol section.

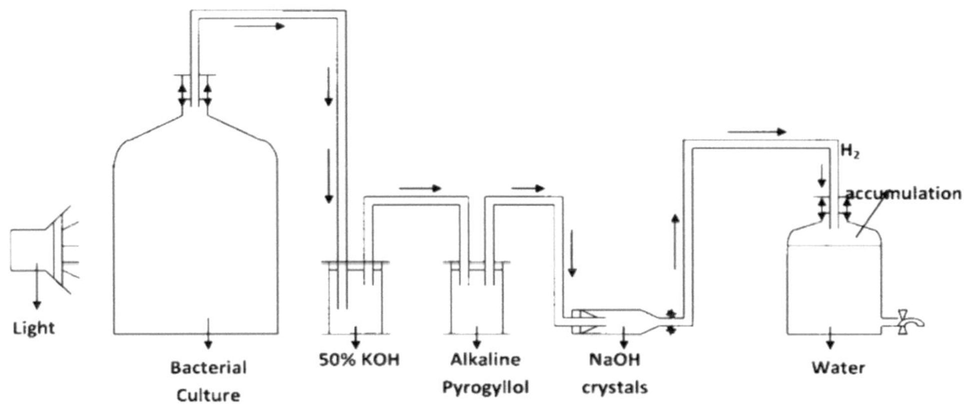


Fig. 3. Schematic Description of Anaerobic Bioreactor for Hydrogen Production

## 11. Conclusions & recommendations

Bacterial and yeast isolates capable of simultaneous saccharification and ethanol fermentation of agro-industrial wastes such as sugarcane bagasse, an industrial cellulosic waste material, may be exploited to convert the very stable carbohydrate material to its monomeric constituents and the biofuel. The consolidated processing of the substrate (sugarcane bagasse) employing mono as well as co-cultures of the microorganisms can be escalated by identifying a suitable low-cost or preferably value-less or even having a negative value substrate that can act as nutritional supplement. Fortification of the bagasse suspensions with processed yogurt whey gave fruitful results in terms of the magnificent microbial growths and ethanol yields (Chaudhary, 2008). Further the ferment remnants by virtue of their relative higher protein/carbohydrate ratio as compared to the intact substrate may find their involvement in raising animal feed from otherwise, an agro waste and an environmental issue. Identifying the nature and mode(s) of generation of fermentation inhibitors following different pretreatments of lignocellulosic biomass appear to highlight hindrances at the industrial level bioconversion of such substrates to biofuel. Removal or developing the inhibitors detoxification agents, both microbial and/or chemical appears imperative. Testing the reported bacterial and yeast isolates for their detoxification potential and tolerance levels for the inhibitors formed during various pretreatments of the bagasse would be major attribute defining the process development.

Albeit the largest renewable resource, the lignocellulosic matter, is recycled in the nature, the process, which directly and/or indirectly supports its own regeneration too within the biosphere. However, owing to slow turn over rates of the huge and abundantly found piled up reservoirs encountered both in natural ecosystems and agro-industrial sectors which represent a pollutant in select situations allow and demand its biotechnological utilizations, respectively. The former notion reminds, in a real sense, using water at human scale from an ocean and fearing decrease in the water to a level that may disturb the ocean.

Whilst the saccharification and ethanol fermentation of various lignocellulosic materials have been and are being reported by different workers from different regions of the world. The research in this area should be further promoted and encouraged. As more diverse data are required to extract valuable information, which hopefully would be employed to make the dream, obtaining biofuel from lignocellulosic material, at commercial level, a reality.

Likewise, biotechnological exploitation of purple non-sulfur bacteria for obtaining $H_2$ following their cultivation in agro/food industrial waste waters appears promising. Hetertrophic nature of such microbes dictates simultaneous provision of $H_2$ source as well as removal of organic load from agro-industrial effluents which otherwise are the source and non-source pollutants of water bodies in many country. Well illuminated land areas throughout the year are also promising for development of photon based bioreactors in an cost-effective way. Biological hydrogen production a clean/rich energy source represents an exciting new area of technology development for bioenergy generation. Low cost, environmental acceptability, renewability and evolution of no $CO_2$ following combustion are attractive features of $H_2$ as compared to hydrocarbon fuels .

In addition to hydrogen production purple non-sulfur bacteria are applicable for treatment of wastewater for lowering the level of volatile fatty acids, organic matter and Biological oxygen demand. Cell mass of purple non-sulfur bacteria may be used as good alternate of manure, fish feed or agricultural supplement because of its richness in proteins and vitamins.

## 12. References


Abraham, S., 2002. *Toward a more secure and cleaner energy future for America: national hydrogen energy road map: production, delivery, storage, conversion, applications, public education and outreach.* U.S. Department of energy, Washington D.C.

Adams, M.W.W., 1990. The structure and mechanism of iron hydrogenases. *Biochimica et Biophysica Acta,*1020: 115-145.

Ahlam, A.W.,2005. *Microbiological conversion of waste fruits and vegetables into ethanol.*Ph.D.thesis Univ.Punjab.Lahore Pakistan.

Ahn, Y., Park, E.J., Oh, Y.K., Park, S., Webster, G. & Weightman, A.J., 2005. Biofilm microbial community of a thermophilic trickling biofilter used for continuous biohydrogen production. *FEMS Microbiology Letters,* 249: 31-38.

Alonso, P.W., Garzone, P. & Cornacchia, G., 2007. Agro-industry sugarcane residues disposal: the trends of their conversion into energy carriers in Cuba. *Waste Manag.,* 27: 869-885.

Anderson, A.J. & Dawes, E.A., 1990. Occurrence, metabolism, metabolic role, and industrial uses of bacterial poly hydroxy alkanolates. *Microbiological Review,*54: 450-472.

Anderson, W.F., Peterson, J., Akin, D.E. & Morrison, W.H., 2005. Enzyme pretreatment of grass lignocellulose for potential high-value co-products and an improved fermentable substrate. *Appl. Biochem. Biotechnol.*, 124: 303-310.

Aoyama, K., Miyake, M. & Yamada, J., 1996. Application of the vector, pKE4-9 carrying strong promoter to the expression of foreign proteins in *Synechoccocus* PCC7942. *Journal of Marine Biotechnology*, 4(1): 64-67.

Archana, A., Sasikala, Ch., Ramana, Ch. V. & Arunasri, K., 2004. "Paraffin wax-overlay of pour plate", a method for the isolation and enumeration of purple non-sulfur bacteria. *Journal of Microbiological Methods*,59(3): 423-425.

Arik, T., Gunduz, U., Yucel, M., Turker, L., Sediroglu, V. & Eroglu, I., 1996. Photoproduction of hydrogen by *Rhodobacter sphaeroides* O.U.001. *Proceedings of the 11th World Hydrogen Energy Conference, Stuttgart, Germany*,3: 2417-2424.

Armor, J.N., 1999. The multiple roles for catalysis in the production of $H_2$. *Applied Catal A: Gen*, 176: 159-176.

Arooj, M.F., Han, S.K., Kim, S.H., Kim, D.H. & Shin, H.S., 2007. Sludge characteristics in anaerobic SBR system producing hydrogen gas. *Water Research*, 41: 1177-1184.

Asada, Y. & Miyake, J., 1999. Photobiological hydrogen production. *Journal of Bioscience and Bioengineering*, 88(1): 1-6.

Asada, Y., Tomizuka, N. & Kawamura, S., 1985. Prolonged hydrogen evolution by a cyanobacterium (blue-green alga), *Anabaena* sp. Journal of Fermentation Technology, 63: 85-90.

Atiyeh, H. & Duvnjak, Z., 2002. Production of fructose and ethanol from sugar beet molasses using *Saccharomyces serevisiae* ATCC 36858. *Biotechnol. Prog.*, 18: 234-239.

Bahl, H., Andersch, W., Braun, K. & Gottschalk, G., 1982. Effect of pH and butyrate concentration on the production of acetone and butanol by *Clostridium acetobutylicum* grown in continuous culture. *Applied Microbiology and Biotechnology*,14: 17-20.

Ballesteros, I., Oliva, J.M., Saez, F. & Ballesteros, M., 2001. Ethanol production from lignocellulosic byproducts of olive oil extraction. *Appl. Biochem. Biotechnol.*, 93: 237-252

Bao, S.W.G.C.X., 2002. Hydrogen production from acetate by *Rhodopseudomonas palustris*. *Chinese Journal of Biotechnology*, 18(4): 486-91.

Barbosa, M.J., Rocha, J.M.S., Tramper, J. &Wijffels, R.H., 2001. Acetate as a carbon source for hydrogen production by photosynthetic bacteria. *Journal ofBiotechnology*, 85(1): 25-33.

Basak, N. & Das, D., 2007. The prospect of purple non-sulfur (PNS) photosynthetic bacteria for hydrogen production: The present state of art. *World Journal of Microbial Biotechnology*,23: 31-42.

Benemann, 1996. Partial characteristics of hydrogen production by fermentative hydrogen producing bacterial strain B49. *Journal of nature biotechnology*, 14: 1101-1103.

Benemann, J.R. & Weare, N.M., 1974. Hydrogen Evolution by Nitrogen-Fixing Anabaena cylindrica Cultures. *Science*, 184: 174-175.

Benemann, J.R., 1997. Feasibility analysis of photobiological hydrogen production. *International Journal of Hydrogen Energy*,22: 979-87.

Bernstein, Winkler and Zierdt–Warshaw, 1996. *Addison – Wesley Environmental science. Ecology and human impact*. Addison - Wesley publishing Co. New York. USA.

Bicho, P.A., Runnals, P.L., Cunnid, J.D. & Lee, H., 1988. Inducttion of xylose reductase and xylitol dehydrogenate activities in *Pachysolen cannothilus* and *Pichia stipitis* on mixed sugar. *Appl. Environ. Microbiol.* Washington, 1: 50-51.

Biebl, H. & Drews, G., 1969. Das in-vivo-pektrum als taxonomisches Merkmal bei Untersuchun en zur Verbreitung. von Athiorhodaceae. *Zentralbl Bakteriol Parasitenkd. Infektionskr. Hyg. Abt.* 2 (123): 425-452.

Bockris, JO'M., 1981. The economics of hydrogen as a fuel. *International Journal of Hydrogen Energy*,6: 223-41.

Bon, E.P.S., 1996. Ethanol production via enzymatic hydrolysis of sugarcane bagasse and straw. *Brazilian bioethanol project* Financial support. Ministry of science and technology/FINEP.

Brock, T.D. & Madigan, M.T., 1991. *Biology of microrogaisms.* 6th Ed. Prentice Hall, New York.

Bryant, D.A. and Frigaard, N.U., 2006.Prokaryotic photosynthesis and phototrophy illuminated. *Trends in Microbiology*,14 (11): 488.

Budden, J.R. & Spencer, P.A., 1993. Tolerance to temperature and water quality for bacterial oxidation: the benefits of BacTech's moderately thermophilic culture. *FEMS Microbiology Reviews* 11: 191-196.

Buranakarl, L., Cheng-Ying, F., Ito, K., Izaki, K. & Takahashi, H., 1985. Production of molecular hydrogen by photosynthetic bacteria with raw starch. *AgriculturalandBiologicalChemistry*, 49 (11): 3339-3341.

Buranakarl, L., Ito, K., Izaki, K. & Takahashi, H., 1988. Purification and characterization of raw starch digestion amylase from non-sulfur purple photosynthetic bacteria. *Agricultural and Biological Chemistry*, 10(3): 173-9.

Bustos, G., Ramirez, J.A., Garrote, G. & Vazquez, M., 2003.Modelling of the hydrolysis of sugarcane bagasse with hydrochloric acid. *Appl. Biochem. Biotechnol.*, 104: 51-68.

Byrom, D., 1987. Polymer synthesis by microorganisms; Technology and economics. *Trend in Biotechnology*,5: 246-250.

Casper, M.S., 1978. *Hydrogen manufacture by electrolysis, thermal decomposition andunusual techniques*, NJ: Naves Data Corp, Park Ridge.

Cecchi, F., Marcomini, A., Pavan, P., Fazzini, G. & Mata-Alvarez, J., 1989. Mesophilic digestion of the refuse organic fraction sorted by plant. Performance and kinetic. *Waste Management Research*, 8: 33-44.

Chandrakant, P. & Bisaria, V.S., 1998. Simultaneous bioconversion of cellulose and hemicellulosic to ethanol. *Crit. Rev. Biotechnol.*, 18: 295-331.

Chaudhary, N. & Qazi, J.I., 2006a. Microbiological saccharification and ethanol production from sugarcane bagasse. *Biotechnol.*, 5: 517-521.

Chaudhary, N. & Qazi, J.I., 2007. Producing ethanol from cellulosic material - an economic strategy. 27th Pakistan Congress of Zoology, held at B.Z. University Multan, Pakistan (February 27 to March 1, 2007). Abstract pp. 158.

Chaudhary, N. & Sharma, C.B., 2005. Production of citric acid and single cell protein from agrowaste. *National Academy Science Letter*, 28: 189-193.

Chaudhary, N. 2008. Designing microbial consortium for ethanol production from sugarcane bagasse. *Ph.D Thesis* University of the Punjab, Lahore, Pakistan.

Chen, C.C., Lin, C.Y. & Chang, J.S., 2001. Kinetics of hydrogen production with continuous anaerobic cultures utilizing sucrose as the limiting substrate. *Appl. Microbiol. Biotechnol.*, 57: 56-64.

Cheng, K.K., Ge, J.P., Zhang, J.A., Ling, H.Z., Zhou, Y.J., Yang, M.D. & Xy, J.M., 2007. Fermentation of pretreated sugarcane bagasse hemicellulose hydrolysate to ethanol by *Pachysolen tannophilus*. *Biotechnol. Lett.*, 29: 1051-1055.

Chenlin, L. & Herbert, F., 2007. Fermentative hydrogen production from wastewater and solid wastes by mixed cultures. *Critical Reviews in Environmental Science and Technology*, 37(1): 1-39.

Choorit, W., Thanakoset, P., Thongpradistha, J., Sasaki, K. & Noparatnaraporn, N., 2002. Identification and cultivation of photosynthetic bacteria in wastewater from a concentrated latex processing factory. *Letters of Biotechnology*, 24: 1055-1058.

Christen, P., Villegas, E. & Revah, S., 1994. Growth and aroma production by *Ceratocystis fimbriata* in various fermentation media. *Biotechnol. Lett.*, 16: 1183-1188.

Claassen, P.A.M., Mars, A.E., Budde, M.A.W., Lai, M., van Niel, E.W.J. & de Vrije, T., 2006. Biological hydrogen production from biomass by thermophilic bacteria. *WHEC*, 16:13-16.

Claassen, P.A.M., van Lier, J.B., Lopez, C.A.M., van Niel, E.W.J., Sijtsma, L.,& Stams, A.J.M., 1999. Utilization of biomass for the supply of energy carriers. *Applied Microbiology and Biotechnology*, 52: 741-755.

Cohen, A., Distel, B., van Deursen, A. & van Andel, J.G., 1985. Role of anaerobic spore-forming bacteria in the acidogenesis of glucose changes induced by discontinuous or low-rate fed supply. *A. van Leeuw Journal of Microbiology*, 51(2): 179-192.

Cooper, P.E., Rands, M.B. & Woo, C-P., 1975. Sulphide reduction in felmongery effluents by red sulphur bacteria. *Journal of Water Pollution and Control Federation*,47: 2088-2099.

Cox, K. E.,& Williamson, Jr. K. D., 1979. *Hydrogen: its technology and implications*. Boca Raton, Chemical Rubber Company Press, Florida.

Coyne, M.S., 1999. *Soil Microbiology*. Delmar publishers 3 Columbia Circle, Box 15015 Albany, New York.

Czerink, S., French, R., Feik, C. & Chorenet, E., 2000. Production of hydrogen from biomass-derived liquids. *Proceedings of the 2000 DOE hydrogen production Review*, Nrel/CP-570-28890.

Czernik, S., French, R., Feik, C. & Chornet, E., 2002. Hydrogen by catalytic steam reforming of liquid byproducts from biomass thermoconversion processes. *Industrial& Engineering Chemistry Research*, 41: 4209-4215.

Czuppon, T. A., Knez, S. A. & Newsome, D. S., 1996. *Othmer Encyclopedia ofChemical Technology*, Wiley, New York:. P.884.

Das, D. & Veziroglu, T. N., 2001. Hydrogen production by biological process: a survey of literature. *International Journal of Hydrogen Energy*,26: 13-28.

De-Castro, H.F, Oliveira, S.C. & Furlan, S.A., 2003. Alternative approach for utilization of pentose stream from sugarcane bagasse by an induced flocculant *Pichia stipitis*. *Appl. Biochem. Biotechnol.*, 105-108:547 – 555.

Dickinson, R.E. & Cicerone, R.J., 1986. Future global warming from atmospheric trace gases. *Nature*, 319: 109-135.

Dimmling, W. & Seipenbusch, R., 1978. Raw materials for the production of SCP. *Proc. Biochem.*, 13: 9-15.

Dominguez, J.M., Gong, C.S. & Tsao, G.T., 1996. Pretreatment of sugarcane bagasse hemicellulose hydrolysate for xylitol production by yeast. *Appl. Biochem. Biotechnol.*, 57-58: 49-56.

Doremus, M.G., Linden, J.C. & Moreira, A.R., 1985. Agitation and pressure effects on acetone and butanol fermentation. *Biotechnology and Bioengineering*, 27: 852-860.

Du Toit, P.J., Olivier, P.L. & Bilijon, V., 1984. Sugarcane bagasse as a possible source of fermentable carbohydrates. I. Characterization of bagasse with regard to monosachride, hemicellulose and amino acid composition. *Biotechnol. Bioeng.*,26: 1071-1078.

Dunn, S., 2001. *Hydrogen futures*: Towards a sustainable energy system paper Worldwatch Institute, Washington, DC. 157.

Enger, E.D. & Smith, B.F., 2002. *Environmental Science:* A study of interrelationships. McGraw Hill Co. Inc. New York NY 10020.

Eroglu, E., Eroglu, I., Gündüz, U. & Yücel, M., 2009. Comparison of physicochemical characteristics and photofermentative hydrogen production potential of wastewaters produced from different olive oil mills in Western-Anatolia, Turkey. *Biomass and Bioenergy*, 33: 706-711.

Evans, G., 2005. *Biowaste and biological waste treatment*. James & James Ltd. 35-37 William road London NWI 3ER, UK.

Fabiano, B. & Perego, P., 2002. Thermodynamic study and optimization of hydrogen production by *Enterobacter aerogenes*. International Journal of hydrogen Energy, 27: 149-56.

Fang, H. H. P. & Liu, Y., 2000. Anaerobic wastewater treatment in (sub-) tropical regions. In: Proceedings of the Symposium on Establishment and Evaluation of Advanced water Treatment Technology Systems Using Functions of Complex Microbial Community, Tokyo, March 6-8, Japan, 109-118.

Fang, H., Li, C. & Zhang, T., 2006. Acidophilic biohydrogen production from rice slurry. *International Journal of Hydrogen Energy*,31(6): 683-692.

Fang, H.H.P. & Liu, H., 2002. Effect of pH on hydrogen production from glucose by a mixed culture. *Bioresource Technology*, 82: 87-93.

Fang, H.H.P., Liu, H. & Zhang, T., 2002a. Characterization of a hydrogen producing granular sludge. *Biotechnology Bioengineering*, 78(1): 44-52.

Fang, H.H.P., Zhang, T. & Liu, H., 2002b. Microbial diversity of a mesophilic hydrogen-producing sludge. *Applied Microbiology and Biotechnology*, 58(1): 112-118.

Fardeau, M.L., Salinas, M.B., L'Haridon, S., Jeanthon, C., Verhe, F., Cayol, J.L., Patel, B.K.C., Garcia, J.L. & Ollivier, B., 2004. Isolation from oil reservoirs of novel thermophilic anaerobes phylogenetically related to *Thermoanaerobacter subterraneus* reassignment of *T. subterraneus, Thermoanaerobacter yonseiensis, Thermoanaerobacter tengcongensis* and *Carboxydidrachium pacificum* to *Caldanaerobacter substerraneus* gen. nov., sp. Nov., comb. Nov. as four novel subspecies. *International Journal of Evolutionary Microbiology*, 54: 467-474.

Farrell, A.E., Plevin, R.J., Turner, B.T., Jones, A.D., O'Hare, M. & Kammen, D.M., 2006. Ethanol can contribute to energy and environmental goals. *Science*, 311: 506-508.

Fascetti, E. & Todini, O., 1995. *Rhodobacter sphaeroides* RV cultivation and hydrogen production in one and two-stage chemostat. *Applied Microbiology and Biotechnology*, 44: 330-305.

Federov, A.S., Tsygankov, A.A., Rao, K.K. & Hall, D.O., 1998. Hydrogen photoproduction by *Rhodobacter sphaeroides* immobilized on polyurethane foam. *Biotechnology Letters*, 20: 1007-1009.

Fennell, D.E. & Gossett, J.M., 1998. Modeling the production of and competition for hydrogen in a dechlorinated culture. *Environmental Science and Technology*, 32: 2450-2460.

Florin, L., Tsokoglou, A. & Happe, T., 2001. A novel type of iron hydrogenase in the green alga *Scenedesmus obliquus* is linked to the photosynthetic electron transport chain. *J. Biol. Chem.*, 276: 6125-6132.

Fond, O., Matta-Ammouri, G., Petitdemange, H. & Engasser, J. M., 1985. The role acids on the production of acetone and butanol by *clostridium acetobutylicum*. *Applied Science and Technology*, 22: 195-200.

Fujita, Y., Ito, J., Ueda, M., Fukuda, H. & Kondo, A., 2004. Synergistic saccharification and direct fermentatio to ethanol of amorphous cellulose by use of an engineered yeast strain co-displaying three types of cellulolytic enzyme. *Environ. Microbiol.* 70: 1207-1212.

Gaffron, H. & Rubin, J., 1942. Fermentative and photochemical production of hydrogen in algae. *Journal of General Physiology*, 26: 219-240.

Gest, H. & Kamen, M. D. 1949b. Studies on the metabolism of photosynthetic bacteria IV. Photochemical production of molecular hydrogen by growing cultures of photosynthetic bacteria. *Journal of Bacteriology*, 58: 239-245.

Gest, H. & Kamen, M. D., 1949a. Photoproduction of molecular hydrogen by *Rhodospirillum rubrum*. *Science*,109: 558-559.

Gest, H., Kamen, M.D. & Bregoff, H.M., 1950. Studies on the metabolism of photosynthetic bacteria. Photoproduction of hydrogen and nitrogen fixation by *rhodospirillum rubrum*. *Journal of Biology and Chemistry*, 182: 153-170.

Getha, K., Chong, V.C. & Vikineswary, S., 1998. Potential use of the phototrophic bacteria, *Rhodopseudomonas palustris* as an aquaculture feed. *Asian Fish Science*, 10: 223-232.

Ghirardi, M.L., King, P.W., Posewitz, M. C., Ching M.P., Fedorov, A., Kim, K., Cohen, J., Schulten, K. & Seibert, M., 2005. Approaches to developing biological $H_2$-photoproducing organisms and processes. *Biochemical Society transactions*,33(1): 70-72.

Goldemberg, J. & Macedo, I.C., 1994. Brazilian alcohol program – an overview. *Energy for Sustainable Development*, 1: 17-22.

Goldemberg, J., 2007. Ethanol for a sustainable energy future. *Science*,315: 808-810.

Gorman, J., 2002. A Cheap, unlimited hydrogen production method for the third world discovered. *Biological Production of Hydrogen fuel*. Science News, 12 October.

Gottschalk, G., 1986. *Bacterial metabolism*. 2nd edition. Springer: New York.

Govindaswamy, S. & Vane, L.M., 2006. Kinetics of growth and ethanol production on different carbon substrates using genetically engineered xylose-fermenting yeast. *Bioresour. Technol.* [Epub ahead of print].

Gray, K.A., Ahao, L. & Emptage, M., 2006. Bioethanol. *Curr. Opin. Chem. Biol.*, 10: 141-146.

Greenbaum, E., 1990. Hydrogen production by photosynthetic water splitting. In: Hydrogen energy progress VIII. Proceedings 8th *World Hydrogen Energy Conference*, (eds, T. N. Veziroglu., P. K. Takashashi), Pergamon Press, Hawaii, New York, 743-54.

Greer, D., 2005. Creating cellulosic ethanol: spinning straw into fuel. *Cellulosic Ethanol*. [Epub ahead of print].

Gregoire-Padro, C. E., 1998. Hydrogen: the once and future fuel. *Energy Fuels*, 12:1-2.

Grupe, H. & Gottschalk, G., 1992. Physiological events in *Clostridium acetobutylicum* during the shift from acidogenesis to solventogenesis in continuous culture and presentation of a model for shift induction. *Applied and Environmental Microbiology,*58: 3896-3902.

Hai, T., Ahlers, H., Gorenflo, V. & Steinbuchel, A., 2000. Axenic cultivation of anoxygenic photrophic bacteria, cynobacteria and microalgae in new closed tubular glass photobioreactor. *Applied Microbiology and Biotechnology,* 53: 383-389.

Hanada, S., 2003. Filamentous anoxygenic phototrophes in hot springs. *Microbes and Environment,* 18: 51-61.

Hansel, A. & Lindblad, P., 1998. Towards optimization of *Cyanobacteria* as biotechnological relevant producers of molecular hydrogen a clean and renewable energy source. *Applied Journal of Microbiology and Biotechnology,*50: 153-160.

Hart, D., 1997. *Hydrogen Power:* The Commercial Future of the Ultimate Fuel. Financial Times Energy Publishing, London.

Hawkes, F.R., Dinsdale, R., Hawkes, D.L. & Hussy, I., 2002. Sustainable fermentative hydrogen production: challenges for process optimization. *International Journal ofHydrogen Energy,* 27: 1339-1347.

Hawkes, F.R., Hussy, I., Kyazze, G., Dinsdale, R. & Hawkes, D.L., 2007. Continuous dark fermentative hydrogen production by mesophilic microflora: principles and progress. *International Journal of Hydrogen Energy,* 32(2): 172-184.

He, D., Bultel, Y., Magnin., J. P., Roux, C. & Willison, J. C., 2005. Hydrogen photosynthesis by *Rhodobacter capsulatus* and its coupling to PEM fuel cell. *Journal of Power Sources,* 141: 19-23.

Hill, J., Nelson, E., Tilman, D., Polasky, S. & Tiffany, D., 2006. Environmental, economic and energetic costs and benefits of biodiesel and ethanol biofuels. *Proc. Natl. Acad. Sci. U.S.A.,* 103(30): 11206-11210.

Hillmer, P. & Gest, H., 1976. $H_2$ metabolism in the photosynthetic bacterium *Rhodopseudomonas capsulata*: $H_2$ production by growing cultures. *Journal ofBacteriology,* 129(2): 724-731.

Hiraishi, A. & Kitamura, H., 1984. Distribution of phototrophic purple non sulfur bacteria in activated sludge systems and other aquatic environment. *Bulletin of the Japanese Society of Scientific Fisheries,*50(11): 1929-1937.

Hiraishi, A. & Ueda, Y., 1994. *Rhodoplanes* gen. nov., a new genus of phototrophic bacteria including *Rhodopseudomonas rosea* as *Rhodoplanesroseus* comb. Nov.and *Rhodoplanes elegans* sp. Nov. *International Journal of Systematic Bacteriology,* 44: 665-673.

Hiraishi, A. Hoshino, Y. & Kitamura, H., 1984. Isoprenoid quinine composition in the classification of *Rhodospirillaceae. Journal of General and Applied Microbiology,* 30: 197-210.

Hiraishi, A., Morishima, Y. & Kitamura, H., 1991. Use of isoprenoid quinine profiles to study the bacterial community structure and population dynamics in the photosynthetic sludge system. *Water and Science Technology,*23: 937-945.

Hiraishi, A., Shi, J. L. & Kitamura, H., 1989. Effects of organic nutrient strength on purple non sulfur bacterial content and metabolic activity of photosynthetic sludge for waste water treatment. *Journal of Fermentation and Bioengineering,* 68: 269-276.

Hisada, T., Okamura, K. & Hiraishi, A., 2007. Isolation and characterization of phototrophic purple nonsulfur bacteria from *Chloroflexus* and Cyanobacterial mats in hot springs. *Microbes and Environment,*22(4): 405-411.

Ho, N. W., Chen, Z., Brainard, A.P. & Sedlak, M., 1999. Successful design and development of genetically engineered *Saccharomyces* yeasts for effective co-fermentation of glucose and xylose from cellulosic biomass to fuel ethanol. *Adv. Biochem. Eng. Biotechnol.* 65: 163-192.

Hofbauer, H., 2007. Conversion technologies: gasification overview. In: *Proceedings of the 15th European Biomass Conference and Exhibition.* Berlin (Germany).

Holm, H.W. & Vennes, I.W., 1970. Occurrence of purple sulphur bacteria in sewage treatment lagoon. *Applied journal of Microbiology,*19: 988-996.

Holt, G. J., Krieg, N. R., Sneath, P.H., Staley, J.T. & Williams, S.T., 1994. *Bergey's Manual of Determinative Bacteriology,* 9th ed. Williams and Wikins Co Baltimore.P. 787.

Honda, R., Fukushi, K. & Yamamoto, K., 2006. Optimization of wastewater feeding for single-cell protein production in an anaerobic wastewater treatment process utilizing purple non-sulfur bacteria in mixed culture condition. *Journal of Biotechnology,*125 (4): 565-573.

Hongpattarakere, T. & Kittikun, A.H., 1995. Optimization of single-cell-protein production from cassava starch using *Schwanniomyces castellii. Biomedical and Life Science,* 11: 607-609.

Hulshoff Pol, L. W. & Lettinga, G., 1986. New technologies for anaerobic wastewater treatment. *Water Science and technology,*18(12): 41-53.

Hussy, I., Hawkes, F.R., Dinsdale, R. & Hawkes, D.L., 2003. Continuous fermentative hydrogen production from a wheat starch co-product by mixed microflora. *Biotechnology and Bioengineering,* 84: 33-44.

Iglesias, J. R., Castrillon, L., Maranon, E. & Sastre, H. 1998. Solid state anaerobic digestion of unsorted municipal solid waste in a pilot plant scale digester. *Bioresource Technology,*63(1): 29-35.

Ike, A., Saimura, C., Hirata, K. & Miyamoto, K., 1996. Environmentally friendly production of $H_2$ inocrporating microalgal $CO_2$ fixation. *Journal of Marine Biotechnology,* 4: 47-51

Ike, A., Yoshihara, K., Nagase, h., Hirata, K. & Miyamoto, K., 1998. Algal $CO_2$ fixation and $H_2$ photoproduction,. In: *BioHydrogen* (ed. Zaborsky, O.R.), Plenum Press, London.p. 265-271.

Imhoff, J.F. & Truper, H.G., 1992. The genus *Rhodospirillum* and related genera, In: *The prokaryotes, 2nd ed.* (eds. Balows, A., Truper, H.G., Dworkin, M., Harder, W. and Schleifer, K.H.), Springer-Verlag. New York, N.Y.2141-2155.

Imhoff, J.F. & Truper, H.G., 1998. The purple nonsulfur bacteria in Bergey's Manual of Systematic Bacteriology, Vol 3, (eds. Staley, J.T., Bryant, M.P., Pfennig, N. and Holt, J.G.), Williams and Wilkins Co. Baltimore.pp. 1658-1661.

Imhoff, J.F., Hirashi, A. & Suling, J., 2005. Anoxygenic phototrophic purple bacteria. In: *Bergey' s manual of systematic bacteriology* (eds., Brenner, D.J., Krieg, N.R., Staley, J.T. and Garrity, G.M.), 2nd ed., vol. 2. The proteobacteria, part A, introductory essays. Springer-Verlag, New York, N. Y. 119-132.

Jansen, P.H. & Harfoot, C.G., 1991. *Rhodopseudomonas rosea* sp. nov., a new purple nonsulfur bacterium. *International Journal of Systematic Bacteriology,* 41(1): 26-30.

Jones, B.E., Grant, W.D., Duckworth, A.W. & Owenson, G.G., 1998. Microbial diversity of soda lakes. *Extremophiles,* 2: 191-200.

Jones, D. T. & Woods, D. R., 1986. Acetone-butanol fermentation revisited. *Microbiology Reviews,* 50: 484-524.

Jung, G. Y., Jung, H. O. Kim, J. R., Ahn, Y. & Park, S., 1999. Isolation and characterization of *Rhodopseudomonas palustris* P4 which utilizes CO with the production of $H_2$. *Biotechnology Letters,* 21: 525-529.

Kaiser, D., 1966. Contribution an l' etude de L'ecologic des bacteries photosynthetique. *Annales de l'Institut Pasteur / Virologie,* 111: 732-749.

Kalia, V., Lal, S., Ghai, R., Mandal, M. & Chauhan, A., 2003. Mining genomic databases to identify novel hydrogen producers. *Trends in Biotechnology,* 21(4): 152-156.

Kamel, B.S., 1979. Utilization of date carbohydrate as substrate in microbial fermentation. *Proc. Biochem.,* 14: 95-100.

Kapdan, I.K. & Kargi, F., 2006. Bio-hydrogen production from waste materials, .*Enzyme and Microbial Technology,* 38: 569-582.

Karhumaa, K., Weidemann, B., Hagerdal, H.B., Boles, E. & Grauslund, G.M.F., 2006. Co-utilization of L-arabinose and D-xylose by laboratory and industrial *Sacchromyces cerevisiae* strains. *Microb. Cell Fact.,* 5:18.

Karr, E.A., Sattley, M.W., Jung, D.O., Madigan, M.T. & Achenbach, L.A., 2003. Remarkable diversity of phototrophic purple bacteria in a permanently frozen Antarctic lake. *Applied and Environmental Microbiology,* 69(8): 4910-4914.

Katsuda, T., Ooshima, H., Azuma, M. & Kato, J., 2006. New detection method for hydrogen gas for screening hydrogen-producing microorganisms using water-soluble wilkinson's catalyst derivatives. *Journal of Bioscience and Bioengineering,*102(3):220-226.

Katzen, R. & Fowler, D.E., 1994. Ethanol from lingo-cellulosic wastes with utilization of recombinant bacteria. *Appl. Biochem. Biotechnol.,* 45: 697 – 707.

Khanal, S. K., Chen, W. H., Li, L. & Sung, S., 2004. Biological hydrogen production: effects of pH and intermediate products. *International Journal Hydrogen Energy,* 29: 1123-31.

Kim, B.C., Lee, S.C., Lee, S.Y., Chang, H.N., Chang, Y.K. & Woo, S.I., 1994. Production of poly (3- hydroxybutyric acid) by fed-batch culture of *Alcaligenes eutrophus* with glucose concentration control. *Biotechnology and Bioengineering,* 43: 892-898.

Kim, J. S., Ito, K. & Takahashi, H., 1980. The relationship between nitrogenase activity and hydrogen evolution in *Rhodopseudomonas palustris. Agricultural Biology Chemistry,* 44(4): 827-833.

Kim, J.S., Ito, K. & Takahashi, H., 1981. Production of molecular hydrogen by *Rhodopseudomonas* sp. *Journal of Fermentation Technology,* 59(3): 185-90.

Kim, J.S., Ito, K., Izaki, K. & Takahashi, H., 1987. Production of Molecular hydrogen by a semi-continuous outdoor culture of *Rhodopseudomonas sphaeroides. Agricultural and BiologicalChemistry,* 51(4): 1173-1174.

Kim, M.K., Choi, K-M., Yin, C-R., Lee, K-Y., Im, W-T., Lim, J. H. & Lee, S-T., 2004. Odorous swine wastewater treatment by purple non-sulfur bacteria, *Rhodopseudomonas palustris,* isolated from eutrophic ponds. *Biotechnology letters,* 26: 819-822.

Kirk, R. E., Othmer, D. F., Grayson, M. & Eckroth, D., 1985. *In Concise Encyclopediaof Chemical Technology XIII.* Wiley-Inter-science, New York, 838-893.

Klasson, K.T., Ackerson, M.D., Clausen, E. C. & Gaddy, J.L., 1992. Bioconversion of synthesis gas into liquid or gaseous fuels. *Enzyme andMicrobial Technology*,14: 602-608.

Klasson, K.T., Cowger, J.P., Ko, C.W., Vega, J.L., Clausen, E.C. & Gaddy, J.L., 1990. Methane production of synthesis gas using mixed culture of *R. rubrum, M. barkeri* and *M. formicum*. *Applied BiochemistryandBiotechnology*, (24/25): 2273-2289.

Kleihues, L., Lenz, O., Bernhard, M., Buhrke, T. & Friedrich, B., 2000. The $H_2$ sensor of Ralstonia eutropha is a member of the subclass of regulatory [NiFe] hydrogenase. *Journal of Bacteriology*,182: 2716-2724.

Kobayashi, M. & Tchan, Y. T., 1973. Treatment of industrial waste solutions and production of useful by-products using a photosynthetic bacterial method. *Water.Resource*, 7: 1219-1224.

Koku, H., Eroglu, I., Gunduz, U., Yucel., M. & Turker, L., 2002. Aspects of the metabolism of hydrogen production by *Rhodobacter sphaeroides*. *International Journal ofHydrogen Energy*, 27: 1315-1329.

Koku, H., Eroglu, I., Gunduz, U., Yucel., M. & Turker, L., 2003. Kinetics of biological hydrogen production by the photoproduction bacterium *Rhodobacter sphaeroides* O. U. 001. *International Journal of Hydrogen Energy*, 28: 381-388.

Kondo, T., Arakawa, M., Hiral, T., Wakayama, T., Hara, M., & Miyake, J., 2002. Enhancement of hydrogen, production by a photosynthetic bacterium mutant with reduced pigment. *Journal of Bioscience bioengineering*, 93: 145-150.

Kotsopoulos, T., Zeng, R. & Angelidaki, I., 2006. Biohydrogen production in granular up-flow anaerobic sludge blanket (UASB) reactors with mixed cultures under hyper-thermophilic temperature (70°C). *Biotechnology and Bioengineering*, 94(2): 296-302.

Lamed, R.J. & Zeikus, J.G., 1980. Ethanol production by thermophilic bacteria: relationship between fermentation product yields of and catabolic enzyme activities in *Clostridium thermocellum* and *Thermoanaerobium brockii*. *J. Bacteriol.*, 144: 569-578.

Laser, M., Schulman, D., Allen, S.G, Lichwa, J., Antal, M.J. Jr. & Lynd, L.R., 2002. Comparison of liquid hot water and steam pretreatments of sugarcane bagasse for biconversion to ethanol. *Bioresour. Technol.*, 81:33 – 44.

Lavarack, B.P., Griffin, G.J. & Rodman, D., 2002. The acid hydrolysis of sugarcane bagasse hemicellulose to produce xylose, arabinose, glucose and other products. *Biomass and Bioenergy*, 23: 367-380.

Lawford, H.G. & Rousseau, J.D., 2003. Cellulosic ethanol: alternative fermentation process designs with wild type and recombinant *Zymmonas mobilis*. *Appl. Biochem. Biotechnol.*,105-108: 457-69.

Lawford, H.G., Rousseau, J.D. & Tolan J.S., 2001. Comparative ethanol productivities of different Zymomonas recombinants fermenting oat hull hydrolysate. *Appl. Biochem. Biotechnol.*, 93: 133-146.

Lay, J. J.; Lee, Y. J. & Noike, T., 2000. Feasibility of biological hydrogen production from organic fraction of municipal solid waste. *WaterResearch*, 33: 2579-2586.

Lay, J.J., 2001. Modelling and optimization of anaerobic digested sludge converting starch to hydrogen. *Biotechnology and Bioengineering*, 68(3): 269-278.

Lee, I. C., Kim, M. S. & Ihm, S. K., 2002. Gasification of glucose in supercritical water. *Industrial& Engineering Chemistry Research*,41: 1182-1188.

Levin, D.B., Pitt, L. & Love, M., 2004. Biohydrogen production: prospects and limitations to practical application. *International Journal of Hydrogen Energy*, 29: 173-185.

Li, C. & Fang, H.H.P., 2007. Fermentative hydrogen production from wastewater and solid wastes by mixed cultures. *Critical Review of Environment and Science Technology,* 37: 1-39.

Lin, Y. & Tanaka, S., 2006.Ethanol fermentation from biomass resources: current state and prospects. *Appl. Microbiol. Biotechnol.,* 69: 627-642.

Lin. C. Y., Hung, C. H., Chen, C. H., Chung, W. T. & Cheng, L. H., 2006. Effects of initial cultivation pH on fermentative hydrogen production from xylose using natural mixed cultures. *Process Biochemistry,* 41: 1383-1390.

Liu, C.F., Ren, J.L., Xu, F., Lin, J.J., Sun, J.X. & Sun, R.C., 2006. Isolation and characterization of cellulose obtained from ultrasonic irradiated sugarcane bagasse. *J. Agric. Food Chem.,* 54: 5742-5748.

Liu, C.F., Sun, R.C., Qin, M.H., Zhang, A.P., Ren, J.L., Ye, J., Luo, W. & Cao, Z.N., 2007. Succinoylation of sugarcane bagasse under ultrasound irradiation. *Bioresour. Technol.,* [Epub ahead of print].

Liu, G. & Shen, J., 2004. Effects of culture and medium conditions on hydrogen production from starch using anaerobic bacteria. *Journal of Bioscience and Bioengineering,* 98(4): 251-256.

Lodhi, M. A. K., 1987. Hydrogen production from renewable sources of energy. *International Journal of Hydrogen Energy,*12: 461-8.

Logan, B. E., 2002. Biological hydrogen production measured in batch anaerobic respirometers. *Environmental Science and Technology,* 36(11): 2530-2535.

Lynd, L.R., 1995. Fuel production biological. In: *Molecular Biology and Biotechnology* (Robert, A. Meyers, ed.), pp.331-333. VCH Publishers, Inc. New York, NY 10010, U.S.A.

Lynd, L.R., Weimer, P.J., van Zyl, W.H. & Pretorius, I.S., 2002. Microbial cellulose utilization: fundamentals and biotechnology. *Microbiol. Mol. Biol. Rev.,* 66: 506-577.

Macler, B., Pelroy, R., Bassham, J., 1979. Hydrogen formation in nearly stoichiometric amounts from glucose by a *Rhodopseudomonas sphaeroides* mutant. *Journal of Bacteriology,* 138: 446-452.

Madigan, M.T., 2003. Anoxygenic phototrophic bacteria from extreme environments. *Photosyn. Res.,* 76: 157-171.

Maeda, I., Miyasaka, H., Umeda, F., Kawase, M. & Yagi, K., 2003. Maximization of hydrogen production ability in high-density suspension of *Rhodovulum sulfidophilum* cells using intracellular poly(3-hydroxcbutyrate) as sole substrate. *Biotechnology Bioengineering,* 81: 474-81.

Maness, P.C., Huang, J., Smolinski, S., Tek, V. & Vanzin, G., 2004. Energy generation from the CO Oxidation-Hydrogen Production pathway in *Rubrivivax gelatinosus. Applied and Environmental Microbiology,* 71(6): 2870-2874.

Mao, X.Y., Miyake, J. & Kawamura, S., 1986. Screening photosynthetic bacteria for hydrogen production from organic acid. *Journal of Fermentation Technology,* 64: 245-249.

Markov, S.A. & Weaver, P.F., 2008. Bioreactors for $H_2$ production by purple non sulfur bacteria. *Applied Biochemistry and Biotechnology,* 145(1-3): 79-86.

Martin, C., Galbe, M., Nilvebrant, N.I. & Jönsson, L.J., 2002. Comparison of the fermentability of enymatic hydrolyzates of sugarcane bagasse pretreated by steam explosion using different impregnating agents. *Appl. Biochem. Biotechnol.,* 98-100: 699-716.

Martinez, A., Rodriguez, M.E., York, S.W., Preston, J.F. & Ingram, L.O., 2000. Effects of $Ca(OH)_2$ treatments "overliming" on the composition and toxicity of bagasse hemicellulose hydrolysates. *Biotechnol. Bioeng.*, 69: 526-536.

Matsunaga, T., Hatano, T., Yamada, A. & Matsumoto, M., 2000. Microaerobic hydrogen production by photosynthetic bacteria in a double phase photobioreactor. *Biotechnology and Bioengineering*,68(6): 647-651.

Mehrabi, S., Ekanemesang, U. M., Aikhionbare, F. O., Kimbro, K. S. and Bender, J., 2001. Identification and characterization of *Rhodopseudomonas spp.*, a purple non-sulfur bacterium from microbial mats. *Biomolecular Engineering*, 18: 49-56.

Melis, T., 2005. Integrated biological hydrogen production. *Proceeding International Hydrogen Energy Congress and Exhibition IHEC 2005.* Istanbul, Turkey.

Meunchang, S., Panichsakpatana, S. & Weaver, R.W., 2005. Inoculation of sugar mill by-products compost with $N_2$-fixing bacteria. *Plant and Soil*, 271: 219-225.

Miller, G.T. Jr., 2003. *Environmental science working with the earth.* Brooks/Cole-Thomson Learning 511 Forest Lodge Road Pacific Grove, CA 93950, USA.

Miyake, J. & Kawamura, S., 1987. Efficiency of light energy conversion to hydrogen by the photosynthetic bacterium *Rhodobacter sphaeroides. Inrternational Journal of Hydrogen Energy*, 39: 147-149.

Miyake, J., Asada, Y. & Kawamura, S., 1989. Nitrogenase,. In: Biomass hand-book (eds. Kitani, O. and Hall, C.W.), Gordon and Breach Science Publishers, New york.362-370.

Miyake, J., Igarashi, Y. & Roegner, M., 2004. Biohydrogen III. Renewable energy system by biological solar energy conversion. London: Elsevier, 1-187.

Miyake, J., Matsunaga, T., & San Pietro, A., 2001. Biohydrogen II. An approach to environmentally acceptable technology London: Pergamon-Elsevier, 1-273.

Miyake, J., Miyake, M. & Asada, Y., 1999. Biotechnological hydrogen production: research for efficient light energy conversion. *Journal of Bacteriology*, 70(1-3): 89-101.

Miyake, J., Tomizuk, N. & Kamibayashi, A., 1982. Prolonged photohydrogen production by *Rhodospirillum rubrum. Journal of Fermentation Technology*, 60: 199-203.

Miyake, M. & Asada, Y., 1996. Efficient transformation of a thermophilic cyanobacterium, *Synechococcus elongates. Journal of Marine Biotechnology*, 4: 113-116.

Miyake, M., Kotani, H. & Asada, Y., 1992. Isolation and identification of restriction endonuclease, Sell from a cyanobacterium, *Synechococcus elongates. Nuclear Acids Research*, 20: 2605.

Miyake, M., Yamada, J., Aoyama, K., Uemura, I., Hoshino, T., Miyake, J. & Asada, Y., 1996. Strong expression of foreign protein in *Synechococcus* sp. PCC7942. *Journal of Marine Biotehcnology*, 4: 64-67.

Miyamoto, K., 1993. Hydrogen production by photosynthetic bacteria and microalgae, p. In: *Recombinant microbes for industrial and agricultural applications* (eds. Murooka, Y. and Imanaka, T.). Marcel Dekker, Basel and Hong Kong,Inc., New York, 771-786.

Miyamoto, K., Benemanu, J.R. & Hallenbeck, P.C., 1979. Nitrogen fixation by thermophilic blue-green algae (cyanobacteria): temperature characteristics and potential use in biophotolysis. *Applied Environmental Microbiology*, 37: 454-458.

Miyoshi, M., 1997. Studien uber die Schwefelrasenbildung und die Schwefelbakterien der Thermen von Yumoto bei Nikko. *Zentralbl. Bakteriol. Parasiten Infek. Abt.*,2: 526-527.

Mizuno, O., Dinsdale, R., Hawkes, F.R., Hawkes, D.L. & Noike, T., 2000. Enhancement of hydrogen production from glucose by nitrogen gas sparging. *Bioresource Technology,* 73(1): 59-65.

Moat,A.G., Foster,J.W. & Spector, M.P., 2004. *Microbial physiology.* Willey-Liss A John Wiley & Sons, Inc. publication replica press pvt. Ltd.,India.

Molina, J.W.F., Ripoli, T.C., Geraldi, R.N. & Amaral, J.R., 1995. Aspectos econômicos e Operacionais do Enfardamento de Residuo de Colheita de Cana-de-Acúcar para Aproveitamento Energético. *STAB-Acúcar, Alcool e Subprodutos,* 13: 28.

Molina, O.E., Perotti de Gálvez, N.I., Frigerio, C.I. & Córdoba, P.R., 1984. Planta piloto de Procesos Industrials microbiologicos, PROIMI, Avda Belgrano y Pje Caseros, 4000 S.M. de Tucumán, Argentina. [Epub ahead of print].

Montgomery, R., 2004. Development of biobased products. *Bioresource Technology,* 91: 1-29.

Morita, T.A. & Silva, S.S., 2000. Inhibition of microbial xylitol production by acetic acid and its relation with fermentative parameters. *Appl. Biochem. Biotechnol.,* 84-86: 801-808.

Moucha, T., Linek, V. & Prokopova, E., 2003. Gas hold-up, mixing time and gas liquid volumetric mass transfer coefficient of various multiple- impeller configuration: Rushton turbine, pitched blade and techmix impeller and their combination. *ChemicalEngineering Science,*58(9): 1839-1896.

Nagadomi, H., Kitamura, T., Watanabe, M. & Sasaki, K., 2000. Simultaneous removal of chemical oxygen demand (COD) phosphate, nitrate and $H_2S$ in the synthetic sewage wastewater using porous ceramic immobilized photosynthetic bacteria. *Letters ofBiotechnology,*22: 1369-1374.

Najafpour, G. D., Basu, R., Clausen,E. C. & Gaddy, G. L., 1995. Water-gas shift reaction for synthesis gas: a biological route. *International Journal of EngineeringScience,*6: 39-49.

Najafpour, G., Younesi, H. & Mohamed, A.R., 2003. Continuous hydrogen production via fermentation of synthesis gas. *International Journal of Petrol Process.: Petrochemical. CoalProcess,*45(3-4): 154-158.

Najafpour, G.D., Younesi, H. & Mohamed, A.R., 2004. Effect of organic substrate on hydrogen production from synthesis gas using *Rhodospirillum rubrum,* in batch culture. *Journal of Biochemical Engineering,* 21 (2): 123–130.

Najafpour, G.D., Younesi, H. & Mohamed, A.R., 2006. A survey on various carbon sources for biological hydrogen production via water-gas reaction using photosynthetic bacterium *Rhodospirillum rubrum. Energy sources,*28: 1013-1026.

Nakada, E., Asada, Y., Arai, T. & Miyake, J., 1995. Light penetration into cell suspensions of photosynthetic bacteria and relation to hydrogen production. *Journal ofFermentation and Bioengineering,* 80(1):53-57.

Nakadomi, H., Hiromatsu, T., Takeno, K., Watanabe, M. & Sasaki, K., 1999. Treatment of aquarium water by denitrifying photosynthetic bacteria using immobilized polyvinyl alcohol beads. *Journal of Bioscience and Bioengineering,* 87: 625-629.

Nakashimada, Y., Rachman, M.A., Kakizono, T. & Nishio, N., 2002. Hydrogen production of *Enterobacter aerogene* altered by extracellular and intracellular redox states. *International Journal of Hydrogen Energy,* 27: 1399-1405.

Nandi, R. & Sengupta, S., 1998. Microbial Production of Hydrogen: An Overview. *Critical Reviews in Microbiology,* 24(1): 61-84.

Nath, K. & Das, D., 2004. Improvement of fermentative hydrogen production: various approaches. *Applied Microbiology and Biotechnology,* 65(5): 520-529.

Nath, S. & Shukla, V., 1997. A new approach to evaluate kinetic parameters and mass transfer coefficients in continuous stirred tank reactor. Application to antibiotic separations. *Chemical and Engineering Journal*, 65: 63-69.

Negro, M.J., Manzanares, P., Ballesteros, I., Oliva, J.M., Cabanas, A. & Ballesteros, M., 2003. Hydrothermal pretreatment conditions to enhance ethanol production from popular biomass. *Appl. Biochem. Biotechnol.*, 108: 87-100.

Nelson, D.L. & Cox, M.M., 2000. *Lehninger Principles of Biochemistry*. Macmillan worth publishers ,41 Madison avenue New York .Ny10010 USA.

Nianzhi, J., Sieracki, M.E., Yao, Z. & Hailian, D., 2003. Aerobic anoxygenic phototrophic bacteria and their roles in marine ecosystems. *Chinese Science Bulletin,* 48(11): 1064-1068.

Niklowitz, W. & Drews, G., 1955. Zur Elektronenmikroskopischen Darstellung der feinstruktur von *Rhodospirillum rubrum.Archives ofMicrobiology,*23: 123-129.

Noda, N., Zorin, N. A., Nakamura, C., Miyake, M., Gogotov, I. N., Asada, Y., Akutsu, H. & Miyake, J. 1998. Langmuir-Blodgett film of hydrogenase for electrochemical hydrogen production. *Thin Solid Films*, 327-329: 639-642.

Noparatnaraporn, N., Nishizawa, Y., Hayashi, M. & Nagai, S., 1983. Single cell protein production from cassava starch by *Rhodopseudomonas gelatinosa. Journal of Fermentation Technology,* 61(5): 515-519.

Noparatnaraporn, N., Sasaki, K., Nishizawa, Y., Hayashi, M. & Nagai, S., 1982. *International union of Microbiological Societies, 13th International Congress of Microbiology*, Boston, Mass., U. S. A. 150.

Novak, T.N., Gritzer, R.F., Leadbetter, E.R. & Godchaux, W., 2004. Photrophic utilization of taurine by the purple nonsulfur bacteria *Rhodopseudomonas palustris* and *Rhodobacter sphaeroides. Microbiology,* 150: 1881-1891.

Oda, Y., Star, B., Huisman, L. A., Gottschal, J. C. & Forney, L. J., 2003. Biogeography of the purple non-sulfur bacterium *Rhodopseudomonas palustris. Applied and Environmental Microbiology,* 69(9): 5186-5191.

Oh, Y. K., Kim, S. H., Kim, M. & Park, S., 2004. Thermophilic biohydrogen production from glucose with trickling biofilters. *Biotechnology Bioengineering,* 88: 690-8

Oh, Y., Seol, E., Lee, E. & Park, S., 2002. Fermentative hydrogen production by a new Chemohetrotrophic bacterium *Rhodopseudomonas palustris* P4. *International Journal of Hydrogen Energy,*27: 1373-1379.

Okubo, Y., Futamata. H. & Hiraishi, A., 2006. Distribution and capacity for utilization of lower fatty acids of phototrophic purple nonsulfur bacteria in wastewater environments. *Microbes and Environment,*20: 135-143.

Oleszkiewicz, J. & Poggi-Varaldo, H.M., 1997. High solids anaerobic digestion of mixed municipal and industrial waste. *Journal of Environment and Engineering,* 123: 1087-1092.

Onodera, H., Miyahara, T. & Noike, T., 1999. Influence of ammonia concentration on hydrogen transformation of sucrose. Asian waterquality'99, 8th IAWQ *Asian –Pacific Regional conference,* 2: 1139-1144.

Ormerod, J.G., Ormerod, K.S. and Gest, H., 1961. Light dependent utilization of organic compounds and photoproduction of molecular hydrogen by photosynthetic bacteria: relationships with nitrogen metabolism. *Archives of Biochemistry Biophysics,* 94: 449-463.

Oser, B.L., 1965 *Physiological chemistry*. The Blakiston Division McGraw-Hill Book C.New York , U.S.A.

Overmann, J., 2001. Diversity and ecology of phototrophic sulfur bacteria. *Microbiology Today*, 28: 116-118.

Padree, A.B., Schachman, H.K. & Stanier, R.Y., 1952. Chromatophores of *Rhodospirillum rubrum*. *Nature*, 169: 282-283.

Page, W. J., 1989. Production of poly-β-hydroxybutyrate by *Azotobacter vinelandii* strain UWD during growth on molasses and other complex carbon source. *Applied Journal of Microbiology and Biotechnology*,31: 329-33.

Pandey, A., Soccol, C. R., Nigam, P. & Soccol, V. T., 2000. Biotechnological potential of agroindustrial residues.1: Sugarcane bagasse. *Bioresource Technol.*,74: 69-80.

Pauss, A., Andre, G., Perrier, M. & Guiot, S.R., 1990. Liquid to gas mass transfer in anaerobic processes: inevitable transfer limitations of methane and hydrogen in the biomethanation process. *Applied and Environmental Microbiology*, 56: 1636-1644.

Pereira, A.A., Martins, G.F., Antunes, P.A., Conrrado, R., Pasquini, D., Job, A.E., Curvelo, A.A., Ferreira, M., Riul, A.Jr. & Constantino, C.J., 2007. Lignin from sugarcane bagasse: extraction, fabrication of nanostructured films and application. *Langmuir*, 23: 6652-6659.

Perez, J.M., Rodriguez, F., Alonso, M.V., Oliet, M. & Echeverria, J.M., 2002. Characterization of a novolac resin substituting phenol by ammonium lignosulfonate as filler or extender. *Bio. Res.*, 2: 270-283.

Pfennig, N., 1987. Van Niel Remembered. *Features*, 53 (2): 75-77.

Pfenning, N. & Truper, H.G., 1974. Bergey's Manual of Determinative Bacteriology, In: *Order 1.Rhodospirilleaes*, 8th ed., (eds., Buchanan, R. E. and Gibbons, N. E), Williams and Wilkins Co., Baltimore, 25–6974.

Pfenning, N., 1977. Phototrophic green and purple bacteria: a comparative, systematic survey. *Annual Review of Microbiology*, 31: 275-290.

Poggi-Varaldo, H.M., Gomez-Cisneros, E., Rodriguez-Vazquez, T., Trejo-Espino, J., Rinderknecht-Seijas, N., Rodriquez-Vazquez, R., Fermandez-Villagomez, G. & Esparza-Garcia, F., 2002. Unsuitability of anaerobic compost from solid substrate anaerobic digestion as soil amendment. In: *Microbiology of composting Heidelberg*, (eds., Insam, H., Riddech, N. and Klammer, S.), Springer, Germany, p. 287-297.

Poggi-Varaldo, H.M., Rodriquez-Vazquez, R., Fermandez-Villagomez, G. & Esparza-Garcia, F., 1997. Inhibition of mesophilic solid substrate anaerobic digestion by ammonia nitrogen. *Applied Microbiology and Biotechnology*, 47: 284-291.

Poggi-Varaldo, H.M., Trejo-Espino, J., Fermandez-Villagomez, G., Esparza-Garcia, F., Caffarel-Mendez, S. & Rinderknecht-Seijas, N., 1999. Quality of anaerobic compost for soil amendment. *Water Science Technology*, 40(11-12): 179-186.

Preuss, V.P, Duke, G. & Rogers, J., 1998. A *sustainable earth*. Cambridge university press. The Edinburgh building Cambridge CB2 2RU, UK.

Prince, R.C. & Kheshgi, H.S., 2005. The photobiological production of hydrogen: potential efficiency and effectiveness as a renewable fuel. *Critical Review of Microbiology*, 31: 19-31.

Proctor, L.M., 1997. Nitrogen-fixing, photosynthetic, anaerobic bacteria associated with pelagic copepods. *Aquatic Microbial Ecology*,12: 105-113.

Qian, D.J., Nakamura, C., Wenk, S.O., Ishikawa, H., Zorin, N. & Miyake, J., 2002. A hydrogen biosensor made of clay, poly (butylviologen), and hydrogenase sandwiched on a glass carbon electrode. *Biosensors and Bioelectronics*, 17: 789-796.

Ramachandran, R., & Menon, P. K., 1998. An overview of industrial uses of hydrogen. *Internatioal Journal of hydrogen energy*, 23: 593-8.

Rey, F.E., Heiniger, E.K. & Harwood, C.S., 2007. Redirection of Metabolism for biological hydrogen production. *Applied and Environmental Microbiology*, 73(5): 1665-1671.

Robert, L.U. & Wolfe, R.S., 1970. Anaerobic growth of purple non sulfur bacteria under dark conditions. *Journal of Bacteriology*,104(1): 462-472.

Rocha, J.S., Barbosa, M.J. & Wijffels, R.H., 2001. Hydrogen production by photosynthetic bacteria: culture media, yields and efficiencies. In: *Biohydrogen II: An approach to environmentally acceptable technology* (eds, Miyake, J., Matsunaga, T., San Pietro, A.), Pergamon, London, p 3-32.

Rodrigues, R.C.L.B., Felipe, M.G.A., Silva, J.B.A., Vitolo, M. & Gomez, P.V., 2001. The influence of pH, temperature and hydrolyzate concentration on the removal of volatile and nonvolatile compounds from sugarcane bagasse hemicellulosic hydrolyzate treated with activated charcoal before or after vaccum evaporation. *Braz. J. Chem. Eng.*, 18: [Epub ahead of print].

Rodriguez, H. & Gallardo, R., 1993. Single cell protein from bagasse pith by a mixed bacterial culture. *Acta Biotechnologica*, 13: 141-149.

Rodriguez, H., Alvarez, R. & Enriquez, A., 1993. Evaluation of different alkali treatments of bagasse pith for cultivation of *Cellulomonas* sp. *World Journal of Microbiology and Biotechnology.*, 9: 213-215.

Rosen, M. A. & Scott, D. S., 1998. Comparative efficiency assessments for a range of hydrogen production processes. *International Journal of Hydrogen Energy*,23: 653-9.

Saeed, A.A.W.S., 2005. Microbiological conversion of waste fruits and vegetables into ethanol. *Department of Zoology, University of the Punjab, Lahore.*

Sahlstrom, L., 2003. A review of survival of pathogenic bacteria in organic waste used in biogas plants. *Bioresource Technology*, 87(2): 161-166.

Sasaki, K. & Nagai, S., 1979. The optimum pH and temperature for the aerobic growth of *Rhodopseudomonas gelatinosa*, and vitamin $B_{12}$ and ubiquinone formation on a starch medium. *JournalofFermentation Technology*,57: 383-386.

Sasaki, M., Adschiri, T. & Arai, K., 2003. Fractionation of sugarcane bagasse by hydrothermal treatment. *Bioresour. Technol.*, 86: 301-304.

Sasikala, C. & Ramana, C. V., 1995. Biotechnological potentials of anoxygenic phototrophic bacteria. 1. Production of single-cell protein, vitamins, ubiquinones, hormones, and enzymes and use in waste treatment. *Advances in Applied Microbiology*, 41: 173-226.

Sasikala, K., Ramana C., Rao, P.R. & Kovacs, K.L., 1993. Anoxygenic phototrophic bacteria: physiology and advances in hydrogen production technology. *Advance Applied Microbiology*, 38: 211-295.

Sastri, M. V. C., 1989. India' hydrogen energy program- a status report. *International Journal of Hydrogen Energy*,14: 507-13.

Schachman, H.K., Padree, A.B. & Stanier, R.Y., 1952. Studies on the macromolecular organization of microbial cells. *Archives of Biochemistry and Biophysics*, 38: 245-260.

Sedlak, M. and Ho, N.W., 2004. Characteriztion of the effectiveness of hexose transporters for transporting xylose during glucose and xylose co-fermentation by a recombinant *Saccharomyces* yeast.*Yeast,* 21:671-84.

Segers, L. & Verstraeta, W., 1983. Conversion of organic acids to hydrogen by Rhodospirillaceae growth with glutamate or dinitrogen as nitrogen source. *Biotech. Bioeng.,* 25: 2843-2853.

Shi, X. Y. & Yu, H. Q., 2005. Response surface analysis on the effect of cell concentration and light intensity on hydrogen production by *Rhodopseudomonas capsulate. Process Biochemistry,* 40: 2475-81.

Shin, H.S., Youn, J.H. & Kim, S.H., 2005. Hydrogen production from food waste in aneaerobic mesophilic and thermophilic acidogenesis. *International Journal of Hydrogen Energy,* 29: 1355-1363.

Sindhu, M.S. & Sandhu, D.K., 1980. Single-cell protein production by trichoderma longibrachiatum on treated sugarcane bagasse. *Biotechnol. Bioeng.,* 22: 689-692.

Singleton, P. & Sainbury, D., 2001. *Dictionary of microbiology and molecular biology.* John Wiley and Sons Ltd., Baffins Lane ,Chichester ,West Sussex po19 IUD,UK.

Slapack, G.E., Russell, I. & Stewart, G.G., 1987. Thermophilic microbes in ethanol production. *CRC press, Inc. Boca Raton, Florida,* 129-134.

Smith, J.E., 1996. Biotechnology, Cambridge University Press, Cambridge.

Sojka, G. A. & Gest, H., 1968. Integration of energy conversion and biosynthesis in the photosynthetic bacterium *Rhodopseudomonas capsulate. Proc. Nat. Acad. Sci. (Wash.)*61: 1486-1493.

Sonderegger, M., Jeppsson, M., Larsson, C., Grauslund, C.M.F., Boles, E., Olsson, L., Martins, S.I., Hagerdal, H.B. & Sauer, U., 2004. Fermentation performance of engineered and evolved xylose-fermenting *Sacchromyces cervisiae* strains. *Biotechnol. Bioeng.,* 1: 90-98.

Sree, K., Sridhar, M., Rao, L.V. & Pandey, A., 1999. Ethanol production in solid substrate fermentation using thermotolerant yeast. *Proc. Biochem.,* 34: 115-119.

Stabnikova, O., Wang, J.Y., Ding, H.B. & Tay, J.H., 2005. Biotransformation of vegetable and fruit processing waste with yeast biomass enriched with selenium. *Bioresour. Technol.,* 96: 747-751.

Stenberg, K., Bollok, M., Reczey, K., Galbe, M. & Zacchi, G., 2000. Effect of substrate and cellulase concentration on simultaneous saccharification and fermentation of steam-pretreated softwood for ethanol production. *Biotechnol. Bioeng.,* 68: 204-210.

Stryer, L., 1995. *Biochemistry.* W.H. Freeman Co. New York.

Sun, Y. & Cheng, J., 2002. Hydrolysis of lignocellulosic materials for ethanol production: a review. *Bioresour. Technol.,* 83:1 – 11.

Sunita, M. & Mitra, C.K., 1993. Photoproduction of hydrogen by photosynthetic bacteria from sewage and wastewater . *Journal of Biosciences,*18 (1): 155-160.

Suzuki, Y. 1982. On hydrogen as fuel gas. *International Journal of Hydrogen Energy,*7: 227-30.

Swoagar, W. C. & Linderstrom, E. S., 1971. Isolation and counting of Athiorhodaceae with membrane filters. *Applied journal of Microbiology,*22 683-687.

Taguchi, F., Mikukami, N., Saito-Taki, T. & Hasegawa, K., 1995. Hydrogen production from continuous fermentation of xylose during growth of *Clostridium* sp. Strain No. 2. *Canadian Journal of Microbiology,* 41: 536-540.

Taherzadeh, M. J. & Karimi, K., 2007. Acid-based hydrolysis processes for ethanol from lignocellulosic materials: a review. *Bioethanol review. Bioresources,* 2: 472-499.

Takabatake, H., Suzuki, K., Ko, I. B. & Noike, T., 2004. Characteristics of anaerobic ammonia removal by mixed culture of hydrogen producing photosynthetic bacteria. *Bioresource Technology,* 95:151-8.

Tanisho, S. & Ishiwata, Y., 1995. Continuous hydrogen production form molasses by fermentation using urethane foam as a support of flocks. *International Journal of Hydrogen Energy,* 20: 541-544.

Tanisho, S., Wakao, N. & Kokakov, Y., 1983. Biological hydrogen production by *Enterobacter aerogenes. Journal of Chemical Engineering, Japan.,* 16: 529-530.

Teixeira, L.C., Linden, J.C. & Schroeder, H.A., 2000. Simultaneous saccharification and co-fermentation of peracetic acid pretreated biomass. *Appl. Biochem. Biotechnol.,* 84-86: 111-127.

Thomas, K.N.G., Bassat, A.B. & Zeikus, J.G., 1981. Ethanol production by thermophilic bacteria: fermentation of cellulosic substrates by co-cultures of *Clostridium thermocellum* and *Clostridum thermohydrosulfuricum. Appl. Environ. Microbiol.,* 41: 1337-1343.

Toivari, M.H., Aristidou, A., Ruohonene, L. & Penttila, M., 2001. Conversion of xylose to ethanol by recombinant *Sacchromyces cervisiae*: importance of xylulokinase (XKSI) and oxygen availability. *Metab. Eng.*3: 236-249.

Tsygankov, A. A., Fedorov, A. S., Laurinavichene, T. V., Gogotov, I. N., Rao, N. N. & Hall, D. O., 1998. Actual and potential rates of hydrogen photoproduction by continuous culture of the purple non-sulfur bacterium *Rhodobacter capsulatus. Applied Journal of Microbial Biotechnology,* 49: 102-107.

Ueno, R., Urano, N. & Kimura S., 2002. effect of temperature and cell density on ethanol fermentation by a thermotolerant aquatic yeast strain isolated from a hot spring environment. *Fisheries Sci.,* 68: 571-578.

Ueno, Y., Haruta, S., Ishii, M. & Igarashi, Y., 2001b. Microbial community in anaerobic hydrogen-producing microflora enriched from sludge compost. *Applied Microbiology and Biotechnology,* 57: 555-562.

Ueno, Y., Otsuka, S. & Morimoto, M., 1995. Hydrogen production from industrial wastewater by anaerobic microflora in chemostat culture. *Journal of Fermentation and Bioengineering,* 79: 395-397.

Valdez-Vazquez, I., Rios-Leal, E., Esparza-Garcia, F., Cecchi, F. & Poggi-Varaldo, H.M., 2005. Semi-continuous solid substrate anaerobic reactors for $H_2$ production from organic waste: Mesophilic versus thermophilic regime. *International Journal of Hydrogen Energy,* 30: 1383-1391.

Van Ginkel, S. & Sung, S., 2001. Biohydrogen production as a Function of pH and Substrate Concentration. *Journal of Environmental Science and Technology,* 35: 4726-4730.

Van Groenestijn, J., Hazewinkel, O. & Bakker, R., 2006. Pretreatment of lignocellulose withbiological acid recycling (Biosulfurol process). *Zuckerindustrie,* 131: 639-641.

Van Haandel, A.C., 2005. Integrated energy production and reduction of the environmental impact at alcohol distillery plants. *Water Sci. Technol.,* 52: 49-57.

Van Niel, C.B., 1929. Photosynthesis of bacteria. In: contributions to marine biology, Standford Univ. Press, 161-169.

Van Niel, C.B., 1931. On the morphology and physiology of the purple and green sulphur bacteria. *Archives ofMicrobiology,*3: 1-112.

Van Niel, C.B., 1944. The culture, general physiology, morphology, and classification of the non-sulfur purple and brown bacteria. *Microbiology and Molecular Biology Reviews,* 8(1):1-120.

Van Niel, E., Claassen, P. & Stams, A., 2003. Substrate and product inhibition of hydrogen production by the extreme thermophile, *Caldicellulosiruptor saccharolyticus. Biotechnology and Bioengineering,* 81(3): 255-262.

Voet, D., Voet, J.G. & Pratt, C.W., 1999. *Fundamentals of Biochemistry.* John Wiley and sons, Inc. New York, USA.

Von, S.M., Zacchi, G., Olsson, L. & Hahn-Hagerdal, B., 1994. Cost analysis of ethanol production from willow using recombinant *Escherichia coli.Biotechnol. Prog.,* 10: 555-560.

Wang, C.H., Lu, W. B. & Chang, J.S., 2007. Feasibility study on fermentative conversion of raw and hydrolyzed starch to hydrogen using anaerobic mixed microflora. *International Journal of Hydrogen Energy,* 32: 3849-3859.

Watanabe, T., Kim, J.S., Ito, K., Buranakarl, L., Kampee, T. & Takahashi, H., 1979. Production of hydrogen by photosynthetic baceria isolated from Thailand. *Microbial Utilization of Renewable Resources,* 1: 181-185.

Weaver, P.F., Lien, S. & Seibert, M., 1979. *Improved hydrogen photoproduction from photosynthetic bacteria and green algae.* Solar Energy Research Institute 21 pp.; NREL Report No. TP-332-409 Colorado, U.S.

Weaver, P.F., Wall, J.D. & Gest, H., 1975. Characterization of *Rhodopseudomonas capsulata.Archives of Microbiology,*105: 207-216.

Weissman, J.C. & Benemann, J.R., 1977. Hydrogen production by nitrogen-fixing cultures of *Anabaean cylindrical. Applied Environmental Microbiology,* 33: 123-131.

Wheals, A.E., Basso, L.c., Alves, D.M. & Amorim, H.V., 1999. Fuel ethanol after 25 years. *Trends biotechnol.,* 17: 482-487.

Wilke, C., Yang, R., Sciamanna, A. & Friedas, R., 1981. Raw material evaluation and process development studies for conversion of biomass to sugars and ethanol. *Biotechnol. Bioeng.,* 23: 163-183.

Wolfrum, E.J. & Watt, A., 2002. Bioreactor design studies for a hydrogen producing bacterium. *Applied Biochemistry and Biotechnology,* 98(100): 611-625.

Wolk, C.P., Ernst, A. & Elhai, J., 1994. Heterocyst metabolism and development, p. 769-823. In: *The molecular biology of cyanobacteria* (ed. Bryant, D.A.). Kluwer Academic Publishers, Dordrecht.

Woodward, J., Orr, M., Cordray, K. & Greenbaum, E., 1996. Enzymatic production of biohydrogen. *Nature,*405: 1014-1015.

Wu, M., Wu, Y. & Wang, M., 2006. Energy and emission benefits of alternative transportation liquid fuels derived from switchgrass: a fuel life cycle assessment. *Biotechnol. Prog.,* 22: 1012-1024.

Wu, S.Y., Hung, C.H., Lin, C.N., Chen, H.W., Lee, A.S. & Chang, J.S., 2006. Fermentative hydrogen production and bacterial community structure in high-rate anaerobic bioreactors containing silicone-immobilized and self-flocculated sludge. *Biotechnology and Bioengineering,* 93(5): 934-946.

Yamagishi, K., 1995. *Interim Evaluation Report of Development of Environmentally Friendly Technology for the Production of Hydrogen*. New Energy and Industrial Technology Development Organization, Japan.

Yang, B., Lu, Y., Gao, K. & Deng, Z., 1997. Study on the hydrolyzate of sugarcane bagasse to ethanol by fermentation. *Chin. J. Biotechnol.*, 13: 253-261.

Yang, H. & Shen, J., 2006. Effect of ferrous iron concentration on anaerobic bio-hydrogen production from starch. *International Journal of Hydrogen Energy*, 31: 2137-2146.

Yasa, I., Cadirci, B.H., Kocyigit, A. & Öztürk, T., 2006. Enrichment and isolation of anoxygenic phototrophic bacteria in Winogradsky Column. *E.U. Journal of Fisheres & Aquatic Science*, 23(1-2): 71-73.

Yetis, M., Gunduz, U., & Eroglu, I., Yucel, M. and Turker, L., 2000. Photoproduction of hydrogen from sugar refinery wastewater by *Rhodobacter sphaeroides* O. U. 001 grown in the wastewater of a sugar refinery. *International Journal of Hydrogen Energy*, 25: 1035-41.

Yetis, M., Gunduz, U., Eroglu, I., Yucel, M. and Turker, L., 1998. Photoproduction of hydrogen from wastewater of a sugar refinery by *Rhodobacter sphaeroides* O.U.001. Paper to be presented in *Marine Bioprocess Engineering*, Noord-wijkerhout, The Netherlands.

Yigit, D.O., Gunduz, U. Turker, L. Yucel, M. & Eroglu, I., 1999. Identification of by-products in hydrogen producing bacteria , *Rhodobacter sphaeroides* O. U. 001 grown in the wastewater of sugar refinery. *Journal of Biotechnology*, 70: 125-131.

Yoch, D.C. & Gotto, J.W., 1982. Effect of light intensity and inhibitors of nitrogen assimilation on $NH^{++}$ inhibition of nitrogenase activity in *Rhodospirillum rubrum* and *Anabaena* sp. *J. Bacteriol.*, 151: 800-806.

Yokoi, H., Saitsu, A., Uchita, H., Hirose, J., Hayashi, S. & Takasaki, Y., 2001. Microbial hydrogen production from sweet potato starch residue. *Journal of Bioscience and Bioengineering*, 91(1): 58-63.

Yokoi, H., Tokushige, T., Hirose, J., Hayashi, S. & Takasai, Y., 1998. $H_2$ production form starch by a mixed culture of *Clostridium butyricum* and *Enterobacter aerogenes.Biotechnology Letters*, 20(2): 143-147

Yokoi, H., Tokushige, T., Hirose, J., Hayashi, S. & Takasaki, Y., 1997. Hydrogen production by immobilized cells of aciduric *Enterobacter aerogenes* strain HO-39. *Journal of Fermentation and Bioengineering* , 83: 484-491.

Yokoyama, H., Moriya, N., Ohmori, H., Waki, M., Ogino, A. & Tanaka, Y., 2007. Community analysis of hydrogen-production extreme thermophilic anerobic microflora enriched from cow manure with five substrates. *Applied Microbiology and Biotechnology*, 77: 213-222.

Younesi, H., Najafpour, G., Ismail, K.S.K., Mohamed, A.R. & Kamaruddin, A.H., 2008. Biohydrogen production in a continuous stirred tank bioreactor from synthesis gas by anaerobic photosynthetic bacterium: *Rhodospirillum rubrum*. *Bioresource Technology*, 99: 2612–2619.

Zaborsky, O.R., 1998. Biohydrogen: *Plenum*, New York.1-552.

Zanin, G.M., Santana, C.C., Bon, E.P., Giordano, R.C., de, Moraes, F.F., Andrietta, S.R., De Carvalho, Neto. C.C., Macedo, I.C., Ramos, L.P.O. & Fontana, J.D., 2000. Brazilian bioethanol program. *Appl. Biochem. Biotechnol.*, 84-86: 1147-61.

Zhang, T., Liu, H. and &, H., 2003. Biohydrogen production from starch in wastewater under thermophilic condition. *Journal of Environmental Management,* 69(2): 149-156.

Zhang, Y. & Shen, J., 2006. Effect of temperature and iron concentration on the growth and hydrogen production of mixed bacteria. *International Journal of Hydrogen Energy,* 31(4): 441-446.

Zhang, Z., Show, K., Tay, J., Liang, D. & Lee, D., 2008. Biohydrogen production with anaerobic fluidized bed reactors—a comparison of biofilm-based and granule-based systems. *International Journal of Hydrogen Energy,* 33(5): 1559-1564.

Zheng, C., Lei, Y., Yu, Q., Lui, X. & Huan, K., 2002. Enzymatic hydrolysis of waste sugarcane bagasse in water media. *Environ. Technol.,* 23(9): 1009-1016.

Zhu, H. Ueda, S. Asada, Y. & Miyake, J., 2002. Hydrogen production as a novel process of wastewater treatment-studies on tofu wastewater with entrapped *R.sphaeroides* and mutagenesis. *International Journal of Hydrogen Energy,* 27: 1349-57

Zhu, H., Suzuki, T., Tsygankov, A,A., Asada, Y. & Miyake, J., 1999. Hydrogen production from tofu wastewater by *Rhodobacter sphaeroides* immobilized in agar gels. *International Journal of Hydrogen Energy,* 24: 305-10.

Zhu, H., Wakayama, T., Asada, Y. & Miyake, J., 2001. Hydrogen production by four cultures with participation by anoxygenic photosynthetic bacterium and anaerobic bacterium in the presence of $NH_4^+$.*International Journal ofHydrogen Energy,* 26: 1149-1154.

Zhu, H.G., Shi, J.L. & Xu, Y.T., 1995. Highly concentrated alcohol waste water treatment by photosynthetic bacteria method. *Environmental Science Shanghai*, 14: 8-10.

Zurrer, H. & Bachofen, R., 1979. Hydrogen Production by the Photosynthetic Bacterium *Rhodospirillum rubrum. Applied and Environmental Microbiology,* 37: 78.

12

# Kinetic Modelling of Dilute Acid Hydrolysis of Lignocellulosic Biomass

P. Lenihan, A. Orozco, E. O'Neill, M.N.M. Ahmad,
D.W. Rooney, C. Mangwandi and G.M. Walker
*School of Chemistry and Chemical Engineering, Queen's University Belfast*
*Northern Ireland*
*UK*

## 1. Introduction

### 1.1 Dilute acid hydrolysis

Dilute Acid hydrolysis refers to the hydrolysis of hemicellulosic material by acids (typically sulphuric, hydrochloric or phosphoric acid) at concentrations of 1-10% using a moderate temperature (in the range of 100-150$^0$C). But in these relatively moderate operational conditions, it proves less effective in the formation of hexoses[1]. This is mainly due to the decomposition of the monosaccharides into less desirable compounds during hydrolysis. These compounds include furfural, a product of dehydration of pentoses and hydroxymethilfurfural-HMF, a product of the dehydration of hexoses. These compounds along with acetic acid which forms during initial decomposition of the hemicelluloses, as a result of hydrolysis of acetyl groups linked to the sugar, inhibit the later fermentation, leading to reduced ethanol yields[2]. The production of these inhibitors increases when hydrolysis takes place at higher temperatures and higher acid concentrations[3].

Sulphuric and Hydrochloric acids are the most commonly used catalysts for hydrolysis of lignocellulosic residues. In contrast to these acids, phosphoric acid can be more advantageous for hydrolysis. Phosphoric acid is less aggressive than other acids which give solutions with higher concentrations of growth inhibitors of microorganisms such as furfural or acetic acid[2].

Dilute Phosphoric acid, on hydrolysates from sugar cane bagasse, has shown fermentable sugars with 21.4 g of sugar $L^{-1}$ with less than 4 g $L^{-1}$ of inhibitors at operating conditions of 6% acid concentration at 100$^0$C for 300 mins[4]. Similarly on hydrolysates from olive tree pruning, have shown hemicelluloses conversion rates of 77% with glucose and reducing sugar concentrations being observed as 89% of the hemicellulosic sugars contained in the raw material at conditions of 8% acid concentration at 90$^0$C for 240 mins[2].

These hydrolysates obtained after the acid hydrolysis need to be processed if they are going to be used as fermentation media. In general the following operations are needed (in this sequence): concentration, detoxification, neutralization and supplementation with nutrients. This process is illustrated in Figure 1.

The concentration of hydrolysates by evaporation is usual to increase the sugar concentration. In this operation, besides water, small amounts of growth inhibitors such as acetic acid, furfural and HMF are removed[1]. A detoxification operation by adsorption on active carbon in the form of charcoal can remove the growth inhibitors cited. In this operation, phenolic compounds proceeding from lignin can also be removed[1]. In the operation of neutralization, it is usual to add chemicals that neutralize the acids of the hydrolysates, forming salts[5]. These salts have low solubility and are normally removed by filtration. For example, hydrolysates containing sulphuric acid are neutralized with calcium carbonate, forming calcium sulphate[1]. Finally, the processed hydrolysates are supplemented with several nutrients to be a favorable fermentation medium. These nutrients contribute the nitrogen and micronutrients needed for the growth of the microorganisms[6].

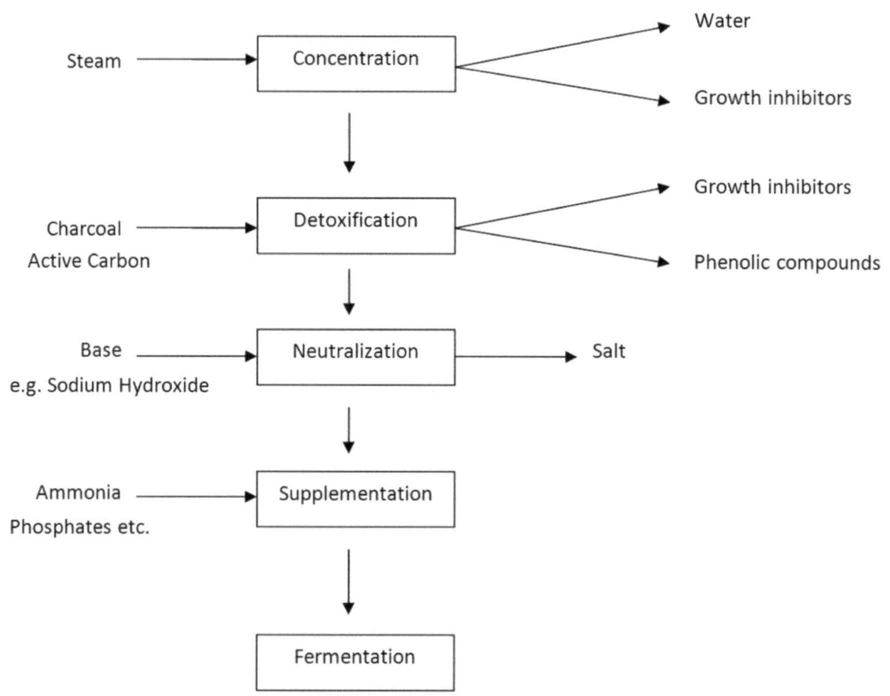


Fig. 1. General operations required after hydrolysis prior to fermentation.

The interest in the use of $H_3PO_4$ is that after neutralization of hydrolysates with NaOH, the salt formed is sodium phosphate. This salt can remain in the hydrolysates because it is used as nutrient by microorganisms. Therefore, an operation of filtration is not needed with the consequent advantage: improve the economics of the process (avoid the filtration to remove the salts and decrease the amount of nutrient needed for fermentation) and is friendly with the environment (the salt formed is not a waste)[1].

### 1.2 Kinetics of acid hydrolysis of cellulose

The Hydrolysis reactions using dilute acid are very complex, mainly because the substrate is in a solid phase and the catalyst in a liquid phase. The reaction rate of hydrolysis depends on a number of variables, such as: temperature, acid concentration, time, substrate concentration and substrate composition. The practical objective of studying the kinetic model is, on a first level, to optimize the process and, on a second level, to obtain Equations useful for economical estimations[4]. The models usually associated with dilute acid hydrolysis were first proposed by Saeman[7], for the hydrolysis of Douglas fir wood using sulphuric acid. The models proposed in the literature use irreversible pseudo-homogeneous first-order reactions. They proposed that hydrolysis of cellulose involves the polymer glucan of cellulose being degraded to monomer glucose which is subsequently converted to decomposition products. This is represented below:

$$\text{glucan(s)} \xrightarrow{K_1} \text{glucose} \xrightarrow{K_2} \text{decomposition products} \quad (1)$$

Where $K_1$ is the rate of conversion of glucan to glucose and $K_2$ is the rate of decomposition of glucose. Both have units of the reciprocal of time ($min^{-1}$). Both reactions were considered to be first order and irreversible. Saeman's model could also be applied to the hydrolysis of hemicellulosic fraction. Therefore this reaction was generalised to:

$$\text{Polymer} \xrightarrow{K_1} \text{Monomer} \xrightarrow{K_2} \text{decomposition products} \quad (2)$$

Where the polymer can be glucose, xylose or araban.

From this reaction model and solving differential Equations, monomer concentration ($M$) as a function of time ($t$) can be represented by[8]:

$$M = \left[\frac{k_1 P_0}{k_2 - k_1}\right](e^{-k_1 t} - e^{-k_2 t}) + M_0 e^{-k_2 t} \quad (3)$$

Where:
$M$ = Monomer Concentration, g $L^{-1}$
$P$ = Polymer Concentration, g $L^{-1}$
$M_0$ = Initial Monomer concentration, g $L^{-1}$
Assuming that the initial monomer concentration to be approximately equal to 0, then Equation (1) can be simplified to[8]:

$$M = \left[\frac{k_1 P_0}{k_2 - k_1}\right](e^{-k_1 t} - e^{-k_2 t}) \quad (4)$$

An alternative model called the two fraction model is often used to describe the reaction kinetics and is tested against the Saeman's model to provide accuracy. This model considers that only a fraction of the polymer reacts. This is called the fast fraction, and the fraction that does not react or reacts slowly is called the slow fraction. The ratio between them is the parameter α. In the case that the slow fraction does not react, the following Equation is used[4]:

$$M = \alpha \left[\frac{k_1 P_0}{k_2 - k_1}\right](e^{-k_1 t} - e^{-k_2 t}) \quad (5)$$

When determining kinetic parameters it is more thorough to apply both models to see if there is deviation of results. If both results returned do not match then it can be concluded that the two fraction model is more accurate. If on the other hand the kinetics reveal similar results, it can be concluded that the reaction is 100% fast fraction with $\alpha = 1$ g $g^{-1}$.
The use of both models has been demonstrated by Gámez et al.[4] in their efforts to hydrolyze sugar cane bagasse into fermentable sugars. The hemicelluloses of sugar cane are primarily xylan. The primary sugar obtained from this process is xylose however there are concentrations of glucose arabinose and furfural obtained also. By carrying out phosphoric acid hydrolysis on the sugar cane bagasse at 100°C at different acid concentrations and reaction times it was found that an optimum yield of 38.6% conversion was obtained at 2% phosphoric acid concentration for 300 mins. Both the two fraction and the Saeman's model

where used to derive reaction constants for the xylose kinetic model and it was found that both were approximately the same showing that the fast reaction was 100%. Therefore the latter model was used because it's simplicity.

## 2. Experimental

### 2.1 Experimental aim

For the purposes of this research the conditions to be varied are temperature and acid concentration. The raw material will be hydrolysed at 135$^0$C, 150$^0$C, 175$^0$C and 200$^0$C using phosphoric acid at 2.5, 5, 7.5 and 10% w/w acid concentrations. The raw material substrate being studied is potato skins.

### 2.2 Materials

#### 2.2.1 Potato peelings

The potato peelings used for this work were obtained from a potato crisp manufacturer, Tayto (NI) Ltd, with a typical composition detailed in Table 1.

| Composition | Proportion |
|---|---|
| Cellulose | 55.25% |
| Hemicellulose | 11.71% |
| Lignin | 14.24% |
| Moisture | 10.0% |
| Ash | 8.8% |

Table 1. The chemical composition of Potato Peel

#### 2.2.2 Cellulose and hemicellulose analysis

The potato peels are first ground to pass a 16 mesh screen. Three 0.3 g samples are weighed into three test tubes and to each is added 3 ml of 85% sulphuric acid that has been cooled to 15$^0$C. The samples are stirred thoroughly before being placed in a water bath at 30$^0$C. This temperature is maintained for 2 hours, stirring the samples every 10 minutes. After a total time of 2 hours the mixture is washed from the vial into an Erlenmeyer flask and made up to 89.11g with distilled water. The dilute solution is autoclaved at 15 pounds steam pressure and 121 $^0$C for 1 hour. At the end of this time the sample is cooled and vacuum filtered to remove unreacted lignin. The filtrate is then syringed through a 0.45 μm filter, before being analysed by HPLC. With 100% conversion assumed the composition of glucose is recognized as cellulose and that of arabinose can be recognized as hemicellulose.

#### 2.2.3 Other analyses

Due to the robust and complex nature of lignin, it is only decomposable through enzyme action, therefore making it virtually insoluble in mineral acids. Having hydrolysed the cellulose and hemicellulose components of the biomass the composition of lignin can be determined quite easily. The process of vacuum filtering the samples results in the separation of the hydrolysate and the remaining solid deposit. This deposit is made up of mainly lignin and ash components. The glass filter crucibles which have been used in the vacuum filter are dried over night in an oven at 110$^0$C before having their weight recorded. They are then placed in a muffle furnace at 550$^0$C for 3 hours to burn off the remaining organic deposits. The weight is then recorded again. The proportion of acid insoluble residue mainly lignin can be calculate using Equation 6 as per the *Standard Test Method for Determination of Acid-Insoluble Residue in Biomass – E1721 - 95* :

$$\text{Percentage of lignin} = \frac{W_2 - W_3}{W_1 \times T_{110}} \times 100 \quad (6)$$

Where,

$W_1$ = Weight of potato peel sample (g)
$W_2$ = Weight of filter crucible after ignition in muffle furnace - Ash sample (g)
$W_3$ = Weight of filter crucible after vacuum filtration - Lignin and Ash (g)
$T_{110}$ = As received sample conversion factor

- Moisture analysis

Moisture content of a sample of potato peel is measured by weighing out a recorded amount of sample and placing it in an oven at 110$^0$C until the dry weight of the sample is constant over a 2 hour period. The sample is then cooled and its weight is recorded. Moisture content is determined by dividing the dry weight by the initial weight.

- Ash analysis

The ash content is calculated by dividing the weight of the filter crucible, after it has been ignited in the muffle furnace $W_2$, by the initial weight of the sample $W_1$ times the conversion factor $T_{110}$.

## 2.3 Experimental procedure

### 2.3.1 Equipment

A 1 Litre continuously stirred pilot batch reactor (Parr reactor) was employed for the experimental programme. The reactor operates at a temperature range of -10 to 350$^0$C up to 130 bar pressure. Operating conditions are modulated by a 4843 controller unit. The total contents of the reactor constitute 700g of which 5% w/w will be the raw material potato peels. The potato peels are dried and milled to 16 mesh or 1mm diameter particles. The remaining 95% w/w content of the reactor is made up of the dilute acid concentration. The acid concentration is not initially added to the reactor but instead is delivered through the acid reservoir during the initialisation of the reaction. For acid concentrations 2.5, 5 and 7.5% w/w this is made by preparing a 70g sample made up of the 85% phosphoric acid required to achieve the desired acid concentration for the reaction and distilled water. The remaining distilled water required to achieve this dilution is mixed with the potato peel and charged to the Parr reactor vessel.[10]

The sample tube is then fitted with a gauze mesh to restrict the solid sample from blocking it. The reactor is secured tightly by 6 bolts to maintain the operating pressure within the vessel during the reaction. The vessel is then attached with the heating jacket and the agitator impellor is connected to begin mixing.

The sample line and acid reservoir are bolted tightly to the reactor. The nitrogen line is then attached to the acid reservoir. Finally the thermocouple which provides feedback to the 4843 controller is inserted and the temperature setpoint is entered. The 4843 controller will then ramp up the jacket heating to achieve and maintain the required operating temperature setpoint. Depending on the whether the temperature output required is 135$^0$C, 150$^0$C, 175$^0$C or 200$^0$C it will take between 30-60 minutes to reach the desired temperature. The impellor is initiated at the same point as the jacket heating element and remains constant for all experiments at the 4843 controller maximum RPM rate of 632. This will ensure that by the time the reaction commences the concentration of potato peels will be constant throughout the vessel. Once the desired temperature setpoint is at steady state the reaction can commence.[10]

### 2.3.2 Reaction procedure

To initialise the reaction the phosphoric acid must be delivered to the reaction vessel from the acid reservoir. The reservoir is first pressurised by opening the nitrogen valve, thus pressurising it to 20 bar. The acid inlet valve is then opened, causing a pressure differential between the reservoir and the reaction vessel which will allow the acid to be delivered to the vessel. The pressure gauge of the vessel is monitored for any increase in pressure, once this is observed it indicates that all the acid has been delivered. At this point the inlet valve is closed and the stop watch is started simultaneously.

Sampling occurs at time intervals of 2, 4, 8, 15, 30, 45, 60, 75 and 90 minutes. A sample tube is secured to the sample line; the sample outlet valve is then opened allowing a maximum of 5 ml of solution to be collected. The sampling procedure is assisted by the elevated pressure within the vessel allowing it to take the briefest amount of time possible. This helps to reduce further reactions of the solution which would spoil the results. The sample tube is coiled through a jug of cold water to further reduce the reaction rate of the solution by rapid cooling. The sample tube must then be cleared using compressed air to prevent contamination of the next sample. Once this has been completed the sample tube is removed sealed and placed in ice to completely cease any possible further reacting. Finally the nitrogen line is opened and the vessel is pressurised slightly to stabilise the reaction vessel, maintain a constant pressure and to clear any blockages in the sample line.[10]

Although contamination of the solution by potato peel particles is severely reduced by the presence of a gauze mesh it is not eliminated, therefore purification by vacuum and syringe filtration must be done in preparation for analysis. On completion of this the samples are sent for analysis in a HPLC.

## 3. Results and discussion

### 3.1 Reaction kinetics

As mentioned previously the reaction which illustrates the production and decomposition of sugars is demonstrated below in Equation (1).

$$\text{Polymer} \xrightarrow{K_1} \text{Monomer} \xrightarrow{K_2} \text{decomposition products} \tag{1}$$

The polymer can be cellulose (glucan) or hemicellulose (araban). Saeman[7] found that a simple two-step reaction model adequately described the production of sugars from hydrolysis. This model assumes an irreversible first-order type reaction. If the reaction is studied in terms of concentration it can be represented by the following Equation (7).

$$\text{Glucan (A)} \xrightarrow{K_1} \text{Glucose (B)} \xrightarrow{K_2} \text{HMF (C)} \tag{7}$$

The formation rate of the product glucose (B) with respet to time is represented by Equation (8):

$$\frac{dC_B}{dt} = k_1 C_A - k_2 C_B \tag{8}$$

By integrating this Equation with respect to time gives an Equation representing the concentration of sugar as a function of time. The Equation (9) below is derived:

$$C_B = k_1 C_{A0}\left(\frac{e^{-k_1 t} - e^{-k_2 t}}{k_2 - k_1}\right) \quad (9)$$

Using this Equation it will be possible to accurately model the reactions kinetics at each of the operating conditions of temperature and acid concentration and therefore determine the reaction constants.

### 3.2 Modelling

The reaction constants $k_1$ and $k_2$ are determined through using the solver function on Microsoft Excel. By minimising the sum of the square of the error between the experimental data and the model data obtained accurate reaction parameters can be found. The solver function operates by attempting to acquire a value of zero error through changing of the $k_1$ and $k_2$ values.

As this is a solid-liquid reaction the initial concentration of A is difficult to determine without going into mass transfer and shrinking particle theory, which is based on the average size of the particles used. On this basis quantitative saccharification is used to determine the concentration of the reactants. By taking into account the solid liquid ratio, the initial concentrations $C_{A0}$ can be established by determining the concentration of their products in an assumed 100% conversion reaction. Through quantitative saccharification the hexosans (glucan) and pentosan (araban) are hydrolysed completely to form hexose (glucose) and pentose (arabinose) respectively. The concentrations, of the sugars produced, which are obtained from analysing the chromatograms are fixed as the initial concentrations of the cellulose and hemicellulose. Therefore the concentration of glucose is assumed that of the $C_{A0}$ of glucan or cellulose, and the concentration of arabinose is assumed that of the $C_{A0}$ of araban or hemicellulose. These values will satisfy the respective parameters within the mathematical model.

The reaction was modelled to determine the kinetics for phosphoric acid concentrations of 2.5, 5.0, 7.5, 10.0% (w/w) and operating temperatures of 135, 150, 175 and 200 °C.

### 3.3 Glucose production

Figure 2 shows the results of the kinetic model constructed to represent glucose at temperatures 135, 150, 175 and 200°C at an acid concentration of 5.0% w/w. The model was constructed through calculating best fit reaction coefficients for $k_1$ and $k_2$.

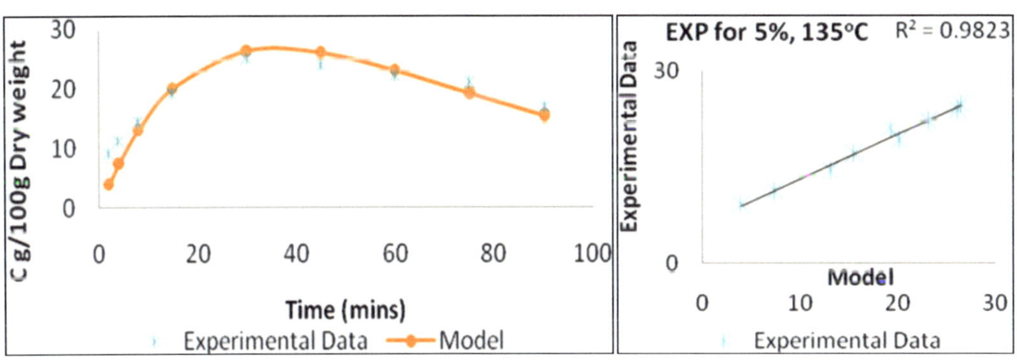


(a.1) Glucose model at 135°C, acid conc. 5.0% (a.2) Fitting Test

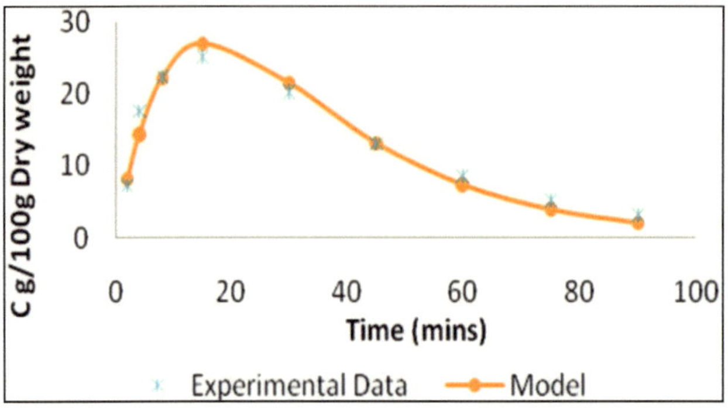


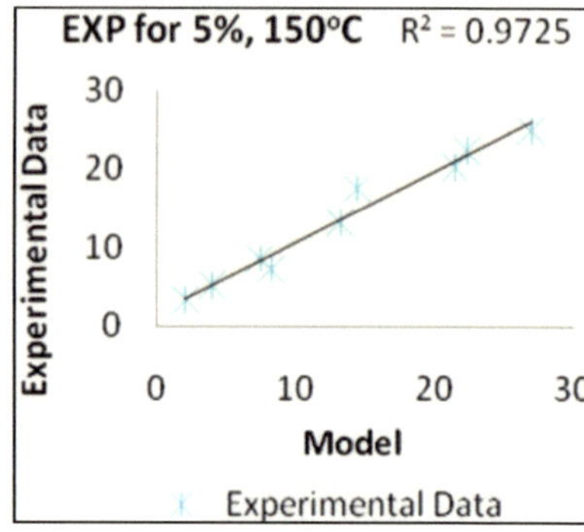


(b.1) Glucose model at 150°C, acid conc. 5.0%

(b.2) Fitting Test

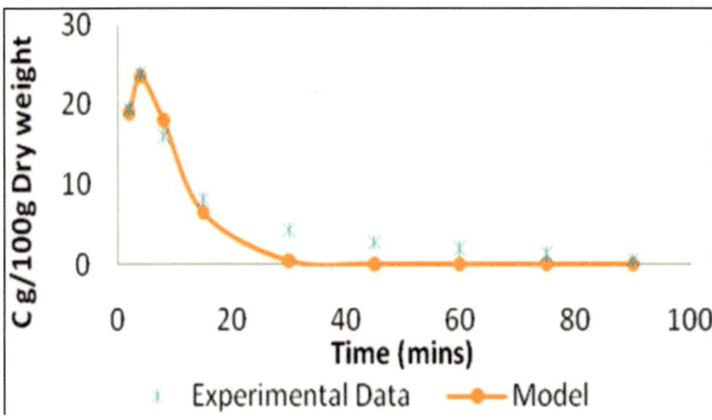


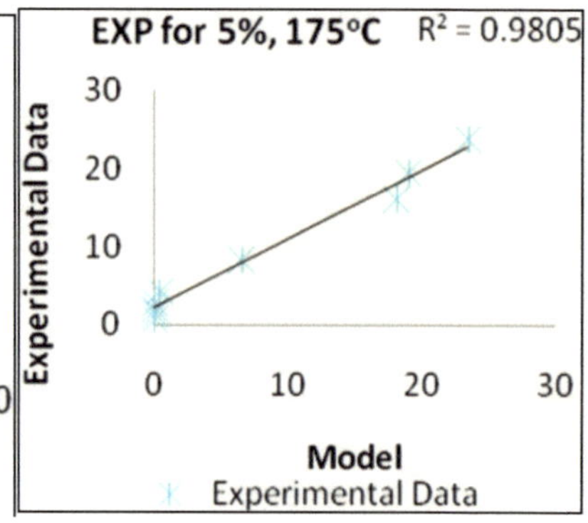


(c.1) Glucose model at 175°C, acid conc. 5.0%

(c.2) Fitting Test

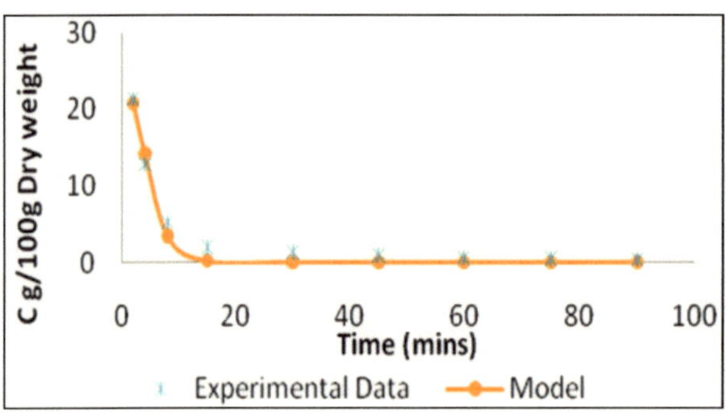


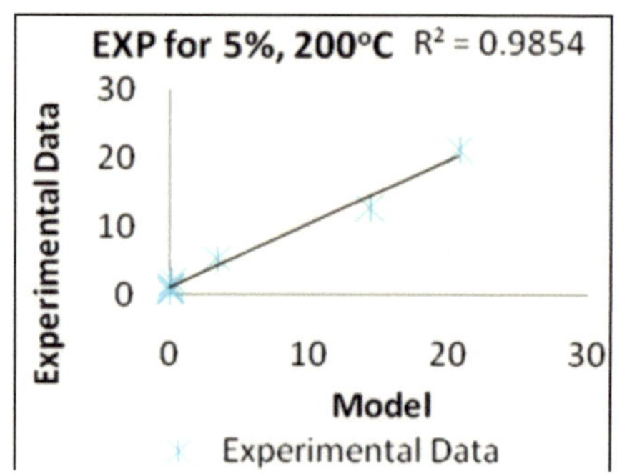


(d.1) Glucose model at 200°C, acid conc. 5.0%

(d.2) Fitting Test

Fig. 2. Glucose model with variation in acid concentration

From observing these graphs it can be seen that the models generated for glucose production at 5.0% w/w acid concentration are highly accurate. Along with the model for the reaction, the experimental values against the model values are constructed to determine if the data fits. The correlations represented by $R^2$ are seen to be well above 0.9 which indicates the high level of accuracy achieved from the model. The kinetic model is of first order principles further demonstrating that the glucose generation reaction is a first order reaction as has been noted in numerous studies.

### 3.4 Arabinose production

Figure 3 shows the results of the kinetic model constructed to represent arabinose at temperatures 135, 150, 175 and 200 °C at an acid concentration of 2.5% w/w. The model was constructed through calculating best fit reaction coefficients for $k_1$ and $k_2$.

From observing these graphs it can be seen that the models generated for arabinose production at 2.5% w/w acid concentration are highly accurate. Along with the model for the reaction the experimental values against the model values are constructed to test that the data fits. The correlations represented by $R^2$ are seen to be well above 0.9 which indicates the high level of accuracy achieved from the model. The kinetic model is of first order principles further demonstrating that the arabinose generation reaction is a first order reaction as has been noted in numerous studies. Due to the relatively low level of arabinose being formed in this hydrolysis reaction there were a number of experiments which registered little or no arabinose formation and therefore could not be modelled accurately. These models were under operating conditions of 7.5 and 10.0% w/w acid concentration at 175 and 200°C. The reasons for these failed models are likely as a result of low concentrations formed which led to indecipherable chromatogram readings from the HPLC.

It can be concluded that the first order models of Glucose and Arabinose, where there was reasonable experimental data, are accurate. Glucose was modelled entirely with all $R^2$ correlations exceeding 0.95, which is acceptable. Due to discrepancies in the data points for a small number of more aggressive reactions, four models of arabinose generation where

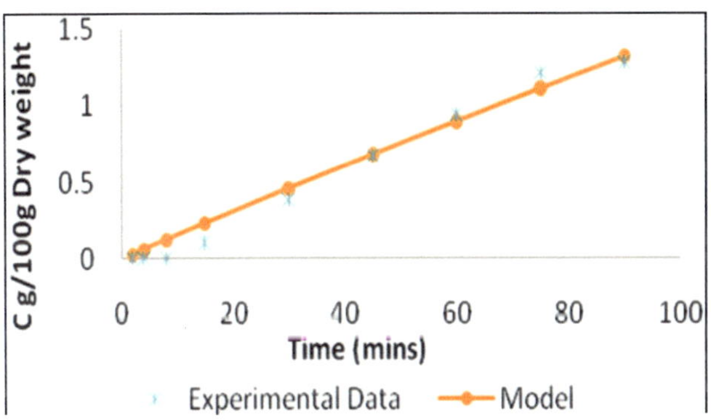


(a.1) Arabinose model at 135°C, acid conc. 2.5%

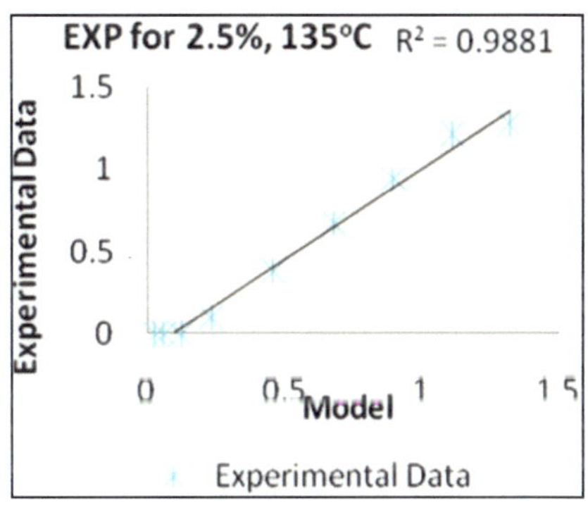


(a.2) Fitting Test

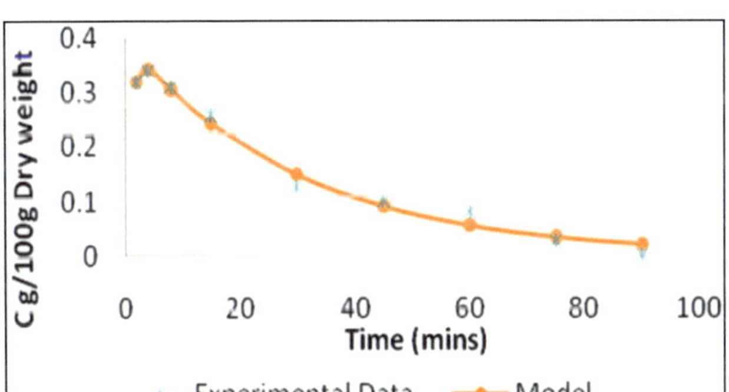


(b.1) Arabinose model at 150°C, acid conc. 2.5%

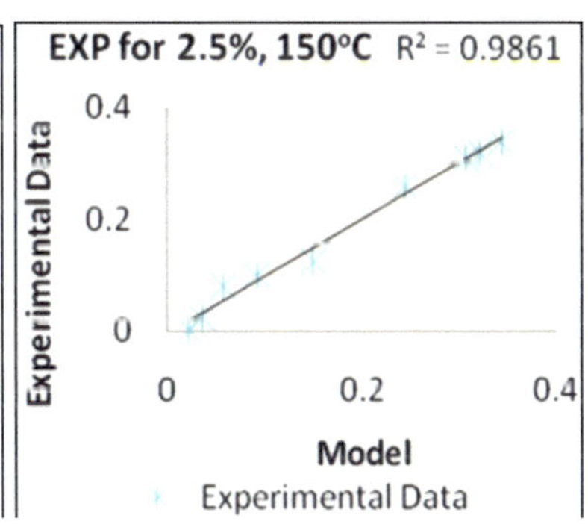


(b.2) Fitting Test

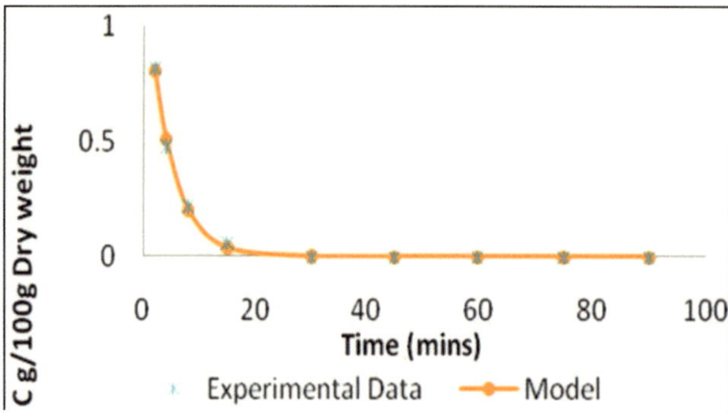


(c.1) Arabinose model at 175°C, acid conc. 2.5%

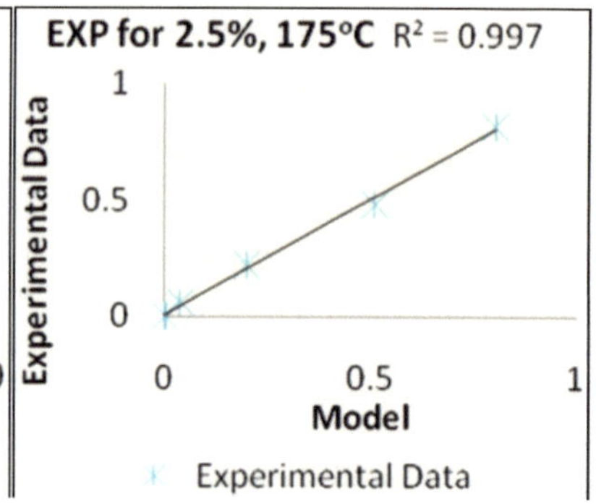


(c.2) Fitting Test

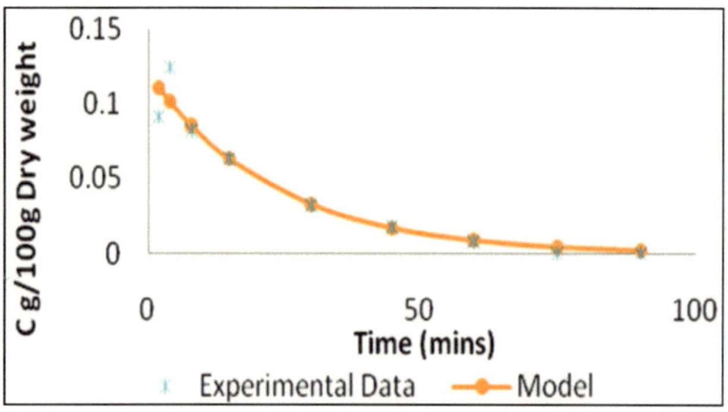


(d.1) Arabinose model at 200°C, acid conc. 2.5%

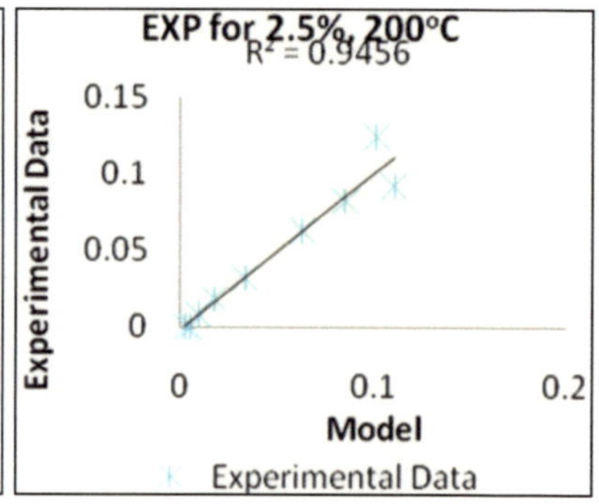


(d.2) Fitting Test

Fig. 3. Arabinose model with variation in reaction temperature

unachievable however the remainder attained $R^2$ correlations exceeding 0.95. To improve the models certain data points were ignored (one or two entries per reaction) when it was judged to be upsetting the model. In all cases this improved the overall yield.

### 3.5 Reaction constants

Having successfully determined the accurate k values for the reactions through the mathematical models, a number of other important kinetic information can be derived. These kinetic coefficients can be correlated with temperature by applying the Arrhenius Equation, and from that additional information, such as the activation energy and the pre-exponential factor, can be obtained. This correlation is demonstrated below[4]:

$$k_i = k_{i0} e^{\frac{-E_a}{RT}} \quad (10)$$

Where,

$k_i$ = Kinetic coefficient ( $i$ = 1 or 2) ($min^{-1}$)

$k_{i0}$ = Pre-exponential factor ( $i$ = 1 or 2) ($min^{-1}$)

$E_a$ = Activation Energy (kJ $mol^{-1}$)

R = Gas Constant, 8.314 (kJ $mol^{-1}$ $K^{-1}$)

T = Temperature (K)

The pre-exponential factor is the link between the temperature and acid concentration variables in the reaction. As can be seen above from the Arrhenius Equation the reaction constant $k_i$ increases exponentially with temperature, to coincide with this the acid concentration affects the kinetic model as per the following Equation 10[4]:

$$k_{i0} = a_i[Ac]^{n_i} \quad (11)$$

Where

$a_i$, $n_i$ = Regression Parameters

Ac = Acid Concentration (% w/w)

Therefore once $k_i$ has been determined, it is possible to derive all the reaction constants associated with the variables of temperature and $H_3PO_4$ concentration. The $k_{10}$ values are unique to the potato peel raw material source and vary greatly between raw materials. These differences can be due to the structure and composition of the material which may neutralize the acid. In order to extract these additional kinetic parameters, both Equations must be rearranged into the form of a linear function; $y = mx + c$, whereby the reaction coefficients are represented by the linear constants $m$ and $c$. Equation 10 and Equation 11 are therefore rearranged as follows[4]:

$$ln\ k_i = -\frac{E_a}{R}\frac{1}{T} + ln\ k_{i0} \quad (12)$$

$$ln\ k_{i0} = n_i ln\ [Ac] + \ln a_i \quad (13)$$

Figure 4 illustrates the linear relationship between the log values of the k coefficients and the inverse of temperature at 5.0% w/w acid concentration. The $R^2$ correlation shows good agreement between the k values, this indicates that the model accurately follows the theoretical Equation (12).

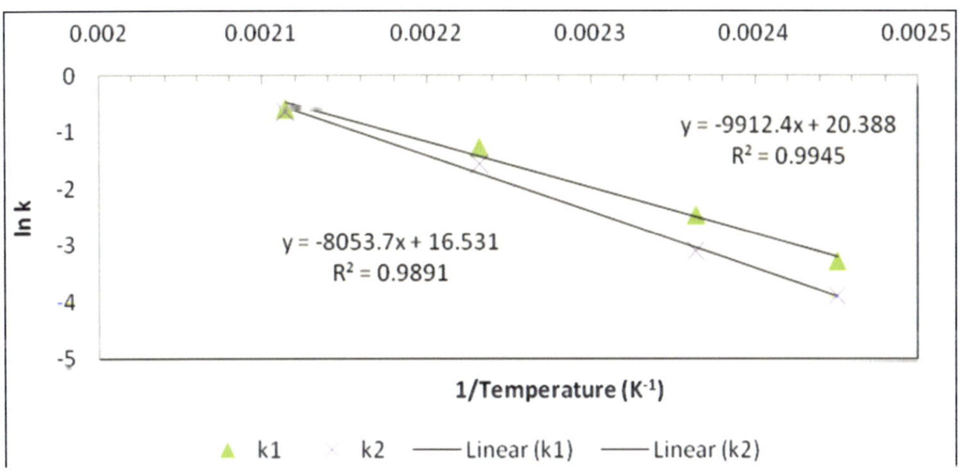


Fig. 4. Relationship between the calculated k values and the temperature of glucose, at 5.0% w/w acid concentration

Similarly with Glucose, Arabinose demonstrates good agreement between the experimental and modelling data. The data extracted from the linear Equation is then used to obtain values for activation energy, $E_a$, while $ln\ k_{i0}$ is used later to establish the mathematical relationship between acid concentration and the rate of reaction. Table 2 shows the reaction constants obtained for hydrolysis of potato peels at temperatures 135, 150, 175 and 200 °C at 5.0% w/w acid concentration. The reaction constants for all other acid concentration reactions are also generated. Having calculated all the $k_{i0}$ values for the reaction constants $k_i$ it is now possible to introduce acid concentration into the reaction model as per the empirical Equation 12. Figure 5 shows that the log values of $k_{i0}$ vary linearly with the log of acid concentration for glucose generation.

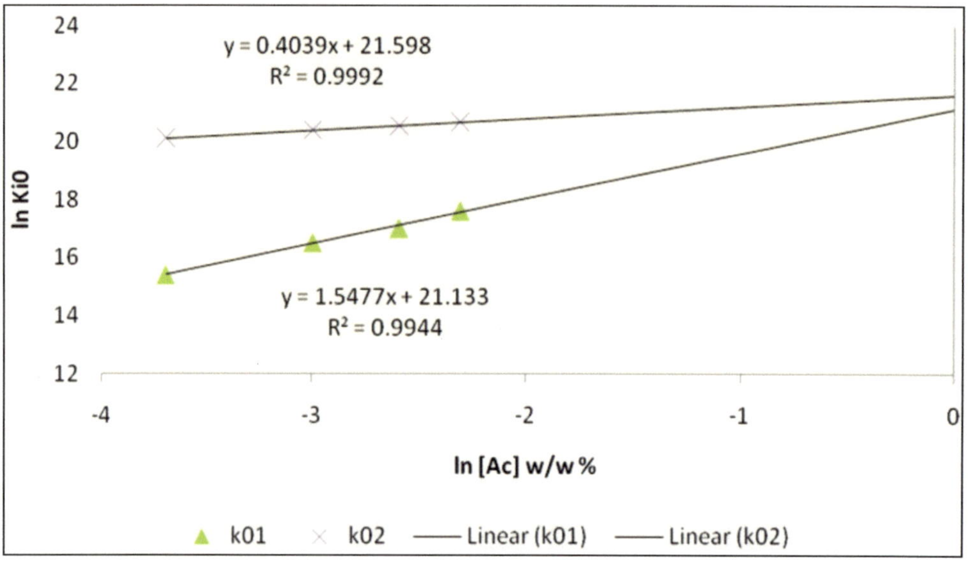


Fig. 5. The relationship between the calculated $k_{i0}$ values and acid concentration for glucose

The model for glucose was found to be more accurate than that of Arabinose due to issues with reliability of Arabinose experimental data obtained brought on by experimental/human error. The linear fit has a high accuracy level, indicated by the $R^2$ correlations obtained. This illustrates a strong agreement between acid concentration and the $k_{i0}$ parameters. Table 3 shows the reaction constants obtained from Equation 12 for both sugars investigated.

| | | 135°C | 150°C | 175°C | 200°C |
|---|---|---|---|---|---|
| (a) Glucose | | | | | |
| $k_1$ (min$^{-1}$) | | 0.03731 | 0.0841 | 0.2772 | 0.5438 |
| $k_2$ (min$^{-1}$) | | 0.02002 | 0.0444 | 0.2054 | 0.5213 |
| Reaction constants | $k_1$ | | $E_a$ = 66.958 kJ, | $k_{10} = 1.51\text{x}10^7$, | $R^2 = 0.9945$ |
| | $k_2$ | | $E_a$ = 82.411 kJ, | $k_{20} = 7.15\text{x}10^8$, | $R^2 = 0.9891$ |
| (a) Arabinose | | | | | |
| $k_1$ (min$^{-1}$) | | 0.0116 | 0.0225 | 0.03903 | 0.041 |
| $k_2$ (min$^{-1}$) | | 0.0683 | 0.2777 | 0.5517 | 3.3665 |
| Reaction constants | $k_1$ | | $E_a$ = 30.939 kJ, | $k_{10}$ = 127.84, | $R^2 = 0.9127$ |
| | $k_2$ | | $E_a$ = 89.226 kJ, | $k_{20} = 2.05\text{x}10^{10}$, | $R^2 = 0.961$ |

Table 2. Reaction constants of sugars released in 5.0% w/w $H_3PO_4$ hydrolysis of Potato Peels

One notable conclusion can be drawn from these results is that the value of $n$ for sugar formation of glucose is markedly higher than that for sugar degradation. This parameter justifies the conclusion that increasing the acid concentration will increase net yield of sugars as the $k_1$ would increase by a magnitude faster than the rate of degradation.
A generalised model for the prediction of sugar production rate from hydrolysis by $H_3PO_4$ has been developed. The kinetic parameters $k_1$ and $k_2$ can be determined at any given temperature and acid concentration, by substituting Equation 10 and Equation 11 to give Equation 14.

$$k_i = a_i e^{\left(\frac{-E_{aj}}{RT}\right)} [Ac]^{n_i} \tag{14}$$

Where,
$i$ = 1 or 2 (depending on whether reaction is sugar formation or degradation)

| | $k_{10}$ | $k_{20}$ |
|---|---|---|
| (a) Glucose | | |
| $a_i$ | $1.51 \times 10^9$ | $2.40 \times 10^9$ |
| $R^2$ | 0.9944 | 0.9992 |
| $n_i$ | 1.5477 | 0.4039 |
| (a) Arabinose | | |
| $a_i$ | $5.54 \times 10^{-18}$ | $4.12 \times 10^{21}$ |
| $R^2$ | 0.99 | 0.988 |
| $n_i$ | -14.89 | 8.733 |

Table 3. Reaction constants of sugars released for varying $H_3PO_4$ concentrations during hydrolysis of Potato Peels

### 3.6 Process optimisation

With a theoretical model in place consisting of accurately determined reaction parameters, the optimum conditions can be examined more closely.
When taking into account the reaction kinetics involved in the formation of sugars through hydrolysis, it is seen that the larger the value of $k_1$ the higher the rate of sugar formation and therefore the lower the cycle time required to maximise yield. In contrast to this the larger the value of $k_2$ the higher the rate of sugar degradation. Hence, the most desirable operating conditions will result in a high value of $k_1$ and a low value of $k_2$.
In practice this scenario is difficult to achieve so the optimum conditions available at the current operating conditions must be investigated. The value of $k_1$ and $k_2$ both increase exponentially with temperature as per the Arrhenius Equation. To increase the temperature would serve to increase the sugar production rate but a rapid decline due to sugar degradation would render the process highly inefficient as was seen earlier. Therefore, to preserve sugars, it is best to limit the temperature of the reaction and to try and manipulate the k values through varying the acid concentration.
As mentioned previously, through detailed analysis of the reaction kinetics it was found that the acid concentration had affected the $k_1$ value by a magnitude greater than it affects $k_2$. By studying the theoretical results of the Arrhenius Equation it can be seen that the values of $k_2$ vary only slightly compared to the variance of $k_1$. This will allow the rate of production to increase at the expense of the rate of degradation allowing for high yields to

be achieved in a shorter time span. One such example is sugar generation at 135°C and 10.0% w/w acid concentration whereby a maximum yield of 55.2g sugar/ 100g dry potato peel is achieved after a short residence time of between 4-8 minutes. As acid degradation is primarily a function of temperature, almost negligible sugar degradation occurs during this period. However using more concentrated acid reactants will require acid recovery techniques which can prove both tricky and expensive to operate.

Baring this in mind a more attractive proposition would be to run the reaction at a more moderate temperature and acid concentration. One possible set of operating condition is 175°C and 7.5% w/w acid concentration. A maximum yield of 38.78 g sugar/ 100g potato peel is observed after 15 minutes residence time and the degradation of this sugar is at a more manageable rate with over 44% sugar retained after 90 minutes. This reaction would be considered easier to control.

If the cost of the process is introduced to the model it will have an effect on the selection of operating conditions. The main aim of a major energy providing company is to maximise profit, therefore the cycle time and throughput of the reaction is paramount. The maximum yield residence time becomes a significant factor in this situation. If a company opts to use a more controllable cost effective reaction model to hydrolyse their feed material such as 135°C and 5.0% w/w acid concentration then a maximum yield of 26.32g sugar/ 100g dry potato peel is achieved after 30 minutes reaction time. However if the reaction conditions are modified slightly to 150°C at the same acid concentration then a maximum yield of 25.97g sugar/ 100g dry potato peel is achieved after 15 minutes. Although the yields are similar the retention time for the reaction is halved by increasing the temperature by 25 deg C. The outcome of this sort of comparison will ultimately be determined by the energy input costs and the costs of treating process wastes related to the reaction.

The limitations of a batch system for this type of reaction are apparent when examining the results. Preparation and cleaning of the reactor are cumbersome processes taking between 2-3 hours to complete from start to finish. Therefore, to achieve and maintain the optimum conditions for this hydrolysis reaction a continuous process with an average retention time is a more desirable option. This is currently the case in bulk manufacturing of biofuel where plug-flow reactors are a viable option.

A summary of the optimum conditions for both glucose and arabinose at each concentration is presented in Table 4.

| Conc. $H_3PO_4$ (% w/w) | Temperature (°C) | Time (min) | Glucose g/ 100g dry Potato Peel |
|---|---|---|---|
| 2.5 | 150 | 8 | 14.2 |
| 5.0 | 150 | 15 | 25.1 |
| 7.5 | 135 | 15 | 36.8 |
| 10.0 | 135 | 4 | 53.8 |

Table 4. (a) Summary of the conditions required for optimum glucose yield at 2.5, 5.0, 7.5 and 10% w/w acid concentration

## 4. Conclusions

It can be seen from Table 5 that a level of 97% glucose conversion has been obtained when compared to the theoretical yield. However an unquantified proportion of this yield is likely to be attributed to residual starch which is present in the feed stock prior to reacting. Starch is a mixture of both aourlose and amlyopectin (usually in 20:80 or 30:70 ratios) which are complex carbohydrate polysaccharides of glucose. Starch readily hydrolyses to form glucose monomers and experimental yields have been found to reach 111% that of the theoretical yield[9]. However the presence of starch should not be considered a contamination of the results as it is common place for a percentage of residual starch to be present after potatoes have been processed therefore it will only increase the attractive quality of potato peels to companies interested in utilising it as a feed material for biofuel production.

| | Maximum recorded yield Dry mass basis (g/100g) | Quantitative saccharification of cellulose Dry mass basis (g/100g) |
|---|---|---|
| Glucose | 53.813 | 55.252 |
| Arabinose | 2.904 | 11.712 |
| Total | 55.217 | 66.96 |

Table 5. Comparing yields with total theoretical yields through Quantitative Saccharification

Although the high conversion of cellulose to glucose is apparent the low level of arabinose conversion is a concern. As mentioned previously arabinose is quite thermally unstable. When reacted at 135°C and 2.5% w/w acid concentration the production rate is quite high and continues to rise until the reaction ends at 90 minutes. If this reaction where to continue unabated the conversion rate for arabinose may reach a more acceptable level. However in an industrial context allowing such a slow reaction is uneconomical but if the temperature is increased by any significant amount, even to 150°C, the thermal instability becomes an issue and degradation of arabinose sets in rapidly thus rendering it a negligible side reaction to the dominant glucose reaction. Having said that even with the unimpressive arabinose conversion, overall sugar yield is 82.5% of the theoretical yield which would be considered quite an acceptable return in the circumstances.

Another conclusion drawn from this study is the difference between the effect of temperature and acid concentration on conversion rates. Both variables have the effect of increasing the overall reaction rate however the production and degradation of sugars is more sensitive to fluctuations in temperature. By increasing either condition the reaction rate increases allowing for a maximum yield to be obtained in a far shorter period of time, however it was found that by increasing the temperature had an overall detrimental effect on the net sugar yield during the reaction. It was found in all cases that as temperature increased the decomposition reaction began to drastically outpace the formation rate which led to rapid declines in the net sugar yield. In contrast to this increases in acid concentration had a less dramatic effect on the production and decomposition rate of sugar. As acid concentrations raised so did the production rate of sugars, but it was seen to affect the sugars in a more stable manner. Although degradation of sugars also increased with increasing acid concentration it was less rapid than was seen with increasing temperature. In conclusion, to run this reaction in the most effective manner temperature should be kept at a reduced level while acid concentration should be the primary reaction dependency.

## 5. Acknowledgement

Dr Walker is currently holder of a Royal Academy of Engineering (UK), Leverhulme Senior Research Fellowship.

## 6. References

[1] Gámez, S., González-Cabriales, J.J., Ramírez, J.A., Garrote, G. and Vázquez, M. Study of the hydrolysis of sugar cane bagasse using phosphoric acid. Journal of food engineering. 2006, 74 (1) pp. 78-88

[2] Romero, I., Moya, M., Sánchez, S., Ruiz , E., Castro, E. and Bravo, V. Ethanolic fermentation of phosphoric acid hydrolysates from olive tree pruning. Industrial Crops and Products. 2007, 25(2) pp. 160-168

[3] Hamelinck, C.N., van Hooijdonk, G., Faaijm, A. Ethanol from lignocellulosic biomass: techno-economic performance in short-, middle- and long-term. Biomass and Bioenergy. 2005, 28(4), pp.384-410

[4] Gámez, S., Ramírez, J.A., Garrote, G. and Vázquez, M. Manufacture of Fermentable Sugar Solutions from Sugar Cane Bagasse Hydrolyzed with Phosphoric Acid at atmospheric Pressure. 2004, 74 pp. 4172-4177

[5] Alves, L.A., Vitolo, M., Felipe, M.G.A., and Silva, J.B.A. Xylose reductase and xylitol dehydrogenase activities of Candida guilliermondii as a function of different treatments of sugarcane bagasse hydrolysate employing experimental design. Applied Biochemistry and Biotechnology. 2002, 98–100 pp. 403–413

[6] Aguilar, R., Ramírez, J. A., Garrote, G., and Vázquez, M. Kinetic study of the acid hydrolysis of sugar cane bagasse. Journal of Food Engineering. 2002, 55 (4) pp. 309-318

[7] Saeman, J.F. Kinetics of wood saccharification. Hydrolysis of cellulose and decomposition of sugars in dilute acid at high temperature. Industrial and Engineering Chemistry. 1945, 37 (43-52)

[8] Rahman, S.H.A., Choudhury, J.P. and Ahmad, A.L. Production of xylose from oil palm empty fruit bunch fiber using sulfuric acid. Biochemical Engineering Journal. 2006, 30(1) pp. 97-103

[9] Kunlan, L., Lixin, X., Jun, L., Jun, P., Guoying, P. and Zuwei, X. Salt-assisted acid hydrolysis of starch to D-glucose under microwave irradiation. Carbohydrate Research. 2001, 331 (1) pp 9-12

[10] Lenihan, P., A. Orozco, E. O'Neill, M.N.M. Ahmad, D. Rooney and G.M. Walker* Dilute acid hydrolysis of Lignocellulosic biomass, Chemical Engineering Journal , Chemical Engineering Journal Volume: 156 Issue: 2, pp 395-403

# 13

# *Trichoderma reesei*: A Fungal Enzyme Producer for Cellulosic Biofuels

Bernhard Seiboth, Christa Ivanova and Verena Seidl-Seiboth
*Vienna University of Technology*
*Austria*

## 1. Introduction

Enzymes are proteins that catalyse chemical reactions by lowering the activation energy needed and thus speed up the reaction itself. The advantage of enzymes is that they can be applied under mild reaction conditions and that they exhibit high substrate specificity, stereoselectivity and less side product formation than conventional chemical reactions, making biotechnological processes often more cost-effective than chemical approaches. Microorganisms, such as bacteria and fungi are widely exploited for the industrial production of numerous enzymes. Filamentous fungi (moulds) can grow on a wide range of substrates and efficiently degrade biopolymers and are thus an attractive resource for new enzymes. The decomposition of cellulosic plant biomass to glucose monomers for biofuel production is a typical example for an application that requires an enzyme-based approach in order to specifically cleave the glycosidic bonds between the glucose monomers of the cellulose chain and release single glucose molecules. The main enzymes necessary to degrade cellulosic plant material are cellulases and hemicellulases. The filamentous fungus *Trichoderma reesei* is today a paradigm for commercial scale production of different cellulases and hemicellulases and is well adapted to fermenter cultivations. Beside well established applications of these enzymes in pulp, paper, food, feed or textile processing industries, these plant cell wall degrading enzymes are nowadays also employed for the saccharification of cellulosic plant biomass to simple sugars for biofuel production (Bouws et al., 2008; Harman and Kubicek, 1998; Kumar et al., 2008). The cellulolytic potential of this pantropical fungus was already recognized during WWII through the deterioration of cotton fabrics of the US Army. Strain QM6a (originally named *T. viride*) was isolated from the cotton canvas of an army tent from Bougainville Island (Solomon Islands). After identification of the fungus as the cause for the massive destruction, it was put under quarantine in the eponymous Quartermaster collection of the US army at Natick. Strain QM6a was later recognized as an own species and named after its principal investigator in those years Elwyn T. Reese (Reese, 1976). It is an important peculiarity that this *T. reesei* strain QM6a is the ancestor of all enzyme producing *T. reesei* strains in commercial use. Later it was found, that, based on DNA-based phylogenetic markers, the asexual fungus *T. reesei* was indistinguishable from the sexually propagating fungus *Hypocrea jecorina*, thus indicating that they belong to the same species (Kuhls et al., 1996). More recent investigations demonstrated that even the original isolate *T. reesei* QM6a, which was for a long time considered to be an asexual clonal line, can be sexually crossed with *H. jecorina* strains when the requirement for a fungal strain with an opposite mating type is considered (Seidl et al., 2009). Thus, although initially classified as different genera, *T. reesei* and *H. jecorina* are manifestations of the same organism and according to the 'holomorph concept' the name of the teleomorph (sexual form) should be used for the species, but due to historical reasons the name of the anamorph (asexual form), *T. reesei,* was - and still is - predominantly used as species name in publications and patents. The genome sequence of *T. reesei* QM6a was published in 2008 (Martinez et al., 2008) and currently two further annotated genome sequences of *Trichoderma* species i.e. the mycoparasitic and biocontrol fungi *T. atroviride* and *T. virens* (Kubicek et al., 2011) are available via the Joint Genome Institute (http:// genome.jgi-psf.org/). A number of species characteristics of *T. reesei* such as its smaller genome size in comparison to *T. atroviride* and *T. virens* or its reduced carbon utilization pattern indicate that *T. reesei* is a rather specialized species and highly adapted to a particular habitat which is, however, still not exactly known (Druzhinina et al., 2010). Beside its long history of safe use in industrial enzyme production which is also recognized by its GRAS status (Generally Recognized as Safe) by the U.S. Food and Drug Administration (FDA), *T. reesei* serves today as a model organism for the regulation and biochemistry of (hemi)cellulose degradation (Aro et al., 2005; Kubicek et al., 2009). For further deployment of enzymes for biofuel production active research efforts in different areas are currently ongoing including consolidated bioprocessing, engineering to improve enzymes by rational design or directed evolution, the generation of even better hyperproducing strains or the expression of enzymes already in the plant itself. Today, *T. reesei* is already the main industrial source for cellulases and hemicellulases (Merino and Cherry, 2007), but despite the fact that industrial strains produce more than 100 g/l of cellulases (Cherry and Fidantsef, 2003), efforts are needed to reduce costs and maximize yield and efficiency of the produced enzyme mixtures (for recent reviews see e.g. Carroll and Somerville, 2009; Wilson, 2009) . In this chapter we will give an overview on *T. reesei* as a cellulase producer and focus on recent genomic and genetic advances that will aid in improving its cellulolytic enzyme mix, enzyme production and performance.

## 2. The cellulose degradation machinery of *T. reesei*

Lignocellulose is a mixture of cellulose associated with hemicellulose, lignin, pectin and other substances in minor amounts. It is the most abundant renewable biomass, produced by the photosynthesis of plants directly from $CO_2$ and accumulates to approximately 200 billion metric tons worldwide. The complex structure of this solid and heterogeneous substrate hampers an efficient conversion to simple sugars and presents a number of technical and economic challenges in bringing cellulosic biofuels to the market. One of the major economical barriers for the production of biofuels is the intrinsic recalcitrance of lignocellulosic plant matter (Himmel et al., 2007). The β-(1,4)-linked glucose polysaccharide cellulose provides structure and rigidity to the plants and makes them more resistant to chemical, microbial or enzymatic degradation. The multiple hydroxyl groups of the glucose monomers are linked to oxygen molecules on neighbouring chains via hydrogen bonds, which results in the formation of microfibrils with high tensile strength. Many properties of cellulose depend also on its chain length or degree of polymerization. In addition to the general low accessibility of the insoluble microcrystalline cellulose fibres for enzymes, the additional presence of lignin and hemicellulose on the cellulose surface interferes with an efficient enzymatic degradation of this biopolymer by cellulases. The recalcitrance of plant biomass used for enzymatic cellulose degradation is usually reduced by a pretreatment step that exposes the cellulose microfibrils. The development of new pretreatment procedures for improved cellulose digestibility by reducing its crystallinity and the optimization of the pretreatment for different plant raw materials are a field of active ongoing research (Pedersen and Meyer, 2010; Wyman et al., 2005a; Wyman et al., 2005b; Zhu et al., 2010). However, even if a satisfactory accessibility of the plant biomass is achieved or even if pure cellulose is used as substrate, cellulose degradation itself is relatively slow and not very effective in comparison to e.g. starch degradation by (gluco)amylases and thus large enzyme loads are required to break down the crystalline cellulose parts to fermentable sugar monomers. This makes the enzyme mixtures a critical cost factor for commercially viable biofuel production (Merino and Cherry, 2007).

### 2.1 *T. reesei* cellulase basics

A number of different types of cellulases are necessary to degrade cellulose, but in recent years also different non-enzymatic proteins were found to add substantially to an efficient degradation. In contrast to some bacteria the cellulase multi-enzyme system of fungi is not assembled in large cellulosome complexes, but the different fungal enzymes activities are produced independently and their concerted attack on the crystalline cellulose results in a synergistic decomposition of the polymer. The cellulolytic system of *T. reesei* can be divided into three major enzyme classes: (i) exoglucanases - in the case of *T. reesei* cellobiohydrolases (CBHs) - liberate the D-glucose dimer cellobiose consecutively from the ends of the cellulose chain (ii) endoglucanases (EGs) randomly cut within the cellulose chain and (iii) β-glucosidases release D-glucose from the soluble oligomeric breakdown products, thereby preventing cellobiose inhibition of the other enzymes (Fig. 1). The whole process occurs simultaneously and the rate limiting step is the depolymerisation of the insoluble cellulose by the CBHs and EGs. Both types of enzymes show considerable synergism in their simultaneous action on crystalline cellulose substrates, which is explained by the endo/exo model for synergistic degradation of cellulose (Beguin and Aubert, 1994; Tomme et al., 1995). In this model EGs hydrolyze internal cellulose bonds randomly, preferentially in the more accessible amorphous regions, thereby creating new free cellulose ends which are then the starting points for CBHs to further degrade the polymer starting from these amorphous regions into the crystalline regions of the cellulose. *T. reesei*, has two CBHs, CEL6A and CEL7A, which both are composed of two structurally and functionally distinct domains, i.e. a catalytic domain and a cellulose binding domain that are linked via a flexible linker. Some but not all of the EGs have also a two domain structure. The cellulose binding domain belongs to the carbohydrate binding modules (CBMs) and is linked either to the N-Terminus or the the C-terminus. It enables efficient attachment of the enzyme to the insoluble substrate and processive cleavage of the cellulose. After successful hydrolytic cleavage of the carbohydrate chain the enzyme does not dissociate from the substrate but stays attached and slides along the chain for the next cleavage to occur. Both CBHs, CEL6A and CEL7A, have been shown to act processively, whereby CEL6A cleaves the cellobiose dimers from the non-reducing end of the cellulose chain and CEL7A from the reducing end (Barr et al., 1996). Classification of the cellulases into exo- and endoglucanases is only a coarse model. It has been shown that some CBHs can also make internal cuts and processive EGs have also been described that produce cellotetraose instead of cellobiose.

In analogy to other enzymes that act on carbohydrate polymers, cellulases have a long substrate-binding groove running along one side of the protein that binds several consecutive sugar residues, generally at least two D-glucose units on each side of the catalytic center. The groove is lined by hydrogen-bonding residues, and in addition aromatic amino acid side chains are often found as hydrophobic platforms that form discrete sugar-binding subsites. EGs generally have an open binding cleft, whereas CBHs have extended loops that surround the cellulose chain so that a substrate-binding tunnel is formed. The first three-dimensional X-ray structure of a cellulase was the catalytic domain of CEL6A (Rouvinen et al., 1990) followed later by CEL7A (Divne et al., 1994). Crystallization of these cellulases for structural analysis was only possible after cleavage of the CBM from the catalytic domain probably due to high flexibility of the linker region. Analysis of the CBMs reveal a wedge shaped structure with one face hydrophilic and the other more hydrophobic. In the CBM of CEL6A three tyrosines form a regular flat surface and are implicated in cellulose binding.

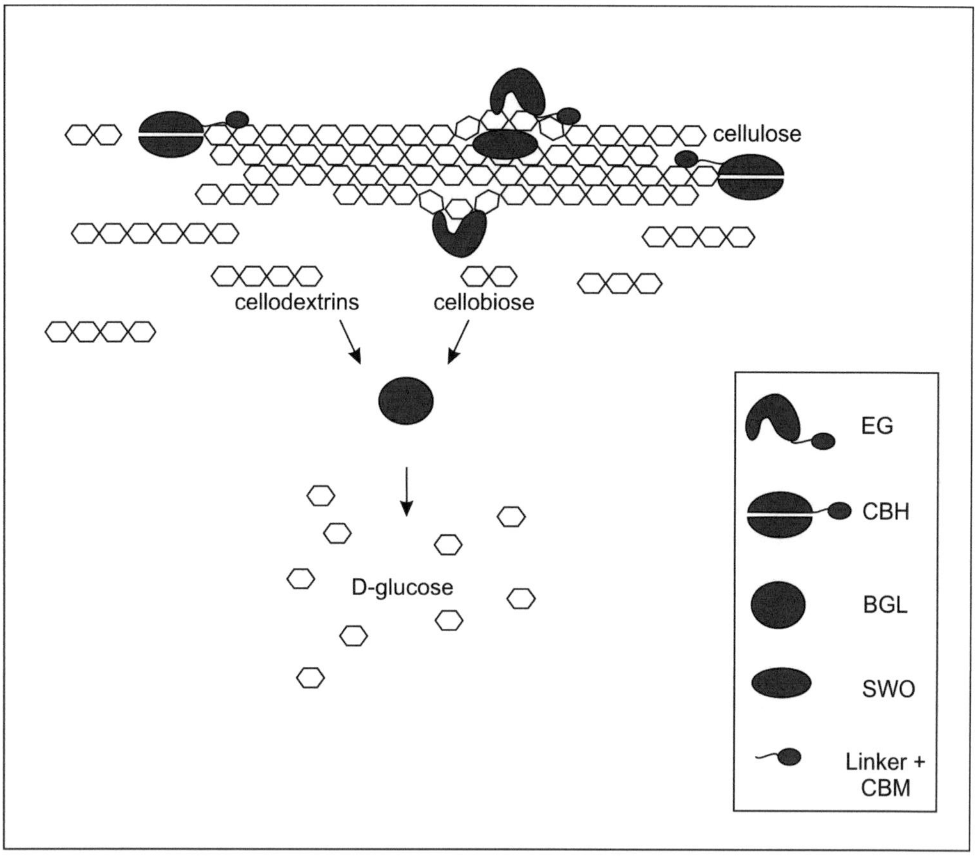


Fig. 1. Degradation of cellulose by cellulases and non-enzymatic proteins of *T. reesei*. Multiple members of the different types of cellulase enzymes - CBHs, EGs and β-glucosidases (BGLs) - degrade the crystalline cellulose synergistically to glucose. CBHs and some of the EGs are composed of a bipartite structure which includes the large catalytic domain and a smaller carbohydrate binding module (CBMs) connected via a flexible linker region. In addition, non-enzymatic proteins such as swollenin (SWO) aid in the degradation of cellulose by disrupting the crystalline structure and thus improving the accessibility of cellulose for enzymatic proteins.

## 2.2 Cellulase classification

A number of different systems exist for the general classification of enzymes based on amino acid sequence similarities or their catalytic mechanism. The classification by the International Union of Biochemistry and Molecular Biology (IUBMB) divides proteins into sub-classes based on the nature of the reactions that they catalyze and uses a numbering code system (EC-numbers) to indicate their enzymatic activity. In this classification the *T. reesei* cellulases belong to the hydrolases and are found in group EC3 with CBHs (EC 3.2.1.91), EGs (EC 3.2.1.4) and β-glucosidases (EC 3.2.1.21). However, this system relies on the biochemical characterisation of a protein and to enable the prediction and classification of new enzymes *in silico*, a sequence-based classification was established that assigns sequences to various families according to their amino acid similarities. Carbohydrate-*a*ctive enz*y*mes, or CAZymes, are categorized into different families in the CAZy database (http://www.cazy.org, Cantarel et al., 2009). These families describe enzymes that degrade, modify, or create glycosidic bonds. The categories in the CAZy database include glycoside hydrolases (GHs; enzymes that hydrolyse or rearrange glycosidic bonds), glycosyltransferases (GTs; enzymes that form glycosidic bonds), polysaccharide lyases (PLs; enzymes with a non-hydrolytic cleavage mechanism of glycosidic bonds) and carbohydrate esterases (CEs; enzymes that hydrolyse carbohydrate esters). In addition a further section is devoted to carbohydrate binding modules (CBMs). There are several types of CBMs which are involved in the adhesion of the enzymatic module to different carbohydrates: The CBMs of cellulases (CBM family 1) are generally specific for binding to the surface of crystalline cellulose and enhance the activity of the enzyme on crystalline cellulose but not on soluble substrates (Bayer et al., 1998). The hydrolysis (GHs) and formation (GTs) of glycosidic bonds can proceed via retention or inversion of the anomeric carbon (C1) of the substrate after cleavage. In an inverting enzyme the hydrolysis of a β-glycosidic bond creates a product with the α-configuration, and vice versa, whereas in retaining enzymes the β-configuration is preserved. A special case of the retaining mechanism is the 'neighbouring group participation', where a side group of the substrate itself assists in the catalytic mechanism. In GHs in general two amino acids at the opposite sides of the sugar plane are the key factors for the nucleophilic attack, stabilization of the transition state and finally hydrolysis of the glycosidic bond. Often acidic side chains of the two catalytic amino acids are important for the reaction mechanism. GHs with a retaining mechanism have also commonly transglycosylating abilities and are able to form oligosaccharides from their hydrolytic reaction products when they are present in high concentrations. In contrast to that, GTs, for which so far predominantly inverting mechanisms have been reported, use activated carbohydrate monomers (e.g. UDP-activated sugars for cell wall biosynthesis) to create glycosidic bonds and thus long carbohydrate chains. Cellulases can be found in several GH families and they are specific for cleaving the β(1→4) bonds of cellulose. Interestingly both, CBHs and EGs are found within the same family (see also Table 1). The large numbers of three-dimensional structures of GH enzymes that have accumulated during the last 15 years have so far largely verified that members of a GH family share the same overall protein fold, a common catalytic site motif, and the same reaction mechanism, i.e. retaining or inverting. For some families only one enzymatic activity has so far been reported, whereas in other families a multitude of different enzymatic activities can be found.

The CAZy database has been online since 1998 and this resource turned out to be extremely valuable for the annotation of novel genome sequences in order to elucidate the carbohydrate degradation potential of an organism by bioinformatic means. As a consequence this sequence-based classification system has been widely accepted and is applied for all organisms, ranging from archaea to bacteria, fungi, plants, animals and humans. As a further consequence of the general acceptance of the CAZy classification, the *T. reesei* cellulases were renamed. For example the cellobiohydrolase CBH1/CBHI is now CEL7A, named by its activity CEL (for cellulase), 7 for the GH family and A denoting that this was the first reported family 7 cellulase from this organism. Table 1 summarizes the corresponding "old" and "CAZy" designations for the components of the *T. reesei* cellulase system.

| GH family | CAZy nomenclature | Previous designation | Cellulase type | Size in amino acids | Position of CBM | Stereo-selectivity |
|---|---|---|---|---|---|---|
| 1 | CEL1A | BGL2 | β-glucosidase | 466 | - | Retaining |
| 1 | CEL1B | | β-glucosidase⁺ | 484 | - | Retaining |
| 3 | CEL3A | BGL1 | β-glucosidase | 744 | - | Retaining |
| 3 | CEL3B | | β-glucosidase⁺ | 874 | - | Retaining |
| 3 | CEL3C | | β-glucosidase⁺ | 833 | - | Retaining |
| 3 | CEL3D | | β-glucosidase⁺ | 700 | - | Retaining |
| 3 | CEL3E | | β-glucosidase⁺ | 765 | - | Retaining |
| 5 | CEL5A | EG2 | endoglucanase | 397 | N | Retaining |
| 5 | CEL5B | | endoglucanase⁺ | 438 | GPI anchor | Retaining |
| 6 | CEL6A | CBH2 | cellobiohydrolase | 447 | N | Inverting |
| 7 | CEL7A | CBH1 | cellobiohydrolase | 497 | C | Retaining |
| 7 | CEL7B | EG1 | endoglucanase | 436 | C | Retaining |
| 12 | CEL12A | EG3 | endoglucanase | 218 | - | Retaining |
| 45 | CEL45A | EG5 | endoglucanase | 270 | C | Inverting |
| 61 | CEL61A | EG4 | endoglucanase | 344 | C | Not known |
| 61 | CEL61B | | Endoglucanase (?)# | 249 | - | Not known |
| 74 | CEL74A | EG6 | endoglucanase/ xyloglucanase* | 818 | C | Inverting |

Table 1. The cellulose degrading enzyme system of *T. reesei*
⁺ according to GH family prediction; # no hydrolytic activity (Harris et al., 2010);
* has probably xyloglucanase activity according to GH family prediction

### 2.3 A comparative genomic view on *T. reesei* cellulases and related enzymes

Efforts to clone different cellulolytic components of *T. reesei* have in the last decades resulted in the identification of the major components of its cellulase system including the two CBHs, different EGs and ß-glucosidases (Table 1). However, a survey of the full potential of *T. reesei* as a biomass-degrading enzyme producer became only possible by sequencing the genome of strain QM6a. Unexpectedly, in view of the fact that *T. reesei* is an efficient plant polysaccharide degrader and important model for microbial cellulose and hemicellulose degradation, analysis of the genome sequence of *T. reesei* QM6a revealed that this fungus has actually a relatively small set of cellulases - especially the limited number of CBHs and

EGs was surprising- , hemicellulases and other plant cell wall degrading enzymes in comparison to other filamentous fungi (Martinez et al., 2008). These findings indicated that at least part of the secret of *T. reesei's* success lies in its efficient induction system of cellulase gene transcription and in its extremely high cellulase production and secretion capacities. The so far described components of the *T. reesei* cellulolytic system include: two CBHs (CEL6A and CEL7A), and eight EGs (CEL5A, CEL5B, CEL7B, CEL12A, CEL45A, CEL61A, CEL61B and CEL74A). In addition, seven β-glucosidases have so far been reported in *T. reesei* (CEL1A, CEL1B, CEL3A, CEL3B, CEL3C, CEL3D, CEL3E) and according to the predictions in the CAZy database several further candidate β-glucosidases can be found in the *T. reesei* genome.

With a total of ca. 200 GH encoding genes, *T. reesei* has already fewer GH than phytopathogenic fungi (*Magnaporthe grisea* and *Fusarium graminearum*) and in general, it has the fewest cellulases and the smallest set of hemicellulases and pectin degrading enzymes of all species able to degrade plant cell walls by a hydrolytic attack. In addition, the number of CBMs, which are involved in substrate recognition and binding of enzymes involved in plant polysaccharide depolymerisation, is the smallest among the Sordariomycetes. The CAZomes which beside GH also comprise the other groups of CAZymes are summarized for *T. reesei* and two further sequenced *Trichoderma* species in comparison to other fungi in Table 2.

| Fungus | GH | GT | PL | CE | CBM |
|---|---|---|---|---|---|
| *Trichoderma reesei* | 193 | 93 | 5 | 17 | 41 |
| *Trichoderma atroviride* | 242 | 96 | 8 | 25 | 62 |
| *Trichoderma virens* | 256 | 98 | 6 | 25 | 87 |
| *Aspergillus nidulans* | 251 | 91 | 21 | 31 | 41 |
| *Aspergillus niger* | 243 | 110 | 8 | 24 | 40 |
| *Gibberella zeae* | 247 | 102 | 21 | 44 | 64 |
| *Magnaporthe grisea* | 232 | 92 | 5 | 48 | 65 |
| *Neurospora crassa* | 173 | 76 | 4 | 23 | 12 |
| *Penicillium chrysogenum* | 216 | 101 | 9 | 22 | 49 |
| *Postia placenta* | 248 | 102 | 8 | 25 | 34 |
| *Saccharomyces cerevisiae* | 46 | 68 | 0 | 3 | 12 |

Table 2. CAZome comparison of *T. reesei* with other fungi (data adapted from Kubicek et al., 2011). GH: glycoside hydrolases, GT: glycosyltransferases, PL: polysaccharide lyasas, CE: carbohydrate esterases, CBM: carbohydrate binding modules.

The analysis of the genomes of the two mycoparasitic *Trichoderma* species, *T. atroviride* and *T. virens* (Kubicek et al., 2011), showed that the number of identified cellulases is not significantly increased in these fungi either, suggesting that the observed low variety of cellulases is a common feature of the genus *Trichoderma* (Tables 3 and 4). *T. atroviride* and *T. virens* have slightly more cellulases than *T. reesei* with additional one and two, respectively, members of family GH12. *T. virens* has in addition another GH45 cellulase containing a CBM1 at the C-terminus. Although only two GH61 proteins have so far received attention in cellulose degradation, all three *Trichoderma* spp. possess another GH61 protein, which could also be involved in plant biomass degradation, but has not been characterized so far. The

GH5 cellulase CEL5B has in all three investigated *Trichoderma* species a putative GPI (Glycophosphatidylinositol)-anchor at the C-terminus, which would render this protein bound to the plasma membrane/fungal cell wall. Families GH5 and GH3 contain, besides cellulases and β-glucosidases, also many proteins with other enzymatic activities (e.g. β-1,6 glucanases, β-xylosidases, mannanases, etc.) and a relatively high number of these enzymes is found in the three analysed *Trichoderma* spp. (11-17 members/GH family/species). A more in-depth analysis and biochemical characterizations will be necessary to reveal the specific activities of members of these families. In contrast to that, the enzymatic characteristics of other GH families are very specific for cellulose degradation, e.g. only two GH7 proteins are found in all three sequenced *Trichoderma* spp. and both of them are prominent members of the cellulase-system, one being CEL7A and the other CEL7B.

| Fungus | GH5* | GH6 | GH7 | GH12 | GH45 | GH61 | Total |
|---|---|---|---|---|---|---|---|
| *T. reesei* | 2 | 1 | 2 | 2 | 1 | 3 | **11** |
| *T. atroviride* | 2 | 1 | 2 | 3 | 1 | 3 | **12** |
| *T. virens* | 2 | 1 | 2 | 4 | 2 | 3 | **14** |

Table 3. Comparison of CBHs and EGs in the three sequenced *Trichoderma* species based on genome annotation data (data adapted from Kubicek et al., 2011). * Numbers indicate only putative cellulases from family GH5.

Beside the insight in the celluloytic potential of the three *Trichoderma* spp, the genome comparison provides also valuable information about the other plant cell wall degrading enzymes and lists different possibilities to improve the plant cell wall degrading potential of *T. reesei* with enzymes from these two closely related species. The general low pectin degrading potential of all three fungi indicates that they are not equipped with enzymes which are necessary to attack living or intact plants but are better adapted to decompose dead and decaying plant matter. Within the group of hemicellulases, the number of xylanases in *T. reesei* is similar to the two mycoparasites, although *T. virens* has two GH10 proteins whereas *T. reesei* and *T. atroviride* have only one, and further one of the GH11 members of *T. atroviride* has a CBM1 at the C-terminus, but no CBMs are found in the respective proteins of *T. reesei* and *T. virens*. Furthermore, *T. atroviride* and *T. virens* have slightly more α-L-arabinofuranosidases (in families GH43, GH51 and GH62), α-glucuronidases (GH67) than *T. reesei* and posses in addition also endo-acting α-arabinanases (GH93).

A relatively low capacity and flexibility to decompose pectin is observed in all three sequenced *Trichoderma* species based on the identified enzymes of families GH28, 78, 88, and 105, families PL1, 3 and 4, and families CE8 and 12, all directly acting on polygalacturonan and rhamnogalacturonan or their respective degradation products. Equally, the number and nature of pectin side-chain degrading enzymes (GH43, 51, 53, and 54) is lower than in other fungi; e.g. *Aspergillus niger* has already at least 39 genes encoding enzymes involved in the depolymerisation of the backbone of pectin (Martens-Uzunova and Schaap, 2009) and in addition several genes encoding enzymatic activities required for the degradation of the arabinan and arabinogalactan side chains. In contrast to that, *T. reesei* has only a total of 6 pectinolytic enzymes, whereas *T. virens* and *T. atroviride* have a slightly better pectin degradation machinery and possess at least 15 pectinolytic enzymes. Thus, the numbers of pectin degrading enzymes (GH78, 88, 105, PL1, CE8) are slightly enriched in these two fungi in comparison to *T. reesei*. These additional proteins could constitute a small albeit important difference facilitating the attachment to roots or the displacement of hyphae in the pectin-rich intercellular space of plants, thereby deepening fungal-plant interactions of *T. virens* and *T. atroviride.*

| GH family | 10 | 11 | 26 | 29 | 39 | 43* | 51* | 53* | 54* | 62 | 67 | 74 | 93 | Total |
|---|---|---|---|---|---|---|---|---|---|---|---|---|---|---|
| *T. reesei* | 1 | 4 | 0 | 0 | 1 | 2 | 0 | 0 | 2 | 1 | 1 | 1 | 0 | **13** |
| *T. atroviride* | 1 | 4 | 0 | 0 | 2 | 5 | 1 | 0 | 2 | 2 | 2 | 1 | 3 | **23** |
| *T. virens* | 2 | 4 | 2 | 0 | 1 | 3 | 0 | 0 | 2 | 3 | 2 | 1 | 1 | **21** |

Table 4. Comparison of hemicellulases in the three sequenced *Trichoderma* species based on genome annotation data (data adapted from Kubicek et al., 2011). * Members of these GH families are also involved in pectin side chain degradation.

### 2.4 New players in cellulose degradation

In addition to the classical cellulases, a number of non enzymatic proteins substantially aid in cellulose degradation. One example for such a non-enzymatic protein is swollenin (SWO1) that modifies cellulose by disrupting its crystalline structures and thereby enhances its hydrolysis (Saloheimo et al., 2002). Swollenin has in addition to an N-terminal CBM1 an expansin domain. These domains are usually found in plant proteins that aid in loosening up the cell wall structure of plants and it has been shown that SWO1 is important for the fungal cellulose degradation process because it is involved in opening up the rigid structure of crystalline cellulose and thus makes the substrate more accessible for cellulases (see also Fig. 1). *Trichoderma* spp. have additional 3-4 proteins with expansin-like domains though these proteins have no CBMs. In addition, no transcriptional upregulation of the respective genes was found under cellulase-inducing conditions indicating that they might have other roles (Verbeke et al., 2009).

There certainly are several uncovered enzymatic or protein activities still hidden in the genome of *T. reesei*. The genes encoding CIP1 and CIP2 are strongly expressed under cellulase-inducing conditions. CIP2 was only recently described to encode a CE15 glucuronoyl esterase, probably cleaving the ester linkage between 4-O-methyl-D-glucuronic acid of glucuronoxylan and lignin alcohols (Li et al., 2007). The function of CIP1 is not known yet. BLAST searches of CIP1 in public databases show that it is similar to glycosyl hydrolases found in bacteria but which are not yet characterized in detail, indicating that CIP1 could have a novel undescribed enzymatic activity or mechanism.

The recalcitrance and crystallinity of the biomass is a major obstacle for the cellulolytic enzyme set of *T. reesei*. In its natural environment biotic and abiotic factors such as other microorganisms, changes in ambient temperature and humidity aid in loosening up the rigid structure of cellulosic material to render it more accessible for *T. reesei's* cellulases and hemicellulases. However, for biotechnological processes this obstacle needs to be overcome by other means. This problem can be approached from two sides: (a) by improving the enzyme mix and (b) by optimizing the pretreatment of the cellulosic biomass. Strong research efforts are ongoing in both of these fields to overcome the recalcitrance of cellulosic biomass and it will be important to combine them in their most effective way. Advanced modeling and microscopy techniques are applied to increase the understanding of enzyme-substrate interactions and to improve the enzymatic activity of existing cellulases by protein engineering approaches with respect to thermostability and activity on insoluble substrates. However, in addition to these important aspects, there is a large enzymatic potential still hidden within the diversity of the fungal kingdom and the ongoing fungal genome sequencing efforts are just beginning to unravel the yet undiscovered potential of so far unknown enzymes that could aid in strongly improving the enzyme mix.

The classification of glycoside hydrolase families is based on amino acid similarities, which implies structural relationship and thus similar enzymatic mechanisms. However, detailed investigation of the sequence and 3D structure of some proteins can reveal that they might not be catalytically active - or at least not use the same catalytic mechanism as other proteins in the same GH family - as was recently reported for GH 61 proteins (Harris et al., 2010; Karkehabadi et al., 2008). While endoglucanase activity was reported for some GH 61 members, e.g. CEL61A from *T. reesei* and CEL61A from *Aspergillus kawachii*, analysis of the crystal structure of another GH61 member, CEL61B from *T. reesei* showed that there is no easily identifiable carbohydrate-binding cleft or pocket or catalytic centre of the types normally seen in GHs. Further, GH61 proteins of the thermophilic ascomycete *Thielavia terrestris* were shown to be capable of enhancing the activity of cellulases from *T. reesei* on pretreated corn stover. The respective GH61 proteins lacked measurable hydrolytic activity by themselves but in the presence of various divalent metal ions they significantly reduced the total protein loading required to hydrolyze lignocellulosic biomass. Structural analysis of the *T. terrestris* GH61E protein further supported the lack of a hydrolytic mechanism, as no evidence for clustering of conserved catalytic acidic residues that are present in almost all known glycoside hydrolases was found. *T. terrestris* GH61E has an important metal ion-binding site, which was investigated by mutagenizing those residues directly or indirectly involved in metal binding. Mutation of the directly interacting His-1 to Asn or His-68 to Ala resulted in a completely inactive protein. A structural comparison search showed that the known structure most similar to *T. reesei* Cel61B is that of CBP21 from the Gram-negative soil bacterium *Serratia marcescens*. For *T. terrestris* GH61E also significant structural similarity was found. *S. marcescens* CBP21 is a member of the CBM family 33 proteins. A polar surface patch which is highly conserved in that structural family has been identified in CBP21 and was shown to be involved in chitin binding (Vaaje-Kolstad et al., 2004). Chitin is a cellulose derivative where the 2-hydroxy group has been substituted with an acetamido group and is the second most abundant biopolymer after cellulose. The crystallinity and insolubility of cellulose and chitin in water are major obstacles for the efficient enzymatic degradation of these abundant biopolymers. Most interestingly, CBP21 was recently shown not only to possess chitin-binding activities, but also to be enzymatically active on chitin. CBP21 has a new enzymatic mechanism that introduces chain breaks by generating oxidized ends and thus opening up the rigid structure of crystalline chitin and making it accessible polysaccharide material for hydrolysis by normal glycoside hydrolases (Vaaje-Kolstad et al., 2010). The products of this reaction were identified as chitin oligosaccharides with a normal sugar at the nonreducing end and an oxidized sugar at the reducing end. In the presence of a reductant such as ascorbic acid, CBP21 boosted chitinase efficiency on crystalline β-chitin to much higher levels than previously observed. Thus, the reaction catalyzed by CBP21 comprises a hydrolytic step and an oxidation step and the authors suggested naming CBP21 a "chitin oxidohydrolase." CBP21 catalysis was found to be inhibited by EDTA, and activity could be restored by adding divalent cations such as $Mg^{2+}$ or $Zn^{2+}$, which may bind to the conserved histidine motif in analogy to the metal-ion binding site in GH61 protein. This shows that, with more protein crystal structures to be solved in the next years, structural comparison might provide unprecedented, novel clues and unravel so far unknown links between different protein families.

### 2.5 Cellulase production and engineering

*T. reesei* is an efficient producer of cellulases and industrial production exceeds 100 g cellulases per liter. The highest producer in the public domain and widely used as a cellulase hyperproducer is RUT-C30 which produces 30 g/l. A number of industrial strains are derived from this mutant. The highest amount of cellulases reported for a certain strain was the industrial strain CL487. Interestingly, we could recently show that in contrast to the published genealogy of this mutant, the strain is in fact a progeny of RUT-C30 since it shares a number of mutations also found in RUT-C30 (Portnoy et al., 2011). Since most of the cellulase genes are regulated in a coordinated way, only the relative ratio of their expression differs in higher production mutants (Foreman et al., 2003). The CBHs CEL6A and CEL7A are the two most abundantly secreted proteins in *T. reesei* under cellulase-inducing growth conditions. These proteins are known to account for 70 to 80% of the total *T. reesei* cellulases whereby CEL7A typically makes up around 60% of the total secreted cellulases (Nummi et al., 1983). Other abundant components of the *T. reesei* cellulase complex are CEL6A (15-20%), CEL7B and CEL5A with up to 10% each.

CBHs are recognized as being the most important enzyme components for cellulose conversion and considerable attention is paid to the mechanistic action of *T. reesei* CEL6A and CEL7A (Barr et al., 1996) with the ultimate goal of improving their performance (Zhang and Lynd, 2006). The relatively low turnover rates of cellulases, e.g. one to four per second for CEL7A (Jalak and Valjamae, 2010) present major challenges for cost-effective production of biofuels. Thus the substrate binding and catalytic residues of the CBHs have already been investigated in considerable detail (for review see Schülein, 2000), but efforts to increase the thermostability and performance of CBHs are still ongoing (Lantz et al., 2010). The action of the catalytic and carbohydrate-binding domains on crystalline cellulose was investigated with atomic force microscopy to visualize and define the role of the two domains in this process (Igarashi et al., 2009). The complete CEL7A protein moved with a velocity of ca. 3.5 nm/sec on crystalline cellulose and interestingly, the average velocity of the sliding catalytic domain alone without CBM was similar to that of the intact CEL7A. This indicates that adsorption on crystalline cellulose through the CBM is not essential for the movement of CEL7A, and the catalytic domain alone seems to be sufficient for the sliding on the substrate. In another study, it was reported that a size reduction apparently occurred only in the width of the cellulose crystal, whereas the height remained relatively constant, indicating that CEL7A selectively hydrolyzes the hydrophobic faces of cellulose. The authors suggested that the limited accessibility of the hydrophobic faces in native cellulose may contribute significantly to the rate-limiting slowness of cellulose hydrolysis (Liu et al., 2011).

Another parameter that has an impact on cellulase performance is protein glycosylation, which can influence the function and stability of proteins. Most enzymes secreted by *T. reesei* are *O*- and *N*-linked glycoproteins. These glycosides are attached to the enzymes during the protein maturation and secretion process and are further modified by trimming and addition of new glycosides within the different compartments of the secretory pathway before the protein is finally exported into the surrounding medium. Glycosylation and other co- and posttranslational modifications (e.g. phosphorylation, proteolytic processing, disulfide bridge formation or attachment of a GPI-anchor) and their potential effects on enzyme performance have to be taken into account when *T. reesei* cellulases are expressed in heterologous host systems. This is especially important when cellulases are expressed *in planta* or more dramatically in bacterial hosts which are in general not able to accomplish posttranslational modifications and usually are used for the intracellular production of enzymes. It is still a challenge to understand the effect of protein glycosylation on cellulase activity and to express active cellulases successfully in heterologous hosts. Expression of CEL7A was already evaluated in different classical protein expression systems, ranging from *Escherichia coli* to the yeasts *Saccharomyces cerevisiae* and *Pichia pastoris* and to other filamentous fungi such as *Aspergillus niger var. awamori*, which often are used to produce large quantities of enzymes and proteins for industrial processes (Adney et al., 2003). However, successful expression of a functional cellulase is limited due to different co- and posttranslational requirements including correct glycosylation but also the formation of disulfide bridges has to be considered. CEL7A expression in *A. niger var. awamori* resulted in a recombinant CEL7A which was overglycosylated in comparison to the native *T. reesei* enzyme. As a consequence of the *N*-overglycosylation a reduced activity and increased non-productive binding on cellulose was observed (Jeoh et al., 2008). Even when produced by *T. reesei*, the glycosylation pattern is not uniform and depends on the fungal strain, the fermentation medium and pH, and the secretion of other carbohydrate modifying activities into the medium (Stals et al., 2004a; Stals et al., 2004b). In addition to the impact of glycosylation in the catalytic domain on enzyme activity, the *O*-glycosylation of the linker region connecting catalytic domain and CBM was examined. In this simulation study, the *O*-glycosylation does not change the stiffness of the linker, but rather provides a 1.6 nm extension, thus expanding the operating range of CEL7A (Beckham et al., 2010). Understanding the expression, posttranslational modifications and even more details of the protein architecture in the context of enzyme performance of CBHs and EGs is important for the development of fuels from cellulosic biomass.

When *T. reesei* is used as a production host for heterologous proteins or large amounts of proteins, improvements in the efficiency of protein folding, quality control and transport in the secretory pathway will be necessary to maintain high level enzyme production. For the recombinant production of homologous and heterologous proteins and enzymes the corresponding genes are usually expressed under the promoter and terminator region of the major cellulases *cel7a* or other cellulases to guarantee a high production under cellulase inducing conditions. Attempts to produce cellulases on other carbon sources such as glucose resulted generally in lower amounts of protein produced even when the strongest available promoters were used (Nakari-Setälä and Penttilä, 1995). Various mechanisms sense the accumulation of proteins within the secretory pathway including the unfolded protein response (UPR) in the ER. UPR is mediated by the proteinkinase/endoribonuclease IRE1 which activates HAC1, a transcription factor which in turn upregulates genes involved in protein folding, glycosylation, transport, and in the degradation of misfolded proteins. UPR can also be stimulated by different chemical agents that prevent protein folding or transport. Using such agents, a novel kind of secretion stress termed repression under secretion stress (RESS) was detected in *T. reesei* which downregulates the transcript levels of genes encoding cellulases but also of secreted proteins in general and thereby reduces both the expression of the cellulases and the recombinant protein driven by cellulase promoters (Pakula et al., 2003).

## 3. Regulation of cellulase expression

The quest of *T. reesei* to most efficiently detect cellulose in the environment, to degrade the insoluble substrate by producing different cellulases, transport the soluble break-down products from the surrounding through the cytoplasmic membrane into the cell and subsequently assimilate these sugars, is an essential process to survive. However, to be able to compete with other microorganisms, *T. reesei* has also to be flexible and immediately respond and adapt to changes in the nutrient composition of the environment. When exposed to a mixture of carbon sources, *T. reesei* will utilize the best carbon and energy source available and downregulate the expression of genes which are involved in the degradation of less favourable and complex carbon sources such as cellulose. Degradation of individual carbon sources present in complex mixtures follows a mainly energy driven hierarchy, although adaptation of e.g. saprobic or plant pathogenic fungi to their habitats has resulted in species specific carbon source priorities. Especially genes encoding enzymes involved in polysaccharide degradation are tightly regulated since their synthesis, co- and posttranslational modifications and export through the secretory pathway requires a high energy input by the fungus. Sophisticated mechanisms including carbon sensing and signalling ensure an assimilation hierarchy and two major control circuits i.e. specific induction and general carbon catabolite repression (CCR) control cellulase gene expression and enable an economic production of cellulases. Since *T. reesei* and most of the other microorganisms are particularly adapted to glucose, the most abundant monosaccharide in nature, this CCR is often termed simply glucose repression although other carbon sources are also able to provoke CCR.

The cellulolytic system of *T. reesei* is subject to multiple levels of control whereby most of the regulation happens at the level of transcription. Another characteristic of cellulase expression is that the different cellulase encoding genes are co-regulated, which means that they are expressed under all conditions at the same relative amounts (Foreman et al., 2003; Ilmen et al., 1997). As the prime function of cellulases is the hydrolysis of cellulose and thus to provide the fungal cells with soluble oligomers as carbon source and for energy production, these enzymes are specifically formed in the presence of cellulose and soluble derivatives thereof. The additional presence of other more easily metabolisable monosaccharides leads to cellulase repression even in the presence of an inducer. In *T. reesei* and other fungi CCR is ultimately mediated by the transcription factor CRE1 (Fig. 2). The CRE1 binding motif in the promoter regions of the cellulase and xylanase genes is 5′-SYGGRG-3′ but its functionality depends also on the context within the promoter region. The *in vivo* functionality of the CRE1 binding sites have been demonstrated for the cellulase *cel7a* promoter and the xylanase *xyn1* promoter. Removal of the CRE1 binding sites led to a basal level of cellulase expression on D-glucose indicating that these genes are now carbon catabolite derepressed (Ilmen et al., 1996a; Mach et al., 1996). A number of these functional CRE1 binding sites are characterized by two closely spaced 5`-SYGGRG-3` motifs and it is assumed that direct CRE1 repression occurs mainly through this double motif. CCR can act on cellulase gene expression on different levels either directly by CRE1 (Dowzer and Kelly, 1991; Ilmen et al., 1996b) which represses individual cellulase expression or transcription of its activators, or by inducer exclusion. Inducer exclusion is a regulatory phenomenon whereby other carbohydrates inhibit uptake of the inducing carbon source. This can be a consequence of the inhibition of inducer uptake through a specific permease by a structurally related sugar or by CRE1-mediated repression of the gene encoding a specific permease of the inducible system. Another possibility would be that formation of the inducer is inhibited within the cell which also prevents induction.

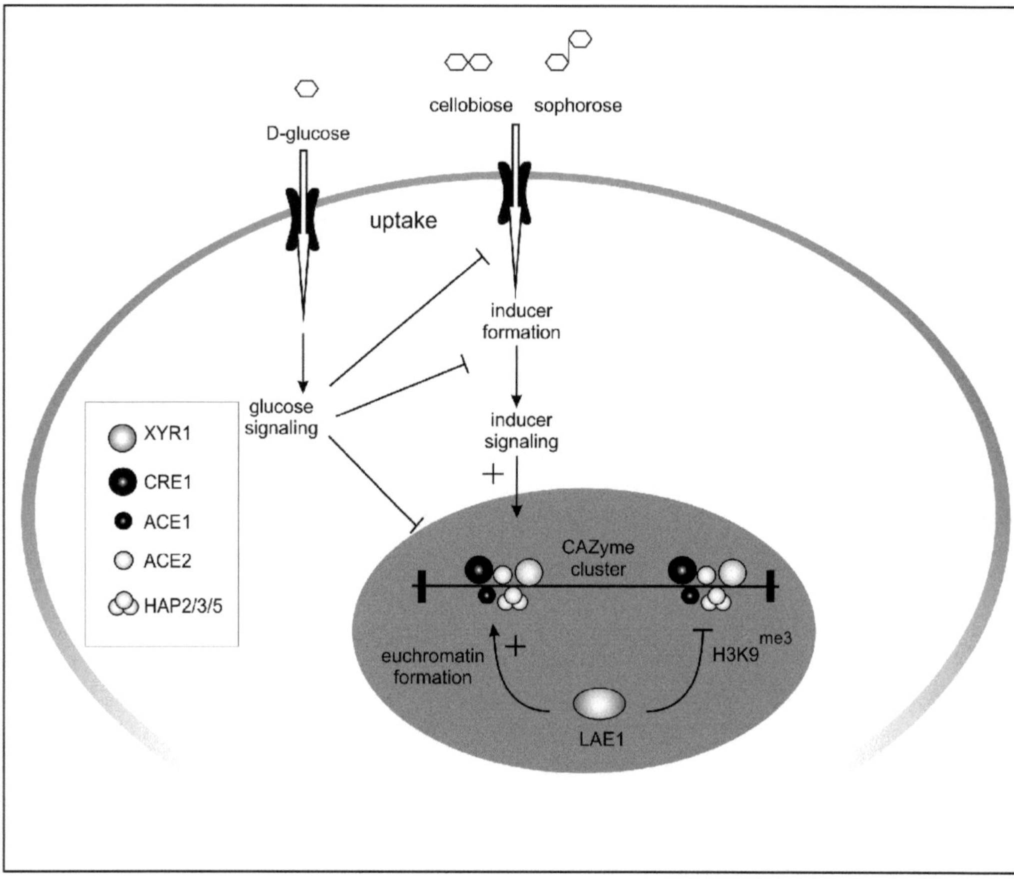


Fig. 2. Regulation of cellulase gene expression in *T. reesei*. Regulation of cellulase genes is accomplished on the transcriptional level by carbon catabolite repression (indicated here by D-glucose) or specific induction by cellulose breakdown (cellobiose) or transglycosylation (sophorose) products. CCR can act on different levels including inducer uptake, inducer formation, and direct repression of cellulase and transcriptional activator gene expression. Positive regulation is accomplished via the transcriptional activators XYR1, ACE2 and the HAP complex and negative regulation via CRE1 and ACE1. In addition cellulase expression in the CAZyme clusters is also regulated on the chromatin level by the putative proteinmethyltransferase LAE1. Potential targets are euchromatin formation and H3K9 methylation.

Since the expression of a vast majority of the cellulase genes in *T. reesei* does not occur during growth on glucose, one of the earliest attempts to improve cellulase production was towards a release from carbon catabolite repression (see the following section). One of the best cellulase producers in the public domain, the carbon catabolite derepressed strain RUT C30, indeed contains a truncation in the *cre1* gene (Ilmen et al., 1996b; Seidl et al., 2008), thus indicating the importance of carbon catabolite derepression for cellulase formation. However, relief from carbon catabolite repression alone is not sufficient for high cellulase production, as cultivation on glucose or other carbon sources results only in low cellulase levels, which do not match cellulase levels under inducing conditions, as confirmed by the deletion of *cre1* in wild-type and different mutant strains. These results establish *cre1* as one of the main targets for strain improvement but also indicate that cellulase hyperproduction is still inducer dependent (Nakari-Setälä et al., 2009). The regulatory pathways by which the presence of D-glucose triggers carbon catabolite repression are still only poorly understood in filamentous fungi. In the yeast *S. cerevisiae*, D-glucose and D-fructose phosphorylating enzymes are involved in D-glucose sensing and carbon catabolite repression, whereby hexokinase 2 contributes to glucose repression. The situation seems to be different in filamentous fungi. In *A. nidulans* mutations in both the gluco- and hexokinase are necessary for a CreA (*A. nidulans* CRE1 homologue)-mediated carbon catabolite derepression on D-glucose which suggests a compensatory effect of these two proteins in single mutants (Flipphi et al., 2003). Similarly, in the *T. reesei* glucokinase and hexokinase deletion strains carbon catabolite repression on D-glucose is maintained and only the double deletion strains are derepressed.

Since carbon catabolite derepression alone is not sufficient for high cellulase production, it is of fundamental interest for high enzyme production to elucidate the molecular basis on how *T. reesei* senses the presence of cellulose and initiates cellulase induction. High cellulase formation requires the presence of an inducing carbon source, which can be derived from cellulose, e.g. the two disaccharides sophorose and cellobiose. For the initial inducer formation it is assumed that a low basal level of cellulases is produced which attack the insoluble cellulose and form the inducer or inducer precursors by releasing small amounts of di- and oligosaccharides. These oligosaccharides are taken up and can then induce further cellulase biosynthesis (Carle-Urioste et al., 1997). CEL6A was found to be the predominant cellulase on *T. reesei* conidia and a *cel6a* knock out strain exhibited a pronounced delay in initiating growth on cellulose and cellulase transcription (Seiboth et al., 1992). The additional deletion of the second CBH-encoding gene *cel7a* led to strains which could not initiate growth on cellulose (Seiboth et al., 1997) without the addition of a soluble inducer for cellulase induction. These studies provide significant support for the role of conidia-bound cellulases for the primary attack on cellulose to provide soluble cellulose break-down products which initiate growth and inducer formation.

Sophorose (2-O-β-D-glucopyranosyl-α-D-glucose) is the strongest cellulase-inducing sugar and is formed by *T. reesei* during cellulose hydrolysis by the cellulase system through transglycosylation reactions (Gritzali and Brown, 1979; Vaheri et al., 1979). It is assumed that sophorose formation requires the action of β-glucosidase(s). In addition to this cellulose-derived inducer a few other inducers including the disaccharide lactose (β-D-galactopyranosyl-(1→4)-D-glucose) and L-sorbose, the only monomeric sugar, are known. L-sorbose induces high cellulase transcript levels but it cannot be effectively applied in industrial fermentations since it also severely affects cell growth and is hardly metabolized by the cell. Due to the fact that lactose is soluble and by far cheaper than the other two inducing disaccharides cellobiose and sophorose, it is often used as a soluble carbon source in *T. reesei* cellulase and recombinant protein fermentations which has been reviewed by us previously (Kubicek et al., 2009; Seiboth et al., 2007). The obligatory presence of an inducer for cellulase gene expression implies tight regulation of their promoter regions. The different cellulase genes are generally regulated in a coordinated way and at least three transcriptional activators including XYR1, ACE2 the HAP2/3/5 complex, as well as the two repressors CRE1 and ACE1 are involved in their regulation (Fig. 2, Aro et al., 2005, Kubicek et al., 2009). Amongst these, XYR1 is clearly the major cellulase activator and indispensable for cellulase expression. It is an orthologue of the *xlnR* gene of *A. niger* and consensus sequences for XYR1 have been found in all inducible *T. reesei* (hemi)cellulase promoters investigated so far. XYR1 is a zinc binuclear cluster protein which binds to a GGCTAA-motif. Deletion of *xyr1* eliminates cellulase induction on cellulose and sophorose, thus proving its essential role in the induction process. In addition it regulates xylanases, β-xylosidase and also some α-L-arabinofuranosidases (Akel et al., 2009; Stricker et al., 2006). ACE2 is a coactivator whose influence is specific for growth on cellulose: *ace2* deletion reduces cellulase activity to 30-70% during growth on cellulose but does not affect induction by sophorose. Expression from the *cel6a* promoter is dependent on a CCAAT box bound by the HAP2/3/5 protein complex and it is assumed that this CCAAT sequence plays a role in the opening of the chromatin structure necessary for high level transcriptional activation. ACE1 is a repressor of cellulase gene transcription and binds to the sequence 5'-AGGCA-3'. Deletion of *ace1* increased the expression of cellulase and xylanase genes in sophorose and cellulose induced culture. ACE1 represses also transcription of *xyr1* (Mach-Aigner et al., 2008). A strain deleted in both *ace1* and *ace2* expressed cellulases and xylanases similar to the Δ*ace1* strain. An interesting difference to the regulation of cellulases and hemicellulases to *A. niger* by XlnR is that D-xylose induces only xylanases but not cellulase in *T. reesei* (Ilmen et al., 1997). This would implicate that the substrate-unspecific activator XYR1 is fine tuned by more specific transcriptional regulators or other signals.

One of the most interesting results from the analysis of the genome sequence of *T. reesei* QM6a was that 130 of the 316 CAZyme genes, including e.g. the two CBH-encoding genes, are located in 25 distinct clusters (Martinez et al., 2008). These CAZyme clusters were found in non-conserved genome regions to other closely related fungi. The precise preservation of the order of genes on a chromosome of related species is described by the term synteny. During species evolution this synteny is often lost due to rearrangements of the gene order in the chromosomes or between chromosomes, resulting in large gaps between syntenic blocks. In the interjacent non-syntenic regions often species-specific characteristics are found which are important for the adaption of the species to its environment. These non-syntenic CAZyme clusters contain also a number of genes encoding proteins involved in secondary metabolism such as polyketide synthase (PKS) or nonribosomal peptide synthase (NRPS). One conclusion from this gene clustering would be that *T. reesei* is able to regulate GHs and secondary metabolites in a coordinate manner and would enable *T. reesei* to fend off competition for nutrients during polysaccharide degradation by producing toxic substances. It is long established that cellulases and hemicellulases are expressed in a coordinate manner and the question arises if their accessibility on a transcriptional level could be enhanced even further by wide domain regulators, which control the accessibility of these regions by chromatin remodelling for the activation of more specific gene regulators such as e.g. the cellulase- and xylanase-regulator XYR1. Clustering of genes encoding enzymes involved in the biosynthesis of secondary metabolites is often found in fungi including *Aspergillus* spp. and regulated beside pathway-specific transcription factors also at an upper hierarchic level by global epigenetic control mechanisms. It was recently established that some of these gene clusters are regulated by a putative protein methyltransferase LaeA which is considered as a master regulator of secondary metabolism in Aspergilli and other species (Palmer and Keller, 2010). Although the exact role of LaeA is unclear, it is suggested that LaeA might control the accessibility of binding factors to chromatin regions of secondary metabolite clusters because the S-adenosyl methionine binding site of LaeA prevents heterochromatin maintenance of some clusters. Although the mechanism leading to these secondary metabolite gene clusters in Aspergilli and other fungi might be different from the CAZyme clusters, the role of the orthologous LAE1 in *T. reesei* was tested. The results show that the presence of LAE1 is required for the expression of the major cellulase and hemicellulase genes, and that its absence resulted in the inability to grow on cellulose. Expression of *lae1* under a strong constitutive promoter resulted in significantly increased cellulase gene transcription and cellulase formation. These data provide an experiment-based explanation of the advantage for clustering of cellulases in the genome of *T. reesei* and illustrate that chromatin regulation is a suitable target for strain improvement (Seiboth et al., ms submitted). In the case of secondary metabolite biosynthetic genes, clustering has been suggested to reflect their evolutionary history (Keller et al., 2005; Zhang et al., 2004). In the"selfish cluster" hypothesis it is suggested that selection occurs by promoting the maintenance of the cluster as a unit, by e.g. horizontal transfer events (Walton, 2000). However, there is no evidence or indication for horizontal gene transfer for the major cellulase and hemicellulases found in the CAZyme clusters and it is therefore likely that these CAZyme clusters are maintained by the operation of coregulation mechanisms which is essentially what is observed for cellulases and hemicellulases which are coinduced by similar substrates and controlled by a set of similar transcriptional regulators. Interestingly, *xyr1* itself is located in one of the CAZyme clusters, and downregulated in the *Δlae1* strain, which indicates that LAE1 controls both, accessibility of the cellulase genes and their major transcriptional activator. The signalling pathway by which the presence of the cellulase inducer is communicated to the *T. reesei* transcriptional machinery is not known yet. Its identification may also shed light on how LAE1 receives the signal so that cellulase gene transcription can be activated.

## 4. Strain improvement in *T. reesei*

One of the key issues in fungal strain breeding is the generation of improved producer strains in terms of enzyme yield. In the past this has traditionally been achieved by a combination of classical mutagenesis during which the fungus was exposed to different mutagens such as X-rays, γ-rays, UV rays, or chemicals, including *N*-methyl-*N'*-nitro-*N*-nitrosoguanidine (NTG) and ethyl methanesulfonate (EMS), in combination with different screening procedures to isolate cellulase-overexpressing strains. Classical mutagenesis is a common tool that has been successfully applied for many microorganisms, including fungi, to improve the production of various industrial enzymes (glucoamylase, lipase or cellulase). A prerequisite in such strain improvement programs is, of course, that the microorganism already produces the enzyme(s) of choice. It is a peculiarity of *T. reesei* that all these cellulase high-producing strains generated by classical mutagenesis and used today in research or by the major enzyme manufacturers are derived from strain QM6a from the Solomon Islands. However, the availability of sophisticated gene manipulation methods and the recent decoding of the genome sequence of *T. reesei* enable us to introduce molecular genetic tools into such programs. One approach to better understand the biology underlying cellulase hyperproduction is the analysis of improved producer strains derived from classical mutagenesis programs. Until recently, only a few data were available on the genomic alterations that occurred in the strains undergoing strain improvement procedures. With the publication of the genome sequence of strain QM6a and the development of high-throughput methods such as massively parallel DNA sequencing and comparative genomic hybridization (CGH), it is now possible to identify the genomic changes that occurred in the different mutant lines.

### 4.1 A short history of the main *T. reesei* strain improvement programs

Over the past decades, academic and industrial research programs have produced many different *T. reesei* strains by random mutagenesis in combination with appropriate selection regimes (reviewed by Harman and Kubicek, 1998) whose cellulase production exceeds by far the level of the original isolate QM6a. By 1979, classical mutagenesis programmes had led to the development of mutant strains with an up to 20 times higher productivity compared to the original isolate, and nowadays, improved producer strains are reported to secrete more than 100 grams of cellulases per litre in industrial fermentations. The first reported cellulase mutant strain was isolated at Natick Laboratories by irradiating conidia of QM6a in a linear particle accelerator (Fig. 3). QM9123 produced twice as much cellulases as its parental strain on cellulose-containing media. Further irradiation resulted in strain QM9414 which produced up to four times higher levels of cellulases than QM6a. QM9414 is an early cellulase mutant which is used today in research and was also often used as the parental strain in improvement programs. At Rutgers University a separate line of high-producing mutants was generated. Among these strains, RUT-C30 has become the most frequently used cellulase hyperproducer in the public domain. Montenecourt and Eveleigh prepared two mutant lines which led to the hypercellulolytic strains RUT-C30 and RL-P37 (Montenecourt and Eveleigh, 1977; Montenecourt and Eveleigh, 1979), RUT-C30 was generated by three rounds of mutagenesis and selection with the ultimate goal of isolating a carbon catabolite derepressed strain. In the first step, UV mutagenesis and selection for the ability to hydrolyze cellulose under carbon catabolite repressing conditions led to strain M7 which was further mutagenised by an NTG treatment under carbon catabolite repression conditions. This led to the isolation of strain NG14, which already showed a strong increase in secreted protein and cellulase activity but still showed considerable catabolite repression. NG14 was subjected to another round of UV mutagenesis and screening for elevated cellulose hydrolysis levels and resistance to 2-deoxyglucose to eliminate carbon catabolite repression leading to RUT-C30. 2-deoxyglucose is a non-metabolizable analogue of glucose in which the 2-hydroxyl group is replaced by a hydrogen atom. This compound is taken up by the cell but cannot undergo further glycolysis and the hexokinase enzyme traps this substance in most cells producing 2-deoxyglucose-6-phosphate which exhibits CCR. The resulting strain RUT-C30 produces twice as much extracellular protein as its parental strain NG14, reaching more than 30 g/L production in industrial fermentations and is carbon catabolite derepressed.

In the last years three major genetic changes in RUT-C30 were revealed. The first discovered mutation is a truncation in the *cre1* gene, the key transcription regulator of carbon catabolite repression in fungi (Ilmen et al., 1996b; Seidl et al., 2008). This is not surprising, since the elimination of carbon catabolite repression was one of the main targets during the mutagenesis program. It was recently shown that the deletion of either the complete *cre1* gene or its replacement with the truncated version of RUT-C30 has largely the same physiological effect, indicating that the truncated *cre1* gene is practically a null allele (Nakari-Setälä et al., 2009). The second mutation detected in RUT-C30 was a frame-shift mutation in the gene *gls2a*, encoding the glucosidase α-subunit, resulting in a truncated gene product (Geysens et al., 2005). This protein is involved in the structural modification of *N*-linked oligosaccharides present on glycoproteins and in the quality control mechanisms of polypeptides such as cellulases which are traversing the endoplasmic reticulum. The third mutation is a large genomic lesion of 85 kb, affecting 29 genes (Seidl et al., 2008), including transcription factors and enzymes of the primary metabolism. Some of the genes lacking in

RUT-C30 could be correlated with pronounced alterations in its phenotype, such as poor growth on α-linked oligosaccharides. This deletion of 85 kb is also present in NG14.

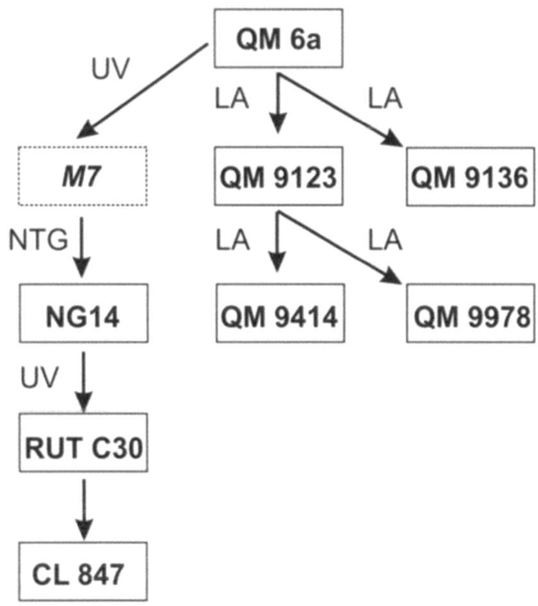


Fig. 3. Pedigree of important *T. reesei* strains derived from the original isolate QM6a by classical mutagenesis. The mutagens used are either irradiation by linear particle accelerators (LA), UV light (UV) or NTG. All the strains except CL487 are sequenced by the Joint Genome Institute. Strain M7 is indicated in italics as it is not available any more. QM9136 and QM9978 are deficient in cellulase production.

#### 4.2 Genome wide approaches to understand cellulase hyperproduction

The *T. reesei* QM6a derived mutant series form a unique pedigree of strains in which each mutant shows an improved ability to produce cellulases compared to its parental strain. Besides these strains there are also mutants available with defects in cellulase expression (Fig. 3). From these mutagenesis programmes it is also evident that several mutations and not just one or two gene modifications are necessary to make a significantly improved cellulase producer strain. It can be expected that a large number of gene alterations are found in these strains affecting the protein synthesis and the secretion ability. The analysis of mutations in high-producing *T. reesei* strains would increase our understanding of the mechanisms underlying efficient cellulase production and their analysis might reveal new key players in cellulase expression. Since most of the improved producer strains (e.g. RUT C30) were subsequently also used in industrial strain improvement programs, the additional mutations found in these industrial hyperproducers could give even further leads on how to improve protein production. The knowledge of the gene alterations would allow targeted genetic engineering for improved production of cellulases and proteins in general.

With the genome sequence of strain QM6a as a reference, it is now possible to use high-throughput methods such as massively parallel DNA sequencing or comparative genomic hybridization (CGH) to detect mutations in these strains. Using massive parallel sequencing numerous mutations were identified in the mutant line consisting of NG14 and RUT-C30 (Le Crom et al., 2009). This analysis revealed 223 single nucleotide variants (SNVs) and 15 small indels (insertions and deletions), and 18 larger deletions which led to the loss of more than 100 kb of genomic DNA in the cellulase overproducing strains. The SNVs resulted in mutations in more than 50 genes which are surprisingly found only in a small group of functions including nuclear transport, transcription, mRNA stability, protein secretion and vacuolar targeting. In both strains, NG14 and RUT-C30, a number of genes belonging to these functional categories are mutated. Interestingly, with the exception of CRE1, none of the transcriptional regulators of cellulase expression (XYR1, ACE1, ACE2 or HAP complex) or the proteinmethyltransferase LAE1 were affected in these strain lines by the different rounds of mutagenesis. These results highlight additional mechanisms which can be of importance for cellulase hyperproduction but they also underscore the complexity of the problem: it will be part of the future research to find out which mutations are detrimental, neutral or beneficial for cellulase production. In a recent study the analysis of genomic alterations was extended to QM9414 and QM9123 (Vitikainen et al., 2010) by array comparative genomic hybridization (aCGH), which allows the analysis of smaller genomes with oligonucleotide tiling arrays at a single nucleotide resolution on a single chip. Again, a large number of genomic alterations were detected in all four cellulase mutant strains. In strains QM9123 and QM9414 a total of 44 novel mutations including 4 translocation breakpoints were detected while in the RUT-C30 line additional 17 mutations to the ones detected by Le Crom et al. (2009) were found. Further, also a number of chromosomal translocation breakpoints were detected in RUT-C30. These results confirm earlier conclusion from electrophoretic karyotyping which indicated differences in the size of the RUT-C30 chromosomes when compared to QM9414, probably due to chromosomal rearrangements. Gene mapping by hybridization had also already pointed to additional rearrangements in the gene order between these strain lines (Carter et al., 1992; Mantyla et al., 1992). It is possible that these chromosomal rearrangements also play a role in cellulase regulation since wide domain regulation as found for LAE1 might be affected in such strains.

Based on the results from these genomic studies, new hypotheses about the mechanisms underlying cellulase and hemicellulase production and secretion in *T. reesei* were developed and novel areas for strain improvement such as nuclear transport, mRNA turnover and vacuolar protein trafficking will now be tested. Till now only two new deletions identified in RUT-C30 (the deletion of an 85 kb region and the deletion of a gene encoding a transcription factor) were tested for cellulase production by constructing knock-out strains, but both deletions did not affect cellulase production. It will be interesting for future research, to see if there is a synergism between the mutations and if some mutations might only become operative in the right genomic background.

## 5. Advanced tool boxes for genetic engineering and strain breeding

Results from genome-wide approaches including the *T. reesei* genome sequencing and annotation project, high-throughput sequencing of improved *T. reesei* strains and microarray analysis under cellulase inducing conditions, provided researchers with long lists of potential candidate genes for further investigations to better understand and further improve cellulase production. Although some of these genes encode proteins with well-characterized orthologues in other fungi and thus their general functions might be predictable, their relevance for cellulase production still needs to be evaluated. Further, many of the newly discovered genes are annotated as 'hypothetical proteins', which indicates that it is difficult to predict any function at all for these proteins. A key technique to assess gene functions and to alter the characteristics of fungal strains is the inactivation of genes by targeted deletion or knock-out. With such a technique, organisms are genetically engineered in such a way that a particular gene (target gene) is completely absent from the organism (it has been "knocked out" in the organism). To facilitate gene knockouts, individual cells are genetically transformed with a DNA construct, the gene deletion fragment. This construct consists of the promoter and terminator region of a selected gene, but the gene's coding region (encoding the protein) is replaced by a marker gene. When this gene deletion cassette is introduced into the organism by transformation, it is able to recombine with the target gene in the organism's genome. Recombination occurs between the promoter and terminator region of the gene deletion fragment and the target gene, resulting in its replacement by the marker in the genome.

The marker gene is also essential to discriminate between fungal cells which were transformed by the gene deletion fragment and those which did not incorporate the DNA. Marker genes can either confer antibiotic resistance (dominant marker) or complement an auxotrophy (auxotrophic marker) and thus enable selection of transformed cells. Auxotrophy is the inability of an organism to synthesize a particular organic compound required for its growth. Functional genetic studies on a large scale basis depend on the construction of well defined gene deletion strains with an efficient gene targeting system. So far the generation of gene deletion strains in filamentous fungi such as *T. reesei* was a time consuming process and the established working procedures were not suitable for carrying out gene deletions on a high-throughput basis. Therefore adequate molecular tool boxes were developed in the last years to overcome these bottlenecks to be able to perform such gene deletions on a large scale basis.

The first limiting step was the construction of the gene deletion fragment itself. Usually this fragment was assembled in a time consuming process based on different genetic engineering techniques including several rounds of DNA restriction digests, DNA ligations and transformations in *E. coli* followed by the verification of the construct before it could be used for fungal transformation and gene deletion. Different methods to facilitate the large scale production of deletion vector constructs were developed which now enable us to bypass traditional vector construction methods. One method to construct deletion fragments relies on the polymerase chain reaction (PCR) and assembles the three fragments simply by PCR bypassing ligation and restriction digests. Another method was successfully tested for *N. crassa* genome project (Colot et al., 2006). This method takes advantage of recombinational cloning in *S. cerevisiae* to rapidly create knockout cassettes (Fig. 4). The different fragments of the deletion construct are individually amplified by PCR with additional short, overlapping ends. The PCR fragments are then transformed into yeast cells and are there assembled to the complete gene deletion fragment by the endogenous recombination machinery via the overlapping ends of the PCR fragments. Once recombination has taken place, the DNA vector can be extracted from the yeast cells and the complete deletion construct can be amplified by PCR. A similar strategy was recently also developed for *T. reesei* (unpublished results).

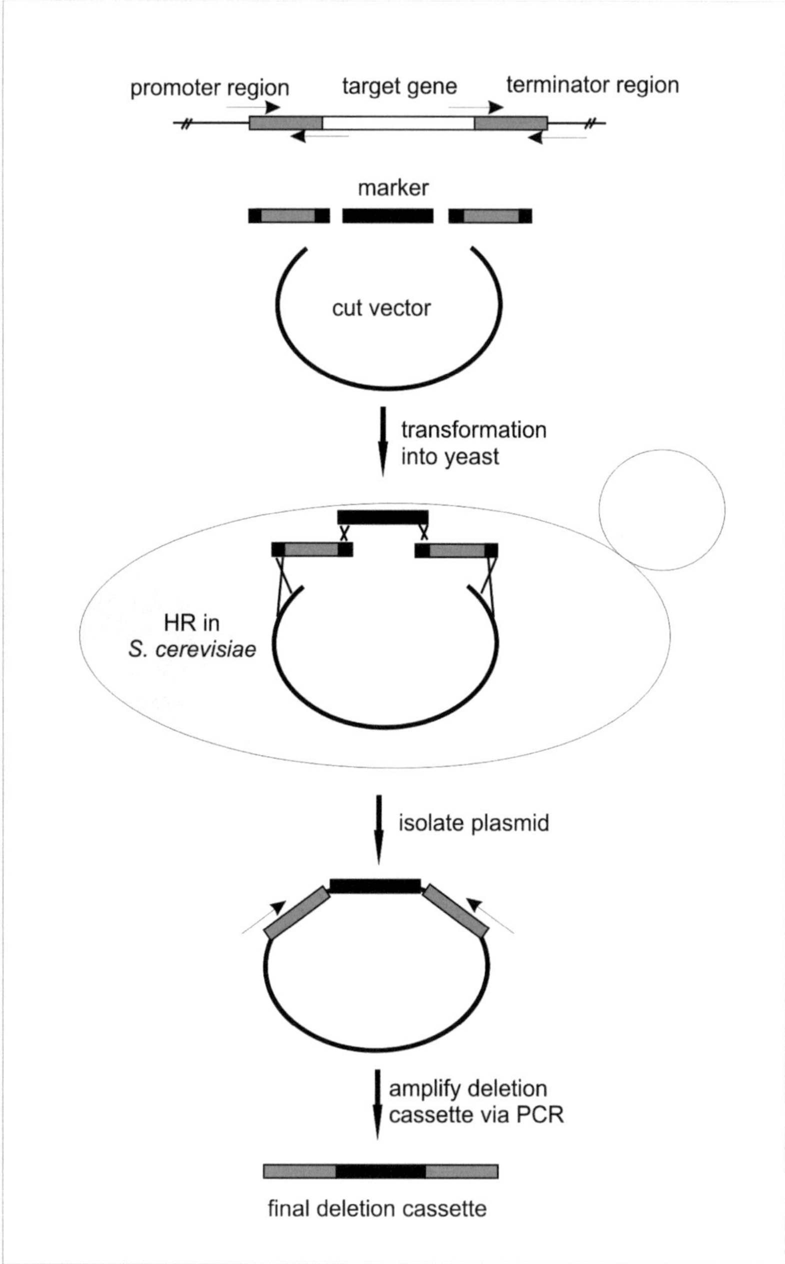


Fig. 4. High-throughput construction of gene deletion fragments by yeast recombinational cloning. Promoter and terminator regions (in grey) of the target gene are amplified from genomic DNA with homologous overlapping ends to the marker gene (white) and to the vector sequence. The flanking regions, the marker gene and the restricted vector are then introduced by cotransformation into *S. cerevisiae* where the deletion vector is assembled by homologous recombination. Subsequently, the deletion vector is extracted and the cassette amplified by PCR.

Another major obstacle for functional genetic studies is the general low frequency of homologous recombination (HR) of filamentous fungi in comparison to the yeast *S. cerevisiae*. This HR frequency is typically ca. 10 % in *T. reesei*. This means that only in one of ten fungal cells that contain the gene deletion fragment the target gene has been deleted while in the other nine the DNA fragment is inserted somewhere else in the genome (= ectopic integration). HR depends on the genomic locus, the genetic marker and the size of the flanking regions. As a consequence of the predominant ectopic insertion of the deletion fragment during genetic transformations, numerous transformants have to be screened to detect the desired ones where a HR between the target gene and the deletion fragment had occurred. Integration of DNA fragments in the genome requires the action of double strand repair mechanisms and is mediated by two main pathways (Fig. 5A), the homologous recombination pathway (HR) and the nonhomologous end joining pathway (NHEJ). The NHEJ pathway seems to be the dominant mode of DNA integration in most fungi including *T. reesei* leading to a ligation of strands without sequence similarity. To improve HR frequencies it is necessary to eliminate this ectopic (random) integration of the deletion cassette or to improve HR frequencies. Impairment of NHEJ seems to be the preferred solution and is usually achieved by inactivation of key components of the NHEJ pathway such as the DNA binding KU70 or 80 proteins (Krappmann, 2007). *T. reesei* strains in which the corresponding *tku70* gene was deleted show improved gene targeting (90-100%) and allow gene knockout studies on a high throughput basis using 1 kb homologous flanking regions (Guangtao et al., 2009). Although this system represents a considerable breakthrough for the study of gene functions in filamentous fungi, studies on the potential side effect resulting from an impaired NHEJ pathway on genome stability and integrity are not available yet. Potentially deleterious effects of this *tku70* deletion can be minimized by reintroduction of the *tku70* gene via transformation or via sexual recombination in sexually propagating fungi, or by only transiently inactivation of the *tku70* gene.

In *T. reesei,* DNA mediated transformation relies on a small number of dominant and auxotrophic markers, which generally limit the number of genetic manipulations. In addition, the presence of dominant markers is often undesired in industrial strains and might be an obstacle in product approval processes. As a solution for all of these problems recyclable markers were developed. Different systems are available that are suitable for removing the marker gene after each transformation step and allowing the reapplication of the same marker in the next round of transformation. The "blaster cassette" is one option to facilitate the removal of the marker (Fig. 5B). Here the marker gene is located between two direct repeats of about 500 bp, which favours excision of the marker by HR between these repeats. This blaster cassette can then be reused for multiple rounds of gene deletions as was already exemplified in *T. reesei*, where we consecutively deleted two hexokinase genes with a blaster cassette (Hartl and Seiboth, 2005). The blaster approach relies on naturally occurring recombination between the two direct repeats flanking the marker gene, which occurs at a relatively low frequency (ca. 1: $10^4$ $10^6$) and relies on the use of a bidirectional marker (see below).

Other recombination systems introduce a specific recombinase (Fig. 5C), which detects short stretches of DNA with a specific sequence and recombines these sites, thereby removing the DNA between the sites. The advantage of this approach is a relatively high recombination frequency, but it requires an additional gene - encoding the recombinase - which has to be regulated in such a way that the marker gene is not removed during the initial fungal transformation but only at a later stage when the removal of the marker gene is desired.

Examples for such systems are the Cre/*loxP* system from *E. coli* bacteriophage P1 or the *S. cerevisiae* FLP/*FRT* system. Both depend on a site-specific recombination mediated by the Cre or FLP recombinase and recognition sites that serve as specific recombinase binding sites (*loxP* or *FRT*).

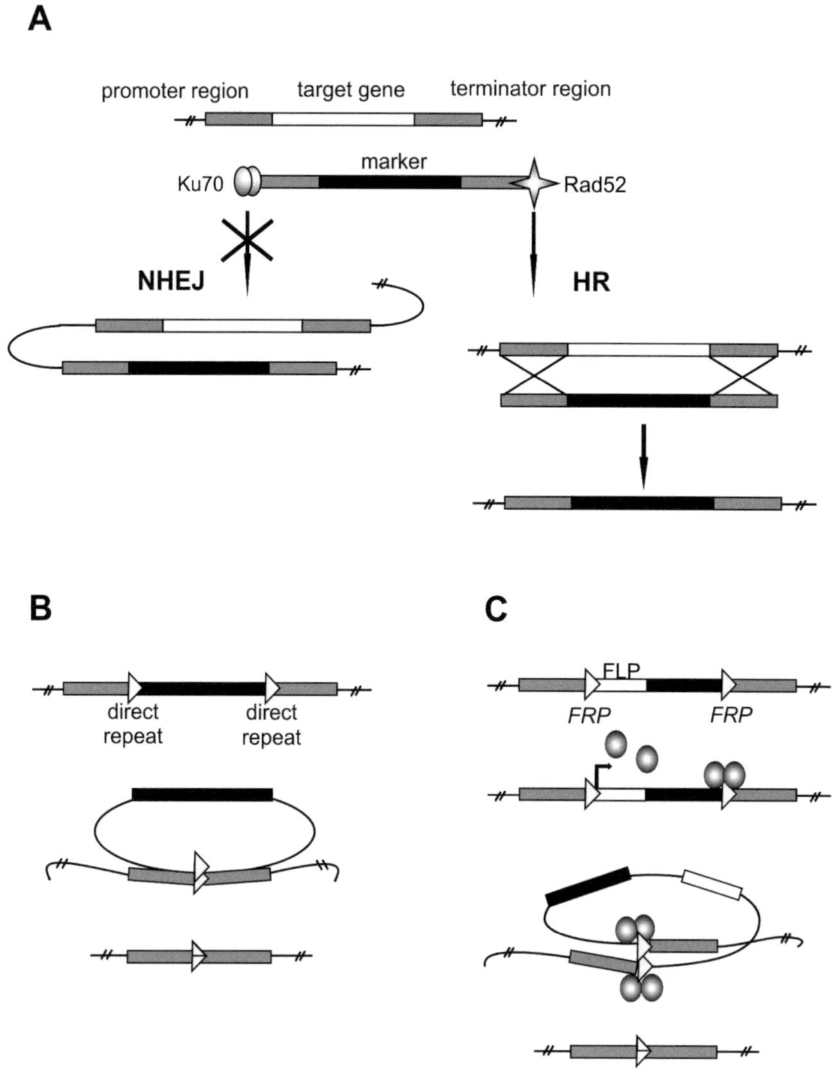


Fig. 5. Homologous DNA integration and different marker removal systems. (A) The HR and NHEJ pathway mediate the integration of DNA into the genome. Inactivation of Ku70 or Ku80 genes or the amplification of Rad52 improve HR frequencies. Marker recycling by a blaster cassette approach (B) or by a FLP/*FRT* recombination system (C). Blaster cassettes are composed of a marker cassette flanked by two direct repeats are responsible for the excision of the blaster cassette by homologous recombination after transformation. In the FLP mediated system the FLP recombinase is introduced with the marker gene and mediates the recombination between two *FRT* sequences (Kopke et al., 2010).

Bidirectional genetic markers, which can be used to select for the absence and presence of the marker, are essential for the application of blaster cassettes but are also advantageous in the recombinase system instead of the normal (unidirectional) markers. Bidirectional selection is accomplished by first selecting for the presence of the marker during the fungal transformation (=positive selection). For example, in the case of a dominant positive selection, the presence of the marker gene results in the resistance of the transformant to a certain antibiotic. Afterwards the loss/absence of the marker is tested (= negative selection). In case the marker gene is still present, the transformant will be sensitive to a certain toxic substance. In the absence of the marker this substance cannot be metabolized and is therefore not toxic anymore. The *T. reesei pyr4* gene (encoding orotidine 5'-phosphate carboxylase) is an example for a bidirectional marker. The *pyr4* gene confers uridine/uracil prototrophy, it is therefore possible to select for the presence or the absence of the marker. Transformants where *pyr4* has integrated into the genomic DNA are selected via *pyr4* function. Such transformants are able to grow on medium without uridine supplementation. These transformants, however, are sensitive to 5-FOA (5-fluorotic acid) and thus removal of the *pyr4* marker by one of the above described methods results in strains which are insensitive to 5-FOA. Similar bidirectional marker cassettes were constructed by fusing a dominant positive marker with a negative marker, e.g. a phleomycin resistance gene was fused in-frame with the herpes simplex virus type 1 thymidine kinase gene. This gene fusion confers phleomycin resistance for dominant positive selection and ganciclovir sensitivity for negative selection (Krappmann et al., 2005).
Although considerable progress has been made in the last decades to understand cellulase production in *T. reesei* by making use of gene deletion strains and other molecular biological tools, the cellulase secretion levels of strains such as RUT C30 or CL487 that were derived from random mutagenesis programs could not be reached. Thus, as described in section 4.2, the interest in the complete set of mutations of these strains was renewed in the past few years by the availability of genome-wide approaches. The results yielded long lists with the total numbers of mutations and most of them have not been functionally analyzed yet. While some of them probably are the cause for cellulase hyperproduction, others are certainly deleterious and it is desirable to repair them in order to improve the fitness of the strain and thus possibly increase cellulase production. Due to the limitations of genetic engineering outlined above, another way to speed up the analysis of these mutations would be the meiotic recombination through sexual crossing. Interestingly, industrially applied filamentous fungal species have so far been considered to be largely asexual. This was also the case for *T. reesei* for more than 50 years, but analysis of the genome sequence of *T. reesei* QM6a provided evidence for its sexual cycle and showed that this is a so-called heterothallic species, which needs another strain of the opposite mating type for sexual propagation. Genome analysis of QM6a revealed an intact mating type locus and using appropriate *H. jecorina* wild-type isolates with the correct opposite mating type enabled us recently to successfully cross *T. reesei* QM6a for the first time under laboratory conditions (Seidl et al., 2009). In other fungi, for example the model organisms *N. crassa* and *A. nidulans,* sexual crossing is, a commonly used technique to combine fungal strain properties. This approach enables the isolation and characterization of auxotrophic mutants and is the reason, why transformation strategies that involve the conversion of auxotrophic mutants to prototrophy are so far only poorly developed in *T. reesei*. Sexual propagation would also be important for repairing detrimental mutations in *T. reesei* production strain lines. Another important application would be the back-crossing of deleted genes, for example the *tku70* gene that

aids in generating gene deletion strains but in the long term might be important for chromosome stability of the fungus. Although successful sexual crossing of *T. reesei* QM6a and notably also of other cellulase hyperproducing mutants including RUT-C30, has been a milestone for strain development in *T. reesei*, several drawbacks still need to be overcome. So far *T. reesei* strains can only be crossed with *H. jecorina* wild-type strains and genetic analysis indicated that the *T. reesei* QM6a strain line has a genetic defect and is not able to develop female fruiting bodies. Therefore, QM6a can only serve as the male mating partner during sexual propagation. Further research will be necessary to elucidate the reasons for this deficiency and to fully establish a system similar to the model fungi *N. crassa* or *A. nidulans*.

## 6. Conclusion

The analysis of the complete genome sequence of *T. reesei* has resulted in the surprising finding that this industrial (hemi)cellulase producer has only a relatively small set of enzymes involved in plant cell wall degradation. This indicates that the efficiency of the enzyme mixtures of *T. reesei* can be further improved by the addition of other cellulases and non-enzymatic proteins. However, even with its apparently limited set of cellulolytic enzymes *T. reesei* gained significant industrial importance over the past decades due to its still yet unmatched enzyme production capacities. The large numbers of mutations detected in different production strain lines by recent advances in genome-wide analysis approaches underscore our limited understanding of the regulatory processes underlying cellulase production and suggest yet unexploited potentials to further improve enzyme production and strain breeding including nuclear transport, transcription, mRNA stability, protein secretion and vacuolar targeting. The relevance of all these alterations can in the next years be tested by using the recently developed toolboxes that enable fungal strain improvement on a high-throughput basis.

## 7. Acknowledgment

The FWF (Austrian Science Fund) supported BS by grant P19421 and VS by P20559 and a Hertha-Firnberg grant (T390).

## 8. References

Adney, W.S.; Chou, Y.C.; Decker, S.R.; Ding, S.Y.; Baker, J.O.; Kunkel, G.; Vinzant, T.B. & Himmel, M.E. (2003). Heterologous expression of *Trichoderma reesei* 1,4-beta-D-glucan cellobiohydrolase (Cel 7A). In, *Applications of Enzymes to Lignocellulosics* Amer Chemical Soc., ISBN pp. 403-437, Washington, USA

Aro, N.; Pakula, T. & Penttilä, M. (2005). Transcriptional regulation of plant cell wall degradation by filamentous fungi. *FEMS Microbiology Reviews*, Vol.29, No. 4, pp. 719-739, ISSN 0168-6445

Barr, B.K.; Hsieh, Y.L.; Ganem, B. & Wilson, D.B. (1996). Identification of two functionally different classes of exocellulases. *Biochemistry*, Vol.35, No. 2, pp. 586-592, ISSN 0006-2960

Bayer, E.A.; Chanzy, H.; Lamed, R. & Shoham, Y. (1998). Cellulose, cellulases and cellulosomes. *Current Opinion in Structural Biology*, Vol.8, No. 5, pp. 548-557, ISSN 0959-440X

Beckham, G.T.; Bomble, Y.J.; Matthews, J.F.; Taylor, C.B.; Resch, M.G.; Yarbrough, J.M.; Decker, S.R.; Bu, L.; Zhao, X.; McCabe, C.; Wohlert, J.; Bergenstrahle, M.; Brady, J.W.; Adney, W.S.; Himmel, M.E. & Crowley, M.F. (2010). The O-glycosylated linker from the *Trichoderma reesei* Family 7 cellulase is a flexible, disordered protein. *Biophysical journal*, Vol.99, No. 11, pp. 3773-3781, ISSN 1542-0086

Beguin, P. & Aubert, J.P. (1994). The biological degradation of cellulose. *FEMS Microbiology Reviews*, Vol.13, No. 1, pp. 25-58, ISSN 0168-6445

Bouws, H.; Wattenberg, A. & Zorn, H. (2008). Fungal secretomes--nature's toolbox for white biotechnology. *Applied Microbiology and Biotechnology*, Vol.80, No. 3, pp. 381-388, ISSN 1432-0614

Cantarel, B.L.; Coutinho, P.M.; Rancurel, C.; Bernard, T.; Lombard, V. & Henrissat, B. (2009). The Carbohydrate-Active EnZymes database (CAZy): an expert resource for Glycogenomics. *Nucleic Acids Research*, Vol.37, No. Database issue, pp. D233-238, ISSN 1362-4962

Carle-Urioste, J.C.; Escobar-Vera, J.; El-Gogary, S.; Henrique-Silva, F.; Torigoi, E.; Crivellaro, O.; Herrera-Estrella, A. & El-Dorry, H. (1997). Cellulase induction in *Trichoderma reesei* by cellulose requires its own basal expression. *Journal of Biological Chemistry*, Vol.272, No. 15, pp. 10169-10174, ISSN 0021-9258

Carroll, A. & Somerville, C. (2009). Cellulosic biofuels. *Annual Revies in Plant Biology*, Vol.60, No., pp. 165-182, ISSN 1545-2123

Carter, G.L.; Allison, D.; Rey, M.W. & Dunn-Coleman, N.S. (1992). Chromosomal and genetic analysis of the electrophoretic karyotype of *Trichoderma reesei*: mapping of the cellulase and xylanase genes. *Molecular microbiology*, Vol.6, No. 15, pp. 2167-2174, ISSN 0950-382X

Cherry, J.R. & Fidantsef, A.L. (2003). Directed evolution of industrial enzymes: an update. *Current Opinion in Biotechnology*, Vol.14, No. 4, pp. 438-443, ISSN 0958- 1669

Colot, H.V.; Park, G.; Turner, G.E.; Ringelberg, C.; Crew, C.M.; Litvinkova, L.; Weiss, R.L.; Borkovich, K.A. & Dunlap, J.C. (2006). A high-throughput gene knockout procedure for *Neurospora* reveals functions for multiple transcription factors. *Proceedings of the National Academy of Science U S A*, Vol.103, No. 27, pp. 10352-10357, ISSN 0027-8424

Divne, C.; Stahlberg, J.; Reinikainen, T.; Ruohonen, L.; Pettersson, G.; Knowles, J.K.; Teeri, T.T. & Jones, T.A. (1994). The three-dimensional crystal structure of the catalytic core of cellobiohydrolase I from *Trichoderma reesei*. *Science*, Vol.265, No. 5171, pp. 524-528., ISSN 0036-8075

Druzhinina, I.S.; Komon-Zelazowska, M.; Atanasova, L.; Seidl, V. & Kubicek, C.P. (2010). Evolution and ecophysiology of the industrial producer *Hypocrea jecorina* (Anamorph *Trichoderma reesei*) and a new sympatric agamospecies related to it. *PLoS One*, Vol.5, No. 2, pp. e9191, ISSN 1932-6203

Flipphi, M.; van de Vondervoort, P.J.; Ruijter, G.J.; Visser, J.; Arst, H.N., Jr. & Felenbok, B. (2003). Onset of carbon catabolite repression in *Aspergillus nidulans*. Parallel involvement of hexokinase and glucokinase in sugar signaling. *The Journal of Biological Chemistry*, Vol.278, No. 14, pp. 11849-11857, ISSN 0021-9258

Foreman, P.K.; Brown, D.; Dankmeyer, L.; Dean, R.; Diener, S.; Dunn-Coleman, N.S.; Goedegebuur, F.; Houfek, T.D.; England, G.J.; Kelley, A.S.; Meerman, H.J.; Mitchell, T.; Mitchinson, C.; Olivares, H.A.; Teunissen, P.J.; Yao, J. & Ward, M. (2003).

Transcriptional regulation of biomass-degrading enzymes in the filamentous fungus *Trichoderma reesei*. *The Journal of Biological Chemistry* Vol.278, No. 34, pp. 31988-31997, ISSN 0021-9258

Geysens, S.; Pakula, T.; Uusitalo, J.; Dewerte, I.; Penttilä, M. & Contreras, R. (2005). Cloning and characterization of the glucosidase II alpha subunit gene of *Trichoderma reesei*: a frameshift mutation results in the aberrant glycosylation profile of the hypercellulolytic strain Rut-C30. *Applied and Environmental Microbiology*, Vol.71, No. 6, pp. 2910-2924, ISSN 0099- 2240

Gritzali, M. & Brown, R.D.J. (1979). The cellulase system of *Trichoderma*: relationship betweeen purified extracellular enzymes from induced or cellulose-grown cells. *Advances in Chemistry Series* Vol.181, No., pp. 237-260, ISSN 0065-2393

Guangtao, Z.; Hartl, L.; Schuster, A.; Polak, S.; Schmoll, M.; Wang, T.; Seidl, V. & Seiboth, B. (2009). Gene targeting in a nonhomologous end joining deficient *Hypocrea jecorina*. *Journal of Biotechnology*, Vol.139, No. 2, pp. 146-151, ISSN 0168-1656

Harman, G.E. & Kubicek, C.P. (1998). *Trichoderma and Gliocladium. Enzymes, biological control and commercial applications*, 2, Taylor & Francis Ltd, ISBN 0748408061, London, Great Britain

Harris, P.V.; Welner, D.; McFarland, K.C.; Re, E.; Navarro Poulsen, J.C.; Brown, K.; Salbo, R.; Ding, H.; Vlasenko, E.; Merino, S.; Xu, F.; Cherry, J.; Larsen, S. & Lo Leggio, L. (2010). Stimulation of lignocellulosic biomass hydrolysis by proteins of glycoside hydrolase family 61: structure and function of a large, enigmatic family. *Biochemistry*, Vol.49, No. 15, pp. 3305-3316, ISSN 1520-4995

Hartl, L. & Seiboth, B. (2005). Sequential gene deletions in *Hypocrea jecorina* using a single blaster cassette. *Current Genetics*, Vol.48, No. 3, pp. 204-211, ISSN 0172-8083

Himmel, M.E.; Ding, S.Y.; Johnson, D.K.; Adney, W.S.; Nimlos, M.R.; Brady, J.W. & Foust, T.D. (2007). Biomass recalcitrance: engineering plants and enzymes for biofuels production. *Science*, Vol.315, No. 5813, pp. 804-807, ISSN 1095-9203

Igarashi, K.; Koivula, A.; Wada, M.; Kimura, S.; Penttilä, M. & Samejima, M. (2009). High speed atomic force microscopy visualizes processive movement of *Trichoderma reesei* cellobiohydrolase I on crystalline cellulose. *The Journal of Biological Chemistry*, Vol.284, No. 52, pp. 36186-36190, ISSN 1083-351X

Ilmen, M.; Saloheimo, A.; Onnela, M.L. & Penttilä, M.E. (1997). Regulation of cellulase gene expression in the filamentous fungus *Trichoderma reesei*. *Applied and Environmental Microbiology*, Vol.63, No. 4, pp. 1298-1306., ISSN 0099- 2240

Ilmen, M.; Thrane, C. & Penttilä, M. (1996). The glucose repressor gene *cre1* of *Trichoderma*: isolation and expression of a full-length and a truncated mutant form. *Molecular and General Genetics*, Vol.251, No. 4, pp. 451-460, ISSN 0026-8925

Jalak, J. & Valjamae, P. (2010). Mechanism of initial rapid rate retardation in cellobiohydrolase catalyzed cellulose hydrolysis. *Biotechnology and Bioengineering*, Vol.106, No. 6, pp. 871-883, ISSN 1097-0290

Jeoh, T.; Michener, W.; Himmel, M.E.; Decker, S.R. & Adney, W.S. (2008). Implications of cellobiohydrolase glycosylation for use in biomass conversion. *Biotechnology for Biofuels*, Vol.1, No. 1, pp. 10, ISSN 1754-6834

Karkehabadi, S.; Hansson, H.; Kim, S.; Piens, K.; Mitchinson, C. & Sandgren, M. (2008). The first structure of a glycoside hydrolase family 61 member, Cel61B from *Hypocrea jecorina*, at 1.6 A resolution. *Journal of Molecular Biology*, Vol.383, No. 1, pp. 144-154, ISSN 1089-8638

Keller, N.P.; Turner, G. & Bennett, J.W. (2005). Fungal secondary metabolism - from biochemistry to genomics. *Nature Reviews in Microbiology*, Vol.3, No. 12, pp. 937-947, ISSN 1740-1526

Kopke, K.; Hoff, B. & Kuck, U. (2010). Application of the Saccharomyces cerevisiae FLP/FRT recombination system in filamentous fungi for marker recycling and construction of knockout strains devoid of heterologous genes. *Applied and Environmental Microbiology*, Vol.76, No. 14, pp. 4664-4674, ISSN 1098-5336

Krappmann, S. (2007). Gene targeting in filamentous fungi: the benefits of impaired repair. *Fungal Biology Reviews*, Vol.21, No. 1, pp. 25-29, ISSN 1749-4613

Krappmann, S.; Bayram, O. & Braus, G.H. (2005). Deletion and allelic exchange of the *Aspergillus fumigatus veA* locus via a novel recyclable marker module. *Eukaryotic Cell*, Vol.4, No. 7, pp. 1298-1307, ISSN 1535-9778

Kubicek, C.P.; Mikus, M.; Schuster, A.; Schmoll, M. & Seiboth, B. (2009). Metabolic engineering strategies for the improvement of cellulase production by *Hypocrea jecorina*. *Biotechnol Biofuels*, Vol.2, No., pp. 19, ISSN 1754-6834

Kubicek, C.P.; Seidl, V.; Zeilinger, S.; Kredics, L.; Nevalainen, H.; Deshpande, N.; Monte, E.; Wiest, A.; McCluskey, K.; Henrissat, B.; Coutinho, P.M.; Casas, S.; Misra, M.; Schmoll, M.; Antal, Z.; Tisch, D.; Seiboth, B.; Shapiro, H.; Zhang, M.; Challacombe, J.; Salamov, A.; Lindquist, E.; Lucas, S.; Mukherjee, M.; Hermosa, R.; Horwitz, B.; von Döhren, H.; Herrera-Estrella, A.; Baker, S.; Kenerley, C.M. & Grigoriev, I.V. (2011). Genome sequence comparison of three species of *Trichoderma* reveals the evolutionary changes from saprophytic to mycoparasitic lifestyle. *Genome Biology*, in press. ISSN 1465-6906

Kumar, R.; Singh, S. & Singh, O.V. (2008). Bioconversion of lignocellulosic biomass: biochemical and molecular perspectives. *Journal of Indian Microbiology and Biotechnology*, Vol.35, No. 5, pp. 377-391, ISSN 1367-5435

Lantz, S.E.; Goedegebuur, F.; Hommes, R.; Kaper, T.; Kelemen, B.R.; Mitchinson, C.; Wallace, L.; Stahlberg, J. & Larenas, E.A. (2010). *Hypocrea jecorina* CEL6A protein engineering. *Biotechnology for Biofuels*, Vol.3, No., pp. 20, ISSN 1754-6834

Le Crom, S.; Schackwitz, W.; Pennacchio, L.; Magnuson, J.K.; Culley, D.E.; Collett, J.R.; Martin, J.; Druzhinina, I.S.; Mathis, H.; Monot, F.; Seiboth, B.; Cherry, B.; Rey, M.; Berka, R.; Kubicek, C.P.; Baker, S.E. & Margeot, A. (2009). Tracking the roots of cellulase hyperproduction by the fungus *Trichoderma reesei* using massively parallel DNA sequencing. *Proceedings of the National Academy of Science U S A*, Vol.106, No. 38, pp. 16151-16156, ISSN 1091-6490

Li, X.L.; Spanikova, S.; de Vries, R.P. & Biely, P. (2007). Identification of genes encoding microbial glucuronoyl esterases. *FEBS Letters*, Vol.581, No. 21, pp. 4029-4035, ISSN 0014-5793

Liu, Y.S.; Baker, J.O.; Zeng, Y.; Himmel, M.E.; Haas, T. & Ding, S.Y. (2011). Cellobiohydrolase hydrolyzes crystalline cellulose on hydrophobic faces. *The Journal of Biological Chemistry*, No., ISSN 1083-351X

Mach-Aigner, A.R.; Pucher, M.E.; Steiger, M.G.; Bauer, G.E.; Preis, S.J. & Mach, R.L. (2008). Transcriptional regulation of *xyr1*, encoding the main regulator of the xylanolytic and cellulolytic enzyme system in *Hypocrea jecorina*. *Applied and Environmental Microbiology*, Vol.74, No. 21, pp. 6554-6562, ISSN 1098-5336

Mach, R.L.; Strauss, J.; Zeilinger, S.; Schindler, M. & Kubicek, C.P. (1996). Carbon catabolite repression of xylanase I (*xyn1*) gene expression in *Trichoderma reesei*. *Molecular Microbiology*, Vol.21, No. 6, pp. 1273-1281., ISSN 0950-382X

Mantyla, A.L.; Rossi, K.H.; Vanhanen, S.A.; Penttilä, M.E.; Suominen, P.L. & Nevalainen, K.M. (1992). Electrophoretic karyotyping of wild-type and mutant *Trichoderma longibrachiatum* (*reesei*) strains. *Current Genetics*, Vol.21, No. 6, pp. 471-477, ISSN 0172-8083

Martens-Uzunova, E.S. & Schaap, P.J. (2009). Assessment of the pectin degrading enzyme network of *Aspergillus niger* by functional genomics. *Fungal Genetics and Biology*, Vol.46 Suppl 1, No., pp. S170-S179, ISSN 1096-0937

Martinez, D.; Berka, R.M.; Henrissat, B.; Saloheimo, M.; Arvas, M.; Baker, S.E.; Chapman, J.; Chertkov, O.; Coutinho, P.M.; Cullen, D.; Danchin, E.G.; Grigoriev, I.V.; Harris, P.; Jackson, M.; Kubicek, C.P.; Han, C.S.; Ho, I.; Larrondo, L.F.; de Leon, A.L.; Magnuson, J.K.; Merino, S.; Misra, M.; Nelson, B.; Putnam, N.; Robbertse, B.; Salamov, A.A.; Schmoll, M.; Terry, A.; Thayer, N.; Westerholm-Parvinen, A.; Schoch, C.L.; Yao, J.; Barabote, R.; Nelson, M.A.; Detter, C.; Bruce, D.; Kuske, C.R.; Xie, G.; Richardson, P.; Rokhsar, D.S.; Lucas, S.M.; Rubin, E.M.; Dunn-Coleman, N.; Ward, M. & Brettin, T.S. (2008). Genome sequencing and analysis of the biomass-degrading fungus *Trichoderma reesei* (syn. *Hypocrea jecorina*). *Nature Biotechnology*, Vol.26, No. 5, pp. 553-560, ISSN 1546-1696

Merino, S.T. & Cherry, J. (2007). Progress and challenges in enzyme development for biomass utilization. *Advances in Biochemistry andEnginering/Biotechnology*, Vol.108, No., pp. 95-120, ISSN 0724-6145

Montenecourt, B.S. & Eveleigh, D.E. (1977). Preparation of mutants of *Trichoderma reesei* with enhanced cellulase production. *Applied and Environmental Microbiology*, Vol.34, No. 6, pp. 777-782, ISSN 0099-2240

Montenecourt, B.S. & Eveleigh, D.E. (1979). Selective Screening Methods for the Isolation of High Yielding Cellulase Mutants of *Trichoderma reesei*, In, *Hydrolysis of Cellulose: Mechanisms of Enzymatic and Acid Catalysis*, Advances in Chemistry. American Chemical Society, ISBN pp. 289-301

Nakari-Setälä, T.; Paloheimo, M.; Kallio, J.; Vehmaanperä, J.; Penttilä, M. & Saloheimo, M. (2009). Genetic modification of carbon catabolite repression in *Trichoderma reesei* for improved protein production. *Applied and Environmental Microbiology*, Vol.75, No. 14, pp. 4853-4860, ISSN 1098-5336

Nakari-Setälä, T. & Penttilä, M. (1995). Production of *Trichoderma reesei* cellulases on glucose-containing media. *Applied and Environmental Microbiology*, Vol.61, No. 10, pp. 3650-3655, ISSN 0099-2240

Nummi, M.; Niku-Paavola, M.L.; Lappalainen, A.; Enari, T.M. & Raunio, V. (1983). Cellobiohydrolase from *Trichoderma reesei*. *Biochemical Journal*, Vol.215, No. 3, pp. 677-683, ISSN 0264-6021

Pakula, T.M.; Laxell, M.; Huuskonen, A.; Uusitalo, J.; Saloheimo, M. & Penttilä, M. (2003). The effects of drugs inhibiting protein secretion in the filamentous fungus *Trichoderma reesei*. Evidence for down-regulation of genes that encode secreted proteins in the stressed cells. *The Journal of Biological Chemistry*, Vol.278, No. 45, pp. 45011-45020, ISSN 0021-9258

Palmer, J.M. & Keller, N.P. (2010). Secondary metabolism in fungi: does chromosomal location matter? *Current Opinion in Microbiology*, Vol.13, No. 4, pp. 431-436, ISSN 1879-0364

Portnoy, T.; Margeot, A.; Seidl-Seiboth, V.; Le Crom, S.; Ben Chaabane, F.; Linke, R.; Seiboth, B. & Kubicek, C.P. (2011). Differential regulation of the cellulase transcription factors XYR1, ACE2, and ACE1 in *Trichoderma reesei* strains producing high and low levels of cellulase. *Eukaryotic Cell*, Vol.10, No. 2, pp. 262-271, ISSN 1535-9786

Reese, E.T. (1976). History of the cellulase program at the U.S. Natick Development Center. *Biotechnology Bioengineering Symposium*, Vol.6, No., pp. 9-20

Rouvinen, J.; Bergfors, T.; Teeri, T.; Knowles, J.K. & Jones, T.A. (1990). Three-dimensional structure of cellobiohydrolase II from *Trichoderma reesei*. *Science*, Vol.249, No. 4967, pp. 380-386, ISSN 1095-9203

Saloheimo, M.; Paloheimo, M.; Hakola, S.; Pere, J.; Swanson, B.; Nyyssonen, E.; Bhatia, A.; Ward, M. & Penttilä, M. (2002). Swollenin, a *Trichoderma reesei* protein with sequence similarity to the plant expansins, exhibits disruption activity on cellulosic materials. *European Journal of Biochemistry*, Vol.269, No. 17, pp. 4202-4211, ISSN 0014-2956

Schülein, M. (2000). Protein engineering of cellulases. *Biochimica et Biophysica Acta*, Vol.1543, No. 2, pp. 239-252, ISSN 0006-3002

Seiboth, B.; Hakola, S.; Mach, R.L.; Suominen, P.L. & Kubicek, C.P. (1997). Role of four major cellulases in triggering of cellulase gene expression by cellulose in *Trichoderma reesei*. *Journal of Bacteriology*, Vol.179, No. 17, pp. 5318-5320.,

Seiboth, B.; Messner, R.; Gruber, F. & Kubicek, C.P. (1992). Disruption of the *Trichoderma reesei cbh2* gene coding for cellobiohydrolase II leads to a delay in the triggering of cellulase formation by cellulose *Journal of General Microbiology*, Vol.138, No., pp. 1259-1264, ISSN 0022-1287

Seiboth, B.; Pakdaman, B.S.; Hartl, L. & Kubicek, C.P. (2007). Lactose metabolism in filamentous fungi: how to deal with an unknown substrate. *Fungal Biology Reviews*, Vol.21, No., pp. 42-48, ISSN 1749-4613

Seidl, V.; Gamauf, C.; Druzhinina, I.S.; Seiboth, B.; Hartl, L. & Kubicek, C.P. (2008). The *Hypocrea jecorina* (*Trichoderma reesei*) hypercellulolytic mutant RUT C30 lacks a 85 kb (29 gene-encoding) region of the wild-type genome. *BMC Genomics*, Vol.9, No., pp. 327, ISSN 1471-2164

Seidl, V.; Seibel, C.; Kubicek, C.P. & Schmoll, M. (2009). Sexual development in the industrial workhorse *Trichoderma reesei*. *Proceedings of the National Academy of Science U S A*, Vol.106, No. 33, pp. 13909-13914, ISSN 1091-6490

Stals, I.; Sandra, K.; Devreese, B.; Van Beeumen, J. & Claeyssens, M. (2004a). Factors influencing glycosylation of *Trichoderma reesei* cellulases. II: N glycosylation of Cel7A core protein isolated from different strains. *Glycobiology*, Vol.14, No. 8, pp. 725-737, ISSN 0959-6658

Stals, I.; Sandra, K.; Geysens, S.; Contreras, R.; Van Beeumen, J. & Claeyssens, M. (2004b). Factors influencing glycosylation of *Trichoderma reesei* cellulases. I: Postsecretorial changes of the O- and N-glycosylation pattern of Cel7A. *Glycobiology*, Vol.14, No. 8, pp. 713-724, ISSN 0959-6658

Tomme, P.; Warren, R.A. & Gilkes, N.R. (1995). Cellulose hydrolysis by bacteria and fungi. *Advances in Microbial Physiology*, Vol.37, No., pp. 1-81, ISSN 0065-2911

Vaaje-Kolstad, G.; Vasella, A.; Peter, M.G.; Netter, C.; Houston, D.R.; Westereng, B.; Synstad, B.; Eijsink, V.G. & van Aalten, D.M. (2004). Interactions of a family 18 chitinase with the designed inhibitor HM508 and its degradation product, chitobiono-delta-lactone. *The Journal of Biological Chemistry*, Vol.279, No. 5, pp. 3612-3619, ISSN 0021-9258

Vaaje-Kolstad, G.; Westereng, B.; Horn, S.J.; Liu, Z.; Zhai, H.; Sorlie, M. & Eijsink, V.G. (2010). An oxidative enzyme boosting the enzymatic conversion of recalcitrant polysaccharides. *Science*, Vol.330, No. 6001, pp. 219-222, ISSN 1095-9203

Vaheri, M.; Leisola, M. & Kaupinnen, V. (1979). Transglycosylation products of the cellulase system of *Trichoderma reesei*. *Biotechnology Letters*, Vol.1, No., pp. 41-46, ISSN 0141-5492.

Verbeke, J.; Coutinho, P.; Mathis, H.; Quenot, A.; Record, E.; Asther, M. & Heiss-Blanquet, S. (2009). Transcriptional profiling of cellulase and expansin-related genes in a hypercellulolytic Trichoderma reesei. *Biotechnology Letters*, Vol.31, No. 9, pp. 1399-1405, ISSN 1573-6776

Vitikainen, M.; Arvas, M.; Pakula, T.; Oja, M.; Penttilä, M. & Saloheimo, M. (2010). Array comparative genomic hybridization analysis of *Trichoderma reesei* strains with enhanced cellulase production properties. *BMC Genomics*, Vol.11, No., pp. 441, ISSN 1471-2164

Wilson, D.B. (2009). Cellulases and biofuels. *Current Opinion in Biotechnology*, Vol.20, No. 3, pp. 295-299, ISSN 1879-0429

Zhang, Y.H. & Lynd, L.R. (2006). A functionally based model for hydrolysis of cellulose by fungal cellulase. *Biotechnology and Bioengineering*, Vol.94, No. 5, pp. 888-898, ISSN 0006-3592

# Part 4

## Biofuels Generalities

# 14

# Application of Response Surface Methodology to Optimize Alkali Concentration, Corn Stover Particle Size, and Extruder Parameters for Maximum Sugar Recovery

Chinnadurai Karunanithy and Kasiviswanathan Muthukumarappan
*Agricultural and Biosystems Engineering, South Dakota State University*
*South Dakota*
*USA*

## 1. Introduction

The National Research Council (2000) has set a goal for the biobased industry of providing at least 10% of liquid fuels by the year 2020 and providing 50% of liquid fuels by the year 2050. The 2007 Energy Act mandates the production of 21 billion gallons of biofuels from non-corn starch materials by 2022. Brazil and the US produce about 60% of the world's ethanol, exploiting sugarcane and corn, respectively. Economics and limitation in grain supply lead to search for alternative resources. Lignocellulosic materials are the most abundant renewable resources on earth (Lynd et al., 2005) and cheaper than corn. Among the crop residues, corn and wheat are the most abundant in the US, roughly 96% of the total biomass (Little, 2001). Corn stover is being considered as one of the main renewable feedstocks for conversion into fuels and chemicals. According to Kadam and McMillon (2003), about 80-100 dry tons of corn stover/year can be utilized for ethanol production. It has been estimated that approximately 256 million dry tons of corn stover will be available in the year 2030 due to collection technologies improvement and a steady yield increase (Perlack et al., 2005). Moreover, corn stover is projected as the feedstock in two of the six commercial-scale lignocellulosic bio-refineries supported by the US Department of Energy (Service, 2007).

Ethanol production from biomass is quite different from the process used for corn grain, because the carbohydrates in biomass are more difficult for hydrolytic enzymes to access than the starch in grain (Gibbons et al., 1986). Unlike corn grain, biomass is composed of 40–50% cellulose, 25–35% hemicellulose, and 15–20% lignin (Saha & Bothast, 1997). Because of the complex structure of biomass and its recalcitrant nature, an additional step called pretreatment is required for ethanol production from biomass in addition to the steps involved in corn ethanol production. The purposes of pretreatment are to open up the biomass structure, to increase accessible surface area, to reduce the cellulose crystallinity, and to increase the porosity, pore size, and pore volume. Extensive pretreatment effort has been made using several methods on different biomasses with varying degrees of success. Acid, alkali, hydrothermal (steam, steam explosion, hot water, pH controlled hot water), and ammonia fiber expansion (AFEX) are a few well recognized pretreatment methods.

Despite biomass pretreatment research of more than three decades, no perfect conversion technology has been established for biofuels production from biomass on a commercial scale (de Leon & Coors, 2008).

Extrusion is a well known technology in the processed food, feed, and plastic industries. An extruder has the ability to provide high shear, rapid heat transfer, effective and rapid mixing in a short residence time, as well as adaptability to many different processes - all in a continuous process. A few extrusion pretreatments (Dale et al., 1999; de Virje et al., 2002; Karunanithy et al., 2008; Karunanithy & Muthukumarappan, 2010a, 2010b, 2010c, 2011a, 2011b, 2011c; Lee et al., 2009; Muthukumarappan & Julson, 2007 ) showed a significant improvement on sugar recovery from corn stover, switchgrass, miscanthus, prairie cord grass, big bluestem, and Douglas fir through enzymatic hydrolysis. Potential fermentation inhibitors such as furfural and HMF were not reported in any of the above studies. Karunanithy and Muthukumarappan (2011a) achieved 85.7, 87.5, and 86.3% of glucose, xylose, and combined sugar recovery, respectively for the optimized pretreatment condition of 180°C barrel temperature, 155 rpm screw speed, 20% wb moisture content, and corn stover particle size 8 mm. The literatures report a sugar recovery of more than 90% or near quantitative using dilute acid (Lloyd & Wyman, 2005; Yan et al., 2009; Zhu et al., 2004, 2005), lime (Kim & Holtzapple, 2005 ), compressed hot water (Liu & Wyman, 2005), steam pretreatment (Bura et al., 2009), steam explosion (Elander et al., 2009; Tucker et al., 2003), a combination of acid and alkali (Varga et al., 2002), ammonia recycle process (Kim et al., 2003) and AFEX (Chundawat et al., 2007). These results show that there still is room to improve sugar recovery from corn stover when pretreated in extrusion in combination with other pretreatment methods.

In general, alkali pretreatment results in less degradation of the sugar compared to acid pretreatments. Considering the construction material of the extruder, addition of acid would lead to corrosion problem; therefore, extruder screws and barrel should be fabricated using acid-resistant stainless steel alloy such as AL6XN (Miller & Hester, 2007). Alkali pretreatment can be as simple as soaking the biomass in NaOH at room temperature or as complicated as treating the biomass in AFEX. Among sodium, calcium, potassium, and ammonium hydroxides, sodium hydroxide is the most studied alkali in biomass pretreatment (Elshafei et al., 1991; MacDonald et al., 1983) and effective also (Keshwani, 2009). MacDonald et al (1983) obtained an overall yield of 77.5% from dilute NaOH pretreatment at a high temperature, whereas Elshafei et al (1991) achieved a theoretical maximum yield of cellulose when corn stover was soaked in 1.0 M NaOH for 24 h at room temperature. Recently, Gupta (2008) reported about 94% glucose digestibility when corn stover was pretreated with 1.5% NaOH at 60°C for 24 h.

Barrel temperature and screw speed are important extruder parameters, which can affect sugar recovery. Biomass size reduction has become an integral part of biomass pretreatment. Studies have shown that the particle size influences the diffusion kinetics (Kim & Lee, 2002), the effectiveness of pretreatment (Chundawat et al., 2007), the enzymatic hydrolysis rate, the rheological properties (Chundawat et al., 2007; Desari & Bersin, 2007), lignin removal (Hu et al., 2008), the sugar yield (Chang et al., 2001; Hu et al., 2008; Yang et al., 2008), acetic acid formation (Guo et al., 2008), and the power requirement for size reduction (Cadoche & Lopez, 1989; Mani et al., 2004; van Walsum et al., 1996). It is a well known that alkali acts as delignification agent at low concentration without degrading the carbohydrates. Hence, the extruder barrel temperature, screw speed, corn stover particle size, and alkali (NaOH) concentration are the independent variables selected for this study.

Optimization of pretreatment conditions is one of the most important stages in the development of an efficient and economic pretreatment method. The traditional one-factor-at-a-time approach is time consuming; moreover, the interactions between independent variables are not considered. Response surface methodology (RSM) is an effective optimization tool wherein many factors and their interactions affecting the response can be identified with fewer experimental trials than one-factor-at-a-time experiment. RSM has been widely used in various fields ranging from food process operations including extrusion (Altan et al., 2008; Jorge et al., 2006), food product development, media composition in biotechnology to bioprocessing such as enzymatic hydrolysis and fermentation. Recently, RSM has been successfully applied to biomass pretreatment by many researchers (Canettieri et al., 2007; Kim & Mazza, 2008; Lu et al., 2007; Neureiter et al., 2002; Rahman et al., 2007; Xin & Saka, 2008). Earlier extrusion pretreatment studies conducted by the authors yielded encouraging results and however, the extrusion factors including alkali concentration were not optimized. The following are the objectives of the present study: 1) to understand and optimize the effect of extruder parameters such as barrel temperature and screw speed, biomass particle size, and alkali (NaOH) concentration for maximum sugar recovery using RSM and adopting a central composite rotatable design (CCRD), and 2) to propose a mathematical model to predict glucose, xylose, and combined sugar recovery from corn stover.

## 2. Materials and methods

### 2.1 Experimental design

A central composite rotatable design (CCRD) with four independent variables was used to study the response pattern and to determine the optimum combination of temperature, screw speed, alkali concentration, and particle size for maximizing the sugar recovery from corn stover. The CCRD combines the vertices of the hypercube whose coordinates are given by a $2^n$ factorial design with star points. The star points provide the estimation of curvature of the nonlinear response surface. The experimental design was developed using Design Expert 7.1.6 (2002), which resulted in 30 runs, in addition 6 more center points were added to allow for the estimation of the pure error sum squares. The 36 experiments (16 factorial, 8 star, and 12 center points) were randomized to maximize the effects of unexplained variability in the observed responses due to extraneous factors. Independent variable levels were selected based on a previous and one-factor-at-a-time experiment. The independent variables were coded according to the following equation

$$x_i = (X_i - X_0) / \Delta X_i \qquad (1)$$

where $x_i$ and $X_i$ are the dimensionless and actual values of the independent variable i, $X_0$ is the actual value of the independent variable at the center point, and $\Delta X_i$ is the step change of $X_i$ corresponding to a unit variation of the dimensionless value. The variables optimized included barrel temperature (45 to 225°C), screw speed (20 to 200 rpm), alkali concentration (0.5 to 2.5%), and particle size (2 to10 mm) each at five levels. -2, -1, 0, 1, and 2 and shown in Table 2.

### 2.2 Biomass preparation

Corn stover obtained from a local farm was ground in a hammer mill (Speedy King, Winona Attrition Mill Co, MN) using 2, 4, 6, 8, and 10 mm sieves to understand the influence of

particle size on sugar recovery. Compositional analysis of corn stover such as glucose, xylose, mannose, arabinose, lignin, and ash was done following Sluiter et al (2008a, 2008b) and reported in the mass balance diagram.
Alkali (NaOH) solutions of different concentrations (0.5, 1.0, 1.5, 2.0, and 2.5 % w/v) were prepared. It was found that corn stover required nine times of solution than that of corn stover weight for complete soaking, which is equivalent to 10% solids loading rate. The different particle size of corn stover was soaked in different alkali concentrations as given in the experimental design for 30 min at room temperature. The black liquid was drained out using cheese cloth followed by manual squeezing to remove excess moisture. Moisture content of the biomass samples was determined as described by Sluiter et al (2008c). The moisture content of alkali soaked samples was in the range of 75-78% wb.

### 2.3 Extrusion pretreatment

Extrusion was performed using a single screw extruder (Brabender Plasti-corder Extruder Model PL 2000, Hackensack, NJ), which had a barrel length to screw diameter ratio (l/d) of 20:1. In order to have a smooth biomass (plug) flow into the die section, the screw discharge end was fitted with conical metal piece. A screw with 3:1compression ratio was selected based on a previous study (Karunanithy & Muthukumarappan, 2010a). The single screw extruder was fitted to a 7.5 hp motor, which had a provision to adjust the screw speed from 0 to 210 rpm. The extruder barrel had provisions to control the temperature of the feed and transition zone in both barrel and die section. The extruder barrel temperature and the screw speed were controlled by a computer connected to the extruder. Extruder feeding was done manually. Compressed air was supplied as a cooling agent along the barrel length. Once the barrel temperature stabilized, about 500 g of biomass was extruded under each pretreatment condition, divided into two batches accounting for variations due to extruder operation, and considered replicates. The mean residence time varied between 30 and 90 sec depending upon the screw speed.

### 2.4 Enzymatic hydrolysis

Enzymatic hydrolysis of pretreated samples (0.3 g in 10 mL hydrolysis volume) was carried out using 0.1M, pH 4.8 sodium citrate buffer for 72 h at 50°C and 150 rpm as described by Selig et al (2008). Based on the literature survey and earlier study, the amount of cellulase (Celluclast 1.5L, activity 70 FPU/g) enzyme was decided to be 15 FPU/g of dry matter. The ratio of cellulase to β- glucosidase (Novo-188, activity 250 CBU/g) was maintained at 1:4 based on earlier study (Karunanithy & Muthukumarappan, 2010b). All these enzymes were provided by Novozyme (Krogshoejvej, Denmark). After hydrolysis, the samples were kept in boiling water for 10 min to inactivate the enzyme action. The supernatant was centrifuged with 16060 g force (13000 rpm for 15 min) and then frozen twice before injecting into the HPLC to remove the impurities which contribute to the pressure increase in the HPLC system. Soluble sugar was quantified using HPLC (Agilent Technologies, Santa Clara, CA; Bio-Rad Aminex 87H column, Hercules, CA) with a mobile phase of 0.005M $H_2SO_4$ at a flow rate of 0.6 mL/min at 65°C and a sample volume of 20 µL as mentioned by Sluiter et al (2008d). The sugar concentration obtained from chromatogram was divided by dry weight of biomass taken for enzymatic hydrolysis in order to know the percentage of different sugars with respect to total biomass. Glucose and xylose are the major sugars present in the biomass as compared to arabinose. Instead of reporting arabinose separately, it was added with glucose and xylose and reported as combined sugar. The sugar recovery reported in this chapter was after enzymatic hydrolysis of the pretreated samples.

$$Y_i,\% = \frac{S_{ip}}{S_{ir}} * 100 \quad (2)$$

$$Y_c,\% = \frac{\sum S_{ip}}{\sum S_{ir}} * 100 \quad (3)$$

Where
$Y_i$- individual sugar recovery, %
$Y_c$- combined sugar recovery, %
$S_{ip}$- individual sugar obtained from hydrolyzate of pretreated samples through HPLC
$S_{ir}$ – individual sugar from raw material

### 2.5 Statistical analysis

The second order polynomial equation was used to describe the effect of independent variables in terms of linear, quadratic, and interactions. The proposed model for the response ($Y_i$) was:

$$Y_i = b_0 + \sum_{i=1}^{4} b_i X_i + \sum_{i=1}^{4} b_{ii} X_i^2 + \sum_{i=1}^{3} \sum_{j=i+1}^{4} b_{ij} X_i X_j + \varepsilon \quad (4)$$

where $Y_i$ is the predicted response; $b_0$ is the interception coefficient; $b_i$, $b_{ii}$ ,and $b_{ij}$ are linear, quadratic, and interaction terms; ε is the random error, and $X_i$ is the independent variables studied. Design Expert 7.1.6 software was used for regression and graphical analysis of the data obtained. Statistical analysis of the model was performed to evaluate the analysis of variance.

The quality of fit of second order equation was expressed by the coefficient of determination $R^2$ and its statistical significance ($\alpha = 0.05$) was determined by the F test. The individual effect of each variable and also the effects of the interaction were determined. Optimization (maximizing sugar recovery) of the fitted polynomial was determined using numerical optimization contained in the Design Expert 7.1.6. After optimizing the pretreatment conditions using RSM, validation was done by extruding corn stover at two different optimum conditions of barrel temperature, screw speed, alkali concentration, and particle size from the numerical solution, depending on the particle size, due to availability of the standard sieve size.

## 3. Results and discussion

### 3.1 Solid loss and washing of the pretreated corn stover

Many researchers (Dawson & Boopathy, 2007; Kaar & Holtzapple, 2000; Titgemeyer et al., 1996) have reported washing of the biomass with different medium after alkali pretreatment; however, it may not be necessary if the alkali concentration is low enough. According to Novozymes biomass kit, most of the enzymes have optimum activity between a pH of 4.5 to 6.5 at 45-70°C. When different biomasses soaked with different alkali concentration, the sample prepared for enzymatic hydrolysis (after adding citrate buffer, DI water, and enzymes) had a pH of 4.8 to 5.4, which is well within the range of Novozyme's recommendations. Depending upon the alkali concentration and particle size, a sugar recovery of 35-45% was recorded. However, the washed samples had lower sugar recovery (due to loss of sugar about 5-7%) than that of unwashed samples.

Considering the solid loss (15%) and delignification (40%) during alkali soaking, the sugar recovery was calculated and shown in the mass balance diagram. The literature values of solid loss and delignification varied from 10.0-67.3 and 27.7-96.0%, respectively, depending upon the alkali usage and the pretreatment conditions employed on corn stover as listed in Table 1. The reason for lower solid loss and delignification is due to room temperature when compared to most of methods mentioned in the table.

### 3.2 Effect of independent variables on sugar recoveries

The experimental glucose, xylose, and combined sugar recoveries from different treatment combinations are presented in Table 2. The proposed quadratic models in terms of actual variables are given below for glucose ($Y_G$), xylose ($Y_X$), and combined sugar ($Y_C$) recovery, where $X_1$, $X_2$, $X_3$, and $X_4$ represent barrel temperature (°C), screw speed (rpm), alkali concentration (% w/v), and particle size (mm) of corn stover, respectively. Similar equations were reported for acid hydrolysis of sugarcane bagasse (Neureiter et al., 2002), oil palm empty fruit bunch (Rahman et al., 2007), eucalyptus (Canettieri et al., 2007), for concentrated acid pretreatment of pine wood (Miller & Hester, 2007 ), acid catalyzed fractionation and enzymatic hydrolysis of flax shives (Kim & Mazza, 2008), hot-compressed water pretreatment of Japanese beech hydrolysis (Xin & Saka, 2008), and extrusion pretreatment of corn stover (Karunanithy & Muthukumarappan, 2011a). Those equations predict the responses well with high $R^2$ and low probability values.

$$\begin{aligned} Y_G = &-206.7 + 0.9664X_1 + 0.4352X_2 + 192.8X_3 + 17.76X_4 + 0.0024X_1X_2 - 0.0172X_1X_3 \\ &+0.0178X_1X_4 - 0.3881X_2X_3 - 0.0190X_2X_4 + 3.8485X_3X_4 - 0.0045X_1^2 - 0.0009X_2^2 \\ &-50.503X_3^2 - 1.7979X_4^2 \end{aligned} \quad (5)$$

$$\begin{aligned} Y_X = &-141.13 + 0.6135X_1 + 0.4986X_2 + 168.87X_3 + 8.895X_4 + 0.0014X_1X_2 - 0.0803X_1X_3 \\ &+0.0245X_1X_4 - 0.1693X_2X_3 + 1.6732X_3X_4 - 0.0029X_1^2 - 0.0021X_2^2 - 46.837X_3^2 - 1.094X_4^2 \end{aligned} \quad (6)$$

$$\begin{aligned} Y_C = &-190.54 + 0.8375X_1 + 0.5593X_2 + 186.72X_3 + 15.47X_4 + 0.0021X_1X_2 - 0.0962X_1X_3 \\ &+0.0228X_1X_4 - 0.3206X_2X_3 - 0.0239X_2X_4 + 3.041X_3X_4 - 0.0039X_1^2 - 0.0015X_2^2 \\ &-48.92X_3^2 - 1.5658X_4^2 \end{aligned} \quad (7)$$

The regression coefficient, standard error, F, and p values are shown in Table 3. All the independent variables had a significant influence on sugar recoveries, as evident from their p values in Table 3. All the independent variables had a positive influence on all the sugar recoveries, as evident from the proposed model equations. The magnitude of the terms indicates the order of influence on sugar recoveries i.e., alkali concentration, particle size, barrel temperature, and screw speed. Not only the linear terms of independent variables but also their quadratic terms contributed to glucose, xylose, and combined sugar recovery, as evident from equation 5 and 7. Again, the difference in magnitude of the quadratic terms explains which variable was dominant for sugar recoveries.

| Pretreatment | Alkali used | Pretreatment conditions | Biomass size, mm | Solid loss, % | Delignification, % | Glucose, % | Xylose, % | Reference |
|---|---|---|---|---|---|---|---|---|
| Alkali | 1-10% NaOH | 1h autoclaving | ~ 3.0 | 56.2-67.3 | 91.0-95.9 | 79.4 | | Varga et al., 2002 |
| Alkali | 10% NaOH | 2h autoclaving | 2.0 | NR | NR | 0.36g/g DM | | Crofcheck & Montross, 2004 |
| Alkali | 1-5% NaOH | 60 C, 24 h | 0.5-2.2 | 26.2-47.6 | 64.4-81.4 | 82-99.8+ | 59.5-71.2+ | Gupta, 2008 |
| Alkali | 1% NaOH | 25 C, 24 h | 0.5-2.2 | 16 | 41.8 | 65.3+ | 49.3+ | Gupta, 2008 |
| Alkali | 1% NaOH+5%$H_2O_2$ | 25 C, 24 h | | 10 | 27.7 | 49.1+ | 31.8+ | Gupta, 2008 |
| Alkali (NaOH) | 2% | | < 2.0 | 41.4 | 73.9 | 81.2 | | Chen et al., 2009 |
| Bayer process sand pretreatment | 0.053-1.007 g NaOH equiv/g | ; 20- | <1.0 | 14.5-27.0 | 4-66.7 | 93.2 | 94.5 | Rodgers et al., 2009 |
| Lime | 0.4 g $Ca(OH)_2$/g | 0.093 g/g 0- and 0.093 g/g, $_2$O/g | <2.0 | 32.3 | 34.8 | 50 | | Chen et al., 2009 |
| Lime | 0.5 g $Ca(OH)_2$/g | weeks, aeration | 6.0 | NR | 43.6-47.7 | 91.3 | 51.8 | Kim & Holtzapple, 2005 |
| Lime | 0.075 g $Ca(OH)_2$/g | 120 C, 4 h, 5 g $H_2O$/g | <0.18->0.84 | 33.0 | 39.5 | 88.0 | 87.7 | Teramoto et al., 2009 |
| Wet oxidation | 2 g/L $Na_2CO_3$ | bar $O_2$ | ~ 3.0 | 24.6-51.4 | 49.4-59.7 | 71.0-85.0 | | Kim & Lee, 2005a |
| ARP | 15 wt% $NH_3$ | 170 C, 1.5 h, 5 ml/min, 2.3 MPa | 0.5-2.2 | 38.4-46.4 | 70.0-85.0 | 92.5-99 | | Kim et al., 2003 |
| ARP | 15 wt% $NH_3$ | 170 C, 1.5 h, 5 ml/min, 2.3/2.5 MPa | 0.5-2.2 | 38.4-46.4<br>47.0-57.0 | 67.9-84.7<br>75.0-81.0 | 71.7-93.4 | | Kim & Lee, 2005a, 2006 |
| ARP | 1 solid:3.3 liquid 15 wt% $NH_3$ | 170 C, 10 min, | 0.5-2.2 | | 70.6 | 98.6 | 48.1 | Gupta et al., 2007 |
| Low liquid ammonia recycle percolation | 3.3 mL of 15 wt% $NH_3$ /g | | 0.5-2.2 | 41.0-47.0 | 59.0-70.0 | 86-95 | 71-86 | Kim et al., 2006 |
| Aqueous ammonia soaking | $NH_3$ | Room temperature, 10 days | 0.5-2.2 | NR | 55.0-74.0 | 86-92 | 72-84 | Kim & Lee, 2005b |
| Aqueous ammonia soaking | 1 solid:6 liquid 15 wt% $NH_3$, | | 0.5-2.2 | 22.6-32.7 | 50.0-77.0 | 85 | 78 | Kim & Lee, 2007 |
| Aqueous ammonia soaking | 1 solid:8liquid 15 wt % | | 0.5-2.2 | NR | 64.7 | 86.7 | 35.1 | Kim & Lee, 2007 |
| AFEX | 0.86-2 g $NH_3$/g | Twin screw extruder, 10:1 | <2.0 | NR | NR | 63-77* | | Dale et al., 1999 |
| AFEX | 1 corn stover:1 $NH_3$ | db | 6.0 | | NR | 75-90 | 50-70 | Teymouri et al., 2005 |

* ruminant digestibility + enzymatic digestibility

Table 1. Different alkali pretreatment methods employed on corn stover and their results as reported in literature

| | | | | | Glucose, % | | | Xylose, % | | | Combined sugar, % | | |
|---|---|---|---|---|---|---|---|---|---|---|---|---|---|
| Treat | Temp | Speed | Alkali | PS | Obsd | Pred | Resl | Obsd | Pred | Resl | Obsd | Pred | Resl |
| 1 | 0(135) | 0(110) | 0(1.5) | 0(6) | 91.3 | 89.0 | 2.3 | 86.5 | 87.0 | -0.5 | 92.0 | 89.2 | 2.8 |
| 2 | 1(180) | -1(65) | 1(2.0) | -1(4) | 61.6 | 61.7 | -0.1 | 64.4 | 63.1 | 1.3 | 60.4 | 59.8 | 0.6 |
| 3 | 0(135) | 0(110) | 0(1.5) | 0(6) | 86.3 | 89.0 | -2.7 | 84.1 | 87.0 | -2.9 | 88.9 | 89.2 | -0.3 |
| 4 | 1(180) | -1(65) | 1(2.0) | 1(8) | 87.0 | 85.1 | 1.9 | 75.0 | 74.8 | 0.2 | 83.0 | 81.1 | 1.9 |
| 5 | 0(135) | 0(110) | 0(1.5) | 0(6) | 90.0 | 89.0 | 1.0 | 87.2 | 87.0 | 0.2 | 87.9 | 89.2 | -1.3 |
| 6 | 0(135) | 0(110) | 0(1.5) | 0(6) | 87.8 | 89.0 | -1.2 | 85.1 | 87.0 | -1.9 | 87.4 | 89.2 | -1.8 |
| 7 | 1(180) | 1(155) | 1(2.0) | -1(4) | 43.7 | 44.3 | -0.6 | 56.8 | 56.3 | 0.5 | 48.1 | 47.9 | 0.2 |
| 8 | -1(90) | -1(65) | -1(1.0) | -1(4) | 52.8 | 51.9 | 0.9 | 57.8 | 57.9 | -0.1 | 55.3 | 54.5 | 0.8 |
| 9 | 0(135) | 0(110) | 0(1.5) | 0(6) | 88.1 | 89.0 | -0.9 | 90.0 | 87.0 | 3.0 | 90.3 | 89.2 | 1.1 |
| 10 | -1(90) | 1(155) | -1(1.0) | -1(4) | 48.9 | 50.0 | -1.1 | 57.0 | 54.8 | 2.2 | 54.3 | 54.3 | 0 |
| 11 | 0(135) | 0(110) | 0(1.5) | 2(10) | 68.3 | 69.3 | -1.0 | 70.5 | 71.7 | -1.2 | 69.7 | 71.0 | -1.3 |
| 12 | 0(135) | 2(200) | 0(1.5) | 0(6) | 67.9 | 68.2 | -0.3 | 62.5 | 63.7 | -1.2 | 65.8 | 66.6 | -0.8 |
| 13 | 1(180) | 1(155) | -1(1.0) | 1(8) | 61.8 | 61.7 | 0.1 | 66.0 | 63.6 | 2.4 | 65.4 | 63.3 | 2.1 |
| 14 | -1(90) | -1(65) | 1(2.0) | -1(4) | 77.7 | 77.0 | 0.7 | 74.8 | 74.7 | 0.1 | 76.9 | 77.1 | -0.2 |
| 15 | 0(135) | 0(110) | 0(1.5) | 0(6) | 88.0 | 89.0 | -1.0 | 90.1 | 87.0 | 3.1 | 89.3 | 89.2 | 0.1 |
| 16 | 0(135) | 0(110) | 2(2.5) | 0(6) | 49.2 | 50.7 | -1.5 | 47.0 | 49.1 | -2.1 | 48.3 | 50.3 | -2.0 |
| 17 | 0(135) | 0(110) | 0(1.5) | 0(6) | 89.2 | 89.0 | 0.2 | 85.9 | 87.0 | -1.1 | 89.4 | 89.2 | 0.2 |
| 18 | 0(135) | 0(110) | 0(1.5) | 0(6) | 87.8 | 89.0 | -1.2 | 85.5 | 87.0 | -1.5 | 86.9 | 89.2 | -2.3 |
| 19 | -2(45) | 0(110) | 0(1.5) | 0(6) | 49.6 | 51.6 | -2.0 | 59.1 | 61.6 | -2.5 | 55.6 | 57.6 | -2.0 |
| 20 | 0(135) | 0(110) | 0(1.5) | 0(6) | 90.8 | 89.0 | 1.8 | 87.8 | 87.0 | 0.8 | 91.1 | 89.2 | 1.9 |
| 21 | -1(90) | 1(155) | -1(1.0) | 1(8) | 46.6 | 44.7 | 1.9 | 47.1 | 47.6 | -0.5 | 47.1 | 46.6 | 0.5 |
| 22 | 1(180) | -1(65) | -1(1.0) | 1(8) | 49.8 | 51.0 | -1.2 | 57.0 | 58.5 | -1.5 | 53.1 | 55.0 | -1.9 |
| 23 | -1(90) | -1(65) | 1(2.0) | 1(8) | 95.4 | 94.1 | 1.3 | 80.1 | 77.5 | 2.6 | 92.1 | 90.2 | 1.9 |
| 24 | 0(135) | -2(20) | 0(1.5) | 0(6) | 92.2 | 94.4 | -2.2 | 74.7 | 76.9 | -2.2 | 85.0 | 87.2 | -2.2 |
| 25 | 0(135) | 0(110) | 0(1.5) | 0(6) | 89.7 | 89.0 | 0.7 | 87.0 | 87.0 | 0 | 89.3 | 89.2 | 0.1 |
| 26 | 0(135) | 0(110) | 0(1.5) | 0(6) | 89.7 | 89.0 | 0.7 | 86.6 | 87.0 | -0.4 | 88.2 | 89.2 | -1.0 |
| 27 | 1(180) | -1(65) | -1(1.0) | -1(4) | 45.3 | 43.0 | 2.3 | 56.3 | 53.5 | 2.8 | 48.4 | 45.8 | 2.6 |
| 28 | 0(135) | 0(110) | -2(0.5) | 0(6) | 25.2 | 26.3 | -1.1 | 30.1 | 31.2 | -1.1 | 29.4 | 30.4 | -1.0 |
| 29 | 0(135) | 0(110) | 0(1.5) | -2(2) | 49.7 | 51.2 | -1.5 | 65.2 | 67.3 | -2.1 | 55.7 | 57.4 | -1.7 |
| 30 | -1(90) | 1(155) | 1(2.0) | -1(4) | 43.2 | 40.2 | 3.0 | 58.7 | 56.4 | 2.3 | 51.0 | 48.1 | 2.9 |
| 31 | -1(90) | -1(65) | -1(1.0) | 1(8) | 54.8 | 53.5 | 1.3 | 56.1 | 54.0 | 2.1 | 57.0 | 55.4 | 1.7 |
| 32 | 1(180) | 1(155) | -1(1.0) | -1(4) | 61.0 | 60.5 | 0.5 | 60.2 | 62.0 | -1.8 | 62.0 | 62.8 | -0.8 |
| 33 | -1(90) | 1(155) | 1(2.0) | 1(8) | 48.8 | 50.3 | -1.5 | 55.5 | 55.9 | -0.4 | 51.9 | 52.6 | -0.7 |
| 34 | 2(225) | 0(110) | 0(1.5) | 0(6) | 52.6 | 53.2 | -0.6 | 65.3 | 66.1 | -0.8 | 56.0 | 57.0 | -1.0 |
| 35 | 0(135) | 0(110) | 0(1.5) | 0(6) | 89.4 | 89.0 | 0.4 | 88.3 | 87.0 | 1.3 | 90.3 | 89.2 | 1.1 |
| 36 | 1(180) | 1(155) | 1(2.0) | 1(8) | 61.7 | 60.9 | 0.8 | 65.5 | 64.7 | 0.8 | 60.8 | 60.6 | 0.2 |

Table 2. Experimental design showing both coded and actual values of variables, observed and predicted responses

| | Glucose | | | | Xylose | | | | Combined sugar | | | |
|---|---|---|---|---|---|---|---|---|---|---|---|---|
| Factor | Coefft | Std error | F value | P value | Coefft | Std error | F value | P value | Coefft | Std error | F value | P value |
| Temp | -91.63 | 26.20 | 12.22 | 0.0021 | -101.86 | 31.82 | 10.24 | 0.0043 | -123.48 | 28.03 | 19.40 | 0.0002 |
| SS | -504.27 | 26.20 | 370.35 | < 0.0001 | -220.44 | 31.82 | 47.98 | < 0.0001 | -416.36 | 28.03 | 220.59 | < 0.0001 |
| AC | -28664.90 | 741.11 | 1496.0 | < 0.0001 | -26607.8 | 900.03 | 873.97 | < 0.0001 | -27785.4 | 792.8 | 1228.0 | < 0.0001 |
| PS | 223.90 | 26.20 | 73.01 | < 0.0001 | 96.48 | 31.82 | 9.19 | 0.0063 | 176.72 | 28.03 | 39.74 | < 0.0001 |
| Temp*SS | 4.86 | 0.45 | 112.00 | < 0.0001 | 2.88 | 0.55 | 26.77 | < 0.0001 | 4.27 | 0.49 | 75.64 | < 0.0001 |
| Temp*AC | -32.29 | 9.19 | 12.33 | 0.0021 | -36.13 | 11.16 | 10.47 | 0.0040 | -43.27 | 9.83 | 19.36 | 0.0002 |
| Temp*PS | 1.60 | 0.45 | 12.21 | 0.0022 | 2.21 | 0.55 | 15.64 | 0.0007 | 2.05 | 0.49 | 17.48 | 0.0004 |
| SS*AC | -174.63 | 9.19 | 360.86 | < 0.0001 | -76.19 | 11.16 | 46.57 | < 0.0001 | -144.28 | 9.83 | 215.20 | < 0.0001 |
| SS*PS | -1.71 | 0.45 | 13.90 | 0.0012 | -0.84 | 0.55 | 2.26 | 0.1473 | -2.14 | 0.49 | 19.08 | 0.0003 |
| AC*PS | 76.96 | 9.19 | 70.09 | < 0.0001 | 33.46 | 11.16 | 8.98 | 0.0069 | 60.81 | 9.83 | 38.23 | < 0.0001 |
| $Temp^2$ | -9.15 | 0.32 | 794.07 | < 0.0001 | -5.79 | 0.39 | 215.42 | < 0.0001 | -7.99 | 0.34 | 528.65 | < 0.0001 |
| $SS^2$ | -1.92 | 0.32 | 34.97 | < 0.0001 | -4.18 | 0.39 | 112.24 | < 0.0001 | -3.08 | 0.34 | 78.64 | < 0.0001 |
| $AC^2$ | -5050.28 | 130.01 | 1508.9 | < 0.0001 | -4683.70 | 157.89 | 879.93 | < 0.0001 | -4892.07 | 139.0 | 1237.0 | < 0.0001 |
| $PS^2$ | -7.19 | 0.32 | 489.56 | < 0.0001 | -4.37 | 0.39 | 122.99 | < 0.0001 | -6.26 | 0.34 | 324.43 | < 0.0001 |

SS- screw speed AC- alkali concentration PS - particle size

Table 3. Coefficient values of the fitted model for different responses

As mentioned earlier, alkali soaked samples had moisture content in the range of 75-78% (wb). Friction is the main mode of material conveyance in a single screw extruder (Yeh & Jaw, 1998). Because water acts as lubricant in the extruder (Hayasi et al., 1992), an increase in moisture content resulted in decrease in the friction between the material, screw shaft, and barrel (Chen et al., 2009) resulted in less disturbance to cell wall of the corn stover. An increase in temperature and screw speed will introduce more energy to the material in the barrel, which would enhance the moisture evaporation at the exit (Yu et al., 2009); thereby, the disturbance to cell wall structure of the corn stover was high. An increase in corn stover particle size increased the glucose, xylose, and combined sugar recovery. Similar trends have been reported for corn stover pretreated with lime (Chang et al., 2001), hot water (Zeng et al., 2007) and wheat straw pretreated in wet oxidation ( Pedersen & Meyer, 2009). It has been reported that sugar yield was more pronounced for larger particles (0.42-1.00 mm) than smaller particles (0.05-0.15 mm) due to topological changes of biomass in lime, hot water, and wet oxidation pretreatment (Chang et al., 2001; Pedersen & Meyer, 2009; Teramoto et al., 2009). An increase of sugar yield with decrease in particle size was reported for corn stover pretreated in AFEX (Chundawat et al., 2007) and wheat straw irradiated with 500 kGy (Yang et al., 2008). Kaar and Holtzapple (2000) found that particle size (<0.8-0.84 mm) had no effect on sugar yield from lime pretreatment of corn stover; more than 95% enzymatic hydrolysis yield was reported from SFEC pretreatment at 180/200°C irrespective of eucalyptus flour size (<2 or < 5 mm) by Teramoto et al (2009). Similarly Guo et al (2008) reported that feedstock size had no impact on the performance of dilute acid pretreatment at an input size below 1x 5 cm.

Not only linear and quadratic terms but also interactions terms were contributed to sugar recoveries as evident from Table 3 and equations 5-7. In order to visualize the interaction effects for glucose recovery, significant interaction response surfaces are shown in Figure 1a-f. As noted in the figure, all the possible interactions had a significant effect on glucose recovery. An increase in screw speed showed a clear negative trend on glucose recovery at low temperature, whereas the glucose recovery was same across screw speeds, as evident from the interaction of temperature with screw speed (Fig. 1a). The glucose recovery of 92% can be achieved with the barrel temperature between 130-140°C along with 1.5-1.8% alkali concentration as seen from the interaction of the temperature and alkali concentration (Fig. 1b). The effect of temperature was prominent with a larger particle size (8 mm) than that of a smaller particle (4 mm), as seen from their interaction surface plot (Fig.1c). The glucose recovery increased with an increase in particle size regardless of barrel temperature as it was clear from their interaction. As noted from the dome shape surface plot that the maximum glucose recovery can be obtained at the barrel temperature between 130 to140°C. These results suggested that the temperature of 130-140°C would be sufficient to remove the moisture through vaporization; further increase in temperature may result in thermal softening of corn stover.

The alkali interaction with screw speed indicated that the alkali concentration of 1.5 to 1.7% would be good enough to obtain a glucose recovery of 95%. An alkali concentration up to 1.7% increased glucose recovery, a further increase in concentration of alkali resulted in decrease on glucose recovery, as evident from the interaction of alkali concentration with screw speed (Fig. 1d). The screw speed had a minimum effect on glucose recovery at a 1% alkali concentration; however, the screw speed effect was reversed at a 2% alkali concentration. The increase in screw speed showed a negative influence on glucose recovery

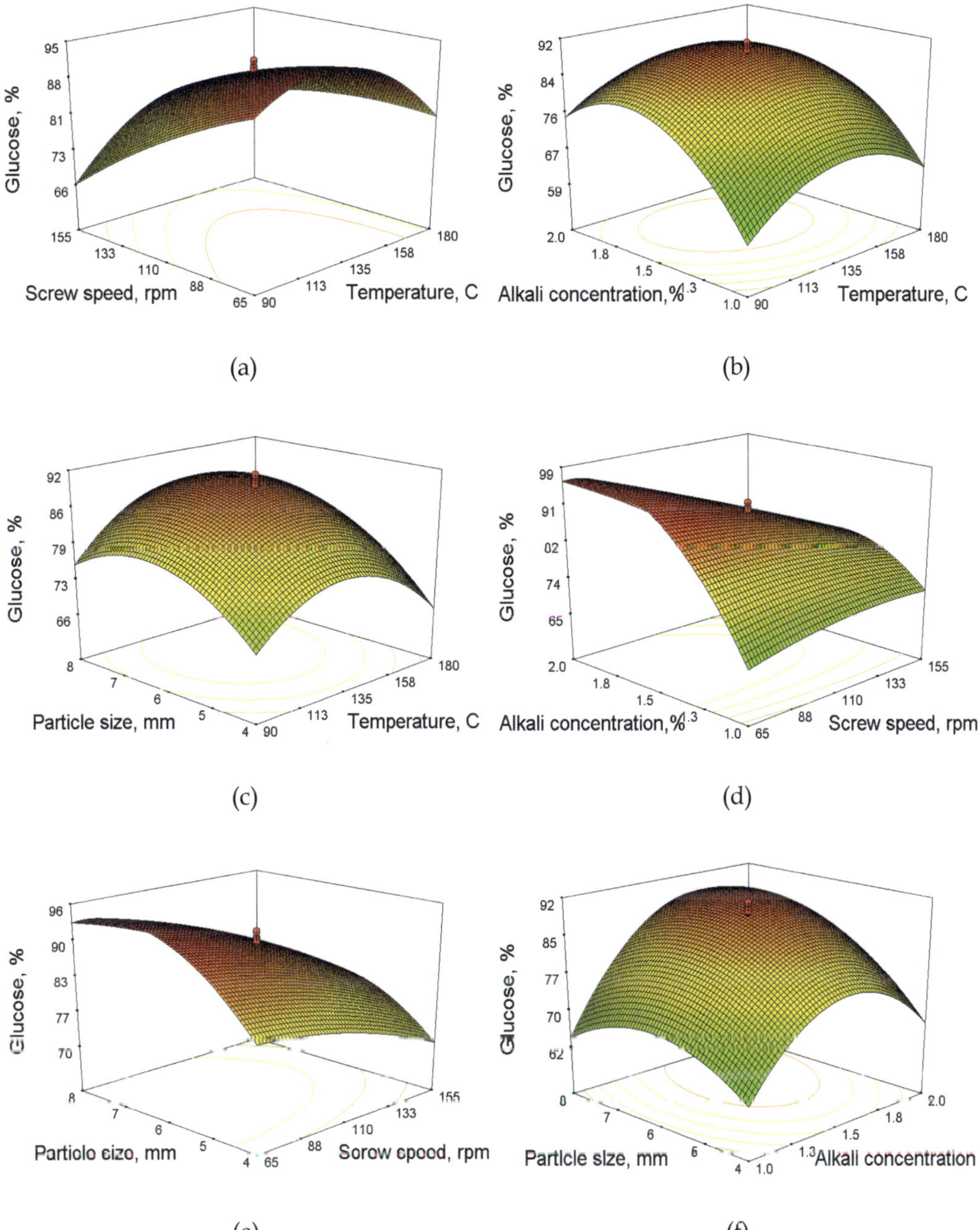


Fig. 1. Interaction effect of two independent variables on glucose recovery from corn stover (when other factors fixed at the center point: 135°C, 110, rpm, 1.5% w/w, and 6 mm )

regardless of the particle size as seen from their interaction (Fig. 1e). The particle size impact on glucose recovery was prominent at a lower screw (65 rpm) speed than that of higher screw speed, as observed in Fig. 1e. The glucose recovery increased with an increase in particle size as evident from the interaction of particle size with temperature or screw speed; however, when particle size interacting with alkali concentration, the glucose recovery increase was noticed till 7 mm only (Fig. 1f). Again, the alkali concentration of 1.5-1.7% would give the maximum of 92% glucose recovery as seen from the alkali concentration interaction with particle size. These results showed that more than 90% glucose recovery was possible at a low screw speed, alkali concentration between 1.5 to 1.7% with particle size of 6-7 mm.

Xylose response surface obtained for the significant interactions of the independent variables through model prediction are shown in Fig. 2a-e. Except screw speed interaction with particle size, all other interactions had an impact on xylose recovery from corn stover. The maximum xylose recovery was predicted at a low screw speed (65 rpm) and the barrel temperature between 130 and 140°C, as evident from the dome shape surface plot (Fig. 2a). The screw speed showed a prominent effect on xylose recovery at a lower screw speed (65 rpm) as compared to a higher screw speed (155 rpm) might be due to more residence time, as seen in Fig. 2a. As observed from the interaction of temperature with alkali concentration, an increase in alkali concentration and temperature had a positive influence on xylose recovery (Fig. 2b). However, the xylose recovery increase was significant till 1.7% alkali concentration and the barrel temperature between 130 and 140°C. The interaction of temperature and particle size indicated that the increase in particle size up to 7 mm exhibited a direct relation with xylose recovery (Fig. 2c). Again, this interaction confirmed that the barrel temperature of 130-140°C would result in a xylose recovery of 91%. Interaction of screw speed and alkali concentration for xylose recovery was similar to glucose recovery; alkali concentration had more prominent effect than that of screw speed. An increase in screw speed had a minimum effect on xylose recovery at an alkali concentration of 1%; however, its effect was clear at a 2% alkali concentration (Fig. 2d). The xylose recovery increased with an increase in alkali concentration till 1.7% and then the increase was negligible as evident from the surface plot (Fig. 2e). The effect of particle size on xylose recovery was similar across the alkali concentrations. The xylose recovery of 91% was predicted with alkali concentration of 1.7% regardless of the particle size. The barrel temperature of 130-150°C, low screw speed, 1.5 to 1.7% alkali concentration, and 6-7 mm corn stover particle size would result in a higher xylose recovery.

The predicted combined sugar response surfaces for the interactions among the independent variables are depicted in Fig. 3a-f. It can be noted that all the possible interactions had contributed for the combined sugar recovery. In general, the screw speed had a negative effect on combined sugar recovery, whereas the particle size had a positive effect, and this trend was also observed in glucose and xylose recovery. This trend might be attributed to a high mean residence time at a low screw speed and a greater resistance offered by a larger particle. The barrel temperature and alkali concentration were somewhere middle of the range results in a higher combined sugar recovery as similar to glucose and xylose recovery. Since the combined sugar is the addition of glucose, xylose, and arabinose, arabinose is being small amount and followed the same trend. The combined sugar recovery of more than 93% was possible depending upon the interaction of independent variables, which was similar to glucose and xylose recoveries.

### 3.3 Comparison of alkali soaking-extrusion results with other pretreatment methods

The maximum glucose (91.3%), xylose (86.5%), and combined sugar (92.0%) recovery was recorded for the treatment combination of 130°C, 110 rpm, 1.5% alkali concentration, and a 6 mm particle size. These pretreatment conditions differed from the maximum sugar recovery conditions (180°C, 155 rpm, 20% moisture content, and 8 mm) reported by Karunanithy and Muthukumarappan (2011b). These authors reported a glucose, xylose, and combined sugar recovery of 88, 90, and 90%, respectively, for optimum pretreated corn stover with only extrusion. The results were comparable to each other, indicating that extrusion alone is good enough to obtain about 90% sugar recovery. This might be due to loss of hemicellulose during alkali soaking; otherwise, the sugar recovery would have reached a near quantitative. Recently, in another study authors (2010a) have reported about 90% sugar recoveries for the pretreatment conditions (150°C, 150 rpm, 4 mm corn stover particle size with a 15% moisture content using 3:1 screw compression ratio). This might be due to a difference in the pretreatment conditions employed and possibly sugar loss during alkali soaking.

The present results were higher than the literature values for different biomasses. de Vrije et al (2002) reported 77% delignification, 69% glucose and 38% of xylose and arabinose conversion from a combined pretreatment of miscanthus in a twin screw extruder (100 rpm and 100°C) and alkali (NaOH 12% and 70°C). It is a fact that higher alkali concentration not only removes the lignin but also degrades the carbohydrates. Hence, the low sugar recovery reported for miscanthus might be due to degradation of carbohydrates and the inherent characteristics of biomass. The retention of carbohydrates depends on the feedstock composition as evident from the ammonia pretreatment of corn stover and poplar (Gupta et al., 2007). Recently, Lee et al (2009) extruded Douglas fir using a twin screw extruder at 50 rpm and 40°C and reported cellulose to glucose conversion of 62.4% when ethylene glycol was added as a cellulose affinity additive. The difference in glucose recovery might be due to delignification, type of extruder, pretreatment conditions (as screw speed, temperature, and particle size), and the inherent characteristics of biomasses. Jung et al (1992) achieved 1.85, 1.73, and 1.58 times higher glucose, xylose, and arabinose recovery, respectively, when maize stalk was pretreated with 1M NaOH at 39°C for 24 h followed by 72 h *in vitro* degradability compared to control sample. These authors reported a delignification of 62% for the above pretreatment condition.

A comprehensive comparison of various pretreatment methods employed on corn stover is listed in Table 4. It could be observed from the table that dilute acid (Yan et al., 2009), maleic acid (Lu & Mosier, 2008), lactic and/ acetic acid (Xu et al., 2009), controlled pH hot water (Mosier et al., 2005), steam (Bura et al., 2009), and steam explosion (Elander et al., 2009) had comparable yield of this extrusion pretreatment. The present results were higher than dilute sulfuric acid (Chen et al., 2009; Zhu et al., 2009), formic acid (Xu et al., 2009a), soaking in ethanol and aqueous ammonia (Kim et al., 2009), and steam explosion (Mosier et al., 2005); however, it was lower than dilute acid (Zhu et al., 2005), inorganic salt-$FeCl_3$ (Liu et al., 2009), cellulose solvent and organic solvent based lignocelluloses fractionation (Zhu et al., 2009). It could be noted that the particle size of corn stover used in most of the pretreatment listed in the table was lower than the extrusion pretreatment, and the higher enzyme dose was employed. Moreover, extrusion has an added advantage as a continuous process, wherein additions of chemicals are easy as demonstrated by by Dale et al (1999), de Vrije et al (2002), and Lee et al (2009).

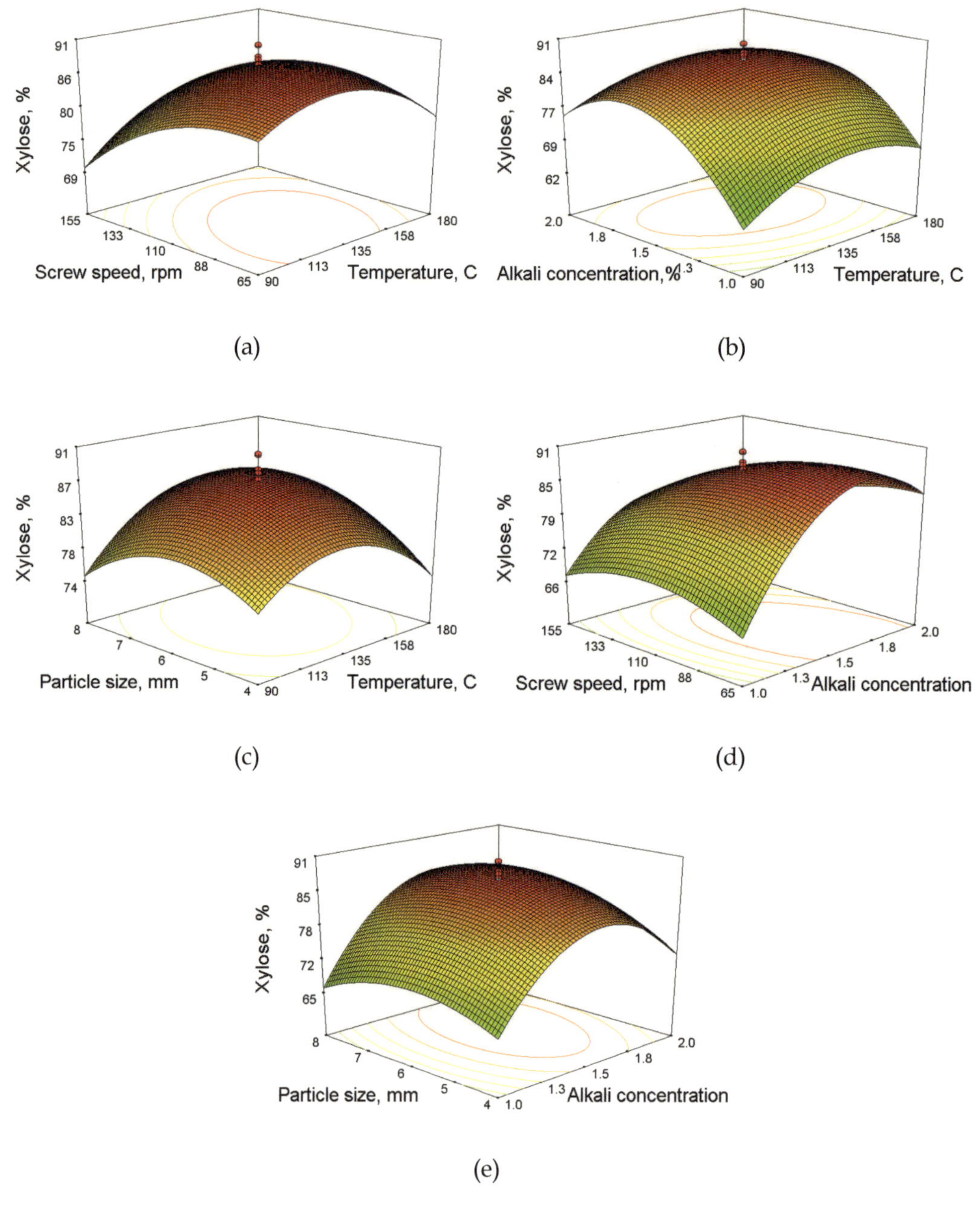


Fig. 2. Interaction effect of two independent variables on xylose recovery from corn stover (when other factors fixed at the center point: 135°C, 110, rpm, 1.5% w/w, and 6 mm)

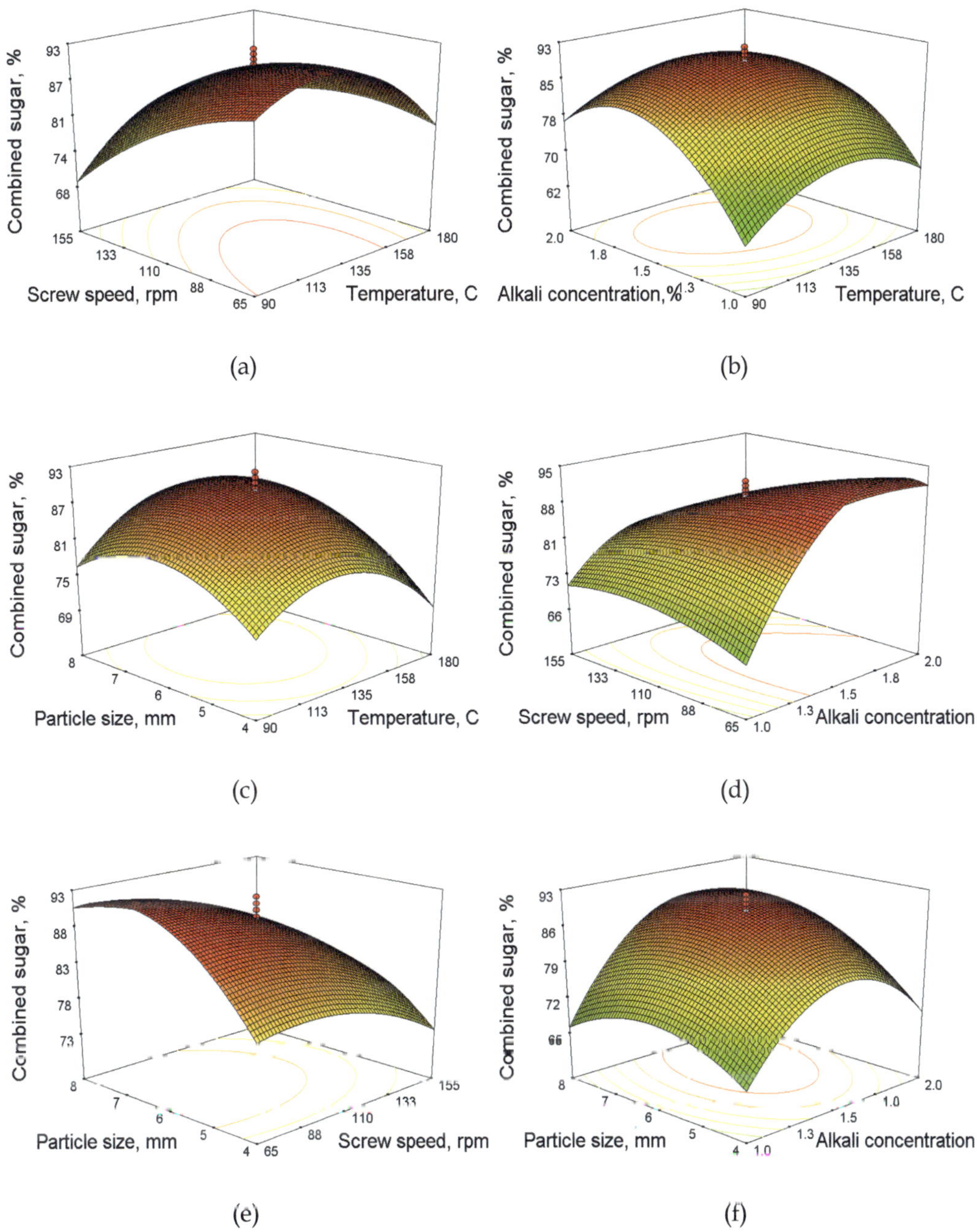


Fig. 3. Interaction effect of two independent variables on combined sugar recovery from corn stover (when other factors fixed at the center point: 135ºC, 110, rpm, 1.5% w/w, and 6 mm)

Corn stover was subjected to several pretreatment methods such as dilute acid, alkali, wet oxidation, hot water, steam, steam explosion, ammonia recycle percolation, and AFEX as found in literature. Table 1 shows different alkalis employed on corn stover and their pretreatment conditions along with glucose and xylose recovery. Although large corn stover particles was used in the present study, the glucose and xylose recovery obtained in this study was comparable with most of the alkali pretreatments listed in the table. However, the result of present study was higher than the values reported for alkali (Varga et al., 2002), wet oxidation (Varga et al., 2003), aqueous ammonia soaking pretreatment (Kim & Lee, 2007), lime and alkali pretreatment (Chen et al., 2009). Rodgers et al (2009) achieved higher glucose and xylose yield than the current study.

The sugar recovery obtained in the present study was comparable with that of the 88-93% glucose reported for dilute acid pretreatment at 140°C with 0.98% $H_2SO_4$ for 40 min (Lloyd & Wyman, 2005), more than 95% glucose and 77% xylose from the dilute acid percolation process at 180°C with 1% acid (w/w) at a flow rate of 10 mL/min for 3 min, followed by $N_2$ through quenching (Zhu et al., 2004). Varga et al (2002) obtained 95.7% glucose when corn stover was pretreated with 1% NaOH for a day followed by a 60 min autoclaving with 1% $H_2SO_4$. A similar result was reported from compressed hot water pretreatment of corn stover at 200°C with a flow rate of 10 mL/ min for 24 min (Liu & Wyman, 2005). The difference in sugar recovery might be due to the mechanisms of different pretreatment methods, the pretreatment conditions employed, and the composition of raw corn stover.

### 3.4 Response surface model evaluation

The predicted and observed responses along with coded and actual variables are presented in Table 2. The closeness of the predicted and observed responses reflects the goodness of fit. The analysis of variance of the observed data, p value ($\alpha$ = 0.05) and the coefficient of determination ($R^2$) of the regression model using CCRD are presented in Table 5. The F value for the glucose, xylose, and combined sugar were very high compared to the tabular $F_{14, 21}$ value of 2.19 indicates that the model was highly significant. As noted from the table that the regression model was significant, whereas the lack of fit was not significant, suggested that proposed model is good. The fact that the coefficient of determination was also close to one reflects the adequacy of the model to represent relationship among the barrel temperature, screw speed, alkali concentration, and particle size on sugar recovery. However, a large value of $R^2$ does not always imply that the regression model is a good one because $R^2$ will increase when a variable is added regardless of whether the additional variable is statistically significant or not (Xin & Saka, 2008). Hence, adjusted and predicted $R^2$ were calculated to check the model adequacy.

The predicted determination coefficient was in reasonable agreement with the adjusted determination coefficient and it also confirms the fitness of the model. The proposed models explain more than 90% of the variations in sugar recoveries. Coefficient of variation (CV) is the ratio of standard error estimate to the mean values expressed as percentage and is another measure to evaluate the goodness of the model. As a general rule, the CV should not be greater than 10% (Cocharan & Cox, 1957; Linko et al., 1984; Vainionpaa & Malkki, 1987). Considering the general rule, a low value of CV (2.68-3.20%) shows that the experiments conducted are precise and reliable. "Adeq Precision" measures the signal-to-noise ratio. The larger the ratio, better the prediction/optimization; in general, a ratio greater than 4 is desirable. The ratio of 38.69-57.38 indicates an adequate signal thus, the model can be used to navigate the design space (Liu et al., 2010).

| Pretreatment | Pretreatment conditions | Biomass size, mm | Hydrolysis conditions | Optimum conditions | Yield | Reference |
|---|---|---|---|---|---|---|
| Dilute acid | 0.2-1% $H_2SO_4$, 160- mL/min, 4.4 min | 0.84-2.0 | 5-15 FPU, 30 CBU/g glucan, 96 h | 15 FPU and 30 CBU/g glucan | >98% enzymatic digestibility | Zhu et al., 2005 |
| Dilute acid | acid), 1 min , 30% slr | 0.25-0.42 | 15 FPU, 30 CBU/g glucan; 5 FPU, 30 CBU/g glucan, 72 h | | 60 and 74% digestibility for 5 and 15 FPU/g glucan | Zhu et al., 2009 |
| Dilute acid | 1.5% $H_2SO_4$ | <2 | 20 FPU, 10 CBU/g DM, 48 h, 8% slr | | 39% hydrolysis yield, 12.2% lignin removal | Chen et al., 2009 |
| Dilute sulfuric acid cycle spray flow-through | 1-3% w/v $H_2SO_4$, 8 L/min, 90 min, -120 min, 8 L/min, 2% w/v $H_2SO_4$ -15 L/min, 90 min, 2% w/v $H_2SO_4$ - L/min, 90 min, 2% w/v $H_2SO_4$ | < 0.42 | 60 FPU, 15 CBU/g glucan, 72 h, 1% glucan loading | 8 L/min | 90-95% glucose, 90-93% xylose, and 70-75% lignin removal | Yan et al., 2009 |
| Maleic acid | 0.05-0.2 M, 150- -60 min, | 0.42 | | 0.2 M, 100- | 80-90% xylose yield | Lu & Mosier, 2008 |
| Formic acid | 15 min, no formic acid; | 2 | 30 FPU/g DM, 2% slr, 24 h; SSF 168 h | | 75.0 and 73.2, 66.4 and 50.8% glucose and xylose yield from without and with formic acid, 76.5 and 69.6% ethanol from with and without formic acid pretreatment | Xu et al., 2009a |
| Lactic and/ acetic acid pretreatment | DM, 41.1 g lactic acid and 15.73 g acetic acid/kg DM, 40.21 g acetic acid/kg DM | 2 | 30 FPU/g DM, 24 h, 2% slr | | 83.9-95.7% glucose, 58.9-81.1% xylose, no inhibitors | Xu et al., 2009b |
| Alkali (NaOH) | 0-0.8% (0-0.058 g NaOH/g), soaking at room temperature, 2 h | < 2 | 46 FPU/g substrate, 65 h | 0.8% NaOH | 50-95 and 65-95% glucose and xylose depending on anatomical fractions | Duguid et al., 2009 |
| $FeCl_3$ pretreatment | 0.1 M/L FeCl3, 140- | 0.18-0.84 | 15-60 FPU, 26.25-105 CBU/g cellulose, 72 h | 105 CBU/g cellulose | 98% hydrolysis yield | Liu et al., 2009 |
| Soaking in ethanol and aqueous ammonia | 1-49 wt% ethanol , 1 solid:9 liquid (15 wt% $NH_3$ 24 h | 0.5-2.0 | 30 FPU, 30 CBU/g glucan, 96 h, 1% glucan loading | 20% ethanol | 100 and 89.6% glucose and xylose retention, 70.1% lignin removal; 80% glucose and 60% xylose | Kim et al., 2009 |
| AFEX | 1 stover:1 $NH_3$ moisture db | < 0.125- >0.841 | 15 FPU, 64 pNPGU/g glucan, xylanase 10% of cellulose, 168 h, 1% glucan loading, washing | | >98% enzymatic hydrolysis, size reduction , washing, and xylanase addition improved the yield | Chundawat et al., 2007 |
| Controlled pH | 160- -30 min | 5.4 | 15 FPU, 30 CBU/g glucan, 72 h, 1% slr | stover slurry | 86.9-87.2% total sugar | Elander et al., 2009; Mosier et al., 2005 |
| Steam pretreatment | 3% $SO_2$, 170 C, 9 min,; 3% $SO_2$, 190 C, 5 min, (0% $SO_2$, 190 C, 5 min); 210 C, 7.8 min, 3% $SO_2$ | 0.5-2.0 | 15 FPU, 30 IU/g cellulose, 72 h, 1% slr | | 58, 88, and 98% glucose; 63, 86, and 98% xylose | Bura et al., 2009 |
| Steam explosion | 0-3% $H_2SO_4$; 180- | 2-3 cm | 15 FPU and 20 IU/g DM; 48 h; 2% slr | | 85% glucose yield | Zimbardi et al., 2007 |
| $SO_2$ catalyzed | 0.5-3% $SO_2$ 170- -10 min | NF | 15 FPU, 30 CBU/g glucan, 72 h | | 93.7% total sugar | Elander et al., |

slr- solids loading rate DM- dry matter

Table 4. Pretreatment and hydrolysis conditions employed on corn stover pretreated in different pretreatment methods

| Response | Source | df | Sum of squares | Mean squares | F value | P value | $R^2$/ Adj $R^2$ /Pred.$R^2$ | CV(%) / Adeq Precision |
|---|---|---|---|---|---|---|---|---|
| Glucose | Regression | 14 | 13943.27 | 995.94 | 294.6 | < 0.0001 | 0.99/ 0.99/ 0.97 | 2.68/ 57.38 |
| | Lack of fit | 10 | 48.42 | 4.84 | 2.36 | 0.0875 | | |
| | Pure error | 11 | 22.56 | 2.05 | | | | |
| | Residual | 21 | 70.99 | 3.38 | | | | |
| | Total | 35 | 14014.27 | | | | | |
| Xylose | Regression | 14 | 7984.55 | 570.32 | 114.38 | < 0.0001 | 0.98/ 0.97/ 0.94 | 3.20/ 38.69 |
| | Lack of fit | 10 | 68.05 | 6.80 | 2.04 | 0.1286 | | |
| | Pure error | 11 | 36.65 | 3.33 | | | | |
| | Residual | 21 | 104.70 | 4.98 | | | | |
| | Total | 35 | 8089.26 | | | | | |
| Combined | Regression | 14 | 11393.72 | 813.83 | 210.33 | < 0.0001 | 0.99/ 0.98/ 0.97 | 2.83/ 47.07 |
| | Lack of fit | 10 | 55.81 | 5.58 | 2.41 | 0.0822 | | |
| | Pure error | 11 | 25.43 | 2.31 | | | | |
| | Residual | 21 | 81.25 | 3.86 | | | | |
| | Total | 35 | 11474.98 | | | | | |

Table 5. Analysis of variance of the fitted model for different responses

### 3.5 Optimization and validation

The interactions discussed in the previous section were for individual sugar recovery. Maximum glucose, xylose, and combined sugar recovery were the desirable responses considered for optimization. Hence, an overlay contour plots superimposing glucose, xylose, and combined sugar recovery responses were depicted in Fig. 4. The shaded region gave wide range of options to select the barrel temperature (90-180°C), screw speed (50-155 rpm), alkali concentration (1.25-2.0%), and particle size (4-9 mm) for maximum glucose (80-95%), xylose (80-95%), and combined sugar (80-90%) recovery from corn stover. Based on the models, numerical optimization was carried out in Design Expert. Considering each response, three solutions were found as shown in Table 6. In order to confirm the predicted responses, corn stover was extruded at a barrel temperature of 133°C, a screw speed of 85 rpm and 1.65% alkali concentration with two different particle sizes since the optimum particle size was 6.45 mm, which is in between 6 and 8 mm. The extruded samples (Fig. 5) were subjected to the enzymatic hydrolysis and sugar measurement as described in the Materials and Methods.

The glucose, xylose, and combined sugar obtained were 91.8, 82.3, and 90.0%, respectively; the values were very close to the predicted values; which was 5.0, 2.5, and 4.3 times higher than control sample. The sugar recovery comparison of control, alkali soaked, and alkali soaked - extruded corn stover is shown in Fig. 6 for better understanding. The alkali soaked corn stover had a double times sugar recovery of control, whereas the alkali soaked-extrusion pretreated corn stover had double times that of alkali soaked corn stover as evident from Fig 6. A mass balance diagram is shown in Fig. 7 for better understanding

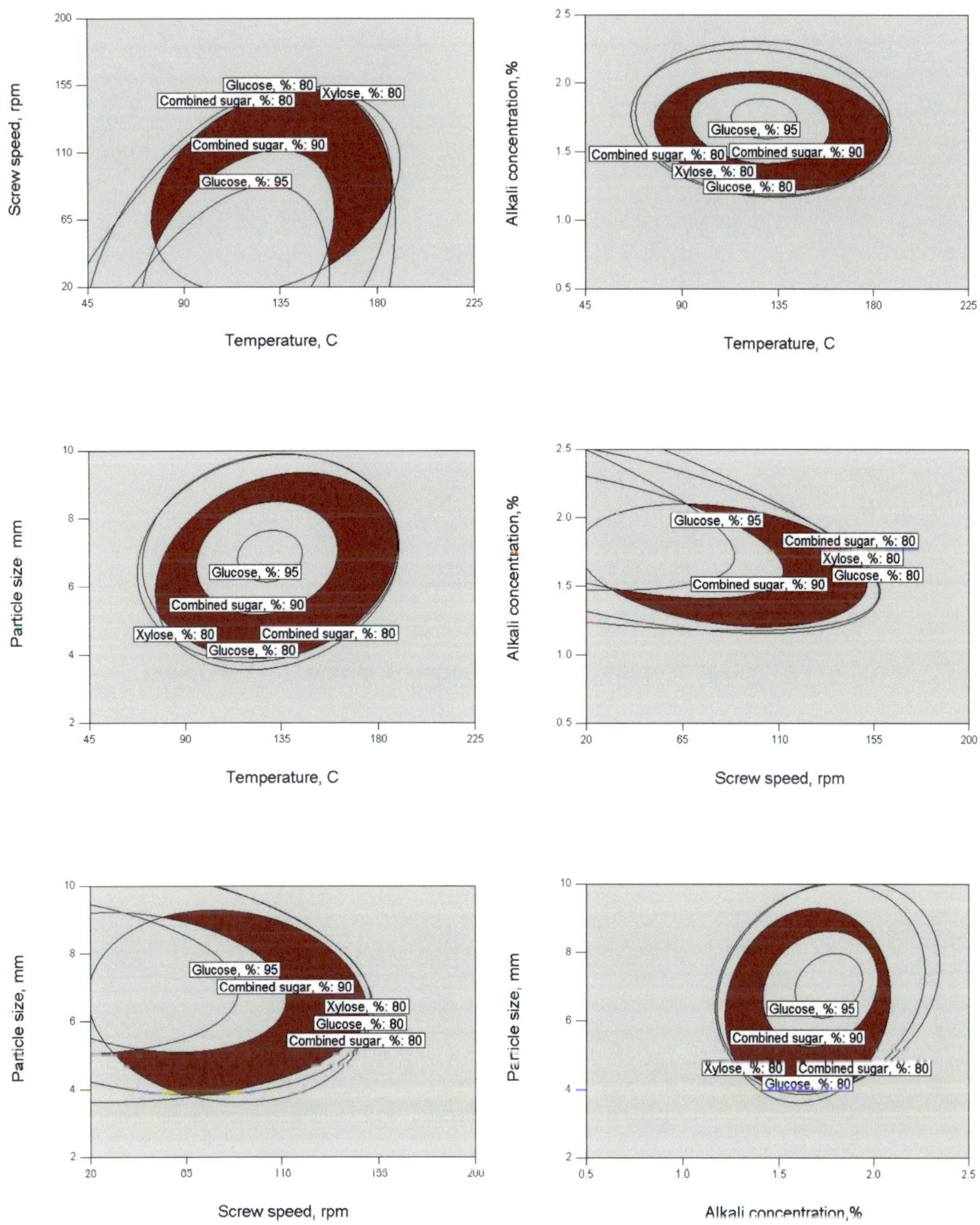


Fig. 4. Superimposed contours for sugar recovery responses as a function of temperature, screw speed, alkali concentration, and particle size of corn stover

Fig. 5. Corn stover extruded at optimum pretreatment conditions for validation

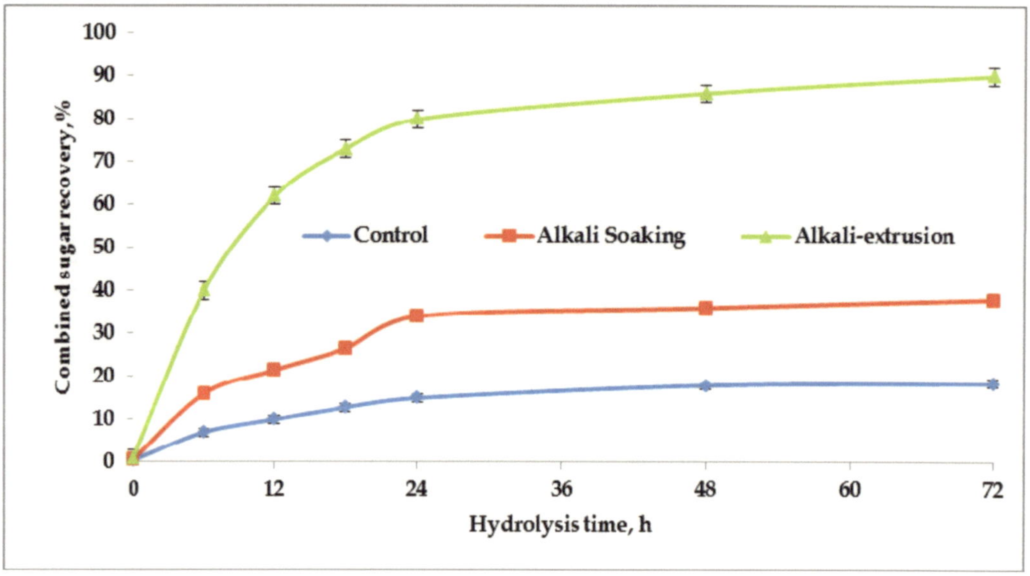


Fig. 6. Comparison of sugar recovery profile from control, alkali soaked, and alkali soaked-extruded corn stover

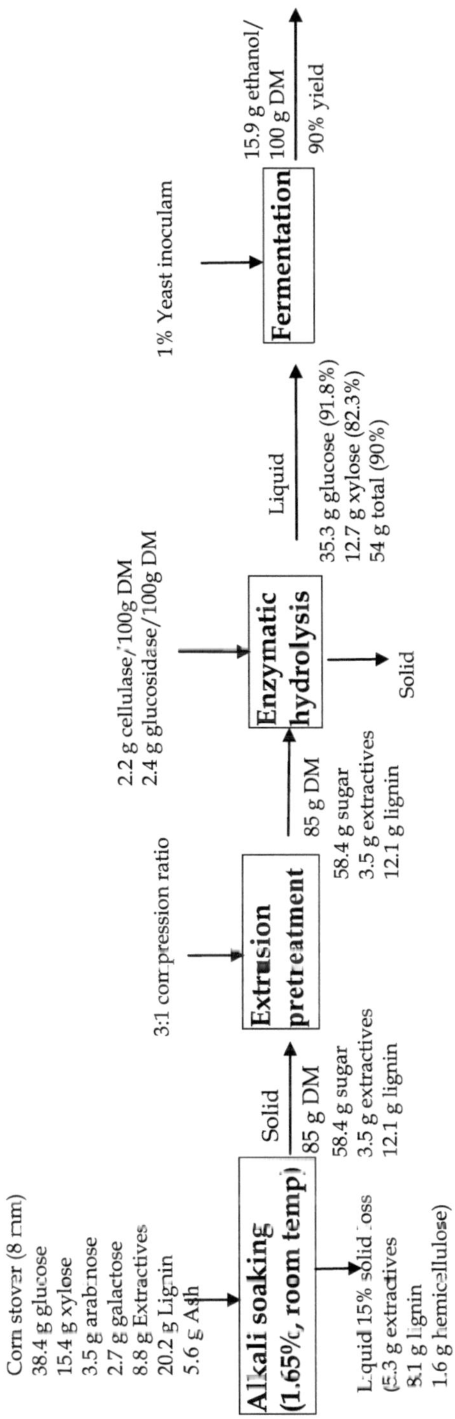


Fig. 7. Mass balance diagram of alkali soaking-extrusion pretreatment of prairie cord grass followed by fermentation

| Solution # | Temperature, °C | Screw speed, rpm | Alkali conc, % | Particle size, mm | Glucose, % | Xylose, % | Combined sugar, % |
|---|---|---|---|---|---|---|---|
| 1 | 133 | 84 | 1.65 | 6.45 | 95.61 | 88.73 | 93.89 |
| 2 | 133 | 85 | 1.65 | 6.46 | 95.52 | 88.73 | 93.84 |
| 3 | 133 | 84 | 1.65 | 6.44 | 95.68 | 88.73 | 93.94 |
| Validation | | | | | | | |
| | 133 | 85 | 1.65 | 6 | 92.65 | 86.59 | 90.53 |
| | 133 | 85 | 1.65 | 8 | 91.76 | 86.26 | 90.08 |

Table 6. Solutions for optimal and validation conditions

(assumption a thumb rule is 50% of the glucose will be converted into ethanol during fermentation with an efficiency of 90%).This optimization study revealed that a larger particle size (8 mm) could be used for biofuels production; thereby the biomass size reduction energy cost can be saved to a greater extent.

Although the optimum pretreatment condition differed from corn stover, similar sugar recovery was reported for switchgrass and prairie cord grass (Karunanithy & Muthukumarappan, 2011b, 2011c). The biomass digestibility depends on the specific types of phenolic acids that constitute the non-core lignin of lignocellulosic biomass (Jung & Deetz, 1993). These phenolic acids are involved in the ester linkages between hemicellulose and lignin. The major non-core lignin phenolic acids are *p*coumaric acid (CA) and ferulic acid (FA). According to Burritt et al (Burritt et al., 1984) and Jung (1989), the ratio of CA to FA present in non-core lignin has a strong negative impact on biomass digestibility. The difference in alkali concentration might be attributed to their pcoumaric to ferulic acids ratio apart from lignin content. The glucose, xylose, and combined sugar recovery of 90.5, 81.5, and 88%, respectively, were reported for the optimum pretreatment conditions of 180°C barrel temperature, 118 rpm screw speed, 2% alkali concentration, and 6 mm switchgrass particle size. The optimum pretreatment conditions of 114°C barrel temperature, 122 rpm screw speed, 1.70% alkali concentration, and 8 mm particle size of prairie cord grass resulted in a glucose, xylose, and combined sugar recovery of 87, 85, and 82%, respectively. The difference in sugar recovery and optimum conditions might be attributed to the inherent nature of biomasses including their chemical compositions.

### 3.6 Byproducts formation

In general, furfural, HMF, and acetic acid are the fermentation inhibitors found in most of the pretreatment at different extent for various feedstocks as listed in Table 7. Pretreatment temperature, residence time, and acid concentration are the important factors influence the degradation process; moreover the degradation is proportional to the pretreatment severities (Bustos et al., 2003; Hodge et al., 2008; Rodríguez et al., 2009; Sassner et al., 2008; Yu et al., 2010; Zeng et al., 2007). When side chains of acetyl group present in hemicellulose are released, acetic acid is generated. Various researchers reported a range of 1.9-7.3% acetyl group for corn stover (Balan et al., 2009; Kim & Lee, 2005b; Torget et al., 1991; Weiss et al., 2009). Acetic acid was the only byproduct found in most of the pretreated corn stover samples in the range of 0.060-0.168 g/L. The highest acetic acid (0.168 g/L) resulted at a

| Biomass | Pretreatment | Conditions | Acetic acid, g/L | HMF, g/L | Furfural, g/L | Reference |
|---|---|---|---|---|---|---|
| Corn stover | Dilute acid | 0.5 1.4% $H_2SO_4$, 165- -12 min, 20% slr | NR | NR | 6-31%# | Schell et al., 2008 |
| Corn stover | Dilute acid | 0.5 1.4% $H_2SO_4$, 165- -12 min | 15.5 g | 4 | 3.9 | Hodge et al., 2008 |
| Corn stover | Dilute acid | 1.5% $H_2SO_4$ | 0.35 | NR | 0.01 | Chen et al., 2009 |
| Corn stover | Dilute acid | 1.5-1.6% $H_2SO_4$, 180- -105 s | NR | 0.1-1%+ | 2-11%+ | Lu & Mosier at al., 2007 |
| Corn stover | Maleic acid | -15 min, 40 and 150 g/L | NR | NR | 1.8 | Lu & Mosier at al., 2007 |
| Corn stover | Formic acid | | 1.18 | 0.28 | 1.3 | Xu et al., 2009a |
| Corn stover | Lime | $_2O/g$ | 0.22 | NR | ND | Chen et al., 2009 |
| Corn stover | Alkali | | 0.19 | NR | ND | Chen et al., 2009 |
| Corn stover | Aqueous ammonia/acid | 10% $NH_3$ - | 0.26 | NR | ND | Chen et al., 2009 |
| Corn stover | Hot water | | NR | 0.01 | 0.09-0.1 | Zeng et al., 2007 |
| Corn stover | Steam | 2% $SO_2$ | 2-4 | < 2 | <2 | |
| Corn stover | Steam | $SO_2$ | NR | 0.18 | 1.26 | 7 |
| Corn stover | Steam explosion | -30% slr | 1.4-4.5 | NR | 0 0.06-0.18 | Lu et al., 2010 |
| Rice straw | Compressed hot water | 140- - | 0.25-3.25 | 0-0.7 | 0-3 | Yu et al., 2010 |
| Rice straw | Hydrothermal | 150- -20 min, liquid/solid ratio 6-8 | 0.4-2.03 | NR | NR | et al., 2009 |
| Sugar cane bagasse | Dilute acid | 150- -20 min,0.05-0.045 mol/L $H_2SO_4$, 4-20% DM | 3.29* | 0.35* | 4.52* | Neureiter et al., 2002 |
| Sugarcane bagasse | Hydrochloric acid | 2-6% HCl, 100- -300 min | NR | NR | 8 | Bustos et al., 2003 |
| Sugarcane bagasse | Wet oxidation | 185- -15 min, 2g $Na_2CO_3$ or 36.5%w/w $H_2SO_4$ pH 3-10 | 0.7-3.0* | 0-0.07 * | 0-0.53* | Martin et al., 2007 |
| Wheat straw | Dilute acid | | NR | NR | 2.5 | Kabel et al., 2007 |
| Wheat straw | Steam explosion | 0.9% $H_2SO_4$ soaking, 18 h, 45 C; 160-200 C, 5-20 min | NR | 0.03-1.51* | 0.03-0.24* | Ballesteros et al., 2006 |
| Wheat straw | Steam explosion | | 5.1 | 0.1 | 1.4 | - |
| Rye straw | Liquid hot water | 170-230 C, 4g/min, 10% slr | NR | 0.2-4.4% | 0-3.6% | Ingram et al., 2009 |
| Barely straw | Steam | 0.2-2.0% $H_2SO_4$, 5 min, 190- | | 0.1-0.3* | 0.6-1.8* | Linde et al., 2006 |
| Oil palm fruit bunch | Dilute acid | 6% $H_2SO_4$ | 2-5 | NR | 1-4 | Rahman et al., 2007 |
| Eucalyptus | Dilute acid | 0.65% $H_2SO_4$ | 3.10 | 0.20 | 1.23 | Canettieri et al., 2007 |
| Aspen | Steam explosion | 2 | 0.74-4.33* | 0.009-0.2* | 0.29-2.01* | De Bari et al., 2007 |
| Salix | Steam | 0.25% or 0.5%$H_2SO_4$, 180- -12 min | NR | 0.1-0.6 | 0.4-2.4 | Sassner et al., 2008 |

# furfural yield *g/100 g + loss of respective sugar slr- solids loading rate NR- not reported ND- not detected DM- dry matter

Table 7. Comparison of inhibitor formation from different feedstocks produced through various pretreatments

barrel temperature of 135°C and a screw speed of 110 rpm, alkali concentration of 1.5% and a particle size of 6 mm. It was noticed that acetic acid concentration was 0.143 g/L for the optimized conditions (133°C, 85 rpm, 1.65%, and 8 mm). The acetic acid formation and its concentration confirmed that the deacetylzation process, however the concentration was below the inhibition limit (Taherzadeh et al., 1997). The particle size of rice straw and silvergrass inversely influenced the acetic acid formation in dilute acid pretreatment (Guo et al., 2008); however, it was not confirmed in this study.
The acetic acid concentration was lower than dilute acid (Chen et al., 2009; Hodge et al., 2008), formic acid (Xu et al., 2009a), lime (Chen et al., 2009), alkali (Chen et al., 2009), hydrothermal pretreatment (Öhgren et al., 2006; Lu et al., 2010) of corn stover; compared to other feedstocks listed in the table. NaOH has higher reactivity with hemicellulose than cellulose due to amorphous characteristics of hemicellulose (Lai, 2001). In addition, Gupta (2008) reported that minimum loss of hemicellulose is unavoidable for corn stover pretreated with dilute NaOH (1-5%) at 60°C for 24 h. However, the mild pretreatment condition (25°C, 1% NaOH, 24 h) minimized the sugar loss from corn stover (Gupta, 2008). No glycerol was found in any of the pretreatment combinations in contrary to the results reported by Karunanithy and Muthukumarappan (Karunanithy & Muthukumarappan, 2010c, 2011a, 2011b, 2011c, 2010a, 2010b).In addition, no furfural and HMF were found in any of the pretreated corn stover samples. The possible reason could be short residence time and no acidic conditions during pretreatment as compared most of the pretreatment methods listed in the table.

## 4. Conclusions

Corn stover was extruded using a single screw extruder at various conditions based on a central composite rotatable design to obtain maximum glucose, xylose, arabinose, and combined sugar recovery. Statistical analyses confirmed that extruder barrel temperature, screw speed, alkali concentration, and particle size had a significant effect on sugar recovery. Response surface methodology was adopted to optimize alkali concentration, corn stover particle size, and extruder parameters for maximum sugar recovery. The proposed quadratic model to predict the sugar recovery had high F and $R^2$ values with low p value represents an adequate relationship among the independent variables studied on sugar recovery from corn stover. It was found that under optimum condition such as barrel temperature of 133°C, screw speed of 85 rpm, alkali concentration of 1.65% w/v with 8 mm particle size, the glucose, xylose, and combined sugar recovery were 91.8, 82.3, and 90.1%, respectively, and also confirmed through validation. This optimization study revealed that larger corn stover particle (8 mm) can be used for biofuels production, which could result in saving of size reduction energy cost.

## 5. Acknowledgements

This research was supported by funding from the Agricultural Experiment Station and North Central Sun Grant Center at South Dakota State University through a grant provided by the US Department of Transportation, Office of the Secretary, Grant No.DTOS59-07-G-00054. Also, enzymes supplied by Novozymes, Inc for conducting this study were greatly appreciated.

## 6. References

Altan, A., McCarthy, KL., & Maskan, M. (2008). Twin-screw extrusion of barley–grape pomace blends: Extrudate characteristics and determination of optimum processing conditions. *J Food Eng.*, 89, 24-32, 0260-8774

Balan, V., da Costa Sousa, L., Chundawat, SPS., Marshall, D., Sharma, LN., Chambliss, CK., & Dale, BE. (2009). Enzymatic digestibility and pretreatment degradation products of AFEX-treated hardwoods (*Populus nigra*). *Biotechnol. Prog.*, 25, 2, 365-375, 1520-6033

Ballesteros, I., Negro, MJ., Oliva, JM., Cabañas, A., Manzanares, P., & Ballesteros, M. (2006). Ethanol production from steam-explosion pretreated wheat straw. *Appl. Biochem. Biotechnol.*, 130, 1-3, 496-508, *0273-2289*

Bura, R., Chandra, R., & Saddler, J. (2009). Influence of xylan on the enzymatic hydrolysis of steam-pretreated corn stover and hybrid poplar. *Biotechnol. Prog.* 25, 2, 315-322, 1520-6033

Burritt, EA., Bittner, AS., Street, JC., & Anderson, MJ. (1984). Correlations of phenolic-acids and xylose content of cell-wall with in vitro dry-matter digestibility of 3 maturing grasses. *J Dairy Sci.*, 67, 6, 1209-1213, *0022-0302*

Bustos, G., Ramírez, JA., Garrote, G., & Vázquez, M. (2003). Modeling of the hydrolysis of sugar cane bagasse with hydrochloric acid. *Appl. Biochem. Biotechnol.*, 104, 52-69, *0273-2289*

Cadoche, L., & Lopez, GD. (1989). Assessment of size reduction as a preliminary step in the production of ethanol from lignocellulosic wastes. *Biolog. Wastes*, 30, 153-157, 0269-7483

Canettieri, EV., Rocha, GJM., Carvalho Jr, JA., & Silva, JBA. (2007). Optimization of acid hydrolysis from the hemicellulosic fraction of Eucalyptus grandis residue using response surface methodology. *Bioresour. Technol.*, 98, 422-428, 0960-8524

Chang, VS., Nagwani, M., Kim, C-H., & Holtzapple, MT. (2001). Oxidative lime pretreatment of high-lignin biomass poplar wood and newspaper. *Appl. Biochem. Biotechnol.*, 94,1-28, *0273-2289*

Chen, FL., Wei, YM., Zhang, B., & Ojokoh, AO. (2010). System parameters and product properties response of soybean protein extruded at wide moisture range. *J Food Eng.*, 96, 208-213, 0260-8774

Chen, M., Zhao, J., & Xia, L. (2009). Comparison of four different chemical pretreatments of corn stover for enhancing enzymatic digestibility. *Biomass Bioenerg.*, 33, 10,1381-1385, *0961-9534*

Chundawat, SPS., Venkatesh, B., & Dale, BE. (2007). Effect of particle size based separation of milled corn stover on AFEX pretreatment and enzymatic digestibility. *Biotechnol. Bioeng.*, 96, 2, 219-231, 1097-0290

Cocharan, WG., & Cox, GM. (1957). In *Experimental Designs*, Bradley RA, Kendall DG, Hunter JS, Watson GS (eds). Wiley: New York, pp. 335-375

Crofcheck, CL., & Montross, MD. (2004). Effect of stover fraction on glucose production using enzymatic hydrolysis. *Transactions of the ASAE*, 47, 3, 841-844, *0001-2351*

Dale, BE., Weaver, J., & Byers ,FM. (1999). Extrusion processing for ammonia fiber explosion (AFEX). *Appl. Biochem. Biotechnol.*, 77,35-45, *0273-2289*

Dawson, L., & Boopathy, R. (2007). Use of post-harvest sugarcane residue for ethanol production. *Bioresour. Technol.*, 98, 9, 1695-1699, 0960-8524

De Bari, I., Nanna, F., & Braccio, G. (2007). $SO_2$-catalyzed steam fractionation of aspen chips for bioethanol production: Optimization of the catalyst impregnation. *Ind. Eng. Chem. Res.*, 46, 23, 7711-7720, 0088- 5885

de Leon, N., & Coors, JG. (2008). Chapter 7: Genetic improvement of corn for lignocellulosic feedstock production, In: *Genetic Improvement of Bioenergy Crops*, W. Vermerris (ed.), pp.185-210. Springer Science, 978-0-387-70804-1, New York

Desari, RK., & Bersin, RE. (2007). The effect of particle size on hydrolysis reaction rates and rheological properties in cellulosic slurries. *Appl. Biochem. Biotechnol.*, 136–140, 289-299, *0273-2289*

Design Expert 7.1.6. Statease, Minneapolis, MN, 2002.

de Vrije, T., de Haas, GG., Tan, GB., Keijsers, ERP., & Claassen, PAM. (2002). Pretreatment of miscanthus for hydrogen production by *thermotoga elfii*. *Inter. J Hydrogen Energy*, 27, 1381-1390, 0360-3199

Duguid, KB., Montross, MD., Radtke, CW., Crofcheck, CL., Wendt, LM., & Shearer SA. (2009). Effect of anatomical fractionation on the enzymatic hydrolysis of acid and alkaline pretreated corn stover. *Bioresour. Technol.*, 100, 21, 5189-5195, 0960-8524

Elander, RT., Dale, BE., Holtzapple, M., Ladisch, MR., Lee, YY., Mitchinson, C., Saddler, JN., & Wyman, CE. (2009). Summary of findings from the Biomass Refining Consortium for Applied Fundamentals and Innovation (CAFI): Corn stover pretreatment. *Cellulose*, 16, 4, 649-659, 1572-882X

Elshafei, AM., Vega ,JL., Klasson, KT., Clausen, E., & Gaddy, JL.(1991). The saccharification of corn stover by cellulase from *Penicillinfuniculosum*. *Bioresour. Technol.*, 35, 73-80, 0960-8524

Gibbons, WR., Westby, CA., & Dobbs, TL. (1986). Intermediate-scale, semicontinuous solid-phase fermentation process for production of fuel ethanol from sweet sorghum. *Appl. Environ. Microbiol.*, 51,1, 115-122, 0099-2240

Guo, G-L., Chen, W-H., Chen, W-H., Men, L-C., & Hwang, W-S. (2008). Characterization of dilute acid pretreatment of silvergrass for ethanol production. *Bioresour. Technol.*, 99, 14, 6046-6053, 0960-8524

Gupta, R. (2008).Alkaline pretreatment of biomass for ethanol production and understanding the factors influencing the cellulose hydrolysis. Ph.D Dissertation, Auburn University, Alabama

Gupta, R., Kim, TH., & Lee, YY. (2007). Substrate dependency and effect of xylanase supplementation on enzymatic hydrolysis of ammonia-treated biomass. *Appl. Biochem. Biotechnol.*, 148,1-3, 59-70, *0273-2289*

Hayashi, N., Hayakawa, I., & Fujio, Y. (1992). Hydration of heat-treated soy protein isolate and its effect on the molten flow properties at an elevated temperature. *Inter. J Food Sci. Technol.*, 27, 565-571, 1365-2621

Hodge, DB., Karim, MN., Schell, DJ., & McMillan, JD. (2008). Soluble and insoluble solids contributions to high-solids enzymatic hydrolysis of lignocelluloses. *Bioresour. Technol.*, 99, 18, 8940-8948, 0960-8524

Hu, Z., Wang, Y., & Wen, Z.(2008). Alkali (NaOH) pretreatment of switchgrass by radio frequency-based dielectric heating. *Appl. Biochem. Biotechnol.* 148, 71-81, *0273-2289*

Ingram, T., Rogalinski, T., Bockemühl, V., Antranikian, G., & Brunner, G. (2009). Semi-continuous liquid hot water pretreatment of rye straw. *J Supercritical Fluids*, 48, 3, 238-246, 0896-8446

Jorge, M., Roberto, G., Karina, PJX., Oliva, CE., Benjamin, R., & Cuauhte´mocc, R. (2006).The optimization of the extrusion process when using maize flour with a modified amino acid profile for making tortillas. *Inter. J Food Sci. Technol.* 41, 727-736, 1365-2621

Jung , HG. (1989). Forage lignins and their effects on fiber digestibility. *Agron. J.*, 81, 1, 33-38, *0002-1962*

Jung, HG., & Deetz, DA. (1993). Cell wall lignification and degradability. p. 315–346. *In* H.G. Jung et al (ed.) Forage cell wall structure and digestibility. ASA, CSSA, and SSSA, Madison, WI, 0891181156

Jung, HG., Valdez, FR., Hatfield, RD., & Blanchette, RA.(1992). Cell wall composition and degradability of forage stems following chemical and biological delignification. *J Sci. Food Agric.*, 58, 347-355, 1097-0010

Kaar, W., & Holtzapple, M. (2000). Using lime pretreatment to facilitate the enzymatic hydrolysis of corn stover. *Biomass Bioenerg.*, 18, 3, 189-199, *0961-9534*

Kabel, MA., Bos, G., Zeevalking, J., Voragen, AGJ., & Schols, HA. (2007). Effect of pretreatment severity on xylan solubility and enzymatic breakdown of the remaining cellulose from wheat straw. *Bioresour. Technol.*, 98, 10, 2034-2042, 0960-8524

Kadam, KL., & McMillan, JD. (2003). Availability of corn stover as a sustainable feedstock for bio-ethanol production. *Bioresour. Technol.*, 88, 17-25, 0960-8524

Karunanithy, C., Muthukumarappan, K., & Julson, JL. (2008). Influence of high shear bioreactor parameters on carbohydrate release from different biomasses. ASABE Paper No. 084109. St. Joseph, Mich.: ASABE

Karunanithy, C., & Muthukumarappan, K. (2010a).Effect of extruder parameters and moisture content of corn stover and big bluestem on sugar recovery from enzymatic hydrolysis. *Biolog Engg.*, 2, 2,91-113, 1934-2799

Karunanithy, C., & Muthukumarappan, K. (2010b). Influence of extruder temperature and screw speed on pretreatment of corn stover while varying enzymes and their ratios. *Appl. Biochem. Biotechnol.*, 162, 264-279, *0273-2289*

Karunanithy, C., & Muthukumarappan, K. 2010c. Effect of extruder parameters and moisture content of switchgrass, prairie cord grass on sugar recovery from enzymatic hydrolysis. Appl. Biochem. Biotechnol. 162, 1785–1803, 0273-2289

Karunanithy, C., & Muthukumarappan, K. (2011a). Optimization of corn stover and extruder parameters for enzymatic hydrolysis using respons surface methodology. Biological Engg, 3(2), 73-95, 1934-2799

Karunanithy, C., & Muthukumarappan, K. (2011b). Optimization of alkali concentration, switchgrass particle size and extruder parameters for maximum sugar recovery using response surface methodology. *Chemical Engg & Tech.*,(Accepted), 1521-4125

Karunanithy, C., & Muthukumarappan, K. (2011c). Optimization of alkali concentration and extruder parameters for maximum sugar recovery from prairie cord grass using response surface methodology. *Biochemical Engg J.*, DOI:10.1016/j.bej.2011.02.001, 1369-703X

Keshwani, DR. (2009). Microwave pretreatment of switchgrass for bioethanol production. PhD dissertation, North Carolina State University, Raleigh, North Carolina

Kim, JW., & Mazza, G. (2008). Optimization of phosphoric acid catalyzed fractionation and enzymatic digestibility of flax shives. *Ind. Crops Prod.*, 28, 346-355, 0926-6690

Kim, S., & Holtzapple, MT. (2005). Lime pretreatment and enzymatic hydrolysis of corn stover. *Bioresour. Technol.*, 96,18, 1994-2006, 0960-8524

Kim, SB., & Lee, YY. (2002). Diffusion of sulfuric acid within lignocellulosic biomass particles and its impact on dilute acid treatment. *Bioresour. Technol.*, 83, 165-171, 0960-8524

Kim, TH., & Lee, YY. (2005a). Pretreatment and fractionation of corn stover by ammonia recycle percolation process. *Bioresour. Technol.*, 96,18, 2007-2013, 0960-8524

Kim, TH., & Lee, YY. (2005b). Pretreatment of corn stover by soaking in aqueous ammonia. *Appl. Biochem. Biotechnol.*, 121-124, 1119-1131, *0273-2289*

Kim, TH,, &Lee, YY. (2006). Fractionation of corn stover by hot-water and aqueous ammonia treatment. *Bioresour. Technol.*, 97, 2, 224-232 , 0960-8524

Kim, TH., & Lee, YY. (2007). Pretreatment of corn stover by soaking in aqueous ammonia at moderate temperatures. *Appl. Biochem. Biotechnol.*, 136-140, 81-92, *0273-2289*

Kim, TH., Kim, JS., Sunwoo, C., & Lee, YY. (2003). Pretreatment of corn stover by aqueous ammonia. *Bioresour. Technol.* 90,39-47, 0960-8524

Kim, TH., Lee, YY., Sunwoo, C., & Kim JS. (2006). Pretreatment of corn stover by low liquid ammonia percolation process. *Appl. Biochem. Biotechnol.*, 133, 41-57, *0273-2289*

Kim, TH., Nghiem, NP., & Hicks, KB. (2009). Pretreatment and fractionation of corn stover by soaking in ethanol and aqueous ammonia. *Appl. Biochem. Biotechnol.,* 153, 1-3,171-179, *0273-2289*

Lai, YZ. (2001). Chemical degradation. Chapter 10 In: Wood and cellulosic chemistry (David N.-S. Hon and Nobuo Shiraishi), Marcel Dekker Inc., New York- Basel, pp .443-512, 08247-00244

Lee, SH., Teramoto, Y., & Endo, T. (2009). Enzymatic saccharification of woody biomass micro/nanofibrillated by continuous extrusion process I - Effect of additives with cellulose affinity. *Bioresour. Technol.*, 100, 1, 275-279, 0960-8524

Linde, M., Galbe, M., & Zacchi, G. (2006). Steam pretreatment of acid-sprayed and acid-soaked barley straw for production of ethanol. *Appl. Biochem. Biotechnol.*, 130, 1-3, 546-562, *0273-2289*

Linko, P., Mattson, C., Linko, YY., & Antila, J. (1984). Production of fat bread by continuous extrusion cooking from high a-amylase. *J Cereal Sci.*, 2, 43-51, *0733-5210*

Little, AD. (2001). Aggressive growth in the use of bioderived energy and products in the United States by 2010. Final Report. www.adltechnology.com

Liu, C., & Wyman, C. (2005). Partial flow of compressed-hot water through corn stover to enhance hemicellulose sugar recovery and enzymatic digestibility of cellulose. *Bioresour. Technol.*, 96,1978-1985, 0960-8524

Liu, L., Sun, J., Li, M., Wang, S., Pei, H., & Zhang, J. (2009). Enhanced enzymatic hydrolysis and structural features of corn stover by $FeCl_3$ pretreatment. *Bioresour. Technol.*, 100, 23, 5853-5858, 0960-8524

Liu, Q., Cheng, K., Zhang, J., Li, J., & Wang, G. (2010). Statistical optimization of recycled-paper enzymatic hydrolysis for simultaneous saccharification and fermentation via central composite design. *Appl. Biochem. Biotechnol.* 160, 604-612, *0273-2289*

Lloyd, TA., & Wyman, CE. (2005). Combined sugar yields for dilute sulfuric acid pretreatment of corn stover followed by enzymatic hydrolysis of the remaining solids. *Bioresour. Technol.*, 96, 1967-1977, 0960-8524

Lu, XB., Zhang, YM., Yang, J., & Liang, Y. (2007). Enzymatic hydrolysis of corn stover after pretreatment with dilute sulfuric acid. *Chem. Eng. Technol.* 30, 7, 938-944, 0930-7516

Lu, Y., & Mosier, N. (2008). Kinetic modeling analysis of maleic acid catalyzed hemicellulose hydrolysis in corn stover. *Biotechnol. Bioeng.*, 101, 6, 1170-1181, 1097-0290

Lu, Y., Wang, Y., Xu, G., Chu, J., Zhuang, Y., & Zhang, S. (2010). Influence of high solid concentration on enzymatic hydrolysis and fermentation of steam-exploded corn stover biomass. *Appl. Biochem. Biotechnol.*, 160, 360-369, *0273-2289*

Lu, YL., & Mosier, NS. (2007). Biomimetic catalysis for hemicellulose hydrolysis in corn stover. *Biotechnol. Prog.*, 23, 1, 116-123, 1520-6033

Lynd, LR., van Zyl, WH., McBride, JE., & Laser, M. (2005). Consolidated bioprocessing of cellulosic biomass: an update. *Current Opin. Biotechnol.*, 16, 577-583, 0958-1669

MacDonald, DG., Bakhshi, NN., Mathews, JF., Roychowdhury, A., Bajpai, P., & Moo-Young, M. (1983). Alkali treatment of corn stover to improve sugar production by enzymatic hydrolysis. *Biotechnol. Bioeng.*, 25, 2067-2076

Mani, S., Tabil, LG., & Sokhansanj, S. (2004). Grinding performance and physical properties of wheat and barley straw, corn stover, and switchgrass. *Biomass Bioenerg.* 27, 339-352, *0961-9534*

Martin, C., Klinke, HB., & Thomsen, AB. (2007). Wet oxidation as a pretreatment method for enhancing the enzymatic convertibility of sugarcane bagasse. *Enz. Micro. Technol.*, 40, 3, 426-432, 0141- 0229

Miller, S., & Hester, R. (2007). Concentrated acid conversion of pine softwood to sugars Part I: Use of a twin-screw reactor for hydrolysis pretreatment. *Chem. Eng. Comm.*, 194, 1, 185-102, 1563-5201

Mosier, N., Hendrickson, R., Ho, N., Sedlak, M., & Ladisch, MR. (2005). Optimization of pH controlled liquid hot water pretreatment of corn stover. *Bioresour. Technol.*, 96, 18,1986-1993, 0960-8524

Muthukumarappan, K., & Julson, JL. (2007). Pretreatment of biomass using a novel extrusion process. 15th European Biomass Conference & Exhibition from Research to Market Development, pp583-586. Florence, Italy: ETA srl 7-11 May 2007

National Research Council. (2000). Biobased industrial products- priorities of research and commercialization. National academy press, Washington, DC

Neureiter, M., Danner, H., Thomasser, C., Saidi, B., & Braun, R. (2002).Dilute acid hydrolysis of sugarcane bagasse at varying conditions. *Appl. Biochem. Biotechnol.*, 98-100, 49-58, *0273-2289*

Öhgren, K., Rudolf, A., Galbe, M., & Zacchi, G. (2006). Fuel ethanol production from steam-pretreated corn stover using SSF at higher dry matter content. *Biomass Bioenerg.*, 30, 10, 863-869, *0961-9534*

Öhgren, K., Bura, R., Saddler, J., & Zacchi, G. (2007). Effect of hemicellulose and lignin removal on enzymatic hydrolysis of steam pretreated corn stover. *Bioresour. Technol.*, 98, 13, 2503-2510, 0960-8524

Pedersen, M., & Meyer, AS. (2009). Influence of substrate particle size and wet oxidation on physical surface structures and enzymatic hydrolysis of wheat straw. *Biotechnol. Prog.* 25, 2, 399-408, 1520-6033

Perlack ,RD., Wright, LL., Turhollow, AF., Graham, RL., Stokes, BJ., & Erbach, DC. (2005). Biomass as feedstock for a bioenergy and bioproduct industry: The technical feasibility of a billion-ton annual supply, U.S. DOE and USDA, April 2005

Rahman, SHA., Choudhury, JP., Ahmad, AL., & Kamaruddin, AH. (2007). Optimization studies on acid hydrolysis of oil palm empty fruit bunch fiber for production of xylose. *Bioresour. Technol.*, 98, 3, 554-559, 0960-8524

Rodgers, M., Hu, Z., & Zhan, X. (2009). Enhancing enzymatic hydrolysis of maize stover by bayer process sand pretreatment. *Energy & Fuels*, 23, 4, 2284-2289, 1520-5029

Rodríguez, A., Moral, A., Sánchez, R., Requejo, A., & Jiménez, L. (2009). Influence of variables in the hydrothermal treatment of rice straw on the composition of the resulting fractions. *Bioresour. Technol.*, 100, 20, 4863-4866, 0960-8524

Saha, B., & Bothast, R. (1997). Enzymes in lignocellulosic biomass conversion. In: Fuels and Chemicals from Biomass. *ACS Symposium Series*, 666,46-56, *0097-6156*

Sassner, P., Martensson, C-G., Galbe, M., & Zacchi, G. (2008). Steam pretreatment of $H_2SO_4$-impregnated Salix for the production of bioethanol. *Bioresour. Technol.*, 99, 1, 137-145, 0960-8524

Schell, DJ., Farmer, J., Newman, M., & Mcmillan, JD. (2003). Dilute–sulfuric acid pretreatment of corn stover in pilot-scale reactor investigation of yields, kinetics, and enzymatic digestibilities of solids. *Appl. Biochem. Biotechnol.*, 105-108, 69-85, *0273-2289*

Selig, M., Weiss, N., Ji, Y. (2008). Enzymatic saccharification of lignocellulosic biomass. NREL/TP-510-42629. National Renewable Energy Laboratory, Golden, CO

Service, RF. (2007). Cellulosic ethanol: Biofuel researchers prepare to reap a new harvest. *Sci.* 315,1488-1491.

Sluiter, A., Hames, B., Ruiz, R., Scarlata, C., Sluiter, J., Templeton, D., & Crocker, D.(2008a) Determination of structural carbohydrates and lignin in biomass. NREL/TP-510-42618. National Renewable Energy Laboratory, Golden, CO

Sluiter, A., Hames, B., Ruiz, R., Scarlata, C., Sluiter, J., & Templeton, D. (2008b). Determination of ash in biomass. NREL/TP-510-42622. National Renewable Energy Laboratory, Golden, CO

Sluiter, A., Hames, B., Hyman, D., Payne, C., Ruiz, R., Scarlata, C., Sluiter, J., Templeton, D., & Wolfe, J. (2008c).Determination of total solids in biomass and total dissolved solids in liquid process samples. NREL/TP-510-42621. National Renewable Energy Laboratory, Golden, CO

Sluiter, A., Hames, B., Ruiz, R., Scarlata, C., Sluiter, J., & Templeton, D. (2008d). Determination of sugars, byproducts, and degradation products in liquid fraction process samples. NREL/TP-510-42623. National Renewable Energy Laboratory, Golden, CO

Taherzadeh, MJ., Niklasson, C., & Lidén G. (1997). Acetic acid - Friend or foe in anaerobic batch conversion of glucose to ethanol by *Saccharomyces cerevisiae? Chem. Eng. Sci.*, 52, 15, 2653-2659, 0009-2509

Teramoto, Y., Lee, SH., & Endo, T. (2009). Cost reduction and feedstock diversity for sulfuric acid-free ethanol cooking of lignocellulosic biomass as a pretreatment to enzymatic saccharification. *Bioresour. Technol.*, 100, 20, 4783-4789,

Teymouri, F., Laureano-Perez, L., Alizadeh, H., & Dale, BE. (2005). Optimization of the ammonia fiber explosion (AFEX) treatment parameters for enzymatic hydrolysis of corn stover. *Bioresour. Technol.* 96, 18, 2014-2018, 0960-8524

Titgemeyer, EC., Cochran, RC., Towne, EG., Armendariz, CK., & Olson, KC.(1996). Elucidation of factors associated with the maturity-related decline in degradability of big bluestem cell wall. *J Animal Sci.*, 74, 3, 648-657, 0021-8812

Toma´s-Pejo´, E., Oliva, JM., Ballesteros, M., & Olsson, L. (2008). Comparison of SHF and SSF processes from steam-exploded wheat straw for ethanol production by xylose-fermenting and robust glucose-fermenting *Saccharomyces cerevisiae* strains. *Biotechnol. Bioeng.*, 100, 6, 1122-1131, 1097-0290

Torget, R., Walter, P., Himmel, M., & Grohmann, K. (1991). Dilute acid pretreatment of corn residues and short-rotation woody crops. *Appl. Biochem. Biotechnol.*, 28/29, 75-86, *0273-2289*

Tucker, MP., Kim, KH., Newman, MM., & Nguyen, QA. (2003). Effects of temperature and moisture on dilute-acid steam explosion pretreatment of corn stover and cellulase enzyme digestibility. *Appl. Biochem. Biotechnol.* 105-108, 165-177, *0273-2289*

Vainionpaa, J., & Malkki, Y. (1987). Extrusion cooking in the manufacture of intermediate moisture food and feed. In *7th World Congress of Food Science and Technology*, Ghee AH (ed.). Singapore Institute of Food Science and Technology: Singapore,1987, pp. 338- 345

van Walsum, G., Allen, SG., Spencer, MJ., Laser, MS., Antal, MJ., & Lynd, LR. (1996). Conversion of lignocellulosics pretreated with hot compressed liquid water to ethanol. *Appl. Biochem. Biotechnol.*, 57/58, 157-170, *0273-2289*

Varga, E., Szengyel, Z., & Reczey, K.(2002). Chemical pretreatments of corn stover for enhancing enzymatic digestibility. *Appl. Biochem. Biotechnol.*, 98-100, 73-87, *0273-2289*

Varga, E., Schmidt, AS., Reczey, K., & Thomsen, AB. (2003). Pretreatment of corn stover using wet oxidation to enhance enzymatic digestibility. *Appl. Biochem. Biotechnol.*, 104, 37-50, *0273-2289*

Weiss, ND., Nagle, NJ., Tucker, MP., & Elander, RT. (2009). High xylose yields from dilute acid pretreatment of corn stover under process-relevant conditions. *Appl. Biochem. Biotechnol.*, 155, 1-3, 418-428, *0273-2289*

Xin, L., & Saka, S.(2008). Optimization of Japanese beech hydrolysis treated with batch hot-compressed water by response surface methodology. *Inter. J Agric. Bio. Eng.*, 1, 2, 239-245, 1934-6352

Xu, J., Thomsen, MH., & Thomsen, AB. (2009a). Pretreatment on corn stover with low concentration of formic acid. *J Microbiol. Biotechnol.*, 19, 8, 845-850, 1738-8872

Xu, J., Thomsen, MH., & Thomsen, AB. (2009b). Enzymatic hydrolysis and fermentability of corn stover pretreated by lactic acid and/or acetic acid. *J Biotechnol.*, 139, 4, 300-305, *0168-1656*

Yan, L., Zhang, H., Chen, J., Lin, Z., Jin, Q., Ji, H., & Huang, H. (2009) Dilute sulfuric acid cycle spray flow-through pretreatment of corn stover for enhancement of sugar recovery. *Bioresour. Technol.*, 100, 5, 1803-1808, 0960-8524

Yang, C., Shen, Z., Yu, G., & Wang, J. (2008). Effect and aftereffect of γ- radiation pretreatment on enzymatic hydrolysis of wheat straw. *Bioresour. Technol.*, 99, 14, 6240 6245, 0960-8524

Yeh, A-I., & Jaw, Y-M. (1998).Modeling residence time distributions for single screw extrusion process. *J Food Eng.*, 35, 211-232, 0260-8774

Yu, L., Ramaswamy, HS., & Boye, J. (2009). Twin-screw extrusion of corn flour and soy protein isolate (spi) blends: A response surface analysis. *Food Bioprocess Technol.*, DOI 10.1007/s11947-009-0294-8, 1935-5149

Yu, G., Yano, S., Inoue, H., Inoue, S., Endo, T., & Sawayama, S. (2010). Pretreatment of rice straw by a hot-compressed water process for enzymatic hydrolysis. *Appl. Biochem. Biotechnol.*, 160, 2, 539-551, *0273-2289*

Zeng, M., Mosier, NS., Huang, C-P., Sherman, DM., & Ladisch, MR. (2007). Microscopic examination of changes of plant cell structure in corn stover due to hot water pretreatment and enzymatic hydrolysis. *Biotechnol. Bioeng.*, 97, 265-278, 1097-0290

Zhu, Y., Lee, YY., & Elander, RT. (2004). Dilute-acid pretreatment of corn stover using a high-solids percolation reactor. *Appl. Biochem. Biotechnol.*, 117, 103-114, *0273-2289*

Zhu, Y., Lee, YY., & Elander, RT. (2005). Optimization of dilute-acid pretreatment of corn stover using a high-solids percolation reactor. *Appl. Biochem. Biotechnol.*, 124, 1-3, 1045-1054, *0273-2289*

Zhu, Z., Sathitsuksanoh, N., Vinzant, T., Schell, DJ., McMillan, JD., & Zhang, Y-HP. (2009). Comparative study of corn stover pretreated by dilute acid and cellulose solvent-based lignocellulose fractionation: Enzymatic hydrolysis, supramolecular structure, and substrate accessibility. *Biotechnol. Bioeng.*, 103, 4, 715-724, 1097-0290

Zimbardi, F., Viola, E., Nanna, F., Larocca, E., Cardinale, M., & Barisano, D. (2007). Acid impregnation and steam explosion of corn stover in batch processes. *Ind. Crops Prod.*, 26, 2,195-206, 0926-6690

15

# Innovative Biological Solutions to Challenges in Sustainable Biofuels Production

Xiaohan Yang[1,2] et al.*
[1]*Biosciences Division, Oak Ridge National Laboratory*
[2]*BioEnergy Science Center, Oak Ridge National Laboratory, Oak Ridge*
*USA*

## 1. Introduction

The rising prices, declining supplies, and concerns about environmental safety and energy security associated with the use of fossil fuels are driving the development and use of biofuels (Gonzalez-Garcia *et al.*, 2010; Markevicius *et al.*, 2010; Singh *et al.*, 2010; Sahin, 2011). Biofuels in general can be defined as liquid, gas and solid fuels predominantly produced from biomass (Demirbas, 2008). In this chapter, we will specifically focus on liquid biofuels which have attracted world-wide attention due to their renewability, sustainability, common availability, reduction of greenhouse gas (GHG) emissions, and biodegradability (Demirbas, 2009; Gonzalez-Garcia *et al.*, 2010; Balat, 2011). Currently there are two major types of liquid biofuels, bioalcohol and biodiesel, as alternatives to gasoline and diesel fuel, respectively. Among the various bioalcohols, bioethanol is currently the most widely used and biobutanol has great growth potential in the future due to its significant properties including high energy content, hydrophobicity, blending ability, compatibility with combustion engines, and octane rating (Kumar & Gayen, 2011). To date, liquid biofuels have been mainly produced in the U.S., Brazil and several European countries (Fig. 1A). Furthermore, there is a regional difference in the preference for biofuels types, with bioethanol preferentially produced in the American and Asian countries (e.g., U.S., Brazil, China, and Canada) while biodiesel is preferentially produced in European countries (e.g., Germany, France) (Fig. 1B).

Bioethanol can be produced from three categories of raw materials: simple sugars, starch, and lignocelluloses (Balat, 2011). Biomass feedstock for biodiesel production is under active development worldwide, with rapeseed and sunflower oils predominating in Europe, palm oil in tropical countries, and soybean oil and animal fats in the United States; and development of additional feedstocks such as *Jatropha* oil and algae for biodiesel is also underway (Dyer *et al.*, 2008; Knothe *et al.*, 2009). In particular, microalgal oil is one of the major renewable biofuels with great potential for replacing petroleum-based liquid fuels (Cooper *et al.*, 2010).

* Ting Li[1], David Weston[1], Abhijit Karve[1], Jessy L. Labbe[1], Lee E. Gunter[1,2], Poornima Sukumar[1], Anne Borland[3], Jin-Gui Chen[1], Stan D. Wullschleger[4], Timothy J. Tschaplinski[1,2] and Gerald A. Tuskan[1,2]
[1]*Biosciences Division, Oak Ridge National Laboratory, USA*
[2]*BioEnergy Science Center, Oak Ridge National Laboratory, Oak Ridge, USA*
[3]*Newcastle Institute for Research on Sustainability, Newcastle University, Newcastle Upon Tyne, UK*
[4]*Environmental Sciences Division, Oak Ridge National Laboratory, USA*

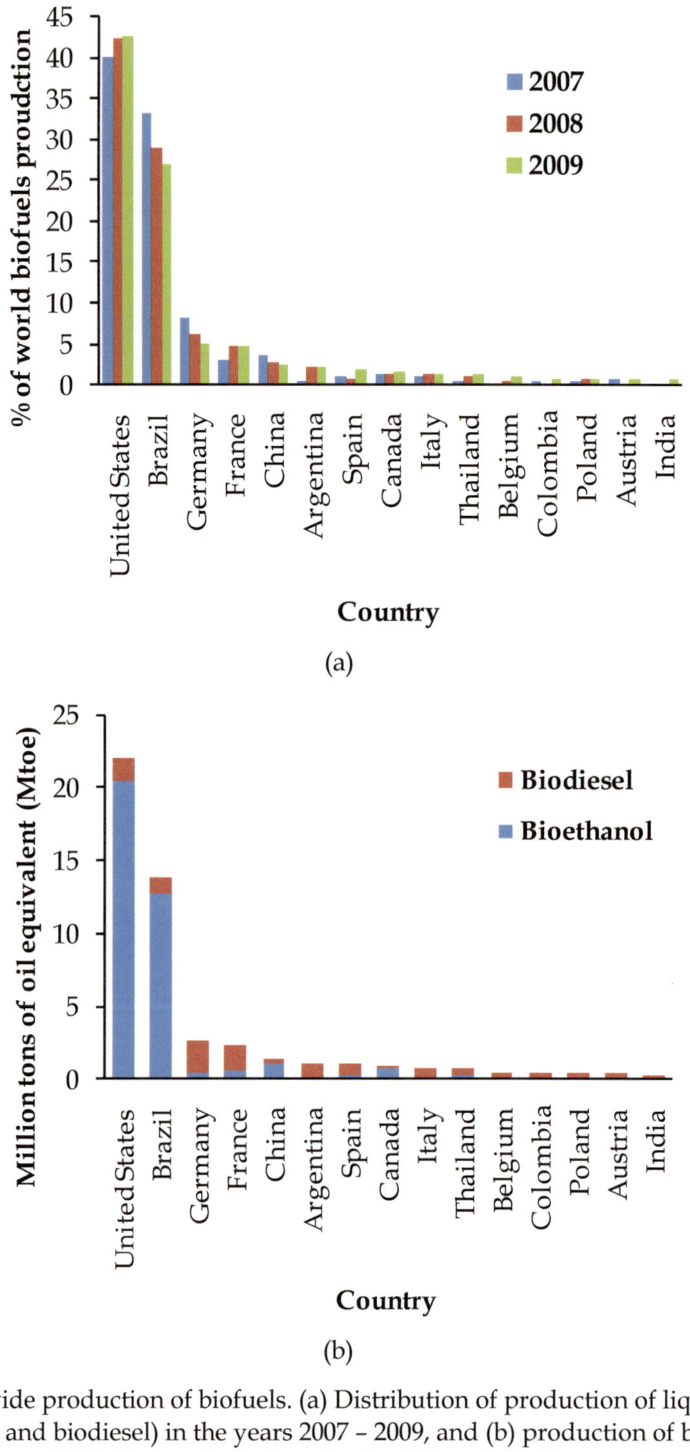


Fig. 1. World-wide production of biofuels. (a) Distribution of production of liquid biofuels (i.e., bioethanol and biodiesel) in the years 2007 - 2009, and (b) production of bioethanol and biodiesel in the year 2009. Drawn from data obtained from http://www.plateforme-biocarburants.ch

Although biofuels have advantages over fossil fuels, the use of biomass does not automatically imply that its production, conversion, and utilization are sustainable given the potential conflict between land use for food versus fuel (Markevicius *et al.*, 2010; Payne, 2010). In this chapter, we will first describe the challenges in the sustainable production of liquid biofuels and then discuss the novel biological approaches for solving these challenges.

## 2. Challenges in sustainable biofuel production

Currently sustainable biofuel production faces several major challenges: 1) Biofuel versus food competition, 2) limited biomass production, 3) recalcitrance of biomass for biofuel production, and 4) less-than-ideal physical properties of biofuels. We will discuss each of these challenges below.

### 2.1 Biofuel versus food competition

Biofuel crops are generally planted on agricultural land and most of the current bioenergy crops are also used as food or animal feed. Such dual-use crops include barley, maize, rice, rye, sorghum, wheat, cassava, potato, sugar beet, sugarcane, rapeseed, and soybean (Gerbens-Leenes *et al.*, 2009; Sahin, 2011). To date, almost all bioethanol has been produced from food sources such as grain or sugarcane (Mussatto *et al.*, 2010; Somerville *et al.*, 2010) and expanding biofuel production from such feedstocks is likely to exacerbate food insecurity and political instability (Payne, 2010). If terrestrial biofuels are to replace ~90 EJ (= $90x10^{18}$ J) mineral oil-derived transport fuels, large areas of good agricultural land will be required: about $5x10^{8}$ ha in the case of biofuels from sugarcane or oil palm and at least $1.8\text{-}3.6x10^{9}$ ha in the case of ethanol from wheat, corn, or sugar beet, an area that is equivalent to the current worldwide cropland (~$1.8x10^{9}$ ha) (Reijnders, 2009). Moreover, bioenergy crops will potentially compete with food crops for inputs such as water and nutrients. Agriculture accounts for ~70% of all the world's freshwater withdrawals (Rosegrant *et al.*, 2009) and a decline in water availability is already a major constraint on agricultural productivity and global food security (de Fraiture *et al.*, 2008). Thus, sustainable production of biofuel feedstocks requires the use of land that is not required or is not suitable for food production (Marko *et al.*, 2009; Reijnders, 2009; Fritsche *et al.*, 2010). Development of new capabilities for biomass production on marginal or abandoned land with minimized water and nitrogen supply would be the best strategy to avoid the biofuel versus food competition. We will discuss several specific approaches to implement this strategy, such as introduction of new crops (see Section 3.1) and transgenic crops (see Section 3.2) that have high water use efficiency (WUE) and nitrogen use efficiency (NUE).

### 2.2 Limited supply of biomass for biofuel production

A major constraint on bioethanol production is the availability of biomass feedstock (Balat, 2011). Currently biofuel production accounts only for a small portion (~2%) of the 1,200 billion liters of annual gasoline consumption worldwide (de Fraiture *et al.*, 2008) and the contribution of biodiesel to global transportation fuel consumption is only 0.14% (Courchesne *et al.*, 2009). Assuming that 50% of the energy content of the feedstock can be recovered as liquid biofuels, the potential of global woody biomass is predicted to produce 73.8 million tonnes (3.1 EJ) of liquid biofuels in the year 2020, accounting for only 2.6% of the global forecasted transportation fuel consumption (117 EJ) (Asikainen, 2010). The production of biofuels from lignocellulose is limited by the amount of plant biomass, as demonstrated by the estimation that lignocellulosic biomass harvested from all switchgrass, hybrid poplar, corn stover, and wheat straw in the United States could produce 10.31 billion gallons of ethanol or 8.27 billion gallons of butanol, which could replace 6.97 or 7.55 billion gallons of gasoline, respectively, leaving a significant gap from the target of 21 billion gallons of biofuels per year (Swana *et al.*, 2011). The major economic factor affecting the input costs of biodiesel production is the feedstock, which is about 75-80% of the total operating cost (Demirbas, 2010). Likewise, the biggest challenge for meeting current and future targets in biodiesel production is the limited supply of feedstocks, which necessitates an increase in the efficiency of plant oil production (Durrett *et al.*, 2008; Li *et al.*, 2010). Limitations in biomass quantity may be attributed to environmental and biochemical constraints on net photosynthetic productivity (Schaub & Vetter, 2008). We will discuss specific approaches for increasing biomass supply for biofuel production, such as the selection of feedstocks for biomass production on marginal land (see Section 3.1), genetic improvement in biofuel yield (see Sections 3.2), and utilization of beneficial microorganisms to increase the yield of bioenergy crops (see Sections 3.4).

### 2.3 Recalcitrance of biomass for biofuel production

Developing non-food, "next-generation" feedstocks such as lignocellulosic biomass has the potential to meet most of the global transportation fuel needs without impacting negatively on food security (Abramson *et al.*, 2010). A major bottleneck for conversion of lignocellulosic biomass to simple sugars (saccharification), to be subsequently converted by microorganisms into ethanol or other products, is the recalcitrance to enzymatic saccharification (Chen & Dixon, 2007; Lionetti *et al.*, 2010). Recalcitrance is mainly due to the heterogeneity and molecular structure of lignocellulose where cellulose is arranged into a network of tight, inter-chain hydrogen bonds that form a crystalline core of microfibrils, embedded in a matrix of hemicellulosic polysaccharides that are covalently linked to lignin, a highly complex aromatic polymer (Vega-Sanchez & Ronald, 2010). Lignin contributes to biomass recalcitrance and consequently increases the costs associated with conversion (Simmons *et al.*, 2010; Vega-Sanchez & Ronald, 2010). Lignins are complex aromatic biopolymers, consisting of (mainly) syringyl (S), guaiacyl (G), and p-hydroxyphenyl (H) units (Simmons *et al.*, 2010). Variations in lignin content and its S-G monomer composition is directly associated with the yield of fermentable sugars (Lee & Voit, 2010). Pectin that embeds the cellulose-hemicellulose network affects the exposure of cellulose to enzymes and consequently the process of saccharification (Lionetti *et al.*, 2010). The lack of efficient biocatalysts and microorganisms to convert lignocellulosic raw materials into liquid fuels is a further bottleneck for sustainable adoption of next-generation feedstocks (Liu & Khosla, 2010). We will discuss several approaches to address the biomass recalcitrance issue, including genetic modification of cell walls (see Section 3.2) and engineering of microorganisms for biomass conversion (see Section 3.3).

### 2.4 Less-than-ideal physical properties of biofuels

The physical properties of current liquid biofuels including bioalcohol and biodiesel are less-than-ideal for applications in transportation. Although bioethanol currently dominates the biofuel market, some of its inherent physical properties, such as low energy content and incompatibility with existing fuel distribution and storage infrastructure, limit its economic use (Peralta-Yahya & Keasling, 2010). Biobutanol is a viable alternative to bioethanol because it has a higher energy content and lower solubility in water, can be transported through existing pipelines, and can be used to supplement both gasoline and diesel fuels (Fortman *et al.*, 2008). However, biobutanol has its own shortcomings: it is produced at a lower titer, is much more toxic than ethanol, and requires more energy than ethanol for distillation-based purification from fermentation broth, due to its high boiling-point (Fortman *et al.*, 2008). For example, the energy yield of n-butanol is about half that of ethanol from corn or switchgrass using current acetone-butanol-ethanol (ABE) technology and the low yield increases n-butanol's life-cycle greenhouse gas emission for the same amount of lower heating value (LHV) compared to ethanol (Pfromm *et al.*, 2010). Also, the net energy (6.53 MJ/L) generated during corn-to-biobutanol conversion is greater than that (0.40 MJ/L) of the corn-derived bioethanol (Swana *et al.*, 2011). Although biodiesel obtained from some oil crops, such as *Calophyllum inophyllum, Azadirachta indica, Terminalia catappa, Madhuca indica, Pongamia pinnata*, and *Jatropha curcas* oils meet current biodiesel standards in both the European Union (EN 14214) and the United States (ASTM D 6751 02), none of the current biodiesel products can be considered to be the "ideal" alternative that matches all of the key fuel properties that ensure the best diesel engine performance (Pinzi *et al.*, 2009). Plant oils are mostly composed of long-chain (C16 and C18) fatty acids (FAs) such as palmitate (16:0), stearate (18:0), oleate (18:1), linoleate (18:2), and linolenate (18:3), and these FAs differ from each other in terms of acyl chain length and number of double bonds, leading to different physical properties (Durrett *et al.*, 2008). One of the major problems associated with biodiesel properties is the poor flow at low temperatures due to the predominant components of long-chain (C16 and C18) FAs in oil produced from biomass feedstock such as oil seeds and algae (Knothe *et al.*, 2009). For example, the cloud point (i.e., below the cloud point, the formation of crystals clogs the diesel injection) of bio-oil is higher than that of fossil diesel, particularly for oil obtained from some major tropical bioenergy crops such as palm (Abolle *et al.*, 2009a; Abolle *et al.*, 2009b). The presence of saturated methyl esters longer than C12 significantly increases the cloud point, even when blended with conventional diesel fuel (Durrett *et al.*, 2008). Therefore, the current forms of pure biodiesel are not suitable for use in colder climates. We will discuss genetic improvement of biofuel quality as a possible strategy to address the limitations in physical properties of liquid biofuels (see Section 3.2.2).

## 3. Biological solutions

### 3.1 Development of new crops for biomass production on marginal lands

To address the two challenges "biofuel versus food competition (Section 2.1)" and "limited supply of biomass for biofuel production (Section 2.2)", it is crucial to find ways to produce biomass on marginal lands that are not useful for food production. For many locations around the world, marginal lands represent a valuable resource that could prove to be a viable option for bioenergy crop production. However, crops will need to be tailored to such water-limited and degraded regions, as current biomass crops (e.g., poplar, sugarcane) are poorly suited for biomass production on such lands without irrigation and proper fertilization. Therefore, land-based biofuel crops with high WUE, drought tolerance, and NUE, as well as aquatic biofuel crops, such as microalgae, have great potential for biofuel production on non-agricultural lands.

### 3.1.1 Land-based biofuel crops with high WUE and drought tolerance

Several emerging or potential bioenergy crops such as *Agave*, sweet sorghum, and *Jatropha* are suitable for production on marginal land because of their high drought tolerance and/or WUE. Succulent species of the genus *Agave* have been cultivated for centuries as sources of alcohol and fibres from rain-fed semi-arid lands. Certain species have been reported to display annual above ground productivities that are comparable to those of the most water-use efficient C3 or C4 crops but with only 20% of the water required for cultivation (Borland *et al.*, 1999). Such characteristics have provoked interest in the potential of *Agave* as a sustainable source of bioenergy feedstock that will not compete with food and fodder production, whilst offering potential for carbon sequestration on marginal and degraded land (Davis *et al.*, 2011). The desirable traits of high productivity and water conservation in *Agave* can be attributed to the operation of crassulacean acid metabolism (CAM), a specialized mode of photosynthetic $CO_2$ acquisition (Fig. 2). CAM is expressed on a background of Rubisco-mediated $CO_2$ fixation via the engagement of nocturnal $CO_2$ uptake catalysed by phosphoenolpyruvate carboxylase (PEPC) and subsequent day-time decarboxylation processes. In CAM plants like *Agave*, stomata open at night when evapo-transpiration rates are low and atmospheric plus respiratory $CO_2$ is fixed in the cytosol by PEPC. The 3-C substrate phosphoenolpyruvate (PEP) is formed from the glycolytic breakdown of carbohydrates. The final 4-C product, malic acid, is stored in a large central vacuole. During the day, malate exits the vacuole and is decarboxylated through the single or combined action of three enzymes (depending on plant species): NADP malic enzyme (NADP-ME), NAD-ME, and phosphoenolpyruvate carboxykinase (PEPCK). In addition to the 3-C products PEP or pyruvate, $CO_2$ is released at a high internal partial pressure ($pCO_2$). This is accompanied by stomatal closure and transpirational water loss is curtailed. By opening their stomata during the cooler night time, CAM plants lose far less water than they would during the warmer day time, and thus *Agave* spp. have lower seasonal water requirements than other bioenergy crops such as corn, sugarcane, Miscanthus, and poplar (Somerville *et al.*, 2010). *Agave* avoids dehydration via structural adaptations such as leaf succulence, and shrinkage of the root cortex (hydraulic isolation) can occur at modest soil deficits with cavitation of the root xylem, curtailing water loss from storage tissues to a drying soil (North *et al.*, 2004). Besides having relatively low requirements for water and nutrients, species such as *A. tequilana*, *A. mapisaga* and *A. salmiana* can provide high yield and high quality biomass for biofuel production. The typically low rates of transpiration in *Agave* leaves obviate the requirement for a highly lignified xylem and so lignin contents are relatively low (3–15% by dry weight) whilst cellulose content is relatively high (up to 68%) (Davis *et al.*, 2011). *Agave* biomass can be harvested year-round, producing up to 500 metric tons (green) of biomass per hectare annually (Austin, 2010a; Austin, 2010b). Some *Agave* cultivars possess higher sugar content than sugarcane in Brazil, higher cellulose content than the fastest-growing *Eucalyptus*, and more dry biomass than the genetically-modified poplar trees (Austin, 2010b). Therefore, *Agave* has the potential to become a new bioenergy crop due to its high water use efficiency (3 - 6 fold higher than C4 or C3 plants, respectively) (Borland *et al.*, 2009), drought tolerance, high yield, and high quality of biomass. One major limitation in the development of *Agave* into an important biomass feedstock is that there is essentially no genomics-based knowledge to inform improvement strategies for bioenergy purposes. Recently, we initiated an *Agave* genomics project at Oak Ridge National Laboratory (USA) to obtain a genomic and biochemical-based understanding of CAM in *Agave* necessary for its consideration as a biofuel feedstock. Several other *Agave* transcriptome sequencing projects have been initiated in the United Kingdom (J Hartwell, personal communication) and Mexico (Simpson *et al.*, 2011).

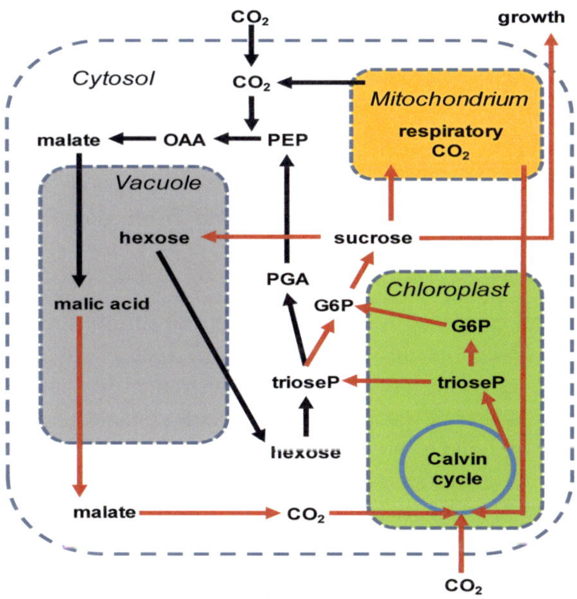


Fig. 2. The CAM pathway. G6P, glucose 6-phosphate; PEP, phosphoenolpyruvate; OAA, oxaloacetate. Red and black arrows represent light-and dark-period reactions, respectively. Adapted from Holtum et al. (2005), Borland et al. (2009), and Wild et al. (2010).

Sweet sorghum is a potential feedstock for bioalcohol production, with advantages in hot and dry climatic conditions over alternatives, such as sugarcane or maize (Raghuwanshi & Birch, 2010) because it has higher tolerance to salt and drought compared to sugarcane and corn that are currently used for biofuel production. Moreover, the high carbohydrate content of sweet sorghum stalk is similar to sugarcane, but its water and fertilizer requirements are much lower than sugarcane (Almodares & Hadi, 2009).

Jatropha (*J. curcas* L.) has gained much attention for biodiesel production in tropical and sub-tropical countries because of its hardiness, ease of propagation, drought tolerance, high oil content, rapid growth, adaptation to wide agro-climatic conditions, and multiple uses of the plant as a whole (Divakara *et al.*, 2010). Jatropha, known commonly as physic nut, is native or naturalized to parts of Asia, Africa and Central/South America, and has been identified as a multipurpose species with many attributes that give it considerable potential as a biodiesel crop in different parts of the world (Gubitz *et al.*, 1999). It has been shown that the seed kernel of this member of the Euphorbiaceae or spurge family contains 40-60% (w/w) oil deemed unsuitable for cooking due to the presence of toxic esters (Shah *et al.*, 2004). The seed oil of Jatropha was used as a diesel fuel substitute during World War II (Agarwal, 2007), and in more recent years the unmodified Jatropha oil and blends with diesel fuel (Banerji *et al.*, 1985; Jones & Miller, 1991) and transesterified oil esters were tested as an alternative fuel for Thailand (Takeda, 1982; Ishii & Takeuchi, 1987). Despite the growing interest in Jatropha as a biofuel feedstock, it lacks improved germplasm and, until recently, active breeding programs had been lacking. Major germplasm collections for

Jatropha are now found in India (Kaushik *et al.*, 2007; Sunil *et al.*, 2008), Africa, and the Philippines. Information on genetic diversity in Jatropha is still limited since most studies have concentrated on accessions from India where the shrub was brought by the Portuguese. Due to its relatively small genome (2C value of 0.85 pg, in the same size range as that of rice) (Carvalho *et al.*, 2008), Jatropha could become a model woody crop for biodiesel production. Genetic and genomic resources for this emerging biofuels crop are becoming available including a transformation system (Li *et al.*, 2008), a 100x coverage of the *J. curcas* genome sequence (http://www.lifetechnologies.com/news-gallery/press-releases/2010/life-techologies-ad-sg-biofuels-complete-sequece-of-jatropha-geo.html), and a growing library of expressed sequence tags (ESTs) from developing and germinating Jatropha seeds (Costa *et al.*, 2010).

### 3.1.2 Land-based biofuel crops with high nitrogen use efficiency

Nitrogen use efficiency (NUE) is dependent on many factors including soil nitrogen (N) availability, uptake and assimilation, and carbon-nitrogen flux, and is one of the major limiting factors in increasing crop productivity (Pathak *et al.*, 2008; Raghuram *et al.*, 2008). Although NUE can be calculated in a number of ways (Good *et al.*, 2004), a simple yet useful metric is yield per unit of available N in the soil (Kant *et al.*, 2011). Kant and colleagues (2011) suggest that plant N use can be divided into two general stages. The first stage is characterized by N uptake, assimilation into organic compounds (e.g., amino acids), and storage. All of these processes contribute to biomass accumulation. The second stage represents the proportion of N that is allocated to the final yield product (e.g., grain, fruit, and biomass). Relative to traditional agronomic crops, both stages must be considered when assessing next generation bioenergy feedstocks (e.g., lignocellulosic crops). For example, the current land use strategy is to relegate bioenergy crops to marginal lands thereby lessening competition with food crops for limiting arable soils. This would have a direct impact on available N and subsequent plant N uptake and assimilation. In regard to the second stage, lignocellulosic bioenergy crops are often perennial with a biomass yield component. By contrast, traditional agronomic crops are often annual with yield components consisting of grain or fruit. Therefore, allocation within a life-cycle context will be an important component and target for NUE improvement of bioenergy feedstocks. Here, we will discuss NUE in the context of next generation bioenergy crops with a focus on N uptake and assimilation, allocation in a life-cycle and growth habit context, and the interaction of N uptake and allocation driven by genetic controls on root architecture.

#### 3.1.2.1 Nitrogen uptake and assimilation

Stage one of the above NUE model is driven by N uptake and assimilation. In agricultural soils, the predominant form of N is nitrate and to a lesser extent ammonium (Crawford & Forde, 2002). Both high- and low-affinity transporters mediate nitrate uptake and transport. In *Arabidopsis*, for example, there are three main classes of nitrate transporters represented by over 67 genes (Kant *et al.*, 2011). After entering the cell, nitrate is reduced to nitrite by nitrate reductase, and nitrite is further reduced to ammonium in plastids by nitrite reductase (Crawford & Forde, 2002). Ammonium is then assimilated into amino acids through the GOGAT (glutamine synthetase/glutamate synthase) cycle. A number of studies have attempted to increase NUE through the expression of genes associated with N uptake and assimilation. For example, Fraisier et al. (2000) constitutively expressed a high-affinity transporter in *Nicotiana*. Although nitrate influx was enhanced, there was no phenotypic difference or measurable change in NUE. Similar results were obtained with genetic approaches to alter the expression of nitrate and nitrite reductase N assimilation genes. In these studies, enzyme abundance was increased but complex regulatory feedbacks resulted in no detectable phenotypic improvement (Good *et al.*, 2004). There has been some success with the overexpression of glutamine synthetase, where a 30% increase in kernel number was reported (Hirel *et al.*, 2006). However, no successful commercial lines have been developed using this approach (Kant *et al.*, 2011), which highlights the challenge in transferring laboratory results to field-based applications.

Given that the predominant form of N is nitrate in agriculture soils, we often overlook the potential for organic N source (e.g., amino acids, peptides, etc.) to contribute to overall plant nutritional status. To date, all plant species tested have the ability to acquire amino acids (Lipson & Nasholm, 2001; Nasholm *et al.*, 2009). This includes species that interact with all major mycorrhizal types and non-mycorrhizal types as well. Numerous studies suggest that organic N is an important mineral substrate in the arctic, boreal, temperate, Mediterranean shrubland, and alpine ecosystems (Nasholm *et al.*, 2009). Our understanding of the mechanism by which organic N enters plant cells and is assimilated is quite limited relative to uptake of nitrate and ammonium. There are numerous amino acid transporters belonging to multiple families (Rentsch *et al.*, 2007), yet few have been functionally characterized. Only a handful of studies have investigated how acquired amino acids are assimilated into the N pathway (Schmidt & Stewart, 1999; Thornton & Robinson, 2005; Persson *et al.*, 2006). Based on their results, it appears that amino acids are more likely to be transaminated rather than deaminated and are able to move into shoots. Mycorrhizal associations are known to facilitate proteolysis of soil nitrogenous compounds and enhance the uptake of organic N to plant hosts (see Section 3.4). For sustainable production of bioenergy feedstocks on marginal lands, strategies for increasing NUE through improvement of organic uptake and assimilation should be considered. Possible strategies include a greater understanding and thus modification of the organic N assimilation pathway, and directed plant-microbe interactions (see Section 3.4).

#### 3.1.2.2 Carbon allocation and NUE in annual versus perennial crops

A key challenge for the production of next generation bioenergy feedstocks is increasing yields while maintaining sustainability. As mentioned previously, the existing agricultural concept of NUE relates N uptake to yield (Moll *et al.*, 1982), generally in terms of grain production, and thus has severe limitations in comparing annual to perennial crops. In ecological studies, NUE is associated with whole-plant physiology, the assimilation of N, and other nutrients that are necessary for carbon fixation into sugars and carbon allocation into tissues forming stems, leaves and roots. For bioenergy crops, an assessment of the growth habit and life cycle of the crop is necessary in order to compare NUE of seed or oil crops to lignocellulosic energy. In addition, it is clear that NUE should be calculated from harvestable rather than total biomass (Weih *et al.*, 2011). In general, NUE for bioenergy crops is not well studied or characterized, and most studies do not address integration of processes. Whereas annuals depend more on acquired nutrients for growth (Chapin *et al.*, 1990), perennial crops have an advantage with traits such as rapid spring regrowth from existing carbon reserves and generally higher NUE (Jorgensen & Schelde, 2001). Lignocellulosic crops such as poplar, willow, Eucalyptus, and Miscanthus have higher NUE than traditional annual cereal crops in part due to differences in harvest time or multiple year rotations which allow higher rates of translocation of N to storage organs like stems and roots (Jorgensen & Schelde, 2001). Ecological studies suggest that NUE is the product of mean retention time (MRT), defined as the length of time a unit of N is present in a population, which is representative of N carryover from annual to perennial plant parts (Berendse & Aerts, 1987; Aerts & Chapin, 2000; Weih *et al.*, 2011). Thus, perennials may compensate for lower N acquisition capabilities by having higher N retention due to a lower total biomass turnover rate (Aerts & Chapin, 2000). A high NUE does not necessarily indicate that the system as a whole is more efficient (Jorgensen & Schelde, 2001). One of the criticisms leveled at bioenergy crops is an increased use of N fertilizers derived from fossil fuels and associated greenhouse gas (GHG) emissions (Scharlemann & Laurance, 2008; Erisman *et al.*, 2010). Most of the major industrialized areas of the world, including the United States, European Union, and China have proposed increasing sustainable energy sources through the development of bioenergy crops. However, there have been few discussions over the environmental impacts of changes in the N cycle as a result of increasing biomass production. Thus, improvements in NUE of bioenergy crops will be crucial for mitigation of GHG associated with the production of biofuels (Erisman *et al.*, 2010). NUE of perennial biofuel crops can be improved through a combination of optimizing soil, fertilizer and water interactions, as well as through improvement in traits associated with the physiology of N uptake and assimilation. Development of higher yield bioenergy crops with increased NUE and decreased or neutral soil and atmospheric N losses is critical in order to create a sustainable source of energy for increasing world energy consumption (Erisman *et al.*, 2010).

#### 3.1.2.3 Root architecture

Plants rely on roots and their dynamic architecture for water and nutrient uptake from soil. It is a dilemma, especially under nutrient restricted conditions, for plants to allocate their limited N resources to root growth for foraging of additional nutrients or to shoot development and reproductive structures. Therefore, it is important to understand the changes associated with root growth and development regulated by nutrients especially in the context of nitrogen. Roots have been shown to absorb various forms of N including inorganic nitrate ions and ammonium ions, and organic amino acids, with the help of membrane localized transporters (Nasholm *et al.*, 2009; Masclaux-Daubresse *et al.*, 2010). Nitrogen availability in soil can modify root architecture dynamically. Moreover, the type of N available can also influence root growth (Walch-Liu & Forde, 2008). High nitrate concentrations can reduce primary and lateral root growth, while low nitrate content can enhance outgrowth of laterals (Walch-Liu *et al.*, 2006). Additionally, lateral root development was reduced in *Arabidopsis* plants grown in high sucrose to nitrate ratio (Malamy & Ryan, 2001). Even though high accumulation of nitrates can cause a decrease in root elongation, localized nitrate supply can induce the elongation of lateral roots. In *Arabidopsis*, within species variation was observed in root growth responses as an adaptive mechanism to N availability (Walch-Liu & Forde, 2008). The influence of N content on root growth has been attributed to NRT2.1, a nitrate transporter, although contradicting reports suggest that this protein could act positively or negatively in regulating lateral root growth (Kant *et al.*, 2011). A recent study has revealed a role for the nitrate transporter NRT1.1 in modulating lateral root development under variable nitrate availabilities. This is accomplished by functioning as a plant hormone (auxin) transporter and by regulating auxin accumulation that is necessary for primordia development (Krouk *et al.*, 2010). There are co-localized QTLs for root architectural traits and N uptake traits (Coque *et al.*, 2008).

More studies are needed to dissect the complex interactions between N content regulation, root architectural modifications, and the genetic control of these structural and functional traits associated with nutrient acquisition.

N allocation is a key component related to growth, development, and yield in plants. The N management of plants varies across growth stages. In the early stage, developing shoots and roots act as a sink for N, with assimilated N being used for production of proteins required for structure as well as other regulatory functions (Hirel *et al.*, 2007). At a later stage, roots and shoots serve as a source for N for developing reproductive and storage organs. N remobilization from senescing tissues to young and developing tissues occur at both stages of growth and reproduction (Masclaux-Daubresse *et al.*, 2010). Additional cycling of N can occur through assimilatory and photorespiratory fluxes throughout the life cycle of plants (Hirel *et al.*, 2007). Under high nutrient conditions and at later stages of plant development, root to shoot ratio is low (Garnett *et al.*, 2009). In soils where leaching loss of nutrients are high, a root system with dynamic growth is relevant in N uptake, rather than having high root/shoot ratio (Garnett *et al.*, 2009). Under low N conditions, there is a negative relation between root number and yield, possibly due to competition for limiting resources between shoot and root (Hirel *et al.*, 2007). There is variation among species in the involvement of root architecture for N uptake before and after flowering. In some species such as maize, grain yield was correlated to root architecture when grown under low and high levels of N (Garnett *et al.*, 2009). Additional regulation comes at the level of nitrate transport components during different stages of root and shoot development, which would directly regulate adaptive responses to various environmental conditions. Root growth and architecture, thus, are important in understanding N uptake efficiency under various soil conditions.

Improving NUE by altering root growth is an important aspect to maximize plant growth and yield. Various aspects of root architecture such as root length, density of lateral roots, age of roots, and root hairs can affect N uptake depending on environmental conditions and N availability. Additionally, mycorrhizal and arbuscular microbial associations in plants have also been shown to enhance N uptake (Hawkins *et al.*, 2000; Parniske, 2008). The duration of N uptake is also relevant. Continuance of N uptake through flowering and early grain development was associated with increased NUE in maize (Worku *et al.*, 2007). Deeper root systems are advantageous in soils where N resources diffuse deep down into the soil profile. Not only the total length, but the root length per volume (root length density) positively correlates with increased NUE, depending on the soil type and species of plants (Garnett *et al.*, 2009). This is due to an increase in root surface area to acquire nutrients from soil, especially in acquiring ammonium ions that are less mobile in soils. However, this is not applicable in soils that have high nutrient content and/or have low leaching, as N levels are saturating and increased surface area due to root hairs is not beneficial (Garnett *et al.*, 2009). A modeling study looking at the relation between N availability and root architecture has shown that the dependence on root morphology in N uptake occurs at low N concentration. Addition of root hairs to the model further reduced the limit of root morphology dependent N concentrations. Moreover, increasing root diameter had no effect on assimilation of nitrate and ammonium ions in the model (Robinson & Rorison, 1983). Within a root system, uptake rates of nitrate ions differ between young and older roots. The older roots could continue to uptake N, even though the rate of uptake might go down, possibly helping improve NUE (Garnett *et al.*, 2009). In an inbred maize line, greater N acquisition was associated with a more responsive root system to low N, a larger and longer root system, and a greater root/shoot ratio (Liu *et al.*, 2009). Proteolytic enzymes from root exudates can help in degrading proteins, which then can be taken up by plants (Garnett *et al.*, 2009). In parallel, certain factors could negatively affect NUE. Efflux of nitrate and ammonium ions from roots can decrease the net uptake, thereby reducing NUE. In addition, the down-regulation of high affinity N transporters when N is not limited reduces net uptake of N. Environmental factors, such as low light levels and low temperature, limit the net uptake of N (Glass, 2003). Understanding root traits that improve NUE could be used to select plants using breeding or genetic modification techniques for enhanced N utilization capacities.

### 3.1.3 Aquatic biofuel crops

Biofuels derived from aquatic microbial oxygenic photoautotrophs (AMOPs) including cyanobacteria, algae, and diatoms offer a number of environmental and economic benefits over terrestrial biofuel feedstocks. AMOPs are inherently more efficient solar collectors than terrestrial plants, use less or no land, can be converted to liquid fuels using simpler technologies than those required to break down cellulose, and offer secondary uses that fossil fuels do not provide (Dismukes *et al.*, 2008). Algae in particular have great potential for the renewable production of several bioenergy carriers such as starches for bioalcohols and lipids for biodiesel (Beer *et al.*, 2009). Compared with terrestrial biofuel feedstocks, algae have higher photosynthetic efficiencies for conversion of solar energy into fuels, higher productivities, use of otherwise nonproductive land, reuse and recovery of waste nutrients, less water consumption, use of saline or brackish waters, year-round production, daily harvesting, and reuse of $CO_2$ from power-plant flue gas or similar sources (Schenk *et al.*, 2008; Beer *et al.*, 2009; Brune *et al.*, 2009; Gouveia & Oliveira, 2009; Posten & Schaub, 2009). The oil yield from microalgae (20,000 to 80,000 liters per acre per year) is 7-31 times greater than the next best terrestrial crop, palm oil (Demirbas & Demirbas, 2011). Among the various microalgae (e.g., *Chlorella vulgaris, Spirulina maxima, Nannochloropsis* sp., *Neochloris oleoabundans, Scenedesmus obliquus* and *Dunaliella tertiolecta*) recently tested, *Neochloris oleoabundans* (fresh water microalga) and *Nannochloropsis* sp. (marine microalga) are suitable for biofuel production due to their high oil content (29.0% and 28.7%, respectively), with a substantial increase (50%) in oil quantity when grown under low nitrogen (Gouveia & Oliveira, 2009). The high productivity of algae suggests that much of the US transportation fuel needs could be met by algal biofuels at a production cost competitive with the cost of petroleum seen during the early part of 2008 (Pienkos & Darzins, 2009). One major limitation is that the current practice used to cultivate, harvest, and process algae for biofuels production is too expensive to make algal biofuel cost-competitive with fossil fuels (van Beilen, 2010).

Cyanobacteria are excellent organisms for biofuel production for a number of reasons: their genomes are relatively easy to manipulate; they are efficient at converting solar energy into biofuels; and they can be grown on non-arable land using photobioreactors (Rittmann, 2008; Liu & Curtiss, 2009; Kumar *et al.*, 2011). An attractive feature of cyanobacteria as a candidate for biofuel-producing microbial systems is that they incorporate the favorable characteristics of both plants and prokaryotics. Unlike the generally utilized biofuel-producing microbes (e.g.,*Escherichia coli, Zymomonas mobilis, Saccharomyces cerevisiae,* etc.) that have been exploited to make biofuels from glucose produced from polysaccharides through fermentation (Lu, 2010), cyanobacteria can absorb solar energy and fix carbon dioxide (thereby contributing to C sequestration) and are more efficient in converting solar energy and carbon dioxide into useable substrates for biofuels as compared to terrestrial plants. Cyanobacterial cultures can have better water conservation than terrestrial plant feedstocks, and many cyanobacterial strains are tolerant of marine, brackish, or industrial waste waters, and might effectively utilize water resources that are not suitable for terrestrial crops (Ducat *et al.*, 2011). In general, compared to plants and eukaryotic microalgae, cyanobacteria are more amenable to genetic manipulation for installing biofuel-producing chemical pathways, as demonstrated by the successful reconstruction of metabolic network in *Synechocystis* sp. PCC 6803 (Knoop *et al.*, 2010; Lu, 2010). Cyanobacterial species have been engineered for the production of biofuels (e.g., alcohols, alkanes and hydrogen) (Ducat *et al.*, 2011) and have been tested as a feedstock for biodiesel production by simultaneous extraction and conversion of total lipids (Wahlen *et al.*, 2011). One limitation for biofuel production is that there is inadequate knowledge of cyanobacterial biology and genetic tools in cyanobacteria are less developed in comparison to traditional bioindustrial workhorse organisms, such as *E. coli* and yeast (Ducat *et al.*, 2011).

### 3.2 Genetic improvement of current bioenergy crops

For sustainable bioenergy production, the crop should be high yielding, fast growing, have low lignin content, and require relatively low energy inputs for its growth and harvest on nonprime agricultural land (Waclawovsky *et al.*, 2010). Genetic engineering can be used to improve bioenergy crops in various aspects such as reducing biomass recalcitrance, enhancing water and nitrogen use efficiency, increasing biofuel yield, and modifying properties of biodiesel. Efficient transformation systems are now available for some biofuel feedstock crops, such as *Camelina sativa* (Lu & Kang, 2008), *J. curcas* (Li *et al.*, 2008; Kumar *et al.*, 2010; Pan *et al.*, 2010), *Panicum virgatum* (Xi *et al.*, 2009), and *Populus* (Song *et al.*, 2006; Cseke *et al.*, 2007; Yang *et al.*, 2009; Yevtushenko & Misra, 2010), making genetic engineering feasible in these crops. Also, genetic diversity in natural or breeding populations has been exploited to develop superior lines for biofuel production. The successful examples of genetic improvement of bioenergy crops are listed in Table 1.

#### 3.2.1 Genetic improvement of biofuel yield

Genes involved in cell wall biogenesis and organization are promising targets for genetic manipulation to overcome the biomass recalcitrance that limits biofuel yields from second generation feedstocks (Yang *et al.*, 2011; Ye *et al.*, 2011). Lignin is one of the most important factors determining cell wall recalcitrance (Simmons *et al.*, 2010; Vanholme *et al.*, 2010). Genetic reduction of lignin content could effectively overcome cell wall recalcitrance to bioconversion, as demonstrated in transgenic alfalfa with down-regulated lignin biosynthetic genes, such as cinnamate 4-hydroxylase (C4H), hydroxycinnamoyl CoA:shikimate hydroxycinnamoyl transferase (HCT), coumaroyl shikimate 3-hydroxylase (C3H), caffeic acid 3-O-methyltransferase (COMT), cinnamoyl CoA reductase (CCR) and cinnamyl alcohol dehydrogenase (CAD) (Chen & Dixon, 2007; Jackson *et al.*, 2008). Switchgrass (*Panicum virgatum* L.) is a leading dedicated bioenergy feedstock in the United States and down-regulation of the switchgrass COMT gene decreases lignin content modestly, reduces the syringyl:guaiacyl lignin monomer ratio, and consequently increases the ethanol yield by up to 38%, using conventional biomass fermentation processes (Fu *et al.*, 2011). Genetic engineering of biofuel crops with transcription factors involved in the regulation of the phenylpropanoid pathway is another efficient approach to modify lignin biosynthesis. For example, the maize (*Zea mays*) R2R3-MYB factor ZmMYB31 down-regulates several genes involved in the synthesis of monolignols and transgenic *Arabidopsis* plants over-expressing ZmMYB31 show a significantly reduced lignin content with unaltered polymer composition, and consequently increased cell wall degradability of the transgenic plants (Fornale *et al.*, 2010). An alternative approach to address the lignin issue is to replace monolignols with compounds containing easily cleavable chemical linkages, such as ester and amide bonds, avoiding the undesirable developmental and structural phenotypes associated with the down-regulation of lignin biosynthetic enzymes in transgenic plants (Vega-Sanchez & Ronald, 2010). Inclusion of monolignol substitutes, such as feruloylquinic acid, methyl caffeate, or caffeoylquinic acid with normal monolignols could considerably suppress lignin formation and substantially improve cell wall hydrolysis and fermentation (Grabber *et al.*, 2010).

Besides lignin, hemicellulose (including xylan, glucuronoxylan, arabinoxylan, glucomannan, and xyloglucan) also contributes to plant cell wall recalcitrance (Vega-Sanchez & Ronald, 2010). It has been demonstrated that modification of hemicellulose could help overcome biomass recalcitrance. For example, loosening hemicellulose by over-expressing xyloglucanase and xylanase in transgenic poplar accelerates the enzymatic degradation of cellulose in wood (Kaida *et al.*, 2009), and lowering hemicellulose in transgenic poplar by under-expressing *PoGT47C*, a glycosyltransferase gene involved in glucuronoxylan biosynthesis, reduced the recalcitrance of wood to cellulase digestion (Lee *et al.*, 2009). As one of the most abundant polysaccharides on Earth, xylan will provide more than one third of the sugars for lignocellulosic biofuel production when using grass or hardwood feedstocks. Genetic mutations can be generated to remove branches from xylan and consequently simplify lignocellulosic biomass, requiring fewer enzymes for complete hydrolysis (Mortimer *et al.*, 2010). Another possible approach for improving saccharification of plant biomass is to modify pectin in the cell wall. For example, reduction of de-methyl-esterified homogalacturonan (HGA) in both *Arabidopsis* and tobacco plants through the expression of a fungal polygalacturonase (PG) or an inhibitor of pectin methylesterase (PMEI) increased the efficiency of enzymatic saccharification (Lionetti *et al.*, 2010).

Biodiesel is produced by the transesterification of triacylglycerol (TAG) to generate fatty acid methyl esters (FAMEs) (Vega-Sanchez & Ronald, 2010). Biodiesel yield can be improved by genetic manipulation of key genes in the TAG biosynthesis pathway. The final and the only committed step in the biosynthesis of TAG is catalyzed by diacylglycerol acyltransferase (DGAT) enzymes. DGAT is a target for genetic manipulation for enhancing TAG production. For example, expressing a codon-optimized version of a DGAT gene from the soil fungus *Umbelopsis ramanniana* in soybean resulted in 1.5% (by weight) increase in seed oil (Lardizabal *et al.*, 2008). Furthermore, transcription factors that regulate the biosynthetic pathways at the transcriptional level can be utilized for increasing lipid production. For example, two soybean Dof-type transcription factor genes, *GmDof4* and *GmDof11*, enhance lipid content in the seeds of transgenic *Arabidopsis* seeds, indicating that *GmDof* genes may augment the lipid content of soybean seeds by up-regulating genes that are associated with the biosynthesis of fatty acids (Wang *et al.*, 2007). On the other hand, glycerol-3-phosphate supply limits oil accumulation in developing seeds and over-expression of a yeast gene encoding cytosolic glycerol-3-phosphate dehydrogenase (*GPD1*) under the control of a seed-specific promoter resulted in 40% increase in seed oil content in oil-seed rape (*Brassica napus*) (Vigeolas *et al.*, 2007). Although TAG is mainly produced in the seeds of oil crop species, plants can also accumulate small amounts of TAG in the vegetative tissues such as leaves, and leaf TAG levels in the model plant *Arabidopsis* can be increased by up to 20 fold by blocking fatty acid breakdown (Slocombe *et al.*, 2009), expanding the scope of biomass feedstock for biodiesel production. This new route to biodiesel production is further demonstrated by the fact that transferring of an *Arabidopsis DGAT* gene into tobacco resulted in up to a 20-fold increase in TAG accumulation in tobacco leaves (Andrianov *et al.*, 2010). The full potential of *J. curcas* for biodiesel production is limited by the lack of high yielding varieties with high oil content, and recent research has been conducted to explore existing diversity for yield and oil content by direct selection, hybridization, and creation of diversity by mutation and biotechnological interventions (Divakara *et al.*, 2010).

Directing photosynthetic carbon partitioning from starch to TAG synthesis may represent a more effective strategy than direct manipulation of the lipid synthesis pathway to increase biodiesel production. For example, inactivation of ADP-glucose pyrophosphorylase in a *Chlamydomonas* starchless mutant led to a 10-fold increase in TAG (Li *et al.*, 2010). The model green alga *Chlamydomonas reinhardtii* accumulates triacylglycerols and forms lipid droplets during nitrogen deprivation, and suppression of the expression of the green algal specific major lipid droplet protein (MLDP) gene using an RNA interference approach led to increased lipid droplet size, but no change in TAG content or metabolism (Moellering & Benning, 2010).

Oil harvesting is a major factor limiting the final yield of biodiesel generated from aquatic biomass. To address the harvesting problem in biodiesel production from liquid culture of algae and cyanobacteria, a controllable inducing lysis system, based on integration of bacteriophage-derived lysis genes, into the *Synechocystis* sp. PCC 6803 genome downstream of a nickel-inducible signal transduction system, can be utilized to facilitate extracting lipids for biofuel production and consequently eliminate the need for mechanical or chemical cell breakage and facilitate recovery of biofuel from cyanobacteria (Liu & Curtiss, 2009).

### 3.2.2 Genetic improvement of biofuel quality

As mentioned in Section 2.4, the physical properties of biofuels need to be improved to match the quality of fossil fuels. A lot of research efforts have been devoted to improve the quality of biodiesel. The polyunsaturated fatty acids linoleic acid (18:2) and alpha-linolenic acid (18:3) are major factors affecting the quality of plant oils for biofuels (Lu *et al.*, 2009). Two approaches can be used to address the issue of biodiesel quality. The first approach is to reduce the levels of both saturated and polyunsaturated fatty acids while increasing the amount of monounsaturated fatty acids, such as oleate (C18:1) or palmitoleate (C16:1) (Durrett *et al.*, 2008; Pinzi *et al.*, 2009; Vega-Sanchez & Ronald, 2010). For example, simultaneous down-regulation of two embryo-specific genes in soybean, Delta-12 fatty acid desaturase *FAD2-1* gene and the *FatB* gene encoding a palmitoyl-thioesterase, increased oleic acid levels to greater than 85% compared with less than 18% in wild-type, and lowered saturated fatty acids levels to less than 6% (Buhr *et al.*, 2002). Phosphatidylcholine:diacylglycerol cholinephosphotransferase (PDCT), encoded by the *Arabidopsis ROD1* gene, is an enzyme for the transfer of 18:1 into the membrane lipid phosphatidylcholine (PC) for desaturation and also for the reverse transfer of 18:2 and 18:3 into the TAG synthesis pathway; and mutation in *ROD1* reduced 18:2 and 18:3 accumulation in seed TAG by 40% (Lu *et al.*, 2009). The second approach is to produce biodiesel comprising medium-chain (C8 and C10) FAs. Currently, *Cuphea* is the only plant source found to produce high levels of medium-chain (C8 and C10) FAs (Fig. 3); and the properties of *Cupea* methyl esters (CuME) meet or exceed the current industrial standard of biodiesel (e.g., CuME displayed a cloud point of -9 to -10°C and a pour point in the range of -21 to -22°C) (Knothe *et al.*, 2009). Understanding the molecular mechanism underlying the accumulation of medium-chain FAs in *Cuphea* and transferring this mechanism to other biomass feedstocks would have great potential for improving biodiesel quality.

| Species | Gene | Biofuel type | References |
|---|---|---|---|
| *Arabidopsis thaliana* (*Arabidopsis*) | fungal polygalacturonase (PG) | Bioalcohol | (Lionetti *et al.*, 2010) |
| *Arabidopsis thaliana* (*Arabidopsis*) | maize R2R3-MYB factor ZmMYB31 | Bioalcohol | (Fornale *et al.*, 2010) |
| *Medicago sativa* (Alfalfa) | cinnamate 4-hydroxylase (*C4H*) | Bioalcohol | (Chen & Dixon, 2007) |
| *Medicago sativa* (Alfalfa) | hydroxycinnamoyl CoA:shikimate hydroxycinnamoyl transferase (*HCT*) | Bioalcohol | (Chen & Dixon, 2007) |
| *Medicago sativa* (Alfalfa) | coumaroyl shikimate 3-hydroxylase (*C3H*) | Bioalcohol | (Chen & Dixon, 2007) |
| *Medicago sativa* (Alfalfa) | caffeic acid 3-O-methyltransferase (*COMT*) | Bioalcohol | (Chen & Dixon, 2007) |
| *Medicago sativa* (Alfalfa) | cinnamoyl CoA reductase (*CCR*) | Bioalcohol | (Jackson *et al.*, 2008) |
| *Medicago sativa* (Alfalfa) | cinnamyl alcohol dehydrogenase (*CAD*) | Bioalcohol | (Jackson *et al.*, 2008) |
| *Panicum virgatum* (Switchgrass) | caffeic acid O-methyltransferase (*COMT*) | Bioalcohol | (Fu *et al.*, 2011) |
| *Populus alba* x *tremula* (Poplar) | *PoGT47C* glycosyltransferase | Bioalcohol | (Lee *et al.*, 2009) |
| *Populus* (Poplar) | Xyloglucanase (AaXEG2) from *Aspergillus* | Bioalcohol | (Kaida *et al.*, 2009) |
| *Populus* (Poplar) | xylanase (HvXYL1) | Bioalcohol | (Kaida *et al.*, 2009) |
| *Populus* (Poplar) | Cellulase (AtCel1) from *Arabidopsis* | Bioalcohol | (Kaida *et al.*, 2009) |
| *Zea mays* (Corn) | R2R3-MYB factor ZmMYB31 | Bioalcohol | (Fornale *et al.*, 2010) |
| *Arabidopsis thaliana* (*Arabidopsis*) | Dof-type transcription factor genes, *GmDof4* and *GmDof11* from soybean | Biodiesel | (Wang *et al.*, 2007) |
| *Arabidopsis thaliana* (*Arabidopsis*) | ROD1 gene (mutation) | Biodiesel | (Lu *et al.*, 2009) |
| *Brassica napus* (Oil-seed rape) | glycerol-3-phosphate dehydrogenase (*GPD1*) gene from yeast | Biodiesel | (Vigeolas *et al.*, 2007) |
| *Glycine max* (Soybean) | Delta-12 fatty acid desaturase (*FAD2-1*) and *FatB* gene encoding a palmitoyl-thioesterase | Biodiesel | (Buhr *et al.*, 2002) |
| *Glycine max* (Soybean) | Diacylglycerol acyltransferase (*DGAT2A*) gene from the soil fungus | Biodiesel | (Lardizabal *et al.*, 2008) |
| *Nicotiana tabacum* (Tobacco) | Diacylglycerol acyltransferase (*DGAT*) gene from *Arabidopsis thaliana* | Biodiesel | (Andrianov *et al.*, 2010) |
| *Chlamydomonas* | ADP-glucose pyrophosphorylase | Biodiesel | (Li *et al.*, 2010) |
| *Chlamydomonas reinhardtii* (green alga) | Major lipid droplet protein (MLDP) | Biodiesel | (Moellering & Benning, 2010) |
| *Synechocystis* sp. PCC 6803 | Bacteriophage-derived lysis genes | Biodiesel | (Liu & Curtiss, 2009) |

Table 1. Improvement of bioenergy crops using transgenic and mutational approaches.

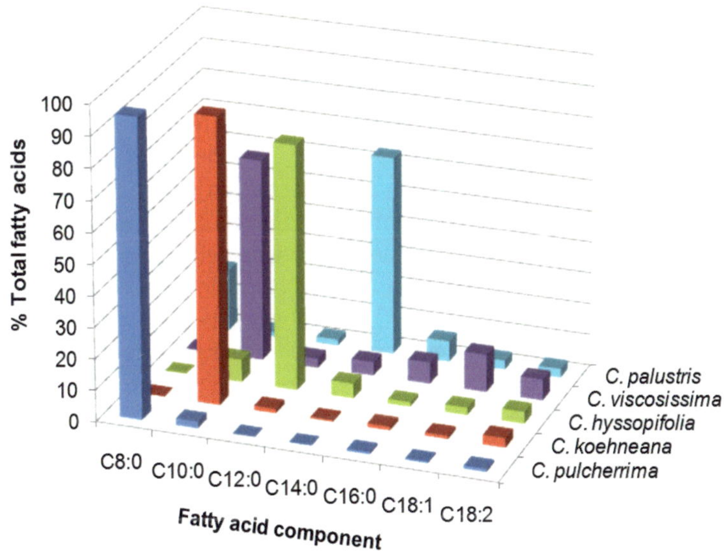


Fig. 3. Variation in fatty acid composition among some *Cuphea* species. Drawn with data from Dehesh (2001) and Knothe et al. (2009).

### 3.3 Improvement of microorganisms in biomass conversion

#### 3.3.1 Metabolic improvement and genetic engineering of microorganisms for biofuel production

As mentioned in Section 2.3, the lack of efficient microorganisms to convert biomass into liquid fuels is a big challenge in biofuel production using non-food lignocellulosic feedstock which has the potential to meet most of the global transportation fuel needs in a sustainable manner. The desirable traits of microorganisms for biofuel production include high substrate utilization and processing capacities, fast and deregulated pathways for sugar transport, good tolerance to inhibitors and product, and high metabolic fluxes (Alper & Stephanopoulos, 2009). With beneficial traits for biofuel-related applications, some native microorganisms, such as *Clostridium acetobutylicum* for the ABE process, have become the unambiguous organisms of choice for biofuel production in industry (Inui *et al.*, 2008; Alper & Stephanopoulos, 2009; Roberts *et al.*, 2010). However, since the properties required for industrial processing are very different from the features evolved in the native biomes, the transformation from an innate capacity of environmental isolates into an industrially relevant performance can sometimes be strenuous (Alper & Stephanopoulos, 2009). For instance, the current mainstream process of bioethanol production makes use of the basic yeast *S. cerevisiae*. This model organism has a proven track record in industrial applications, has superior conversion yields of ethanol from glucose, can tolerate ethanol, and has been the organism of choice for hundreds of years in fermentations to produce wine and other spirits. However, native strains of *S. cerevisiae* have not been exposed to the high concentrations of sugars, aromatic components, and adverse conditions that typically arise in the industrial conversion of lignocellulose to ethanol (Alper & Stephanopoulos, 2009). The same situation exists in the production of butanol using *C. acetobutylicum* that converts acetyl-coA into a mixture of butonal, acetone, and ethanol, and has limited tolerance to the produced solvents (Alper & Stephanopoulos, 2009; Mao *et al.*, 2010). Despite the difficulties in the utilization of these native microorganisms, which are derived from environmental isolates, the innate capacity of these cells to use recalcitrant substrates is immense. With the advent of modern genetic tools and synthetic biology approaches, we are capable of harnessing the commonly used industrial microorganisms (e.g., *E. coli* and *S. cerevisiae*) for biofuel production (Alper & Stephanopoulos, 2009; Clomburg & Gonzalez, 2010; Sommer *et al.*, 2010). Global transcription machinery engineering, in which transcription factors are adapted to industrial needs by creating mutant libraries and searching for dominant mutations, has proved successful, being able to enhance cellular traits in *E. coli* and yeast species (Liu *et al.*, 2010). Recently, Atsumi et al. (2008) cloned the genes involved in an alternative butanol pathway into *E. coli*, endowing it with the ability to produce reasonable amounts of isobutanol and other alcohols, such as isopropanol. This application, gene transfer along with global transcription machinery engineering, offers the prospect of a desired combination of a high biofuel production and a genetically tractable host. The industrial application of several native and model microorganisms is described as follows.

### 3.3.2 Industrial application of several representative microorganisms

#### 3.3.2.1 Yeast

As mentioned in the Introduction section, bioethanol is currently the most widely used liquid biofuel, with the global market dominated by Brazil and the United States. The Brazilian system is based on sucrose obtained from sugarcane, which can be converted to bioethanol directly by yeast species without enzymatic pre-treatment, allowing this system to produce an energy surplus estimated at about eightfold (Goldemberg, 2007; Robertson *et al.*, 2008; Argueso *et al.*, 2009). Yeast is a well-established fermenting microorganism in existing commercial-scale ethanol industries. PE-2 is one of the most widely adopted yeast strains for the sugarcane fermentation process, used in about 30% of Brazilian distilleries, generating roughly 10% of the world's bioethanol supply (Argueso *et al.*, 2009). The generation and conversion of fermentable sugars from lignocellulosic materials to ethanol is strongly dependent on the feedstock pretreatment and strain selection (Lau & Dale, 2009). Fermentation of hydrolysates derived from pretreated lignocellulosic biomass is often preceded by washing, nutrient supplementation, and detoxification, which are very costly processes. Recently, a promising technology, known as consolidated bioprocessing (CBP), was developed for biofuel production from lignocellulosic biomass. It involves the use of a single microorganism to convert pretreated lignocellulosic biomass to ethanol by combining cellulase production, cellulose hydrolysis, and sugar fermentation into a single step (Linger *et al.*, 2010; Wen *et al.*, 2010). Although yeast is utilized to ferment sugars derived from cornstarch or sugarcane into ethanol, it cannot ferment the cellodextrins naturally released from lignocellulosic biomass by cellulases and requires multiple enzymes, including β-glucosidases, to quantitatively produce fermentable glucose (Sun & Cheng, 2002; Galazka *et al.*, 2010; Chundawat *et al.*, 2011). Several promising yeast strains have been created, such as 424A(LNH-ST) that exhibits excellent co-fermentation of glucose and xylose (Lau & Dale, 2009). Contrary to yeast, cellulolytic fungi such as *Neurospora crassa* grow well on cellodextrins. Engineering of the *N. crassa* cellodextrin transport system into *S. cerevisiae* promotes efficient growth of this yeast on cellodextrins, and the engineered yeast strains more rapidly convert cellulose to ethanol when compared with yeast lacking this system in simultaneous fermentation experiments (Galazka *et al.*, 2010). An alternative engineering strategy to construct CBP-enabling yeast species is to endow *S. cerevisiae* with the ability to utilize cellulose by heterologously expressing a functional cellulase system (Wen *et al.*, 2010). Nature has provided two ways of designing such yeast strains: noncomplexed cellulase systems and complexed cellulase systems (i.e., cellulosomes) (Wen *et al.*, 2010; Chundawat *et al.*, 2011). By mimicking the noncomplexed cellulase system, several groups successfully constructed cellulolytic *S. cerevisiae* strains that directly ferment amorphous cellulose to ethanol, although the titer and yield were relatively low (Fujita *et al.*, 2004; Den Haan *et al.*, 2007; Wen *et al.*, 2010). Compared to the noncomplexed cellulase systems, the cellulosome could provide a "quantum leap" in the development of biofuel technology thanks to its highly ordered structural organization that enables enzyme proximity synergy and enzyme-substrate-microbe complex synergy (Bayer *et al.*, 2007). To date, the trifunctional minicellulosomes have been successfully assembled *in vivo* in *S. cerevisiae*, and the resulting recombinant strain could simultaneously hydrolyze and ferment amorphous cellulose to ethanol, providing a relatively convenient engineering platform (Wen *et al.*, 2010).

In the post-genomic era, the availability of rich genomic, proteomic, and metabolomic information provides a solid foundation for yeast strain improvement and engineering. In 1996, the *S. cerevisiae* laboratory strain S288c became the first eukaryote to have its genome completely sequenced (Bayer *et al.*, 2007; Argueso *et al.*, 2009). Since then, other haploid strains from diverse backgrounds have been sequenced (RM11-1a, YJM789, M22, YPS163, and AWRI1631; http://www.broad.mit.edu/), followed by a large-scale effort to determine the genome sequences of many others (Bayer *et al.*, 2007; Wei *et al.*, 2007; Doniger *et al.*, 2008; Argueso *et al.*, 2009). Extensive analyses have been conducted to examine the nucleotide sequence diversity between these strains and the results from these studies provide valuable insights for synthetic biology and artificial biology to create efficient and robust yeast strains.

#### 3.3.2.2 Clostridium

*C. thermocellum* is a Gram-positive bacterium that is able to ferment cellulose to ethanol, acetic acid, lactic acid, formic acid, hydrogen, and $CO_2$. As mentioned earlier, *C. thermocellum* is naturally capable of producing butanol. Biobutanol is an attractive fuel as it possesses better energy properties than ethanol, including higher energy content per volume, lower water absorption, and better blending ability. Additionally, *C. thermocellum* appears to be a cellulose-utilizing specialist (Freier *et al.*, 1988; Demain *et al.*, 2005; Tripathi *et al.*, 2010) and produces cellulosomes, a multienzyme cellulose-solubilizing complex (Bayer *et al.*, 1985; Bayer *et al.*, 2004; Gold & Martin, 2007; Tripathi *et al.*, 2010). Because of the exemplary capacity of *C. thermocellum* to convert cellulosic biomass without the addition of purified cellulose or hemicellulase enzymes, the CBP platform using *C. thermocellum* provides a promising means for low cost production of renewable biofuels. Metabolic engineering is required in order to increase the yield of ethanol or other desired products and decrease the rate of mixed-product fermentations carried out by wild type *C. thermocellum*. Unfortunately, reliable genetic tractability has been elusive for *Clostridium* species, in terms of transformation efficiency and screenable genetic marker development (Tripathi *et al.*, 2010). The transformation protocol remains complex and cumbersome in *Clostridium* species, such as *C. acetobutylicum*, *C. perfringens*, *C. septicum*, and *C. thermocellum*, and the efficiency does not compare with that of typical model organisms. When it comes to the selectable or screenable phenotypes, comprehensive work has been carried out with genetically tractable model organisms, such as *E. coli*, but not in *Clostridium*. Several studies have been performed to transfer these selectable markers into *Clostridium* species. One prominent system transferred to *Clostridium* involves the genes encoding the enzyme orotidine 5-phosphate decarboxylase (PyrF) (Boeke *et al.*, 1984; Haas *et al.*, 1990; Tripathi *et al.*, 2010). Many more studies are being undertaken to develop more efficient genetic improvement and engineering approaches for *Clostridium* species.

#### 3.3.2.3 Zymomonas mobilis

Gram-negative fermentative bacterium *Z. mobilis* has been studied for its exceptionally high ethanol production rate and tolerance to the toxicity of the final product and has become a particularly attractive microbial candidate for the CBP platforms (Skotnicki *et al.*, 1983; Linger *et al.*, 2010). *Z. mobilis* is capable of fermenting sugars at low pH and has a naturally high tolerance to many inhibitory compounds existing in hydrolysates derived from lignocellulosic biomass (Zhang *et al.*, 1995; Linger *et al.*, 2010). Additionally, the native Entner-Doudoroff pathway in *Z. mobilis* allows it to reach the near-theoretical maximum ethanol yields during fermentation while achieving relatively low biomass formation (Swings & De Ley, 1977; Linger *et al.*, 2010). To establish *Z. mobilis* as a CBP host, a necessary prerequisite is that *Z. mobilis* must have high levels of cellulolytic enzyme expression. However, achieving high-level expression of cellulases is not the only hurdle to overcome. It is imperative that these enzymes must be translocated to the extracellular space and contact the lignocellulosic substrate directly (Linger *et al.*, 2010). The most obvious means to achieve this translocation is by harnessing the host's protein secretion apparatus. It has been reported that several *Z. mobilis* strains natively produce an endogenous activity against carboxymethyl cellulose and that this activity can be detected extracellularly, which can be adapted to secrete cellulolytic enzymes (Linger *et al.*, 2010). All these results suggest that *Z. mobilis* may be adept at producing cellulases, and as this attribute is essential for an industrial application, *Z. mobilis* serves as an ideal candidate for CBP. To date, *Z. mobilis* has shown successful records in CBP and has been successfully engineered to ferment the pentose ($C_5$) sugars, xylose, and arabinose (Zhang *et al.*, 1995; Deanda *et al.*, 1996; Linger *et al.*, 2010).

#### 3.3.2.4 *Trichoderma reesei*

*T. reesei* (syn. *Hypocrea jecorina*) is a mesophilic soft-rot ascomycete fungus (Mandels & Reese, 1957; Martinez *et al.*, 2008). This biomass-degrading fungus represents a paradigm for the production of bioethanol and a range of key biochemical building blocks, such as aspartic acid, glucaric acid, glutamic acid, glycerol, sorbitol, and hydroxybutyrolactone, because it naturally possesses enzymes that hydrolyze lignocellulosic polysaccharides (Martinez *et al.*, 2008; Alper & Stephanopoulos, 2009). It has enjoyed a long history of safe use in industrial enzyme production and is currently widely used as a source of cellulases and hemicellulases for the hydrolysis of plant cell wall polysaccharides (Nevalainen *et al.*, 1994; Martinez *et al.*, 2008). Although genetic engineering techniques, gene knockout protocols, and DNA-mediated transformation systems have improved the performance of industrial *T. reesei* strains (Martinez *et al.*, 2008), further studies are needed to expand its extraordinary potential for biofuel production.

### 3.4 Utilization of beneficial microorganisms to increase the yield of bioenergy crops

All plant-associated microenvironments, especially the rhizosphere, are colonized by the microbes in high abundance (Berg *et al.*, 2005). Soil microorganisms including bacteria and mycorrhizal fungi promote plant growth either directly by acting as biofertilizers, phytostimulators, rhizoremediators or indirectly as biocontrol agents. The controlled use of microbes has emerged as a promising solution for the sustainable production of agronomically important crops. This is important as the production of bioenergy feedstocks has the potential to place additional burden to already constrained natural resources such as land, water and nutrients. In this section we discuss how the partnerships between plants and their microbial associates can be used to bolster biomass production of bioenergy feedstocks in an environmentally-conscious fashion.

The population density of the bacteria in the plant rhizosphere is high, with estimates ranging from $10^5$-$10^7$ CFU $g^{-1}$ fresh weight of bacteria (Bais *et al.*, 2006). Although rhizobacteria may be neutral or antagonistic to host plant growth and productivity, most (about two thirds) are reputed as beneficial (Furnkranz *et al.*, 2009). This has been demonstrated in several studies with rhizobacteria. For example, different isolates of *Methylobactrium* have been shown to improve germination, growth and yield of sugarcane (Madhaiyan *et al.*, 2005), and *Enterobactor* sp. 638 has been shown to have a pronounced influence on growth and development of poplar cuttings in marginal soils (van der Lelie *et al.*, 2009). As described earlier (Section 2.1), one way of avoiding competition between food and bioenergy crops is to modify bioenergy feedstocks for growth on marginal lands. These marginal lands are comprised of soil that lack one or more essential nutrient, are water limited or are contaminated by pollutants such as heavy metals. Plant-associated bacteria can be used for the economic production of biofuels by enabling the cultivation of bioenergy crops on these otherwise unsuitable marginal lands. For example, several greenhouse and field studies have demonstrated the efficiency of non-nodule forming nitrogen fixing bacteria on different host plant species including sugarcane, soybean and rice (Boddey *et al.*, 1995; Mano & Morisaki, 2008; Mishra *et al.*, 2009). In switchgrass, inoculation of the seedlings by a consortium of different rhizosphere microbes increased N-uptake up to 6-fold (Brejda *et al.*, 1998). In poplar and willow, there is a role for endophytes in fixing atmospheric nitrogen (Doty *et al.*, 2009). Several genera of bacteria including *Bacillus, Enterobactor, Pseudomonas* and *Azotobactor* have been shown to mineralize or solubilize phosphate in the rhizosphere making it available to the plant (Vassilev *et al.*, 2006 and references therein).

The ability by which plants acclimate and tolerate abiotic stress can be enhanced by their microbial associates. With plant-rhizobacteria interactions, for example, the bacteria produce compounds including phytohormones (e.g., auxin and ethylene), which in turn modulate plant growth and can improve host plant stress tolerance and fitness. The bacteria *Azotobactor* and *Azospirillium* were originally thought to improve host plant growth through fixed nitrogen, but additional studies have identified multiple mechanisms including the production of hormones such as Indole-3-acetic acid, Gibberellins, and cytokinins (Okon *et al.*, 1998). Many root associated bacteria are known to produce auxin derivatives (e.g., indole-3-acetic acid) and such bacteria can modify root architecture, which in turn influences water and nutrient uptake (see Section 3.1.2). In poplar, inoculation of rooted cuttings with auxin-producing endophytic bacteria improved growth by up to 60% (Taghavi *et al.*, 2009). Rhizobacteria also modulate ethylene levels in plants either through the auxin they produce or with the activity of bacterial 1-aminocyclopropane-1-carboxylate (ACC) deaminase. Bacteria possessing this enzyme can use ACC as an immediate precursor of ethylene, thereby reducing plant ethylene levels that leads to increased root growth. This is important given that ethylene plays a key role in stress signal transduction pathways. In addition to auxin, ethylene and gibberellin producing bacteria have been isolated from pine (Bent *et al.*, 2001), rapeseed (Noel *et al.*, 1996), lettuce (Noel *et al.*, 1996), and soybean (Garcia de Salamone *et al.*, 2001). Some of these bacteria stimulate plant growth by gibberellin biosynthesis (Gutierrez Manero *et al.*, 2001). Although our current understanding of the role of soil bacteria in improving host plant abiotic stress tolerance is limited, a few studies have shown some promise with bioenergy feedstocks using this approach. One notable example is from Ye *et al.* (2005), where inoculation of *Miscanthus* with a consortium of soil bacterial enhanced tolerance to salinity.

Some bioenergy feedstocks such as poplar and willow have been used for remediation of groundwater and soil contaminants such as BTEX (benzene, toluene, ethylbenzene and the xylene isomers), TCE (trichloroethylene), and diesel. In poplar, selective enrichment of the *rhizospheric* and endophytic bacteria has been observed in the presence of the contaminants (Barac *et al.*, 2009). Use of recombinant bacteria modified to contain specific degradation pathways has emerged as a novel tool for growing plants on the contaminated soil (van der Lelie *et al.*, 2009). Inorganic pollutants such as heavy metals induce oxidative stress by enhancing ethylene production which in turn reduces biomass productivity (Arshad *et al.*, 2007). Inoculation of plants with bacteria harboring ACC deaminase can be used to enhance plant growth and improve metal tolerance. However, further experimentation is required to test this possibility.

In addition to their role in plant nutrition and rhizoremediation, management of plant-microbe interactions can be used in low-cost integrated disease management strategies. Many soil bacteria produce anti-microbial compounds which prevent the growth of harmful soil born fungi. This strategy has shown some promise in bioenergy crops. For instance, in Eucalyptus, a strain of *Pseudomonas fulva* has been shown to reduce *Cylindrocladium candelabrum* growth by 33%, which causes mini-cutting rot in Eucalyptus and several other tree species. A study by Fucikovsky *et al.* (2006) has shown some promise for this approach in controlling bacterial infection of *Agave*, an emerging bioenergy feedstock plant. In addition to their anti-microbial activity, soil microbes and endophytes have also been used to activate plant defense systems against pathogens and herbivory. This phenomenon known as induced systemic resistance (ISR) is largely dependent on the ethylene and jasmonic acid signaling in the plant (van Loon, 2007). On the microbial side, several compounds secreted by the soil bacteria such as salicylic acid, Acyl homoserine lactones, acetoin, and 2,3-butanediol have been shown to induce ISR (Ryu *et al.*, 2003; Shuhegge *et al.*, 2006; van Loon, 2007). Interestingly, unlike other biocontrol associations ISR does not require an extensive colonization of the host plant (Kamilova *et al.*, 2005). However, due to the complexity of the bacterial communities in the soil, a more comprehensive understanding of their genomes and secretomes is necessary before we further explore the use of soil bacteria as biocontrol agents.

The mycorrhizal symbiosis between soil fungi and plant roots represents the most widespread association between plants and microbes. Mycorrhizal symbioses are prevalent in all major terrestrial biomes (Smith *et al.*, 1997). Currently we face many global challenges to our energy supply (see Section 2), and soil functioning through plant-mycorrhiza interactions could play an important role in helping us address these challenges. Specifically, plant-mycorrhiza interactions may 1) enhance carbon sequestration in terrestrial ecosystems to stabilize the atmospheric $CO_2$ concentration, 2) increase the production of food and bioenergy crops by increasing nutrient availability, 3) remediate degraded, polluted or desertified soils, and 4) develop sustainable cropping systems aimed at improving WUE and soil properties to minimize erosion, water pollution, and eutrophication (Schreiner *et al.*, 2003). All of these aspects make plant-mycorrhiza interactions an excellent approach for improving the sustainability of bioenergy feedstock productivity.

Mycorrhizal fungi are an important soil carbon sink and often constitute 20-30% of total soil microbial biomass (Leake *et al.*, 2004). They can reduce soil carbon loss by immobilizing carbon in their mycelium, by extending root lifespan, and by improving carbon sequestration in soil aggregates (Langley *et al.*, 2006; Rillig & Mummey, 2006). Bacteria and fungi play distinct roles because of their inherent stoichiometry, especially of C and N. The average C : N ratio in bacteria is about 4 and in fungi about 10, and fungi generally respire less, resulting in higher carbon use efficiency (CUE) relative to bacteria (Six *et al.*, 2006). Recent studies, however, found considerable overlap in CUE-values of bacteria and fungi that is dependent on a number of factors including species and functional group identity, quantity and quality of substrates, and abiotic factors (Six *et al.*, 2006). Mycorrhizal fungi may have higher CUE than saprophytic fungi and bacteria (Wallander *et al.*, 2003). Furthermore, fungal mycelia are more recalcitrant in soil relative to bacteria. Mycelia are comprised of complex nutrient-poor carbon forms such as chitin and melanin, while bacterial membranes mainly consist of phospholipids that are quickly re-assimilated by soil biota. Although, the mechanisms of microbial contribution to soil organic carbon sequestration are poorly understood *in situ*, an overall increase in fungal-dominance is typically associated with high organic-matter content and low substrate quality, i.e. high C:N ratio (Bardgett, 2005; van der Heijden *et al.*, 2008). The effect of mycorrhizal fungi on soil carbon sequestration may be highly specific to the combination of plant and symbiont species (Kiers & van der Heijden, 2006) and soil fertility (Allen *et al.*, 2003). These underlying traits need further elucidation, yet it appears that across ecosystems, different types of mycorrhizal fungi prevail and are related to particular plant traits and growth limiting nutrients (Cornelissen *et al.*, 2001; Read & Perez-Moreno, 2003).

So far, mycorrhizal application has shown a substantial increase in the yield properties such as aboveground biomass (Sramek *et al.*, 2000). Although no clear mechanism other than an improvement in the nutritional status has been proposed (Toussaint, 2007), beneficial fungus–plant interactions has shown enhancement in productivity of crops by synthesizing a number of active compounds such as alkaloids, oils, resins, tannins, natural rubber, gums, waxes, dyes, flavors and fragrances, pharmaceuticals, and pesticides (Rai *et al.*, 2001). For example, the suitable selection of host plant–fungus genotype led to an altered accumulation of essential oil levels in arbuscular mycorrhiza-colonized plants of *Mentha arvensis* (Freitas *et al.*, 2004) and sweet basil *Ocimum basilicum* L. (Copetta *et al.*, 2006; Copetta *et al.*, 2007; Toussaint, 2007).

Colonization with mycorrhizal fungi results in improvements in plant fitness and nutrition (Smith *et al.*, 1997). The network of extrametrical hyphae facilitate acquisition and transport of many ions to roots, particularly mobile ions such as P, N, K, S, Ca, and Zn. In addition, mycorrhizal fungi enhance the reabsorption of nutrients lost through root exudation and contribute to the soil fertility (Hamel, 2004; Rillig, 2004). A functional specialization is recognized according to the type of the mycorrhizal fungi, arbuscular mycorrhiza (AM) or ectomycorrhiza (EM). The most important function of AM for plant growth is increasing uptake of P. There has been strong evidence that supports the role of AM mycelia in mineralization and uptake of organic P (Tarafdar & Marschner, 1994; Koide & Kabir, 2000). The rapid linear extension rates and narrow diameters of AM hyphal networks along with the wall-bound extracellular phosphatase enzymes (Joner *et al.*, 2000) enable the enzymes to reach in soil pores that are otherwise inaccessible due to their small size and distance from the root. It is well established that many EM fungi are active producers of phytase and phosphatase enzymes (Leake & Read, 1997), and some can obtain both P and N from a range of organic sources, including partially decayed tree litter, pollen, and nematodes (Read & Perez-Moreno, 2003). In soil microcosms, between 35% and 40% of the total P content of partially decayed tree litter was removed by colonizing EM mycelium, with the majority of this P being mobilized from organic compounds. In the absence of EM mycelium, moist and non-sterile partially decayed tree litter releases inorganic P slowly (Bending & Read, 1995). It was reported that 15% of P and 12% of N supplied to trees in boreal forest ecosystems may come from EM derived associations (Read & Perez-Moreno, 2003). Furthermore, some EM fungi are toxic to fungal-feeding micro-arthropods such as collembola and significant amounts of N can be obtained by mycorrhizal fungi digesting of dead collembola (Klironomos & Hart, 2001). In addition, mycorrhizal fungi appear to be able to acquire P from a range of inorganic P sources, including some calcium and aluminium phosphates that have extremely low solubility (Yao *et al.*, 2001), but it is not known whether the fungi are directly involved in their solubilization. Uptake of insoluble P sources by AM may be facilitated by P-solubilizing bacteria, and there may be mutualistic interactions between these two groups of organisms (Villegas & Fortin, 2001). EM mycelia have also been shown to obtain P from a range of sparingly soluble mineral sources such as aluminium phosphate (Cumming & Weinstein, 1990), and their production of organic chelators such as citric and oxalic acids, together with hydroxamate siderophores, are implicated in major mineral weathering processes and podsolization (van Breemen *et al.*, 2000). These findings are of importance for biogeochemistry and processes of soil maturation. Besides their roles in P nutrition, both AM and EM fungi play a major role in the uptake of N by plants. Based on the studies of monoxenic fungal cultures, AM mycelium has been shown to have a role in the uptake of ammonium, nitrate, glycine, and glutamine. AM fungi increase decomposition and subsequent capture of inorganic N from complex organic materials such as plant litter (Hodge *et al.*, 2001). These kinds of responses have been considered characteristic of EM but not AM fungi (Leake & Read, 1997). Furthermore, ectomycorrhizal fungi have high-affinity amino acid uptake systems (Wallenda *et al.*, 2000) and highly developed proteolytic capabilities enabling them to directly access macromolecular N (Abuzinadah & Read, 1989). Although use of mycorrhizal fungi for improving crop production has been limited to medicine or food production, studies are ongoing to explore their roles in bioenergy production.

## 4. Conclusion and perspectives

Declining availability and political instability in the supply of fossil fuels have focused efforts on developing liquid biofuels to meet our ever-increasing energy requirements. However, a huge gap remains between biofuel production and future energy needs, as reflected by the fact that current biomass generated on agricultural lands cannot support sustainable biofuel production, and the physical properties of both bioethanol and biodiesel are less than ideal for application in transportation. In this chapter, we have described four major challenges in sustainable biofuel production and discussed biological innovations for solving these challenges. Currently, biofuels are commercially produced mostly from the so-called first generation bioenergy biomass (e.g., corn and soybean), and worldwide efforts have been undertaken to realize the potential of next-generation bioenergy crops (e.g., switchgrass, *Populus*, Jatropha, and algae). With the availability of increasing numbers of sequenced plant genomes (http://www.phytozome.net/) across a large evolutionary space, a better understanding of the gene networks regulating the biological pathways relevant to biomass composition, productivity and resource use efficiency will be obtained. Such knowledge can subsequently be exploited to design effective strategies for the genetic improvement of bioenergy crops that will include overcoming the recalcitrance of lignocellulose to enzymatic saccharification.

CAM species such as *Agave* show considerable promise as a biofuel crop for the future due to their high water-use efficiency, tolerance to abiotic stress (e.g., drought and high temperatures), and potential for high biomass production on marginal lands (Borland *et al.*, 2009; Jaradat, 2010; Somerville *et al.*, 2010). Further research is needed to establish the relationship between CAM and nutrient uptake and assimilation in order to further enhance the significance of using *Agave* as a biofuel feedstock. Reported discrepancies on how the water-conserving CAM pathway impacts on the use and allocation of N need to be resolved in order to fully exploit the sustainable farming of *Agave* for biomass by reducing dependence on commercial nutrients, minimising the cost of production and diminishing environmental pollution.

The newly-developed synthetic biology (i.e., the ability to design and chemically synthesize genetic sequences imported into host cells) could expand our capacity to construct and improve pathway performance, enabling diversification of the biofuel-type molecules produced in standard model organisms (Alper & Stephanopoulos, 2009). For producing biofuels identical or similar to petroleum-derived transportation fuels, synthetic approaches have been used to engineer microbes to synthesize biofuels, such as butanol and fatty acid- or isoprenoid-based fuels, which are nearly identical to gasoline and diesel (Ghim *et al.*, 2010). Furthermore, the recent introduction of artificial biology, fuelled by the capacity to synthesize large pieces of DNA, has made it possible to construct cellular systems *de novo* (Alper & Stephanopoulos, 2009; Biello & Harmon, 2010; Bornscheuer, 2010; Noskov *et al.*, 2011) and thus has created a new efficient strategy for sustainable production of biofuels with ideal quality and in commercial quantities.

A better understanding of the soil microorganisms and their interactions with the host plants in their ecosystem will ensure an opportunity for the use of bacteria and mycorrhizal fungi to enhance sustainable bioenergy crop production. Thus, in properly managed agricultural systems, microbial symbioses can act as biofertilizer, biocontrol agent, and soil improver, likely being one of the key solutions to the problems associated with sustainable biofuel production. Recent genome sequencing efforts for the plant-associated microbes have been increasing our knowledge about these organisms and the way they interact with the plants (Martin *et al.*, 2008; Taghavi *et al.*, 2009). We still need to find better ways to inoculate and identify suitable vectors for introducing these beneficial microbes in the plant ecosystem. The increasing amount of genomic data and the systems biology studies will help us find the most suitable consortia of microbes for inoculation in the coming years.

## 5. Acknowledgment

We would like to thank Wellington Muchero for thoughtful and insightful comments on the manuscript. This work was supported by the Laboratory Directed Research and Development Program of Oak Ridge National Laboratory, the BioEnergy Science Center, and the Genomic Science Program, U.S. Department of Energy, Office of Science, Biological and Environmental Research. The BioEnergy Science Center is a U.S. Department of Energy Bioenergy Research Center supported by the Office of Biological and Environmental Research in the DOE Office of Science. Oak Ridge National Laboratory is managed by UT-Battelle, LLC for the U.S. Department of Energy under Contract Number DE-AC05-00OR22725.

## 6. References

Abolle, A.; Kouakou L. & Planche H. (2009a). The viscosity of diesel oil and mixtures with straight vegetable oils: Palm, cabbage palm, cotton, groundnut, copra and sunflower. *Biomass Bioenerg,* Vol.33, No.9, pp. 1116-1121.

Abolle, A.; Loukou K. & Henri P. (2009b). The density and cloud point of diesel oil mixtures with the straight vegetable oils (SVO): Palm, cabbage palm, cotton, groundnut, copra and sunflower. *Biomass Bioenerg,* Vol.33, No.12, pp. 1653-1659.

Abramson, M.; Shoseyov O. & Shani Z. (2010). Plant cell wall reconstruction toward improved lignocellulosic production and processability. *Plant Sci,* Vol.178, No.2, pp. 61-72.

Abuzinadah, R. A. & Read D. J. (1989). The role of proteins in the nitrogen nutrition of ectomycorrhizal plants IV. The utilization of peptides by birch *Betula pendula* Roth. Infected with different mycorrhizal fungi. *New Phytol,* Vol.112, pp. 55-60.

Aerts, R. & Chapin F. S. (2000). The mineral nutrition of wild plants revisited: A re-evaluation of processes and patterns. *Adv Ecol Res,* Vol.30, pp. 1-67.

Agarwal, A. K. (2007). Biofuels (alcohols and biodiesel) applications as fuels for internal combustion engines. *Prog Energ Combust,* Vol.33, No.3, pp. 233-271.

Allen, M. F.; Swenson W.; Querejeta J. I.; Egerton-Warburton L. M. & Treseder K. K. (2003). Ecology of mycorrhizae: a conceptual framework for complex interactions among plants and fungi. *Annu Rev Phytopathol,* Vol.41, pp. 271-303.

Almodares, A. & Hadi M. R. (2009). Production of bioethanol from sweet sorghum: A review. *Afr J Agr Res,* Vol.4, No.9, pp. 772-780.

Alper, H. & Stephanopoulos G. (2009). Engineering for biofuels: exploiting innate microbial capacity or importing biosynthetic potential? *Nat Rev Microbiol,* Vol.7, No.10, pp. 715-723.

Andrianov, V.; Borisjuk N.; Pogrebnyak N.; Brinker A.; Dixon J.; Spitsin S.; Flynn J.; Matyszczuk P.; Andryszak K.; Laurelli M. et al. (2010). Tobacco as a production platform for biofuel: overexpression of *Arabidopsis DGAT* and *LEC2* genes increases accumulation and shifts the composition of lipids in green biomass. *Plant Biotechnol J,* Vol.8, No.3, pp. 277-287.

Argueso, J. L.; Carazzolle M. F.; Mieczkowski P. A.; Duarte F. M.; Netto O. V. C.; Missawa S. K.; Galzerani F.; Costa G. G. L.; Vidal R. O.; Noronha M. F. et al. (2009). Genome structure of a *Saccharomyces cerevisiae* strain widely used in bioethanol production. *Genome Res,* Vol.19, No.12, pp. 2258-2270.

Arshad, M.; Saleem M. & Hussain S. (2007). Perspectives of bacterial ACC deaminase in phytoremediation. *Trends Biotechnol,* Vol.25, No.8, pp. 356-362.

Asikainen, A. (2010). Availability of wood biomass for biorefining. *Cell Chem Technol,* Vol.44, No.4-6, pp. 111-115.

Atsumi, S.; Cann A. F.; Connor M. R.; Shen C. R.; Smith K. M.; Brynildsen M. P.; Chou K. J. Y.; Hanai T. & Liao J. C. (2008). Metabolic engineering of *Escherichia coli* for 1-butanol production. *Metab Eng,* Vol.10, No.6, pp. 305-311.

Austin, A. (2010a). Agave shows potential as biofuel feedstock. *Biomass magzine,* Vol.February.

Austin, A. (2010b). Avant-garde Agave. *Biomass magzine,* Vol.May.

Bais, H. P.; Weir T. L.; Perry L. G.; Gilroy S. & Vivanco J. M. (2006). The role of root exudates in rhizosphere interactions with plants and other organisms. *Annual Review of Plant Biology,* Vol.57, No.1, pp. 233-266.

Balat, M. (2011). Production of bioethanol from lignocellulosic materials via the biochemical pathway: A review. *Energ Convers Manage,* Vol.52, No.2, pp. 858-875.

Banerji, R.; Chowdhury A. R.; Misra G.; Sudarsanam G.; Verma S. C. & Srivastava G. S. (1985). Jatropha seed oils for energy. *Biomass,* Vol.8, No.4, pp. 277-282.

Barac, T.; Weyens N.; Oeyen L.; Taghavi S.; van der Lelie D.; Dubin D.; Split M. & Vangronsveld J. (2009). Application of poplar and its associated microorganisms for the in situ remediation of a BTEX contaminated groundwater plume. *Int J Phytoremediation,* Vol.11, pp. 416-424.

Bardgett, R. D. (2005). *The biology of soil: A community and ecosystem approach.* Oxford University Press, Oxford.

Bayer, E. A.; Belaich J. P.; Shoham Y. & Lamed R. (2004). The cellulosomes: multienzyme machines for degradation of plant cell wall polysaccharides. *Annu Rev Microbiol,* Vol.58, pp. 521-554.

Bayer, E. A.; Lamed R. & Himmel M. E. (2007). The potential of cellulases and cellulosomes for cellulosic waste management. *Curr Opin Biotechnol,* Vol.18, No.3, pp. 237-245.

Bayer, E. A.; Setter E. & Lamed R. (1985). Organization and distribution of the cellulosome in *Clostridium thermocellum*. *J Bacteriol,* Vol.163, No.2, pp. 552-559.

Beer, L. L.; Boyd E. S.; Peters J. W. & Posewitz M. C. (2009). Engineering algae for biohydrogen and biofuel production. *Curr Opin Biotech,* Vol.20, No.3, pp. 264-271.

Bending, G. D. & Read D. J. (1995). The structure and function of the vegetative mycelium of ectomycorrhizal plants. V. The foraging behaviour of ectomycorrhizal mycelium and the translocation of nutrients from exploited litter. *New Phytol,* Vol.130, pp. 401-409.

Bent, E.; Tuzun S.; Chanway C. P. & Enebak S. (2001). Alterations in plant growth and in root hormone levels of lodgepole pines inoculated with rhizobacteria. *Can J Microbiol,* Vol.47, No.9, pp. 793-800.

Berendse, F. & Aerts R. (1987). Nitrogen use efficiency: a biological meaningful definition? *Functional Ecology,* Vol.1, pp. 293-296.

Berg, G.; Eberl L. & Hartmann A. (2005). The rhizosphere as a reservoir for opportunistic human pathogenic bacteria. *Environ Microbiol,* Vol.71, pp. 4203-4213.

Biello, D. & Harmon K. (2010). Tools for Life The ability to make cells with artificial genomes bodes well for basic biology. *Sci Am,* Vol.303, No.2, pp. 17-17.

Boddey, R.; Oliveira O. C.; Urquiaga S.; Reis V.; Olivares F. L.; Baldani V. & Döbereiner J. (1995). Biological nitrogen fixation associated with sugar cane and rice: contributions and prospects for improvement. *Plant Soil,* Vol.174, No.1, pp. 195-209.

Boeke, J. D.; LaCroute F. & Fink G. R. (1984). A positive selection for mutants lacking orotidine-5'-phosphate decarboxylase activity in yeast: 5-fluoro-orotic acid resistance. *Mol Gen Genet,* Vol.197, No.2, pp. 345-346.

Borland, A. M.; Griffiths H.; Hartwell J. & Smith J. A. C. (2009). Exploiting the potential of plants with crassulacean acid metabolism for bioenergy production on marginal lands. *J Exp Bot,* Vol.60, No.10, pp. 2879-2896.

Borland, A. M.; Hartwell J.; Jenkins G. I.; Wilkins M. B. & Nimmo H. G. (1999). Metabolite control overrides circadian regulation of phosphoenolpyruvate carboxylase kinase and $CO_2$ fixation in Crassulacean acid metabolism. *Plant Physiol,* Vol.121, No.3, pp. 889-896.

Bornscheuer, U. T. (2010). The first artificial cell-a revolutionary step in synthetic biology? *Angew Chem Int Edit,* Vol.49, No.31, pp. 5228-5230.

Brejda, J. J.; Moser L. E. & Vogel K. P. (1998). Evaluation of switchgrass rhizosphere microflora for enhancing seedling yield and nutrient uptake. Agron J, Vol.90, No.6, pp. 753-758.

Brune, D. E.; Lundquist T. J. & Benemann J. R. (2009). Microalgal biomass for greenhouse gas reductions: Potential for replacement of fossil fuels and animal feeds. *Journal of Environmental Engineering,* Vol.135, No.11, pp. 1136-1144.

Buhr, T.; Sato S.; Ebrahim F.; Xing A. Q.; Zhou Y.; Mathiesen M.; Schweiger B.; Kinney A.; Staswick P. & Clemente T. (2002). Ribozyme termination of RNA transcripts down-regulate seed fatty acid genes in transgenic soybean. *Plant J,* Vol.30, No.2, pp. 155-163.

Carvalho, C. R.; Clarindo W. R.; Praca M. M.; Araujo F. S. & Carels N. (2008). Genome size, base composition and karyotype of *Jatropha curcas* L., an important biofuel plant. *Plant Sci,* Vol.174, No.6, pp. 613-617.

Chapin, F. S.; Schulze E. D. & Mooney H. A. (1990). The ecology and economics of storage in plants. *Annu Rev Ecol Syst,* Vol.21, pp. 423-447.

Chen, F. & Dixon R. A. (2007). Lignin modification improves fermentable sugar yields for biofuel production. *Nat Biotechnol,* Vol.25, No.7, pp. 759-761.

Chundawat, S. P. S.; Beckham G. T.; Himmel M. E. & Dale B. E. (2011). Deconstruction of lignocellulosic biomass to fuels and chemicals. *Annu Rev Chem Biomol Eng,* Vol.2, pp. 6.1-6.25.

Clomburg, J. M. & Gonzalez R. (2010). Biofuel production in *Escherichia coli*: the role of metabolic engineering and synthetic biology. *Appl Microbiol Biot,* Vol.86, No.2, pp. 419-434.

Cooper, M. S.; Hardin W. R.; Petersen T. W. & Cattolico R. A. (2010). Visualizing "green oil" in live algal cells. *J Biosci Bioeng,* Vol.109, No.2, pp. 198-201.

Copetta, A.; Lingua G.; Bardi L.; Masoero G. & Berta G. (2007). Influence of arbuscular mycorrhizal fungi on growth and essential oil composition in *Ocimum basilicum* var. Genovese. *Caryologia,* Vol.60, pp. 106-110.

Copetta, A.; Lingua G. & Berta G. (2006). Effects of three AM fungi on growth, distribution of glandular hairs, and essential oil production in *Ocimum basilicum* L. var. Genovese. *Mycorrhiza,* Vol.16, pp. 485-494.

Coque, M.; Martin A.; Veyrieras J. B.; Hirel B. & Gallais A. (2008). Genetic variation for N-remobilization and postsilking N-uptake in a set of maize recombinant inbred lines. 3. QTL detection and coincidences. *Theor Appl Genet,* Vol.117, No.5, pp. 729-747.

Cornelissen, J. H. C.; Aerts R.; Cerabolini B.; Werger M. J. A. & van der Heijden M. G. A. (2001). Carbon cycling traits of plant species are linked with mycorrhizal strategy. *Oecologia,* Vol.129, pp. 611-619.

Costa, G. G. L.; Cardoso K. C.; Del Bem L. E. V.; Lima A. C.; Cunha M. A. S.; de Campos-Leite L.; Vicentini R.; Papes F.; Moreira R. C.; Yunes J. A. et al. (2010). Transcriptome analysis of the oil-rich seed of the bioenergy crop *Jatropha curcas* L. *BMC Genomics,* Vol.11, pp. 462.

Courchesne, N. M. D.; Parisien A.; Wang B. & Lan C. Q. (2009). Enhancement of lipid production using biochemical, genetic and transcription factor engineering approaches. *J Biotechnol,* Vol.141, No.1-2, pp. 31-41.

Crawford, N. M. & Forde B. G. (2002). Molecular and developmental biology of inorganic nitrogen nutrition. In *The Arabidopsis book,* (ed. EM Meyerowitz). American Society of Plant Biologists, Rockville, MD.

Cseke, L. J.; Cseke S. B. & Podila G. K. (2007). High efficiency poplar transformation. *Plant Cell Rep,* Vol.26, No.9, pp. 1529-1538.

Cumming, J. R. & Weinstein L. H. (1990). Utilization of AlPO4 as a phosphorus source by ectomycorrhizal *Pinus rigida. New Phyto,* Vol.116, pp. 99-106.

Davis, S. C.; Dohleman F. G. & Long S. P. (2011). The global potential for Agave as a biofuel feedstock. *GCB Bioenergy,* Vol.3, No.1, pp. 68-78.

de Fraiture, C.; Giordano M. & Liao Y. S. (2008). Biofuels and implications for agricultural water use: blue impacts of green energy. *Water Policy,* Vol.10, pp. 67-81.

Deanda, K.; Zhang M.; Eddy C. & Picataggio S. (1996). Development of an arabinose-fermenting *Zymomonas mobilis* strain by metabolic pathway engineering. *Appl Environ Microbiol,* Vol.62, No.12, pp. 4465-4470.

Dehesh, K. (2001). How can we genetically engineer oilseed crops to produce high levels of medium-chain fatty acids? *Eur J Lipid Sci Technol,* Vol.103, No.10, pp. 688-697.

Demain, A. L.; Newcomb M. & Wu J. H. (2005). Cellulase, clostridia, and ethanol. *Microbiol Mol Biol Rev,* Vol.69, No.1, pp. 124-154.

Demirbas, A. (2008). Biofuels sources, biofuel policy, biofuel economy and global biofuel projections. *Energ Convers Manage,* Vol.49, No.8, pp. 2106-2116.

Demirbas, A. (2009). Biofuels securing the planet's future energy needs. *Energ Convers Manage,* Vol.50, No.9, pp. 2239-2249.

Demirbas, A. (2010). Social, economic, environmental and policy aspects of biofuels. *Ener Educ Sci Tech-B,* Vol.2, No.1-2, pp. 75-109.

Demirbas, A. & Demirbas M. F. (2011). Importance of algae oil as a source of biodiesel. *Energ Convers Manage,* Vol.52, No.1, pp. 163-170.

Den Haan, R.; Rose S. H.; Lynd L. R. & van Zyl W. H. (2007). Hydrolysis and fermentation of amorphous cellulose by recombinant *Saccharomyces cerevisiae. Metab Eng,* Vol.9, No.1, pp. 87-94.

Dismukes, G. C.; Carrieri D.; Bennette N.; Ananyev G. M. & Posewitz M. C. (2008). Aquatic phototrophs: efficient alternatives to land-based crops for biofuels. *Curr Opin Biotech,* Vol.19, No.3, pp. 235-240.

Divakara, B. N.; Upadhyaya H. D., Wani S. P. & Gowda C. L. L. (2010). Biology and genetic improvement of *Jatropha curcas* L.: A review. *Appl Energ,* Vol.87, No.3, pp. 732-742.

Doniger, S. W.; Kim H. S.; Swain D.; Corcuera D.; Williams M.; Yang S. P. & Fay J. C. (2008). A catalog of neutral and deleterious polymorphism in yeast. *Plos Genetics,* Vol.4, No.8, pp. e1000183.

Doty, S. L.; Oakley B.; Xin G.; Kang J. W.; Singleton G.; Khan Z.; Vajzovic A. & Staley J. T. (2009). Diazotrophic endophytes of native black cottonwood and willow. *Symbiosis,* Vol.47, No.1, pp. 23-33.

Ducat, D. C.; Way J. C. & Silver P. A. (2011). Engineering cyanobacteria to generate high-value products. *Trends Biotechnol,* Vol.29, No.2, pp. 95-103.

Durrett, T. P.; Benning C. & Ohlrogge J. (2008). Plant triacylglycerols as feedstocks for the production of biofuels. *Plant J,* Vol.54, No.4, pp. 593-607.

Dyer, J. M.; Stymne S.; Green A. G. & Carlsson A. S. (2008). High-value oils from plants. *Plant J,* Vol.54, No.4, pp. 640-655.

Erisman, J. W.; van Grinsven H.; Leip A.; Mosier A. & Bleeker A. (2010). Nitrogen and biofuels; an overview of the current state of knowledge. *Nutr Cycl Agroecosys,* Vol.86, No.2, pp. 211-223.

Fornale, S.; Shi X. H.; Chai C. L.; Encina A.; Irar S.; Capellades M.; Fuguet E.; Torres J. L.; Rovira P.; Puigdomenech P. et al. (2010). ZmMYB31 directly represses maize lignin genes and redirects the phenylpropanoid metabolic flux. *Plant J,* Vol.64, No.4, pp. 633-644.

Fortman, J. L.; Chhabra S.; Mukhopadhyay A.; Chou H.; Lee T. S.; Steen E. & Keasling J. D. (2008). Biofuel alternatives to ethanol: pumping the microbial well. *Trends Biotechnol,* Vol.26, No.7, pp. 375-381.

Fraisier, V.; Gojon A.; Tillard P. & Daniel-Vedele F. (2000). Constitutive expression of a putative high-affinity nitrate transporter in *Nicotiana plumbaginifolia*: evidence for post-transcriptional regulation by a reduced nitrogen source. *Plant J,* Vol.23, No.4, pp. 489-496.

Freier, D.; Mothershed C. P. & Wiegel J. (1988). Characterization of *Clostridium thermocellum* JW20. *Appl Environ Microbiol,* Vol.54, No.1, pp. 204-211.

Freitas, M. S. M.; Martins M. A. & Vieira E. I. J. C. (2004). Yield and quality of essential oils of *Mentha arvensis* in response to inoculation with arbuscular mycorrhizal fungi. *Pesqui Agropecu Bras,* Vol.39, pp. 887-894.

Fritsche, U. R.; Sims R. E. H. & Monti A. (2010). Direct and indirect land-use competition issues for energy crops and their sustainable production - an overview. *Biofuel Bioprod Bior,* Vol.4, No.6, pp. 692-704.

Fu, C.; Mielenz J. R.; Xiao X.; Ge Y.; Hamilton C. Y.; Rodriguez M., Jr.; Chen F.; Foston M.; Ragauskas A.; Bouton J. et al. (2011). Genetic manipulation of lignin reduces recalcitrance and improves ethanol production from switchgrass. *P Natl Acad Sci USA,* Vol.108, No.9, pp. 3803-3808.

Fucikovsky, L. (2006). Antagonistic coccus type bacteria killing a coryneform type bacterium, the pathogen of *Agave tequilana*. *Proceedings of the 1st International Symposium on Biological Control of Bacterial Plant Diseases,* pp. 67.

Fujita, Y.; Ito J.; Ueda M.; Fukuda H. & Kondo A. (2004). Synergistic saccharification, and direct fermentation to ethanol, of amorphous cellulose by use of an engineered yeast strain codisplaying three types of cellulolytic enzyme. *Appl Environ Microbiol,* Vol.70, No.2, pp. 1207-1212.

Furnkranz, M.; Muller H. & Berg G. (2009). Characterization of plant growth promoting bacteria from crops in Bolivia. *J Plant Dis Protect,* Vol.116, No.4, pp. 149-155.

Galazka, J. M.; Tian C. G.; Beeson W. T.; Martinez B.; Glass N. L. & Cate J. H. D. (2010). Cellodextrin transport in yeast for improved biofuel production. *Science,* Vol.330, No.6000, pp. 84-86.

Garcia de Salamone, I. E.; Hynes R. K. & Nelson L. M. (2001). Cytokinin production by plant growth promoting rhizobacteria and selected mutants. *Can J Microbiol,* Vol.47, No.5, pp. 404-411.

Garnett, T.; Conn V. & Kaiser B. N. (2009). Root based approaches to improving nitrogen use efficiency in plants. *Plant Cell Environ,* Vol.32, No.9, pp. 1272-1283.

Gerbens-Leenes, W.; Hoekstra A. Y. & van der Meer T. H. (2009). The water footprint of bioenergy. *P Natl Acad Sci USA,* Vol.106, No.25, pp. 10219-10223.

Ghim, C. M.; Kim T.; Mitchell R. J. & Lee S. K. (2010). Synthetic biology for biofuels: Building designer microbes from the scratch. *Biotechnol Bioproc E,* Vol.15, No.1, pp. 11-21.

Glass, A. D. M. (2003). Nitrogen use efficiency of crop plants: Physiological constraints upon nitrogen absorption. *Crit Rev Plant Sci,* Vol.22, No.5, pp. 453-470.

Gold, N. D. & Martin V. J. (2007). Global view of the *Clostridium thermocellum* cellulosome revealed by quantitative proteomic analysis. *J Bacteriol,* Vol.189, No.19, pp. 6787-6795.

Goldemberg, J. (2007). Ethanol for a sustainable energy future. *Science,* Vol.315, No.5813, pp. 808-810.

Gonzalez-Garcia, S.; Moreira M. T. & Feijoo G. (2010). Comparative environmental performance of lignocellulosic ethanol from different feedstocks. *Renew Sust Energ Rev,* Vol.14, No.7, pp. 2077-2085.

Good, A. G.; Shrawat A. K. & Muench D. G. (2004). Can less yield more? Is reducing nutrient input into the environment compatible with maintaining crop production? *Trends Plant Sci,* Vol.9, No.12, pp. 597-605.

Gouveia, L. & Oliveira A. C. (2009). Microalgae as a raw material for biofuels production. *J Ind Microbiol Biot,* Vol.36, No.2, pp. 269-271.

Grabber, J. H.; Schatz P. F.; Kim H.; Lu F. C. & Ralph J. (2010). Identifying new lignin bioengineering targets: 1. Monolignol-substitute impacts on lignin formation and cell wall fermentability. *BMC Plant Biol,* Vol.10, pp. 114.

Gubitz, G. M.; Mittelbach M. & Trabi M. (1999). Exploitation of the tropical oil seed plant *Jatropha curcas* L. *Bioresource Technol,* Vol.67, No.1, pp. 73-82.

Gutierrez Manero, F. J.; Ramos Solano B. & Probanza A. (2001). The plant growth promoting rhizobacteria *Bacillus pumilus* and *Bacillus licheniformis* produce high amounts of physiologically active gibberellins. *Physiologia Plantarum,* Vol.111, No.2, pp. 206 211.

Haas, L. O.; Cregg J. M. & Gleeson M. A. (1990). Development of an integrative DNA transformation system for the yeast *Candida tropicalis. Journal of Bacteriology,* Vol.172, No.8, pp. 4571-4577.

Hamel, C. (2004). Impact of arbuscular mycorrhiza fungi on N and P cycling in the root zone. *Canadian Journal of Soil Science,* Vol.84, No.4, pp. 383-395.

Hawkins, H. J.; Johansen A. & George E. (2000). Uptake and transport of organic and inorganic nitrogen by arbuscular mycorrhizal fungi. *Plant Soil,* Vol.226, No.2, pp. 275-285.

Hirel, B.; Le Gouis J.; Ney B. & Gallais A. (2007). The challenge of improving nitrogen use efficiency in crop plants: towards a more central role for genetic variability and quantitative genetics within integrated approaches. *J Exp Bot,* Vol.58, No.9, pp. 2369-2387.

Hirel, B.; Martin A.; Lee J.; Kichey T.; Gerentes D.; Zivy M.; Tatout C.; Dubois F.; Balliau T.; Valot B. et al. (2006). Two cytosolic glutamine synthetase isoforms of maize are specifically involved in the control of grain production. *Plant Cell,* Vol.18, No.11, pp. 3252-3274.

Hodge, A.; Campbell C. D. & Fitter A. H. (2001). An arbuscular mycorrhizal fungus accelerates decomposition and acquires nitrogen directly from organic material. *Nature (Lond),* Vol.413, pp. 297-299.

Holtum, J. A. M.; Smith J. A. C. & Neuhaus H. E. (2005). Intracellular transport and pathways of carbon flow in plants with crassulacean acid metabolism. *Funct Plant Biol,* Vol.32, No.5, pp. 429-449.

Inui, M.; Suda M.; Kimura S.; Yasuda K.; Suzuki H.; Toda H.; Yamamoto S.; Okino S.; Suzuki N. & Yukawa H. (2008). Expression of *Clostridium acetobutylicum* butanol synthetic genes in *Escherichia coli. Appl Microbiol Biot,* Vol.77, No.6, pp. 1305-1316.

Ishii, Y. & Takeuchi R. (1987). Transesterified curcas oil blends for farm diesel engines. *Transactions of the American Society of Agricultural Engineers,* Vol.30, No.3, pp. 605-609.

Jackson, L. A.; Shadle G. L.; Zhou R.; Nakashima J.; Chen F. & Dixon R. A. (2008). Improving saccharification efficiency of alfalfa stems through modification of the terminal stages of monolignol biosynthesis. *Bioenerg Res,* Vol.1, No.3-4, pp. 180-192.

Jaradat, A. A. (2010). Genetic resources of energy crops: Biological systems to combat climate change. *Aust J Crop Sci,* Vol.4, No.5, pp. 309-323.

Joner, E. J.; van Aarle I. M. & Vosátka M. (2000). Phosphatase activity of extra-radical arbuscular mycorrhizal hyphae: a review. *Plant Soil,* Vol.226, pp. 199-210.

Jones, N. & Miller J. H. (1991). *Jatropha curcas*: A multipurpose species for problematic sites. *Land Resources Series,* Vol.1, pp. 1–12

Jorgensen, U. & Schelde K. (2001). Energy crop water and nutrient use efficiency. (ed. SRC IEA Bionenergy Task 17), pp. 1-36. International Energy Agency.

Kaida, R.; Kaku T.; Baba K.; Oyadomari M.; Watanabe T.; Nishida K.; Kanaya T.; Shani Z.; Shoseyov O. & Hayashi T. (2009). Loosening xyloglucan accelerates the enzymatic degradation of cellulose in wood. *Mol Plant,* Vol.2, No.5, pp. 904-909.

Kamilova, F.; Validov S.; Azarova T.; Mulders I. & Lugtenberg B. (2005). Enrichment for enhanced competitive plant root tip colonizers selects for a new class of biocontrol bacteria. *Environ Microbiol,* Vol.7, pp. 1809-1817.

Kant, S.; Bi Y. M. & Rothstein S. J. (2011). Understanding plant response to nitrogen limitation for the improvement of crop nitrogen use efficiency. *J Exp Bot,* Vol.62, No.4, pp. 1499-1509.

Kaushik, N.; Kumar K. & Kumar S. (2007). Potential of *Jatropha curcas* for biofuels. *J Biobased Mater Bio,* Vol.1, No.3, pp. 301-314.

Kiers, E. T. & van der Heijden M. G. A. (2006). Mutualistic stability in the arbuscular mycorrhizal symbiosis: exploring hypotheses of evolutionary cooperation. *Ecology,* Vol.87, pp. 1627-1636.

Klironomos, J. N. & Hart M. M. (2001). Animal nitrogen swap for plant carbon. *Nature,* Vol.410, pp. 651-652.

Knoop, H.; Zilliges Y.; Lockau W. & Steuer R. (2010). The metabolic network of *Synechocystis* sp. PCC 6803: Systemic properties of autotrophic growth. *Plant Physiol,* Vol.154, No.1, pp. 410-422.

Knothe, G.; Cermak S. C. & Evangelista R. L. (2009). *Cuphea* oil as source of biodiesel with improved fuel properties caused by high content of methyl decanoate. *Energ Fuel,* Vol.23, pp. 1743-1747.

Koide, R. T. & Kabir Z. (2000). Extraradical hyphae of the mycorrhizal fungus *Glomus intraradices* can hydrolyse organic phosphate. *New PhytoL,* Vol.148, pp. 511-517.

Krouk, G.; Lacombe B.; Bielach A.; Perrine-Walker F.; Malinska K.; Mounier E.; Hoyerova K.; Tillard P.; Leon S.; Ljung K. et al. (2010). Nitrate-regulated auxin transport by NRT1.1 defines a mechanism for nutrient sensing in plants. *Dev Cell,* Vol.18, No.6, pp. 927-937.

Kumar, K.; Dasgupta C. N.; Nayak B.; Lindblad P. & Das D. (2011). Development of suitable photobioreactors for $CO_2$ sequestration addressing global warming using green algae and cyanobacteria. *Bioresource Technol,* Vol.102, No.8, pp. 4945-4953

Kumar, M. & Gayen K. (2011). Developments in biobutanol production: New insights. *Appl Energ,* Vol.88, No.6, pp. 1999-2012

Kumar, N.; Anand K. G. V.; Pamidimarri D. V. N. S.; Sarkar T.; Reddy M. P.; Radhakrishnan T.; Kaul T.; Reddy M. K. & Sopori S. K. (2010). Stable genetic transformation of *Jatropha curcas* via *Agrobacterium tumefaciens*-mediated gene transfer using leaf explants. *Ind Crop Prod,* Vol.32, No.1, pp. 41-47.

Langley, J. A.; Chapman S. K. & Hungate B. A. (2006). Ectomycorrhizal colonization slows root decomposition: the postmortem fungal legacy. *Ecol Lett,* Vol.9, pp. 955-959.

Lardizabal, K.; Effertz R.; Levering C.; Mai J.; Pedroso M. C.; Jury T.; Aasen E.; Gruys K. & Bennett K. (2008). Expression of *Umbelopsis ramanniana DGAT2A* in seed increases oil in soybean. *Plant Physiol,* Vol.148, No.1, pp. 89-96.

Lau, M. W. & Dale B. E. (2009). Cellulosic ethanol production from AFEX-treated corn stover using *Saccharomyces cerevisiae* 424A(LNH-ST). *P Natl Acad Sci USA,* Vol.106, No.5, pp. 1368-1373.

Leake, J.; Johnson D.; Donnelly D.; Muckle G.; Boddy L. & Read D. (2004). Networks of power and influence: the role of mycorrhizal mycelium in controlling plant communities and agroecosystem functioning. *Can J Bot,* Vol.82, pp. 1016-1045.

Leake, J. R. & Read D. J. (1997). *Mycorrhizal fungi in terrestrial habitats.* Springer-Verlag, Heidelberg.

Lee, C. H.; Teng Q.; Huang W. L.; Zhong R. Q. & Ye Z. H. (2009). Down-regulation of PoGT47C expression in poplar results in a reduced glucuronoxylan content and an increased wood digestibility by cellulase. *Plant Cell Physiol,* Vol.50, No.6, pp. 1075-1089.

Lee, Y. & Voit E. O. (2010). Mathematical modeling of monolignol biosynthesis in *Populus* xylem. *Math Biosci,* Vol.228, No.1, pp. 78-89.

Li, M. R.; Li H. Q.; Jiang H. W.; Pan X. P. & Wu G. J. (2008). Establishment of an *Agrobacteriuim*-mediated cotyledon disc transformation method for *Jatropha curcas*. *Plant Cell Tiss Org*, Vol.92, No.2, pp. 173-181.

Li, Y. T.; Han D. X.; Hu G. R.; Dauvillee D.; Sommerfeld M.; Ball S. & Hua Q. (2010). *Chlamydomonas* starchless mutant defective in ADP-glucose pyrophosphorylase hyper-accumulates triacylglycerol. *Metab Eng*, Vol.12, No.4, pp. 387-391.

Linger, J. G.; Adney W. S. & Darzins A. (2010). Heterologous expression and extracellular secretion of cellulolytic enzymes by *Zymomonas mobilis*. *Appl Environ Microb*, Vol.76, No.19, pp. 6360-6369.

Lionetti, V.; Francocci F.; Ferrari S.; Volpi C.; Bellincampi D.; Galletti R.; D'Ovidio R.; De Lorenzo G. & Cervone F. (2010). Engineering the cell wall by reducing de-methyl-esterified homogalacturonan improves saccharification of plant tissues for bioconversion. *P Natl Acad Sci USA*, Vol.107, No.2, pp. 616-621.

Lipson, D. & Nasholm T. (2001). The unexpected versatility of plants: organic nitrogen use and availability in terrestrial ecosystems. *Oecologia*, Vol.128, No.3, pp. 305-316.

Liu, H. M.; Yan M.; Lai C. G.; Xu L. & Ouyang P. K. (2010). gTME for improved xylose fermentation of *Saccharomyces cerevisiae*. *Appl Biochem Biotech*, Vol.160, No.2, pp. 574-582.

Liu, J. X.; Chen F. J.; Olokhnuud C.; Glass A. D. M.; Tong Y. P.; Zhang F. S. & Mi G. H. (2009). Root size and nitrogen-uptake activity in two maize (*Zea mays*) inbred lines differing in nitrogen-use efficiency. *Journal of Plant Nutrition and Soil Science*, Vol.172, No.2, pp. 230-236.

Liu, T. G. & Khosla C. (2010). Genetic engineering of *Escherichia coli* for biofuel production. *Annu Rev Genet*, Vol.44, pp. 53-69.

Liu, X. Y. & Curtiss R. (2009). Nickel-inducible lysis system in *Synechocystis* sp PCC 6803. *P Natl Acad Sci USA*, Vol.106, No.51, pp. 21550-21554.

Lu, C. F. & Kang J. L. (2008). Generation of transgenic plants of a potential oilseed crop *Camelina sativa* by *Agrobacterium*-mediated transformation. *Plant Cell Rep*, Vol.27, No.2, pp. 273-278.

Lu, C. F.; Xin Z. G.; Ren Z. H.; Miquel M. & Browse J. (2009). An enzyme regulating triacylglycerol composition is encoded by the *ROD1* gene of *Arabidopsis*. *P Natl Acad Sci USA*, Vol.106, No.44, pp. 18837-18842.

Lu, X. F. (2010). A perspective: Photosynthetic production of fatty acid-based biofuels in genetically engineered cyanobacteria. *Biotechnol Adv*, Vol.28, No.6, pp. 742-746.

Madhaiyan, M.; Poonguzhali S.; Lee H. S.; Hari K.; Sundaram S. P. & Sa T. M. (2005). Pink-pigmented facultative methylotrophic bacteria accelerate germination, growth and yield of sugarcane clone. *Biology and Fertility of Soils*, Vol.41, No.5, pp. 350-358.

Malamy, J. E. & Ryan K. S. (2001). Environmental regulation of lateral root initiation in *Arabidopsis*. *Plant Physiol*, Vol.127, No.3, pp. 899-909.

Mandels, M. & Reese E. T. (1957). Induction of cellulase in *Trichoderma viride* as influenced by carbon sources and metals. *Journal of Bacteriology*, Vol.73, No.2, pp. 269-278.

Mano, H. & Morisaki H. (2008). Endophytic bacteria in the rice plant. *Microbes Environ*, Vol.23, No.2, pp. 109-117.

Mao, S. M.; Luo Y. A. M.; Zhang T. R.; Li J. S.; Bao G. A. H.; Zhu Y.; Chen Z. G.; Zhang Y. P.; Li Y. & Ma Y. H. (2010). Proteome reference map and comparative proteomic analysis between a wild type *Clostridium acetobutylicum* DSM 1731 and its mutant with enhanced butanol tolerance and butanol yield. *J Proteome Res,* Vol.9, No.6, pp. 3046-3061.

Markevicius, A.; Katinas V.; Perednis E. & Tamasauskiene M. (2010). Trends and sustainability criteria of the production and use of liquid biofuels. *Renew Sust Energ Rev,* Vol.14, No.9, pp. 3226-3231.

Marko, B.; Popp J. & Somogyi A. (2009). Sustainable land use for food, feed and biofuels. *Cereal Res Commun,* Vol.37, pp. 643-646.

Martin, F.; Aerts A.; Ahren D.; Brun A.; Danchin E. G. J.; Duchaussoy F.; Gibon J.; Kohler A.; Lindquist E.; Pereda V. et al. (2008). The genome of *Laccaria bicolor* provides insights into mycorrhizal symbiosis. *Nature,* Vol.452, No.7183, pp. 88-U87.

Martinez, D.; Berka R. M.; Henrissat B.; Saloheimo M.; Arvas M.; Baker S. E.; Chapman J.; Chertkov O.; Coutinho P. M.; Cullen D. et al. (2008). Genome sequencing and analysis of the biomass-degrading fungus *Trichoderma reesei* (syn. *Hypocrea jecorina*). *Nat Biotechnol,* Vol.26, No.5, pp. 553-560.

Masclaux-Daubresse, C.; Daniel-Vedele F.; Dechorgnat J.; Chardon F.; Gaufichon L. & Suzuki A. (2010). Nitrogen uptake, assimilation and remobilization in plants: challenges for sustainable and productive agriculture. *Ann Bot-London,* Vol.105, No.7, pp. 1141-1157.

Mishra, P.; Mishra S.; Selvakumar G.; Kundu S. & Gupta S. (2009). Enhanced soybean (*Glycine max* L.) plant growth and nodulation by *Bradyrhizobium japonicum*-SB1 in presence of *Bacillus thuringiensis*-KR1. *Acta Agriculturae Scandinavica, B,* Vol.59, No.2, pp. 189-196.

Moellering, E. R. & Benning C. (2010). RNA interference silencing of a major lipid droplet protein affects lipid droplet size in *Chlamydomonas reinhardtii. Eukaryot Cell,* Vol.9, No.1, pp. 97-106.

Moll, R. H.; Kamprath E. J. & Jackson W. A. (1982). Analysis and interpretation of factors which contribute to efficiency of nitrogen-utilization. *Agron J,* Vol.74, No.3, pp. 562-564.

Mortimer, J. C.; Miles G. P.; Brown D. M.; Zhang Z. N.; Segura M. P.; Weimar T.; Yu X. L.; Seffen K. A.; Stephens E.; Turner S. R. et al. (2010). Absence of branches from xylan in *Arabidopsis* gux mutants reveals potential for simplification of lignocellulosic biomass. *P Natl Acad Sci USA,* Vol.107, No.40, pp. 17409-17414.

Mussatto, S. I.; Dragone G.; Guimaraes P. M. R.; Silva J. P. A.; Carneiro L. M.; Roberto I. C.; Vicente A.; Domingues L. & Teixeira J. A. (2010). Technological trends, global market, and challenges of bio-ethanol production. *Biotechnol Adv,* Vol.28, No.6, pp. 817-830.

Nasholm, T.; Kielland K. & Ganeteg U. (2009). Uptake of organic nitrogen by plants. *New Phytol,* Vol.182, No.1, pp. 31-48.

Nevalainen, H.; Suominen P. & Taimisto K. (1994). On the safety of *Trichoderma reesei. J Biotechnol,* Vol.37, No.3, pp. 193-200.

Noel, T.; Sheng C.; Yost C.; Pharis R. & Hynes M. (1996). *Rhizobium leguminosarum* as a plant growth-promoting rhizobacterium: direct growth promotion of canola and lettuce. *Can J Microbiol,* Vol.42, No.3, pp. 279-283.

North, G. B.; Martre P. & Nobel P. S. (2004). Aquaporins account for variations in hydraulic conductance for metabolically active root regions of *Agave deserti* in wet, dry, and rewetted soil. *Plant Cell Environ,* Vol.27, No.2, pp. 219-228.

Noskov, V. N.; Chuang R. Y.; Gibson D. G.; Leem S. H.; Larionov V. & Kouprina N. (2011). Isolation of circular yeast artificial chromosomes for synthetic biology and functional genomics studies. *Nature Protocols,* Vol.6, No.1, pp. 89-96.

Okon, Y.; Bloemberg G. V. & Lugtenberg B. J. J. (1998). *Biotechnology of biofertilization and phytostimulation*. Marcel Dekker, Inc., New York.

Pan, J. L.; Fu Q. T. & Xu Z. F. (2010). *Agrobacterium tumefaciens*-mediated transformation of biofuel plant *Jatropha curcas* using kanamycin selection. *Afr J Biotechnol,* Vol.9, No.39, pp. 6477-6481.

Parniske, M. (2008). Arbuscular mycorrhiza: the mother of plant root endosymbioses. *Nat Rev Microbiol,* Vol.6, No.10, pp. 763-775.

Pathak, R. R.; Ahmad A.; Lochab S. & Raghuram N. (2008). Molecular physiology of plant nitrogen use efficiency and biotechnological options for its enhancement. *Curr Sci India,* Vol.94, No.11, pp. 1394-1403.

Payne, W. A. (2010). Are biofuels antithetic to long-term sustainability of soil and water resources? *Advances in Agronomy,* Vol.105, pp. 1-46.

Peralta-Yahya, P. P. & Keasling J. D. (2010). Advanced biofuel production in microbes. *Biotechnol J,* Vol.5, No.2, pp. 147-162.

Persson, J.; Gardestrom P. & Nasholm T. (2006). Uptake, metabolism and distribution of organic and inorganic nitrogen sources by *Pinus sylvestris*. *J Exp Bot,* Vol.57, No.11, pp. 2651-2659.

Pfromm, P. H.; Amanor-Boadu V.; Nelson R.; Vadlani P. & Madl R. (2010). Bio-butanol vs. bio-ethanol: A technical and economic assessment for corn and switchgrass fermented by yeast or *Clostridium acetobutylicum*. *Biomass Bioenerg,* Vol.34, No.4, pp. 515-524.

Pienkos, P. T. & Darzins A. (2009). The promise and challenges of microalgal-derived biofuels. *Biofuel Bioprod Bior,* Vol.3, No.4, pp. 431-440.

Pinzi, S.; Garcia I. L.; Lopez-Gimenez F. J.; de Castro M. D. L.; Dorado G. & Dorado M. P. (2009). The ideal vegetable oil-based biodiesel composition: A review of social, economical and technical implications. *Energ Fuel,* Vol.23, pp. 2325-2341.

Posten, C. & Schaub G. (2009). Microalgae and terrestrial biomass as source for fuels-A process view. *J Biotechnol,* Vol.142, No.1, pp. 64-69.

Raghuram, N.; Pathak R. R.; Ahmad A. & Lochab S. (2008). Molecular physiology of plant nitrogen use efficiency and biotechnological options for its enhancement. *Curr Sci India,* Vol.94, No.11, pp. 1394-1403.

Raghuwanshi, A. & Birch R. G. (2010). Genetic transformation of sweet sorghum. *Plant Cell Rep,* Vol.29, No.9, pp. 997-1005.

Rai, M.; Acharya D.; Singh A. & Varma A. (2001). Positive growth responses of the medicinal plants *Spilanthes calva* and *Withania somnifera* to inoculation by *Piriformospora indica* in a field trial. *Mycorrhiza,* Vol.11, pp. 123-128.

Read, D. J. & Perez-Moreno J. (2003). Mycorrhizas and nutrient cycling in ecosystems - a journey towards relevance? *New Phytol,* Vol.157, pp. 475-492.

Reijnders, L. (2009). Microalgal and terrestrial transport biofuels to displace fossil fuels. *Energies,* Vol.2, No.1, pp. 48-56.

Rentsch, D.; Schmidt S. & Tegeder M. (2007). Transporters for uptake and allocation of organic nitrogen compounds in plants. *Febs Lett,* Vol.581, No.12, pp. 2281-2289.

Rillig, M. C. (2004). Arbuscular mycorrhizae, glomalin, and soil aggregation. *Can J Soil Sci,* Vol.84, pp. 355-363.

Rillig, M. C. & Mummey D. L. (2006). Mycorrhizas and soil structure. *New Phyto,* Vol.171, pp. 41-53.

Rittmann, B. E. (2008). Opportunities for renewable bioenergy using microorganisms. *Biotechnol Bioeng,* Vol.100, No.2, pp. 203-212.

Roberts, S. B.; Gowen C. M.; Brooks J. P. & Fong S. S. (2010). Genome-scale metabolic analysis of *Clostridium thermocellum* for bioethanol production. *BMC Syst Biol,* Vol.4, pp. 31.

Robertson, G. P.; Dale V. H.; Doering O. C.; Hamburg S. P.; Melillo J. M.; Wander M. M.; Parton W. J.; Adler P. R.; Barney J. N.; Cruse R. M. et al. (2008). Agriculture. Sustainable biofuels redux. *Science,* Vol.322, No.5898, pp. 49-50.

Robinson, D. & Rorison I. H. (1983). Relationships between root morphology and nitrogen availability in a recent theoretical-model describing nitrogen uptake from soil. *Plant Cell Environ,* Vol.6, No.8, pp. 641-647.

Rosegrant, M. W.; Ringler C. & Zhu T. J. (2009). Water for agriculture: maintaining food security under growing scarcity. *Annu Rev Env Resour,* Vol.34, pp. 205-222.

Ryu, C. M.; Farag M. A.; Hu C. H.; Reddy M. S.; Wei H. X.; Pare P. W. & Kloepper J. W. (2003). Bacterial volatiles promote growth in *Arabidopsis. P Natl Acad Sci USA,* Vol.100, No.8, pp. 4927-4932.

Sahin, Y. (2011). Environmental impacts of biofuels. *Ener Educ Sci Tech-A,* Vol.26, No.2, pp. 129-142.

Scharlemann, J. P. W. & Laurance W. F. (2008). Environmental science - How green are biofuels? *Science,* Vol.319, No.5859, pp. 43-44.

Schaub, G. & Vetter A. (2008). Biofuels for automobiles - An overview. *Chem Eng Technol,* Vol.31, No.5, pp. 721-729.

Schenk, P. M.; Thomas-Hall S. R.; Stephens E.; Marx U. C.; Mussgnug J. H.; Posten C.; Kruse O. & Hankamer B. (2008). Second generation biofuels: High-efficiency microalgae for biodiesel production. *Bioenerg Res,* Vol.1, No.1, pp. 20-43.

Schmidt, S. & Stewart G. R. (1999). Glycine metabolism by plant roots and its occurrence in Australian plant communities. *Aust J Plant Physiol,* Vol.26, No.3, pp. 253-264.

Schreiner, R. P.; Mihara K. L.; McDaniel K. L. & Benthlenfalvay G. J. (2003). Mycorrhizal fungi influence plant and soil functions and interactions. *Plant Soil,* Vol.188, pp. 199-209.

Shah, S.; Sharma S. & Gupta M. N. (2004). Biodiesel preparation by lipase-catalyzed transesterification of *Jatropha* oil. *Energ Fuel,* Vol.18, No.1, pp. 154-159.

Shuhegge, R.; Ihring A.; Gantner S.; Bahnweg G.; Knaooe C. & al. e. (2006). Induction of systemic resistance in tomato by N-acyl-L-homoserine lactone-producing rhizosphere bacteria. *Plant Cell Environ,* Vol.29, pp. 909-918.

Simmons, B. A.; Logue D. & Ralph J. (2010). Advances in modifying lignin for enhanced biofuel production. *Curr Opin Plant Biol,* Vol.13, No.3, pp. 313-320.

Simpson, J.; Martinez Hernandez A.; Jazmin Abraham Juarez M.; Delgado Sandoval S.; Sanchez Villarreal A. & Cortes R. (2011). Genomic resources and transcriptome mining in *Agave tequilana. GCB Bioenergy,* Vol.3, No.1, pp. 25-36.

Singh, A.; Pant D.; Korres N. E.; Nizami A. S.; Prasad S. & Murphy J. D. (2010). Key issues in life cycle assessment of ethanol production from lignocellulosic biomass: Challenges and perspectives. *Bioresource Technol,* Vol.101, No.13, pp. 5003-5012.

Six, J.; Frey S. D.; Thiet R. K. & Batten K. M. (2006). Bacterial and fungal contributions to carbon sequestration in agroecosystems. *Soil Sci Soc Am J,* Vol.70, pp. 555-569.

Skotnicki, M. L.; Warr R. G.; Goodman A. E.; Lee K. J. & Rogers P. L. (1983). High-productivity alcohol fermentations using *Zymomonas mobilis. Biochem Soc Symp,* Vol.48, pp. 53-86.

Slocombe, S. P.; Cornah J.; Pinfield-Wells H.; Soady K.; Zhang Q. Y.; Gilday A.; Dyer J. M. & Graham I. A. (2009). Oil accumulation in leaves directed by modification of fatty acid breakdown and lipid synthesis pathways. *Plant Biotechnol J,* Vol.7, No.7, pp. 694-703.

Smith, S. E.; Read D. J. & Harley J. L. (1997). *Mycorrhizal symbiosis.* Academic Press, San Diego, Calif.

Somerville, C.; Youngs H.; Taylor C.; Davis S. C. & Long S. P. (2010). Feedstocks for lignocellulosic biofuels. *Science,* Vol.329, No.5993, pp. 790-792.

Sommer, M. O. A.; Church G. M. & Dantas G. (2010). A functional metagenomic approach for expanding the synthetic biology toolbox for biomass conversion. *Mol Syst Biol,* Vol.6, pp. 360.

Song, J. Y.; Lu S. F.; Chen Z. Z.; Lourenco R. & Chiang V. L. (2006). Genetic transformation of *Populus trichocarpa* genotype Nisqually-1: A functional genomic tool for woody plants. *Plant Cell Physiol,* Vol.47, No.11, pp. 1582-1589.

Sramek, F.; Dubsky M. & Vosa´tka M. (2000). Effect of arbuscular mycorrhizal fungi and *Trichoderma harzianum* on three species of balcony plants. *Plant Prod Sci,* Vol.46, pp. 127-131.

Sun, Y. & Cheng J. (2002). Hydrolysis of lignocellulosic materials for ethanol production: a review. *Bioresour Technol,* Vol.83, No.1, pp. 1-11.

Sunil, N.; Varaprasad K. S.; Sivaraj N.; Kumar T. S.; Abraham B. & Prasad R. B. N. (2008). Assessing *Jatropha curcas* L. germplasm *in-situ* - A case study. *Biomass Bioenerg,* Vol.32, No.3, pp. 198-202.

Swana, J.; Yang Y.; Behnam M. & Thompson R. (2011). An analysis of net energy production and feedstock availability for biobutanol and bioethanol. *Bioresource Technol,* Vol.102, No.2, pp. 2112-2117.

Swings, J. & De Ley J. (1977). The biology of *Zymomonas. Bacteriol Rev,* Vol.41, No.1, pp. 1-46.

Taghavi, S.; Garafola C.; Monchy S.; Newman L.; Hoffman A.; Weyens N.; Barac T.; Vangronsveld J. & Van Der Lelie D. (2009). Genome survey and characterization of endophytic bacteria exhibiting a beneficial effect on growth and development of poplar trees. *Appl Environ Microb,* Vol.75, No.3, pp. 748.

Takeda, Y. (1982). Development study on *Jatropha curcas* (Sabu Dum) oil as a substitute for diesel engine oil in Thailand. *J Agr Assoc China,* No.120, pp. 1-8.

Tarafdar, J. C. & Marschner H. (1994). Phosphatase activity in the rhizosphere and hyphosphere of VA mycorrhizal wheat supplied with inorganic and organic phosphorus. *Soil Biol Biochem,* Vol.26, pp. 387-395.

Thornton, B. & Robinson D. (2005). Uptake and assimilation of nitrogen from solutions containing multiple N sources. *Plant Cell Environ,* Vol.28, No.6, pp. 813-821.

Toussaint, J. P. (2007). Investigating physiological changes in the aerial parts of AM plants: what do we know and where should we be heading? *Mycorrhiza,* Vol.17, pp. 349-353.

Tripathi, S. A.; Olson D. G.; Argyros D. A.; Miller B. B.; Barrett T. F.; Murphy D. M.; McCool J. D.; Warner A. K.; Rajgarhia V. B.; Lynd L. R. et al. (2010). Development of pyrF-based genetic system for targeted gene deletion in *Clostridium thermocellum* and creation of a pta mutant. *Appl Environ Microb,* Vol.76, No.19, pp. 6591-6599.

van Beilen, J. B. (2010). Why microalgal biofuels won't save the internal combustion machine. *Biofuel Bioprod Bior,* Vol.4, No.1, pp. 41-52.

van Breemen, N.; Lundström U. S. & Jongmans A. G. (2000). Do plants drive podzolization via rock-eating mycorrhizal fungi? *Geoderma,* Vol.94, pp. 163-171.

van der Heijden, M. G. A.; Bardgett R. D. & van Straalen N. M. (2008). The unseen majority: soil microbes as drivers of plant diversity and productivity in terrestrial ecosystems. *Ecol Lett,* Vol.11, pp. 296-310.

van der Lelie, D.; Taghavi S.; Monchy S.; Schwender J.; Miller L.; Ferrieri R.; Rogers A.; Wu X.; Zhu W. & Weyens N. (2009). Poplar and its bacterial endophytes: coexistence and harmony. *Crit Rev Plant Sci,* Vol.28, No.5, pp. 346-358.

van Loon, L. C. (2007). Plant responses to plant growth-promoting bacteria. *Eur J Plant Pathol,* Vol.119, pp. 243-254.

Vanholme, R.; Van Acker R. & Boerjan W. (2010). Potential of *Arabidopsis* systems biology to advance the biofuel field. *Trends Biotechnol,* Vol.28, No.11, pp. 543-547.

Vassilev, N.; Vassileva M. & Nikolaeva I. (2006). Simultaneous P-solubilizing and biocontrol activity of microorganisms: potentials and future trends. *Appl Microbiol Biot,* Vol.71, No.2, pp. 137-144.

Vega-Sanchez, M. E. & Ronald P. C. (2010). Genetic and biotechnological approaches for biofuel crop improvement. *Curr Opin Biotech,* Vol.21, No.2, pp. 218-224.

Vigeolas, H.; Waldeck P.; Zank T. & Geigenberger P. (2007). Increasing seed oil content in oil-seed rape (*Brassica napus* L.) by over-expression of a yeast glycerol-3-phosphate dehydrogenase under the control of a seed-specific promoter. *Plant Biotechnol J,* Vol.5, No.3, pp. 431-441.

Villegas, J. & Fortin J. A. (2001). Phosphorus solubilization and pH changes as a result of the interactions between soil bacteria and arbuscular mycorrhizal fungi on a medium containing NH4+ as nitrogen source. *Can J Bot,* Vol.79, pp. 865-870.

Waclawovsky, A. J.; Sato P. M.; Lembke C. G.; Moore P. H. & Souza G. M. (2010). Sugarcane for bioenergy production: an assessment of yield and regulation of sucrose content. *Plant Biotechnol J,* Vol.8, No.3, pp. 263-276.

Wahlen, B. D.; Willis R. M. & Seefeldt L. C. (2011). Biodiesel production by simultaneous extraction and conversion of total lipids from microalgae, cyanobacteria, and wild mixed-cultures. *Bioresource Technol,* Vol.102, No.3, pp. 2724-2730.

Walch-Liu, P. & Forde B. G. (2008). Nitrate signalling mediated by the NRT1.1 nitrate transporter antagonises L-glutamate-induced changes in root architecture. *Plant J,* Vol.54, No.5, pp. 820-828.

Walch-Liu, P.; Ivanov I. I.; Filleur S.; Gan Y. B.; Remans T. & Forde B. G. (2006). Nitrogen regulation of root branching. *Ann Bot-London,* Vol.97, No.5, pp. 875-881.

Wallander, H.; Mahmood S.; Hagerberg D.; Johansson L. & Pallon J. (2003). Elemental composition of ectomycorrhizal mycelia identified by PCR-RFLP analysis and grown in contact with apatite or wood ash in forest soil. *FEMS Microbiol Ecol,* Vol.44, pp. 57-65.

Wallenda, T.; Stober C.; Höbom L.; Schinkel H.; George E.; P. H. & Read D. J. (2000). *Nitrogen uptake processes in roots and mycorrhizas*. Springer-Verlag, Heidelberg.

Wang, H. W.; Zhang B.; Hao Y. J.; Huang J.; Tian A. G.; Liao Y.; Zhang J. S. & Chen S. Y. (2007). The soybean Dof-type transcription factor genes, *GmDof4* and *GmDof11*, enhance lipid content in the seeds of transgenic *Arabidopsis* plants. *Plant J*, Vol.52, No.4, pp. 716-729.

Wei, W.; McCusker J. H.; Hyman R. W.; Jones T.; Ning Y.; Cao Z.; Gu Z.; Bruno D.; Miranda M.; Nguyen M. et al. (2007). Genome sequencing and comparative analysis of *Saccharomyces cerevisiae* strain YJM789. *P Natl Acad Sci USA*, Vol.104, No.31, pp. 12825-12830.

Weih, M.; Asplund L. & Bergkvist G. (2011). Assessment of nutrient use in annual and perennial crops: A functional concept for analyzing nitrogen use efficiency. *Plant Soil*, Vol.339, No.1-2, pp. 513-520.

Wen, F.; Sun J. & Zhao H. M. (2010). Yeast surface display of trifunctional minicellulosomes for simultaneous saccharification and fermentation of cellulose to ethanol. *Appl Environ Microb*, Vol.76, No.4, pp. 1251-1260.

Wild, B.; Wanek W.; Postl W. & Richter A. (2010). Contribution of carbon fixed by Rubisco and PEPC to phloem export in the Crassulacean acid metabolism plant *Kalanchoe daigremontiana*. *J Exp Bot*, Vol.61, No.5, pp. 1375-1383.

Worku, M.; Banziger M.; Erley G. S. A.; Friesen D.; Diallo A. O. & Horst W. J. (2007). Nitrogen uptake and utilization in contrasting nitrogen efficient tropical maize hybrids. *Crop Sci*, Vol.47, No.2, pp. 519-528.

Xi, Y. J.; Fu C. X.; Ge Y. X.; Nandakumar R.; Hisano H.; Bouton J. & Wang Z. Y. (2009). *Agrobacterium*-mediated transformation of switchgrass and inheritance of the transgenes. *Bioenerg Res*, Vol.2, No.4, pp. 275-283.

Yang, X.; Kalluri U. C.; DiFazio S. P.; Wullschleger S. D.; Tschaplinski T. J.; Cheng Z. M. & Tuskan G. A. (2009). Poplar genomics: State of the science. *Crit Rev Plant Sci*, Vol.28, No.5, pp. 285-308.

Yang, X.; Ye C.-y.; Bisaria A.; Tuskan G. A. & Kalluri U. C. (2011). Identification of candidate genes in *Populus* cell wall biosynthesis using text-mining, co-expression network and comparative genomics. *Plant Science* (doi:10.1016/j.plantsci.2011.01.020).

Yao, Q.; Li X.; Feng G. & Christie P. (2001). Mobilization of sparingly soluble inorganic phosphates by the external mycelium of an arbuscular mycorrhizal fungus. *Plant Soil*, Vol.230, pp. 279-285.

Ye, B.; Saito A. & Minamisawa K. (2005). Effect of inoculation with anaerobic nitrogen fixing consortium on salt tolerance of *Miscanthus sinensis*. *Soil Science & Plant Nutrition*, Vol.51, No.2, pp. 243-249.

Ye, C.-Y.; Li T.; Tuskan G. A.; Tschaplinski T. J. & Yang X. (2011). Comparative analysis of GT14/GT14-like gene family in *Arabidopsis*, *Oryza*, *Populus*, *Sorghum* and *Vitis*. *Plant Science* (doi:10.1016/j.plantsci.2011.01.021).

Yevtushenko, D. P. & Misra S. (2010). Efficient *Agrobacterium*-mediated transformation of commercial hybrid poplar *Populus nigra* L. x *P. maximowiczii* A. Henry. *Plant Cell Rep*, Vol.29, No.3, pp. 211-221.

Zhang, M.; Eddy C.; Deanda K.; Finkelstein M. & Picataggio S. (1995). Metabolic engineering of a pentose metabolism pathway in ethanologenic *Zymomonas mobilis*. *Science*, Vol.267, No.5195, pp. 240-243.

16

# Microwave-Assisted Synthesis of Biofuels

Armando T. Quitain[1,2], Shunsaku Katoh[2] and Motonobu Goto[3]
[1]*Graduate School of Science and Technology, Kumamoto University*
[2]*RIST Kagawa, Kagawa Industry Support Foundation*
[3]*Bioelectrics Research Center, Kumamoto University*
*Japan*

## 1. Introduction

Environmentally benign and economically viable alternatives to fossil derived fuels are seriously being explored due to increasing global demand for energy, coupled with the threats posed by the recent climate changes. As potential alternatives, biodiesel and biomass-derivable ethyl *tert*-butyl ether (ETBE) are being pursued. Research for the development of efficient and energy-saving methods for the production of these two promising biofuels had gained significant momentum over the past few years.

Literature searches on published articles having "biodiesel" on its title using *JST Plus/JMEDPlus/JST17580* database resulted to heavy turnouts. The trend follows the graph shown in Fig. 1 in terms of annual publication of related articles. Similar searches including those appearing in abstracts and keywords using SCOPUS database, resulted to more than 6,800 hits. The data clearly indicates a dramatically increasing global interest on research, development and analysis related to biodiesel production, especially during the past decade, due primarily to the motivation of reducing fossil-derived carbon dioxide emissions to the atmosphere.

Biodiesel is a mixture of fatty acid methyl esters produced from the transesterification of plant oils or animal fats with methanol over alkali or acid catalysts as shown in Fig. 2. As reported in the book edited by Saka (2006), the conventional method for its production utilizes homogeneous alkali catalysts, such as NaOH and KOH, in a batch mode (Fig. 3). Post treatment procedures after reaction require neutralization of catalysts and their removal from the products utilizing enormous amount of water. Alternative methods to avoid the problems and high costs of treating wastewater associated with the process are being explored. Moreover, the demand for biodiesel is highly expected to increase, and a more efficient continuous process is being sought with the purpose of reducing capital or production costs.

Several review articles have already been published discussing various alternative production methods for biodiesel. The most noteworthy is the review article published by Ma and Hanna (1999) focusing on the transesterification process, its mechanism, kinetics and effects of reaction parameters such as moisture and free fatty acid contents, molar ratio, reaction time and temperature among many others. Moreover, the source of raw materials and manufacturing costs take the major hurdle in the commercialization of biodiesel, thus alternative sources such as the use of waste cooking oil has long been considered. Kulkarni and Dalai (2006) had reviewed published articles related to the utilization of waste cooking

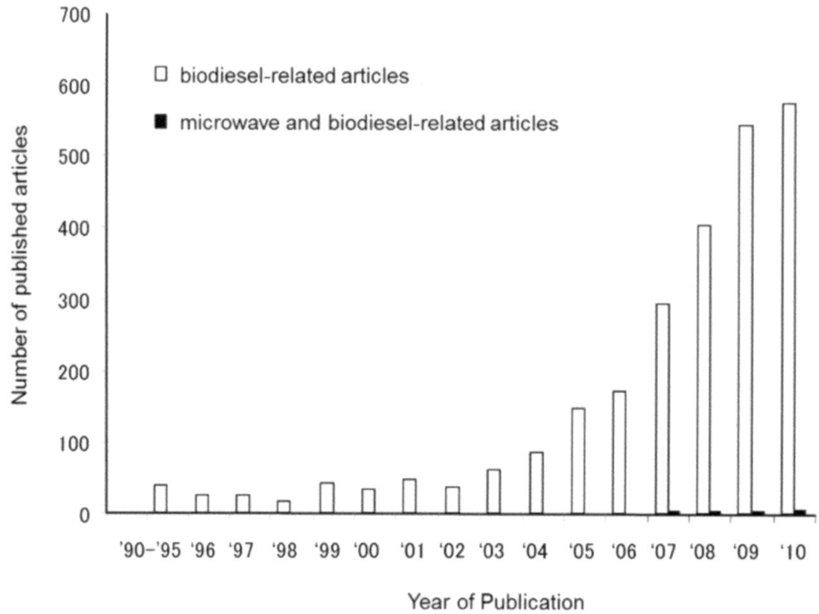


Fig. 1. Trends of biodiesel and microwave related publications based on searches using JST Plus/JMED/JST17580 database

oil as an economic source for biodiesel, showing different methods for the transesterification of oil and the performance of obtained biodiesel in a commercial diesel engine. They concluded that the biodiesel obtained from waste cooking oil gives better engine performance and less emission when tested on commercial diesel engines. Among the many techniques presented, including the use of enzymes, two-step method (acid-catalyzed followed by alkaline-catalyzed step) and supercritical methanol, no particular method was recommended to be superior. In addition, similar reviews on different techniques for the production of biodiesel from waste vegetable oil have been written and summarized by Refaat et al. (2010). Each technique presented has its advantages and drawbacks, and the choice of suitable method depends primarily on its economic viability.

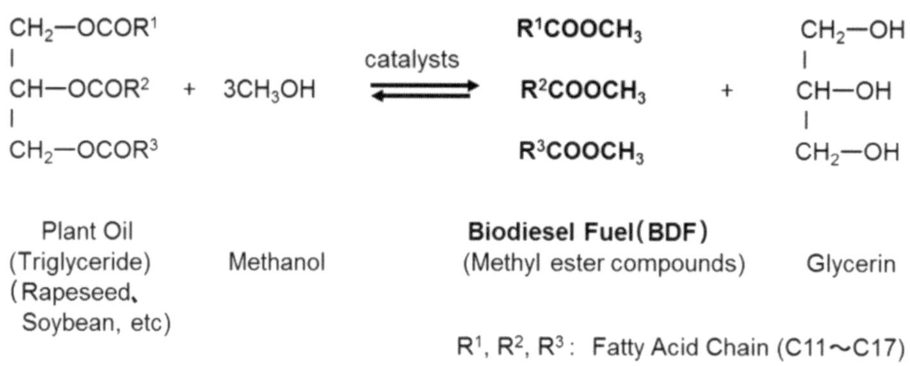


Fig. 2. Reaction mechanism of transesterification of plant oil for biodiesel production

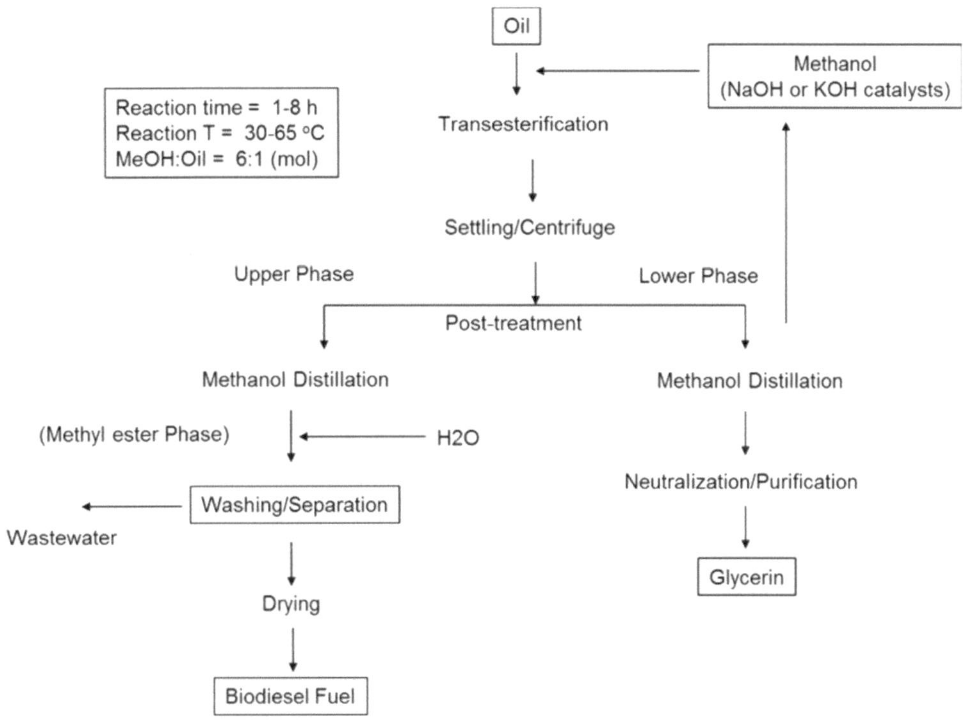


Fig. 3. Conventional alkaline-wash water process for production of biodiesel (Saka, 1996)

Aside from biodiesel, ETBE, which had been previously considered as a replacement for lead as gasoline octane booster, has also gained popularity over ethanol as biofuel due to its superior properties which blend well with gasoline. ETBE also outranks MTBE as an octane enhancer due to its low blending Reid vapor pressure, beside, ETBE is a better option because it is derived from ethanol (EtOH) which can be obtained from renewable resources like biomass. ETBE is produced from the reaction of isobutene (IB) and EtOH, however, the supply of IB, which is mostly derived from non-renewable crude oil, may become limited, and for this reason, alternative routes for its synthesis are also currently being explored. *tert*-Butyl alcohol (TBA), a major byproduct of propylene oxide production from isobutane and propylene, can be employed instead of IB as a reactant (Yang and Goto, 1997). With suitable enzyme catalysts, TBA can also be possibly produced from biomass sources similar to that used for bioethanol production. The prospects are high for the use of these two biomass-derivable lower alcohols for the production of suitable fuel for conventional engines running on gasoline.

The synthesis route for ETBE production employing *tert*-butyl alcohol (TBA) instead of IB has long been investigated by Norris and Rigby (1932) using concentrated sulfuric acid as catalyst. Recently, Habenicht et al. (1995) investigated the reaction at elevated temperatures and pressures. Yin et al. (1995) introduced the use of heterogeneous catalysts such as ion-exchange resin and heteropoly acid. Matouq et al. (1996) applied an advanced method of reactive distillation, utilizing low-grade alcohol catalyzed by potassium hydrogen sulfate ($KHSO_4$). Results showed that ETBE could be produced from the reaction, and that the reactive distillation column was a good choice to separate ETBE from the reacting mixture. In subsequent works, pervaporation technique was incorporated at the bottom of

the column to remove the water byproducts, shifting the equilibrium forward, resulting to better yields (Matouq et al., 1997, Yang & Goto, 1997). The works were further extended on the utilization of ethanol at a concentration as low as that obtained from the fermentation of biomass (about 2.67 mol% in aqueous solution) (Roukas et al., 1995). Almost complete conversion of TBA was obtained, with ETBE selectivity of about 36% (Quitain et al., 1999).
In this chapter, works on the application of microwave irradiation to the syntheses of these two most promising biofuels by microwave irradiation will be discussed and summarized, focusing on our recent studies on microwave-assisted heterogeneously catalyzed processes.

## 2. Fundamentals of microwave irradiation

For rapid synthesis of the two abovementioned biofuels, the application of microwave technology has been proposed. Microwave technology relies on the use of electromagnetic waves to generate heat by the oscillation of molecules upon microwave absorption. The electromagnetic spectrum for microwaves is in between infrared radiation and radiofrequencies of 30 GHz to 300 MHz, respectively, corresponding to wavelengths of 1cm to 1 m. Domestic and industrial microwave systems are required to operate at either 12.2 cm (2.45 GHz) or 33.3 cm (900 MHz) in order not to interfere with the wavelength ranges being utilized for RADAR transmissions and telecommunications (Mingos & Baghurst, 1997).
In microwave-assisted heating, unlike the conventional methods, the heat is generated within the material, thus rapid heating occurs. As a result of this rapid heating, many microwave-assisted organic reactions are accelerated, incomparable with those obtained using the conventional methods. Thus, higher yields and selectivity of target compounds can be obtained at shorter reaction times. In addition, many reactions not possible using the conventional heating methods, had been reported to occur under microwave heating. Some very useful information on the fundamentals of microwave-enhanced chemistry, its sample preparation and applications are well presented in the book edited by Kingston and Haswell (1997).
Other than the advantages of rapid heating, microwave non-thermal effects on reaction likely occur, obtaining dramatic increase in the yield even at milder conditions. Although doubts are cast on the true existence of non-thermal effects, some evidences had been reported and postulates had also been made by several researchers. These were summarized in a review article published by de la Hoz et al. (2005) comparing them with the thermal effects. The review of Jacob et al. (1995) on thermal and non-thermal interaction of microwaves with materials attributed some interesting results on specific microwave effects. Evidences on reaction rate enhancement due to some reasons other than the thermal effects such as "hotspots" or localized heating, molecular agitation, improved transport properties were discussed. They suggested that due to the interaction of microwave with the materials, heating cannot be simply treated as that similar to the conventional methods as there are a lot of possible mechanisms of activation of materials that might possibly occur.
The abovementioned thermal and non-thermal effects of microwave irradiation offer enormous benefits to the synthesis of biofuels including energy efficiency, development of a compact process, rapid heating and instant on-off process (instant heating-cooling process), among many other possible advantages.

## 3. Previous studies on the application of microwave irradiation to the synthesis of biofuels

Due to the benefits and advantages accompanying the use of microwave irradiation, its applications to organic synthesis increased significantly in recent years. However, unlike the momentum that biodiesel research has gained over the past decade, the application of microwave for its production is still in its infancy. The pioneering work on this topic was reported by Breccia et al. (1999), on the use of a domestic microwave apparatus for the synthesis of biodiesel by reaction between methanol and commercial seed oils. In this work, they found that the reaction was practically complete in less than 2 min under microwave irradiation. Activities of several catalysts such as sodium methylate, sodium hydroxide, sodium carbonate, sulfuric acid, benzensulfonic acid and boron carbide were briefly discussed. Boron carbide, which actively absorbs microwave, was reported to be the most effective and promising catalyst for the transesterification. Results of preliminary experiments using a laboratory scale plant for continuous process was also reported, and based on a few obtained data, they concluded that the application of MW both in continuous and batch-wise process was proven to be practical on an industrial scale.

Upon the introduction of scientific microwave apparatus in the market, its use for the investigation of biodiesel synthesis started in 2004, based on the works of Mazzocchia et al. (2004) on the application of heterogeneous catalysts. The catalysts used were mostly zeolites, and the reaction was allowed to proceed at 170 °C for 2h in a sealed vessel. However, only moderate conversions were obtained using this technique.

Research on the use of scientific microwave apparatus was then conducted by the group of Leadbeater and Stencel of the University of Connecticut in 2006 (Leadbeater & Stencel, 2006). They used a 3-kg scale reactor apparatus allowing the reaction to proceed under atmospheric conditions in few minutes. Homogeneous catalysts such as KOH and NaOH were used to accelerate the reaction. The work was extended by the same group to a continuous flow method at flowrates up to 7.2 L/min using a 4L reaction vessel. They also found out that the continuous-flow microwave method was more energy-efficient than the conventional heating methods based on rudimentary energy consumption calculations.

Similar works had been performed by the group of Hernando et al. in 2007 using homogeneous catalysts, and they were able to obtain yield above 95% in 1 min of reaction time. They even used additives such as methyl *tert*-butyl ether (MTBE) to enhance the solubility of the reactants.

Interests on the techniques spread worldwide, and several works then followed mostly on the application to various oil feedstocks. In Thailand, experiments on the use of microwave for the production of biodiesel from waste frying palm oil were reported (Lertsathapornsuk et al., 2008). In this work, domestic microwave apparatus was modified for continuous transesterification. In New Mexico, the group of Patil et al. (2009) tried the techniques on Camelina Sativa oil. In Chicago (US), Majewski et al. (2009), experimented on the transesterification of corn and soybean oil. In China, Zhang et al. (2010) worked on yellow horn oil, Yuan et al. (2009) on castor oil using sulfated activated carbon as microwave absorption catalyst, and in Taiwan, the use of nanopowder calcium oxide to the transesterification of soybean oil was reported. In the Philippines, works using Kenaf seed oil has also been reported (Rathana et al., 2010).

Recently, Leadbeater et al. (2008) applied microwave heating for both batch and continuous flow process for production of biodiesel utilizing butanol, an alcohol that can be generated from agriculture feedstocks similar to that used for ethanol production encouraging utilization of totally renewable based feedstocks. The work was extended by collaborative research groups in Europe (2008) under supercritical conditions for a microwave-assisted catalyst-free transesterification of triglycerides (Geuens et al., 2008). Researchers from Brazil have also tried applying microwave for the activation of enzymatic catalysts used for biodiesel production (Nogueira et al., 2010).

To date, several homogeneous, heterogeneous metal oxide and metal salt catalysts have been evaluated for the microwave-assisted synthesis of biodiesel (Breccia et al., 1999). Among the many catalysts investigated, homogeneous basic catalysts such as KOH and NaOH are the most preferred, and commonly used in the conventional process of transesterification because of its high activity even at low concentration. The production of methyl esters, with methanol as the reactant, proceeds at very high yields even under mild conditions, and reaction generally takes about an hour to complete. For the treatment of free fatty acids present in the oil feedstocks, the use of sulfuric acid is widely considered.

However, there are drawbacks on the use of these homogeneous catalysts including the tedious post treatment procedures of neutralization and washing of products resulting into enormous amount of wastewater produced in the process. Thus, the use of inexpensive heterogeneous catalysts suitable for microwave irradiation is being explored. Our group had been working on this topic since 2006, and had successfully completed application of two related Japanese patents as a result of our extensive works. The results have also been presented in various domestic and international conferences related to microwave application and biomass energy conversions (Quitain et al., 2008; Quitain et al., 2009).

In addition, for ETBE production, Tokyo Electric Co. has applied two patents using microwave irradiation. However, the reported maximum conversion at atmospheric conditions is too low at around 28%. Results of our recent research utilizing the same techniques yielded similar results. However, better conversion closed to 90% was obtained upon the application of constant microwave power, and allowing the reaction temperature to reach solvothermal conditions in a sealed vessel.

## 4. Advantages of heterogeneous catalysis for the synthesis of biofuels

Conventional methods of producing biodiesel normally utilize homogeneous catalysts to accelerate the reaction. At the end of the reaction, the catalyst is neutralized and removed from the products requiring enormous amount of water, which is usually about 80% of produced amount of biodiesel in mass basis. Alternative methods to avoid the problems and high costs of treating wastewater associated with the process are being explored. The use of heterogeneous catalysts offers much benefits as this would eliminate the tedious post treatment procedures of dealing with the wastewater. Besides, the use of solid catalysts accompanies easier product separation resulting to a more economical process.

Several heterogeneous catalysts including basic, acidic, acid-base and enzymes that are suitable for biodiesel production had been reviewed recently by Semwal et al. (2011). The review is very useful in the selection of suitable catalysts and the corresponding optimum conditions. Several solid catalysts have been investigated for biodiesel synthesis but their applications were limited due to lower reaction rates and unfavorable side reactions. Basic heterogeneous catalysts have also been investigated, and the catalytic activity was found to be affected by the presence of water.

Among the many reported catalysts, Ca-based solid catalysts such as CaO and $Ca(OH)_2$ had caught our interest and are deemed most promising because of their availability and low cost. Works are still in progress for modifying these types of catalysts to make them more suitable for a wide range of biofuel feedstocks.

## 5. Our recent works on microwave-assisted synthesis of biofuels

### 5.1 Biodiesel production

#### 5.1.1 Experimental procedures

We have been investigating the application of microwave irradiation to the synthesis of biodiesel using the abovementioned Ca-based solid catalysts. In most of the experiments, rapeseed oil commercially available from Nacalai Tesque (Japan) was used. The average molecular weight of the oil was assumed to be 806 (Kusdiana & Saka, 2001). Methanol (HPLC grade), $Ca(OH)_2$ and CaO (99.9%) were purchased from Wako (Japan), while other catalysts were purchased from Sigma-Aldrich (Japan). In some experiments, commercial slaked lime in pellet form supplied by Inoue Lime Industrial Company (Kochi, Japan) was also used.

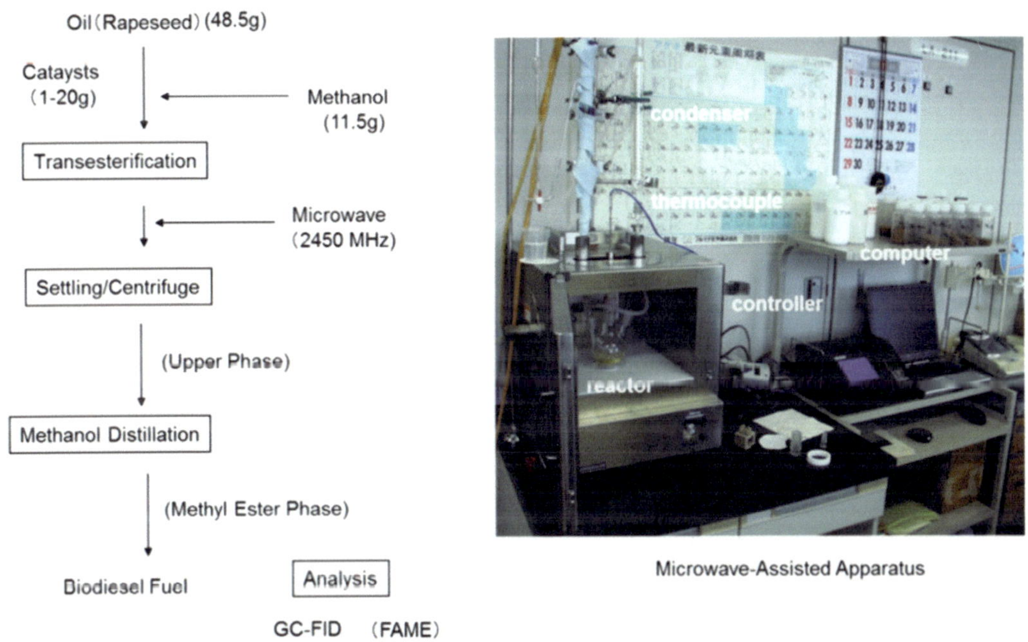


Fig. 4. Outline of experimental procedures and microwave-assisted apparatus used for batch experiments

All microwave-assisted batch experiments were performed using an in-house microwave apparatus shown in Fig. 4, working at 2.45 GHz frequency, with a power programmable from 0 to 700W. Temperature could be controlled, and the reactants could be mixed using a magnetic stirrer. Continuous experiments were performed in a similar apparatus (Shikoku Instrumentation Co. Ltd., Kagawa, Japan) design for microwave-assisted drying, but the power can be programmed to a maximum of 1500W.

In a typical batch experiment (also shown in Fig. 4), about 11.5 g methanol and 48.5 g rapeseed oil (MeOH-oil molar ratio = 6) were placed in a three-necked round bottom flask, and heated either in an oil bath or in a microwave apparatus described above. In all runs, the MeOH-oil ratio was fixed at a commonly used molar ratio of 6, which is also the ratio being employed in industrial scale production of biodiesel. The amount of catalyst was varied from 1 to 20 g. The reaction temperature was set at 60 °C, unless otherwise specified. In experiments involving constant microwave heating power, the temperature was not controlled, but the maximum attained temperature was noted.

The products were collected, then centrifuged to separate the catalysts and the glycerin phase. The unreacted MeOH in the products was then removed using a rotary evaporator at 70 °C.

The products were analyzed of its composition by a gas chromatography - flame ionization detector (GC-FID) apparatus (Shimadzu GC-14B) connected to a computer for data collection and analysis. Component separation was made in a 50m x 0.25mm CP Sil 88 capillary column (GL Science, Japan), tailor-made for FAME analysis using helium as a carrier gas. The column, detection and injection temperatures were set to 190, 300 and 270 °C, respectively. The sample injection volume was 5 µl and peak identification was made by comparing the retention time between the sample and the standard compound. FAME quantitative mixtures (GL Science, Japan) were used for peak identification and for quantitative analysis.

### 5.1.2 Evaluation of catalytic activities of various solid catalysts

Preliminary experiments were conducted to evaluate catalytic activities of various solid catalysts such as Amberlyst 15, Amberlite-OH, Amberlite-Acid, zeolite, sulfated zirconia (in powder and pellet forms), $Ca(OH)_2$ and CaO. Among the catalysts investigated, $Ca(OH)_2$ showed to be the most active, while CaO also gave fairly good results as shown in Fig. 5. The use of these two relatively cheap catalysts showed potential for biodiesel production, thus $Ca(OH)_2$ was used in the succeeding experiments unless otherwise specified.

### 5.1.3 Comparison of microwave and conventional heating

Microwave heating for the production of biodiesel in a batch mode was compared with that of the conventional. In case of the conventional method, the oil bath temperature was set at 60°C, and the mixtures of reactants and catalysts were heated for 1 min. Using microwave irradiation, the power was set at 700 W. Heating for 1 min, the bulk temperature of the mixtures did not reach 60°C in all runs.

Fig. 6 shows a remarkable increase in the yield of methyl esters using microwave heating compared to the conventional. The yield, corresponding to the amount of methyl esters in the oil phase, reached above 95% using 20g $Ca(OH)_2$.

Even if the bulk temperature did not reach 60°C, it is likely that localized heating above 60°C occurred at the surface of the catalysts, which brought about a significant increase in reaction rate, resulting into high yield. This is advantageous especially on the viewpoint of equipment design as this entails less provision for heat and pressure-resisting reactor materials.

In this proposed method, the reaction time was reduced to less than 60 sec compared to 1 to 8 h using the conventional method. In addition, the use of solid catalysts avoids the rigors and complexities of dealing with post-reaction treatments (*i. e.* neutralization of homogeneous catalysts and washing of the products with water). Furthermore, with short reaction time, development of a continuous process is highly feasible thus reducing equipment costs.

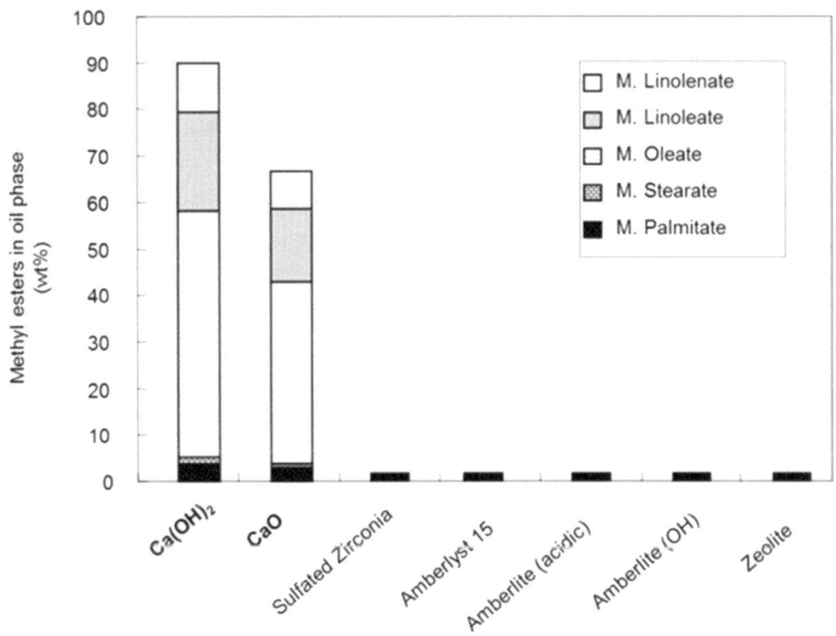


Fig. 5. Comparison of activities of various catalysts under microwave irradiation (t = 1min, catalysts = 10g, MW = 700W)

### 5.1.4 Effect of operating conditions

Fig. 7 shows the effect of microwave irradiation power on the yield at various amounts of catalysts. No significant differences were observed at 140 and 350W, but the yields were comparatively high at 700W especially at 10 and 20g $Ca(OH)_2$.

Using 10g $Ca(OH)_2$, the reaction time was increased to 5 min at the same microwave power of 700W. Results showed that while the temperature increased sharply above 110 °C in just 5 min, the yield decreased to 20%. It is likely that reverse reaction took place brought about by an increase in reaction temperature and the subsequent evaporation of MeOH from the reaction zone. The same results were observed in the works of Hernando et al. (2007) on the batch tests performed with microwaves.

### 5.1.5 Comparison with other vegetable oils

The fatty acid compositions of various oils differ as shown in Table 1. Rapeseed oil contains mostly oleic acid, while soybean and coconut oils are rich in linoleic and lauric acids, respectively. These differences might have an effect on the transesterification of the oils, thus the results obtained using rapeseed oil were compared with that of soybean and coconut oils. Fig. 8 shows that almost similar results were obtained with different types of vegetable oils investigated, with the yield for coconut oil higher than the two other types of oil. This implies that the proposed method can be applied to any types of oil (or fat) feedstocks for biodiesel production.

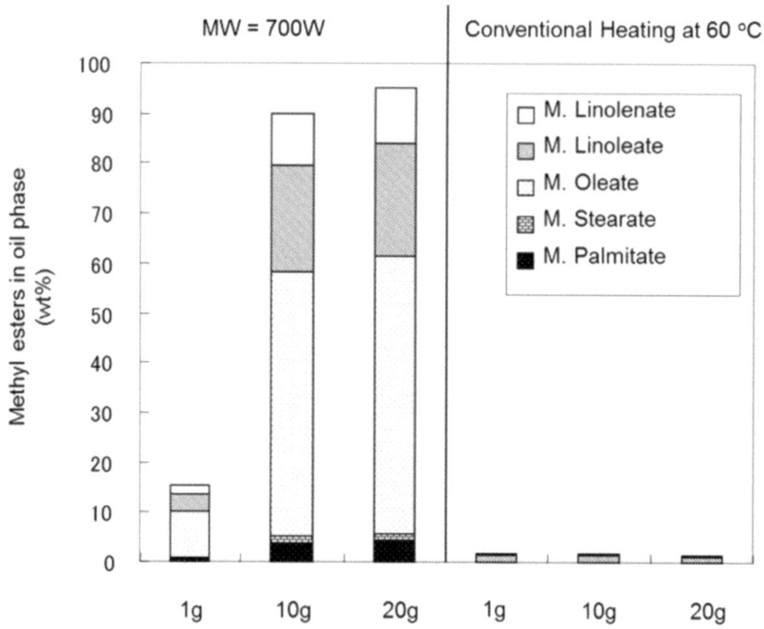


Fig. 6. Comparison of microwave and conventional heating (t = 1min, catalysts: $Ca(OH)_2$, MW = 700W)

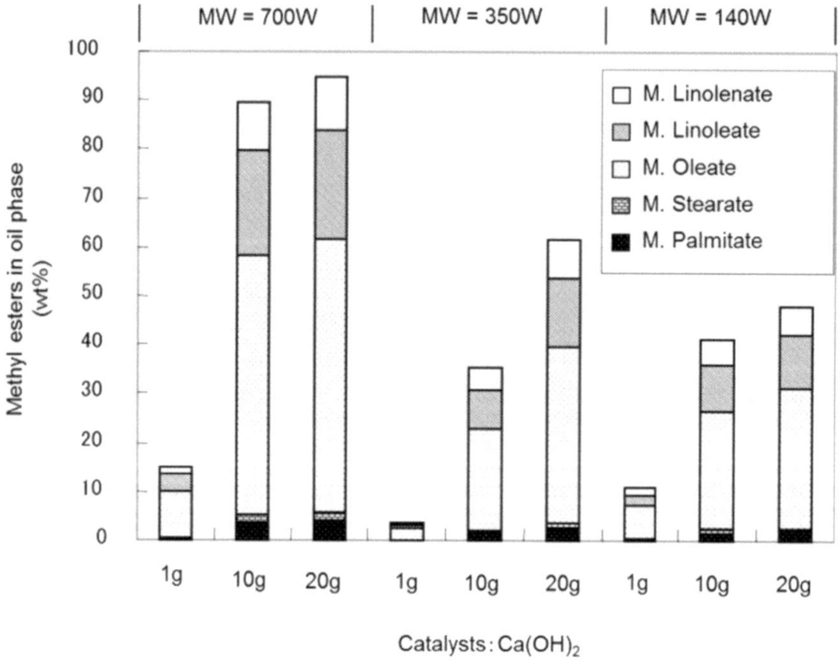


Fig. 7. Dependency of yield on microwave irradiation power at various amounts of $Ca(OH)_2$ catalysts (t=1min)

| Oil/Fats | Fatty Acid Composition (wt%) | | | | | | | | |
|---|---|---|---|---|---|---|---|---|---|
| | Caprylic (C8:0) | Capric (C10:0) | Lauric (C12:0) | Myristic (C14:0) | Palmitic (C16:0) | Stearic (C18:0) | Oleic (C18:1) | Linoleic (C18:2) | Linolenic (C18:3) |
| Rapeseed | | | | | 6 | 1 | **58** | 24 | 11 |
| Soybean | | | | 0.3 | 7.8 | 2.5 | 26 | **51** | 5 |
| Coconut | 8 | 8 | **48** | 16 | 8.5 | 2.5 | 6.5 | 2 | |

Table 1. Fatty acid composition of various oil feedstocks investigated

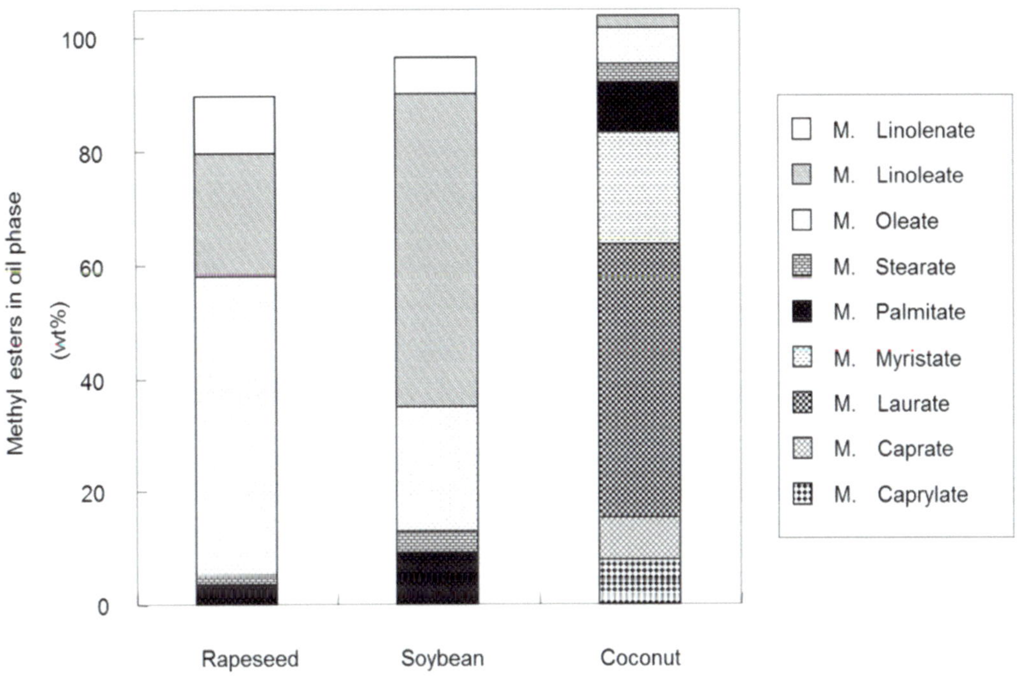


Fig. 8. Comparison of yield using various types of oil feedstocks (t = 1min, Ca(OH)2 = 10g, MW = 700W)

### 5.1.6 Comparison of commercial slaked lime with pure $Ca(OH)_2$ catalysts

For low-cost production of biodiesel, cheap and readily available catalysts for its production are being sought. The use of cheap commercial grade $Ca(OH)_2$ catalysts could be considered. For this purpose, the activity of the commercial slake lime pellets (supplied by Inoue Lime Co. Ltd., Kochi, Japan) containing 60% $Ca(OH)_2$ was compared with that of pure $Ca(OH)_2$. Results in Fig. 9 show that the yield was low at around 30% using the un-pretreated pellets. If pre-dried, the yield increased by more than 10%. After further pulverization, a three-fold increase in the yield was obtained using dried catalysts as a result of the increase in surface contact area. In the succeeding continuous-flow experiments, non-pulverized dried slaked lime pellets were used.

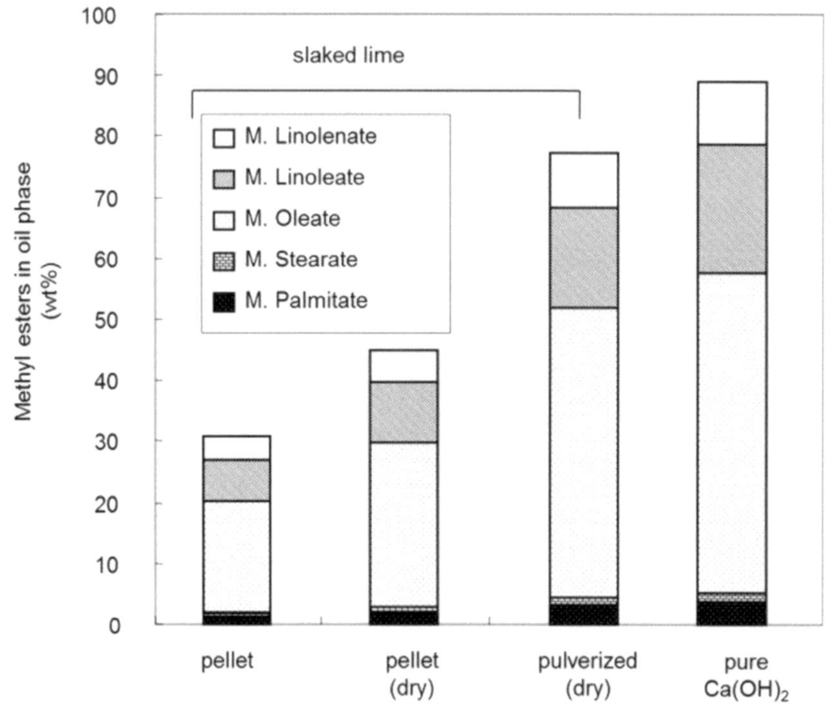


Fig. 9. Experimental results using commercial slaked lime (60% pure) compared to pure $Ca(OH)_2$ catalysts (t=1min, catalysts=10g, MW=700W)

### 5.1.7 Typical continuous-flow experimental methods and results

For continuous-flow experiments, about 120g of slaked lime pellets was placed in a 100-ml glass flask which served as a reactor. Slaked lime was selected based on the results of our previous work on the investigation of activities of various catalysts. Methanol and rapeseed oil (fixed at a molar ratio of 6) were vigorously mixed and allowed to pass through the reactor at various flow rates. The residence time was calculated based on the void space of the reactor after placing the catalysts.

Fig. 10 shows the typical experimental conditions and results of the continuous process for biodiesel production using microwave. The temperature was controlled at 60°C by supplying microwaves at a maximum peak of 30% corresponding to a power of about 300 W. The flow rates were varied from 12 to 50 ml/min corresponding to residence times of 7 to 1.5 min, respectively.

In runs 1 and 2, high yields were observed initially when most of the microwave irradiation were supplied and absorbed by the reactants and catalysts. However, above the set reaction temperature of 60°C, the microwave irradiation automatically ceased, which could possibly cause an intermittent lowering of the yield. If the temperature decreased below 60°C, the system automatically activated, supplying the microwave again to the reactor. The reaction reached steady state after about 8 sampling runs corresponding to a run time of about 8 and 16 mins for runs 1 and 2, respectively. Other than the thermal effects, microwave effects were evident from the experimental results, and thorough investigation would be necessary to further validate this interesting phenomenon.

| | Run 1 | Run 2 | Run 3 |
|---|---|---|---|
| Flowrate (ml/min) : | 50 | 25 | 12 |
| Corresponding residence time(min): | 1.5 | 3.5 | 7 |
| Sampling period (min): | 1 | 2 | 4 |

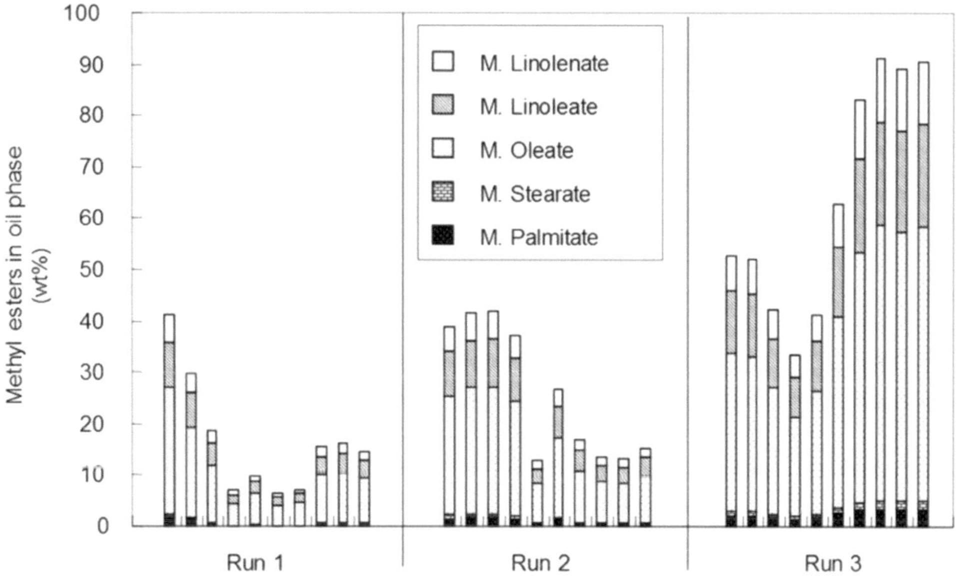


Fig. 10. Typical continuous-flow experimental results for biodiesel production under microwave irradiation

In run 3, steady state was also observed after about 8 sampling runs, and a yield above 90% was obtained in short residence time of 7 min. Mixing of the reactants and catalysts by stirring could improve the homogeneity of microwave absorption and could increase the yield. However, in the case of solid catalyst pellets, recirculation of the reactants or products back to the reactor could be a better alternative to stirring.

### 5.1.8 Combined reaction and separation in a single cavity

The applicability of a combined reaction and separation technique in a single microwave cavity was also investigated using a Soxhlet-extractor-inspired apparatus. In this experiment, the reactants were supplied on top of the glass reactor vessel. Once a predefined level was reached, the products were siphoned down to a distiller right below the reactor to undergo separation of unreacted MeOH from the products and glycerin under microwave irradiation. The time elapsed from the introduction of the reactants to the reactor until the moment it was siphoned to the distiller served as the residence or reaction time.

In a typical run, the temperature was controlled at 60°C by microwave irradiation at a maximum peak of 30% corresponding to about 300 W. The flow rates were varied from 12 to 50 ml/min corresponding to residence times of 20 to 6 min, respectively.

Results in Fig. 11 showed that a combination of reaction and distillation units in a single cavity could be promising for the separation of unreacted MeOH. However, results of the preliminary experiments showed that the yield was lower than those obtained using the batch reactor. One possible reason for this was the difference in the sizes of the two reactors. It was also likely that the supplied microwave power in this experiment was not sufficient for both reaction and distillation processes to occur simultaneously and more efficiently. A rigorous investigation to optimize the process are sought in order to further validate the economic feasibility of this proposed process.

| | Run 1 | Run 2 | Run 3 |
|---|---|---|---|
| Flowrate (ml/min) : | 50 | 25 | 12 |
| Corresponding residence time(min): | 6 | 10 | 20 |
| Sampling period (min): | 4 | 5 | 10 |
| MeOH collected(g): | 132 | 177 | 5 |

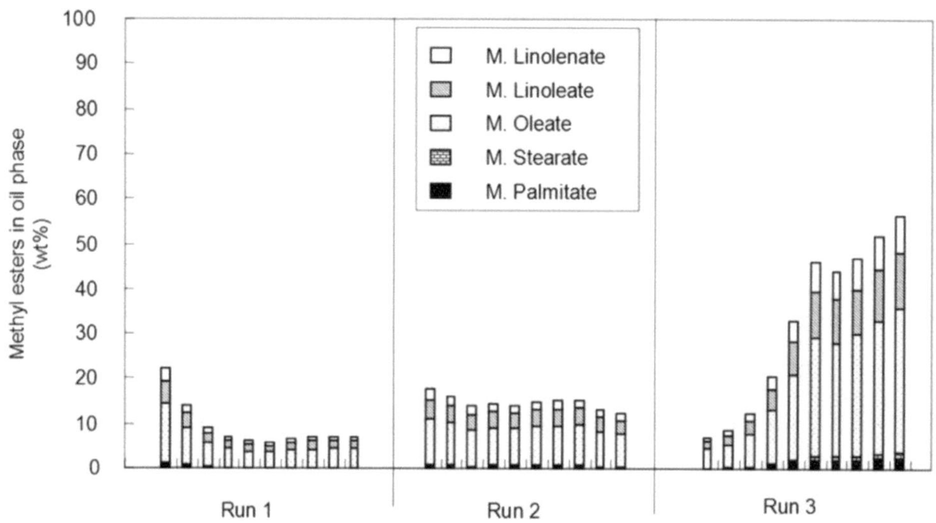


Fig. 11. Results using combined microwave-reactive distillation experiments in a single cavity.

### 5.1.9 Esterification of free fatty acids present in oil feedstock

Most of the oil feedstocks for biodiesel syntheses contain relatively high amount of fatty acids especially the waste cooking oil. This has become a big hurdle for industrialization of the proposed process, because the presence of fatty acids significantly affects the solubility of Ca-based catalysts in the products. Government quality standards for biodiesel require the level of Ca to be below 5 ppm, while the fatty acid content should not exceed 1wt%. Thus, pretreatment of free fatty acids in oil is necessary prior to transesterification of the triglyceride contents. In this regard, microwave irradiation was also applied to convert free fatty acids into biodiesel. Results in Fig. 12 show an 88% conversion of fatty acids in waste cooking oil in 1 min of microwave irradiation at a power of 700W using ion exchange resin as catalysts. With these results, a two-step process shown in Fig. 13 is proposed for the conversion of waste oil, or any type of oil feedstocks containing high amount of fatty acids, to biodiesel fuel. The process consists of a first step of esterification of fatty acids followed by a second step of transesterification of the triglyceride. While the two-step process seems ideal for the treatment of free fatty acids in oil, this also minimizes the solubility of Ca-based catalysts as a result of the reduction of fatty acid contents.

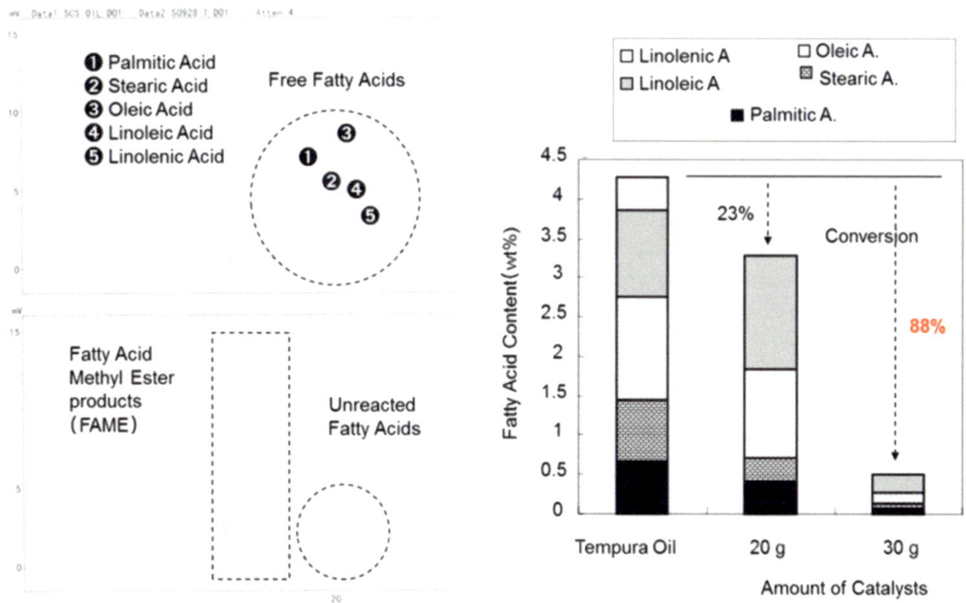


Fig. 12. Results of experiments on esterification of free fatty acids in waste oil using microwave (MWpower=700W, t = 1 min)

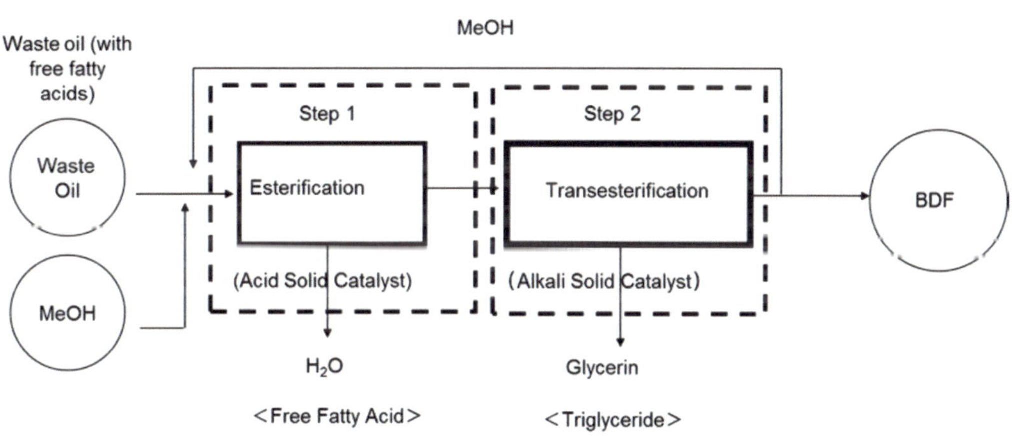


Fig. 13. Proposed two-step process for conversion of waste oil to BDF using solid catalysts under microwave irradiation

### 5.2 ETBE production

The method was also applied to the synthesis of ethyl *tert*-butyl ether (ETBE) from two biomass-derivable alcohols (ethanol and *tert*-butyl alcohol). ETBE, is commonly used as an additive in gasoline to increase the octane number. Recently, EtOH is the most after-sought biofuel replacement for crude-oil-derived gasoline. However, ETBE is thought to offer equal or greater air quality benefits as ethanol, while being technically and logistically less challenging. Unlike ethanol, ETBE does not induce evaporation of gasoline, which is one of the causes of smog, and does not absorb moisture from the atmosphere.

Microwave-assisted experiments were performed using the same microwave apparatus used for biodiesel synthesis, working at 2.45 GHz frequency, with a power programmable from 0 to 1000W. In a typical run, about 0.25mol of EtOH and TBA, and 20 g of catalyst were placed in a reactor vessel, and heated using a microwave apparatus described above. GC-FID was used for the analysis of the products using an internal standard.

Preliminary experiments on the evaluation of catalytic activities of various solid catalysts such as Amberlyst 15JWET, sulfated zirconia, sulfated charcoal and zeolite showed Amberlyst 15JWET to be the most effective as shown in Fig. 14. The yield of ETBE using sulfated charcoal and and zeolite is almost negligible compared to that of Amberlyst 15JWET and sulfated zirconia. Thus, Amberlyst 15JWET was used in the experiments unless otherwise specified.

Experiments at atmospheric pressure using a batch reactor showed that the yield hardly increased beyond the 20% level. The experiments were extended to continuous flow at various conditions, but the yield did not exceed 35% as shown in Fig. 15. Almost similar results were obtained by other researchers (Japanese Patent JP2007-126450), and the lower yield was likely due to the selective dehydration of TBA to IB, a highly volatile compound that could easily escape from the reaction zone. If IB could further react with EtOH to produce ETBE, better yield could be obtained. Thus, experiments using a sealed reactor vessel were conducted using another microwave apparatus (Milestone Ethos). Fig. 16 shows the conversion of EtOH to ETBE using MW at various power and irradiation time. A maximum yield of about 87% was obtained at MW power of 350W at irradiation time of 1 min. At this condition, the attained temperature was around 87°C, higher than the boiling points of the two alcohols, as shown in Fig. 17. The conversion was also found to be dependent on the amount of catalysts, reaction time and microwave power.

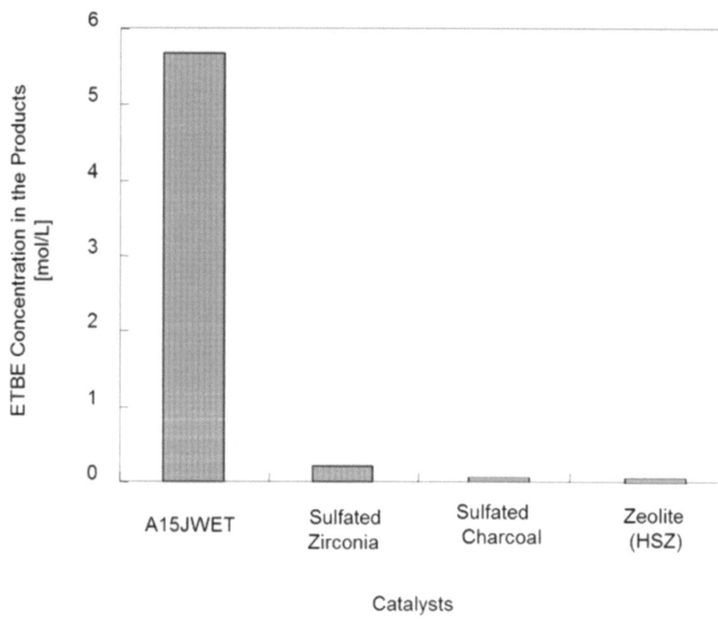


Fig. 14. Comparison of activities of various acidic heterogeneous catalysts for production of ETBE under microwave irradiation (TBA=ETOH=0.25mol, MW=350W, t=1min)

| | Run 1 | Run 2 | Run 3 | Run 4 | Run 5 | Run 6 | Run 7 | Run 8 |
|---|---|---|---|---|---|---|---|---|
| Flowrate (ml/min) : | 100 | 100 | 50 | 50 | 20 | 20 | 10 | 10 |
| Residence time(min): | 1 | 1 | 2 | 2 | 5 | 5 | 10 | 10 |
| Sampling time (min): | 1 | 1 | 2 | 2 | 5 | 5 | 10 | 10 |
| MW duty (%): | 13 | 20 | 13 | 20 | 13 | 20 | 13 | 20 |
| Temperature (°C): | 28 | 67 | 69 | 71 | 70 | 71 | 70 | 71 |

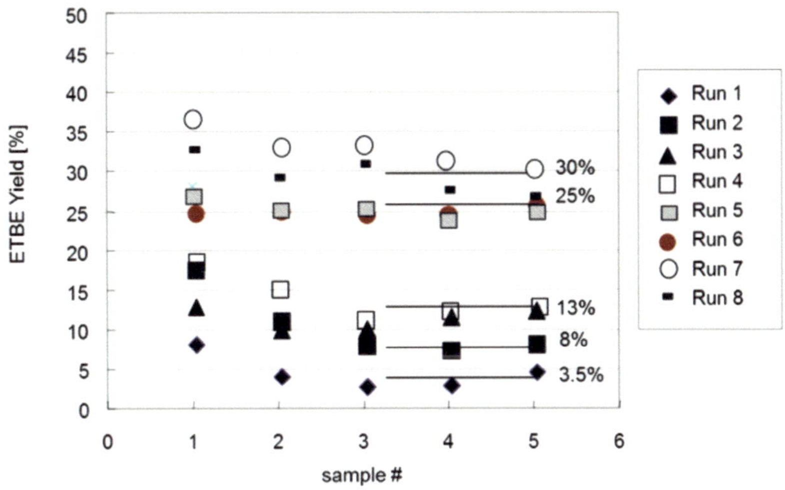


Fig. 15. Typical continuous-flow experimental results for ETBE production using A15JWET ion exchange resin catalysts under microwave irradiation

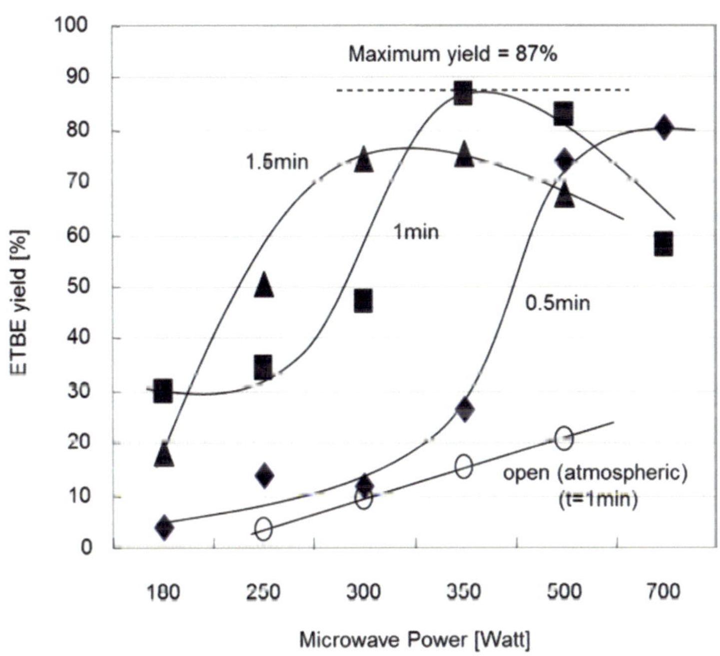


Fig. 16. Yield of ETBE at various microwave irradiation power and reaction time (closed system, TBA=EtOH=0.25mol, A15JWET = 20g )

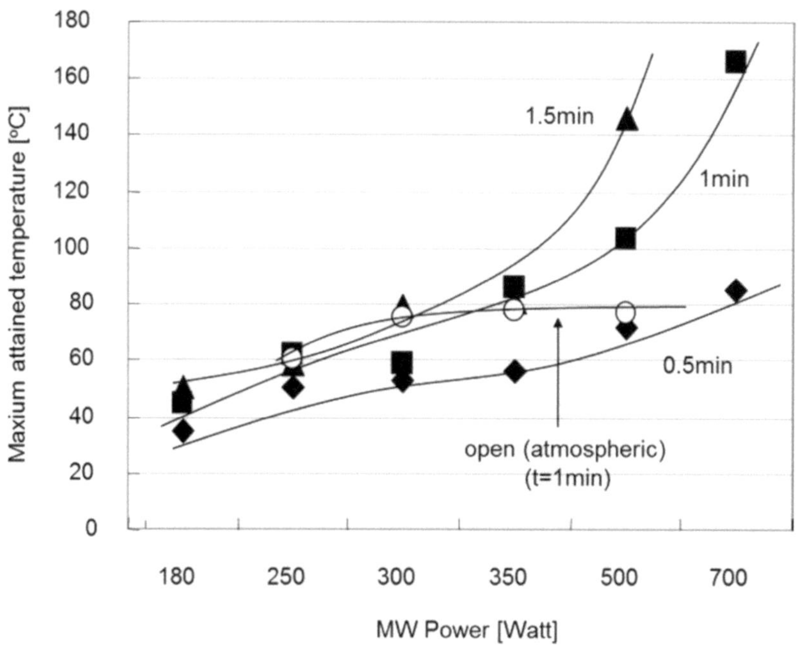


Fig. 17. Maximum attained temperature at various microwave irradiation power and reaction time (closed system, TBA=EtOH=0.25mol, A15JWET = 20g )

| | Yin et al. (1995) | Quitain et al. (1999) | Ito et al. (2007) (JP2007-126450) | This Method |
|---|---|---|---|---|
| Heating Method | Conventional | Conventional | Microwave | Microwave |
| Reaction Type | batch, open | continuous reactive distillation | batch, open | batch, closed |
| Solid Catalyst | A15 | A15 (pellet) | A15JWET | A15JWET |
| Reaction Conditions | T = 65°C, t = 4h | T = 75°C, t = 3h | T = 70°C, t = 0.5-4h | MW=350W, t = 1min (T = ~86°C) |
| g catalyst/ mol EtOH | 10 | ~ | 8 | 80 |
| Stirring | YES | NO | YES | NO |
| Yield(%) | 26 | 38 | 9 ~ 28 | 87 |

Table 2. Comparison of the methods for production of ETBE from TBA and ETOH with the proposed microwave-solid catalyst method

The results of ETBE synthesis using microwave irradiation were summarized in Table 2 in comparison with the methods reported in literature. While several methods reported a conversion in the range of 9 to 38 % using the conventional heating methods, we obtained a maximum conversion of about 87% under microwave irradiation at a power of 350 W for 1 min. At this condition, the bulk temperature reached about 86 °C, which was above the boiling points of the two alcohols. The results imply that solvothermal condition (closed system) is ideal for ETBE synthesis because the IB generated from the dehydration of TBA can further react with EtOH to produce ETBE, resulting to higher conversion.

## 6. Problems associated with the proposed process

Microwave-assisted reactions offer several great advantages to the synthesis of biofuels, however, there are also some drawbacks associated with its use. Microwave could not work well with large quantities of materials, and thus could not be easily converted from laboratory to a multikilogram industrial-scale production. The penetration depth of MW irradiation into the absorbing materials is only a few centimeters, and this significantly limits scale up of the technology.

Microwave irradiation is non-homogeneous and formation of "hotspots" is likely, thus control of reaction is too difficult. Mixing may improve homogeneity, however, with the use of solid catalysts, appropriate methods of mixing remains a challenge.

Safety consideration is another factor for industrial utilization of microwave. The use of batch microwave reactors, for the processing of comparatively large volumes under pressure may not be safe because any malfunction or rupture of a large pressurized reaction vessel, which are usually made of Teflon or glass materials, may result into massive spillage causing significant operational damages to the working place and the environment.

## 7. Outlook and future prospects

As the demand for biofuels continue to increase in the near future, and while the search for an efficient and low-cost production process continues, the global outlook is positive for the use of microwave irradiation to the synthesis of two most promising biofuels - biodiesel and bioETBE.

To overcome the limitations for scaling up microwave-assisted technology for biodiesel production, development of a continuous process is suggested, but still poses several challenges that require detailed investigation. The future also calls for the development of cheap, effective and stable solid catalysts for the synthesis of the abovementioned fuels.

While the use of microwave irradiation offers great benefits with regards to rapid reaction or synthesis, safety is a big factor to consider in designing a large scale production plant. However, this can be avoided if multilayered compact reactors operating under microwave irradiation can be developed instead.

## 8. Conclusion

This chapter has presented syntheses of two most promising biofuels, *i. e.* biodiesel and bioETBE, by microwave-assisted heating. Methods for the production of the biofuels reported in literatures were reviewed, and the advantages of the proposed process of using microwave and heterogeneous catalysts were outlined and discussed. The benefits have been indicated using the results of our recent works, however there are some drawbacks that would require thorough investigation prior to its commercialization. Although the field is in its infancy, the outlook is bright for the proposed methods due to foreseen high global demands for biofuels. The next few years should see development of continuous compact process, along with cheap, effective and stable solid catalysts.

## 9. Acknowledgements

The research works were mostly supported financially by the Japan Science and Technology. Funds for collaborative research between Kagawa Industry Support Foundation and Shodoshima Clean Service, Co. Ltd. for the investigation of the treatment of free fatty acids are also gratefully acknowledged.

## 10. References

Azcan, N. & Danisman, A. (2007). Alkali Catalyzed Transesterification of Cottonseed Oil by Microwave Irradiation. *Fuel*, Vol. 86, pp2639-2644.

Azcan, N. & Danisman, A. (2008). Microwave Assisted Transesterification of Rapeseed Oil. *Fuel* , Vol. 87, pp. 1781-1788.

Barnard, T. M.; Leadbeater, N. E.; Boucher, M. B.; Stencel, L. M. & B. A. Wilhite, B. A. (2007). Continuous-Flow Preparation of Biodiesel Using Microwave Heating. *Energy and Fuels*, Vol. 21, pp. 1777-1781.

Boldor, D.; Kanitkar, A.; Terigar, B. G.; Leonardi, C.; Lima, M. & Breitenbeck, G. A. (2010). Microwave Assisted Extraction of Biodiesel Feedstock from the Seeds of Invasive Chinese Tallow Tree. *Environ. Sci. Technol.* , Vol. 44, pp. 4019-4025.

Breccia, A.; Esposito, B.; Breccia Fratadocchi, G. & Fini, A. (1999). Reaction Between Methanol and Commercial Seeed Oils Under Microwave Irradiation. *Journal of Microwave Power and Electromagnetic Energy*, Vol. 34, No. 1, pp. 3-8.

De la Hoz, A.; Diaz-Ortiz, A. & Moreno, A. (2005). Critical Review: Microwaves in Organic Synthesis. Thermal and Non-Thermal Microwave Effects. *Chem. Soc. Rev.*, Vol. 34, pp. 164-178.

Demirbas, A. (2003). Biodiesel Fuels from Vegetable Oils Via Catalytic and Non-Catalytic Supercritical Alcohol Transesterifications and Other Methods: A Survey. *Energy Conversion and Management*, Vol. 44, pp. 2093-2109.

Geuens, J.; Kremsner, J. M.; Nebel, B. A.; Schober, S.; Dommisse, R. A.; Mittelbach, M.; Tavernier, S.; Kappe, C. O. & Maes, B. U. W. (2008). Microwave-Assisted Catalyst-Free Transesterification of Triglycerides with 1-Butanol under Supercritical Conditions. *Energy and Fuels*, Vol. 22, pp. 643-645.

Groisman, Y. & Gedanken, A. (2008). Continuous Flow, Circulating Microwave System and Its Application in Nanoparticle Fabrication and Biodiesel Synthesis. *J. Phys. Chem. C* , Vol. 112, pp. 8802-8808.

Habenicht, C.; Kam, L. C.; Wilschut, M. J. & Antal, M. J. (1995). Homogeneous Catalysis of Ethyl tert-Butyl Ether Formation from tert-Butyl Alcohol in Hot, Compressed Liquid Ethanol. *Ind. Eng. Chem. Res.*, Vol. 34, pp. 3784-3792.

Hernando, J., Leton, P.; Matia, M. P., Novella, J. L. & Alvarez-Builla, J. (2007). Biodiesel and FAME Synthesis Assisted by Microwaves: Homogeneous Batch and Flow Processes. *Fuel*, Vol. 86, pp. 1641-1644.

Hsiao, M.-C.; Lin, C.-C.; Chang, Y.-H. & Chen, L.-C. (2010). Ultrasonic Mixing and Closed Microwave Irradiation-Assisted Transesterification of Soybean Oil. *Fuel*, Vol. 89, pp. 3618-3622.

Jacob J.; Chia, L. H. L. & Boey, F. Y. C. (1995). Review: Thermal and Non-Thermal Interaction of Microwave Irradiation with Materials. *Journal of Material Science,* Vol. 30, pp. 5321-5327.

Japanese Patent Application: JP2007-126450. ETBE Synthesis Using Microwave.

Japanese Patent Application: JP2008-133250. Production of Alkyl Ether by Microwave.

Kingston, H. M. (Skip) & S. J. Haswell, S. J. (1996). *Microwave-Enhanced Chemistry: Fundamentals, Sample Preparation and Applications,* American Chemical Society, ISBN 0-8412-3375-6, Washington DC, USA.

Kulkarni, M. G. & Dalai, K. A. (2006). Waste Cooking Oil-An Economical Source for Biodiesel: A Review. *Ind. Eng. Chem. Res.* , Vol. 45, pp. 2901-2913.

Kusdiana, D. & Saka, S. (2001). Kinetics of Transesterification in Rapeseed Oil to Biodiesel Fuel as Treated in Supercritical Methanol. *Fuel*, Vol. 80, pp. 693-698.

Leadbeater, N. E. & Stencel, L. M. (2006). Fast, Easy Preparation of Biodiesel Using Microwave Heating. *Energy and Fuels*, Vol. 20, pp. 2281-2283.

Leadbeater, N. E., Barnard, T. M. & Stencel, L. (2008). Batch and Continuous-Flow Preparation of Biodiesel Derived from Butanol and Facilitated by Microwave Heating. *Energy and Fuels*, Vol. 22, pp. 2005-2008.

Lertsathapornsuk, V.; Pairintra, R.; Aryusuk, K. & Krisnangkura, K. (2008). Microwave Assisted in Continuous Biodiesel Production from Waste Frying Palm Oil and Its Performance in a 100 kW Diesel Generator. *Fuel Processing Technology,* Vol. 89, pp. 1330-1336.

Lin, C.-C.; Hsiao, M.-C. & Chang, Y.-H. (2011). Microwave Irradiation-Assisted Transesterification of Soybean Oil to Biodiesel Catalyzed by Nanopowder Calcium Oxide. *Fuel*, in press.

Ma, F. & Hanna, M. A. (1999). Biodiesel production: A Review. *Bioresource Technology*, Vol. 70, pp. 1-15.

Majewski, M. W.; Pollack, S. A. & Curtis-Palmer, V. A. (2009). Diphenylammonium Salt Catalysts for Microwave Assisted Triglyceride Transesterification of Corn and Soybean Oil for Biodiesel Production. *Tetrahedron Letters*, Vol. 50, pp. 5175-5177.

Matouq, M.; Quitain, A. T.; Takahashi, K. & Goto, S. (1996). Reactive Distillation for Synthesizing Ethyl tert-Butyl Ether from Low-Grade Alcohol Catalyzed by Potassium Hydrogen Sulfate. *Ind. Eng. Chem. Res.*, Vol. 35, pp. 982-984.

Mingos, D. M. P. & Baghurst, D. R. (1997). Applications of Microwave Dielectric Heating Effects to Synthetic Problems in Chemistry, In: Microwave-Enhanced Chemistry, H. M. (Skip) Kingston & S. J. Haswell, (Ed.), pp. 3-50, American Chemical Society, ISBN 0-8412-3375-6, Washington DC, USA.

Mazzocchia, C.; Modica, G.; Kaddouri, A. & Nannicini, R. (2004). Fatty Acid Methyl Esters Synthesis from Triglycerides Over Heterogeneous Catalysts in the Presence of Microwaves. *C. R. Chimie*, Vol. 7, pp. 601-605.

Nogueira, B. M.; Carretoni, C.; Cruz, R.; Freitas, S.; Melo Jr., P. A.; Costa-Felix, R.; Pinto, J. C. & Nele, M. (2010). Microwave Activation of Enzymatic Catalysts for Biodiesel Production. *Journal of Molecular Catalysis B: Enzymatic*, Vol. 67, pp. 117-121.

Norris, J. F. & Rigby, G. W. (1932). The Reactivity of Atoms and Groups in Organic Compounds. *J. Am. Chem. Soc.*, Vol. 54, pp. 2088-2100.

Ozturk, G.; Kafadar, A. B.; Zahir Duz, M.; Saydut, A. & Hamamci, C. (2010). Microwave-Assisted Transesterification of Maize (Zeamays L.) Oil as a Biodiesel. *Energy, Exploration and Exploitation*, Vol. 28, No. 1, pp. 47-58.

Patil, P. D.; Gude, V. G.; Camacho, L. M. & Deng, S. (2010). Microwave-Assisted Catalytic Transesterification of Camelina Sativa Oil. *Energy Fuels*, Vol. 24, pp. 1298-1304.

Quitain, A. T.; Itoh, H. & S. Goto, S. (1999). Reactive Distillation for Synthesizing Ethyl tert-Butyl Ether from Bioethanol. *J. Chem. Eng. Japan*, Vol. 32, pp. 280-287.

Quitain, A. T.; Baclayon, D. P.; Chikata, T. & Katoh, S. (2008). Microwave-Assisted Heterogeneous Catalyzed Process for Biodiesel Production, *Proceedings of Global Congress for Microwave Energy Applications (GCMEA) 2008*, Otsu Prince Hotel, Lake Biwa, Otsu, Japan, August 4-8, 2008.

Quitain, A. T.; Chikata, T. & Katoh, S. (2008). Biodiesel Production Using Microwave and Solid Catalysts. Japanese Patent Application: JP 2008-046969.

Quitain, A. T. (2009). Method and Apparatus for Biodiesel Production. Japanese Patent Application: JP 2009-195494.

Quitain, A. T.; Chikata, T. & Katoh, S. (2009). Microwave-Assisted Heterogeneous Catalyzed Process for Highly Efficient Biodiesel Production, *Proceedings of 6th Biomass-Asia Workshop*, Hotel Granvia Hiroshima, Hiroshima, Japan, November 18-20, 2009.

Rathana, Y., S.; Roces, A.; Bacani, F. T.; Tan, R. R.; Kubouchi, M. & Piyachat, Y. (2010). Microwave-Enhanced Alkali Catalyzed Transesterification of Kenaf Seed Oil. *International Journal of Chemical Reactor Engineering* , Vol. 8, pp. S5.

Refaat, A. A. (2010). Different Techniques for the Production of Biodiesel from Waste Vegetable Oil. *Int. J. Environ. Sci. Tech.* , Vol. 7, No. 1, pp. 183-213.

Roukas, T. (1995). Ethanol Production from Carob Pod Extract by Immobilized Saccharomyces Cerivisiae Cells on the Mineral Kissiris. *Food Biotechnology*, Vol. 9, pp. 175-178.

Saka, S. (1996). *All About Biodiesel*, IPC Publishing, ISBN- 9784901493567, Tokyo, Japan.

Semwal, S.; Arora, A. K.; Badoni, R. P. & Tuli, D. K. (2011). Review: Biodiesel Production Using Heterogeneous Catalysts. *Bioresource Technology*, Vol. 102, pp. 2151-2161.

Socha, A. & Sello, J. K. (2010). Efficient Conversion of Triacylglycerols and Fatty Acids to Biodiesel in a Microwave Reactor Using Metal Triflate Catalysts. *Org. Biomol. Chem.*, Vol. 8, pp. 4753-4756.

Yang, B. & Goto, S. (1997). Pervaporation with Reactive Distillation for the Production of Ethyl tert-Butyl Ether. *Sep. Sci. Tech.*, Vol. 32, pp. 971-981.

Yin, X.; Yang, B. & Goto, S. (1995). Kinetics of Liquid-Phase Synthesis of Ethyl tert-Butyl Ether from tert-Butyl Alcohol and Ethanol Catalyzed by Ion-Exchange Resin and Heteropoly Acid. *Int. J. Chem. Kinetics*, Vol. 27, pp. 1065-1074.

Yuan, H.; Yang, B. L. & Zhu, G. L. (2009). Synthesis of Biodiesel Using Microwave Absorption Catalysts. *Energy and Fuels*, Vol. 23, pp. 548-552.

Zhang, S.; Zu, Y.-G.; Fu, Y.-J.; Luo, M.; Zhang, D.-Y. & Efferth, T. (2010). Rapid Microwave-Assisted Transesterification of Yellow Horn Oil to Biodiesel Using a Heteropolyacid Solid Catalyst. *Bioresource Technology*, Vol. 101, pp. 931-936.

# 17

# Fertilizer Potential of Biofuel Byproducts

Amber Moore
*University of Idaho*
*United States of America*

## 1. Introduction

Ethanol and biodiesel production have become important industries worldwide. For example, ethanol production in the United States has increased exponentially from 6.0 billion liters in 2000 to 49 billion liters in 2010, an increase of 800% (RFA, 2011). Similarly, biodiesel production increased in the U.S. from 424 million liters in 2005 to 1,191 million liters in 2010 (NBB, 2011). These booming industries have not only changed how we view our automotive fuel, they have also forced us to consider uses for their valuable byproducts.

The majority of ethanol produced in the U.S. is through the dry grind process. Dried distillers grains (DDGS) is the predominate byproduct of dry-grind ethanol production. It has been estimated that for every liter of ethanol produced, 3.5 kg of DDGS are left over (Rausch and Belyea, 2006). Based on the production estimates listed above, this would have equated to 34.9 million metric tonnes of dried distillers grain produced in the U.S. in 2010.

Soybean oil and canola oil are the most commonly used vegetable oils for biodiesel production in the U.S. The oil is extracted from soybeans and canola seed through a process of cold-pressing, leaving behind valuable seed meal byproducts. An estimated 80% of soybean seed 60% of canola seed is left from the extraction process as seed meal, creating a significant quantity of this important byproduct (NBB, 2008; Herkes, 2010).

The most common use of DDGS, soybean meal, and canola meal is for animal consumption as animal feed. DDGS, soybean meal, and canola meal contain an estimated 26, 47, and 35% crude protein, respectively, therefore these byproducts are considered highly valuable animal feed sources (Table 1). However, issues including over-saturation of DDGS in animal feed markets, animal feed quality issues, and the extremely high costs associated with oilseed meals has forced some biofuel producers to consider other markets for these byproducts.

A potential market for DDGS and oilseed meals that biofuel producers may consider is the fertilizer/agriculture market. DDGS and oilseed meals are rich in plant macronutrients (nitrogen (N), phosphorus (P), potassium (K), calcium (Ca), magnesium (Mg), and sulfur (S)) and plant micronutrients (zinc (Zn), iron (Fe), manganese (Mn), and boron (B)) In addition, these materials have a relatively low carbon-N ratios (ranging from 5:1 to 15:1), therefore these materials are highly decomposable and are able to release organic forms of N to plant available forms of N shortly after field application. This property is of particular interest to organic agriculture markets, where nutrient sources and fertilizers containing enough readily mineralizeable N to impact yield are scarce.

The purpose of this chapter is to discuss byproducts of ethanol and biodiesel production as effective nutrient sources for plants. Specifically, we will cover 1) how these byproducts are made, 2) plant nutrient composition of the byproducts, 3) competition with animal feed markets, 4) situations where biofuel byproducts should be considered as a fertilizer, and 5) a review of field and laboratory research studies that investigate the fertilizer value of these byproducts.

## 2. Common byproducts of biofuel production

In this section, we will provide some background information on how biofuel byproducts of greatest interest to growers are generated during the production of ethanol from dry-grind processing and biodiesel from soybean and canola oils.

### 2.1 DDGS from ethanol production

Over the past 15 years, dry-grind corn processing has become the primary method for producing ethanol in the U.S., favored over wet milling and dry milling (Rausch and Belyea, 2006; RFA, 2011). The simplified steps for making ethanol from corn using the dry-grind method are 1) mixing ground grain corn with water, 2) cooking the mixture and adding amylase enzyme, and 3) liquefying the mixture and adding glucoamylase enzyme and yeast for fermentation. The products of fermentation are ethanol, water, and solids. The solids, also referred to as "whole stillage", are removed from the bottom of the fermentation unit and centrifuged to separate the solids into wet grains and thin stillage. The thin stillage is generally evaporated to form a thick syrup, which is added back to the wet grains. These wet grains are referred to as wet distillers grains with solubles (syrup) added, or WDGS. The final step of this process is to dry the wet grains to form DDGS with solubles (syrup) added, or DDGS (Rausch and Belyea, 2006).

Photo 1. Typical sample of DDGS from ethanol production. Photograph taken by Amber Moore.

## 2.2 Oilseed meals from biodiesel production

The main components used to make biodiesel are soybean or canola oil and an alcohol source, typically methanol. Seed meal byproducts are left after oil is extracted from soybeans and canola oil. While the methods used for extracting oil vary slightly between soybeans and canola seeds, oil is extracted from most oilseeds through a process of cleaning, drying, dehulling, size reduction, flaking, cooking, and tempering (Dunford). The most critical step of this process is cooking/tempering, which is also referred to as conditioning. During this step, oilseeds are cooked or tempered to denature proteins, which releases oil from the seeds and inactivates enzymes. Once they reach a specified temperature, cooked/tempered seeds are pressed to separate oil from the seed. The portion of seed solids remaining after this process is referred to as the meal. The oilseed meal can be used immediately as an animal feed without further treatment.

Photo 2. Typical sample of canola meal from oil-pressed canola seeds. Photograph taken by Amber Moore.

## 3. Plant nutrient composition

As mentioned above, DDCS and oilseed meals are rich in plant macronutrients (N, P, K, Ca, Mg, and S) and plant micronutrients (Zn, Fe, Mn, and B). While all of these nutrients are important for optimal plant growth, N is considered the most critical for all crops. Nitrogen is needed by plants to support vegetative growth and chlorophyll production. Unfortunately, plant available forms of N (ammonium ($NH_4$-N) and nitrate ($NO_3$-N)) are quickly taken up by plants, leached out of the soil root zone, converted to ammonia ($NH_3$) gas, and/or utilized by soil microbes, therefore plant available forms of N do not usually stay in the soil for longer than a single growing season. Also, many organic waste products, such as compost and cattle manure, contain relatively low concentrations of N that may take several years to be converted by microbes to forms that can be used by the plant.

Fortunately, most biofuel byproducts contain high enough concentrations of N to be rapidly available for plant use within the first growing season after application. To quickly and roughly estimate PAN from organic amendments based on total N content, the equation that supports the Oregon State University Organic Fertilizer calculator can be used (Sullivan et al., 2010). To predict full-season PAN for organic materials with less than 6% total N, the calculator uses the equation %PAN = ((-30 +15*(fertilizer total N%)) + 15%). To predict full-season PAN for organic materials with more than 6% total N, the calculator estimates 75% PAN. Applying the value averages from Table 2 and the OSU Organic Fertilizer Calculator, PAN estimates over a growing season for most climates and soils are 50% for DDGS, 75% for soybean meal, 70% for canola meal, and 71% for mustard meals, all varieties. Based on these estimates, the availability of N from biofuel byproducts is only slightly less than most chemical N fertilizers, which are assumed to be 90-100% plant available shortly after application. In this case, it is possible that conventional and organic growers could grow plants near or at optimal yields based on chemical N fertilizer applications.

In addition to N, P and K content of biofuel byproducts is also of great value to growers. As reported by Nelson et al. (2009), the N:P of the DDGS used in the study (5.5:1) was comparable to corn uptake ratios (5.9:1), suggesting that DDGS and other biofuel byproducts with similar N:P ratios could meet both the N and P needs of a crop when applied on a N basis without over- or under-applying P. In 2008, many U.S. growers were unable to afford P or K chemical fertilizers. Based on values listed in table 2, biofuel byproducts may be a viable option for these growers in the future if and when P and K fertilizer prices increase beyond what a grower can afford to pay.

With the addition of a wide variety of other plant macronutrients and micronutrients, biofuel byproducts have the potential to be an all-purpose fertilizer for many growers.

## 4. Competition with use as animal feeds

While the concept of using oilseed meals as a fertilizer is a relatively new one in the U.S., growers in Asia have widely considered rapeseed oil cake as a viable fertilizer source for over 40 years (Abe et al., 2010; Chen and Hsieh, 1972; Kora and Shingte, 1988; Singh and Gurumurt, 1984). The primary reason why biofuel byproducts are not widely used for fertilizer applications is competition with the animal feed industry. DDGS and many oilseed meals hold more value as an animal feed than as a fertilizer source, therefore animal producers are willing to pay for materials based on feed value instead of the lower fertilizer value. For example, one economic analysis showed that feed value of DDGS exceeded the fertilizer value by an estimate \$99 $tonne^{-1}$ (Lory et al., 2008). Using the U.S. price values of 2007 for anhydrous ammonia, diammonium phosphate (DAP), and potassium chloride (KCl) and total nutrient estimates of 50 g N $kg^{-1}$, 8.6 g P $kg^{-1}$, and 12.1 g K $kg^{-1}$ for DDGS, Lory et al. (2008) estimated that the theoretical fertilizer value of the DDGS was \$39 $tonne^{-1}$, in comparison to the 2007 U.S. animal feed value for DDGS of \$172 $tonne^{-1}$. Similar cost comparisons likely exist for soybean and canola meal. At these rates, conventional growers would have a difficult time justifying paying feed value costs for a fertilizer source, which far exceed the value of widely available chemical fertilizers. In addition, ethanol and vegetable oil producers would have an equally difficult challenge in lowering prices to accommodate grower needs.

| Material | C:N | Moisture | N | P | K | Ca | Mg | S | Fe | Mn | Zn | B | Cu | Na | Reference |
|---|---|---|---|---|---|---|---|---|---|---|---|---|---|---|---|
| | | g kg-1 | | | | | | | mg kg-1 | | | | | | |
| DDGS | 12.0 | ---- | 41 | 3 | 4 | 0 | 0.9 | 0.4 | 104 | 6 | 20 | 15 | 2 | ---- | Moore et al., 2010 |
| | ---- | 111 | 48.3* | 8.9 | 9.4 | 6 | 3.3 | 4.7 | 119.8 | 15.8 | 97.5 | ---- | 5.9 | 2400 | Spiehs et al., 2002 |
| | ---- | 108 | 41.5* | 8.2 | 10.4 | 2.4 | 3.2 | 5.7 | 74.5 | 13.7 | 58.0 | ---- | 4.6 | 1700 | Shurson, 2010 |
| | ---- | 102 | 38.2 | 6.9 | 11.5 | ---- | ---- | ---- | 87.0 | 18.8 | 78.3 | ---- | 6.5 | ---- | Nelson et al., 2009 |
| Soybean meal | 5.1 | ---- | 64.3 | ---- | ---- | ---- | ---- | ---- | ---- | ---- | ---- | ---- | ---- | ---- | Rubins and Bear, 1942 |
| | 5.4 | 109 | 87 | ---- | ---- | ---- | ---- | ---- | ---- | ---- | ---- | ---- | ---- | ---- | Van Kessel et al., 2000 |
| Canola meal (B. napus) | 7.6 | ---- | 63 | 14 | 15 | 4.2 | 4.4 | 7 | ---- | ---- | ---- | ---- | ---- | ---- | Snyder et al., 2010 |
| | 8.0 | 25 | 56.6 | ---- | ---- | ---- | ---- | ---- | ---- | ---- | ---- | ---- | ---- | ---- | Gale et al., 2006 |
| | | ---- | 51 | 6.63 | 7.75 | 6.4 | 2.9 | 4.7 | 186 | 41 | 31 | 36 | 5 | ---- | Banuelos and Hanson, 2010 |
| Mustard meal (B. carinata) | 14.4 | ---- | 61 | 6 | 11 | ---- | ---- | 6.0 | ---- | ---- | ---- | ---- | ---- | ---- | Balesh et al., 2005 |
| Mustard meal (Sinapis alba) | 8.5 | ---- | 58 | 3 | 5 | 1 | 1.2 | 0.4 | 54 | 8 | 23 | 15 | 1 | ---- | Moore et al., 2010 |
| | 8.4 | ---- | 58 | 13 | 10 | 4.8 | 3.8 | 18 | ---- | ---- | ---- | ---- | ---- | ---- | Snyder et al., 2010 |
| | ---- | ---- | 49 | 9.62 | 10.9 | 7.7 | 3.7 | 12.6 | 313 | 32 | 49 | 49 | 9 | ---- | Banuelos and Hanson, 2010 |
| Mustard meal (B. juncea) | 7.8 | ---- | 62 | 13 | 10 | 4.3 | 5.0 | 18 | ---- | ---- | ---- | ---- | ---- | ---- | Snyder et al., 2010 |

1

*Based on crude protein value, assume that protein is 16% N

Table 1. Total plant nutrient content and C:N ration of various biofuel byproducts.

## 5. When to consider biofuel byproducts as a fertilizer source

While competition with the animal feed industry has made it difficult for biofuel byproducts to gain recognition as a viable fertilizer source, there are several factors that may cause biofuel producers to consider applications as fertilizers for their byproducts.

### 5.1 Advantages of DDGS in fertilizer markets

#### 5.1.1 Feed market value of DDGS

One concern is that DDGS supply will increase beyond the demand that currently exists for this material as animal feed. In the last three years, ethanol producers have avoided hitting the "feed wall" by exporting DDGS to countries like China, Mexico, and Canada (RFA, 2011). In 2010, DDGS exports from the U.S. were an estimated 8 million metric tons, which is almost 25% of all distillers grain produced in the U.S. (RFA, 2011). While exports have helped producers find markets for the time being, the ethanol industry will have no choice but to reach out to other markets, such as fertilizer markets, if ethanol production continues to increase at its current rate.

Another economic issue for using DDGS as animal feed is competition with other feeds (Table 2). Dietary ingredients are generally the single largest expense in animal production. In addition to DDGS, animal producers select from a wide variety of other products, including corn grain, cottonseed meal, and rice meal. For example, due to a higher protein content, DDGS is typically more expensive than corn. Producers will naturally prefer the most economic dietary ingredients, which can put additional pressure on marketing ethanol products.

| Feed source | Price ($/tonne) |
|---|---|
| Alfalfa meal | 229 |
| Cottonseed meal | 284 |
| DDGS | 204 |
| Rice bran | 146 |
| Soybean meal | 396 |
| Wheat bran | 159 |

Table 2. U.S. byproduct feed prices for February, 2011. Source: ERS, 2011.

#### 5.1.2 Animal feed consumption issues with DDGS

There are a wide variety of concerns for using DDGS as an animal feed source. One of the biggest concerns is the high P concentrations of the grains (Morse et al., 1992). Phosphorus contents of DDGS range between 5.4 and 8.2 g $kg^{-1}$, which exceeds the nutritional requirements of most ruminants, and is also high in comparison to corn grains (NRC, 1980). Excess P in the diet is excreted as manure, which is usually land-applied for disposal. From here, the P can be transported via water and wind to nearby waterways, polluting the waters by stimulating excessive algal growth, depleting dissolved oxygen levels, thus negatively impacting aquatic plant and animal species. It is possible that some animal producers will not purchase high P feeds, such as DDGS, due to a lack a of manure disposal options.

Another issue that producers face is the variability in the nutritional composition of DDGS. It has been estimated that protein content can vary as much as 25-35% from one batch to the next (Belyea et al., 1989). Mineral inconsistencies have also been documented (Arosemena et al., 1995; Belyea et al., 2004). Miscalculations of proteins and mineral content in feed could affect the quality of animal products and animal health.

Other concerns animal consumption issues with DDGS include difficult digestibility due to high fiber content (Rausch and Belyea, 2006), stimulation of thiamine deficiencies due to high sulfur content (Rausch and Belyea, 2006), and carcinogenic aflatoxins concentrated in DDGS from tainted corn grain (Blanco-Canqui et al., 2002).

### 5.1.3 Organic markets for DDGS

In recent years, organic growers in the U.S. have been showing increased interest in using DDGS as a possible fertilizer source for organic production. With low C:N ratios and typical N content ranging between 3.8 and 4.8% N, DDGS is a promising option as a N fertilizer source. This is critical for organic producers, who have limited options for affordable organically certified N fertilizer sources. Nitrogen is generally the most limiting nutrient for plant growth, and therefore can have the greatest impact of all of the plant nutrients on crop yield and quality. Organic N fertilizer sources are generally limited to raw animal manures, composted animal manures, and leguminous and non-leguminous cover crops. While raw animal manures like poultry litter and swine manure can feasibly meet the N requirements of most crops, issues arise with limited availability to specific regions, timing limitations on using raw manures for organic crops, and non-acceptance by some growers and international markets of raw manures as an organically certified practice.

Another reason why organic growers may be interested in DDGS is the potential for weed suppression. A small number of studies of have identified weed suppression characteristics in DDGS, although the mechanism is poorly understood at this time (Boydston et al., 2007; Liu and Christians, 1994). While it has been shown to be only a mildly effective herbicide, organic growers may still be interested in these properties since options for certified organic herbicides are extremely limited.

The approval of using DDGS as a certified organic fertilizer source in the U.S. has been somewhat controversial in recent years. The two major issues have been the approval of corn steep liquor (CSL) and the addition of antibiotics to DDGS. Corn steep liquor is a byproduct of the wet milling process, which differs from the dry-grind process used to produce DDGS. However, the approval of DDGS by individual state organic regulatory agencies has often been directly linked to the approval of CSL. After extensive debate, the Organic Materials Review Institute (OMRI) announced on January 13, 2011 that they would continue to treat CSL as an allowed non-synthetic ingredient for the purpose of product review (OMRI, 2011a). However, it is possible that this ruling could be overturned, so it is advised that organic growers stay up to date with their country's approval of CSL when considering DDGS field applications.

The other concern with using DDGS as an organic certified fertilizer source is the addition of antibiotics to the grains. Antibiotics are often added by ethanol producers during the fermentation process to destroy unwanted bacteria, leaving small concentrations of antibiotic residues in the DDGS. While this does occur, the use of antibiotics in agriculture is difficult to avoid. Antibiotic residues may also be found in manure and compost materials, however the use of antibiotics in animals whose manures are used for organic agriculture is not regulated in the U.S. It should be noted that there is no clear rule that prohibits the use of DDGS containing trace residues of antibiotics as fertilizers. As with this and any other organic practice, we strongly recommend consulting your local organic certifier before using DDGS or any questionable fertilizer amendment.

### 5.2 Advantages of oilseed meals in fertilizer markets

One concern in using oilseeds as animal feed is cost. As listed in Table 1, soybean meal is 1.9 times greater in cost than DDGS, 2.5 greater than wheat bran, and 2.7 times greater than rice bran. As mentioned above, animal feed accounts for the greatest in any animal operation, therefore animal producers are likely to choose the most economical option for their operation.

Similar to DDGS, the high N concentration and plant N availability of oilseed meals is appealing to organic growers who have limited options for fertilizer N sources. Most oilseed meals are currently approved by OMRI for use as an organic fertilizer in the U.S. as "uncomposted plant materials) NOP standard 205.203(c)(3) (NOP, 2011). Soybean meal is currently listed in the Generic Materials List as an allowed Crop Fertilizer and Soil Amendment (OMRI, 2011b). While canola meal, mustard meal, and other oilseed meals are not listed, products containing canola meal and mustard meal are listed under the OMRI materials database. Oilseed meals do not have the approval issues associated with DDGS because the pressing process used for extracting oils from the oilseeds is fairly simple and does not require the addition of chemical solvents or antibiotics during processing. Again, we recommend working with your organic certifier before using oilseed meals as a fertilizer source, even though they are generally approved as an organic fertilizer source. For example, individual certifiers may have concerns if the seeds used were genetically modified organism (GMO).

Mustard meal has unique properties that make it a favorable fertilizer and even herbicide source, yet a very poor option as an animal feed. Mustard meal is considered harmful to animals as a feed source due to high concentrations of erucic acid and glucosinilates (Joseffson, 1970). However, the presence of glucosilinates, which break down to isothiocyanates, can be beneficial for a wide variety of pesticide applications. Mustard seed meals have been shown to control weeds (Boydston et al., 2007; Norsworthy and Meehan, 2005; Rice et al., 2007; Vaughn et al., 2006), insect pests (Elberson et al., 1996, 1997); nematodes (Walker, 1996; Walker, 1997), and pathogens (Chung et al., 2002; Mazzola et al., 2007). Organic markets have taken interest in the pesticide properties of mustard seed meals, especially since effective organically certified pesticide option are limited. Organic growers can therefore benefit from both fertilizer and pesticide benefits of mustard meals. However, growers must be careful of applying the mustard meals too close to planting. Mustard meals can be non-discriminate and can burn emerging crop plants if not enough time is allowed for the isothiocyanate compounds to break down in the soil.

## 6. Biofuel byproducts as fertilizers – research efforts

As described above, the high N content (38 - 87 g N $kg^{-1}$) and low C:N ratio (5.1 - 12.0) strongly suggest that biofuel byproducts are effective N fertilizer sources (Table 1). Nitrogen mineralization studies are commonly used to closely determine plant available N (PAN) in organic amendments, such as biofuel byproducts, as opposed to roughly estimating PAN based on C:N and N content. These studies can account for influences from the organic amendment on the soil microbial populations that convert organic N to plant available forms of N ($NH_4$-N and $NO_3$-N) unrelated to N content.

One of the most common methods for measuring N mineralization is to monitor changes in $NH_4$-N and $NO_3$-N concentrations in a soil amended with a specific organic material over time. PAN measurements listed in Table 3 for various biofuel byproducts all fall in this category (Moore et al. 2010; Rubins and Bear, 1942; Van Kessel et al., 2000; and Gale et al., 2006). Another method for measuring N mineralization is to use $^{15}N$ labeling, where the N in the organic amendment is labeled with the rare $^{15}N$ isotope, and the changes in concentration of $^{15}NH_4$-N and $^{15}NO_3$-N are monitored. This method was used by Snyder et al. (2010) to determine PAN for canola and mustard meals (Table 3). While these methods do differ, the PAN estimates are generally comparable to each other.

Based on N mineralization incubation studies with biofuel byproducts using methods described above, N appears to be quickly available to plants from soybean meal and canola meal sources. Reviewing N mineralization studies of canola meal and soybean incubated at various soil types and incubation temperatures, N mineralization patterns were relatively consistent among studies. These studies showed that between 93 and 100% of N added as either canola meal or soybean meal was in the plant available forms ($NH_4$-N and $NO_3$-N) in the first 28 days of the 40-112 day incubation periods (Table 3) (Rubins and Bear, 1942; Van Kessel et al., 2000; Gale et al. 2006; Snyder et al., 2010). These findings suggest that soybean and canola meal would be excellent as preplant N fertilizer sources, especially for crops able to utilize N within the first month or two of growth. Although both would be considered excellent N sources, soybean meal appears to have at least 20% great PAN values than canola meal (Table 3). This difference can easily be attributed to the lower C:N (4.7-5.4 for soybean meal compared to 7.8 – 8.0 canola meal) and higher concentrations of N (76-87 g N $kg^{-1}$ for soybean meal compared to 57-63 g N $kg^{-1}$ for canola meal).

In contrast to soybean and canola meal, research studies suggest that N from DDGS and mustard meal mineralizes somewhat slowly over a growing season. This effect is not caused by N content, especially for mustard meal, which has a comparable N content of 58 g N $kg^{-1}$ compared to canola meal, which has a N content of 57-63 g N $kg^{-1}$ (Moore et al, 2010; Gale et al., 2006; Snyder et al., 2010). For example, Moore et al. (2010) found that only 56% of N released from DDGS over a growing season was plant available within the first 28 days. The authors speculated that the mild weed-suppressing chemical compounds in DDGS may also suppress nitrifying bacteria populations, although this effect has not been investigated. Similar trends have been seed with mustard meal applications, with two studies showing that a range of 55-82% N released from mustard meal over a growing season was plant available within the first 28 days (Table 3) (Moore et al., 2010; Snyder et al. 2010). Snyder et al. (2010) also recorded a 25% decrease in PAN between day 15 and 45 for Sinapis alba mustard meal, attributing the decrease to a delayed release of isothiocyanate from glucosinilates in the mustard meal.

The effect of mustard meal on nitrifying bacteria populations is better understood than DDGS. Bending and Lincoln (2000) discovered that isothiocyanate compounds found in mustard seed both reduced populations and inhibited the growth of nitrifying bacteria. Snyder et al. (2009) also showed that microbial biomass N concentrations for two varieties of mustard meal treatments (*Brassica juncea* and *Sinapis alba*) were 48 and 67% lower than for canola meal treatments, which the authors attributed to the biocidal properties of the isothiocyanate compounds in the mustard meal. The slow-release N characteristics of DDGS and mustard meal are beneficial for plant growth, supplying N throughout a growing season when the plant is most likely to use it. When all of the N is in plant available form early in the season, as is the case for most mineral N fertilizers, what is not taken up by plants is quickly leached out of the soil and therefore no longer available for plant uptake.

| Amendment | C:N | Total N | PAN (8-14 days) | PAN (15-28 days) | PAN (29-56 days) | PAN (57-126 days) | Reference |
|---|---|---|---|---|---|---|---|
| | | $g\ kg^{-1}$ | --------------------------------%-------------------------------- | | | | |
| **DDGS** | 12 | 41 | 17 | 31 | 46 | 55 | Moore et al., 2010 |
| **Soybean meal** | 4.7 | 76 | ----- | 61 | 65 | ----- | Rubins and Bear, 1942 |
| | 5.4 | 87 | 46 | 62 | 62 | 64 | Van Kessel et al., 2000 |
| **Canola meal** | 8 | 57 | 39 | ----- | 39 | 41 | Gale et al., 2006 |
| | 7.6 | 63 | ----- | 42.5 | 42.5 | ----- | Snyder et al., 2010 |
| **Mustard meal (S. alba)** | 8.5 | 58 | 17 | 33 | 49 | 60 | Moore et al., 2010 |
| | 8.4 | 58 | ----- | 45.8 | 34.1 | ----- | Snyder et al., 2010 |
| **Mustard meal (B. juncea)** | 7.8 | 62 | ----- | 40.5 | 49.3 | ----- | Snyder et al., 2010 |

Table 3. Plant available N (PAN) estimates from the total N pool for various biofuel byproducts.

The majority of peer-reviewed field research conducted has illustrated yield increases with increased rates of biofuel byproducts. Nelson et al. (2009) investigated three field application rates of DDGS in comparison to a slow-release poly coated urea (PCU) and anhydrous ammonia on a corn grain crop. Using application rates of 0, 1.2, 2.4, and 3.6 tonne DDGS $ha^{-1}$ (or 0, 46, 92, and 138 kg N $ha^{-1}$), Nelson et al. (2009) showed a grain yield increase of 1.41 and 1.56 kg grain $ha^{-1}$ for every kg $ha^{-1}$ of DDGS applied in medium and high yielding environments. Balesh et al. (2005) applied mustard meals as a fertilizer for tef at three application rates of 0.24, 0.38, and 0.51 tonne meal $ha^{-1}$ (or 15, 23, and 31 kg N $ha^{-1}$). Over the three rates and two years of field research, Balesh et al. (2005) showed that tef grain yield increases ranged from 2 to 116% in comparison to control treatments that did not receive any N applications. At the highest rate (31 kg N $ha^{-1}$), tef grain yields at least doubled in comparison to the control in both years, and were comparable to the Ethiopian national tef production average of 800 kg $ha^{-1}$. These finding suggests that mustard meal may be used as a viable substitute for chemical fertilizers in tef production. Balesh et al. (2005) also found that a decreased particle size of mustard meal increased yields, with mustard meal powder

applications increasing yields by 39 and 5% in 1994 and 1995, respectively. It should be mentioned that the authors applied mustard meals 20 days prior to planting to avoid toxicity issues with the germinating seed (Balesh et al., 2005). Banuelos and Hanson (2010) studied the application of selenium-enriched mustard and canola meals to strawberries. In this study, fruit yields increased at both rates of canola meal application (4.5 and 13.4 tonne acre$^{-1}$, or 230 and 683 kg N ha$^{-1}$) and at two of the three rates of mustard meal applications (2.2 and 4.5 tonne ha$^{-1}$, or 108 and 220 kg N ha$^{-1}$). An increase in Ca, P, and Mn concentrations in the fruit was observed for canola and mustard meal applications. Over both years, the authors noted that strawberry plants transplanted shortly after mustard meal applications were initially stunted and discolored; however, strawberry survival, growth, and productivity were similar among all treatments. These three studies clearly illustrate that biofuel byproducts can significantly improve crop yields.
Addressing the potential for plant toxicity with mustard meal applications, there are a few studies that have illustrated negative impacts on plant growth with increasing application rates. Snyder et al. (2009) found that mustard meal applications of 1 and 2 tonne ha$^{-1}$ (or approximately 59 and 118 kg N ha$^{-1}$) applied 36 days prior to planting did not affect emergence, while applying the *Sinapis alba* mustard meal only 15 days prior to planting decreased emergence up to 40%. Regardless, total fresh market yield was either comparable to or greater than the control for both years of the two-year study. Strawberry yields decreased by 42% compared to the control treatment when mustard meal was applied at rates of 13.4 tonne acre$^{-1}$ (Banuelos and Hanson, 2010). In both studies, the authors attribute the decrease in plant growth to glucosinilates in the mustard meals. When in contact with water, the glucosinilates break down to form isothiocyanates, which is a know plant growth suppressant.

## 7. Conclusion

DDGS byproducts of ethanol production and oilseed meal byproducts of biodiesel production can be effective fertilizer sources for plants. In addition to containing most macronutrients and micronutrients needed to support plant growth, these byproducts are have low C:N ratios, which means that they are rapidly decomposable and can release nutrients to plants in a timely manner. Competition with animal feed markets has prevented widespread adoption of biofuel byproducts as fertilizers. However, fertilizer markets may become more appealing to biofuel producers due to over-production of DDGS, high costs of oilseed meals, DDGS feed quality issues, and interest from organic growers in alternative nutrient-rich fertilizer sources. Research studies have illustrated that biofuel byproducts can be used as effective fertilizer sources, although growers working with mustard meals should be cautioned to allow enough time to pass between application and planting for phytotoxic glucosinilates to lose their potency. If and when the time ever comes when biofuel byproducts can compete with animal feeds as an economically viable fertilizer source, they will provide a much needed competitive fertilizer source to chemical fertilizers for conventional growers, and will help to produce yields for organic crops comparable to those seen for conventional crops.

## 8. Acknowledgements

I would like to acknowledge Rick Boydston, Harold Collins, and Ashok Alva from the USDA ARS Vegetable and Forage Crops Research Laboratory in Prosser, Washington for inspiring me to investigate the use of DDGS and mustard meals as a N fertilizer source.

## 9. References


Abe, S.S., S. Hashimoto, T. Umezane, T. Yamaguchi, S. Yamamoto, S. Yamada, T. Endo, and N. Nakata. (2010). Agronomic and environmental performance of rapeseed oilcake in the lowland rice farming of Japan. *Communications in Soil Science and Plant Analysis*. Vol. 41, pp. 1944-1958.

Arosemena, A., E.J. Peters, and J.G. Fadel. (1995). Extent of variability in nutrient composition within selected by-product feedstuffs. *Animal Feed Science Technology*. Vol. 54, pp. 103-120.

Balesh T., F. Zapata, and J.B. Aune. (2005). Evaluation of mustard meal as organic fertiliser on tef (*Eragrostis tef (Zucc) trotter*) under field and greenhouse conditions. *Nutrient Cycling in Agroecosystems*. Vol. 73, pp. 49-57.

Banuelos, G.S. and B.D. Hanson. (2010). Use of selenium-enriched mustard and canola seed meals as potential bioherbicides and green fertilizer in strawberry production. *Horticultural Science*. Vol. 45 pp. 1567-1572.

Belyea, R.L., B.J. Steevens, R.R. Restrepo, and A.P. Clubb. 1989. Variation in composition of by-product feeds. *Journal of Dairy Science*. Vol. 72, pp. 2329-2345.

Belyea, R.L., K.D. Rausch, and M.E. Tumbleson. (2004). Composition of corn and distillers dried grains with soluble in dry grind ethanol processing. *Bioresource Technology*. Vol. 94, pp. 293-298.

Bending G.D. and S.D. Lincoln. (2000). Inhibition of soil nitrifying bacteria communities and their activities by glucosinolate hydrolysis products. *Soil Biology & Biochemistry*. Vol. 32, pp. 1261-1269.

Blanco-Canqui, H., C.J. Gantzer, S.H. Anderson, E.E. Alberts, and F. Ghidey. (2002). Saturated hydraulic conductivity and its impact on simulated runoff for claypan soils. *Soil Science Society of America Journal*. Vol. 66, pp. 1596-1602.

Boydston, R., A. McGuire, S. Vaughn, and J. Collins. (2007). Effect of mustard seed meal on early weed emergence in peppermint and potato in sustaining the Pacific Northwest. *Food, Farm, and Natural Resource Systems, Newsletter*. CSANR. Washington State University Extension. Vol. 5, pp. 4-6.

Chen, M. and S.T. Hsieh. (1972). The effects of urea, rape seed oil cake, compound fertilizers and increase phosphorus and potassium on growth and quality of flue-cured tobacco. *Journal of the Agricultural Association of China*. Vol. 78, pp. 50-55.

Chung, W.C., J.W. Huang, H.C. Huang, and J.F. Jen. (2002). Effect of ground *Brassica* seed meal on control of *Rhizoctonia* damping-off of cabbage. *Canadian Journal of Plant Pathology*. Vol. 24, pp. 211-218.

Dunford, N. Oil and oilseed processing I. FAPC-158. *Oklahoma State University Food Technology Fact Sheet*. Oklahoma Cooperative Extension Service. pp. 1-4.

Elberson, L.R., V. Borek, J.P. McCaffry, and M.J. Morra. (1996). Toxicity of rapeseed meal-amended soil to wireworms (*Coleoptera: Elateridae*). *Journal of Agricultural Entomology*. Vol. 13, pp. 323-330.

Elberson, L.R., J.P. McCaffry, and R.R. Tripepi. (1997). Use of rapeseed meal to control black vine weevil larvae infesting potted rhododendron. *Journal of Environmental Horticulture*. Vol. 15, pp. 173-176.

ERS. (2011). Feed grains database. United States Department of Agriculture - Economic Research Service. Access on March 25, 2011. Available from: <http://www.ers.usda.gov/Data/FeedGrains/>

Gale, E.S., D.M. Sullivan, C.G. Cogger, A.I. Bary, D.D. Hemphill, and E.A. Myhre. (2006). Estimating plant-available nitrogen release from manures, composts, and specialty products. *Journal of Environmental Quality*. Vol. 35, pp. 2321-2331.

Joseffson E. (1970). Glucosinolate content and amino acid composition of rapeseed meal as affected by sulphur and nitrogen nutrition. *Journal of the Science of Food and Agriculture*. Vol. 21, pp. 98–102.

Hall, B. and P. Sullivan. (2001). Alternative soil amendments. *ATTRA Horticultural Technical Note.* Accessed on March 25, 2011. Available from: <http://attra.ncat.org/attra-pub/PDF/altsoil.pdf>

Herkes, J. 2010. Rapeseed and canola for biodiesel production. Accessed on March 25, 2011. Available from: < http://www.extension.org/pages/26629/rapeseed-and-canola-for-biodiesel-production>

Kora, P.A. and A.K. Shingte. (1988). The mineralization of nitrogen from non-edible oil cakes in medium black calcareous and lateritic soils. *Agricultural Research Journal of Kerala*. Vol. 26, pp. 215-226.

Liu, D. L. and N. E. Christians. (1994). Isolation and identification of root-inhibiting compounds from corn gluten hydrolysate. *Journal of Plant Growth Regulation*. Vol. 13, pp. 227-230.

Liu, K. (2008). Particle size distribution of distillers dried grains with solubles (DDGS) and relationships to compositional and color properties. *Bioresource Technology*. Vol. 99, pp. 8421-8428.

Lory, J.A., R.E. Massey, C.D. Fulhage, M.C. Shannon, R.L. Belyea, and J.M. Zulovich. (2008). Comparing the feed, fertilizer, and fuel value of distiller's grains. *Crop Management Network*. Accessed on March 25, 2011. Available from: <http://www.plantmanagementnetwork.org/pub/cm/review/2008/distillers/>

Mazzola, M., J. Brown, A.D. Izzo, and M.F. Cohen. (2007). Mechanism of action and efficacy of seed meal-induced pathogen suppression differ in a *Brassicacaeae* species and time-dependent manner. *Phytopathology*. Vol. 97, pp. 454-460.

Moore, A.D., A.K. Alva, H.P. Collins, and R.A. Boydston. (2010). Mineralization of nitrogen from biofuel by-products and animal manures amended to a sandy soil. *Communications in Soil and Plant Analysis*. Vol. 41, pp. 1315-1326.

Morse, D., H.H. Head, C.J. Wilcox, H.H. Van Horn, C.D. Hissem, and B. Harris, Jr. (1992). Effects of concentration of dietary phosphorus on amount and route of excretion. *Journal of Dairy Science*. Vol. 75, pp. 3039-3049.

NBB. (2011). Biodiesel production estimates CY2005-current. National Biodiesel Board. Accessed on March 25, 2011. Available from <http://www.biodiesel.org/resources/fuelfactsheets/>

NBB. (2008). Food vs. fuel: FAQ. National Biodiesel Board. Accessed on March 25, 2011. Available from: < http://biodieselsustainability.com/faq.html>

Nelson, K.A., P.P. Motavelli, and R.L. Smoot. (2009). Utility of dried distillers grain as a fertilizer source for corn. *Journal of Agricultural Science*. Vol. 1 pp. 3-12.

NOP. 2011. Electronic code of federal regulations. United States Department of Agriculture, National Organic Program. Accessed on March 25, 2011. Available from: <http://ecfr.gpoaccess.gov/cgi/t/text/text-idx?c=ecfr&sid=3f34f4c22f9aa8e6d9864cc2683cea02&tpl=/ecfrbrowse/Title07/7cfr205_main_02.tpl>

Norsworthy, J.K., and J.T. Meehan IV. (2005). Herbicidal activity of eight isothiocyanates on Texas panicum (*Panicum texanum*), large crabgrass (*Digitaria sanguinalis*), and sickelpod (*Senna obtusifolia*). *Weed Science*. Vol. 53, pp. 515-520.

NRC. (1980). Phosphorus. In: *Mineral tolerances of domestic animals*. National Research Council, National Academy of Sciences, Washington D.C. pp. 364-377.

OMRI. (2011a). OMRI Position on corn steep liquor. Organic Materials Review Institute. Accessed on March 25, 2011. Available from: <http://www.omri.org/news/96333/omri-position-corn-steep-liquor>

OMRI. (2011b). Generic Materials List. Organic Materials Review Institute. Accessed on March 25, 2011. Available from: <http://www.omri.org/omri-lists>

Peterson, C.L., D.L. Reece, B.L. Hammond, J. Thompson, and S.M. Beck. 1997. Processing characterization, and per

Rausch, K. D., and Belyea, R. L. (2006). The future of coproducts from corn processing. *Applied Biochemistry and Biotechnology*. Vol. 128, pp. 47-86.

RFA. (2011). 2011 Ethanol Industry Outlook. Renewable Fuels Association. Accessed on March 25, 2011. Available from: www.ethanolrfa.org.

Rice, A.R., J.L. Johnson-Maynard, D.C. Thill, and M.J. Morra. (2007). Vegetable crop emergence and weed control following amendment with different *Brassicaceae* seed meals. *Renewable Agriculture and Food Systems*. Vol. 22, pp. 204-212.

Rubins, E.J. and F.E. Bear. (1942). Carbon-nitrogen ratios in organic fertilizer materials in relation the availability of their nitrogen. *Soil Science*. Vol. 54, pp. 411-424.

Shurson, J. (2010). US DDGS comparison tables. Dept. of Animal Sci., Univ. of Minnesota, St. Paul, MN. Accessed on March 25, 2011. Available from: <http://www.ddgs.umn.edu/profiles/us_profile_comparison_tables-Aug2010.pdf>

Singh, U. and K. Gurumurt. (1984). Oil cakes from oil seeds of forest origin - their potential as fertilizer. *Indian Journal of Forestry*. Vol. 7, pp. 12-18.

Snyder, A.J., J.L. Johnson-Maynard, and M.J. Morra. (2010). Nitrogen mineralization in soil incubated with N-15 labeled *Brassicaceae* seed meals. *Applied Soil Ecology*. Vol. 46, pp. 73-80.

Snyder, A.J., J.L. Johnson-Maynard, and M.J. Morra. (2009). Seed meals from Brassicaceae oilseed crops as soil amendments: Influence on carrot growth, microbial biomass, and nitrogen mineralization. *Horticultural Science*. Vol. 44, pp. 354-361.

Spiehs, M.J., M.H. Whitney, and G.C. Shurson. (2002). Nutrient database for distiller's dried grains with soluble produced from new ethanol plants in Minnesota and South Dakota. *Journal of Animal Science*. Vol. 80, pp. 2639-2645.

Sullivan, D.M., N.A. Andrews, J.M. Luna, and J.P.G. McQueen. (2010). Estimating N contribution from organic fertilizers and cover crop residues using online calculators. *Proceedings of the 19th World Congress of Soil Science*. Brisbane, Australia, August, 2010.

Van Kessel, J.S., J.B. Reeves, and J.J. Messinger. (2000). Nitrogen and carbon mineralization of potential manure components. *Journal of Environmental Quality*. Vol. 29, pp. 1669-1677.

Vaughn, S.F., D.E. Palmquist, S.M. Duval, and M.A. Berhow. (2006). Herbicidal activity of glucosinolate-containing seedmeals. *Weed Science*. Vol. 54, pp. 743-748.

Walker, J.T. (1996). Crambe and rapeseed meal as soil amendments: Nematicidal potential and phytotoxic effects. *Crop Protection*. Vol. 15, pp. 433-437.

Walker, G.E. (1997). Effects of *Brassica* residues and other organic amendments on abundance and sex ratio of *Tylenculus semipenetrans* in soil. *Australian Journal of Experimental Agriculture*. Vol. 37, pp. 693-700.

# 18

# The Past, Present, and Future of Biofuels – Biobutanol as Promising Alternative

Köpke Michael, Noack Steffi and Dürre Peter
*Institute of Microbiology and Biotechnology, University of Ulm*
*Germany*

## 1. Introduction

While the worldwide energy consumption is projected to grow by 57 % until 2030 (U. S. Energy Information Administration [EIA], 2007) fossil sources are limited and it is questionable how long they will last. Meanwhile, even unconventional sources such as tar sands and oil shales become economically producible since crude oil prices have reached sustained highs and even surpassed a historic mark of US$ 145 per barrel in June 2008 (EIA, 2011).

Besides concerns about avaibility and prices of fossil fuels as well as the quest for energy independence, there is also an intense discussion about environmental impacts. Burning of fossil fuels leads to a massive increase of the greenhouse gas $CO_2$ in the atmosphere and is thus contributing to global warming (Intergovernmental Panel on Climate Change [IPCC], 2007). This could be counterbalanced by using alternative energy sources. Biofuels are the most promising alternative energy for the transportation sector, which is most rapidly growing (an annual average of 3 % is projected until 2030, especially due to increasing mobility in China and India) and accounts for over 20 % of the worldwide primary energy (EIA, 2010).

Biofuels can either be produced microbially or chemically from renewable biomass and are therefore $CO_2$ neutral. However, only few compounds such as alcohols (ethanol, butanol), alkyl esters of fatty acids (biodiesel), and alkanes (renewable diesel) have the required properties. Today, only bioethanol, biodiesel, and renewable diesel are produced at industrial scale, but several second generation technologies are on path to commercialization. Especially biobutanol fermentation seems to be a promising alternative.

## 2. The past: History of biofuels

The use of biofuels is no novel invention. Fueling up with vegetable oils or ethanol was popular long before the development of the combustion engine. Vegetable and animal oil lamps have been used since the dawn of civilization. Already in 1834, the first US patent for alcohol as a lamp fuel was awarded to S. Casey (Kovarik, 1998). Around 1850, thousands of distilleries produced an estimated 24 million liters (90 million gallons) of "Camphene" (a camphor oil scented blend of turpentine and ethanol) per year (Kovarik, 1998).

Biofuels have also been used since the early days of the car industry. Even the invention of the first combustion engine, the "Otto cycle", was performed with biofuels. German engineer Nikolaus August Otto ran his early engines in the 1860s on ethanol, a fermentation product of yeasts. Interestingly, Otto's initial financing came from Eugen Langen, who owned a sugar refining company having links to the alcohol markets of Europe (Kovarik, 1998). Furthermore, Henry Ford's first prototype automobile, the "Quadricycle", in the 1880s could be operated with ethanol as fuel and his "Model T", the "Tin Lizzie", the most popular car produced between 1908 and 1927, was originally designed to run on pure ethanol. Ford was a big supporter of alcoholic fuels and told a New York Times reporter in 1925 (Ford, 1925): *"The fuel of the future is going to come from fruit like that sumach out by the road, or from apples, weeds, sawdust -- almost anything. There is fuel in every bit of vegetable matter that can be fermented. There's enough alcohol in one year's yield of an acre of potatoes to drive the machinery necessary to cultivate the fields for a hundred years."*

However, due to economic issues, pure ethanol was not able to prevail over gasoline. Before 1906, high taxes were levied on ethanol in the United States and later, gasoline became cheaply available due to the discovery of large oil reserves in Texas and Pennsylvania. Soon however, ethanol was recognized as an effective anti-knocking additive for combustion engines when mixed with gasoline. Blends such as "Agrol" (up to 17 % ethanol fermented from grain) in the United States, "Koolmotor" and "Cleveland Discol" (up to 30 % ethanol fermented from grain) in Britain, "Monopolin" (25 % ethanol fermented from potatoes) in Germany, "Benzalcool" (20 % ethanol) and "Robur" (30 % ethanol and 22 % methanol) in Italy, "Lattbentyl" (25 % ethanol fermented from paper mill wastes) in Sweden, "Moltaco" (20 % ethanol) in Hungary, "Benzolite" (55 % ethanol) in China, "Natalite" (up to 40 % ethanol fermented from sugar cane) in South Africa, "Gasonol" (20 % ethanol fermented from sugar cane) in the Philippines, "Shellkol" (up to 35 % ethanol fermented from molasses) in Australia, and "Espiritu" (20 % ethanol fermented from molasses) in Cuba were common between 1925-1945 (Kovarik, 1998; Giebelhaus, 1980; Finaly, 2004).

Ethanol was not the only biofuel used in the car industry at that time. Rudolph Diesel, inventor of the "Diesel oil-engine", tested his engine with peanut oil at the world's fair "Exposition Universelle Internationale" of 1900 in Paris (Nitske & Wilson, 1965; Knothe, 2001.) In 1912, Diesel published two articles (Diesel, 1912a, 1912b) in which he reflected: *"The fact that fat oils from vegetable sources can be used may seem insignificant to-day, but such oils may perhaps become in course of time of the same importance as some natural mineral oils and the tar products are now. (…) In any case, they make it certain that motor power can still be produced from the heat of the sun, which is always available for agricultural purposes, even when all our natural stores of solid and liquid fuels are exhausted."* During the 1920s however, diesel engine manufacturers altered their engines to petroleum-derived diesel fuel due to cheaper prices and lower viscosity (the viscosity of vegetable oil is about an order of magnitude higher), which led to better atomization of the fuel in the engine's combustion chamber (Knothe, 2001). This problem was solved when the Belgian patent 422,877 was granted on August 31$^{st}$ 1937 to George Chavanne of the University of Brussels (Chavanne, 1937). It describes the use of methyl and ethyl esters of vegetable oil, obtained by acid-catalyzed transesterification, as diesel fuel, being the first report on what is today known as biodiesel (Chavanne, 1943).

During World War II, vegetable oils and alcohols were used as supplementary or emergency fuels in most belligerent nations. For instance, it was reported that the Japanese battleship "Yamato" used refined soybean oil as bunker fuel (Knothe, 2001). Nevertheless, with the outbreak of WW II, virtually all resources were diverted from industrial alcohol production to synthetic rubber or ammunition (Finaly, 2004). After the war, gasoline dominated the market almost completely because of cheap Middle East oil. Only during periods of scarcity such as the oil crises 1973-74 and 1979-80, many countries showed renewed interest in biofuels. The only nation that revived the bioethanol industry permanently was Brazil. Government, farmers, alcohol producers, and car manufactures cooperated in the 1970s to launch the alcohol program "Pró-álcool". Brazil began to produce ethanol by fermentation of sugar cane. Initially, gasoline was blended with 20 % ("E20") or 25 % ("E25") ethanol, and after the second oil crisis pure ethanol ("AEHC" ("Álcool etílico hidratado combustível", hydrated ethyl alcohol fuel) or "E96") was also available as fuel, causing the car industry to implement the necessary engine modifications. The Brazilian bioethanol production increased from 600 million liters (160 million gallons) in 1975 to 13.7 billion liters (3.6 billion gallons) in 1997, by far the highest in the world (International Energy Agency [IEA], 2004).

Not until 2006, the United States surpassed the Brazilian bioethanol production with an annual capacity of 18.4 billion liters (4.9 billion gallons) compared to 17 billion liters (4.5 billion gallons). This was a dramatic increase taking into account that the US had produced only a comparatively low amount of 4.9 billion liters (1.3 billion gallons) bioethanol in 1997 (Figure 1a; Renewable Fuels Association [RFA], 2011). Today, the US doubled this number and produces around 40.1 billion liters (10.6 billion gallons) bioethanol (Figure 1a; RFA, 2011). The major comeback of biofuels in the United States and most other nations was driven by the enormous rise of the crude oil prices since the late 1990s (Figure 1b; EIA, 2011). Meanwhile, oil prices have reached sustained highs of over US$ 80 per barrel and might well continue to increase due to political instability in the Middle East and concerns over the potential oil peak (highest production rate), demonstrated in July 2008 when the oil price surpassed US$ 145 per barrel for a short time.

Other important drivers include the quest to gain energy independence (Schubert, 2006) and growing concerns on the effect of greenhouse gas emissions on the world's climate, which could be counterbalanced by using renewable biomass for biofuel production. Also, MTBE (Methyl *tert*-butyl ether), an oxygenated anti-knocking additive for engines, was restricted in many countries and banned in some US states such as California and New York (which account for app. 45 % of the United States MTBE consumption) in response to environmental and health concerns (EIA, 2003; IEA, 2004). As a suspected carcinogenic agent, MTBE began turning up in significant amounts in ground water, since it is highly soluble in water, binds weakly to soil, and is not readily biodegradable in the environment (Squillance et al., 1997; EIA, 2003; IEA, 2004). This created an additional 10 billion liters (2.7 billion gallon) market for ethanol, which can be used as substitute directly or can be converted to the more environmental friendly ETBE (Ethyl *tert*-butyl ether) (EIA, 2003 & 2007). Moreover, many countries all over the world granted tax exemptions or paid subsidies for biofuels and set mandatory targets for the use of biofuels. The US congress for example established a renewable fuels mandate of 136 billion liter (36 billion gallons) by 2022 (Department of Energy [DOE], 2010), while the European Union agreed to satisfy 10 % of its transport fuel needs from renewable sources, including biofuels, hydrogen, and green electricity (EurActive Network, 2008). As a consequence, research and development of new, second generation biofuels got another push forward.

However, presently only first generation bioethanol and biodiesel are produced at a large industrial scale. While most countries produce mainly bioethanol, some European nations such as Germany, France, Spain and Italy focus on biodiesel, too. In 2009, Brazil and the United States produced app. 85 % of the world's bioethanol, whereas Europe produced about 85 % of all biodiesel (Table 1; RFA, 2010; EIA, 2007; Biofuels platform, 2010; European Biodiesel Board, 2010).

A)

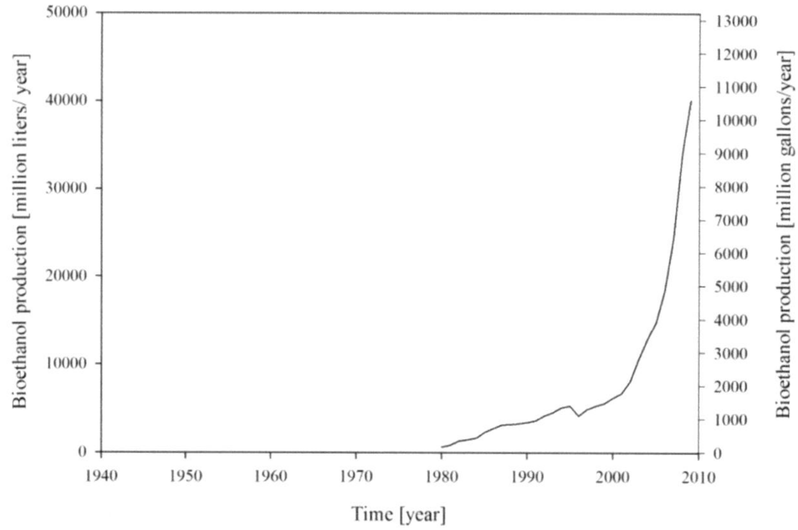


B)

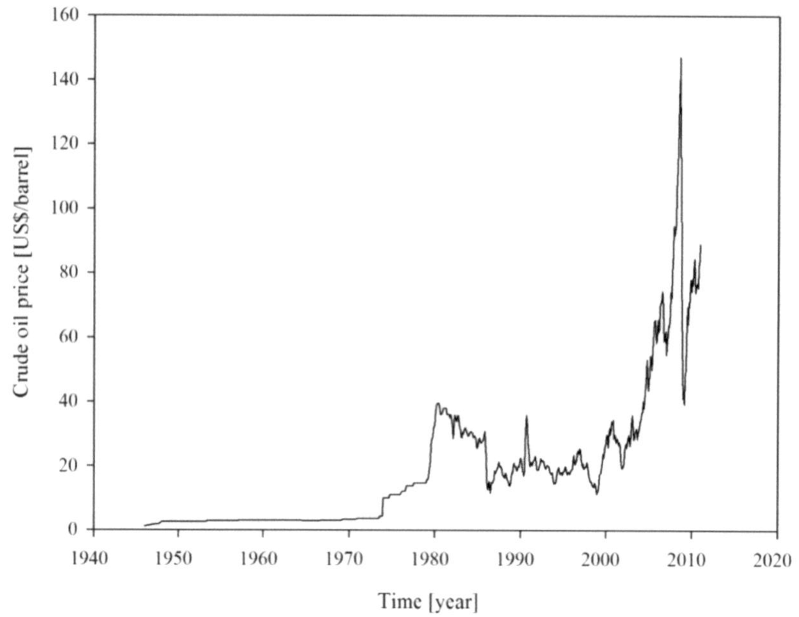


Fig. 1. Bioethanol production and crude oil price. A, development of the bioethanol production in the United States (RFA, 2010). B, development of the crude oil price (West Texas Intermediate) (EIA, 2011).

Biofuels are even used in motor sports nowadays. In 2007, the Scotsman Dario Franchetti won the iconic "Indianapolis 500" race with his 670-horsepower "Indy car" running on pure bioethanol. Moreover, in February 2008 a Virgin Atlantic Boeing 747-400 was the first commercial aircraft testing a 20 % blend of biofuel (a mixture of coconut and babassu nut oils) with 80 % fossil jet fuel (kerosene, $C_9$-$C_{18}$) in one of its four engines on a flight from London Heathrow to Amsterdam Shiphol airport (Virgin Atlantic, 2008). Airplanes in general have problems using ethanol because it freezes at an altitude of about 5 kilometers (16,400 feet). Long chain fatty acids, alkanes and also biobutanol have been identified as potential alternative by the Virgin Group, which announced to invest over US$ 3 billion over the next 10 years in renewable energy initiatives (Ernsting, 2008). One of Virgin Fuels partners, Gevo, Inc. is working on production of this biofuel with metabolic engineered *Escherichia coli* and yeast (Atsumi et al., 2008a, 2008b, Liao et al., 2008; Gunawardena et al., 2008; Buelter et al., 2008; Hawkins et al., 2009).

| Country | Bioethanol | | Biodiesel | | Total production | |
|---|---|---|---|---|---|---|
| | [Ml] | [Mgal] | [Ml] | [Mgal] | [Ml] | [Mgal] |
| US | 40,125 | 10,600 | 1,703 | 450 | 41,828 | 11,050 |
| Brazil | 24,900 | 6,578 | n. d. a. | | 24,900 | 6,578 |
| Germany | 750 | 198 | 3,218 | 850 | 3,968 | 1,048 |
| France | 1,250 | 330 | 2,483 | 656 | 3,733 | 986 |
| China | 2,052 | 542 | n. d. a. | | 2,052 | 542 |
| Thailand | 1,647 | 435 | n. d. a. | | 1,647 | 435 |
| Spain | 465 | 123 | 1,089 | 288 | 1,554 | 410 |
| Canada | 1,102 | 291 | n. d. a. | | 1,102 | 291 |
| Italy | 72 | 19 | 934 | 247 | 1,006 | 266 |
| Belgium | 143 | 38 | 527 | 139 | 670 | 177 |
| Poland | 166 | 44 | 421 | 111 | 587 | 155 |
| Austria | 180 | 48 | 393 | 104 | 573 | 151 |
| Sweden | 175 | 46 | 295 | 78 | 470 | 124 |
| Netherlands | 0 | 0 | 409 | 108 | 409 | 108 |
| India | 348 | 92 | n. d. a. | | 348 | 92 |
| Czech Rep. | 113 | 30 | 208 | 55 | 321 | 85 |
| Portugal | 0 | 0 | 317 | 84 | 317 | 84 |
| Hungary | 150 | 40 | 169 | 45 | 319 | 84 |
| Columbia | 314 | 83 | n. d. a. | | 314 | 83 |
| Finland | 4 | 1 | 279 | 74 | 283 | 75 |
| UK | 70 | 18 | 174 | 46 | 244 | 64 |
| Australia | 216 | 57 | n. d. a. | | 216 | 57 |
| Rest EU | 163 | 43 | 549 | 146 | 712 | 189 |
| Others | 935 | 247 | n. d. a. | | 935 | 247 |
| Total | 75,340 | 19,903 | 13,168 | 3,481 | 88,509 | 23,381 |

n. d. a. = no data available. Data were taken from RFA, 2010; EIA, 2007; Biofuels platform; 2010; European Biodiesel Board, 2010.

Table 1. Production of biofuels in 2009

The use of butanol as biofuel has already been reported in 2005, when David Ramey toured the United States in a 13-year old Buick fueled by pure butanol. Although consumption was 9 % higher, emissions of carbon monoxide (CO), hydrocarbons, and nitrogen oxides ($NO_x$) were decreased enormously. He meanwhile started the company Butyl Fuel, LLC. While this is a fairly small enterprise, two major global players, BP and DuPont, also announced to start fermentative biobutanol production from sugar beet in June 2006 and formed the joint-venture Butamax™Advanced Biofuels, LLC in 2009 with the aim to commercialize biobutanol by 2013. A first commercial plant with a capacity of 420 million liters (111 million gallons) will be built in Saltend, UK (Butamax, 2011a).

About 150 years ago, Louis Pasteur discovered that butanol can be formed by microbes (Pasteur, 1862). The culture he used for his experiments was probably a mixture of different clostridia (strictly or moderately anaerobic, spore-forming, Gram-positive bacteria, unable of dissimilatory sulfate reduction). More detailed studies on butanol producing bacteria were then conducted by Albert Fitz, who finally described the isolation of pure cultures of "*Bacillus butylicus*" from cow feces and hay (Fitz, 1876; Fitz, 1877; Fitz, 1878; Fitz, 1882). Other scientists, e. g. Martinus Beijerinck and Sergei Winogradsky, isolated further solvent-forming bacteria around 1900. These organisms received names such as "*Granulobacter saccharobutyricum*", "*Amylobacter butylicus*", and "*Bacillus orthobutylicus*", which are no longer taxonomically valid (Dürre & Bahl, 1996). Presumably all of them belong to the genus *Clostridium*, which was back then only used as a morphological description, meaning small spindle (Dürre, 2001).

Almost at the same time, considerable interest in synthetic rubber started as a result of the increase in the price of natural material due to its use in automobile tires. In 1910, the British company Strange and Graham, Ltd. launched a project to study butanol formation by microbial fermentation because of its use as precursor of butadiene, the starting material for the synthetic rubber production (as well as of isoamyl alcohol as a precursor of isoprene). The project was pursued by help of Auguste Fernbach and Moïse Schoen from the Institute Pasteur in Paris and William Perkins and Charles Weizmann from Manchester University. Fernbach isolated an acetone-butanol producer in 1911, but Weizmann separated in 1912, continuing his work at Manchester University. He succeeded in isolating an organism, later named *Clostridium acetobutylicum*, which produced significantly larger amounts of acetone and butanol than the strain isolated by Fernbach (McCoy et al., 1926). Patent applications were filed for the Fernbach process in 1911 and 1912 (Fernbach & Strange, 1911ab & 1912) and for the Weizmann process in 1915 (Weizmann, 1915). In 1913, Strange and Graham, Ltd. started production with the so-called ABE fermentation (for acetone-butanol-ethanol) based on the Fernbach process, first at Rainham, UK and later at King's Lynn, UK (Gabriel, 1928; Jones & Woods, 1986; Dürre & Bahl, 1996).

As an irony of fate, natural rubber became available at this time at much cheaper prices and in large quantities, because the new plantations in Asia started to be fully productive. However, the outbreak of World War I led to a sudden and large demand for acetone as a solvent for the production of cordite (smokeless gunpowder). The dominant source for acetone up to this time was calcium acetate imported from Austria, Germany, and the United States. As acetate imports from Austria and Germany were not available during that time and the production capacity in the US was almost negligible compared to the required quantities, Strange and Graham, Ltd. were contracted by the British War Office to supply acetone. However, their production was relatively inefficient, with an average capacity of about 440 kg (970 pounds) of acetone per week. Therefore, a switch to the Weizmann process was requested. Thus, production could be increased to app. 900 kg (2,000 pounds) acetone per week. Hence, the disregarded by-product acetone helped the ABE fermentation process to an international breakthrough, becoming eventually the second largest biotechnological process ever performed (Jones & Woods, 1986; Dürre & Bahl, 1996). Due to the threat by German submarines, grain and corn could not be imported to the United Kingdom in the required quantities any longer. Therefore, the Weizmann process was transferred to Canada and the United States. Plants were built in Toronto in 1916 and Terre Haute, Indiana in 1917 (Gabriel, 1928; Ross, 1961; Jones & Woods, 1986; Dürre & Bahl, 1996). The constant supply of acetone was certainly a decisive factor in winning World War I. Weizmann declined any rewards or personal honors by the British government, but, being a member of the Zionist movement, clarified that his only wish was to see a home established for the Jews in Palestine. There is no doubt that this attitude affected the Balfour declaration of 1917, leading to the foundation of the State of Israel. In succession, Weizmann became its first president (Ross, 1961; Dürre & Bahl, 1996).

At the end of the war in 1918, there was no longer a high demand for acetone and consequently all production plants were closed. During the whole war, butanol (about twice the amount of the produced acetone) was considered a white elephant and simply stored in huge containers (Killeffer, 1927). However, the situation changed in 1920, when the United States implemented the prohibition. As a result no amyl alcohol, obtained as a by-product of the ethanol fermentation, was available for the production of amyl acetate, needed in large amounts by the rapidly growing automobile industry as solvent for lacquers. Butanol and its ester butyl acetate proved to be a well-suited alternative. The Commercial Solvents Corporation (CSC) was founded, obtained the patent rights to the Weizmann process, took over the plant at Terre Haute from the Allied War Board in 1919, and started butanol production in 1920. Despite the general recession of 1920, which forced a shutdown of several months, and a bacteriophage infection in 1923, which cut the yields dramatically, the plant was enlarged. Additionally, a new plant was opened in Peoria, Illinois in 1923, consisting of 32 189,000-liter (50,000-gallon) fermenters and enlarged in 1927 to 96 fermenters (Gabriel, 1928; Gabriel & Crawford, 1930; Ross, 1961; Jones & Woods, 1986; Dürre & Bahl, 1996).

After expiration of the Weizmann patent, many new strains were isolated (McCutchan & Hickey, 1954) and patented and new fermentation plants were built in the United States, Puerto Rico, South Africa, Egypt, the former Soviet Union, India, China, Japan, and Australia. Until 1945, two-thirds of the butanol in the United States was produced by fermentation. During World War II, the focus shifted to acetone production again (Jones & Woods, 1986; Dürre & Bahl, 1996). However, a few years after the end of the war, most of the plants in Western countries were closed because of rising substrate prices and competition by the growing petrochemical industry. The ABE fermentation was only continued in countries that were cut off from international supplies for political or monetary reasons. For instance, the South African apartheid regime ran a plant in Germiston with a capacity of 1,080 $m^3$ (11,625 cubic foot) until 1982 (Jones & Woods, 1986; Jones, 2001). The former USSR operated at least eight plants, some of them up to the late 1980s. Continuous fermentations with lignocellulose hydrolates as substrate and working volumes of more than 2,000 $m^3$ (21,530 cubic foot) were carried out. During the 1960s and 1970s more than 100,000 tons of butanol per year were produced (Zverlov, 2006). China also developed the continuous fermentation process and about 30 plants produced an annual amount of 170,000 tons of solvents at its peak in the 1980s. Afterwards the production decreased successively and the last plant was closed in 2004 (Chiao & Sun, 2007).

Meanwhile, China reopened some of its fermentation plants and constructed new ones, with an expected annual solvent production of up to 1,000,000 tons in the next five years (Chiao & Sun, 2007; Ni & Sun, 2009). New plants were also built or are planned in the United States (Gevo, 2009; Cobalt Technologies, 2010), the United Kingdom (Butamax, 2011a), Brazil (Afschar et al., 1990), France (Marchal et al., 1992; Nimcevic & Gapes, 2000), and Austria (Nimcevic & Gapes, 2000; Gapes, 2000). Thus, the biological butanol production faces bright prospects in the future.

## 3. Specifications for biofuels

Biofuels have to meet defined physical and chemical criteria for the use in modern combustion engines and for the use of the existing distribution infrastructure. First, the state of aggregation is of particular importance. Biofuels for transportation should be liquid at ambient temperature and atmospheric pressure which is true for alcohols, biodiesel, and renewable diesel. Gaseous biofuels such as hydrogen and methane (biogas) will require the development of a new infrastructure and modified engines. Second, biofuels should have the same properties as petrochemical-based fuels (Table 2).

| Fuel | Gasoline | Ethanol | Butanol | Diesel | Biodiesel |
|---|---|---|---|---|---|
| Energy density [MJ/l] | 32-35 | 21.2 | 29.2 | 35-42 | 32-42 |
| Mileage [%] | 100 | 61-66 | 83-91 | 100 | 90-100 |
| Air-fuel ratio | 14.6 | 9.0 | 11.2 | 15.0 | 13.8 |
| Research octane number (RON) | 91-99 | 129 | 96 | - | - |
| Motor octane number (MON) | 81-89 | 102 | 78 | - | - |
| Cetane number (CN) | - | - | - | 50-60 | 45-70 |
| Vapor pressure [hPa] | 35-90 (at 20 °C) | 58 (at 20 °C) | 6.7 (at 20 °C) | - | - |
| Flashpoint [°C] | < -20 | 12 | 35-37 | 55-60 | 100-190 |
| Enthalpy of vaporization [MJ/kg] | 0.36 | 0.92 | 0.43 | - | - |
| Kinematic viscosity [mm²/s] | 0.4-0.8 (at 20 °C) | 1.5 (at 20 °C) | 3.6 (at 20 °C) | 1.2-3.5 (at 40 °C) | 2-9.5 (at 40 °C) |

Table 2. Physical and chemical properties of biofuels

Basically, biodiesel has similar characteristics as regular diesel. However, there are significant differences between biodiesel fuels produced from various vegetable sources (Fukuda et al., 2001), which can lead to damage of diesel engines. Critical points are dilution of motor oil, coking of piston rings, corrosion of hydraulic components, and depositions in the injection system, resulting from the production process and fuel aging. Hence, this is perhaps the biggest problem of biodiesel fuels, causing a couple of automotive manufacturers to refuse the use of biodiesel in some of their models. Newly developed, highly efficient motor technologies require low sulfur and saturated hydrocarbons, without aromatic compounds. A way out of this problem will be the use of renewable diesel and diesel from BtL (biomass to liquid) biofuels.

Similarly, ethanol has a number of disadvantages that can be avoided when using butanol as a biofuel (see section 5.5). The reason why currently so many countries use biodiesel and ethanol despite a number of disadvantages is simply the fact that they are currently available in large quantities. In future, however, superior second generation biofuels such as renewable diesel, butanol, renewable gasoline, and BtL fuels will gain much more importance.

## 4. The present: First generation biofuels

### 4.1 Bioethanol

The bioethanol fermentation is by far the largest biotechnological process worldwide. Most common are batch fermentations with the yeast *Saccharomyces cerevisiae* from corn, sugar cane, cassava, wheat, or rye as substrate. Sugar beet has recently been introduced as well.

During the last few years, industrial production strains have been improved for higher ethanol yields, specific ethanol productivity, inhibitor insensitivity, and product tolerance (up to 20 % ethanol). Meanwhile, the bioethanol fermentation is a mature technology. Worldwide, a total of around 75 billion liters (20 billion gallons) bioethanol were used in 2009 as biofuel (Table 1). The worlds largest bioethanol producer is POET, LLC (www.poet.com/) with over 27 plants producing more than 6 billion liters (1.5 billion gallons) (POET, 2011).

However, the increasing demands for sugar cane and especially corn are in serious competition with the food industry (food-vs-fuel discussion). Mexico has already seen huge demonstrations about the rising costs of the national dish tortillas, which are made from corn. A world bank report blamed the increasing biofuel use as one of the prime causes of raising food prices (Mitchell, 2008), and Oxfam claimed that current biofuel policies will push millions of people into poverty (Oxfam, 2008). The use of large land areas for growing monocultures is criticized as environmentally hazardous as well, and the suger cane fields in Brazil are usually burned before harvest (to kill snakes and make the suger cane easier to handle), releasing green house gases methane and nitrous oxide. Thus, corn and sugar cane ethanol can only be an interim solution, until second generation biofuels are commercialized, which are sustainable and independent of the use of food.

### 4.2 Biodiesel

Biodiesel is a chemically synthesized biofuel. Worldwide, more than 13.1 billion liters (3.5 billion gallons) have been used in 2009 as a pure fuel or fuel additive (Table 1). Biodiesel is defined as monoalkyl esters of fatty acids from vegetable oil or animal fats. It is produced by transesterification of the parent oil with an alcohol and this process is therefore also designated alcoholysis (Figure 2). The alcohol generally used for this process is chemically synthesized methanol because of its cheap price, but it is also possible to use higher alcohols from microbial fermentation such as ethanol or butanol. The resulting products are FAME (fatty acid methyl esters), FAEE (fatty acid ethyl esters), or FABE (fatty acid butyl esters), respectively. The reaction is catalyzed by acids, alkalis, or lipase enzymes (Fukuda et al. 2001; Akoh et al., 2007).

Many different sources can be used as parent oil. Most common are vegetable oils. Primarily important is the oil yield per area, but climate and economics finally determine which vegetable oil is used. While the US rely on soybean oil (400-446 liter oil/ha (43-48 gallons per acre)) as major feedstock, European countries prefer rapeseed (canola) oil (1,100-1,190 liter oil/ha (118-127 gallons per acre)), and tropical countries palm oil (2,400-5,950 liter oil/ha (257-636 gallos per acre)). Sunflower oil (690 liter oil/ha (73 gallons per acre)), peanut/groundnut oil, cottonseed oil, castor oil, and safflower oil are also used commercially (Akoh et al., 2007; Chisti, 2007; Fairless, 2007).

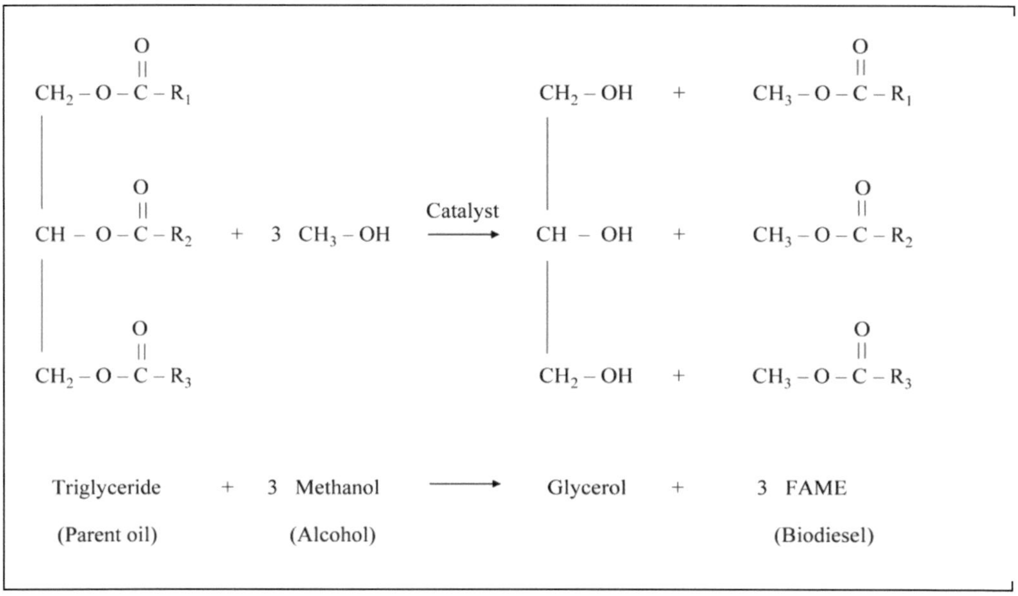


Fig. 2. Chemical reactions leading to biodiesel. R1-R3: saturated and/or unsaturated hydrocarbons of different chain length.

Moreover, the use of jatropha seed oil (1,300-1,892 liter oil/ha (139-202 gallons per acre)) seems to be very attractive because it does not compete with the food industry and the plant *Jatropha curcas* produces seeds containing up to 35 % oil and is resilient to pests and drought. Therefore, it can even grow in dry savanna. Hundreds of thousands hectares (respectively acres) are already in cultivation in South Asia, Africa, Middle and South America, and already in 2007, BP and D1 Oils launched a corresponding project (Fairless, 2007; BP, 2007). Jatropha-based biodiesel is also considered as aviation fuel (Air New Zealand, 2008). Other sources are animal fats (fish oils, blubber, lards, tallow, etc.) and even waste oils (frying oils, soapstocks, etc.).

However, microbial oils can also be used. Especially microalgal oils show great potential and are theoretically $CO_2$ neutral (like vegetable oils). Microalgae grow fast (biomass doubling time of 3.5 hours during exponential growth) and are rich in oil (up to 80 % weight of dry biomass). Photobioreactor experiments demonstrated an oil yield of 136,900 liter oil per hectar (14,635 gallons per acre) (with 70 % oil in biomass). Nevertheless, microalgal biomass production is generally considered more expensive than growing crops, despite having a higher energy yield per area (Schubert, 2006; Chisti, 2007; Chisti, 2008) . Another big challenge is the high demand for water and especially fertilizers. As result, algae biofuels had a worse environmental footprint than corn ethanol in a recent life cycle analysis (Clarens et al., 2010). However, algae oils have big potential as jet fuel. Companies trying to commerzialize algae biofuels include Algenol Biofuels (www.algenolbiofuels.com/), Aquaflow (www.aquaflowgroup.com/), Sapphire Energy, Inc. (www.sapphireenergy.com/), or Solazyme, Inc. (www.solazyme.com/).

In addition, oil production with some yeasts, fungi, or bacteria has also been evaluated (Ratledge, 1993). Metabolically engineered *E. coli* were not only shown to produce free fatty acids, but also biodiesel directly (up to 26 % FAEE of dry cellmass), referred to as microdiesel. The genes for the pyruvate decarboxylase (*pdc*) and alcohol dehydrogenase (*adhB*) of *Z. mobilis* have been introduced in *E. coli* to produce large amounts of ethanol. Additionally, the gene for the acyltransferase (*atfA*) of *Acinetobacter baylyi* strain ADP1 has been subcloned on the same plasmid in *E. coli*. The corresponding enzyme has an extraordinary low substrate specificity and is able to esterify ethanol with the acyl moieties of coenzyme A thioesters of fatty acids. However, supplementation of exogenic fatty acids (oleic acid) was necessary for a substantial FAEE yield, because the acyltransferase did not use the *de novo* synthesized fatty acids properly. Nevertheless, the feasibility of a new microbiological biodiesel production process was demonstrated and can now be further developed (Kalscheuer et al., 2006).

In general, biodiesel shows lower emissions of particulate matter and carbon monoxide (CO) than regular diesel fuel, but slightly enhanced nitrogen oxides ($NO_x$) production. It is also only moderately more mutagenic. However, rapeseed oil used directly as a fuel for a diesel engine showed a strong increase in mutagenicity as measured by the Ames test (Bünger et al., 2006; Bünger et al., 2007). In general, vegetable oil represents currently only a niche application. Although pure fuel costs are lower than those for diesel, modification of engines is required and the higher viscosity leads to cold start problems.

### 4.3 Renewable diesel

The expression "renewable diesel" became meanwhile standard in the US, while "hydrotreated vegetable oil" (HVO) is still more common in Europe. However, HVO is a term lacking precision, as except for vegetable oil also animal fat components such as lard and tallow are used for production. The triglycerides react with hydrogen at a catalyst, thereby forming propane (from the glycerol moiety), $CO_2$, $H_2O$, and hydrocarbons (from the fatty acids by splitting the ester bond and removal of the carboxy group) (Figure 3). Propane can be used as a fuel as well or as feedstock for the chemical industry. Gasoline (in fact: biogasoline; chain lengths from $C_4$ to $C_{12}$) is produced as a side product. The major hydrocarbons of a chain length between $C_{12}$ and $C_{20}$ (well in the diesel range of $C_{10}$ to $C_{25}$) are fully saturated, free of oxygen and aromatic compounds as well as low in sulfur content and producing less nitrous oxide upon burning, thus representing a superior bio-based diesel fuel, chemically equivalent to the crude oil-based product. Renewable diesel is about to enter the market in large quantities. ConocoPhillips (www.conocophillips.com/) started in 2006 to produce renewable diesel commercially, now reaching 150,000 liters (39,600 gallons) per day in its Whitegate refinery in Cork, Ireland (Mason & Ghonasgi, 2008). Neste Oil Corporation (www.nesteoil.com/) announced in May 2008 to produce an annual 170,000 tons of renewable synthetic diesel under the brand name NExBTL in its Provoo refinery, Finland. Feedstocks include palm oil, rapeseed oil, and animal fats (Oja, 2008) .

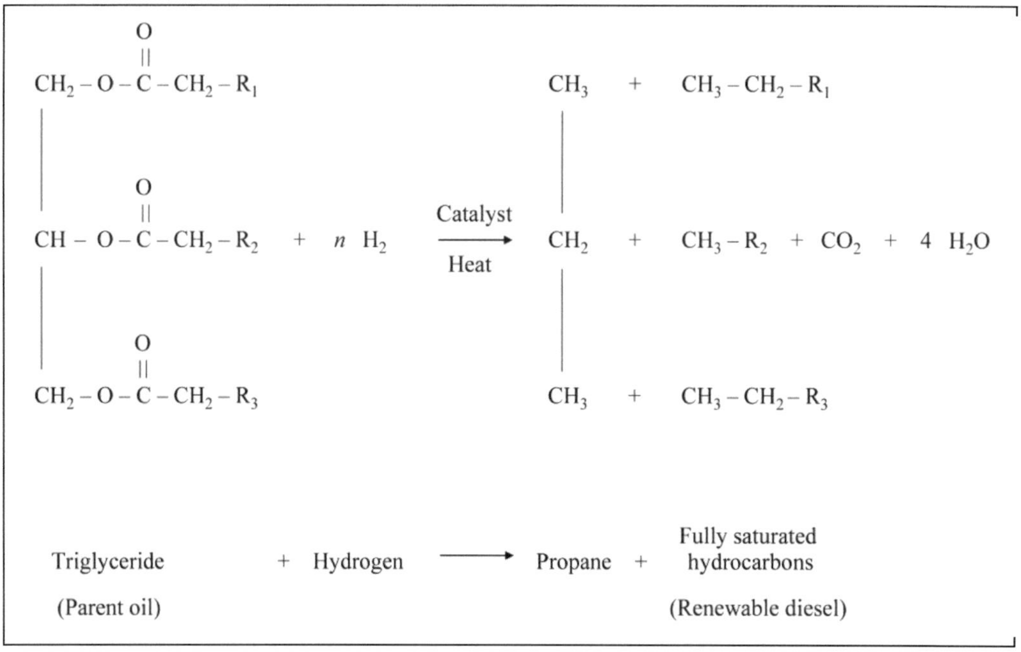


Fig. 3. Chemical reactions leading to renewable diesel. R1-R3: saturated and/or unsaturated hydrocarbons of different chain length.

## 5. The future: Second generation biofuels

### 5.1 Cellulosic biofuels

The use of biomass such as wood, dedicated energy crops, agricultural residues, and municipal solid waste would be a very attractive alternative. Production of up to 4,000 dry tons biomass per $km^2$ (1,545 tons per square mile) and year have been reported in field tries and an annual global biomass average of about 1,000 dry tons per $km^2$ (390 tons per square mile) is conservatively estimated (Ragauskas et al., 2006). The US can app. yield over 1 billion dry tons of biomass and continue to meet food, feed, and export demands (Perlack et al., 2005).

Whereas corn grain consist of starch and sugar cane of sucrose, biomass is composed of lignocellulose (typically 40-50 % cellulose, 25-35 % hemicellulose and 15-20 % lignin), which is the main component of the plant cell wall and therefore very resistant to degradation (Schubert, 2006; Gray et al., 2006). One approach towards degradation is based on thermo-chemical pretreatment and enzymatic hydrolysis of the lignocellulose into fermentable sugars and afterwards into so-called cellulose (or cellulosic) biofuels such as bioethanol.

However, the enzymatic hydrolysis reaction (most commonly with cellulases from the fungus *Trichoderma reesei*) is (still) too expensive (2.5-5 US cent per liter (10-20 US cents per gallon) of ethanol produced) and time-consuming (about 100-fold slower than the average fermentation rate with yeast) (Schubert, 2006; Gray et al., 2006). Many new enzymes from bacteria and fungi have been isolated and characterized in the last few years (Hildén & Johansson, 2004). Recently, the metagenome of the hindgut of the wood-feeding termite *Nasutitermes* has been analyzed with the aim to find novel cellulases (Warnecke et al., 2007). Moreover, well-known enzymes have either been engineered to improve their performance or produced heterologously in an existing system such as *T. reesei* (Warnecke et al., 2007; Viikari, 2007). Companies such as Codexis, Inc. (www.codexis.com/), Genencor® (www.genencor.com/), or Novozymes (www.novozymes.com/) are working on efficient enzyme solutions, while others such as ZeaChem, Inc. (www.zeachem.com/) try to optimize the thermochemical breakdown route.

Cellulose consists exclusively of glucose, hemicellulose contains a complex mixture of carbohydrates with 15-50 % pentoses such as xylose and arabinose (Schubert, 2006; Gray et al., 2006). *S. cerevisiae* as well as the bacterial work horse *Zymomonas mobilis* (which has even a higher ethanol yield on starch than *S. cerevisiae*) are naturally not able to ferment pentose sugars. Therefore, new strains of both organisms have been developed in the last few years by metabolic engineering with additional pentose metabolic pathways (Hahn-Hägerdal, 2007; Jeffries & Yin, 2003; Dien et al., 2003) . Furthermore, bacteria such as *E. coli* or *Klebsiella oxytoca* which can use a wide spectrum of sugars have been genetically modified towards ethanol production (Dien et al., 2003; Jarboe et al., 2007). Nevertheless, the industry is still relying on the yeast *S. cerevisiae* due to its robustness.

Already in 2004, Shell and Iogen (www.iogen.ca/) announced the successful production of cellulosic ethanol in one complete process for commercial use and are currently operating a demonstration plant with an output of 5,000 - 6,000 litres (1400-1600 gallons) of cellulosic ethanol per day (Iogen, 2011; Schubert, 2006). The world leading bioethanol producer POET, LLC is looking into production of cellulosic ethanol too.

Some microorganisms such as *Clostridium cellulolyticum, C. phytofermentans,* or *C. thermocellum* are also capable to ferment cellulose directly into ethanol, in so called consolidated bioprocessing process (CBP) (Lynd et al., 2005). Companies like Mascoma (www.mascoma.com/) and Qteros, Inc (www.qteros.com/) are working to improve the conversion rates to commercial requirements. Heterologous cellulase expression in *Z. mobilis* and *S. cerevisiae* was successfully shown as well, but yields are nonsatisfying so far and there are still some limitations (Lynd et al., 2002; Demain et al., 2005; Schubert, 2006). However, biomass can also be gasified and then converted into a biofuels.

### 5.2 Biomass-to-Liquid (BtL) biofuels

One approach to convert lignocellulosic biomass into a biofuel borrows a technology from the coal industry, the "Fischer-Tropsch process". This method was invented in the petroleum-poor but coal-rich Germany of the 1920s and describes the generation of synthesis gas (or syngas, consisting mainly of carbon monoxide and hydrogen) and its reaction into liquid alkanes, alkenes, and alcohols (Schubert, 2006). The desired product formation can be adjusted by process conditions and the use of appropriate catalysts (usually metals such as iron or cobalt).

While alcohols or diesel produced via such a "CtL" (coal-to-liquid) or "GtL" (natural gas-to-liquid) process is certainly not considered as biofuel, the syngas could also be generated from muncipal solid waste or biomass in a so-called "BtL" (biomass-to-liquid) process. Renewable diesel produced this way is known under the brand names SunFuel® or SunDiesel® (cooperation of Choren Industries, Shell, Volkswagen, and Daimler; www.sunfuel.de/). So far however, only pilot plants started to operate, by a series of companies such as Choren Industries (www.choren.com/), Enerkem (www.enerkem.com/), Fulcrum BioEnerg, Inc. (www.fulcrum-bioenergy.com/), or Rentech (www.rentechinc.com/). In 2008, Range Fuels (www.rangefuels.com/) started to built a first commercial plant with a US$ 76 million grant from the US Department of Energy. However,

instead a proposed goal of an annual 100 million gallon ethanol from wood chips, only 4 million gallons of methanol are produced (Energy Collective, 2011).
Besides huge problems with product specificity, sulfur gases and the accumulation of tar, which leads to consequential poisoning of the noble catalysts, are major issues for these processes. However, some acetogenic bacteria such as *Clostridium ljungdahlii* (Köpke et al., 2010) are also capable to ferment synthesis gas directly into bioethanol via the acetyl-CoA "Wood-Ljungdahl" pathway (Köpke et al., 2011; Tirado-Acevedo, 2010; Henstra, 2007). These bacteria are by far more specific and less affected by sulfur gases and tar (Ragauskas, 2006; Bredwell et al., 1999; Vega et al., 1990). Three companies, Coskata, Inc. (www.coskata.com/), IneosBio (www.ineosbio.com/), and LanzaTech NZ Ltd. (www.lanzatech.co.nz/), have already committed themselves to this technology and aiming to commercialize their process over the next few years. Coskata, Inc. is currently designing a 200 million liter (55 million gallon) per year facility in Greene County, Alabama with a US$ 250 million loan guarantee by the USDA (Coskata, 2011), IneosBio has started to construct a 30 million liter (8 million gallon) per year plant near Vero Beach, Florida (InesBio, 2011), and Lanzatech NZ Ltd. signed a commercial partnership deal towards the construction of a commercial plant at a steel mill in China (LanzaTech, 2010) with the aim to produce an annual 190 million liter (50 million gallons) bioethanol directly from steel mill off gases by 2013.

### 5.3 Methane from biogas

Methane or natural gas is already in use as a fuel for cars. 1 kg (2.2 pounds) of methane is equivalent to app. 1.4 liters (0.264 gallons) of gasoline. Thus, methane derived from biogas (mainly a mixture of $CH_4$ and $CO_2$, plus minor amounts of water, $H_2S$, and other gases) could be directly used without further modifications of the respective engines. However, purification of methane from biogas is required, and this process results in high costs. Also, due to the low percentage of cars currently running on methane, the existing infrastructure of distribution of gaseous fuels does not meet a large demand. Thus, methane from biogas is not expected to enter the market in relevant quantities within the next few years.

### 5.4 Hydrogen from biomass

Hydrogen can serve as a fuel in Otto-type combustion engines and in fuel cells. Thus, single solutions as well as so-called "hybrid cars" (combination of different engine types) will be suited to run on $H_2$. Prototypes of cars with fuel cells already exist from several manufacturers, but mass production is not expected before app. 2015. Several processes have been developed to generate $H_2$ from biomass and respective pilot plants are in operation in e.g. Austria, Germany, and the US (Schindler, 2008). In principle, biomass is converted into synthesis gas (a mixture of mostly CO and $H_2$), from which $H_2$ is further purified and compressed. Other ways of biohydrogen formation include photosynthetic production, fermentative production, and nitrogenase-mediated production (Levin et al., 2004; Nath & Das, 2004; Prince & Kheshgi, 2005). In general, the economics of the different systems are so far not sufficient to allow introduction of a large scale industrial process. A breakthrough at the laboratory level has been achieved recently: Highly efficient hydrogen production from any type of biodegradable organic matter became possible in electrohydrogenic reactors with mixed cultures of bacteria (Cheng & Logan, 2007).

### 5.5 Butanol (biobased-butanol, biobutanol)

While alcoholic fuels have almost the same properties as gasoline (Table 2), butanol provides a number of advantages over ethanol. Primarily, the energy density of butanol is significantly higher, resulting in an increased mileage. In addition, the air-fuel ratio of butanol is higher, which means that it can be run at richer mixtures and therefore produce more power. The octane rating of butanol is lower to that of ethanol, but similar to that of gasoline. Butanol has a lower vapor pressure and is thus safer to handle. Furthermore, the enthalpy of vaporization of ethanol is more than twice of that of butanol, which can lead to insufficient vaporization and cause problems during starts in cold weather. Additionally, ethanol is corrosive and hygroscopic. Especially, aluminum parts are attacked. Hence, ethanol cannot be distributed in pipelines and must be transported by tanker trucks, rail car, or river barge (IEA, 2004). Blending gasoline with ethanol at the production facility or refinery long before distribution is not feasible and must occur shortly before use (IEA, 2004), increasing the risk of a contamination of groundwater in case of spills. Butanol can be blended with gasoline well ahead of distribution and can be transported by the existing infrastructure. While ethanol can only be blended up to 85 % with gasoline, butanol can be blended in any concentration and used in existing car engines without any modification.

While butanol can also be produced by the Fischer-Tropsch process, the more economical synthesis route is by fermentation. This biotechnological procedure has a long-lasting history, as already described. The production organism usually used is *C. acetobutylicum*. During the last decades, intensive investigations led to a significant increase in our knowledge on physiology, metabolic regulation, and genetic manipulation of this organism, which in combination with improved fermentation and downstream processing will allow the revival of a large scale industrial bioprocess.

Next to the fermentative route, an alternative non-fermentative approach has been established by metabolic engineering of the amino acid biosynthesis pathways, which allows for production of 1-butanol and also iso-butanol. While most companies like Butalco GmbH (www.butalco.com/), ButylFuel, LLC (www.butanol.com/), Cobalt Technologies (www.cobalttech.com/), Green Biologics, Ltd. (www.greenbiologics.com), Metabolic Explorer (www.metabolic-explorer.com/), or Tetravitae Bioscience, Inc. (www.tetravitae.com/) are focusing on optimization of the classical ABE fermentation, Butamax™Advanced Biofuels LLC (www.butamax.com/) and Gevo (www.gevo.com/) are persuing the non-fermentative path, too.

#### 5.5.1 Fermentative production of butanol (ABE fermentation)

##### 5.5.1.1 Metabolism and enzymes of *Clostridium acetobutylicum*

*C. acetobutylicum* is a Gram-positive anaerobe organism. Its metabolism is characterized by a biphasic fermentation (Figure 4), starting with the formation of the acids acetate and butyrate. As a consequence of the accumulation of free acids and the resulting pH drop, the essential proton gradient between inside and outside of the cell gets destroyed and *C. acetobutylicum* dies. The strategy of *C. acetobutylicum* to survive is to decrease acid production at the end of exponential growth and to switch the metabolism to solvent production. Acetate and butyrate are taken up from medium into the cell and metabolized into acetone and butanol. By conversion of acids into solvents, the pH increases again. Butanol, however, is also toxic to *C. acetobutylicum* and poses a serious threat to the cells, having damaging effects on membranes and some membrane proteins. Therefore, the

bacteria start with the formation of endospores at the same time. Byproducts of the fermentation are acetoin, ethanol, and lactate. $CO_2$ and $H_2$ are produced, too.
The main carbon sources for *C. acetobutylicum* are starch and sugers. Respective degradative enzymes have been found, purified, and characterized (e. g. an α-amylase from strain ATCC 824) (Paquet et al., 1991; Annous & Blaschek, 1994). *C. acetobutylicum* is unable to feed on cellulose (Lee et al., 1985).
This is curious because the genome of *C. acetobutylicum* contains a number of genes, which code for cellulosome components (Nölling et al., 2001). By their overexpression in *E. coli* or *C. acetobutylicum* it could be shown that part of the respective proteins are functional (Lopez-Contreras et al., 2003; Sabathé & Soucaille, 2003). Phosphotransferase systems perform the uptake of many sugars. Glucose is degraded by glycolysis (Embden-Meyerhof-Parnas pathway) to pyruvate, which again is metabolized into acetyl-CoA by a pyruvate:ferredoxin-oxidoreductase (Nölling et al., 2001). This enzyme consists of an unknown number of 123 kDa subunits and is extremely oxygen-sensitive (Meinecke et al., 1989). Acetate is formed from acetyl-CoA. The latter one is also partly converted to butyrate.

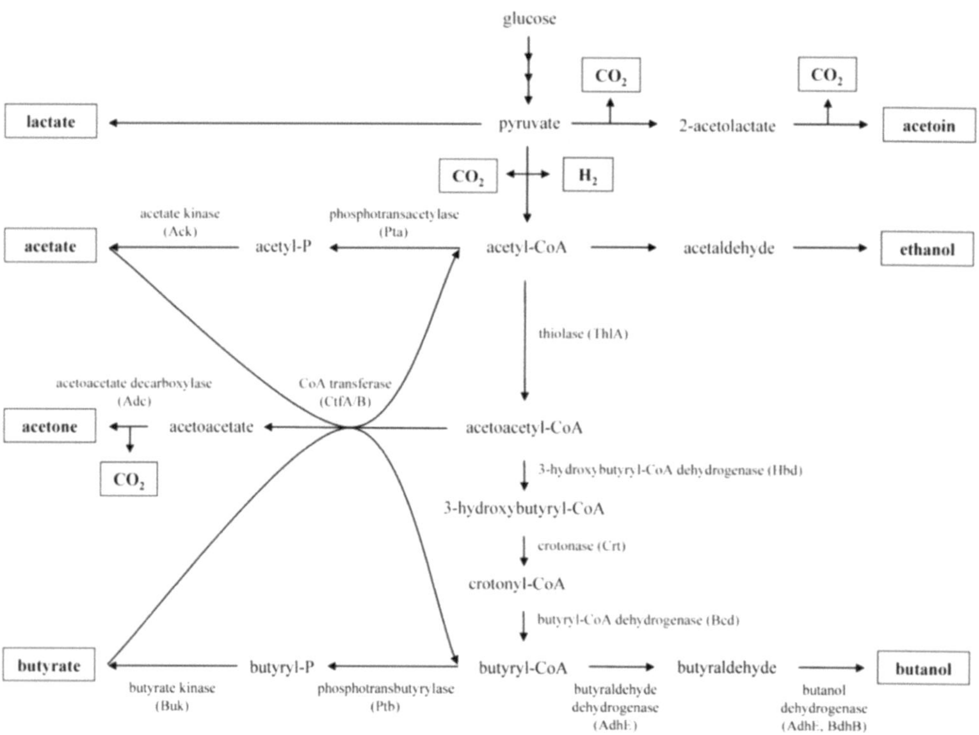


Fig. 4. Catabolic pathways of acid and solvent formation in *C. acetobutylicum*. The single reactions shown do not represent stoichiometric fermentation balances.

Phosphotransacetylase (Pta) and acetate kinase (Ack) (Gavard et al., 1957) are involved in acetate formation. Both are only strongly expressed during the acidogenic growth phase (Andersch et al., 1983). Pta was purified from *C. beijerinckii* and Ack from *C. acetobutylicum*. The former possesses a molecular mass of 56 to 57 kDa (Chen, 1993), the latter is a dimeric enzyme, consisting of identical subunits with a molecular mass of 43 kDa. The enzyme prefers the

substrates ATP and acetate (Winzer et al., 1997). Ack from *C. saccharobutylicum* P262 was also purified. The two subunits have each a molecular mass of 42 kDa (Diez-Gonzalez et al., 1997).
Butyrate formation starts with the condensation of two molecules acetyl-CoA to one molecule acetoacetyl-CoA by a thiolase (ThlA). Acetoacetyl-CoA is reduced to 3-hydroxybutyryl-CoA with NADH by a 3-hydroxybutyryl-CoA-dehydrogenase (Hbd). By dehydration via a crotonase (Crt or Cch), crotonyl-CoA is formed. Crotonyl-CoA is reduced by a butyryl-CoA-dehydrogenase (Bcd) to butyryl-CoA. The formation of butyrate from butyryl-CoA is mediated by two enzymes, a phosphotransbutyrylase (Ptb) and a butyrate kinase (Buk). The thiolase of *C. acetobutylicum* is a tetramer formed of identical subunits, each with a molecular mass of 44 kDa (Wiesenborn et al., 1988). *C. acetobutylicum* and *C. beijerinckii* possess a different gene for a second thiolase (Winzer et al., 2000). The genes for two different thiolases have also been found in *Clostridium pasteuricum* (Berndt & Schlegel, 1975). 3-Hydroxybutyryl-CoA dehydrogenase (Hbd) from *C. beijerinckii* NRRL B593 is a protein consisting of several identical subunits, each with a molecular mass of 31 kDa. Together they compose an enzyme with a total mass of 231 kDa (Colby & Chen, 1992). A crotonase (crotonyl-CoA hydratase) was isolated from a *C. acetobutylicum* strain. This strain was not further specified (Waterson et al., 1972). The purified protein was composed of four identical parts. Each of these subunits possesses a molecular mass of 40 kDa. A remarkable feature of this enzyme is its limited substrate specificity and sensitivity towards high concentrations of crotonyl-CoA. It was found that the enzyme only acts on C4- and C6-enoyl-CoA (Waterson et al., 1972). A butyryl-CoA dehydrogenase has not been purified yet. Phosphotransbutyrylase (Ptb) from both *C. acetobutylicum* strain ATCC 824 and from *C. beijerickii* has also been characterized. The molecular masses determined for the purified proteins were 264 kDa and 205 kDa, respectively. Ptb of *C. acetobutylicum* is composed of subunits with a molecular mass of 33 kDa. The enzyme of *C. beijerickii* is formed from subunits of 31 kDa (Waterson, 1972; Thompson & Chen, 1990). Characterization of a butyrate kinase purified from *C. acetobutylicum* ATCC 824 showed a very low activity with acetate (only 6 % of that with butyrate). That enzyme possesses two subunits of high similarity, each with a molecular mass of 39 kDa (Hartmanis, 1987).
With the beginning of solventogenesis, *C. acetobutylicum* takes up butyrate and acetate. Butyrate, and to a lesser extent acetate, are converted into butyryl-CoA and acetyl-CoA by an acetoacetyl-CoA:acetate/butyrate-coenzyme A transferase (CoA transferase, CtfA/B), while acetoacetyl-CoA is simultaneously converted to acetoacetate. Acetone is formed by the decarboxylation of acetoacetate mediated by an acetoacetate decarboxylase (Laursen & Westheimer, 1966). In some *C. beijerinckii* strains, a further reduction of acetone to 2-propanol is catalyzed by a primary/secondary alcohol dehydrogenase. Sequencing analyses of the *C. acetobutylicum* strain ATCC 824 determined a molecular mass of 22.7 and 23.7 kDa for the two different subunits of the CoA transferase. A molecular mass of 23.6 kDa for both subunits of the protein in DSM 792 (Cary et al., 1990; Gerischer & Dürre, 1990; Petersen et al., 1993; Fischer et al., 1993) was found. Acetoacetate decarboxylase of *C. acetobutylicum* consists of 12 identical subunits, each of them with a molecular mass of 28 kDa, forming an holoenzyme of 330 kDa, while the enzyme from *C. beijerinckii* is only 200-230 kDa (Gerischer & Dürre, 1990; Petersen & Bennett, 1990; Chen, 1993).
In *C. acetobutylicum*, butanol formation is initiated by a bifunctional butyraldehyde/butanol dehydrogenase E (AdhE). Just prior to butanol synthesis the transcription of the respective gene is induced (Sauer & Dürre, 1995; Grimmler et al., 2011). After inactivation of *adhE* significantly less butanol was produced by *C. acetobutylicum*. A solvent-negative mutant

regained the ability to produce butanol after its transformation with *adhE* (Nair & Papoutsakis, 1994; Green & Bennett, 1996). Additionally, a butanol dehydrogenase (BdhB or BdhII) is involved in butanol production (Fischer et al., 1993; Petersen et al., 1991; Nair et al., 1994) This enzyme forms a dimer, consisting of two identical subunits. A molecular mass of 42 kDa was determined for each of them. Investigations of the enzyme activity showed that it is 46-fold higher with butyraldehyde than with acetaldehyde (Welch et al., 1989). The second butanol dehydrogenase is BdhA. Its enzyme activity is only twice as high with butyraldehyde than with acetaldehyde (Walter et al., 1992). Like BdhB it forms a dimer with a subunit size of 42 kDa. Both enzymes were purified from *C. acetobutylicum* ATCC 824.

*C. acetobutylicum* also possesses a second *adhE* gene (*adhE2*), which has a 66 % identity to *adhE* (Fontaine et al., 2002). It is transcribed in continuous culture only under acidogenic conditions (Grimmler et al., 2011) or in alcohologenic cultures (only butanol and ethanol formation) when grown at neutral pH on glycerol (Fontaine et al., 2002).

#### 5.5.1.2 Genomic arrangement of elements required for sugar degradation, acid formation, and solventogenesis

The genome of *C. acetobutylicum* consists of a 3.94 Mbp chromosome and the megaplasmid pSOL1 (192 kbp) (Nölling et al., 2001). Phosphofructokinase (*pfk*) and pyruvate kinase (*pyk*) genes, whose products are involved in degradation of glucose, are arranged in one operon (Belouski et al., 1998) Genes for glyceraldehyde-3-phospate dehydrogenase (*gap*), phosphoglycerate kinase (*pgk*), and triosephosphate isomerase (*tpi*) are grouped together in this order in a common operon. Although *gaf, pgk* and *tpi* are transcribed together, a separate transcript of *tpi* was also found (Schreiber & Dürre, 1999). The transcription start points are located in front of *gap* and *tpi*. Phosphotransacetylase (*pta*) and acetate kinase (*ack*) genes are clustered in one operon, with *ack* downstream of *pta* (Boynton, 1996). Thiolase (*thlA*) forms a monocistronic operon. Transcription of *thlA* is initiated at a typical $\sigma^A$-dependent promoter (Winzer et al., 2000; Stim-Herndon, 1995). An operon consisting of, at least, *thlR-thlB-thlC* includes the gene of a second thiolase (*thlB*), which shows a different transcription pattern in continuous culture than *thlA* (Grimmler et al., 2011). A vegetative sigma-factor dependent promoter seems to be responsible for transcription of *thlB* (Winzer et al., 2000). Crotonase (*crt*), butyryl-CoA dehydrogenase (*bcd*), and 3-hydroxybutyryl-CoA dehydrogenase (*hbd*) form a cluster together with *etfB* and *etfA* (gene products with homology to electron transfer flavoproteins). Except for a putative promoter upstream of *crt*, no further transcription start points were found upstream of the start codons of all these genes. This indicates that *crt-bcd-etfB-etfA-hbd* form one transcription unit, called the *bcs* operon (butyryl-CoA-synthesis) (Boynton et al., 1996). Phosphotransbutyrylase (*ptb*) and butyrate kinase (*buk*) are located in a common operon on the chromosome of *C. acetobutylicum* (Cary et al., 1988; Walter et al., 1993). The respective *ptb* promoter, located 57 bp upstream of *ptb*, initiates transcription of this operon during the acidogenic phase (Tummala et al., 1999; Feustel et al., 2004). Another butyrate kinase gene is also present in *C. acetobutylicum* (Huang et al., 2000).

The gene of the solventogenic enzyme acetoacetate decarboxylase (*adc*) is arranged in a monocistronic operon and is controlled by a $\sigma^A$-dependent promoter. Compared to the other promoters, which regulate transcription of solventogenic enzymes, the *adc* promoter allows highest expression (Feustel et al., 2004; Gerischer & Dürre 1992). The terminator of this operon is formed by a 28-bp stem-loop, 6 bp downstream of an UAA stop codon (Petersen et al., 1993; Gerischer & Dürre, 1992). This UAA is followed by another UAA stop codon. The terminator of *adc* is located next to the *sol* operon which is transcribed by the RNA polymerase in the reverse direction of *adc*. It also terminates transcription of the *sol* operon and thus functions bidirectionally (Figure 5).

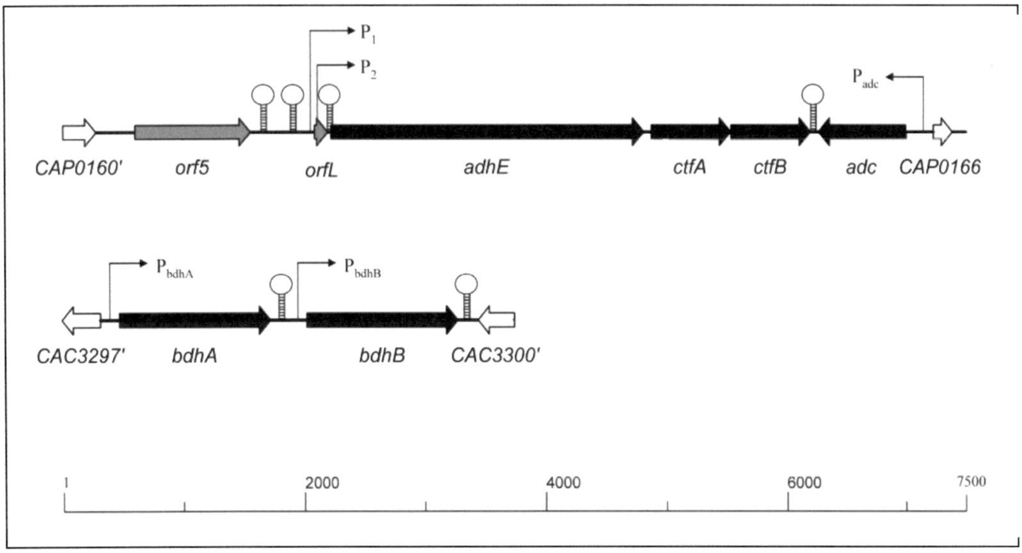


Fig. 5. Organization of megaplasmid and chromosomal gene regions encoding solventogenic enzymes in *C. acetobutylicum*. Promoter positions are indicated by $P_{adc}$, $P_1$, $P_{bdhA}$, and $P_{bdhB}$. $P_2$ represents a mRNA-processing site. Possible stem-loop structures are indicated by hairpin symbols (Thormann et al., 2002).

A small peptide of unknown function (*orfL*), a butyraldehyde/butanol dehydrogenase (*adhE* or *aad*) and the two subunits of the CoA transferase (*ctfA* and *ctfB*) are encoded by this operon in the order *orfL-adhE-ctfA-ctfB* (Petersen et al., 1993; Fischer et al., 1993; Nair et al., 1994). The *sol* operon of other solventogenic clostridia (*C. beijerinckii*, *C. saccharobutylicum*, and *C. saccharoperbutylacetonicum*) includes the *adc* gene and contains an aldehyde dehydrogenase gene (*ald*) instead of *adhE* (Chen & Blaschek, 1999; Kosaka et al., 2007). The product of *adhE* is multifunctional. It has alcohol dehydrogenase activity (C-terminus) as well as aldehyde dehydrogenase activity (N-terminus), like the ethanol-forming *E. coli* enzyme. The latter one acts moreover as pyruvate:formate-lyase deactivase (Goodlove et al., 1989; Kessler et al., 1991).

*adhE2* forms a monocistronic operon. Two promoters $S_1$ and $S_2$ are deduced (Fontaine et al., 2002), but only the distal one ($S_2$) shows a convincing homology to $\sigma^A$-dependent control regions (Dürre, 2004).

Some of the genes responsible for solventogenesis are carried on the megaplasmid pSOL1 (192 kbp) (*adc*, *sol*, and *adhE2*) (Nölling et al., 2001; Cornillot et al., 1997). The genes *bdhA* and *bdhB* for two butanol dehydrogenases have been identified on the chromosome. They form two consecutively located monocistronic operons. Their transcription is induced by $\sigma^A$ dependent promoters and stopped by rho-independent terminators (Petersen et al., 1991; Walter et al., 1992). As already mentioned, the genes for CoA transferase and acetoacetate decarboxylase are arranged in two different operons. The physiological conditions at the onset of solventogenesis might explain the separation of the genes for enzymes, which both catalyze acetate formation. At that time, large amounts of acids are accumulated. To prevent a collapse of the proton gradient across the cytoplasmatic membrane and cell death of the organism, the acids have to be disposed. After their uptake by *C. acetobutylicum*, a CoA transferase catalyzes formation of mainly butyryl-CoA and some acetyl-CoA (Dürre et al., 1995). The butyraldehyde/butanol dehydrogenase E subsequently mediates the conversion of butyryl-CoA into butanol. For this reason, the genetic information for *ctfA/B* and *adhE* is organized in a common transcription unit, the *sol* operon (Figure 5). Before butanol is formed, butyrate must be activated by its transformation into butyryl-CoA. Decarboxylation of acetoacetate is only needed to drive this initial reaction for thermodynamic reasons. Therefore, the *adc* gene is organized as a monocistronic operon.

**5.5.1.3 Control mechanisms**

At the beginning of solventogenesis in *C. acetobutylicum*, the enzymes required for solvent formation are induced or derepressed and the activity of some acidogenic enzymes is decreased (Andersch et al., 1983; Dürre et al., 1987; Hartmanis & Gatenbeck, 1984; Yan et al., 1988). Transcription of the respective genes for solventogenic enzymes starts several hours before solvents are being produced (Grimmler et al., 2011). *adc* is transcribed already with beginning of the exponential growth. Transcription is increased to its maximum in the stationary phase and slowed down afterwards (Sauer & Dürre, 1995; Gerischer & Dürre, 1992). The phosphorylated form of Spo0A initiates transcription of genes responsible for endospore formation and also solvent formation (Dürre & Hollergschwandner, 2004). There are Spo0A~P binding sites upstream of the *adc* and *sol* promoters of *C. acetobutylicum*. DNA-binding studies clearly revealed the participation of the phosphorylated transcription factor in regulation of solvent formation (Ravagnani et al., 2000). Spo0A inactivation reduced expression of genes responsible for solventogenesis (Harris et al., 2002). However, a complete removal of all binding motifs (0A boxes) for Spo0A upstream of the *adc* promoter caused reduction of the transcription of solventogenic enzymes, but did not abolish their expression completely. Consequently, an additional transcription factor must be involved in *adc* regulation (Böhringer, 2002). A surplus of carbon source, a pH below 4.3, limiting phosphate or sulphate concentrations, high concentrations of acetate and butyrate, and a higher temperature stimulate solvent production (Dürre & Bahl 1996; Dürre, 1998; Dürre et al., 2002; Bahl, 1983). Although signals triggering onset of solventogenesis are still unknown, all of the above mentioned factors change the topology of DNA (degree of DNA supercoiling), which influences the binding of regulatory proteins. DNA isolated from *C. acetobutylicum* during solventogenesis is more relaxed than the one extracted during acidogenesis (Wong & Bennett, 1996). Experiments with added novobiocin showed a dramatic increase of *adc* and *sol* transcription, thus supporting the essential influence of DNA topology (Ullmann, 1996; Ullmann & Dürre, 1998).
The induction of the *sol* operon takes place far before the bulk production of solvents (Sauer & Dürre, 1995; Feustel et al., 2004) . The promoter $P_1$ is responsible for the transcription of *sol* (Thormann et al., 2002). *bdhB* is transcribed late during exponential growth when *sol* operon expression is already diminished. The gene product of *bdhB* thus is responsible for most of the butanol production (Sauer & Dürre, 1995). In accordance, upstream of the *bdhB* promoter a 0A box (5´-TGTAGAA) was found (Ravagnani et al., 2000). BdhA, an alcohol dehydrogenase, seems to be responsible for the removal of reducing equivalents (electron sink). This explains the constitutive expression of the respective gene under certain conditions (Sauer & Dürre, 1995).

The onset of solventogenesis is accompanied with the induction of other operons. The enzymes encoded by the *serCAXS* operon are needed for serine biosynthesis. Together with solvent formation, the induction of heat shock proteins such as DnaK, GroEL, and Hsp18 (Sauer & Dürre, 1995; Schaffer et al., 2002; Terracciano et al., 1988; Pich et al., 1990; Sauer & Dürre, 1993; Bahl et al., 1995) and a putative stress protein, PdxY, takes place (Schaffer et al., 2002). By microarray analysis, a lot of pioneering studies with *C. acetobutylicum* have been performed, leading to a better understanding of the regulation of solvent production and sporulation (Tomas et al., 2003a, 2003b; Alsaker et al., 2004; Tummala et al., 2003; Tomas et al., 2004; Borden & Papoutsakis, 2007; Alsaker & Papoutsakis, 2005; Jones et al., 2008; Grimmler et al., 2011).

#### 5.5.1.4 Improvement of the biological production of butanol

The historical fermentative production of butanol was economically inferior to the petrochemically synthesis. However, there is meanwhile a lot of potential to increase the competitiveness of the biotechnological process. Different strategies of metabolic engineering and process optimization offer a way to improve solvent productivity, solvent specifity, butanol tolerance, and substrate utilization:

- Butanol is fermented in a desired production organism such as *E. coli* or yeast. This requires the introduction of the genes for butanol synthesis into the respective organism. *E. coli* or yeast is a suitable host for the production of valuable metabolites as it is easy to manipulate and handle (Farmer & Liao, 2000; Martin et al., 2003; Causey et al., 2004; Kim et al., 2007). Engineering of a synthetic pathway for 1-butanol production was already successfully demonstrated by numerous groups in both organimsms with genes from *C. acetobutylicum* or *C. beijerinckii* (Atsumi et al., 2008a; Inui et al., 2007; Donaldson et al., 2007b; Liao et al., 2008; Gunawardena et al., 2008, Buelter et al., 2008; Steen et al., 2008; Nielsen et al., 2009). While these first attempts resulted only in low butanol titers up to 16.2 mM, a recent study demonstrates butanol production of 30 g/l with a recombinant *E. coli* strain JCL166 (Δ*adhE*, Δ*ldhA*, Δ*frd*) (Shen et al., 2011), which is comparable to and even exceeds the native producer *C. acetobutylicum*. A very promising result was also the successful production of butanol directly from synthesis gas by a metabolically engineered *Clostridium ljungdahlii* strain (Köpke et al., 2010). Although the butanol yield was very low, it shows the potential of sustainable producing a superior biofuel such as butanol from an abundent non-food source.
- Another possibility is to modify *C. acetobutylicum* in a way to achieve higher butanol rates or eliminate undesired byproducts, creating a homobutanol producer (forming only butanol and some $CO_2$ and $H_2$). With a growing number of genetic tools for *Clostridium* becoming available, some remarkable progress has been made with recombinant strains. The production of 238 mM butanol found with a *orf5*-negative strain overexpressing *adhE* represent the highest value ever reported (Harris et al., 2001). Inactivation of acetate and butyrate production of *C. acetobutylicum* can be achieved by mutations in the phosphotransacetylase and phosphotransbutyrylase genes or the corresponding kinase genes, respectively. Harris et al. inactivated the butyrate kinase gene (*buk*) (Harris et al., 2000). This led to solvent superpoduction of 76 mM acetone and 225 mM butanol, when fermentation was carried out at pH 5 or below. It was the first time that the assumed barrier of biological solvent production of 200 mM was exceeded. In addition to *buk* inactivation, the *adhE* gene was overexpressed. This

mutant produced 66 mM acetone and 226 mM butanol. It could be shown that much more solvents are produced by *C. acetobutylicum* after a plasmid containing the *adc, ctfA* and *ctfB* genes were transformed into the bacterium. However, the control plasmid also caused a slight stimulation of the solvent production (Mermelstein et al., 1993). Another possibility of metabolic engineering is the improvement of the solvent tolerance of *C. acetobutylicum* (Tomas et al., 2003b; Tomas et al., 2004; Borden & Papoutsakis, 2007) Butanol tolerance and thus solvent production could be increased by overexpression of the chaperone-encoding *groESL* genes (Tomas et al., 2003b) By overexpression of the cyclopropane fatty acid synthase gene (*cfa*) the lipid composition of the membrane is altered, resulting in an increased butanol resistance. The disadvantage of this method is a significant lower butanol production (Zhao et al., 2003). Further improvements are possible by inactivation of genes leading to acetate, acetoin, acetone, ethanol, and lactate formation.

- Downstream processing is another way to improve economics of butanol fermentation. Distillation of butanol from the fermentation broth is very energy consuming. Alternative recovery methods might be better suited (Dürre, 1998; Santangelo & Dürre, 1996; Ezeji et al., 2004; Ezeji et al., 2007). With regard to the energy consumption, adsorption with molecular sieves (silicate) is much more efficient than gas stripping and pervaporation (Qureshi et al., 2005). During gas stripping, the products are eliminated from the fermentation media and then concentrated by condensation. The advantages of this method are that the microorganisms are not disturbed by the gases as well as the continuous working flow (Ezeji et al., 2004). During pervaporation, the product diffuses selectively across a membrane. Disadvantages of this method are the low selectivity and the incomplete removal of solvents from the fermentation broth. Moreover, the membranes are expensive. Another problem is fouling and clogging of the membranes. During liquid-liquid extraction the desired product is separated from the growth medium by mixing (and following dissolving) in a solvent. This method is only applicable with a solvent nontoxic to the bacteria. During perstraction, a membrane separates culture and extracting solvents, but it suffers from the same problems as pervaporation. Thus, a lot of recovery methods are available that make a more economic butanol purification possible.
- Over the last decades, several alternative carbon sources were evaluated as well. Processes with feedstocks such as apple pomace (fructose, glucose, sucrose) (Voget et al., 1985), whey (lactose) (Maddox et al., 1994), and lignocellulose (xylan and cellulose) (Maddox & Murray, 1983; Yu & Saddler, 1983; Yu et al., 1985; Fond et al., 1983) were developed. The butanol/acetone ratio after fermentation of whey (e.g. 100:1) is superior to that found with starch or molasses (2:1) (Bahl et al., 1986). An increased butanol content simplifies the product recovery process. The disadvantage of whey is its poor nutrient content. Consequently, a much lower productivity is found in comparison to molasses as substrate (Maddox, 1980; Welsh & Veliky, 1984; Ennis & Maddox, 1985; Linden et al., 1986). A fluidized bed reactor of bonechar-immobilized cells was used to improve fermentation with whey. The solvents were removed and concentrated by pervaporation (Maddox et al., 1994). Friedl et al. suggested using immobilized cells of *C. acetobutylicum* to optimize acetone-butanol fermentation (Friedl et al., 1991). The product was removed by pervaporation. With a lactose concentration of 380 mM in the feed solution, a stable high solvent productivity of 47 mM $h^{-1}$ was obtained. Lignocellulose as such cannot be used for fermentation, but pretreatment will release

hexoses and pentoses. Suitable pretreatments of lignocellulose are the Iogen process and methods that have been developed by Green Biologics (Abingdon, Qxfordshire, UK) (Green Biologics, 2007), and Green Sugar GmbH (Dresden, Germany) (Green Sugar GmbH, 2007).

#### 5.5.2 Non-fermentative butanol production

Alternative routes to butanol have also been inspected. By expression of a 2-ketoacid decarboxylase Kivd and an alcohol dehydrogenase Adh from *Saccharomyces cerevisiae* in *E. coli*, 1-butanol and also iso-butanol (along with some other alcohols) could be produced non-fermentatively from intermediates of the amino acid biosynthesis (Atsumi et al., 2008b; Hawkins et al., 2009; Donaldson et al. 2006 & 2007a). Both, Butamax™Advanced Biofuels LLC and Gevo, Inc. developed a respective technology, now having a lawsuit for patent infringement after Butamax's patent was granted (Butamax, 2011b).

### 5.6 Other biofuels

With the fast progress of synthetic biology (Peralta-Yahya & Keasling, 2010), companies such as Amyris, Inc. (www.amyrisbiotech.com/), Codexis, Inc. (www.codexis.com/),LS9, Inc. (www.ls9.com/), or OPX Biotechnologies, Inc. (www.opxbiotechnologies.com/) are trying to develop a range of new biofuels with even superior properties, identical to those of gasoline or suitable as jet fuel.

## 6. Conclusion

Since fossil sources are limited and burning of these fuels leads to massive increase of the greenhouse gas $CO_2$ in the atmosphere, microbial production of biofuels became important again. First generation biofuels, however, have major drawbacks, as they compete with food industry or have unfavourable properties. Several second generation biofuels have been developed over the last few years and are on the way of commercialization, but need to be proven at scale. Biobutanol is one of the most promising second generation biofuels, providing a lot of advantages over bioethanol and has already been successfully used at large scale over decades. The acetone-butanol-ethanol fermentation looks back to a nearly 100 year old history and has already been used industrially at the beginning of the 19th century. Although a lot of fermentation plants were closed after the World War II, research regarding physiology, biochemistry, and genetics of *C. acetobutylicum* was continued. On the basis of these findings, the biological efficiency of solvent production is constantly been improved by metabolic engineering, downstream processing, and alternative substrates as surrogate for sugar. Several plants in China and Brazil are already operating again, and global players such as BP and DuPont dedicated themselves to production of biobutanol. Recently, metabolic engineering efforts demonstrated butanol production in high yields with *E. coli*, or from an alternative carbon sources such as syngas with *C. ljungdahlii*.

## 7. Acknowledgements

Work was supported by grants from the BMBF GenoMik and GenoMikPlus projects (Competence Network Göttingen), and the SysMO projects COSMIC (PtJ-BIO/SysMO/P-D-01-06-13) and COSMIC2 (0315782A), www.sysmo.net.

## 8. References


Afschar, A. S.; Van Rossell, C. E. & Schaller, K. (1990). Bacterial conversion of molasses to acetone and butanol. *Appl Microbiol Biotechnol*, Vol.34, pp. 168-171

Akoh, C. C.; Chang, S.-W., Lee, G.-C. & Shaw, J.-F. (2007). Enzymatic approach to biodiesel production. *J Agric Food Chem*, Vol.55, No.22, pp. 8995-9005

Air New Zealand (2008) Biofuel test flight. Available from http://www.airnewzealand.co.nz/biofuel-test

Alsaker, K. V. & Papoutsakis, E. T. (2005). Transcriptional program of early sporulation and stationary-phase events in *Clostridium acetobutylicum*. *J Bacteriol*, Vol.187, pp. 7103-7118

Alsaker, K. V.; Spitzer, T. R. & Papoutsakis, E. T. (2004). Transcriptional analysis of *spo0A* overexpression in *Clostridium acetobutylicum* and its effect on the cell's response to butanol stress. *J Bacteriol*, Vol.186, pp. 1959-1971

Andersch, W.; Bahl, H. & Gottschalk, G. (1983). Level of enzymes involved in acetate, butyrate, acetone and butanol formation by *Clostridium acetobutylicum*. *Appl Microbiol Biotechnol*, Vol.18, pp. 327-332

Annous, B. A. & Blaschek, H. P. (1994). Regulation of hydrogen metabolism in *Butyribacterium methylotrophicum* by substrate and pH. *J Ind Microbiol*, Vol.13, pp. 10-16

Atsumi, S.; Cann, A. F.; Connor, M. R.; Shen, C. R.; Smith, K. M.; Brynildsen, M. P.; Chou, K. J.; Hanai, T. & Liao, J. C. (2008a). Metabolic engineering of *Escherichia coli* for 1-butanol production. *Metab Eng*, Vol.10, pp. 305-311

Atsumi, S.; Hanai, T. & Liao, J. C. (2008b). Non-fermentative pathways for synthesis of branched-chain higher alcohols as biofuels. *Nature*, Vol.451, pp. 86-89

Bahl, H. (1983). Kontinuierliche Aceton-Butanol-Gärung durch *Clostridium acetobutylicum*, PhD Thesis, University of Göttingen, Germany

Bahl, H.; Gottwald, M.; Kuhn, A.; Rale, V.; Andersch, W. & Gottschalk, G. (1986). Nutritional factors affecting the ratio of solvents produced by *Clostridium acetobutylicum*. *Appl Environ Microbiol*, Vol.52, pp. 169-172

Bahl, H.; Müller, H.; Behrens, S.; Joseph, H. & Narberhaus, F. (1995). Expression of heat shock genes in *Clostridium acetobutylicum*. *FEMS Microbiol Rev*, Vol.17, pp. 341-348

Belouski, E.; Watson, D. E. & Bennett, G. N. (1998). Cloning, sequence, and expression of the phosphofructokinase gene of *Clostridium acetobutylicum* ATCC 824 in *Escherichia coli*. *Curr Microbiol*, Vol.37, pp. 17-22

Berndt, H. & Schlegel, H. G. (1975). Kinetics and properties of β-ketothiolase from *Clostridium pasteurianum*. *Arch Microbiol*, Vol.103, pp. 21-30

Biofuels platform (2010). Bioethanol production in the EU. Available from http://www.biofuels-platform.ch/en/infos/eu-bioethanol.php

Böhringer, M. (2002). Molekularbiologische und enzymatische Untersuchungen zur Regulation des Genes der Acetacetat-Decarboxylase von *Clostridium acetobutylicum*, PhD Thesis, University of Ulm, Germany

Borden, J. R. & Papoutsakis, E. T. (2007). Dynamics of genomic-library enrichment and identification of solvent tolerance genes for *Clostridium acetobutylicum*. *Appl Environ Microbiol*, Vol.73, pp. 3061-3068

Boynton, Z. L.; Bennett, G. N. & Rudolph, F. B. (1996). Cloning, sequencing, and expression of clustered genes encoding β-hydroxybutyryl-coenzyme A (CoA) dehydrogenase, crotonase, and butyryl-CoA dehydrogenase from *Clostridium acetobutylicum* ATCC 824. *Appl Environ Microbiol*, Vol.62, pp. 2758-2766

Boynton, Z. L.; Bennett, G. N. & Rudolph, F. B. (1996). Cloning, sequencing, and expression of genes encoding phosphotransacetylase and acetate kinase from *Clostridium acetobutylicum* ATCC 824. *J Bacteriol*, Vol.178, pp. 3015-3024

BP (2007). Jatropha fact sheet. Available from http://www.bp.com/liveassets/bp_internet/globalbp/STAGING/global_assets/downloads/B/Biofuels_Factsheet_Jatropha.pdf

Bredwell, M. D.; Srivastava, P. & Worden, R. M. (1999). Reactor design issues for synthesis-gas fermentations. *Biotechnol Prog*, Vol.15, No.5, pp. 834-844

Buelter, T; Hawkins, A.; Kersh, K.; Meinhold, P.; Peters, M.; Subbian, E. (2008). International patent WO 2008/143704

Bünger, J.; Krahl, J.; Munack, A.; Ruschel, Y.; Schröder, O.; Emmert, B.; Westphal, G.; Müller, M.; Hallier, E. & Brüning, T. (2007). Strong mutagenic effects of diesel engine emissions using vegetable oil as fuel. *Arch Toxicol*, Vol. 81, pp. 599-603

Bünger, J.; Krahl, J.; Weigel, A.; Schröder, O.; Brüning, T.; Müller, M.; Hallier, E. & Westphal, G. (2006). Influence of fuel properties, nitrogen oxides, and exhaust treatment by an oxidation catalytic converter on the mutagenicity of diesel engine emissions. *Arch Toxicol*, Vol.80, pp. 540-546

Butamax (2011a). Biofuels factsheet. Available from http://www.butamax.com/_assets/pdf/butamax_advanced_biofuels_llc_fact_sheet.pdf

Butamax (2011b). Butamax™ Files Patent Infringement Action Against Gevo, Inc. to Protect Biobutanol Intellectual Property. Available from http://www.butamax.com/_assets/pdf/butamax_press_release_140111.pdf

Cary, J. W.; Petersen, D. J.; Papoutsakis, E. T. & Bennett, G. N. (1990). Cloning and expression of *Clostridium acetobutylicum* ATCC 824 acetoacetyl-coenzyme A:acetate/butyrate:coenzyme A-transferase in *Escherichia coli*. *Appl Environ Microbiol*, Vol.56, pp. 1576-1583

Cary, J. W.; Petersen, D. J.; Papoutsakis, E. T. & Bennett, G.N. (1988). Cloning and expression of *Clostridium acetobutylicum* phosphotransbutyrylase and butyrate kinase genes in *Escherichia coli*. *J Bacteriol*, Vol.170, pp. 4613-4618

Causey, T. B.; Shanmugam, K. T., Yomano, L. P. & Ingram, L. O. (2004). Engineering *Escherichia coli* for efficient conversion of glucose to pyruvate. *Proc Natl Acad Sci USA*, Vol.101, pp. 2235-2240

Chavanne, G. (1937). Belgian patent 422,877

Chavanne, G. (1943). Sur un mode d'utilisation possible de l'huile de palme à la fabrication d'un carburant lourd. *Bull Soc Chim*, Vol.10, pp. 52-58

Chen, C.-K. & Blaschek, H. P. (1999). Effect of acetate on molecular and physiological aspects of *Clostridium beijerinckii* NCIMB 8052 solvent production and strain degeneration. *Appl Environ Microbiol*, Vol.56, pp. 499-505

Chen, J.-S. (1993). Properties of acid- and solvent-forming enzymes of clostridia. In: The Clostridia and Biotechnology, D.R.Woods (Ed.) pp. 51-76,, Butterworth-Heinemann: Stoneham, MA, USA

Cheng, S. & Logan, B. E. (2007). Sustainable and efficient biohydrogen production via electrohydrogenesis. *Proc Natl Acad Sci U S A*, Vol.104, pp. 18871-18873

Chiao, J.-S. & Sun, Z.-H. (2007). History of the acetone-butanol-ethanol fermentation industry in China: development of continuous production technology. *J Mol Microbiol Biotechnol*, Vol.13, pp. 12-14

Chisti, Y. (2007). Biodiesel from microalgae. *Biotechnol Adv*, Vol.25, pp. 294-306

Chisti, Y. (2008). Biodiesel from microalgae beats bioethanol. *Trends Biotechnol*, Vol.26, pp. 126-131

Clarens, A. F.; Resurreccion, E. P.; White, M. A. & Colosi, L. M. (2010). Environmental life cycle comparison of algae to other bioenergy feedstocks. *Environ Sci Technol*, Vol.44, pp. 1813-1819

Cobalt Technologies (2010) Cobalt Technologies Launches Biobutanol Pilot Facility in California. Available from http://www.cobalttech.com/news/news-item/cobalt-technologies-launches-biobutanol-pilot-facility-in-california/

Colby, G. D. & Chen, J.-S. (1992). Purification and properties of 3-hydroxybutyryl-coenzyme A dehydrogenase from *C. beijerinckii* NRRL B593. *Appl Environ Microbiol*, Vol.58, pp. 3297-3302

Cornillot, E.; Nair, R. V.; Papoutsakis, E. T. & Soucaille, P. (1997). The genes for butanol and acetone formation in *C. acetobutylicum* ATCC 824 reside on a large plasmid whose loss leads to degeneration of the strain. *J Bacteriol*, Vol.179, pp. 5442-5447

Coskata (2011) Coskata supported biorefinery selected to receive $250 million loan guarantee from USDA. Available from http://www.coskata.com/media/?story=1039F848-CD91-4401-9511-B507D6A02DEE

Demain, A. L.; Newcomb, M. & Wu, J. H. (2005). Cellulase, clostridia, and ethanol. *Microbiol Mol Biol Rev*, Vol.69, pp. 124-154

Department of Energy (2010). Obama announces steps to boost biofuels, clean coal. Available from http://www.energy.gov/news/8596.htm

Dien, B. S.; Cotta, M. A. & Jeffries, T. W. (2003). Bacteria engineered for fuel ethanol production: current status. *Appl Microbiol Biotechnol*, Vol.63, pp. 258-266

Diesel, R. (1912a). The Diesel oil-engine. *Engineering*, Vol.93, pp. 395-406

Diesel, R. (1912b). The Diesel oil-engine and its industrial importance particularly for Great Britain. *Proc Inst Mech Eng*, pp. 179-280

Diez-Gonzalez, F.; Russell, J. B. & Hunter, J. B. (1997). NAD-independent lactate and butyryl-CoA dehydrogenases of *Clostridium acetobutylicum* P262. *Arch Microbiol*, Vol.166, pp. 418-420

Donaldson, G. K.; Eliot, A. C.; Flint, D.; Maggio-Hall, L. A. & Nagarajan, V. (2006) United States Patent WO 2007/050671

Donaldson, G. K.; Eliot, A. C.; Flint, D.; Maggio-Hall, L. A. & Nagarajan, V. (2007a) International Patent US 7851188

Donaldson, G. K.; Huang L. L., Maggio-Hall L. A., Nagarajan V., Nakamura C. E. & Suh W. (2007b). International Patent.WO 2007/041269

Dürre, P. & Bahl, H. (1996). Microbial production of acetone/butanol/isopropanol. In: Biotechnology, M. Roehr, (Ed.), VCH Verlagsgesellschaft: Weinheim, Vol.6, 2nd Ed., pp. 229-268

Dürre, P. & Hollergschwandner, C. (2004) Initiation of endospore formation in *Clostridium acetobutylicum*. *Anaerobe*, Vol.10, pp. 69-74

Dürre, P. (1998). New insights and novel developments in clostridial acetone/butanol/isopropanol fermentation. *Appl Microbiol Biotechnol*, Vol.49, pp. 639-648

Dürre, P. (2001). From Pandora's Box to Cornucopia: Clostridia – A historical perspective. In: *Clostridia. Biotechnology and Medical Applications*, Bahl H. & Dürre P. (Eds.), pp. 1-17, Wiley-VCH Verlag, Weinheim, Germany

Dürre, P. (2004). Solventogenesis by clostridia. In: Strict and Facultative Anaerobes. Medical and Environmental Aspects, Nakano M.M. & Zuber P.(Eds.), pp. 329-342, Horizon Bioscience, Wymondham, UK

Dürre, P. (2007). Biobutanol: an attractive biofuel. *Biotechnol J*, Vol.2, pp. 1525-1534

Dürre, P.; Böhringer, M.; Nakotte, S.; Schaffer, S.; Thormann, K. & Zickner, B. (2002). Transcriptional regulation of solventogenesis in *Clostridium acetobutylicum*. *J Mol Microbiol Biotechnol*, Vol.4, pp. 295-300

Dürre, P.; Fischer, R.-J.; Kuhn, A.; Lorenz, K.; Schreiber, W.; Stürzenhofecker, B.; Ullmann, S.; Winzer, K. & Sauer, U. (1995). Solventogenic enzymes of *Clostridium acetobutylicum* - catalytic properties, genetic organization, and transcriptional regulation. *FEMS Microbiol Rev*, Vol.17, pp. 251-262

Dürre, P.; Kuhn, A.; Gottwald, M. & Gottschalk, G. (1987). Enzymatic investigations on butanol dehydrogenase and butyraldehyde dehydrogenase in extracts of *Clostridium acetobutylicum*. *Appl Microbiol Biotechnol*, Vol.26, pp. 268-272

Energy Collective (2011) The media's role in the range fuels fiasco. Available from http://theenergycollective.com/robertrapier/52015/media-s-role-range-fuels-fiasco

Ennis, B. M. & Maddox, I. S. (1985). Optimal conditions for long-term stability of acetone-butanol production by continuous cultures of *Clostridium acetobutylicum*. *Biotechnol Lett*, Vol.7, pp. 503-508

Ernsting, A. (2008). Richard Branson and the Virgin Group's biofuel investment: From palm oil and soya biodiesel in trains to synthetic biology for aviation fuels. Available from http://www.biofuelwatch.org.uk/files/virgin_factsheet.pdf

EurActive Network (2008). EU agrees 10% 'green fuel' target in renewables deal. Available from http://www.euractiv.com/en/transport/eu-agrees-10-green-fuel-target-renewables-deal/article-177812

European Biodiesel Board (2011). The EU biodiesel statistics. Available from http://www.ebb-eu.org/stats.php

Ezeji, T. C.; Qureshi, N. & Blaschek, H. P. (2004). Butanol fermentation research: upstream and downstream manipulations. *Chem Record*, Vol.4, pp. 305-314

Ezeji, T. C.; Qureshi, N. & Blaschek, H. P. (2005). Continuous butanol fermentation and feed starch retrogradation: butanol fermentation sustainability using *Clostridium beijerinckii* BA101. *J Biotechnol*, Vol.115, pp. 179-187

Ezeji, T. C.; Qureshi, N. & Blaschek, H. P. (2007). Bioproduction of butanol from biomass. from genes to bioreactors.*Curr Opin Biotechnol*, Vol.18, pp. 220-227

Fairless, D. (2007). The little shrub that could - maybe. *Nature*, Vol.449, pp. 652-655

Farmer, W. R. & Liao, J. C. (2000). Improving lycopene production in *Escherichia coli by* engineering metabolic control. *Nat Biotechnol*, Vol.18, pp. 533-537

Fernbach, A. & Strange, E. H. (1911a). British patent 15,203

Fernbach, A. & Strange, E. H. (1911b). British patent 15,204

Fernbach, A. & Strange, E. H. (1912). British patent 15,073

Feustel, L.; Nakotte, S. & Dürre, P. (2004). Characterization and development of two reporter gene systems for *Clostridium acetobutylicum*. *Appl Environ Microbiol*, Vol.70, pp. 798-803

Finaly M. R. (2004). Old efforts at new uses: a brief history of chemurgy and the American search for biobased materials. *J Ind Ecol*, Vol.7, pp. 33-46

Fischer, R. J.; Helms, J. & Dürre, P. (1993). Cloning, sequencing, and molecular analysis of the *sol* operon of *Clostridium acetobutylicum*, a chromosomal locus involved in solventogenesis. *J Bacteriol*, Vol.175, pp. 6959-6969

Fitz, A. (1876). Ueber die Gährung des Glycerins. *Ber Dtsch Chem Ges*, Vol.9, pp. 1348-1352

Fitz, A. (1877). Ueber Schizomyceten-Gährungen II [Glycerin, Mannit, Stärke, Dextrin]. *Ber Dtsch Chem Ges*, Vol.10, pp. 276-283

Fitz, A. (1878). Ueber Schizomyceten-Gährungen III. *Ber Dtsch Chem Ges*, Vol.11, pp. 42-55

Fitz, A. (1882). Ueber Spaltpilzgährungen. VII. Mittheilung. *Ber Dtsch Chem Ges*, Vol.15, pp. 867-880

Fond, O.; Petitdemange, E.; Petitdemange, H. &Engasser, J. M. (1983). Cellulose fermentation by a coculture of a mesophilic cellulolytic *Clostridium* and *Clostridium acetobutylicum*. *Biotechnol Bioeng Symp*, Vol.13, pp. 217-224

Fontaine, L.; Meynial-Salles, I.; Girbal, L.; Yang, X.; Croux, C. & Soucaille, P. (2002). Molecular characterization and transcriptional analysis of *adhE2*, the gene encoding the NADH-dependent aldehyde/alcohol dehydrogenase responsible for butanol production in alcohologenic cultures of *Clostridium acetobutylicum* ATCC 824. *J Bacteriol*, Vol.184, pp. 821-830

Ford, H. (1925). Ford predicts fuel from vegetation. *The New York Times*, September 20th, p. 24

Friedl, A.; Qureshi, N., & Maddox, I. S. (1991). Continuous acetone-butanol-ethanol (ABE) fermentation using immobilized cells of *Clostridium acetobutylicum* in a packed bed reactor and integration with product removal by pervaporation. *Biotechnol Bioeng*, Vol.38, pp. 518-527

Fukuda, H.; Kondo, A. & Noda, H. J (2001). Biodiesel fuel production by transesterification of oils. *Biosci Bioeng*, Vol.92, pp. 405-416

Gabriel, C. L. & Crawford, F. M. (1930). Development of the butyl-acetonic fermentation industry. *Ind Eng Chem* Vol.22, pp. 1163-1065

Gabriel, C. L. (1928). Butanol fermentation process. *Ind Eng Chem*, Vol.20, pp. 1063-1067

Gapes, J. R. (2000). The history of the acetone-butanol project in Austria. *J Mol Microbiol Biotechnol*, Vol.2, pp. 5-8

Gavard, R.; Hautecoeur, B. & Descourtieux, H. (1957). Phosphotransbutyrylase de *Clostridium acetobutylicum*. *C R Hebd Seances Acad Sci*, Vol.244, pp. 2323-2326

Gerischer, U. & Dürre, P. (1990). Cloning, sequencing, and molecular analysis of the acetoacetate decarboxylase gene region from *Clostridium acetobutylicum*. *J Bacteriol*, Vol.172, pp. 6907-6918

Gerischer, U. & Dürre, P. (1992). mRNA analysis of the *adc* gene region of *Clostridium acetobutylicum* during the shift to solventogenesis. *J Bacteriol*, Vol.174, pp.426-433

Gevo (2009) Gevo announces first retrofit of ethanol demonstration plant to biobutanol for the oil and chemical industries. Available from http://ir.gevo.com/preview/phoenix.zhtml?c=238618&p=irol-newsArticle&ID=1491570&highlight=

Giebelhaus A. W. (1980). Farming for fuel: the alcohol motor fuel movement of the 1930s. *Agric Hist*, Vol.54, pp. 173-184

Goodlove, P. E.; Cunningham, P. R.; Parker, J. & Clark, D. P. (1989). Cloning and sequence analysis of the fermentative alcohol-dehydrogenase-encoding gene of *Escherichia coli*. *Gene*, Vol.85, pp. 209-214

Gray, K. A.; Zhao, L. & Emptage, M. (2006). Bioethanol. *Curr Opin Chem Biol*, Vol.10, pp. 141-146

Green Biologics (2007) Biofuels technology. Available from: http://www.greenbiologics.com/biofuels.asp

Green, E. M. & Bennett, G. N. (1996). Genetic manipulation of acid and solvent formation in *Clostridium acetobutylicum* ATCC 824. *Appl Biochem Biotechnol*, Vol.57/58, pp. 213-221

Grimmler, C.; Janssen, H.; Krauße, D.; Fischer, R.-J.; Bahl, H.; Dürre, P.; Liebl, W. & Ehrenreich, R. (2011). Genome-wide gene expression analysis of the switch between acidogenesis and solventogenesis in continuous cultures of *Clostridium acetobutylicum*. *J Mol Microbiol Biotechnol*, Vol. 20, pp. 1-15

Gunawardena, U.; Meinhold, P.; Peters, M. W.; Urano, J.; Feldman R. M. (2008). International patent WO 2008/010124

Hahn-Hägerdal, B.; Karhumaa, K.; Fonseca, C.; Spencer-Martins, I. & Gorwa-Grauslund, M. F. (2007). Towards industrial pentose-fermenting yeast strains. *Appl Microbiol Biotechnol*, Vol.74, pp. 937-953

Harris, L. M.; Blank, L.; Desai, R. P. & Welker, N. E. (2001). Fermentation characterization and flux analysis of recombinant strains of *Clostridium acetobutylicum* with an inactivated *solR* gene. *J Ind Microbiol Biotechnol*, Vol.27, pp. 322-328

Harris, L. M.; Desai, R. P.; Welker, N. E.; Papoutsakis, E. T. (2000). Characterization of recombinant strains of the *Clostridium acetobutylicum* butyrate kinase inactivation mutant: need for new phenomenological models for solventogenesis and butanol inhibition? *Biotechnol Bioeng*, Vol.67, pp. 1-11

Harris, L. M.; Welker, N. E. & Papoutsakis, E. T. (2002). Northern, morphological, and fermentation analysis of *spo0A* inactivation and overexpression in *Clostridium acetobutylicum* ATCC 824. *J Bacteriol*, Vol.184, pp. 3586-3597

Hartmanis, M. G. N. & Gatenbeck, S. (1984). Intermediary metabolism in *Clostridium acetobutylicum* - levels of enzymes involved in the formation of acetate and butyrate. *Appl Environ Microbiol*, Vol.48, pp. 1277-1283

Hartmanis, M. G. N. (1987). Butyrate kinase from *Clostridium acetobutylicum*. *J Biol Chem*, Vol.262, pp. 617-621

Hawkins, A. W.; Glassner, D. A.; Buelter, T.; Wade, J.; Meinhold, P.; Peters, M. W.; Grueber, P. R.; Evanko, W. A.; Aristidou, A. A. (2009). International patent WO 2009/059253

Henstra, A. M.; Sipma, J.; Rinzema, A. & Stams, A. J. M. (2007). Microbiology of synthesis gas fermentation for biofuel production. *Curr Opin Biotechnol*, Vol.18, pp. 200-206

Hildén, L. & Johansson, G. (2004). Recent developments on cellulases and carbohydrate-binding modules with cellulose affinity. *Biotechnol Lett*, Vol.26, pp. 1683-1693

Huang, K.-X.; Huang, S.; Rudolph, F. B. & Bennett, G. N. (2000). Identification and characterization of a second butyrate kinase from *Clostridium acetobutylicum* ATCC 824. *J Mol Microbiol Biotechnol*, Vol.2, pp. 33-38

IneosBio (2011) INEOS Bio JV receives commitment for $75 million USDA loan guarantee. Available from http://www.ineosbio.com/76-Press_releases-14.htm

Intergovernmental Panel on Climate Change (2007). *Climate Change 2007 - The Physical Science Basis*, Cambridge University Press, ISBN 978 0521 88009-1, New York, NY, USA

International Energy Agency (2004). *Biofuels for transport: An international perspective*, OECD Publishing, ISBN 92-64-01-51-24, Paris, France

Inui, M.; Suda M.; Kimura, S.; Yasuda, K.; Suzuki, H.; Toda, H.; Yamamoto, S.; Okino, S.; Suzuki, N. & Yukawa, H. (2007). Expression of *Clostridium acetobutylicum* butanol synthetic genes in *Escherichia coli*. *Appl Microbiol Biotechnol*, Vol.77, pp. 1305-1316

Iogen (2011) Iogen's cellulosic ethanol demonstration plant. Available from http://www.iogen.ca/company/demo_plant/index.html

Jarboe, L. R.; Grabar, T. B.; Yomano, L. P.; Shanmugan, K. T. & Ingram, L. O. (2007). Development of ethanologenic bacteria. *Adv Biochem Engin/Biotechnol*, Vol.108, pp. 237-261

Jeffries, T. W. & Yin, Y.-S. (2003). Metabolic engineering for improved fermentation of pentoses by yeasts. *Appl Microbiol Biotechnol*, Vol.63, pp. 495-509

Jones, D. T. & Woods, D. R. (1986). Acetone-butanol fermentation revisited. *Microbiol Rev*, Vol.50, pp. 484-524

Jones, D. T. (2001) Applied acetone-butanol fermentation. In: *Clostridia. Biotechnology and Medical Applications*, Bahl H. & Dürre P. (Eds.), pp. 125-168, Wiley-VCH Verlag, Weinheim, Germany

Jones, S. W.; Paredes, C. J.; Tracy, B.; Cheng, N.; Sillers, R.; Senger, R. S. & Papoutsakis, E.T. (2008). The transcriptional program underlying the physiology of clostridial sporulation. *Genome Biol*, Vol.9, pp. R114

Kalscheuer, R.; Stölting, T. & Steinbüchel, A. (2006). Microdiesel: *Escherichia coli* engineered for fuel production. *Microbiology*, Vol.152, pp. 2529-2536

Kessler, D.; Leibrecht, I. & Knappe, J. (1991). Pyruvate-formate-lyase-deactivase and acetyl-CoA reductase activities of *Escherichia coli* reside on a polymeric protein particle encoded by *adhE*. *FEBS Lett*, Vol.281, pp. 59-63

Killeffer, D. H. (1927). Butanol and acetone from corn - A description of the fermentation process. *Ind Eng Chem*, Vol.19, pp. 46-50

Kim, Y.; Ingram, L. O. & Shanmugam, K. T. (2007). Construction of an *Escherichia coli* K-12 mutant for homoethanologenic fermentation of glucose or xylose without foreign genes. *Appl Environ Microbiol*, Vol.73, pp. 1766-1771.

Knothe, G. (2001). Historical perspectives on vegetable oil-based *Diesel* fuels. *inform*, Vol.12, pp. 1103-1107

Köpke, M.; Held, C.; Hujer, S.; Liesegang, H.; Wiezer, A.; Wollherr, A.; Ehrenreich, A.; Liebl, W.; Gottschalk, G. & Dürre, P. (2010). *Clostridium ljungdahlii* represents a microbial production platform based on syngas. *Proc Natl Acad Sci U S A*, Vol.107, pp. 13087-92

Köpke, M.; Mihalcea, C.; Bromley, J. C.; Simpson, S. D. (2011). Fermentative production of ethanol from carbon monoxide. *Curr. Opin. Biotechnol.* doi:10.1016/j.copbio.2011.01.005

Kosaka, T.; Nakayama, S.; Nakaya, K.; Yoshino, S.; Furukawa, K. (2007). Characterization of the *sol* Operon in butanol-hyperproducing *Clostridium saccharoperbutylacetonicum* strain N1-4 and its degeneration mechanism. *Biosci Biotechnol Biochem*, Vol.71, pp. 58-68

Kovarik, B. (1998). Henry Ford, Charles F. Kettering and the fuel of the future. *Automot Hist Rev*, Vol.32, pp. 7-27

LanzaTech (2010) LanzaTech signs first commercial partnership deal. Available from http://www.lanzatech.co.nz/site_resources/library/Expo_2010.pdf

Laursen, R. A. & Westheimer, F. H. (1966). The active site of acetoacetate decarboxylase. *J Am Chem Soc*, Vol.88, pp. 3426-3430

Lee, S. F.; Forsberg, C. W. & Gibbins, L. N. (1985). Cellulolytic activity of *Clostridium acetobutylicum*. *Appl Environ Microbiol*, Vol.50, pp. 220-228

Levin, D. B.; Pitt, L. & Love, M. (2004). Biohydrogen production: prospects and limitations to practical application. *Int J Hydrogen Energy*, Vol.29, pp. 173-185

Liao, J. C.; Atsumi, S.; Brynildsen, M. P.; Cann, A. F.; Chou, K. J.; Shen, C. R.; Smith, K. M.; Hanai, T.; Connor, M. R. (2008). International patent WO 2008/124523

Linden, J. C.; Moreira, A. R.; Lenz, T. G. (1986). Acetone and butanol. In *Comprehensive Biotechnology. The Principles of Biotechnology; Engineering Consideration*; Cooney, C. L. and Humphrey, A. E. (Eds.), pp 915-931; Pergamon Press: Oxford, UK

Lopez-Contreras, A. M.; Martens, A. A.; Szijarto, N.; Mooibroek, H. Claassen, P. A. M.; van der Oost, J. & Willem de Vos M. (2003). Production by *Clostridium acetobutylicum* ATCC 824 of CelG, a cellulosomal glycoside hydrolase belonging to family 9. *Appl Environ Microbiol*, Vol.69, pp. 869-877

Lynd L. R., van Zyl W. H., McBride J. E. & Laser M. (2005). Consolidated bioprocessing of cellulosic biomass: an update. *Curr Opin Biotechnol*, Vol.16, pp. 577-583

Lynd, L. R.; Weimer, P. J.; van Zyl, W. H. & Pretorius, I. S. (2002). Microbial cellulose utilization. *Microbiol Mol Biol Rev*, Vol.66, pp. 506-577

Maddox, I. S. & Murray, A. E. (1983). Production of n-butanol by fermentation of wood hydrolysate. *Biotechnol Lett*, Vol.5, pp. 175-178

Maddox, I. S. (1980). Production of n-butanol from whey filtrate using *Clostridium acetobutylicum* N.C.I.B. 2951. *Biotechnol Lett*, Vol.2, pp. 493-498

Maddox, I. S.; Qureshi, N. & Gutierrez, N. A. (1994). Utilization of whey by clostridia and process technology. In: *The Clostridia and Biotechnology*, Woods D. R. (Ed.), pp. 343-369, Butterworth-Heinemann: Stoneham, MA, USA

Marchal, R.; Ropars, M.; Pourquié, J.; Fayolle, F.& Vandecasteele, J. P. (1992). Large-scale enzymatic hydrolysis of agricultural lignocellulosic biomass. Part 2: Conversion into acetone-butanol. *Bioresour Technol*, Vol.42, pp. 205-217

Martin, V.; Pitera, D. J.; Withers, S. T.; Newman, J. D. & Keasling, J. D. (2003). Engineering a mevalonate pathway in *Escherichia coli* for production of terpenoids. *Nat Biotechnol*, Vol.21, pp. 796-802

Mason, S. & Ghonasgi, D. (2008). Renewable diesel - case study, Symposium New biofuels, Berlin, May 6-7 2008. Available from http://www.fnr-server.de/ftp/pdf/literatur/pdf_332-index.htm

McCoy, E.; Fred, E. B.; Peterson, W. H. & Hastings, E. G. (1926). A cultural study of the acetone butyl alcohol organism. *J Infect Dis*, Vol.39, pp. 457-483

McCutchan, W. N. & Hickey, R. J. (1954) The butanol-acetone fermentations. *Ind Ferment*, Vol.1, pp. 347-388.

Meinecke, B.; Bertram, J. & Gottschalk, G. (1989). Purification and characterization of the pyruvate-ferredoxin oxidoreductase from *Clostridium acetobutylicum*. *Arch Microbiol*, Vol.152, pp. 244-250

Mermelstein, L. D.; Papoutsakis, E. T.; Petersen, D. J. & Bennett, G. N. (1993). Metabolic engineering of *Clostridium acetobutylicum* ATCC 824 for increased solvent production by enhancement of acetone formation enzyme activities using a synthetic acetone operon. *Biotechnol Bioeng*, Vol.42, pp. 1053-1060

Mitchell D. (2008) A note on rising food prices. World bank policy research working paper 4682

Nair, R. V.; Bennett, G. N. & Papoutsakis, E. T. (1994). Molecular characterization of an aldehyde/alcohol dehydrogenase gene from *Clostridium acetobutylicum* ATCC 824. *J Bacteriol*, Vol.176, pp. 871-885

Nair, R.V. & Papoutsakis, E.T. (1994). Expression of plasmid-encoded *aad* in *Clostridium acetobutylicum* M5 restores vigorous butanol production. *J Bacteriol*, Vol.176, pp. 5843-5846

Nath, K. & Das, D. (2004). Improvement of fermentative hydrogen production: Various approaches. *Appl Microbiol Biotechnol*, Vol.65, pp. 520-529

Ni, Y. & Sun, Z. (2009). Recent progress on industrial fermentative production of acetone-butanol-ethanol by *Clostridium acetobutylicum* in China. *Appl Microbiol Biotechnol*, Vol.83, pp. 415-423

Nielsen D. R., Leonard E., Yoon S. H., Tseng H. C., Yuan C. & Prather K. L. (2009). Engineering alternative butanol production platforms in heterologous bacteria. *Metab Eng*, Vol.11, pp. 262-73

Nimcevic, D. & Gapes, J. R. (2000). The acetone-butanol fermentation in pilot plant and pre-industrial scale. *J Mol Microbiol Biotechnol*, Vol. 2, pp. 15-20

Nitske, W. R. & Wilson, C. M. (1965) *Rudolf Diesel, pioneer of the age of power*, University of Oklahoma Press, Norman, OK, USA

Nölling, J.; Breton, G.; Omelchenko, M. V.; Makarova, K. S.; Zeng, Q.; Gibson, R.; Lee, H. M.; Dubois, J. A.; Qiu, D.; Hitti, J.; GTC Sequencing Center Production; Finishing and Bioinformatics Teams; Wolf, Y. I.; Tatusov, R. L.; Sabathe, F.; Doucette-Stamm, L.; Soucaille, P.; Daly, M. J., Bennett, G. N.; Koonin, E. V. & Smith, D. R. (2001). Genome sequence and comparative analysis of the solvent-producing bacterium *Clostridium acetobutylicum*. *J Bacteriol*, Vol.183, pp. 4823-4838

Oja, S. (2008). NExBTL - Next generation renewable diesel, Symposium New biofuels, Berlin, May 6-7 2008. Available from http://www.fnr-server.de/ftp/pdf/Literatur/pdf_332-index.htm

Oxfam (2008) Another Inconvinient Truth. *Oxfam briefing paper*, Vol. 114. Available from http://www.oxfam.org/sites/www.oxfam.org/files/bp114-inconvenient-truth-biofuels-0806.pdf

Paquet, V.; Croux, C.; Goma, G. & Soucaille, P. (1991). Purification and characterization of the extracellular alpha-amylase from *Clostridium acetobutylicum* ATCC 824. *Appl Environ Microbiol*, Vol.57, pp. 212-218

Pasteur, L. (1862). Quelques résultats nouveaux relatifs aux fermentations acétique et butyrique. *Bull Soc Chim Paris*, May, pp. 52-53

Peralta-Yahya P. P. & Keasling J. D. (2010). Advanced biofuel production in microbes. *Biotechnol J.*, Vol.5 pp. 147-162

Perlack, R. D., Wright, L. L., Turhollow, A. F., Graham, R. L., Stokes, B. J. & Erbach, D. C. (2005). Biomass as feedstock for a bioenergy and bioproducts industry: the technical feasibility of a billion-ton annual supply. Available from http://www1.eere.energy.gov/biomass/pdfs/final_billionton_vision_report2.pdf

Petersen, D. J. & Bennett, G. N. (1990). Purification of acetoacetate decarboxylase from *Clostridium acetobutylicum* ATCC 824 and cloning of the acetoacetate decarboxylase gene in *Escherichia coli*. *Appl Environ Microbiol*, Vol.56, pp. 3491-3498

Petersen, D. J.; Cary, J. W.; Vanderleyden, J. & Bennett, G. N. (1993). Sequence and arrangement of genes encoding enzymes of the acetone-production pathway of *Clostridium acetobutylicum*. *Gene*, Vol.123, pp. 93-97

Petersen, D. J.; Welch, R. W.; Rudolph, F. B. & Bennett, G. N. (1991). Molecular cloning of an alcohol (butanol) dehydrogenase gene cluster from *Clostridium acetobutylicum* ATCC 824. *J Bacteriol*, Vol.173, pp. 1831-1834

Pich, A.; Narberhaus, F. & Bahl, H. (1990). Induction of heat shock proteins during initiation of solvent formation in *Clostridium acetobutylicum*. *Appl Microbiol Biotechnol*, Vol.33, pp. 697-704

POET (2011) POET Biorefining Plants. Available from http://www.poet.com/inspiration/plants.asp

Prince, R. C. & Kheshgi, H. S. (2005). The photobiological production of hydrogen: potential efficiency and effectiveness as a renewable fuel. *Crit Rev Microbiol*, Vol.31, pp. 19-31

Qureshi, N.; Hughes, S.; Maddox, I. S. & Cotta, M. A. (2005). Energy-efficient recovery of butanol from model solutions and fermentation broth by adsorption. *Bioprocess Biosyst Eng*, Vol.27, pp. 215-222

Ragauskas, A. J.; Williams, C. K.; Davison, B. H.; Britovsek, G.; Cairney, J.; Eckert, C. A.; Frederick Jr., W. J.; Hallett, J. P.; Leak, D. J.; Liotta, C. L.; Mielenz, J. R.; Murphy, R.; Templer, R. & Tschaplinski, T. (2006). The path forward for biofuels and biomaterials. *Science*, Vol.311, pp. 484-489.

Ratledge, C. (1993). Single cell oils - Have they a biotechnological future? *Trends Biotechnol*, Vol.11, pp. 278-284

Ravagnani, A.; Jennert, K. C. B.; Steiner, E.; Grünberg, R.; Jefferies, J. R.; Wilkinson, S. R.; Young, D. I.; Tidswell, E. C.; Brown, D. P.; Youngman, P.; Morris, J. G. & Young M. (2000) Spo0A directly controls the switch from acid to solvent production in solvent-forming clostridia. *Mol Microbiol*, Vol.37, pp. 1172-1185.

Renewable Fuels Association (2011). Ethanol industry statistics. Available from http://www.ethanolrfa.org/pages/statistics

Ross, D. (1961). The acetone-butanol fermentation. *Prog Ind Microbiol*, Vol.3, pp. 73-85

Sabathé, F. & Soucaille, P. (2003). Characterization of the CipA scaffolding protein and in vivo production of a minicellulosome in *Clostridium acetobutylicum*. *J Bacteriol*, Vol.185, pp. 1092-1096

Santangelo, J. D. & Dürre, P. (1996). Microbial production of acetone and butanol: Can history be repeated? *Chimica oggi/Chemistry today*, Vol.14, pp. 29-35

Sauer, U. & Dürre, P. (1993). Sequence and molecular characterization of a DNA region encoding a small heat shock protein of *Clostridium acetobutylicum*. *J Bacteriol*, Vol.175, pp. 3394-3400

Sauer, U. & Dürre, P. (1995). Differential induction of genes related to solvent formation during the shift from acidogenesis to solventogenesis in continuous culture of *Clostridium acetobutylicum*. *FEMS Microbiol Lett*, Vol.125, pp. 115-120

Schaffer, S., Isci, N., Zickner, B. & Dürre, P. (2002). Changes in protein synthesis and identification of proteins specifically induced during solventogenesis in *Clostridium acetobutylicum*. *Electrophoresis*, Vol.23, pp. 110-121

Schindler, J. (2008). Wasserstoff aus Biomasse. Symposium New biofuels, Berlin, May 6-7 2008. Available from http://www.fnr-server.de/ftp/pdf/Literatur /pdf_332-index.htm

Schreiber, W. & Dürre, P. (1999). The glyceraldehyde-3-phosphate dehydrogenase of *Clostridium acetobutylicum*: isolation and purification of the enzyme, and sequencing and localization of the *gap* gene within a cluster of other glycolytic genes. *Microbiology*, Vol.145, pp. 1839-1847

Schubert, C. (2006). Can biofuels finally take center stage? *Nature Biotechnol*, Vol.24, pp. 777-784

Shen, C. R.; Lan, E. I.; Dekishima, Y.: Baez, A.; Cho, K., M. & Liao, J. C. (2011). High titer anaerobic 1-butanol synthesis in *Escherichia coli* enabled by driving forces. *Appl Environ Microbiol*, Epub. doi:10.1128/AEM.03034-10

Squillance, P. J.; Pankow, F. J.; Korte, N. E. & Zogorski, J. S. (1997). Review of the environmental behavior and fate of methyl *tert*-butyl ether. *Environ Toxicol Chem*, Vol.16, pp. 1836-1844

Steen, E. J.; Chan R.; Prasad N.; Myers S.; Petzold C. J.; Redding A.; Ouellet M. & Keasling J. D. (2008). Metabolic engineering of *Saccharomyces cerevisiae* for the production of n-butanol. *Microb Cell Fact*, Vol.7, pp. 36

Stim-Herndon, K. P.; Petersen, D. J. & Bennett, G. N. (1995). Characterization of an acetyl-CoA C-acetyltransferase (thiolase) gene from *Clostridium acetobutylicum* ATCC 824. *Gene*, Vol.154, pp. 81-85

Terracciano, J. S.; Rapoport, E. & Kashket, E. R. (1988). Intracellular conditions required for initiation of solvent production by *Clostridium acetobutylicum*. *Appl Environ Microbiol*, Vol.54, pp. 1989-1995

Thompson, D. K. & Chen, J.-S. (1990). Purification and properties of an acetoacetyl coenzyme A-reacting phosphotransbutyrylase. *Appl Environ Microbiol*, Vol.56, pp. 607-613

Thormann, K.; Feustel, L.; Lorenz, K.; Nakotte, S. & Dürre, P. (2002). Control of butanol formation in *Clostridium acetobutylicum* by transcriptional activation. *J Bacteriol*, Vol.184 pp. 1966-1973

Tirado-Acevedo O, Chinn MS, Grunden AM (2010). Production of biofuels from synthesis gas using microbial catalysts. *Adv Appl Microbiol*, Vol.70, pp. 57-92.

Tomas, C. A.; Alsaker, K. V.; Bonarius, H. P. & Hendriksen, W. T.; Yang, H.; Beamish, J. A.; Paredes, C. J. & Papoutsakis E. T. (2003a). DNA array-based transcriptional analysis of asporogenous, nonsolventogenic *Clostridium acetobutylicum* strains SKO1 and M5. *J Bacteriol*, Vol.185, pp. 4539-4547

Tomas, C. A.; Beamish, J. A. & Papoutsakis, E. T. (2004). Transcriptional analysis of butanol stress and tolerance in *Clostridium acetobutylicum*. *J Bacteriol*, Vol.186, pp. 2006-2018

Tomas, C. A.; Welker, N. E. & Papoutsakis, E. T. (2003b). Overexpression of *groESL* in *Clostridium acetobutylicum* results in increased solvent production and tolerance, prolonged metabolism, and changes in the cell's transcriptional program. *Appl Environ Microbiol*, Vol.69, pp. 4951-4965

Tummala, S. B., Junne, S. G., Paredes, C. J. & Papoutsakis, E. T. (2003). Transcriptional analysis of product-concentration driven changes in cellular programs of recombinant *Clostridium acetobutylicum* strains.. *Biotechnol Bioeng*, Vol.84, pp. 842-854

Tummala, S. B.; Welker, N. E. & Papoutsakis, E. T. (1999). Development and characterization of a gene expression reporter system for *Clostridium acetobutylicum* ATCC 824.. *Appl Environ Microbiol*, Vol.65, pp. 3793-3799

Ullmann, S. & Dürre, P. (1998). Changes in DNA topology are involved in triggering the onset of solventogenesis in *Clostridium acetobutylicum*. *Recent Res Dev Microbiol*, Vol.2, pp. 281-294

Ullmann, S.; Kuhn, A. & Dürre, P. (1996). DNA topology and gene expression in *Clostridium acetobutylicum*: implications for the regulation of solventogenesis. *Biotechnol Lett*, Vol.18, pp. 1413-1418

U.S. Energy Information Administration (2003). Status and impact of state MTBE bans. Available from http://tonto.eia.doe.gov/ftproot/service/mtbe.pdf

U.S. Energy Information Administration (2007). *International Energy Outlook 2007*, US government printing office, Washington, DC, USA

U.S. Energy Information Administration (2010). *Annual Energy Review 2009*, US government printing office, Washington, DC, USA

U.S. Energy Information Administion (2011). Spot prices for crude oil and petroleum products. Available from: http://www.eia.doe.gov/dnav/pet/pet_pri_spt_s1_d.htm

Vega, J. L.; Klasson, K. T.; Kimmel, D. E.; Clausen, E. C. & Gaddy, J. L. (1990). Sulfur gas tolerance and toxicity of CO-utilizing and methanogenic bacteria. *Appl Biochem Biotechnol*, Vol.24, pp. 329-340

Viikari, L.; Alapuranen, M.; Puranen, T.; Vehmaanperä, J. & Siika-aho, M. (2007). Thermostable enzymes in lignocellulose hydrolysis. *Adv Biochem Engin/Biotechnol*, Vol. 108, pp. 121-145

Virgin Atlantic (2008). Biofuel flight demonstration. Available from http://www.virgin-atlantic.com:80/en/us/allaboutus/environment/biofuel.jsp

Voget, C. E.; Mignone, C. F. & Ertola, R. J. (1985). Butanol production from apple pomace. *Biotechnol Lett*, Vol.7, pp. 43-46

Walter, K. A.; Bennett, G. N. & Papoutsakis, E. T. (1992). Molecular characterization of two *Clostridium acetobutylicum* ATCC 824 butanol dehydrogenase isozyme genes. *J Bacteriol*, Vol.174, pp. 7149-7158

Walter, K. A.; Nair, R. V.; Cary, J. W.; Bennett, G. N. & Papoutsakis, E. T. (1993). Sequence and arrangement of two genes of the butyrate-synthesis pathway of *Clostridium acetobutylicum* ATCC 824. *Gene*, Vol.134, pp. 107-111

Warnecke, F.; Luginbühl, P.; Ivanova, N.; Ghassemian, M.; Richardson, T. H.; Stege, J. T.; Cayouette, M.; McHardy, A. C.; Djordjevic, G.; Aboushadi, N.; Sorek, R.; Tringe, S. G.; Podar, M.; Martin, H. G.; Kunin, V.; Dalevi, D.; Madejska, J.; Kirton, E.; Platt, D.; Szeto, E.; Salamov, A.; Barry, K.; Mikhailova, N.; Kyrpides, N. C.; Matson, E. G.; Ottesen, E. A.; Zhang, X.; Hernández, M.; Murillo, C.; Acosta, L. G.; Rigoutsos, I.; Tamayo, G.; Green, B. D.; Chang, C.; Rubin, E. M.; Mathur, E. J.; Robertson, D. E.;

Hugenholtz, P. & Leadbetter, J. R. (2007). Metagenomic and functional analysis of hindgut microbiota of a wood-feeding higher termite. *Nature*, Vol.450, pp. 560-565

Waterson, R. M.; Castellino, F. J.; Hass, G. M. & Hill, R. L. (1972). Purification and characterization of crotonase from *Clostridium acetobutylicum*. *J Biol Chem*, Vol.247, pp. 5266-5271

Weizmann C. (1915). British patent 4,835

Welch, R. W.; Rudolph, F. B. & Papoutsakis, E. T. (1989). Purification and characterization of the NADH-dependent butanol dehydrogenase from *Clostridium acetobutylicum* (ATCC 824). *Arch Biochem Biophys*, Vol.273, pp. 309-318

Welsh, F. W. & Veliky, I. A. (1984). Production of acetone butanol from acid whey. *Biotechnol Lett*, Vol.6, pp. 61-64

Wiesenborn, D. P.; Rudolph, F. B. & Papoutsakis, E. T. (1988). Thiolase from *Clostridium acetobutylicum* ATCC 824 and its role in the synthesis of acids and solvents. *Appl Environ Microbiol*, Vol.54, pp. 2717-2722

Wiesenborn, D. P.; Rudolph, F. B. & Papoutsakis, E. T. (1989). Phosphotransbutyrylase from *Clostridium acetobutylicum* ATCC 824 and its role in acidogenesis. *Appl Environ Microbiol*, Vol.55, pp. 317-322

Winzer, K.; Lorenz, K. & Dürre, P. (1997). Acetate kinase from *Clostridium acetobutylicum*: a highly specific enzyme that is actively transcribed during acidogenesis and solventogenesis. *Microbiology*, Vol.143, pp. 3279-3286

Winzer, K.; Lorenz, K.; Zickner, B. & Dürre, P. (2000). Differential regulation of two thiolase genes from *Clostridium acetobutylicum* DSM 792. *J Mol Microbiol Biotechnol*, Vol.2, pp. 531-541

Wong, J. & Bennett, G. N. (1996). Recombination-induced variants of *Clostridium acetobutylicum* ATCC 824 with increased solvent production. *J Ind Microbiol*, Vol.16, pp. 354-359

Yan, R.-T.; Zhu, C.-X.; Golemboski, C. & Chen, J.-S. (1988). Expression of solvent-forming enzymes and onset of solvent production in batch cultures of *Clostridium beijerinckii* („*Clostridium butylicum*"). *Appl Environ Microbiol*, Vol.54, pp. 642-648

Yu , E. K. C. & Saddler, J. N. (1983). Enhanced acetone-butanol fermentation by *Clostridium acetobutylicum* grown on D-xylose in the presence of acetic or butyric acid. *FEMS Microbiol Lett*, Vol.18, pp. 103-107

Yu, E. K. C.; Chan, M. K.-H. & Saddler, J. N. (1985). Butanol production from cellulosic substrates by sequential co-culture of *Clostridium thermocellum* and *C. acetobutylicum*. *Biotechnol Lett*, Vol.7, pp. 509-514

Zhao, Y.; Hindorff, L.A.; Chuang, A.; Monroe-Augustus, M.; Lyristis, M.; Harrison, M. L.; Rudolph, F. B. & Bennett, G. N. (2003). Expression of a cloned cyclopropane fatty acid synthase gene reduces solvent formation in *Clostridium acetobutylicum* ATCC 824. *Appl Environ Microbiol*, Vol.69, pp. 2831-2841

Zverlov, V. V.; Berezina, O.; Velikodvorskaya, G. A. & Schwarz, W. H. (2006) Bacterial acetone and butanol production by industrial fermentation in the Soviet Union: use of hydrolyzed agricultural waste for biorefinery. *Appl Microbiol Biotechnol*, Vol.71, pp. 587-597

# 19

# DMF - A New Biofuel Candidate

Guohong Tian, Ritchie Daniel and Hongming Xu
*University of Birmingham and Newcastle University*
*United Kingdom*

## 1. Introduction

Although there are very few biomass-derived fuels which are competitive to fossil fuels, the scientific research community has never ceased to search for renewable biofuel alternatives. For use in the transportation sector, a practical biofuel must:

- be produced economically and efficiently;
- have a relatively high energy density, in both volume and weight, in order to achieve competitive mileage;
- be easy to store and distribute;
- have sufficient reserves, which do not threaten ecosystems;
- be sourced from a variety of bio-mass or bio-waste;
- have suitable properties for use in existing power generation systems without the need for major modifications.

Most potential biofuels meet the majority of these requirements. However, none satisfies them all, not even the most successful gasoline-alternative biofuel, bio-ethanol. Despite its success in the Brazilian market, bio-ethanol requires a large amount of sugar cane, which can put stress on food production and the environment.

Hitherto, little has been able to compete with bio-ethanol, that is until the production techniques of 2,5-dimethylfuran, known as DMF, were significantly improved (Yuriy Roman-Leshkov 2007; Zhao H 2007). These advances have sparked interest in the research community to investigate the adaptability of this new biofuel candidate to be used in power generation systems, particularly in internal combustion engines.

## 2. What is DMF?

DMF, a derivative of furan, is a heterocyclic compound with the formula $(CH_3)_2C_4H_2O$. Its main properties are listed in Table 1. In order to better understand the relative strength of DMF, the corresponding properties for bio-ethanol (referred to as ethanol in this Chapter) and gasoline are also listed in the table.

As shown in the table, DMF has several advantages over ethanol as a gasoline-alternative biofuel. Firstly, the gravimetric energy density, or lower heating value, of DMF is approximately 1/3 higher than ethanol, much closer to that of gasoline, which improves mileage for the same size of fuel tank. In terms of volume, the lower heating value of DMF is less than 10% lower than gasoline and over 40% higher than ethanol. Secondly, unlike ethanol, DMF is insoluble in water, which makes it easier to store. Thirdly, DMF has a

| | | | |
|---|---|---|---|
| Name(s) | 2,5 Dimethylfuran<br>2,5 Dimethyl-furan<br>2,5 Dimethylfurane<br>2,5 Dimethyloxole | Ethanol<br>Ethyl Alcohol<br>Ethyl Hydroxide<br>EtOH | Gasoline<br>Petrol |
| CAS Registry Number | 625-86-5 | 64-17-5 | 8006-61-9<br>86290-81-5 |
| Linear Structure Formula | $(CH_3)_2C_4H_2O$ | $CH_3OCH_3$ | Variable |
| Molecular Formula | $C_6H_8O$ | $C_2H_6O$ | $C_2$ to $C_{14}$ |
| Molecule Schematic | 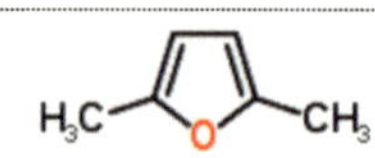 |  | Variable |
| M, Molecular Mass | 96.1289g/mol[1] | 46.069g/mol[1] | 100 – 105g/mol [2] |
| Type of Substance | Heterocyclic | Acyclic | Aliphatic Hydrocarbon Mixture |
| Appearance | Colorless Liquid | Colorless Liquid | Colorless to amber colored liquid |
| Aroma | Spicy, Smokey | Vinous | Petroleum odor |
| Safety | Flammable, Irritant | Flammable, Irritant, CNS effects | Highly Flammable, Irritant |
| Water Solubility[3] | Insoluble<br><1mg/ml @73°F | Highly soluble<br>>=100mg/ml<br>@73°F | Insoluble |
| MP, Boiling Point (1atm) | −62.8°C[3] | −130°C[4] | - |
| BP, Boiling Point (1atm)[1] | 93.0°C | 77.3°C | 96.3°C(FBP) |
| Enthalpy of Vaporization (20°C)[4] | 31.91kJ/mol | 43.2496kJ/mol | - |
| Enthalpy of Formation[5] | −97.9kJ/mol<br>@298.15K | −234.963kJ/mol | - |
| Vapor Pressure | 7.2 kPa (22°C)[6] | 7.869 kPa(25°C)[7] | 72.007 kPa4(25°C) |
| ρ, Density of Liquid | 895.4kg/m³ @20°C | 793.63kg /m³@15°C | - |

1 Beilstein Database April 2008

2 US Department of Energy. Alternative Fuels and Advanced Vehicles Data Centre. October 2008

3 Cameo Chemicals, US Gov. Chemical Data Sheet, 2, 5 Dimethylfuran. October 2008

4 Chemspider.com January 2009 (NB: Prediction Software is utilised by this source when experimental values are unavailable)

5 F M WELLE & Others. Thermochemical Studies for Determination of the Standard Molar Enthalpies of Formation of Alkyl-Substituted Furans and Some Ethers. Structural Chemistry June 1998

6 NTP, 1992

7 http://www.iterasi.net/openviewer.aspx?sqrlitid=qpgkevc0qugba9dhaqnsyw

| | | | |
|---|---|---|---|
| Vapor Density (air=1) | 3.31[7] | 1.59[8] | 3-4[9] |
| Refraction Index (20°C)[4] | 1.443 | 1.361 | - |
| Specific Gravity (4°C, 1atm)[4] | 0.892-0.898 | 0.910 | - |
| Molar Volume (4°C, 1atm)[4] | 104.7cm$^3$ | 59cm$^3$ | - |
| Surface Tension[4] | 25.9dyne cm | 22.3dyne cm | 20dyne cm$^{-1}$ |
| Polar Surface Area[4] | 13.14Å$^2$ | 9.23Å$^2$ | - |
| Viscosity (1atm, 20°C) | 0.65cP[10] | 1.2cP[11] | 0.4-0.5cP[12] |
| Research Octane Number (RON) | - | 110[2] | 95[13] |
| Heat of Vaporization | 332 kJ/kg | 840 kJ/kg | 373 kJ/kg |
| Lower Heating Value | 33.7MJ/kg | 26.9MJ/kg | 42.9MJ/kg |

Table 1. Main properties of DMF, benchmarked with ethanol and gasoline

higher boiling point, which makes it less volatile and more practical as a liquid fuel for transportation. Fourthly, DMF has a similar heat of vaporisation to gasoline, which will help to overcome the difficulty of cold starts seen with ethanol. Finally, and most attractively, DMF is completely uncompetitive with food and potentially more energy efficient to produce.

Although DMF has been used as an octane improver for gasoline, it has served other purposes. Before being considered as an alternative fuel, DMF was used in the food industry with a low production efficiency and high cost. However, new production techniques have been reported by a group of biochemists at the University of Wisconsin-Madison in the US, who published their work in Nature, in 2007 (Yuriy Roman-Leshkov 2007). Here, the authors claimed a new development that allows the mass production of DMF from bio-mass with low energy consumption and high yield. Figure 1 shows the rationale for converting these carbohydrates to DMF. As described in the publication, five oxygen atoms need to be removed from a hexose, for example, fructose to produce DMF. The selective removal can be accomplished in two steps: first, by removing three oxygen atoms through dehydration to produce 5-hydroxymethylfurfural (HMF); and second, by removing two oxygen atoms through hydrogenolysis to produce DMF by the way of the following intermediates: 2-methyl,5-hydroxymethylfuran and 2-methylfuran (4 and 5 in Figure 1)(Yuriy Roman-Leshkov 2007).

8 Material Safety Data Sheet, Ethanol, Absolute. https://fscimage.fishersci.com/msds/89308.htm
9 GASOLINE SAFETY. http://www.burnsurgery.org/Documents/gasoline_safety.doc
10 knovel database
11 http://en.wikipedia.org/wiki/Ethanol
12 http://chemed.chem.purdue.edu/genchem/topicreview/bp/ch14/property.php
13 Delphi. Worldwide Emissions Standards, Passenger Cars & Light Duty Trucks. 2008

Shortly after their publication, this concept was further developed by Zhao and his co-workers, who observed high yields of HMF (the intermediate for DMF; see Figure 1), without the need for acid catalysts (Zhao H 2007). Not only does Zhao's method dramatically reduce the production costs, it also includes glucose as a potential feedstock for HMF. Furthermore, Mascal reported that cellulose itself can be converted into furanic products (Mascal 2008). Such advances have attracted attention towards DMF as a potential gasoline-alternative biofuel (Luque 2008).

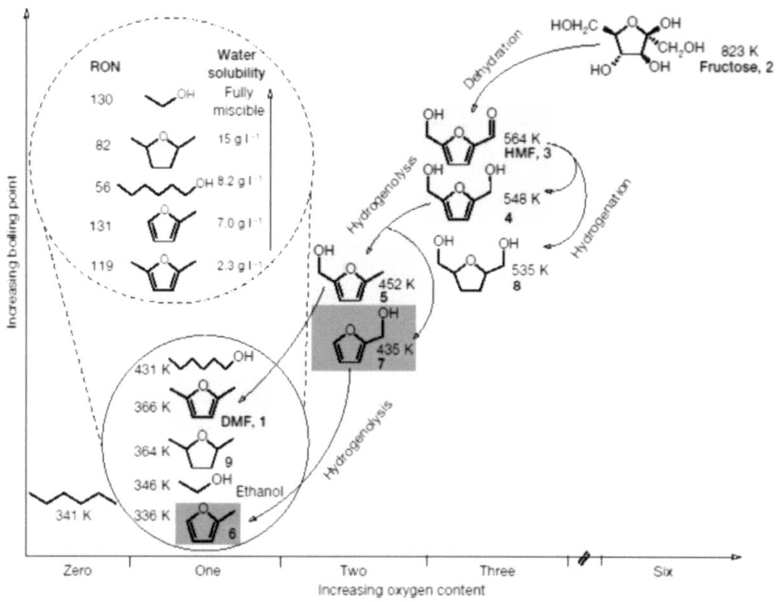


Fig. 1. The rationale for converting carbohydrates to DMF(Yuriy Roman-Leshkov 2007)

The success of the biochemists triggered a quick response from the automotive research community. Engineers were interested in the adaptability of DMF as an alternative automotive fuel. The research was firstly initiated at Birmingham University in the UK to investigate its combustion performance and emissions as an alternative fuel in modern automotive engines. The portfolio of work covers several key aspects of internal combustion engine science. This includes the spray characteristics, laminar burning velocity, engine performance, regulated and unregulated engine emissions and some novel methods for optimising the use of biofuels. Xi'an Jiaotong University in China has also reported a series of studies on the burning behaviour of DMF(Xuesong Wu 2009).

## 3. Spray characteristics

Fuel spray characteristics are essential to engine performance and emissions, particularly for direct-injection engines. The initial factors include the injection pressure, injector design, and fuel properties. They directly influence the fuel-air mixture generation and consequently, combustion behaviour. Optical methods, utilising shadowgraph and high-speed cameras, are often used to measure the main spray characteristics with time, including the spray cone angle and penetration length. For more advanced engine designs, the behaviour of such parameters as the fuel droplet size and velocity distribution are preferred. The information can be measured accurately by advanced laser based measurement techniques, e.g. Phase Doppler Particle Analyser (PDPA, also known as Phase Doppler Anemometry, PDA). This information can be used to understand the environmental impact of the spray development in order to help the design of injection and combustion systems. At the same time, such details can supplement computational simulation models, which can help to refine the engine design process.

A group of shadowgraph spray images from a multi-hole injector at different injection timings is shown in Figure 2. The images were cut in order to focus on one spray. The injection pressure in the fuel rail was 10MPa and the injection duration was set to 2ms. The first spray image was captured 0.8ms after the injection trigger, slightly longer than the 0.7ms solenoid response time for the specific injector used for the test. This method clearly shows the spray development with time, where the penetration length reached over 70mm at 2ms after the start of injection. It also shows a leaner spray at the injection tip due to the evaporation of the fuel droplets. This information is very helpful for the design and optimisation of injectors and combustion chambers, in order to prevent any fuel impingement on the cylinder wall or piston bowl, which is likely to increase the unburned hydrocarbon emissions. However, the information obtained from the image is limited and more comprehensive techniques are needed to provide more details of the injection pattern.

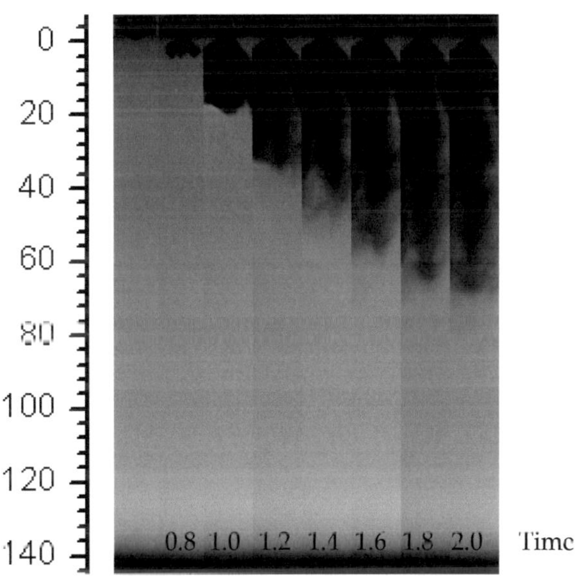


Fig. 2. Shadowgraph images for DMF spray (injection pressure: 100MPa, length unit: mm, time unit: ms)

PDPA is an advanced laser-based technology used to measure liquid droplet size and velocity distribution. It gives statistical information for a high number of droplets (>$10^4$) by obtaining the size and velocity information for each droplet passing through the microscopic measuring point. It normally takes hundreds of injections to gather a sufficient number of validated droplets at one point. With the results from different positions in the spray, a spray map with detailed size and velocity information can be provided to better describe the spray. This helps the comprehensive understanding of the spray characteristics.

A typical PDPA measurement result (velocity) at one test point is illustrated in Figure 3. Each point represents a single measured droplet, where, in this figure, there are over 15,000 droplets. For a clearer comparison between the different fuels, a smooth profile was averaged over 0.05ms time intervals. The curve clearly indicates the spray structure and therefore the trend of velocity development.

The spray structure can be arbitrarily divided into two parts: spray 'head' and 'tail', which are marked in the figure. In the spray head, the droplets have a relatively high velocity with a low deviation. However, in the spray tail, the mean velocity drops quickly until it reaches a minimum. In this case, 1.1ms after the start of injection (ASOI), the spray head reached the measuring point and the droplets were captured. The delay consists of two parts: the solenoid response time, which is approximately 0.7ms (for this injector), and the spray travelling time. The highest velocity of the droplets was 80m/s and the lowest was 20m/s approximately, which gives a wide velocity distribution range. The droplet generation process can explain this phenomenon. The first stage of this process includes the liquid filament breakup (after injection) due to the shearing force between the air and the spray envelope. Resultantly, these droplets slow down and are hit by the ensuing droplets, causing their trajectories to rapidly change. As a result, the real velocity of the ensuing droplets with unchanged trajectories is higher than those with changed trajectories. After the spray head has passed, several droplets with much lower velocity reach the measuring volume (recognized as the 'tail' of the spray). This results in the negative velocity of some droplets which could induce vortices caused by the pressure gradient of the surrounding gas. Several low velocity droplets were captured before and during the spray head. These droplets are believed to be the residual droplets from the previous injection and were ignored due to the negligible quantity.

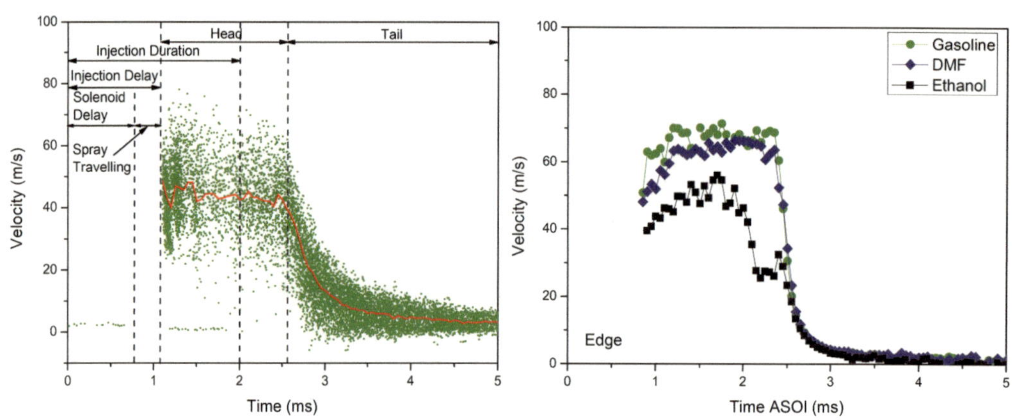


(a) (b)

Fig. 3. (a) Typical PDPA measured velocity distribution (tested fuel: benchmarking gasoline, injection pressure: 10MPa, measure position: 32mm downstream from nozzle at the axle) and (b) mean velocities of tested fuels (measure position: 8mm downstream from nozzle at the edge of spray)

Although the test results shown in Figure 3 are only from one velocity vector, juxtaposing the data of DMF with gasoline and ethanol helps to understand the behaviour of the fuel. These results show the great similarity of DMF and gasoline in terms of droplet velocity.

However, ethanol was observed to have an overall lower velocity distribution at all test points, as highlighted by the example in Figure 3 (b). In fact, DMF and gasoline have very close physical properties, which play a major role in the spray characteristics.

The droplet size distribution is normally more useful than the velocity information, since it gives direct information about how liquid fuels break up, evaporate and generate the desired fuel-air mixture. Statistical values, such as the Sauter Mean Diameter (SMD), are widely used in computational simulation models. The formula to calculate the SMD is as follows:

$$D_{32}=\frac{\sum N_i d_i^3}{\sum N_i d_i^2} \tag{1}$$

As a multiple test point approach, PDPA is capable of giving droplet size information at all test points. The SMD maps for the three test fuels have been drawn out by this means and are shown in Figure 4. As discovered by other researchers, all three test fuels have larger droplets close to the nozzle than downstream and closer to the middle than at the edge.

For all the test fuels, the droplet size decreases rapidly soon after the fuel was injected. After 24mm downstream of the spray, the droplet sizes become stable, which can be explained by the secondary break-up process. The break-up lengths of the three fuels are also similar. Primary break-up cannot be detected by this method due to the high density of droplets which can be considered filaments and not necessarily droplets.

For gasoline, a slight increase in droplet size along the axis can be observed, but not for DMF or ethanol. Further than 40mm from the nozzle, no significant difference was observed in terms of droplet size between the gasoline and DMF sprays. However, the ethanol spray behaves differently. Near the nozzle, the ethanol droplets are the smallest of the three fuels. However, the ethanol spray conversely has the largest droplets downstream. The SMD variation of ethanol along the axis (16.9-20.2μm) is lower than for the other two fuels (16-22.5μm and 14.9-23.4μm, for gasoline and DMF, respectively).

The stable droplet size is mainly affected by three physical properties; vapor density, surface tension and relative velocity. According to Chu's theory (Chu 1986), the stable droplet size is proportional to the surface tension, inversely proportional to the vapor density and square of the relative velocity. DMF, ethanol and gasoline have similar surface tensions (25.9, 22.3 and 20dyne/cm, respectively, see Table 1). However, ethanol has the lowest vapor density at 1.59 (relative to air, see Table 1), which, according to the theory (Chu 1986), is why ethanol has the largest stable droplet size. Nevertheless, this theory cannot explain why ethanol has the smallest droplet size near the nozzle. It can be assumed that ethanol has the shortest primary break-up length, which means at the same position away from the nozzle, more droplets break-up in the ethanol spray than in the other two. However, more research is required to fully support this hypothesis.

Another issue is the effect on the SMD of the injection pressure. As shown in Figure 5, the SMDs of DMF, ethanol and gasoline were measured at the same arbitrary sampling point with injection pressure variations from 50bar to 150bar, thus covering most modern direct-injection spark-ignition applications. Generally, the higher the injection pressure, the smaller the SMD becomes. Within the tested pressure region, the profile shows good linearity. The SMD for gasoline and DMF behave similarly. At 50bar, the SMD for both gasoline and DMF are around 21μm and decrease in unison to 13μm at 150bar. Ethanol, once again, performs differently. With increasing injection pressure, the SMD of the ethanol droplets reduces at a lower rate. At 15MPa injection pressure, Ethanol has an SMD more than 15μm, which is 2μm higher than gasoline.

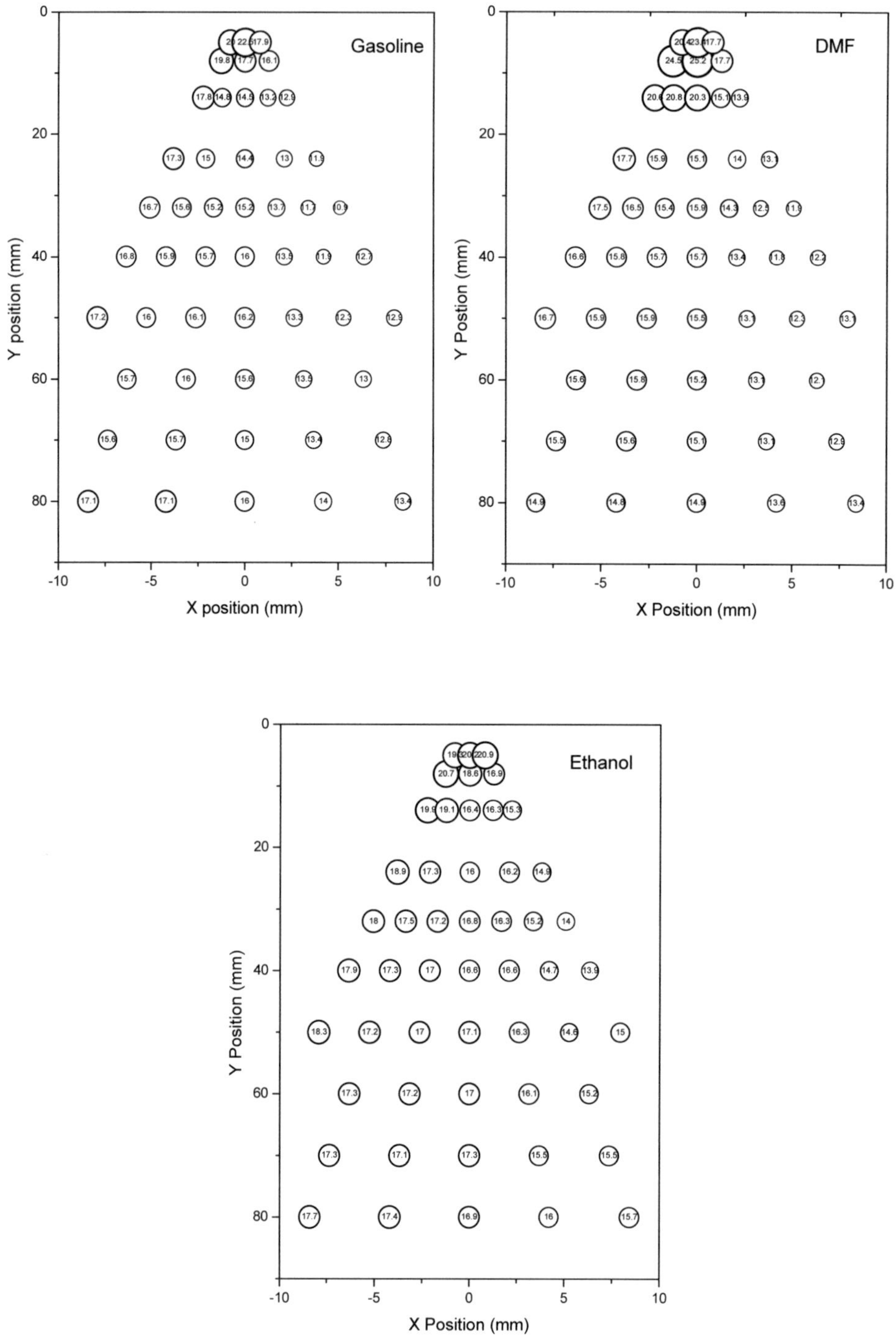


Fig. 4. SMD maps for gasoline, ethanol and DMF (Injection pressure: 10MPa, injection duration: 2mm)

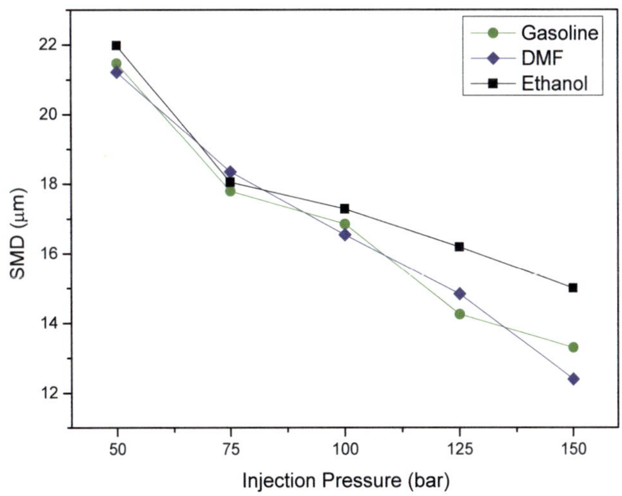


Fig. 5. Effect on injection pressure on SMD (sampling point: 32mm downstream of nozzle on axis)

In summary, the spray characteristics of DMF are particularly favourable. The spray pattern when using DMF is very similar to that with gasoline. Although the DMF spray has a marginally lower velocity than with gasoline, the competing biofuel, ethanol, has a significantly lower velocity in the spray head region and a higher rate of reduction compared to the others. The DMF spray also has a negligibly larger droplet size than the gasoline spray, while ethanol has more abundant droplets. These findings highlight DMF's suitability to match gasoline's behaviour and so would not require major modifications to the fuel injection system.

## 4. Laminar burning velocity

Laminar burning velocity is another important fuel characteristic and is necessary for simulation work. For phenomenological combustion models, a complete understanding of the chemical kinetics of the flame reactions is required. However, due to the complexity and stochastic nature of turbulence, most numerical models rely on the experimental laminar burning velocity to interpret the turbulent component. In fact, the laminar burning velocity directly affects the burn rate and thus the engine performance. Different methodologies are available to study the laminar burning velocity. A widely used technique is to study premixed quiescent combustion, such as that in a constant volume vessel. In this method, a spherical flame develops from initial spark energy and propagates outwardly. The laminar flame speed is then interpreted using high-speed photography. As the product mass in the vessel increases, the pressure and unburned gas temperature upstream of the flame increase as well. However, as this methodology necessitates steady-state ambient conditions, only the data sampled before the increase of pressure and temperature can be used for data analysis (empirically when the spherical flame diameters are smaller than 10% of the vessel inner diameter).

Schlieren techniques are one of the most commonly used optical access methods, due to the cost effectiveness, accuracy and simplicity. The Schlieren method captures images with varying density gradients found at the flame front during combustion. A typical Schlieren test bench plan is presented in Figure 6.

The well established premixed combustion theory will not be discussed here. However, several basic concepts of laminar flame propagation are helpful to understand the following discussion. The concepts will be limited to the case of quiescent premixed combustion with outwardly propagating spherical flames (Law 2006).

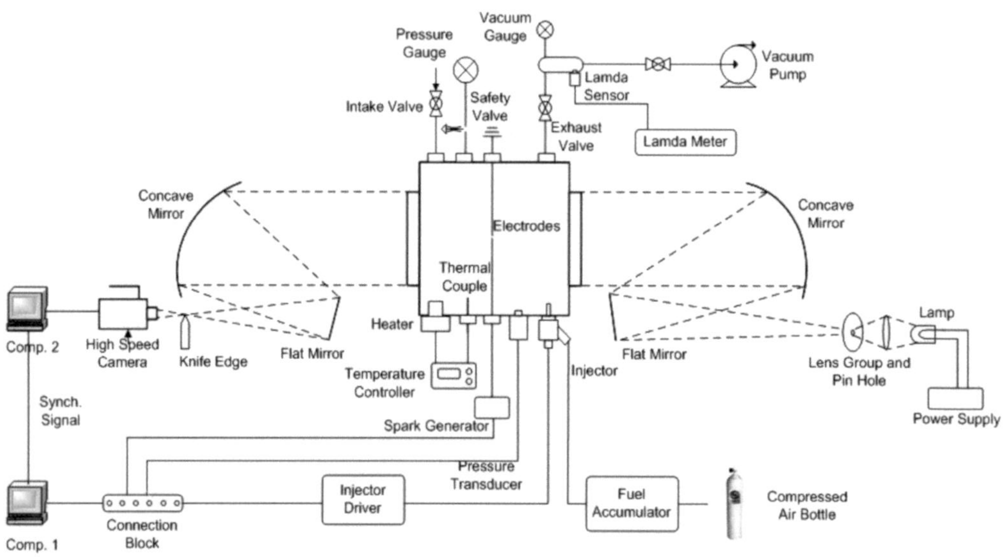


Fig. 6. Schematic of Schlieren test setup

Stretch rate: the flame curvature has a strong influence on the structure and thus the flame speed measurement. The intensity of stretch is represented by a stretch rate,

$$\alpha = \left(\frac{2}{r_f}\right)\left(\frac{dr_f}{dt}\right) \tag{2}$$

with units $s^{-1}$, where $r_f$ is the measured spherical flame radius.
Stretched flame speed: the direct measurement of the flame front propagation speed is defined as:

$$S_n = \frac{dr_f}{dt} \tag{3}$$

Therefore, the stretch rate is:

$$\alpha = \frac{2}{r_f} S_n \tag{4}$$

Unstretched flame speed: the downstream flame speed is significantly affected by the stretched flame and can be shown to be approximately linear to the stretch rate. This linearity allows the extrapolation of the stretched flame speed to the to zero stretch rate, where we can obtain the unstretched flame speed $S_s$.
Markstein length: the Markstein length, $L_b$, is determined by the negative slope of the line when extrapolating the stretched flame speed to the zero stretch rate:

$$S_n = S_s - L_b \cdot \alpha \tag{5}$$

This indicates the influence of stretch rate on the flame propagation speed and characterizes the diffusion-thermal instability. Positive Markstein lengths indicate that the flame is stable to the diffusion-thermal effect, while a negative Markstein length indicates that the flame surface is distorted by the diffusion-thermal effect, which leads to flame speed acceleration and the onset of cellularity in the flame surface. According to the research from Bradley et al. (Bradley 1998), if the Markstein number exceeds 1.5, the flame is initially stable until a critical flame radius is reached.

Laminar burning velocity: the laminar burning velocity $u_l$ is deduced from the unstretched flame speed $S_s$ using the equation:

$$u_l = S_s \frac{\rho_b}{\rho_u} \tag{6}$$

where $\rho_b$ and $\rho_u$ are the burned and unburned gas densities respectively. The pressures are assumed constant in order to find them via the conservation of mass equation:

$$\frac{\rho_b}{\rho_u} = \frac{V_u}{V_b} = \frac{n_u T_u}{n_b T_b} \tag{7}$$

where $n_b$ and $n_u$ are the mole numbers of the products and reactants, and $T_b$ and $T_u$ are the adiabatic flame and initial temperatures, which can be inferred.

Figure 7 juxtaposes three groups of high-speed schlieren images chronologically for DMF, ethanol and gasoline. The images show the stoichiometric (λ=1), premixed combustion under a pressure of 0.1MPa and temperature of 75°C and are indexed by the time after ignition.

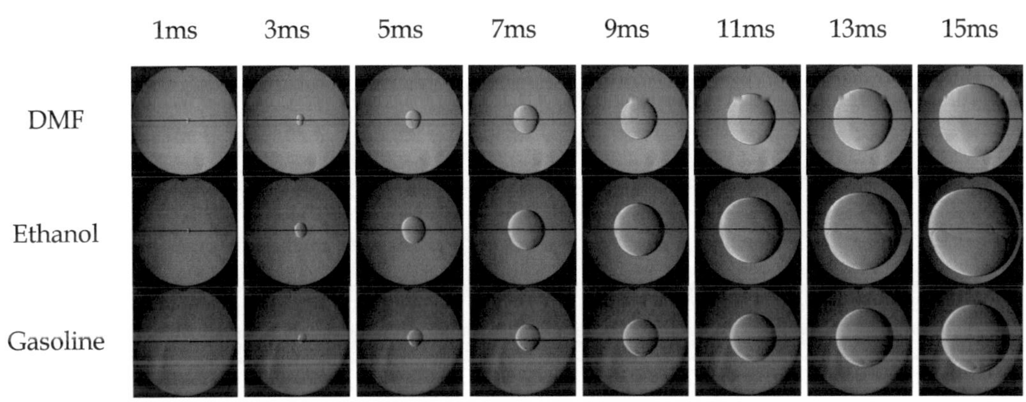


Fig. 7. High speed Schlieren images of DMF, ethanol and gasoline laminar flame propagation

The flame speed when using ethanol is clearly the highest; the relatively advanced flame front, which is established after 3ms, is maintained throughout the expansion. There is very little difference between the cases of DMF and gasoline.

The initiation of the gasoline flame shows the effect of the instability of the spark energy. This is shown between 3 and 5ms, where the flame begins to propagate above the electrode gap. In the DMF and ethanol cases, the centre of the spherical flame is much closer to the gap and produces a more even distribution relative to the window. As a consequence of the quenching at the electrode-flame interface, the lower flame surface of gasoline develops a crack. However, as the flame develops, it grows evenly in all radial directions. Therefore, by restricting the analysis region to radii greater than 6mm (up to 18mm, approximately 10% of the vessel inner diameter), the influence of the spark instability can be ignored.

The DMF flame propagates evenly in all radial directions; the surface appears smooth and is therefore stable. However, as with the other flames, slight wrinkling is observed near the electrode due to the quenching effect, but this does not affect the overall shape. When the spherical flame approaches the edge of the window, the shape becomes distorted. This effect can be observed more clearly on the ethanol flame after 15ms, as it has reached the most advanced stage. The shape of the flame becomes more oval with flatter vertical surfaces. This is believed to be related to the instability of buoyancy and the influence of the internal geometry of the combustion chamber. From the ethanol flame images, a slight crack is also observed after 5ms, which is believed to be caused by the influence of the electrode.

The stretched flame speed is described as the rate of change of the schlieren flame radius. In order to avoid the impact of the spark energy and the pressure increase in the vessel, only specific images (6mm diameter to 1/10 of the vessel inner diameter, as described previously) were used for the analysis. Figure 8 shows the stretched flame speed of DMF under various equivalence ratios, benchmarked with ethanol and gasoline. The stretched flame speeds were plotted against the stretch rate. The further the flame propagates, the greater the radius and the smaller the stretch rate become. Therefore, with respect to time, the points move towards the zero stretch rate. Through linear extrapolation, both the unstretched flame speed, $S_s$, and the Markstein lengths, $L_b$, can be determined at the zero stretch rate using the negative slope of the line, shown in Figure 9.

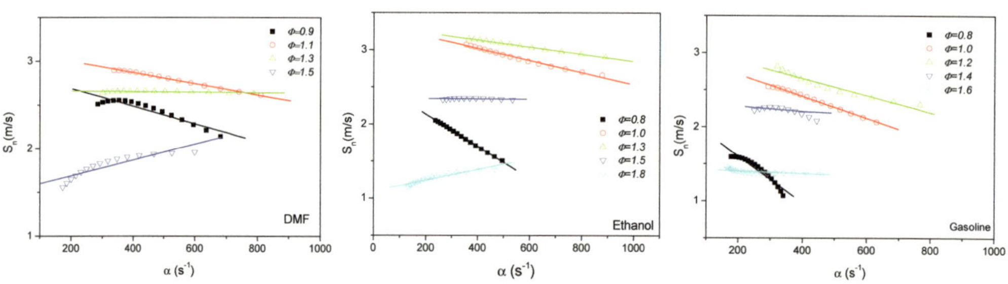


Fig. 8. Stretched flame speed of DMF, benchmarked with ethanol and gasoline (initial temperature: 75°C, pressure: 0.1MPa)

Generally, the Markstein lengths decrease with increasing equivalence ratio, or $\varphi$ ($\lambda^{-1}$) for each initial temperature. This is because the test fuels are heavy hydrocarbon-air mixtures, and the Markstein length depends only on the Lewis number of the deficient reactants (Bechtold 2001).

With regards to temperature, the Markstein lengths have increased at low equivalence ratios (0.8 - 0.9) when the temperature increased. This suggests the flames are much more stable at higher temperatures. However, at high temperature the Markstein lengths then decrease more rapidly from these higher values, suggesting the stability quickly decays. This is also shown by the earlier entry into negative Martkstein lengths, with respect to the equivalence ratio, for all three fuels. This indicates that at higher temperatures, the fuel-air mixtures tend to be more diffusion-thermal unstable and the flame is more susceptible to accelerate with increasing stretch rate. The onset of cellular activity would be quicker and cracks in the flame surface would occur more easily. The rapid decay results in lower Markstein lengths at rich equivalence ratios.

With the exception of 75°C, the ethanol flame appears to be the most stable through the entire range of air-fuel mixtures. However, the difference is insignificant. The Markstein lengths of the DMF flame, on the other hand, are always lower than these of the ethanol flame and partly so for the gasoline flame for this temperature range. This suggests that the DMF flame is slightly more unstable.

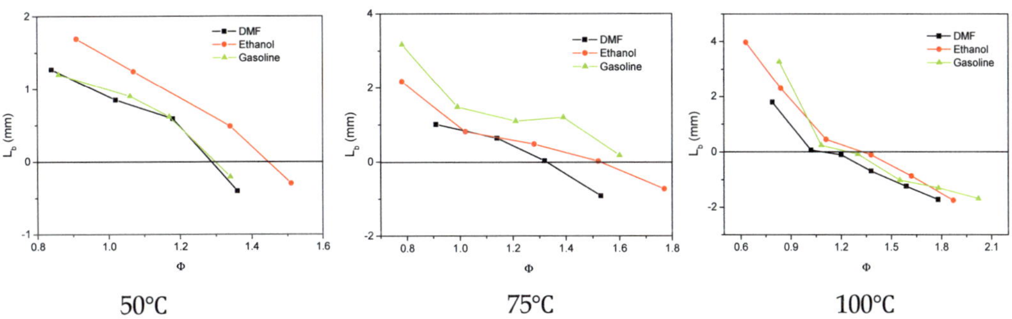


Fig. 9. Markstein lengths of DMF, ethanol and gasoline at various initial temperatures

The Markstein length normally decreases with increasing initial pressure for heavy hydrocarbon fuels. This means at higher initial pressures, the flame propagation is less affected by the flame stretch rate and the flames are less stable, especially for rich mixtures. Figure 10 illustrates the schlieren images at different initial pressures from 0.1MPa to 0.5MPa for a rich mixture (φ=1.2). The fuel used in this test was 20% DMF blended with 80% iso-octane, by volume. For the initial pressures of 0.1 and 0.25MPa, the flame front was smooth without cracks, even when the flame radius reached 30mm. However, at the initial pressure of 0.5MPa, cracks appeared on the flame front when the flame radius was only 10mm. These cracks develop into a cellular structure when the flame radius approaches 30mm under this condition. This is due to the combined effect of diffusional-thermal and hydrodynamic instabilities. The Markstein length and the flame thickness decrease with initial pressure. The increased hydrodynamic instability cannot be reduced by the decreased diffusional-thermal stability.

The aforementioned results can be used to deduce the laminar burning velocities, one of the key indicators of flame behaviour. Figure 11 shows the results for the three fuels for a range of initial temperatures. It is clear that ethanol has the highest burning velocity amongst the three fuels for all the initial temperatures. For instance, at 50°C, the peak laminar burning velocity of ethanol (56cm/s) is 13cm/s higher than that for the gasoline mixture (43cm/s). This difference increases to 17cm/s for 75°C and 15cm/s for 100°C. This reinforces ethanol's superiority, in terms of laminar burning velocity. For the other two fuels, gasoline's laminar burning velocity is more closely matched by DMF, although the laminar burning velocity of DMF is marginally lower. The difference increases as temperature increases and equivalence ratios rise above 1.2. Nevertheless, their profiles are very similar and the curvature is less significant than for ethanol. In real-world engineering applications, e.g. for commercial automotive engines, the equivalence ratios will only slightly deviate around stoichiometric conditions (Φ=0.9-1.2) due to the requirement of the after treatment system to meet emissions legislations. In this region the difference between the laminar burning velocity of DMF and gasoline is less than 10%.

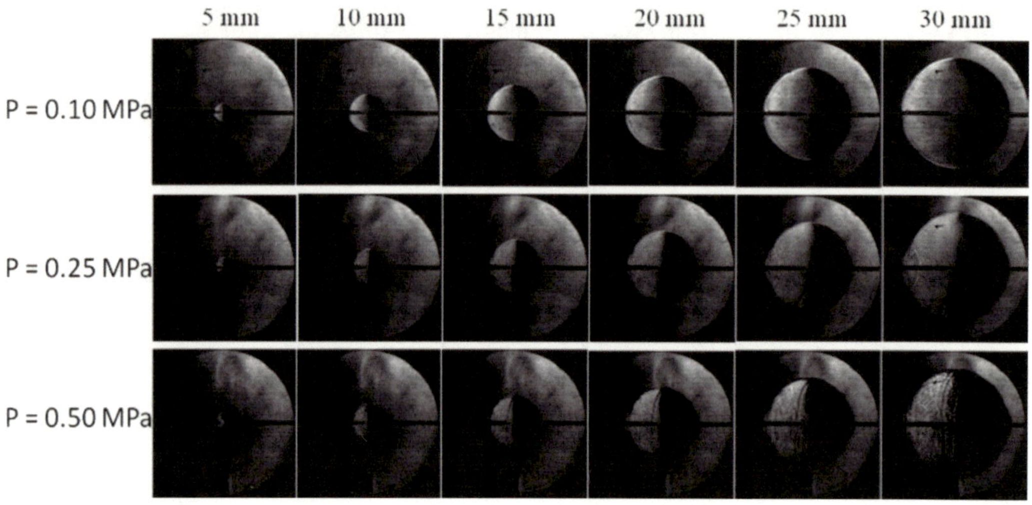


Fig. 10. Schlieren images of flames for DMF/iso-octane-air mixtures at the equivalence ratio of 1.2 and the initial temperature of 393K at three different initial pressures

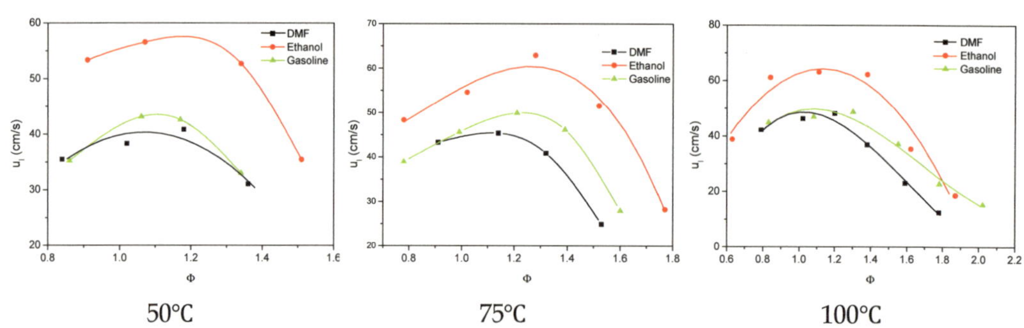


Fig. 11. Laminar burning velocities at various initial temperatures

Ethanol shows high potential in terms of its laminar burning velocity. Although the velocity decreases significantly when the equivalence ratio goes beyond 1.6, the magnitude is very high in the wide near-stoichiometry range. The difference between the laminar burning velocity of DMF and ethanol in a typical engine application region (Φ=0.9-1.2) is almost 30%. The curvatures of the lines within these graphs closely resemble those of other fuels under similar ambient test conditions (temperature and pressure), as documented by Heywood (Heywood 1988). He states that the peak laminar burning velocity of gasoline at 27°C is 36cm/s. In our tests, at 50°C, the peak is 44cm/s, an increase of 8cm/s. This highlights how the initial temperature affects the burning velocity. Increasing the temperature also appears to bring the peak burning velocity for the other equivalence ratios toward a maximum burning velocity. For instance, when the temperature is doubled to 100°C, the maximum remains the same, but the difference between the surrounding equivalence ratio points is

reduced. Therefore, there appears to be a maximum laminar burning velocity. This is certainly apparent for ethanol, where its maximum laminar burning velocity is approximately 62cm/s. For DMF, the maximum is less clear from the data available, but this lies somewhere in the range 41-48cm/s.
In summary, the fundamental spray characteristics and laminar flame propagation tests show how DMF has a high potential for use in spark-ignition engines. Its spray and laminar burning characteristics are very close to those of gasoline, which means it could be adopted in current spark-ignition engines without major modifications.

## 5. Engine performance and regulated emissions

Thus far, the analysis of DMF has shown promising similarities to gasoline. However, the experiments were performed in controlled environments whereas in an engine application, the local temperatures and pressures are varying cyclically. Therefore, the next step is to analyse the combustion characteristics in an engine, in order to understand the influence of the complex combination of various parameters. Within the research community, this can be achieved using single-cylinder engines.
In this section, the engine performance (for instance, optimum spark timing, fuel economy or exhaust temperature) and the regulated engine-out emissions (nitrogen oxides ($NO_x$), carbon monoxide (CO) and total hydrocarbons (THC)) are discussed. Furthermore, as DMF is a new biofuel candidate, it is important to analyse the relative products of combustion. Not only is it necessary to look at solid particles but also the individual hydrocarbons. This allows us to assess the concern for the environment and to humans. Therefore, detailed analysis of the solid particles is made in the next section of this chapter.
Once more, DMF is benchmarked against gasoline and compared to ethanol. As can be expected, the three fuels have different locations of optimum spark timing (referred to as the maximum brake torque or, MBT timing) due to their differing anti-knocking properties, or octane numbers. Figure 12 shows these optimum spark timings at various engine loads for DMF, ethanol and unleaded gasoline (ULG). The indicated mean effective pressure (IMEP) is a widely used scale for engine research and represents the engine load. Unlike engine torque, which is measured from the dynamometer, the IMEP is calculated using the raw crank-angle-resolved in-cylinder pressure trace. This reflects the power directly delivered from the fuel energy to the piston, before any friction losses are considered. This method is more meaningful for engine research because the friction losses vary greatly between engines. For a naturally aspirated spark-ignition engine, the IMEP is usually around 10 bar at wide open throttle (highest load).
Figure 12 clearly shows that at low load (3.5bar IMEP) there is no difference in the MBT location between the three fuels; the spark sweeps generate a relatively flat IMEP curve around 34°bTDC. However, throughout the remaining load range, ethanol allows the most advanced spark timing due to a higher knock resistance and burning velocity. Ethanol also generates greater charge-cooling because of a higher latent heat of vaporization (840kJ/kg for ethanol, compared to 332kJ/kg and 373kJ/kg for DMF and gasoline respectively, see Table 1). This lowers the combustion temperature and reduces the end-gas auto-ignition, or knocking tendency. At the highest load, the MBT spark timing when using ethanol is 11CAD more advanced than with gasoline and 5CAD more than with DMF. Until 6.5bar IMEP, the MBT timing between DMF and ethanol is only separated by 1CAD. Despite this, the maximum IMEP when using DMF is limited by audible knock at 5.5bar IMEP and the

theoretical MBT timing cannot be achieved. Although it is believed that DMF has a higher octane number than ethanol (Yuriy Roman-Leshkov 2007), the MBT timing in the current engine testing is more retarded than that with ethanol, due to this onset of knock. The optimum spark timing for gasoline is clearly the most retarded, once again limited by knock.

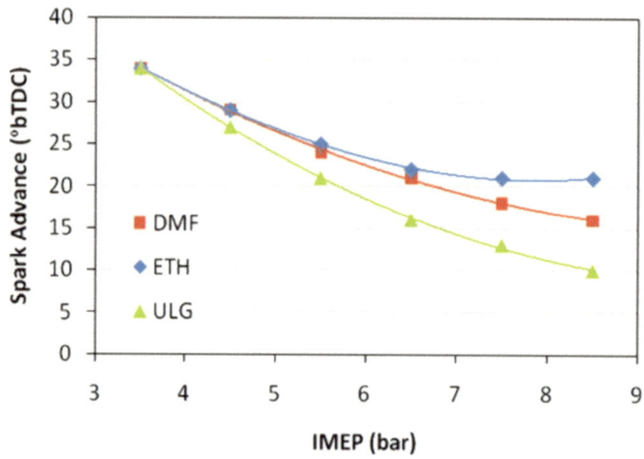


Fig. 12. Optimum Spark Timings at Various Engine Loads for DMF, Ethanol and Gasoline

When using 97 RON gasoline, a knock margin (2CAD retard) was enforced as early at 4.5bar IMEP, which restricted the ability to find the optimum MBT timing. For DMF, however, this safe margin was not enforced until 5.5bar IMEP, as previously mentioned. Although this produces an advantage in efficiency for DMF over 97 RON gasoline, in terms of knock suppression, and therefore spark advance, the anti-knock qualities of DMF are not as proficient as ethanol's due to the lower in-cylinder gas temperature associated with the latter. The studies by Gautam and Martin (Gautam 2000) have shown that the knock suppression capability of oxygen containing fuels is related to the relative oxygen content . Therefore, it is perhaps not surprising for DMF to have poorer anti-knock qualities than ethanol, as it contains less relative oxygen (see Table 1).

Figure 13(a) shows the gravimetric fuel consumption. Although there are differences between the three fuels, whereby DMF is more efficient than ethanol, Figure 13(b) is perhaps more relevant to the end user. The volumetric fuel consumption of DMF is very similar to that of gasoline, due to similar volumetric calorific values. DMF's calorific value is only 6% less than gasoline's, whereas ethanol is 33% less. Ethanol suffers from a low energy density, which would require more re-fuelling for the same volume compared to DMF.

Another fuel conversion efficiency measure is the indicated efficiency, or inversely, the gasoline equivalent indicated specific fuel consumption, or ISFCE. The indicated efficiency is defined as the ratio of the indicated power output to the lower heating value contained in the burning fuel mass. The gasoline equivalent fuel consumption is calculated using the equation shown. First, the numerator converts the fuel consumption rate into an energy rate using the lower heating value of the fuel. Then, once this is divided by the gasoline lower heating value, the fuel rate is converted to a gasoline equivalent. Clearly, the indicated efficiency and gasoline equivalent fuel consumption are inversely proportional. Introducing these two parameters allows a direct comparison of the energy conversion efficiency of the fuel between DMF, gasoline and ethanol. The ISFCE normalizes the lower heating value relative to gasoline and so presents a similar trend and offset as found with the indicated efficiency.

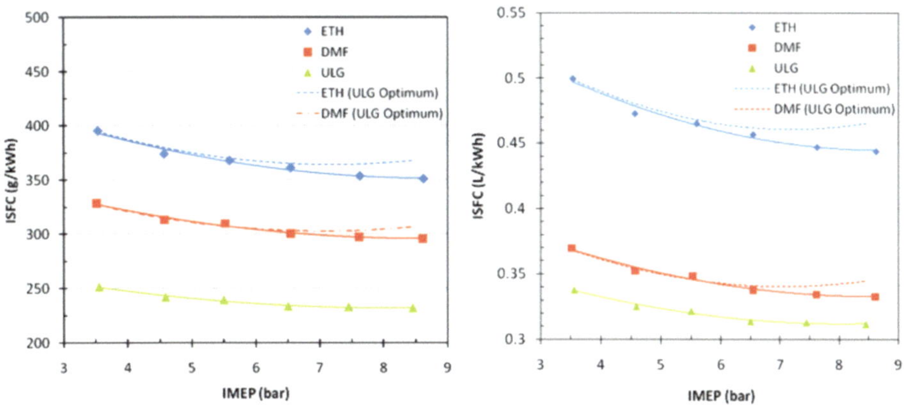


(a) (b)

Fig. 13. (a) Gravimetric and (b) Volumetric Indicated Specific Fuel Consumption

$$ISCE_x = \frac{ISFC_x \times LHV_x}{LHV_{ULG}} \tag{8}$$

Figure 14 highlights the relationship between the indicated efficiency and the ISFCE between the three fuels. Under gasoline MBT timing, as shown by the dashed and dashed-dot lines for ethanol and DMF respectively, the three fuels reach an efficiency peak between 7 and 7.5bar IMEP. Although the curvature is similar, the offset is different. Using advanced modeling software, the authors calculated that DMF burns at the highest combustion temperature, which should lead to a higher thermodynamic efficiency. However, this effect is compensated by the higher heat exchange loss. This is also reflected in the ISFCE. Therefore, this preliminary work suggests that DMF is less efficient than ethanol in transferring the fuel energy into useful engine work when using the fuel-specific MBT timing. Nevertheless, both fuels are competitive to gasoline, especially at high engine loads.

The exhaust gas temperature is another important parameter because it influences the catalyst converter and turbocharger performance (if equipped). Figure 15 shows the exhaust gas temperature 55mm downstream of the cylinder head for the various loads and spark timings between the three fuels. Ethanol combustion clearly results in the lowest exhaust temperature. The reason for this is twofold. Firstly, ethanol has a relatively high latent heat of evaporation, which absorbs more heat in the compression stage. Secondly, the MBT timing for ethanol is much more advanced than for DMF and gasoline, which leads to lower in-cylinder temperatures when the exhaust valves open as the combustion process completes earlier. This also means that more energy is extracted from the burned gas to deliver the useful work, which can explain the higher efficiency seen in Figure 14. When using DMF, the exhaust temperature is approximately 30°C lower than with gasoline at the highest load when using the fuel-specific MBT timing. However, for ethanol, this drop is more than double that seen with DMF (approximately 65°C). As previously mentioned, ethanol has more advanced MBT timing, which advances the combustion and causes a lower exhaust temperature. This effect can be seen when comparing the gasoline specific MBT timing with the fuel-specific MBT timing. When using DMF, for instance, at gasoline MBT timing, the temperature difference between DMF and gasoline is negligible (shown by the dotted line). However, when the spark timing is optimized, the exhaust temperature when using DMF drops significantly. At these temperatures the exhaust gas temperatures are suitable for the efficient performance of three-way catalysts. However, when using a turbocharger, which is likely to become more prominent with engine downsizing, it is more useful to have higher exhaust gas temperatures, so that there is more potential to extract useful energy.

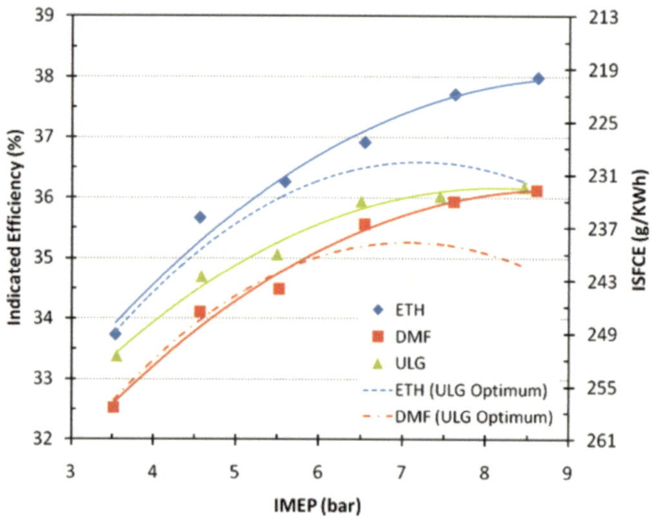


Fig. 14. Indicated Efficiency/Gasoline Equivalent Indicated Specific Fuel Consumption

In spark-ignition engines, the necessity to throttle the intake air causes a restriction to the flow, which leads to pumping losses (not seen in diesel engines). This pumping loss is one of the main reasons why spark-ignition engines have lower efficiency than compression-ignition engines. The volumetric efficiency, which can be thought of as opposite to the pumping loss, is defined as the ratio between the actual intake air mass and the maximum theoretical intake air mass without any losses. The comparison of the pumping losses between DMF and ethanol is shown in Figure 16. This loss directly affects the fuel economy and the differences between the two fuels can be better explained by their physicochemical properties. As shown in Table 1, ethanol contains more oxygen atoms, by weight, which results in a lower stoichiometric air-fuel ratio, or AFR. This reduces the throttle angle demand and as a result increases the net pumping loss and fuel consumption. On initial inspection, this behavior explains the trend of the volumetric efficiency compared to DMF and gasoline (Figure 16). At lower loads (<6bar IMEP), there is little difference in the volumetric efficiency between the oxygen content fuels. However, at higher loads (≥6bar IMEP), the separation is more evident, which demonstrates the effect of ethanol's low stoichiometric AFR. For gasoline, whose stoichiometric AFR is much higher (14.5), the volumetric efficiency is superior because more air is required to compensate for no oxygen in the fuel. However, with closer inspection of the pumping losses, or pumping mean effective pressure (PMEP) below 6bar IMEP, ethanol overcomes the higher throttling requirement. This is largely due to the higher charge-cooling effect of ethanol. At 3.5bar IMEP, the pumping loss, when fuelled with DMF at gasoline MBT timing is 48.9kPa, whereas for ethanol the loss is 47.9kPa. This represents an advantage to ethanol of 1kPa. At 8.5bar IMEP, this advantage shifts in DMF's favor by 2.1kPa (15.7kPa for DMF and 17.8kPa for ethanol). This reduction in pumping loss for ethanol at low load could be attributed to its high heat of vaporization. This effect on volumetric efficiency in a direct-injection spark-ignition engine is well documented by other researchers (Heywood 1988; Stone 1999). Therefore, the relatively high charge-cooling effect created when the ethanol fuel is injected and then evaporated, counteracts the pumping loss due to throttling. As the fuel is injected, the in-cylinder temperature reduces as the fuel vaporizes. This increases the density of the intake air, which allows more air to be consumed. Following the cross-over with DMF at 6bar IMEP, the charge-cooling advantage is then superseded by the much higher throttling losses at higher load.

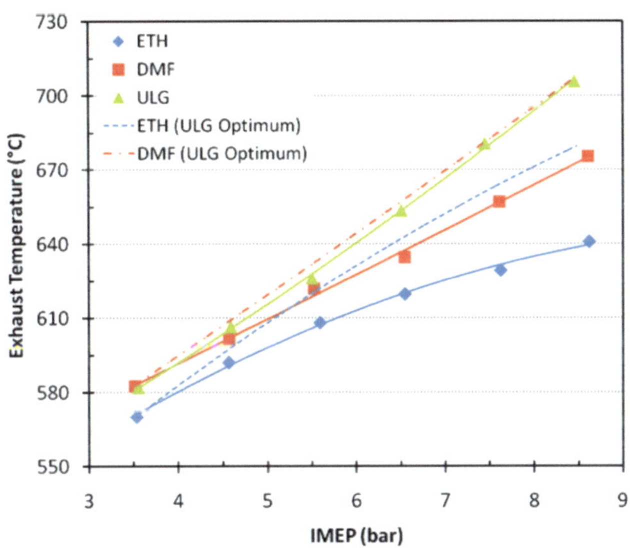


Fig. 15. Exhaust Gas Temperatures

The maximum in-cylinder pressures $P_{max}$, or peak pressure, for the three fuels are shown in Figure 17. At low load, the difference between these three fuels is marginal, but as the load increases, the peak pressures of ethanol and DMF are significantly higher than that with gasoline. At loads higher than 6.5bar IMEP, the peak pressure when using ethanol exceeds that when using DMF. At 8.5bar IMEP, the peak pressure with ethanol is 7bar higher than with DMF and almost 20bar higher than with gasoline. In Figure 17, the peak pressures of ethanol and DMF when using the MBT timing for gasoline are shown with dashed and dashed dot lines respectively. The difference between the three fuels is negligible and is explained by the different combustion speeds. Although the laminar burning velocity data is available from previous research, it is the turbulent burning velocity which dominates the combustion speed. Without going into depth on the fundamentals of turbulent combustion theory, the combustion speed is normally determined by the in-cylinder pressure trace and is defined as the period in crank angle degrees or, CAD, between 10 and 90% of the total energy released, or mass fraction burned (MFB). This combustion duration data for the three fuels is shown in Figure 18.

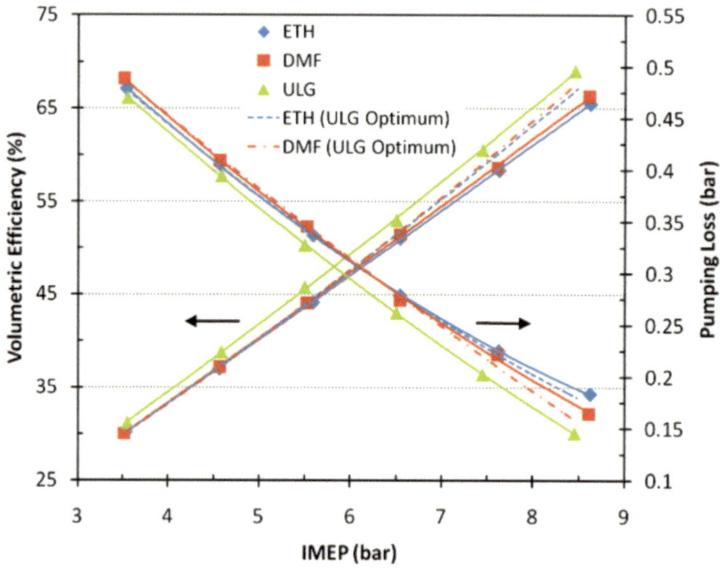


Fig. 16. Volumetric Efficiency and Pumping Losses

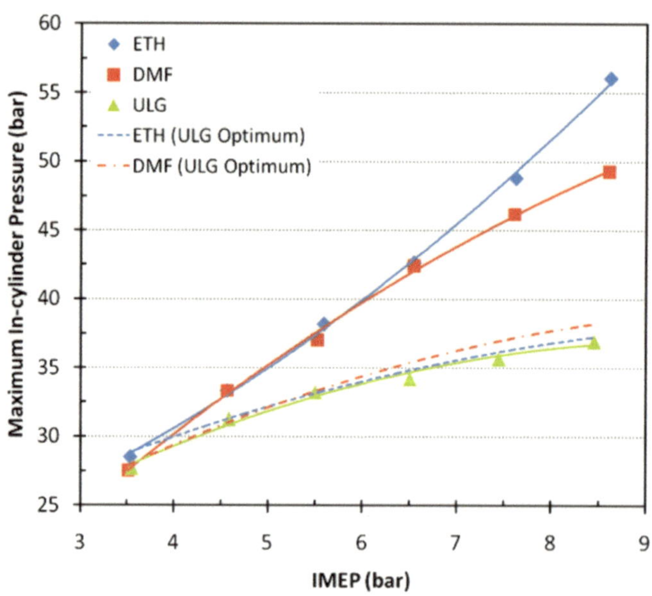


Fig. 17. Maximum In-cylinder Pressure

Evidently, when using the fuel-specific MBT timing, the combustion speeds for both DMF and ethanol are higher than with gasoline. At 8.5bar IMEP, the combustion of the oxygen content fuels ends approximately 4CAD before gasoline. However, between the two oxygen content fuels, the combustion duration when using DMF is marginally lower, than for ethanol. Apart from the low and high loads (3.5bar and 8.5bar IMEP respectively), the combustion duration of DMF is at least 0.35CAD lower, than with ethanol. At fixed gasoline MBT timing, it is again DMF that burns the fastest (apart from 3.5bar IMEP). At 8.5bar IMEP, DMF burns 1CAD quicker than ethanol and 1.3CAD quicker than gasoline. That explains why at this point, DMF has slightly higher peak in-cylinder pressures than ethanol.

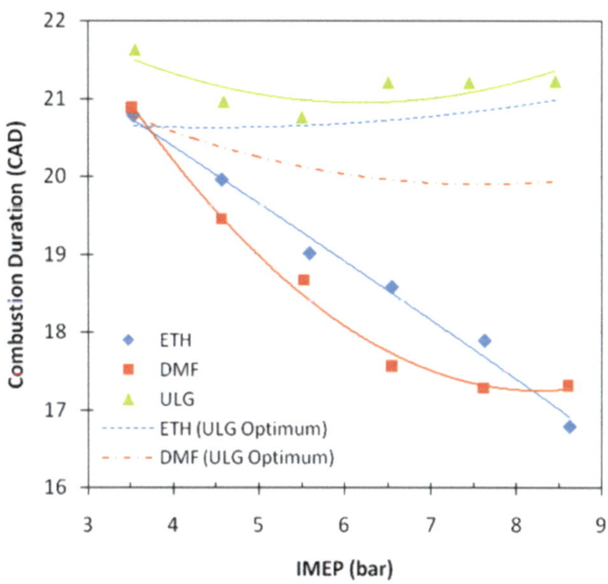


Fig. 18. Combustion Duration (10-90% MFB)

With further analysis of the combustion event, we can study the combustion duration either side of the 50% MFB point, or CA50 when using the gasoline MBT timing. At 8.5bar IMEP, DMF combustion is shown to be consistently quicker than both ethanol and gasoline. This is shown in Figure 19. For all three fuels, the duration before CA50 (10-50% MFB), is slightly higher than that afterwards (50-90% MFB). For DMF, the 10-50% MFB duration is 10.75CAD compared to 9.2CAD for the 50-90% MFB duration, which represents a difference between ethanol of 0.12CAD and 0.83CAD respectively. For DMF and gasoline, the reduction in duration after the CA50 location is 15% and 12% respectively. However, for ethanol this reduction is limited to 8%. Although at this load, ethanol burns more quickly than gasoline before CA50, afterwards gasoline is superior.

These results highlight the advantage of the burning speed of DMF combustion compared to that when using ethanol despite having a lower measured laminar flame speed which was introduced previously. Overall, however, under these test conditions, both oxygen content fuels burn more quickly than gasoline.

The engine-out emissions are compared between the three fuels at the various loads and spark-timing conditions. Firstly, the regulated emissions, which include nitrous oxides

($NO_x$), unburned total hydrocarbons (THC), and carbon monoxide (CO) emissions are evaluated. This is followed by an analysis of the carbon dioxide ($CO_2$) emissions.

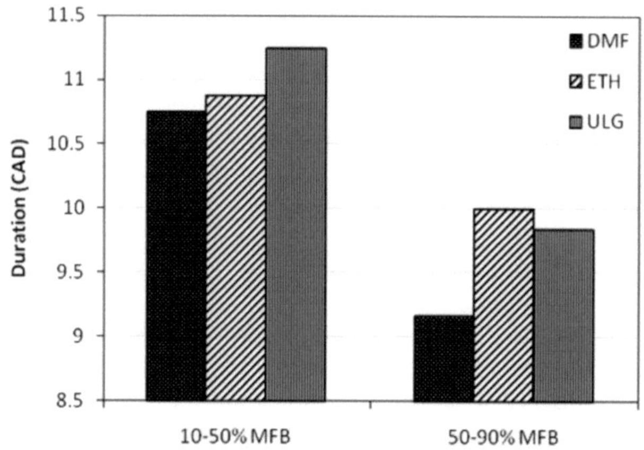


Fig. 19. Combustion Durations either side of CA50 (8.5bar IMEP, Gasoline Optimum Spark Timing)

$NO_x$ is one of most concerning engine-out emissions. It is 20 times more toxic than CO and is the main component to cause photochemical smog. The formation of $NO_x$ mainly occurs when the combustion temperature exceeds 1800K and the mixture is locally rich of oxygen. The indicated specific $NO_x$ (is$NO_x$) emissions are illustrated in Figure 20. The indicated specific emissions shows the mass of emissions gas from the engine with each unit of power produced and has the units of g/kWh. It is clear that the is$NO_x$ production is load dependant and generally increases with load. When using the same spark timing, ethanol produces much lower is$NO_x$ emissions compared to gasoline. This is because ethanol burns at a relatively higher rate and with a lower combustion temperature. Although DMF appears to have a marginally quicker burning rate than ethanol, the is$NO_x$ emissions are more similar to gasoline because the combustion temperatures are much higher. For fuel-specific optimum ignition timing, the production of is$NO_x$ emissions increases. For ethanol, this increase with load is much larger than for DMF. Above 7.5bar IMEP, the emissions are now comparable with gasoline. Optimized MBT timing for ethanol is 11CAD more advanced than the gasoline optimum spark timing, which rapidly increases the in-cylinder pressures. The peak combustion pressures for DMF are similar to ethanol but the temperatures are somewhat higher due to the lower charge-cooling effect. The relative is$NO_x$ emissions can also be attributed to the H/C ratio. Ethanol, which produces the lowest is$NO_x$ emissions, is the highest H/C ratio fuel, whereas DMF produces the highest is$NO_x$ emissions and has the lowest H/C ratio. Therefore, the is$NO_x$ emissions have an inverse relationship to the H/C ratio of the fuel.

The indicated specific total hydrocarbon emissions (isTHC) are much lower for ethanol at all conditions, than for DMF and gasoline, as suggested in Figure 21. Ethanol's oxygen content is much higher than DMF, which, together with it's high combustion efficiency, aids the oxidation of unburned hydrocarbons as oxygen is more readily available. As the load increases from 3.5 to 8.5bar IMEP, the isHC emissions decrease by approximately 30% for all fuels. This is due to increased combustion temperatures and thus combustion efficiency. The

HC and CO emissions reduce due to greater oxidation of the hydrogen and carbon molecules. However, the level of isTHC emissions produced from DMF combustion is between that with gasoline and ethanol.

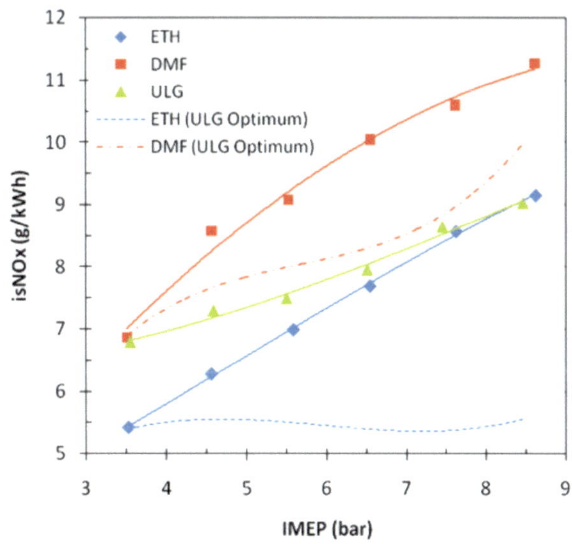


Fig. 20. Indicated Specific $NO_x$ Emissions

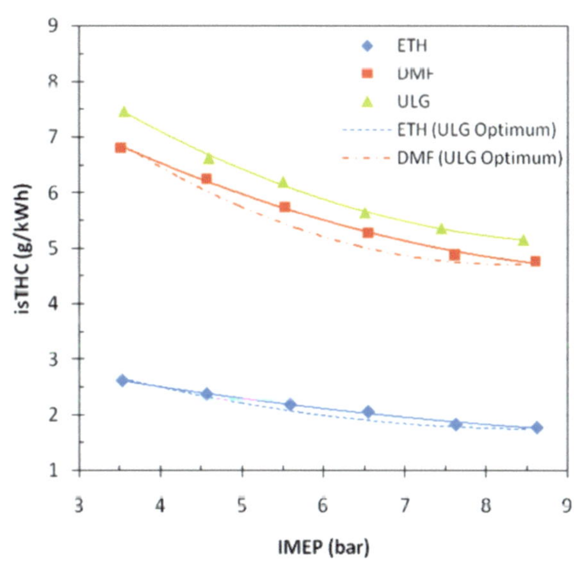


Fig. 21. Indicated Specific Total Hydrocarbon Emissions

The indicated specific carbon monoxide emissions (isCO) comparison between the three fuels is made in Figure 22. Similarly to the isTHC emissions, the isCO emissions generally decrease as load increases. This trend is similar to the inverse of indicated efficiency, where the lowest isCO emissions arise at the highest efficiency. Between the two oxygen content fuels, ethanol consistently produces the lowest isCO emissions for all test conditions. This is due to a higher combustion efficiency and oxygen content. Under gasoline optimum spark timing, the difference increases with load. At 3.5bar IMEP, ethanol is 1g/kWh lower, whereas at 8.5bar IMEP this difference increases to 3g/kWh. Under fuel-specific optimum ignition timing the largest difference is seen at medium loads. For gasoline, the relationship with respect to load is less predictable. The peak at 4.5bar IMEP could be explained by the relatively lower combustion efficiency at this point. At this load, the mixture may be inhomogeneous, resulting in localized pockets of fuel-rich mixture and more incomplete combustion. However, the remaining isCO emissions fluctuate within a similar range as the two biofuels, which all decrease to a minimum around 7.5bar IMEP.

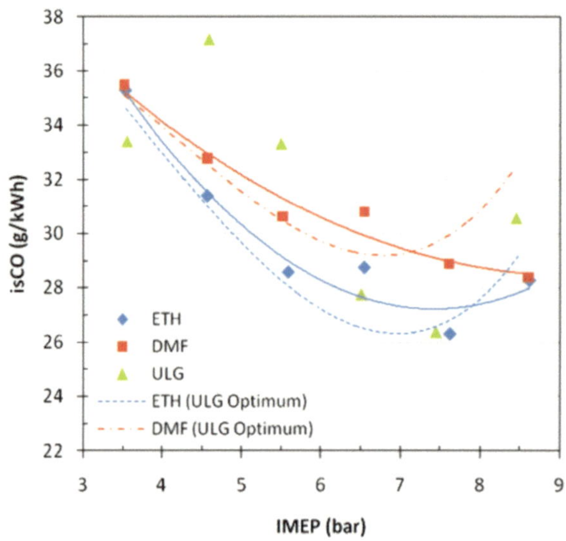


Fig. 22. Indicated Specific Carbon Monoxide Emissions

Although carbon dioxide ($CO_2$) is a non-toxic gas, which is not classified as a pollutant engine emission, it is one of the substances responsible for global warming through the greenhouse effect. Therefore, a consideration of the indicated specific $CO_2$ (is$CO_2$) production is made between the three fuels. This is shown in Figure 23. Here, the is$CO_2$ emissions decrease with increasing load and advancing ignition timing towards the optimum spark timing. The is$CO_2$ emissions are an indication of the completeness of combustion. Therefore, as the load is increased, the combustion is more complete, which is shown by the increase in combustion efficiency. When using gasoline optimum spark timing, DMF and ethanol combustion produces a peak in efficiency between 6 and 7bar IMEP and a minimum in is$CO_2$. Although both biofuels produce higher engine-out is$CO_2$ emissions than gasoline, they both have the added benefit of consuming the $CO_2$ in the atmosphere during their raw production. Therefore, it is fairer to compare the relative lifecycle $CO_2$ emissions.

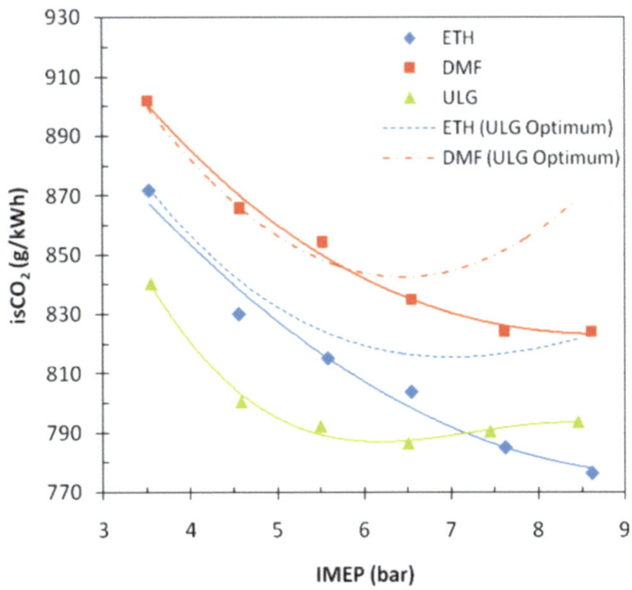


Fig. 23. Indicated Specific Carbon Dioxide Emissions

The engine tests further highlight the competition DMF creates with ethanol in replacing gasoline as a spark-ignition fuel. Due to ethanol's ability to suppress knock, the engine could be run at more advanced spark timing conditions allowing the optimum or MBT timing to be used. The optimized timing for DMF and gasoline, however, was limited by engine knock, and so was retarded. Despite this, DMF was more resistant to knock than gasoline, which suggests that DMF's octane rating is between that of ethanol (approximately 110 RON) and 97 RON gasoline. Despite this, the speed of combustion when fuelled with DMF was faster than that with ethanol. In fact, both DMF and ethanol have lower initial and total combustion durations than gasoline. These lower combustion durations were further reduced when the ignition timing was optimized, highlighting the rapid combustion of these oxygen content fuels. Additionally, the volumetric fuel consumption rate of DMF is similar to that of gasoline. This suggests that a consumer using DMF as a substitute for gasoline, would benefit from a similar range with the same volume of fuel. Although the indicated efficiency of DMF is lower than gasoline, there is potentially more extractable energy in the exhaust gas. Finally, in terms of engine-out emissions, the results for DMF were similar to that with gasoline. The only penalty was seen in the way of $NO_x$ emissions, which would be reduced through the catalytic converter. These $NO_x$ levels were also higher than that for ethanol, mainly due to the higher combustion temperatures. Ethanol's higher latent heat of vaporization offers better charge-cooling. However, this does introduce cold-start issues which would not be experienced when using DMF.

In the next section, an overview of the unregulated emissions is made. This focuses on the solid particles which are produced through DMF combustion. This is then followed by an investigation of a novel approach for the improved use of biofuels in spark-ignition engines through the dual-injection fuelling system.

## 6. Unregulated emissions

Currently, the particulate matter (PM) emissions do not form part of the emissions legislations for gasoline spark-ignition engines in Europe or the US. However, control of these emissions is expected to commence in European and US regulation in 2014. This will require, not only the monitoring of particulate mass, but also the particulate number for all light-duty vehicles. Therefore, an understanding of these emissions will become much more important and it is necessary to understand the implication of the PM emissions when using biofuels. In this section, the PM emissions between the three fuels is studied at low and high load.

The PM size distributions for the three fuels at 3.5bar IMEP using gasoline MBT timing (34°bTDC), are shown in Figure 24 (a). Typically, the PM size distribution consists of two modes: the accumulation and nucleation modes. The former consists of solid carbonaceous species usually greater than 50nm in diameter, whereas the latter consists of liquid particles usually less than 50nm in diameter. The separation between the nucleation and accumulation modes is shown clearly by the peak in the size distributions around 50nm in Figure 24 (a). At this low load, the size distribution shows marginally more accumulation mode particles than nucleation ones. For gasoline, 62.1% of the total particles are accumulation mode particles, whereas for DMF and ethanol, this rises to 64.4 and 67.1%, respectively. The difference between DMF and ethanol is 21,805particles/cm$^3$. This might be caused by DMF's lower viscosity and surface tension, which leads to smaller injected fuel particles. Also, the in-cylinder temperature when using ethanol has been seen to be much lower than that for DMF. This is caused by the greater charge-cooling effect when using ethanol, which counteracts the benefit that the higher oxygen content would have in helping to lower the PM emissions.

The PM size distributions are also compared at 8.5bar IMEP using gasoline MBT timing (10°bTDC). This is shown in Figure 24 (b). Evidently, the effect of load has a significant impact on the PM emissions. The separation between the nucleation and accumulation modes is shown clearly by the inflection in the size distributions around 50nm in Figure 24 (b). Here, the nucleation mode particles dominate the particle size distribution for each fuel, with peaks between 25-30nm. For ethanol and DMF, the accumulation mode now only represents 2.1 and 1.7% of the total particle concentrations, respectively. However, for gasoline, the proportion of accumulation mode particles is much higher (18.3%). In absolute terms, gasoline combustion also produces more accumulation mode particles than when fuelled with the biofuels; both biofuels produce less than 4,000particles/cm$^3$, whereas gasoline produces almost 21,000particles/cm$^3$. This is likely to be caused by the higher droplet velocity of gasoline and relatively high mean droplet diameter, found in previous studies. This could increase the impingement on the piston surface, which would explain the relatively lower combustion efficiency. At this load, the biofuels also burn at higher pressures and temperatures, which helps to promote pyrolysis and further reduce the solid carbonaceous emissions. The opposite is true, however, for gasoline when considering the nucleation mode concentration at 8.5bar IMEP. Now, gasoline combustion produces a similar amount of particles to that of ethanol, and DMF is the worst offender; almost 122,000particles/cm$^3$, or 30% more particles are produced when using DMF. This relative increase in nucleation mode particles for DMF was seen in previous experiments by the authors and is likely to be caused by the incomplete combustion of the ring structure of DMF, which are known to be soot precursors.

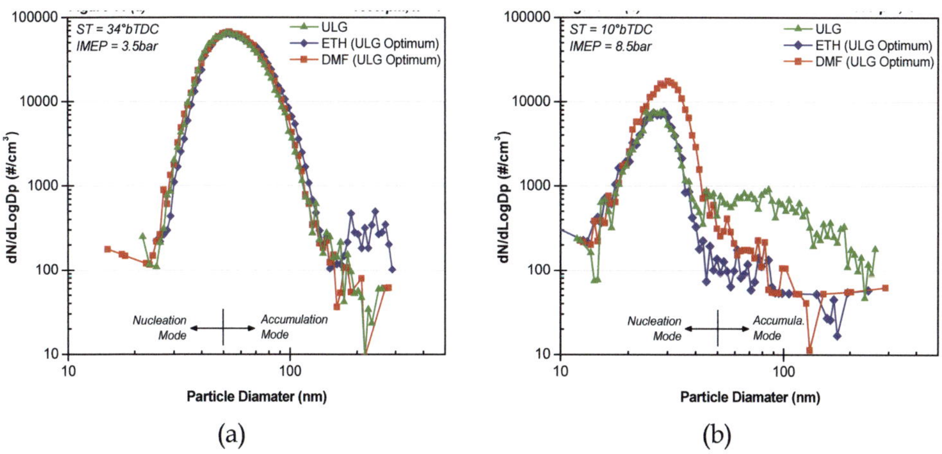


Fig. 24. (a) PM Size Distributions at 3.5bar and (b) 8.5bar IMEP using Gasoline MBT/KL-MBT Timing

In summary, the PM emissions for DMF have been shown to be comparable to that of gasoline at low load. At a higher concentration of load, DMF produces high nucleation mode particles compared to gasoline but a lower concentration of accumulation mode particles. Compared to ethanol, DMF shows similar sized accumulation mode particles at high load, which helps to reduce the average particle size.

The PM emissions work is ongoing and further results are likely to follow in publications by the authors. Furthermore, individual hydrocarbon speciation will show more details of the toxic compounds which are emitted from DMF, ethanol and gasoline and help to better understand the potential effect on the environment and to human health.

## 7. Dual-Injection - a potential method to utilise biofuels

There are many different approaches to partially substitute biofuels for gasoline. The simplest method is to run the engine on a blend of biofuel and gasoline which is supplied at the pump and so externally mixed. This, however, requires a fixed blend ratio to allow the engine to always run in the optimized condition. Alternatively, engines can have flex-fuel capability, with an ability to be smart enough to recognise different blend ratios at any moment and adapt the control strategy to reflect the change in fuel supply. This method is commonly used across Brazil for any blend of ethanol and gasoline. However, another approach is the so-called dual-fuel system. This method, which is referred to as dual-injection in this chapter, also increases system complexity similarly to a flex-fuel vehicle. However, dual-injection offers improved flexibility and control of the instantaneous blend ratio.

In the dual-injection tests, gasoline, is supplied as the main fuel and is delivered through a port-fuel injection (PFI) system. This is then gradually substituted by the supplementary fuel (DMF, ethanol and gasoline) using the direct-injection (DI) system in order to reach the required AFR Both injectors were calibrated to enable the precise control of the blend ratio at any point. For all stable conditions the engine was run at the stoichiometric AFR and constant preset throttle positions (for simplicity). As fractions of PFI gasoline were replaced by the DI fuel, improved volumetric efficiency and load outputs were expected as the throttle was kept constant. Of course, much more calibration work is required to fully realise this method. However, the use for this approach in real production engines is very promising.

The results in this section are shown using stacked graphs to reflect the three initial engine load conditions (represented by initial manifold air pressure, $MAP_i$) using PFI gasoline. For each normalized graph, the vertical axis shows the relative change in each key parameter from the 100% PFI condition. The horizontal axis shows the reduction in the PFI mass fraction, also from the 100% PFI condition, which has been compensated for using DI fuelling to maintain stoichiometry. Error bars have been used to highlight the test repeatability for the three repeats.

Figure 25 shows the relationship between the normalized volumetric air flow rates with varying PFI mass fraction for the three fuels at the three different initial engine loads. It is clear that the volumetric air flow rate increases with decreasing PFI mass fraction regardless of the $MAP_i$ and the DI fuel used. This is mainly because of the improved charge-cooling effect and lower PFI partial pressure. For gasoline, the maximum increase in volumetric air flow rate at the $MAP_i$ of 0.065MPa is 1.1%. The vaporization of the gasoline causes the intake air to cool, which increases its density and thus allows more air to flow into the cylinder. Meanwhile, the partial pressure of the PFI fuel decreases with decreasing PFI mass fraction and further contributes to the improvement in volumetric air flow rate. The increase in volumetric air flow rate for ethanol in DI is much higher than that seen with gasoline and DMF, regardless of the $MAP_i$. This is caused by two reasons. Firstly, ethanol has a much higher latent heat of vaporization, which results in more charge-cooling. Secondly, ethanol has a lower stoichiometric AFR. Therefore, in order to maintain stoichiometry, more ethanol is required, which amplifies the aforementioned reduced partial pressure effect. For DMF, the latent heat of vaporization is marginally lower than that for gasoline. Although this would help with engine cold starts, it produces less charge-cooling when the engine is warm. However, DMF has a lower stoichiometric AFR compared to gasoline, which requires more fuel at the same $MAP_i$. Therefore, the combined effects of the latent heat of vaporization and the stoichiometric AFR make the volumetric air flow rate of DMF fuelling comparative to that of gasoline. Nevertheless, the anticipated improvement with volumetric air flow rate is seen consistently throughout the dual-injection strategy and so offers benefits over the 100% PFI case.

The increase in volumetric air flow rate requires more fuel flow to maintain stoichiometry. Therefore, it is reasonable to expect a higher engine load output (IMEP) as the PFI fraction is reduced. This behaviour is shown in Figure 26. Firstly, it is useful to observe this effect when fuelled with gasoline in PFI and DI. Even at low $MAP_i$, the increase in load was 1%, where at high $MAP_i$ this increases to 5%. Further improvements were obtained using the two biofuels. At high $MAP_i$, 100% ethanol and DMF in DI produced improvements of 12% and 7% respectively, with near-linear improvements as the PFI fraction was reduced. Even at low $MAP_i$, the increases of 6% and 2% are notable, for ethanol and DMF, respectively. For all three fuels, the improvement in engine load is due to the improved volumetric efficiency, as previously highlighted. This effect was enhanced when using ethanol and DMF largely due to the lower stoichiometric air fuel ratios compared to gasoline. This means that more fuel is required to maintain stoichiometry. Despite the slightly lower volumetric energy density of DMF, due to the lower stoichiometric AFR, more fuel energy is supplied to the cylinder when using DMF than with gasoline. The effect of stoichiometric AFR on the volumetric efficiency is further enhanced when using ethanol, due to the higher latent heat of evaporation. This leads to a strong charge-cooling effect and allows even more fuel to be delivered as the density of the intake air increases. The combination of these two factors allows the dual-injection strategy to benefit from the increased energy output.

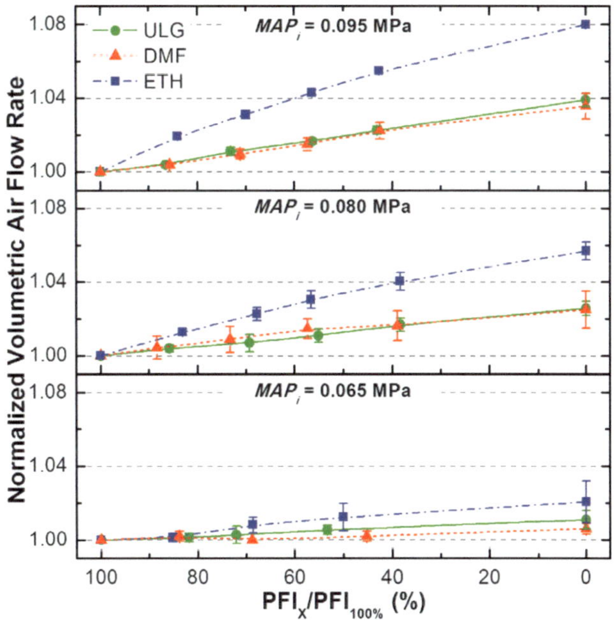


Fig. 25. Normalized Volumetric Air Flow Rate to 100% PFI condition with reduced mass fractions of PFI fuelling using Gasoline, DMF and Ethanol at three different MAPi

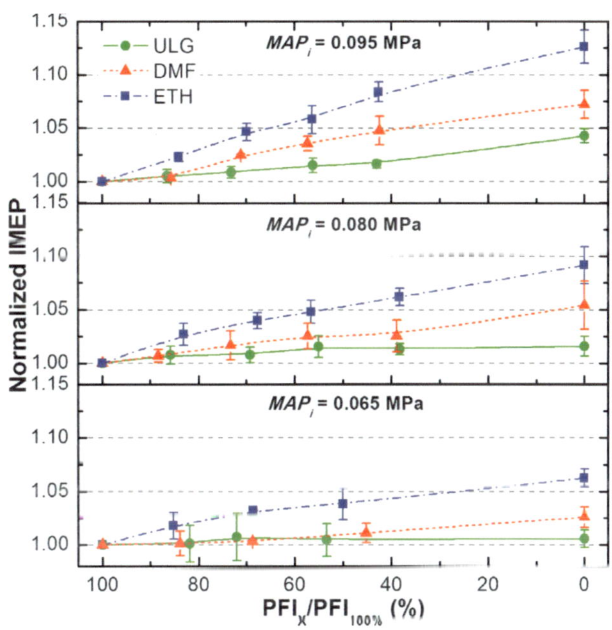


Fig. 26. Normalized IMEP to 100% PFI condition with reduced mass fractions of PFI fuelling using Gasoline, DMF and Ethanol

The indicated thermal efficiency, which is shown in Figure 27, presents varying trends between each DI fuel used. It is ethanol that shows the greatest improvements to efficiency compared to the 100% PFI case. This is apparent from as low as 85% PFI mass fraction. This highlights the high combustion and fuel conversion efficiency of ethanol compared to gasoline and DMF in DI. In Section 5, ethanol was shown to produce 1.2g/kWh lower gasoline equivalent fuel consumption compared to gasoline when using fixed gasoline timing at high load. This gap was increased to 11.2g/kWh when the timing was optimised. This suggests that further improvements to the indicated efficiency can be made through parameter optimization. DMF, on the other hand, produces less indicated efficiency than the other fuels and even 100% PFI in most instances. For most of the test conditions the gap between DMF and gasoline increases with higher mass fractions of DMF in DI. The lower efficiency of DMF, which has been discussed in Section 5 when using pure fuels, seems to have the same effect when using blended fuels. However, both injection timing (which affects mixture generation) and ignition timing were not optimized and the improvement of engine economy is expected when fuelled with DMF with more careful calibration work.

The additional blend control parameter is one of the attractions of the dual-injection system. The engine management system (EMS) would have the flexibility to choose between different fuels and blend ratios at any moment in the engine speed and load map. For instance, when fuel economy becomes a priority, the EMS can decide to run with the lowest fuel consumption strategy. Alternatively, when maximum engine power is the main requirement, the EMS will choose the highest power output strategy. Furthermore, any issues which arise when using the 100% fuel can be overcome. For instance, ethanol is known to suffer from cold starts due to the lower heat of vaporisation. Therefore, on engine warm-up, 100% PFI gasoline could be used until the engine is warm at which point ethanol can begin to be utilised. The availability and cost of the biofuels also play an important role when choosing the current operating strategy in order to optimise the system.

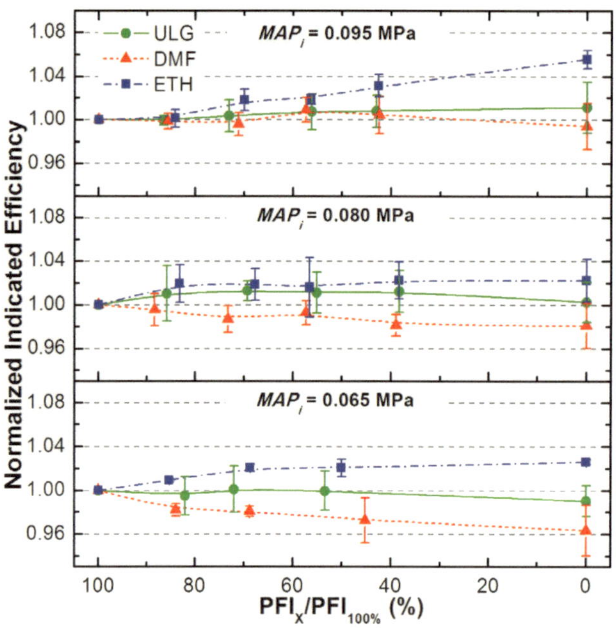


Fig. 27. Normalized Indicated Efficiency (IE) to 100% PFI condition with reduced mass fractions of PFI using Gasoline, DMF and Ethanol

The engine-out emissions produced during the dual-injection strategy are discussed below. Previously, in depth discussions were made concerning regulated, unregulated, gaseous and particulate emissions. Therefore, this section only concerns itself with three major emissions: THC, $NO_x$ and $CO_2$.

Figure 28 illustrates the indicated specific hydrocarbon emissions using the different fuels at the different $MAP_i$. When using gasoline in DI, the isHC emissions marginally increase at low $MAP_i$ as the PFI mass fraction is decreased. However, as the load is increased, the effect of gasoline in DI reduces the isHC emissions. Many researchers have proved the benefit of HC reduction for direct-injection at high load (Zhao 1999), because unlike the fuel evaporation process in the intake manifold for port fuel injection, the higher injection pressure for DI offers an improvement of liquid fuel atomization and a reduction of wall wetting. When using increased fractions of ethanol or DMF in DI, the increase of the direct-injection mass fraction reduces the isHC emissions, regardless of the $MAP_i$. Both DMF and ethanol contain an oxygen atom in their molecular structures, which helps to reduce the HC emissions, as oxygen is more readily available. The higher relative increase in IMEP (Figure 26), together with the more readily available oxygen atoms, improves the oxidation rate of the unburned hydrocarbons.

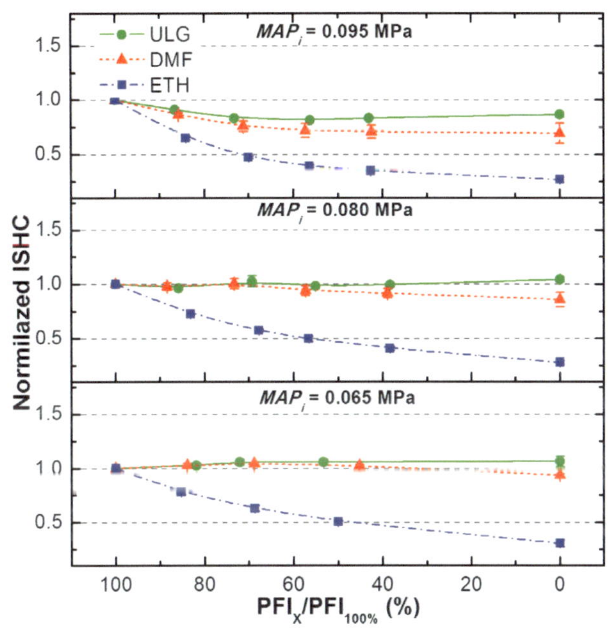


Fig. 28. Normalized Indicated Specific Hydrocarbon Emissions to 100% PFI condition with reduced mass fractions of PFI fuelling using Gasoline, DMF and Ethanol

Figure 29 shows the trend in the normalized indicated specific $NO_x$ ($isNO_x$) emissions between the three fuels. The $NO_x$ emissions are related to the fuel type. For the same excess air ratio, the nitric oxide or NO emissions, which represent the majority of $NO_x$ emissions, decrease with increasing H:C ratio (Harrington 1973). For the three fuel candidates, the H:C ratio increases in the order of DMF, gasoline, and then ethanol. Thus, the $NO_x$ emissions should reflect this order based on the pure fuel test results. As shown in Figure 29, the $isNO_x$ emissions are in the decreasing order of DMF, gasoline, and ethanol, which reflects the H:C ratio order and generally agrees with previous results. In terms of fuel delivery, when reducing the PFI proportion and thus increasing the DI proportion, the charge-cooling effect is increased, which decreases the in-cylinder temperature and helps to suppress $NO_x$

formation. This effect is clearly shown when substituting gasoline from PFI to DI. Although this decrease is moderate for gasoline, it is much more obvious than for ethanol. Ethanol has a higher heat of vaporization and lower stoichimetric air-fuel ratio. Thus, the charge-cooling effect is much more prominent when using ethanol.

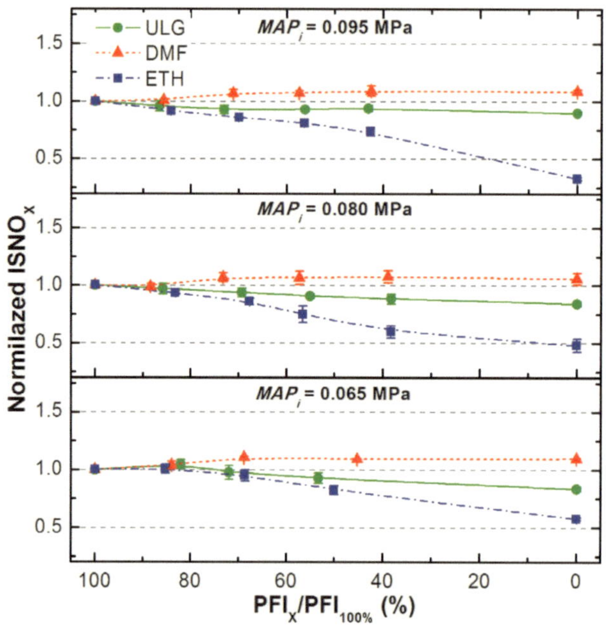


Fig. 29. Normalized Indicated Specific $NO_x$ Emissions to 100% PFI condition with reduced mass fractions of PFI fuelling using Gasoline, DMF and Ethanol

Figure 30 shows the normalized indicated specific carbon dioxide ($isCO_2$) emissions. Carbon dioxide is a non-toxic gas and is not classified as a pollutant engine emission. However, in recent years, the monitoring of $CO_2$ emissions has become more critical because it is considered to contribute to global temperature rises. Thus, the normalized $isCO_2$ emissions are highlighted. For gasoline, the dual-injection strategy helps to reduce the $isCO_2$ emissions at each $MAP_i$. The $CO_2$ emissions give an indication of the combustion efficiency and have been shown to decrease when switching from PFI to DI. The drop in efficiency helps to explain the reduction in $isCO_2$ emissions when using only gasoline. The H:C ratio also affects the $CO_2$ emissions concentration. Therefore, although ethanol has the highest H:C ratio, which helps to reduce the $isCO_2$ emissions, the relatively larger increase in IMEP compromises this benefit. This explains the similar $isCO_2$ performance of ethanol and gasoline. When using DMF in DI, the $isCO_2$ emissions increase at each $MAP_i$. This increase is mainly due to the lower H:C ratio. However, as a biofuel candidate, the lifecycle $CO_2$ emissions for DMF must be considered. DMF consumes $CO_2$ in its production stage, which would help to offset the increase in the engine-out $CO_2$ emissions.

In summary, this section has introduced a novel approach for utilizing biofuels in modern spark-ignition engines. The application of the dual-injection system is obviously beneficial despite a potential increase in cost. The engine power output increases with direct-injection substitution regardless of the alternative fuel used. We believe that, with more careful optimisation of injection and spark timing, the drop in efficiency and increases in $NO_x$ and

$CO_2$ emissions will be compensated for. Nevertheless, the reduction in HC emissions and attraction of blend ratio control and optimisation highlights the benefit of the dual-injection strategy with such biofuels as ethanol and DMF.

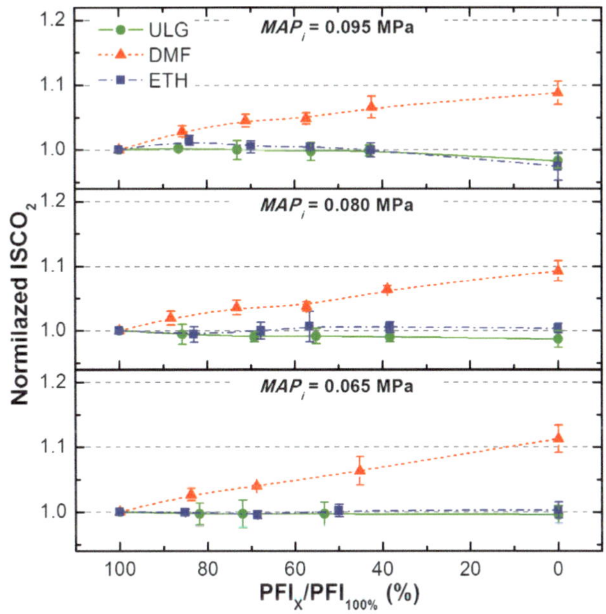


Fig. 30. Normalized Indicated Specific Carbon Dioxide Emissions to 100% PFI condition with reduced mass fractions of PFI fuelling using Gasoline, DMF and Ethanol

## 8. Summary

In this chapter, we introduced 2,5-dimethylfuran, a newly recognised and potential biomass-derived alternative fuel for spark ignition engines. Initial tests have included the fundamental combustion and spray analysis, basic engine tests to examine the effect on the engine performance and regulated emissions and further, more detailed unregulated emissions analysis, including particulate emissions. The chapter concludes with a new approach for the improved utilisation of biofuels and is referred to as the dual-injection system.

The results show how DMF is a promising alternative fuel. It has very similar properties to gasoline with regards to combustion, which means that it can be easily adopted with current spark-ignition engine technologies without the need for major modifications. Compared to the other major gasoline-alternative biofuel, ethanol, DMF is marginally inferior in terms of its laminar burning velocity, engine efficiency and engine-out emissions. However, since these emissions are nowadays combated by very effective aftertreatment systems, the advantages of DMF's higher energy density and potentially better cold-start performance are evident. Furthermore, the production of DMF does not compete with food and so makes it an attractive option.

The PM emissions of DMF (currently an unregulated requirement in Europe and the US) have been shown to be comparable to that of gasoline at low load. This work is ongoing and further results are likely to follow in publications by the authors, including an investigation of the individual hydrocarbons. Such results will show details of the toxic compounds and help to better understand the potential effect on the environment and to human health.

Finally, the dual-injection strategy offers an efficient and flexible use of biofuels, whilst still using gasoline. It offers the high engine performance of a modern DI system, the low $NO_x$, $CO_2$ and HC emissions of a PFI system and the possibility to run the engine using any blend ratio with limited implications to hardware cost.

## 9. Acknowledgement

The authors of this chapter would like to acknowledge the support from Jaguar Cars Ltd, Shell Global Solutions and various research assistants and technicians at the University of Birmingham, UK. They would also like to acknowledge the support from their international collaborators at Tsinghua University, China and Xuesong Wu from Xi'an Jiaotong University, China for his help with the dual-injection system.

## 10. References

Bechtold, J. K., Matalon, M. (2001). "The Dependence of the Markstein Length on Stoichiometry." Combustion and Flame 127: 1906-1913.

Chu, C.-C. C. (1986). One-Dimensional Transient Fluid Model for Fuel-Coolant Interaction Analysis, University of Wisconsin-Madison. PhD Thesis.

Bradley, D., Hicks, R.A., Lawes, M., Sheppard, C.G.W. and Woolley, R. (1998). "The Measurement of Laminar Burning Velocities and Markstein Numbers for Iso-octane–Air and Iso-octane–n-Heptane–Air Mixtures at Elevated Temperatures and Pressures in an Explosion Bomb." Combustion and Flame 115: 126–144.

Gautam, M. and Martin, D.W. (2000). "Combustion characteristics of higher-alcohol/gasoline blends." IMechE 214: 497-511.

Harrington, J.A. and Shishu, R.C. (1973). "A Single-Cylinder Engine Study of the Effects of Fuel Type, Fuel Stoichiometry, and Hydrogen-to-Carbon Ratio and CO, NO, and HC Exhaust Emissions." SAE 730476.

Heywood, J. B. (1988). Internal Combustion Engine Fundamentals, McGraw-Hill.

Law, C. K. (2006). Combustion Physics, Cambridge University Press.

Luque, R., Herrero-Davila, L., Campelo, J.M., Clark, J.H, Hidalgo, J.M., Luna, D., Marinasa, J.M. and Romeroa, A.A. (2008). "Biofuels: a technological perspective." Energy & Environmental Science 1: 542-564.

Mascal, M. and Nikitin, E.B. (2008). "Direct High-Yield Conversion of Cellulose into Biofuel." Angewandte Chemie Internation Edition 47: 7924-7926.

Stone, R. (1999). Introduction to Internal Combustion Engines, Basingstoke: Macmillan Press Ltd.

Wu, X., Huang, Z. and Jin, C. (2009). "Measurements of Laminar Burning Velocities and Markstein Lengths of 2,5-Dimethylfuran-Air-Diluent Premixed Flames." Energy and Fuels 23: 4355-4362.

Roman-Leshkov, R., Barrett, C.J., Liu, Z.Y. and Dumesic, J.A. (2007). "Production of dimethylfuran for liquid fuels from biomass-derived carbohydrates." Nature 447: 982-986.

Zhao, F., Lai, M.C. and Harrington, D.L. (1999). "Automotive spark-ignited direct-injection gasoline engines." Prog. Energy Combust. Sci. 25(5): 437-562.

Zhao, H., Holladay, J.E., Brown, H. and Zhang, Z.C. (2007). "Metal Chlorides in Ionic Liquid Sovlents Convert Sugars to 5-Hydroxymethylfuran." Science 5831: 1597-1600.

# 20

# Biofuels: From Hopes to Reality

Carioca J.O.B.[1,2], Friedrich, H.[3] and Ehrenberger, S.[3]
[1]*Universidade Federal do Ceará, Depto. de Tecnologia de Alimentos Campus do Pici, Fortaleza*
[2]*Centro de Energias Alternativas e Meio ambiente CENEA; Fortaleza*
[3]*Deutsches Zentrum für Luft- und Raumfahrt e.V, Stuttgart*
[1,2]*Brazil*
[3]*Germany*

## 1. Introdution

Recently, the United Nations System addressed to the world a report to create a green economy based in the following directions: *the potential to achieve sustainable development and to eradicate poverty on an unprecedented scale, with speed and effectiveness.* A green economy with low carbon potential to substitute fossil fuels, which addresses climate change, creates decent jobs, and reduces import dependencies. Much more than biofuels, green economies pay attention for market high values bioproducts. According to the United Nations Systems, this potential derives from two concurrent changes. First, there is a change indicating society risks and the necessity of rethinking the economy process considering the necessity of changing material availability; second, there is an increasing recognition that the natural environment forms the basis of our physical assets and must be managed as a source of growth, prosperity, and well-being [1]. Also this program addresses some questions related to important issues and directions concerning research like: *How to manage a smooth and fair transition from a brown economy to a green one at global level? How to ensure that green policies are not used as an excuse for trade protectionism? How to measure progress in the transition to a green economy?* As it can be seen, the United Nations Systems was precisely able to address specific directions related with a sustainable economy and the use of natural resource globally. Certainly the three global conferences on the environment and the UNEP's Green Economy Initiative, launched in late 2008 [2] were important steps to achieve such broad comprehension involving a large number of themes like: agriculture, renewable energy, cities, fisheries, water, forest, wastes, buildings, transports, tourism and manufacturing.

Certainly, these conclusions came out from the long time of understanding related with the indiscriminate use of natural resources and from the importance they have to achieve a sustainable world, as well as, contribute to diminish the environmental problem created by the use of fossil fuels for transport and electricity generation. To exemplify the importance of this report, since the 18th century, wood was the main source of material and energy for mankind. At that time, wood was primarily used for heating, house, ships, bridge constructions and metal processing. These historical uses caused a large deforestation in many areas of Europe. Unfortunately, this kind of process still continues devastating the Amazon forest, as well as other bioresources in Africa and Asia [3], [4]. At the end of the

19th century, the discovered of coal surpassed biomass use as a main source of energy and materials, mainly due to the beginning of the industrial era in England. Since then, the use of coal expanded in large proportion in Europe. As a consequence, it was observed the starting of an extensive use of non-renewable fossil sources of raw materials in the world, which was succeeded by the use of oil since the First World War. Below in Table 1 is shown data on the proportion of fossil fuels use, as well as renewable and nuclear according to the United Nations System report on green economy [5].

| Global energy consumption sharing | % |
|---|---|
| I- Fossil fuels | 78.0 |
| II- Renewable | 19.0 |
| Wind/solar/biomass/geothermal power -generation | 0.7 |
| Biofuels | 0.6 |
| Biomass/solar/geothermal hot water/heating | 1.4 |
| Hydropower | 3.2 |
| Traditional biomass | 13.0 |
| III -Nuclear | 2.8 |

Source: [5]

Table 1. Renewable energy share of global final energy consumption in 2008

Unfortunately, the contribution of fossil fuels is still very high, contributing to increase the greenhouse gas effect, and therefore the increasing of the $CO_2$concentration in the atmosphere. Another worse situation is related with the use of biomass in developing countries, mainly in Sub-Saharan Africa, Latin America and South Eastern Asia. Certainly, it is due to the lack of education and capital investment to access modern technological processes, like power generation using boilers and modern turbines, efficient heat transfer systems, as well as, biofuels production [6].

So, it is urgently necessary to increase the contribution of biofuels, still very low, to diminish environmental impacts. It is interesting to point out the message addressed to the world by Giácomo Ciamician in 1912 [7] known as the father of Photochemistry. At that time he expressed his environmental worries and concerns about the use of coal in Europe during the International Congress of Applied Chemistry in the USA, claiming that biomass should be used for chemical synthesis instead of energy. Fortunately the scientific development related with the agricultural resources, a green revolution is happening since 1970, promoting the intense use of advanced biotechnologies aiming at cell, protoplasts, tissue, and vegetable cultures by the use of genetic engineering to multiply plants and to create new productivity clones rapidly. These extraordinary advances and experiences together with the increase in the photosynthetic efficiency could be applied to increase agricultural productivity of some species to attend new society demands, such as biofuels production. Certainly, the choice of the right cultures in terms of its photosynthetic yields is of primary importance for the success of the biofuels production systems. The photosynthetic capacity of each vegetal depends on the quantity of $CO_2$ absorbed per unit of time on the foliar surface, and also, on the type of mechanisms (C3 or C4) involved in this process according to Melvin Calvin, the 1961 Chemistry Nobel Prize. Normally, it also depends on the radiation intensity, temperature, and $CO_2$ concentration and availability. Unfortunately, the maximum theoretical yield of the photosynthetic process is about 6.5%, but the practical yield is no more than 2.5% for some species like sugar cane and corn depending on the place where the plant is cultivated. Fortunately, there is a large amount of land and water resources available to grow C4 photosynthetic mechanism type of plants which is more appropriate to higher biomass production. The World Energy Outlook 2006 [8] is absolutely emphatic in recognizing that the world use of renewable resources is the only way to invert the actual $CO_2$ concentration in the atmosphere. As part of the photosynthesis process specific type of microalgae may produce vegetable oil or carbohydrates, as well as protein as its main constituents. It can generate much more oil per hectare than other plants used for biofuels production, such as traditional oil seed plants like soya, cotton and castor beans. Microalgae can grow in salt water, freshwater or even in domestic effluents, in ponds or photobioreactors using land not suitable for food production.

So, after the oil price crisis of 1967, which elevated the fossil fuels prices to a new platform much higher than the original one, as well as the environmental problems originated by the increasing of $CO_2$ concentration in the atmosphere, many countries created biofuel programs by themselves to replace fossil fuels. In conclusion, many of these biofuels programs are facing a lot of problems, such as; competition on land use for food and energy, fuels specifications, as well as environmental issues, like effluent disposal. In spite of biomass biodiversity and the renewable character of biofuels, they are still gaining space at regional level and impacting land use. As good examples, ethanol has gained space in Brazil and in the United States of America as well as in Asia. A second generation of technologies is in course to convert cellulose into ethanol through thermal conversion or bio conversion. Thermal processes present better results in terms of yields [9]. Biodiesel from vegetable oils are still receiving large subsidies in many countries where it is produced. Unfortunately it is based on traditional oil plants which have low productivity, and it is non economic under the actual oil prices scenario. Fortunately, a new technology based on yeast synthetic biology is able to produce a new molecule Biofene$^R$ [10] which promises to be an adequate substitute for diesel soon in Brazil. But clearly, no one can predict what kind of biofuel will properly succeed fossil fuels derivates realistically at the international market. So, regulations, standards, and targets are important issues to provide a realistic direction to the biofuel in the market. It is possible to see now some conflicts between countries resources availability to produce certain types of biofuels and the international standard motor specifications, yields and construction materials. It is important to point out that the sugar cane agroindustry development in Brazil is presenting a new phase related with companies association to gain production scale. That is the case of Corsan agribusiness alcohol group and Shell [11] which will be soon the greatest company in the biofuels sector in Brazil. Other international groups as BASF [12] and DOW [13], and a national group BRASQUEM [14] are investing in green biopolymers production like polyethylene based into dehydrated ethanol. This is a new reality of the bioresources use in direction to add-value products. Naturally, it is the right way to achieve green economy goals which will contribute to create jobs in the field improving simultaneously the environmental problem.

## 2. Biofuels and renewable scenarios best option

The studies made by OECD-FAO Agricultural Outlook 2008-2017 [15], as well as the Energy Out look-2006 [8], related to trends, scenarios and policies for fuel production and supply, emphasize that there will be an instability if economies will continue as before. So, instead of using these studies, this work will use the United Nations Green Economy report which offers one of the best options to be taken into consideration in the future analysis of the role of world renewable energy and biofuels production and use, once it comprises several progressive studies made in several areas like; agriculture, renewable energy, cities, fisheries, water, forest, wastes, buildings, transports, tourism and manufacturing. The United Nations green economy report was compiled by UNEP´s Green Economy Initiative, launched in late 2008, in straight collaboration with economists and experts worldwide. It demonstrates that the greening of economies is not generally a drag on growth but rather a new engine of growth; that it is a net generator of decent jobs, and that it is also a vital strategy for the elimination of persistent poverty. The report also seeks to motivate policy makers to create the enabling conditions for increased investments in a transition to a green economy. For its purpose, UNEP has developed a working definition of a green economy as one that results in *improved human well-being and social equity, while significantly reducing environmental risks and ecological scarcities.* In its simplest expression, a green economy can be thought of as one which is low carbon, resource efficient and socially inclusive [2].

Greening the energy sector aims at a renewable and sustainable energy system. This process involves improvements in energy efficiency, a much greater supply of energy from renewable sources and reducing greenhouse gas emissions and pollution. The most direct approach is to reduce the use of fossil fuels - an energy source whose combustion accounts for two thirds of all GHG emissions [16], as mentioned by the chapter on renewable [5]. According to [5] an adequate definition to *renewable energy is that one derived from natural processes that are replenished constantly. In its various forms, it derives directly or indirectly from the sun, or from heat generated deep within the earth. Included in this definition is energy generated from solar, wind, biomass, geothermal, hydropower and ocean resources, and biofuels and hydrogen derived from renewable resource.*

Following the UNEP´s Green Economy Initiative, it will be now set up the basic assumptions related with renewable energy and biofuels:

1. According to Table 1, the share of renewable energy in global final energy consumption in 2008 was 19%, but it could increase to the order of 27% depending of the investments done;
2. World primary energy demand is expected to continue growing. The International Energy Agency's Current Policies Scenario, which assumes no change in policies as of mid-2010, projects a world growth rate of 1.4 per cent per year up to 2035 according to data shown in Table 2.
3. World primary energy mix in the IEA Current Policies Scenario is shown in Table 3. These data show the importance of biomass, traditional and modern, in relation to other sources.
4. per cent at the peak in 2007, and is projected to remain high in the period to 2030 [18];
5. Worldwide investment in renewable energy assets - without large hydropower - grew by a factor of seven from US$17 billion in 2004 to US$126 billion in 2008. For OECD countries the share of renewable has risen from 4.6 per cent in 1973 to 7.7 per cent in 2009 [17];
6. In April 2010, the UN Secretary-General's Advisory Group on Energy and Climate Change-AGECC [19] published a report, which calls on the UN and its Member States to commit themselves to two achievable goals: universal access to modern energy services and a global energy intensity reduction of 40 per cent by 2030;
7. Expenditure on oil alone increased from 1 per cent of global GDP in 1998 to around 4 IPCC (2007) and IEA (2008c) estimate that in order to limit the rise of average global

temperature to 2 degrees Celsius, the concentration of GHG should not exceed 450 parts per million (ppm) CO2-eq. This translates to a peak of global emissions in 2015 and at least a 50 per cent cut in global emissions in 2050, compared with 2005. In 2009, the G8 committed to an 80 per cent cut in their emissions by 2050 in order to contribute to a global 50 per cent cut by 2050, although a precise baseline was not specified. The 80 per cent reduction would yield some space for developing countries to have a less stark reduction trajectory while reaching the global 50 per cent target.

| | Total energy demand [Mtoe] | | Growth rate 2008-2035[a] [%] | Share in total energy demand [per cent] | |
|---|---|---|---|---|---|
| | 2008 | 2035 | | 2008 | 2035 |
| OECD | 5421 | 5877 | 0.3 | 44.2 | 32.6 |
| Non-OECD | 6516 | 11,696 | 2.2 | 53.1 | 64.8 |
| Europe/Eurasia | 1151 | 1470 | 0.9 | 9.4 | 8.1 |
| Asia | 3545 | 7240 | 2.7 | 28.9 | 40.1 |
| China | 2131 | 4215 | 2.6 | 17.4 | 23.4 |
| India | 620 | 1535 | 3.4 | 5.1 | 8.5 |
| Middle East | 596 | 1124 | 2.4 | 4.9 | 6.2 |
| Africa | 655 | 948 | 1.4 | 5.3 | 5.3 |
| Latin America | 569 | 914 | 1.8 | 4.6 | 5.1 |
| World[b] | 12,271 | 18,048 | 1.4 | 100.0 | 100.0 |

a. Compound average annual growth rate. b. World includes international marine and aviation bunkers (not included in regional totals), and some countries/regions excluded here.

Source: [17]

Table 2. Primary energy demand by region in the IEA Current Policies scenario

| | Total energy demand [Mtoe] | | Growth rate 2008-2035[a] [%] | Share in total energy demand [per cent] | |
|---|---|---|---|---|---|
| | 2008 | 2035 | | 2008 | 2035 |
| Coal | 3,315 | 5,281 | 1.7 | 27.0 | 29.3 |
| Oil | 4,059 | 5,026 | 0.8 | 33.1 | 27.8 |
| Gas | 2,596 | 4,039 | 1.7 | 21.2 | 22.4 |
| Nuclear | 712 | 1,081 | 1.6 | 5.8 | 6.0 |
| Hydro | 276 | 439 | 1.7 | 2.2 | 2.4 |
| Biomass and waste[b] | 1,225 | 1,715 | 1.3 | 10.0 | 9.5 |
| Other renewables | 89 | 468 | 6.3 | 0.7 | 2.6 |
| Total | 12,271 | 18,048 | 1.4 | 100.0 | 100.0 |

a. Compound average annual growth rate. b. Includes traditional and modern uses.

Source: [17]

Table 3. World primary energy mix in the IEA Current Policies scenario

8. So, biofuels importance to modernize energy sector and help economy to attend the Millennium Development Goals and its links to energy access as shown in Table 4 [21].

| Millennium Development Goal | | How modern energy will help attain the MDGs |
|---|---|---|
| 1 | Eradicate extreme poverty and hunger by reducing the proportion of people whose income is less than US$1 per day (in US$PPP) | Increases household incomes by improving productivity in terms of time saving, increasing output, and value addition, and diversifying economic activity. Energy for irrigation increases food production and access to nutrition. |
| 2, 3 | Achieve universal primary education and promote gender equality | Provides time for education, facilitating teaching and learning by empowering especially women and children to become educated on health and productive activities, instead of traditional energy related activities. |
| 4, 5, 6 | Reduce child and maternal mortality and reduce disease | Improved health through access to clean water, cleaner cooking fuels, heat for boiling water, and better agricultural yields. Health clinics with modern fuels and electricity can refrigerate vaccines, sterilise equipment, and provide lighting. |
| 7 | Ensure environmental sustainability | Cleaner fuels, renewable energy technologies, and energy efficiency can help mitigate environmental impacts at the local, regional and global levels. |

Source: [21]

Table 4. Millennium development goals and links to energy access

## 3. Biofuels and the green mobility

Transport sector and the electric power generation are the two most important sectors responsible for the actual climate change due to their large contribution for the production of greenhouse gases emissions. At present, transportation sector consumes more than half of global liquid fossil fuels; emits nearly a quarter of the world's energy-related $CO_2$; generates more than 80 per cent of the air pollution in developed countries cities; results in more than 1.27 million fatal traffic accidents per year; and produces chronic traffic congestion in many of the world's urban areas. These costs to society, which can add up to more than 10 per cent of a country's GDP, are likely to grow, primarily because of the expected growth of the global vehicle fleet [22]. According to the green economy report chapter on ***green transport*** is hereby defined as one that supports environmental sustainability through e.g. the protection of the global climate, ecosystems, public health and natural resources. It also supports the other pillars of sustainable development, namely economic (affordable, fair and efficient transport that supports a sustainable competitive economy as well as balanced regional development and the creation of decent jobs) and social (e.g. allowing the basic access and development needs of individuals, companies and society to be met safely and in a manner consistent with human and ecosystem health, and promoting poverty reduction and equity within and between successive generations). This definition was developed through extensive discussions with transport experts including those at UN agencies, and was based on a review of existing and well-acknowledged definitions such as ECMT [23]. Commercial vehicles comprise over of land transportation energy consumption [24]. So the changing in the standard of use **of fossil fuels by biofuels should be the central question**

**to improve efficiency and environmental pollution. In this sense, the introduction of mixtures of biofuels and traditional fuels is a relevant theme to improve environmental impacts and expand biofuels in the market.** Improving vehicles efficiency and biofuels use is a priority to reduce urban air pollution and greenhouse gas emissions.

The executive summary on Transport Technologies and Policy Scenarios to 2050 made by the Energy Council [24] emphasizes important directions concerning biofuels and the importance to increase vehicle efficiency. According to this summary, the following points should be observed:

1. Fuel sustainability is measured in terms of three important points: accessibility, availability and acceptability. Quantitatively, the sustainability can be expressed by the consume reduction, it means, by the vehicle efficiency increase. So, sustainability of fuels is straight related to vehicle technologies;
2. Alternative fuels will also increase steadily in penetration with second generation of biofuels such as synthetic biomass-to-liquid **(BTL) growing significantly by 2035** and synthetic gas-to- liquid (GTL) already expected to grow strongly in the coming decade;
3. Aviation fuels present a particular opportunity for alternative fuels, since they can be produced through synthetic Fisher-Tropsch process, which can use gas, coal, or biomass as a feedstock. In aviation, efficiency could increase up to 30%, through engine and materials technologies and flight management;
4. Assuming economically, environmentally and socially sustainable production, the highest potential lies in biofuels to reduce fossil energy and therefore greenhouse gases. They have potential to **reduce fossil energy up to 90%;**
5. Conventional biofuels like ethanol and biodiesel or hydro-treated vegetable oils from plants can be expected to retain some market share even to 2050;
6. Hybrid vehicle may penetrate in certain applications. Electric power utilization in transport will also increase, in particular in OECD and richer countries;
7. **Hydrogen fuel and fuel cell vehicles are expected to gain a market firmly by 2035** and grow towards 2050. Until 2050, gasoline and diesel fuels will still play a major role, but their biofuel portion will be significant;
8. Hydrogen and fuel calls can contribute significantly in the passenger vehicle sector if the mayor challenge of fuel cells cost, hydrogen storage, hydrogen production and hydrogen delivery can be overcome. In order to make substantial improvements in sustainability of energy for transport sector over the next 43 years, breakthroughs in technology will be necessary.

Another important analysis and contribution for the energy sector was done by LBST in Germany [25]. They pointed out the unsolved dilemma of industrialized society: **economy grows versus the need to reduce emissions to mitigate climate change from the years beyond 2000**. According to them, the oil peak will be around 2005-2010; the gas peak around 2015 – 2025; coal peak around 2010-2040; and nuclear peak will be 2015-2020. So, to achieve a sustainable growth, several types of renewable energy should be introduced beyond the year 2004. This study shows a similar situation like IEA model presented. So, they came to the conclusions that **hydrogen is the adequate carrier for future transport sector due to its potential, yield, storability and efficiency.** Besides that, it has a broad feedstock flexibility to be produced from fossil or renewable. Once using hydrogen, it will facilitate the development of green transport.

### 3.1 Battery and the fuel cell driven cars

The energy conversion via electrochemical cells is not subject to the limitation of Carnot cycle [26] which presents a very low efficiency and dissipates a large amount of heat associated with $CO_2$ emissions. Fuel cell technology offers a better way to convert fuels energy, mainly hydrogen, into electric energy given as reaction product, water, a fact that mitigates $CO_2$ - emissions. This concept was first demonstrated by William Grove, in 1839 when he connected four cells containing platinum electrodes with sulfuric acid, in series at which hydrogen and oxygen were consumed to produce electricity, and the electricity in turn was used to decompose water [27]. However, electrode kinetic has a great influence in the whole electrochemical process efficiency. This pioneer discovery started the enormous research work which resulted in the improvements in fuel cell technology and certainly in the conception of different types of fuel cells like; alkaline, acid, proton exchange membrane, and solid polymer electrolyte, among others [27]. Certainly, the first field of application was associated with the use of fuel cell technology for space vehicles where safety, life of the components, and system launch weight were critical factors in the late fifties. The excellent results obtained in this application encouraged the research and development program for terrestrial uses in the sixties. The mobile application came as natural development in the beginning of seventies. As long as oil derived liquid fuels are available, there was a very little competition for the internal combustion engine or diesel powered vehicles. Several conventional and advanced batteries are being considered for vehicles, which will eventually replace the mechanically powered vehicles. According to the Ragone plot, a very useful illustration for predicting performance characteristics of electrical vehicles is shown in Figure 1. According to these data, the plots of the specific power (W/kg) versus specific energy relations (Wh/kg) are shown for several types of vehicles using batteries, fuel cells.

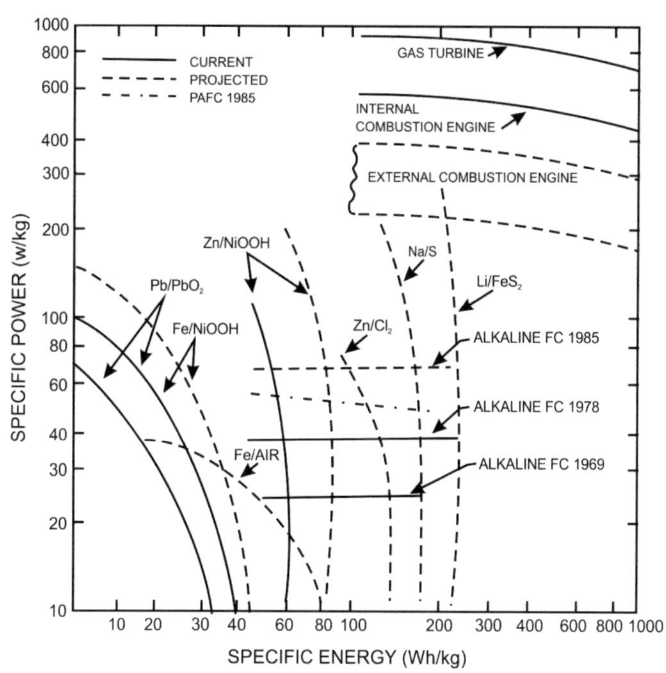


Fig. 1. Ragone Plots: Batteries, Fuel Cells and Heat Engines

They are compared with internal and external combustion engines and gas turbine. In conclusion, the plot demonstrates:

- The mechanical energy converters have the desirable characteristics of high power/weight required for start-up and acceleration, and energy/weight ratio required for range;
- Fuel cell systems have low power/weight ratios, but can attain high energy ratios;
- Batteries can attain high power/weight ratios but have relatively low energy/ weight ratios.

### 3.2 Future scenarios for new vehicles – Vector 21

At the Institute of Vehicle Concepts- which is part of the German Aerospace Centre in Stuttgart, the tool VECTOR 21 [28] has been developed that predicts the distribution of future vehicle fleets according to some boundary conditions, especially those related with potential "Climate Change Protection Scenarios". For this purpose the software combines different customer groups with vehicles which, subject to rational decision, best fits the respective requirement profile. Influences of exogenic scenario boundary conditions such as $CO_2$ restrictions of new vehicle fleets, the price of oil, taxes, subsidies and charges are also taken into account. In the course of this the algorithm accesses an extensive database filled with different technologies. In addition to the type of technology this database contains projections on the influence on the energy consumption of the whole vehicle and the production costs of numerous energy-efficient technologies. All technologies permitted for the respective year are combined in a matrix and a specific $CO_2$ emission determined. Table 5 clarifies the boundary conditions stored in the model using the example of a "Climate Protection" scenario.

| Scenario assumptions | | 2010 | 2020 | 2030 | 2040 |
|---|---|---|---|---|---|
| Oil price | [€/bbl] | 80 | 100 | 130 | |
| CNG taxes | [%] | 20 | from 2018: 100 | | |
| Electricity price | [€ ct/kWh] | 21,5 | 34,1 | 37,3 | 36,4 |
| Hydrogen price | [€ ct/kWh] | 22,3 | 39,0 | 37,6 | 36,5 |
| Hydrogen share from electrolysis | [%] | 100% | | | |
| $CO_2$ intensity of electricity | [g/kWh] | 540 | 510 | 21 (from 2025) | |
| $CO_2$ intensity of hydrogen | [g/kWh] | 648 | 612 | 25 (from 2025) | |
| $CO_2$ target (EU) | [g $CO_2$/km] | 2015: 130 | 95 | 70 | 60 |
| $CO_2$ penalty | [€/ g $CO_2$/km)] | 95 | | | |
| Willingness-to-pay | [%] | 0-20 | | | |
| Vehicle size categories (small /medium / large) | [S/M/L %] | (25/55/20) | (28/50/22) | (30/45/25) | |

Source: [29]

Table 5. Boundary conditions of an exemplary "Climate Protection" scenario

The outcome for the composition of the new vehicle fleet as well as the overall vehicle fleet is shown in Figure 2. Due to severe penalties, alternative vehicle concepts will achieve a considerable market introduction. The ambitious $CO_2$ targets together with the high $CO_2$ reduction potential of hydrogen and electricity will encourage costomers to buy more innovative technologies. In 2040, only 10 % of conventional combustion engines would still be in the vehicle fleet under these conditions. The $CO_2$ emissions of new cars can be reduced by almost 70 % in 2040 compared to 2010. The results show that the number of different propulsion and vehicle concepts on the market will grow. For the introduction of electricity and hydrogen, progressive decisions regarding $CO_2$ targets and energy production have to be taken. Biomass based liquid fuels can contribute to this development as despite of the growing importance of alternative propulsion technologies, conventional vehicles will still play an important part in the mid-term future.

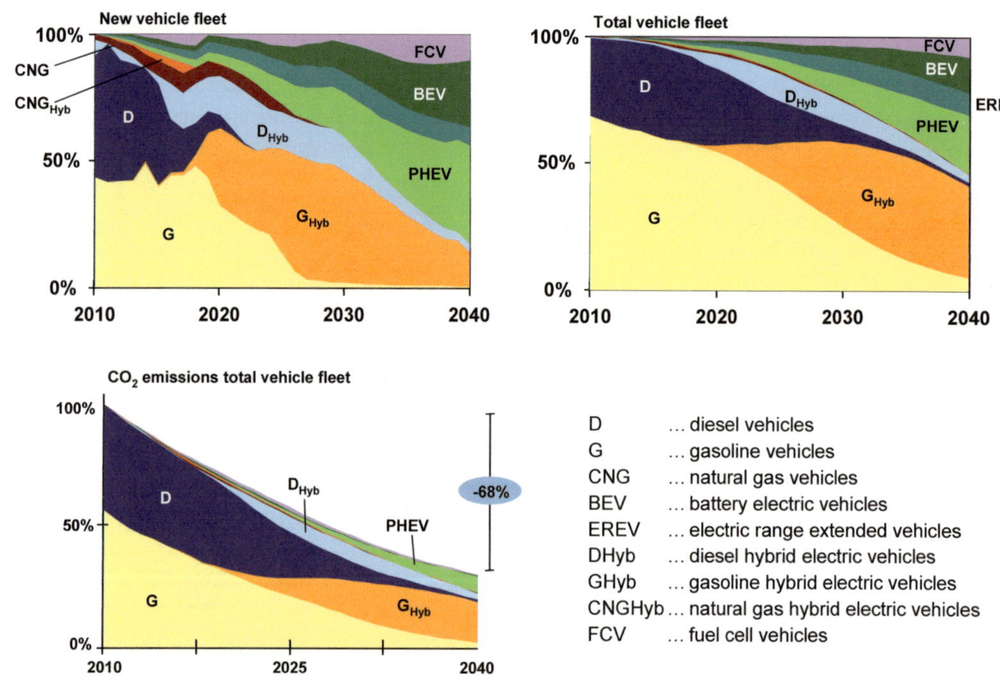


Fig. 2. Vehicle fleet development and $CO_2$ emission reduction for "Climate Protection" scenario

### 3.3 The Brazilian experience with the "flex car technology"

After the economic crisis of the seventies, Brazilian Government decided to implement the National Alcohol Program - *Pro-Álcool* in 1975. It was a program financed by the government to phase out automobile fuels derived from fossil fuels, such as gasoline, in favor of ethanol produced from sugar cane. To achieve this goal, they provided three important initial directions for the ethanol industry: to guarantee purchases by the state-owned oil company Petrobras; to apply low-interest loans for agro-industrial ethanol enterprises and fixed gasoline and ethanol price where hydrous ethanol sold for 95% of the government-set gasoline price at the pump. Concerning to the car industry, initially it was

necessary to develop several important changes in the existing Otto cycle motors related to new materials to avoid corrosion, motor rate of compression, as well as, the injection systems [30]. All of these initial efforts came out with the development of the "flex car technology" which permitted the existing gasoline motor specification to be adapted for using a full range of ethanol-gasoline blends nationally.
Basically, "**flex car technology**" comprises a set of three important devices; a software, an injector and a level sensor, which works integrated with the existing gasoline motor, permitting the use of different mixtures of ethanol and gasoline fuels, including, pure ethanol and pure gasoline. It was the most important improvement done in the injection fuel system since ethanol was nationally launched late in the seventies as a biofuel. Its main purpose was to achieve an adequate fit between the existing gasoline motor specifications and the ethanol characteristics, mainly its specific heat (cal/g). In order to achieve a desirable motor performance, the motor compression rate was established in 11.5:1, an intermediate value between the gasoline motor compression rate 9.5:1 and the rate for alcohol fuel 12:1 [30]. The success of "flex technology" vehicles together with the mandatory mixture (E25) blend throughout the country, have allowed ethanol fuel consumption in the country to achieve a 50% market share of the gasoline-powered fleet by February 2008 [31].
The Brazilian car manufacturing industry introduced the flexible car technology in 2003, reaching 92.3% share of all new cars and light vehicle sales by 2009. In December 2009 they represented 39% of Brazil's registered Otto cycle engine vehicle fleet, and the cumulative production of flex-fuel cars and light commercial vehicles reached the milestone of 10 million vehicles in March 2010. There are around 70 flex models available in the market since 2010, from 11 major car makers [32]. This technology has been extended to bus and now for motorcycles [33].
The Brazilian Government has developed and establish of agreement which allows the technology transfer to some Latin American Countries, Caribbean, Andean Countries and also with the USA, which imposed a U$D 0.54 tariff on every gallon of imported ethanol from Brazil [31].

## 4. Biofuels processing technologies – scenarios of options

Chemical and biochemical industry all over the world is facing an unprecedented change with respect to its basic sources of material [34]. Oil that has supported the society needs in terms of its basic products for such a long time. Now industry is facing a real change towards the use of biomass facing the declining of oil production in the near future. As it was pointed out in the introduction, in the 18th century, wood was used mainly for heating and constructions purposes. At that time, animal oil was used mainly for illumination, as well as horse power was used basically for human transportation. With the advent of coal and oil processing industry in the 19th and 20th centuries, the automobile came to substitute animal power, mainly with the arrival of the vapor machine in the end of the 19th century. Now, at the beginning of the 21st century the world is facing another change towards the use of renewable resources to produce biofuels to fit car specifications.
A broad analysis of the state of art of different biofuels technologies was presented by the author in the publications [3], [35], [36]. In this paper, the main focus will be dedicated to build a realistic vision related to the biofuels introduction in the international market. So, it will be presented possible options of scenarios for biofuels from 2010, from the current status, until the year 2040, when the green mobility should be in course. As it can be seen in

Figure 3 (annexed), the current status of ethanol, biodiesel and biogas production processes demonstrate that these technologies are already in the market in spite of some technical or environmental problems they present. Moreover, Figure 3 shows that some results of the second generation of technologies are coming out until 2015 -2020, and some new results concerning biohydrogen production will permit and contribute for the achievement of the green mobility, that will be available around 2025 which will permit the use of vehicles free of CO2-emissions. This constitutes one important vision of the present work. The data and analysis presented in several papers consolidate this vision. However, it is possible to conclude that it is not possible to foresee what type of biofuel will succeed properly the fossil fuels. Society is still facing several options, perhaps due to the large biodiversity of biomass present in nature at regional level [37]. In practical terms, effort has been done in terms of the scientific knowledge to use these materials to produce different types of biofuels in these regions. But unfortunately, motor technologies are very restrictive in terms of biofuels specifications. However, due to other society needs it is necessary to point out those two categories of plant materials that should be considered to attend sustainability: food and non-food plant materials, residues and wastes. Primarily, the technological efforts need to drive biomass to human food demands in terms of plants with components to supply food purposes using specific land resources. Wastes and non-food material like lignocellulosic (wood, straw, grass, wastes and residues) are present all over the world and technological efforts should be directed to convert them into sustainable energetic sources of biofuels. Acting this way, it is possible to avoid competitiveness with food sector, one of the great problems of the countries involved with biofuels production in our days. Associated with such types of plant materials, its photosynthetic process is straight related with their productivity and land use, as well as, the process requirements to the use of chemical or biochemical catalysts to achieve good processing yields and high productivity. Genetic tools today offer great chances to improve microorganisms and plant species yields and productivity, so, they must be considered essential knowledge to be applied [38]. Finally, process net energy ratio, life cycle analysis and the use of biorefinery concept are today important means to attain process sustainability and economic production scale in general terms [39],[3],[35]. Certainly, science, education and training are the basic requirements to reach technological development and to provide the countries with the support for the best use of their natural resources, which should be guided by sustainable and ethical policies.

### 4.1 Ethanol production from renewable resources

As mentioned in the Figure 3, ethanol production from sugar cane in Brazil is a **current mature technology** in use all over the country. An important and realistic analysis of the results of the sustainability of the sugar cane agroindustry in Brazil is presented in the publication edited by Carvalho [40], who pointed out some interesting themes, which are subdivided in other smaller themes during the discussions of the five main themes mentioned in sequence: impacts of the use of materials resources; environmental impacts; sustainable basis of the agricultural production; impacts of the commercial actions; socioeconomic impacts of the sector. In synthesis, nowadays Brazilian ethanol production is a commercial technology practiced with yields of about 6000 liters of ethanol per hectare, presenting a cost of approximately US$ 0.20 per liter. About 8 Million of hectares are being used for the production of about 22.5 Million of ethanol liters in 2008 [41]. The use of this technology has contributed for the reduction of $CO_2$-emissions in the country, as well as to

create a large number of jobs in the field [40]. Ethanol from starch raw materials is a quite close technology which hydrolyzes amylose and amylopectin carbohydrates into sugars. Nowadays **hydrolysis has been used on industrial scale** to convert grains, potatoes and roots like cassava to produce ethanol in those countries where sugar cane is not available. Actually, the USA is using about 20% of their corn to ethanol, having a total production of about 33.8 million of liters [41]. Corn and cassava have been considered to the **current ethanol production in the USA**, in some Asian and Latin American. Maize (Zea mays L. ssp.), hybrid maize, due to its high grain yield is preferred by farmers over conventional varieties. Cassava or manioc (Manihot esculenta) is a woody shrub of the Euphorbiaceae (spurge family) native from South America extensively cultivated as an annual crop in tropical and subtropical regions for its edible starchy tuberous root. Cassava is the third largest source of carbohydrates for human food in the world, where Africa is the largest center of production. In spite of having a positive energy balance, starchy plants present a lower net energy ratio than sugar cane, a fact that is being considered very critical [42].
The development of integrated biorefineries is a new theme which is being considered relevant with high priority for this sector [3], [43]. The conversion of sugar cane residues, like bagasse, leaves and straw through an enzymatic hydrolysis unit coupled to the ethanol traditional plant is being considered as the second generation technology to increase ethanol production to about 13,000 liters per hectare [9]. Certainly, the development of an **effective pre-treatment process** and components separation is essential to achieve this goal [44]. Lignocellulosic are complex materials and its fractioning is considered a **hard task** [45]. In this paper, there is still a process under development which will possibly be ready for commercial use around the year 2015, as indicated in Figure 3.

### 4.2 Vegetable oils as diesel substitute

Biodiesel from vegetable oils is a **mature technology** which is in use in several countries, but it still has many problems [46]. One of the most important problems it is related to the low productivity of the traditional oleaginous species, like seeds soya, cotton, coconut, rapeseed and castor to support biodiesel production. These plants present C3 mechanism, which is responsible for its low productivity. This fact demands large areas of land, besides the competitiveness with the food products. Glycerin is a byproduct of the reaction between vegetable oils and alcohols (methyl or ethyl) which represent about 12% of the original oil used. Due to its large production, it does not have a definite market. Actually, this technology receives a lot of subsides, fact that is related with their non economic aspect. So, considering land use in medium and long terms, Biodiesel will not achieve the global market even in the actual context of oil price. In our view, this technology could be used in specific cases in order to attend social or energetic ends as it is shown in the scenarios detailed in Figure 3.

#### 4.2.1 HBio – process: A mixture of diesel and vegetable green fuels

In Brazil, Petrobras developed an interesting process denominated Hbio, which is a patented technology [47]. This process is adequate to use vegetable oils in the existing catalytic oil refining process to produce green fuels, as shown below, in the Figure 4. The process improves the quality of fossil diesel due to the good qualities of the vegetable oils derivates obtained in this process mainly in those countries that does not have oil adequate to diesel production. Once there is vegetable oil available, it is possible to produce green

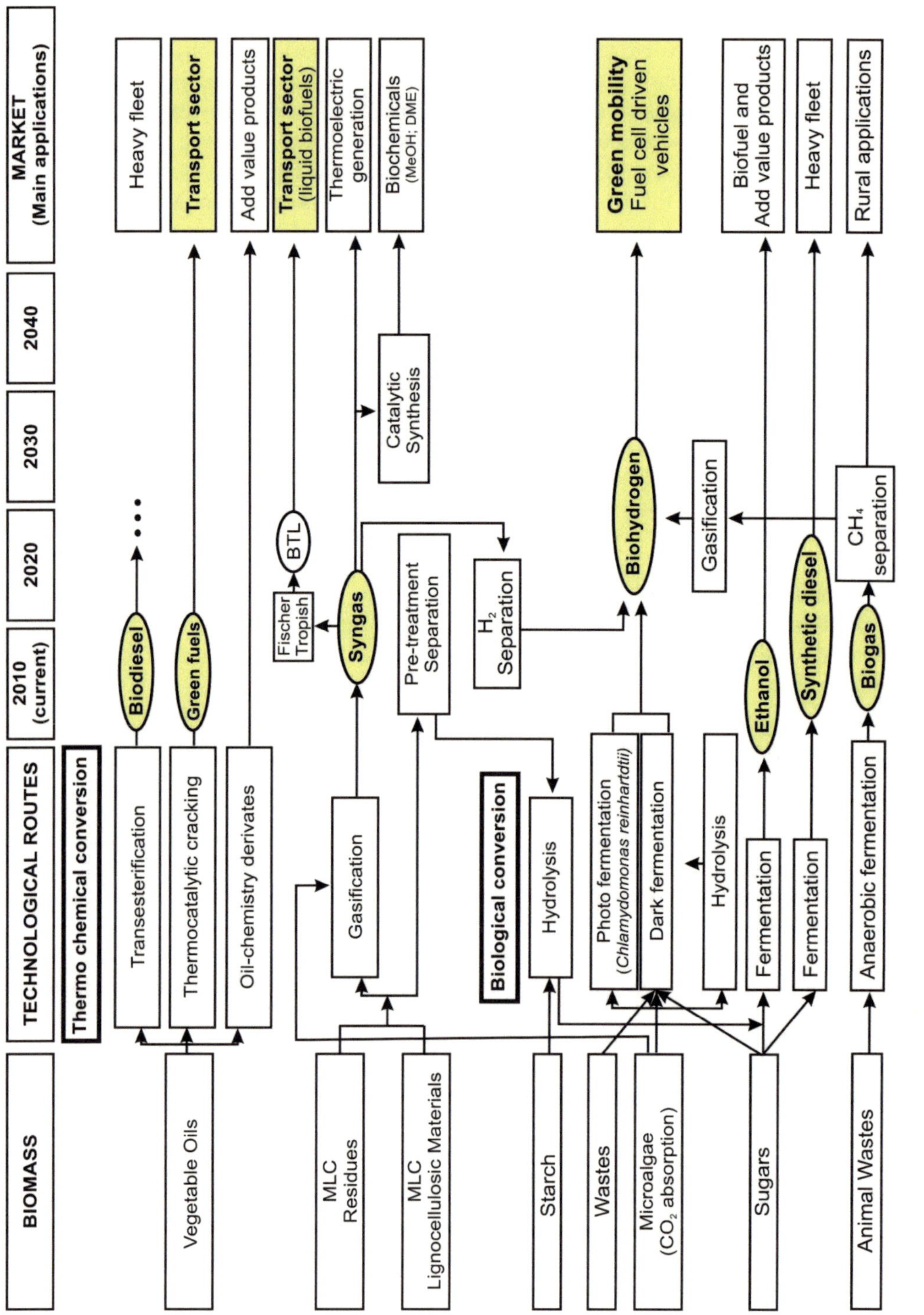


Fig. 3. Biofuels realistic options market scenarios

fuels or mixture of green fuels with diesel. Other oil companies developed equivalent process in the USA. It is important to observe that there is no glycerin production in this process. So, this process is considered an option for future specific applications. This technology is considered ready for implantation in refineries, depending on the availability of vegetable oils, as well as, from the Brazilian Government decision, since Petrobras is a State Company.

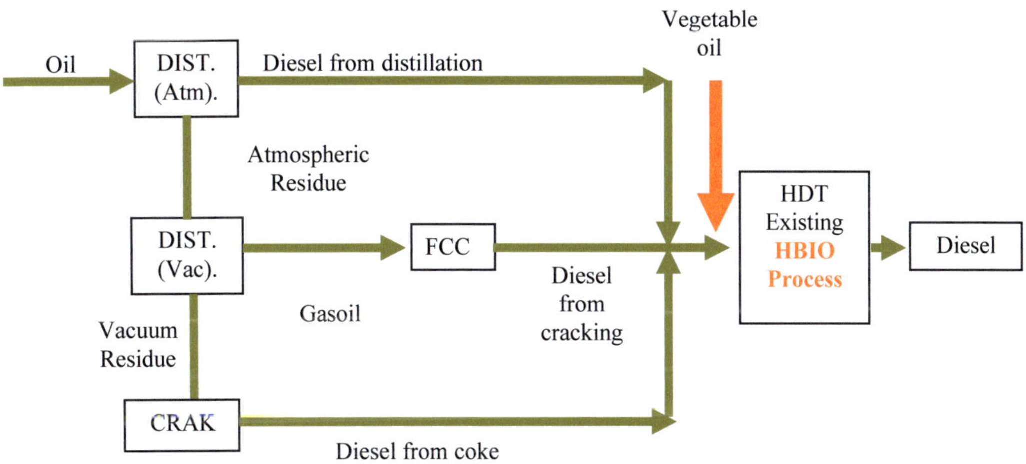


Fig. 4. Flow diagram of the HBio - Process

#### 4.2.2 Synthetic diesel

Finally, Figure 3 mention the work been developed in pilot scale by the Amyris Company in Campinas São Paulo [10] on synthetic biology. This process use a specific engineered strain of *S. cerevisiae* yeast to convert sugar into a hydrocarbon molecule Biofene[R], from which it is produced Farsene, the synthetic diesel. It is a well-known fact that microorganisms can metabolize sugars to produce a large number of isoprenoid compounds [38]. Isoprenoid compounds are commonly known as terpenes and carotenoids. In fact, it is a sustainable product that could be used to substitute diesel oil with a positive impact on the transport sector. It is expected that this technology will enter in commercial scale, in the next decade in those countries that have sugar available in large scale.

### 4.3 Thermal conversion of lignocellulosic materials - LCM into syngas

Two technologies have been considered to produce biofuels from LCM: Biological conversion and thermal conversion as it is shown in Figure 3. Urban LCM fraction also is considered into this broad category of materials. Actually, many countries are engaged in the development of these technologies. Table 6 presents a comparison between three different processes for biomass conversion according to the second generation technologies mentioned. They are: Biomass Integrated Gasification with Gas Turbine-BIG/GT; thermochemical and bioconversion. According to the data presented in Table 6, thermo chemical conversion of biomass into biofuels is considered more attractive.

| Products | BIG/GT | Thermochemical | Biochemical |
|---|---|---|---|
| Ethanol (liters / ton, DM) | -- | 333 | 246 |
| Electricity(kWh/ton, DM) | 1750 | 606 | 226 |
| Total efficiency (%) | 35 | 50 | 33 |

Source: [9]

Table 6. Products and processes efficiencies from second of generation of LCM conversion (It was considered one dry ton of dry biomass)

According to Huber [43] the production of biofuels from lignocellulosic biomass can significantly reduce the external countries dependence on foreign oil, create new jobs, improve rural economy, reduce greenhouse gas emissions, and improve national security. Wayne and collaborators [48] pointed out the importance of the future biomass gasification role concerning biofuels production and conversion through the use of fluidized bed biomass gasifier.

According to the most relevant publications discussed in this paper, thermal conversion of LCM into synthetic gas (Syngas) is a possible candidate technology to be used at international basis, not only because cellulose is present all over the world as urban and plant residues or as straight biomass, but also, due to its flexibility to produce biofuels and chemicals of great interest for the chemical industry like; ethanol, methanol, dimethyl ether (DME), higher alcohols, diesel, gasoline, wax and other chemicals, as shown in Figure 3. Basically, it uses the Fisher-Tropsch process that was developed in Germany in 1920 to produce liquid fuels from coal. Now it is being adapted to use biomass. The Biomass to Liquids (BTL) process produces hydrocarbons via synthetic Fischer-Tropsch technology through synthesis gas derived from biomass. Concerning the gasification process, the main constrain is related to the great variety of types of cellulose residues to be used and their low density, fact that requires large gasifier units. So, it is necessary that a biomass pre-treatment unity linked with an adequate gasifier to produce and adjust the $H_2/CO$ ratio to feed the Fischer- Tropsch process through a shift reactor. A more specific problem to be solved is the cleaning of the synthesis gas to avoid contamination of the catalysts system used in the Fischer-Tropsch process. Basically, two types of liquid products can be manufactured, namely, hydrocarbons and oxygenates, such as methanol. Furthermore, dimethyl ether (DME), which is a high cetane number product, can also be obtained when methanol undergoes dehydration. Because of the hydrocarbons linearity in the product mixtures, synthetic diesel produced presents a high cetane number. In Brazil, the "Centro de Pesquisas da Petrobras-CENPES" [49] is deeply involved with this technological development.

Integrated biorefineries employ various combinations of feedstocks and conversion technologies to produce a variety of products, with the main focus on producing biofuels. Side products can include chemicals (or other materials), heat and power. The renewable feedstocks utilized in integrated biorefineries include, but are not limited to: grain such as corn, wheat sorghum, and barley; energy crops such as switch grass, miscanthus, willow and poplar; forest and industrial residues such as bagasse, stover, straws, sawdust and paper mill waste. Taking into consideration the original biomass, the yield on syngas varies between 75 e 88 % [50].

Another interesting option using microalgae to produce syngas is being developed by Ofelia and collaborators [51], according to Figure 5. In this study, microalgae are produced using a photobioreactor (PBR) using stack gas from capture of $CO_2$ emissions from an NGCC (Natural-Gas-fired Combined Cycle) power plant. The process was simulated in UNISIM Design (Honeywell). The results should be used to the synthesis of chemical products and fuels, in an industrial ecologic arrangement, where $CO_2$ is promoted from waste to feedstock, referred to as Carbon Capture and Industrial Sequestration - CCIS.

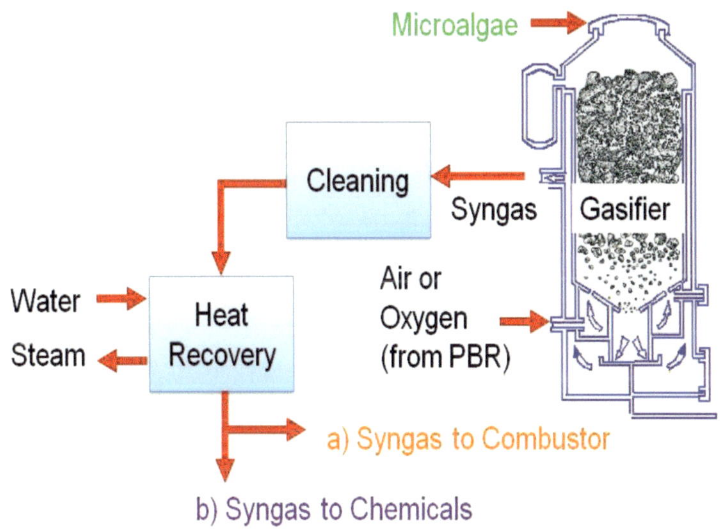


Fig. 5. Integrated microalgae production and gasification

According to these results, it is possible to foresee that thermochemical conversion of LCM present a great potential to produce biofuels and chemicals from non-food plant material, being a strong candidate to succeed fossil fuels. This technological route can produce liquid fuels with a life cycle of greenhouse gases practically zero. This possibility is indicated also in Figure 4 as possible commercial technology around the year 2020-2030. Main restriction is due to microalgae production in large scale.

Microalgae will become the most important biofuel source of biomass in the near future. Besides syngas, it also can be used to produce biogas, or biodiesel using $CO_2$ stack gas from thermoelectric plants, reducing $CO_2$ concentration in the atmosphere. Microalgae are unicellular photosynthetic microorganisms living in saline or freshwater environments that convert sunlight, water and carbon dioxide to algal biomass [52].

## 4.4 Biohydrogen production

### 4.4.1 Introduction

In this article, it will be covered only a discussion on the biohydrogen process, once they are straight connected with the main focus of this article. However, it is necessary to have a clear idea of the existing hydrogen technologies, mainly, its cost, in order to understand the potential for the biological hydrogen production technologies. Also it is important to recognize that hydrogen can be produced straight from biomass, without intensive technology use, including also, decentralized production.

Hydrogen is an energy carrier that could be produced actually by different processes like, natural gas reforming, coal gasification, nuclear, water electrolysis, thermal water splitting and biomass, photo-electrolysis and biological processes [53]. Hydrogen can be used for power generation or used as a transport fuel, mainly in association with fuel cells. Natural gas and coal are the two most current and cheapest sources of $H_2$ production. These processes release $CO_2$, and so, capture and storage are required to reduce the $CO_2$-emissions. Decentralized production of Hydrogen is the best choice for market uptake and for avoiding distribution costs due to the need of infrastructure.

According to IEA - Energy Technology Essentials [53] hydrogen could gain market share in the transport sector if costs of production, distribution and end-use technologies decrease in consonance with the expectations, as well as, if strong policies are placed in order to reduce $CO_2$ emissions. Under adequate circumstances, hydrogen could be entering the market around the year 2020. For that, it should be powering around 700 million fuel cell driven vehicles by 2050, which means, 30% of the projected global fleet [53]. According to IEA report, costs varies, depending of sources and raw materials costs. In summary, hydrogen from electrolysis costs above $30/GJH$_2$, once considering electricity cost at $35/MWh with 80% efficiency; hydrogen from steam methane reforming (SMR) costs ranges between $10-$15/GJH$_2$, considering natural gas price between $6-$9/GJ; coal gasification costs range between $7-$10/GJH$_2$, assuming $1-$1.5/GJ for coal price, and $35-40/MWh for electricity with 45% electrical efficiency; hydrogen from nuclear using its high temperature heat costs about $10-$20/GJ, and from megawatt-scale concentrating solar power systems $20-30/GJ .

### 4.4.2 Biological hydrogen processes

A number of photosynthetic processes for $H_2$ production from water have been proposed and studied for over three decades [54]. These include direct and indirect water-splitting ("bio-photolysis") processes using microalgae that contain the enzymes hydrogenase and nitrogenase. Alternative processes are briefly reviewed for potential practicality and a novel process for photo-biological water splitting is proposed [55].

In the late 1999, Melis [56] discovered that the green algae, *Chlamydomonas reinhartdtii* could be forced to produce straight hydrogen under sulphur-free anaerobic conditions, jointly with researchers of the National Renewable Energy Laboratory-NREL, Golden, Colorado, which developed a preliminary cost analysis of this process [56]. Subjected to this condition, this algae switches from oxygen production (normal photosynthesis), to the hydrogen production. Unfortunately this development still require advances in genetic engineering to improve the efficiency of the photosynthetic process, now understood as performed in two stages.

The biology provides a wide range of approaches to produce hydrogen, including direct and indirect bio-photolysis as well as photo-fermentation and dark-fermentation [57]. Currently these biological technologies are not suitable for solving every day energy problems, but they present high potential for that according to the research perspective. Dark-fermentation is a promising approach to produce hydrogen in a sustainable way and was already examined in lab-scale in many projects [58], [59]. Short hydraulic retention times and high metabolic rates are advantages of the process. The incomplete transformation of the large organic molecules into various organic acids is a disadvantage. So a second process step is required.

Within Europe union strong effort has been made to develop an integrated project denominated by Hyvolution using resources of the 6th Framework Programme on Sustainable Energy Systems aiming at the development a blue print for an industrial scale process for decentralized hydrogen production at small scale from locally produced biomass according to the Figure 6, below [60]. Another important work is being developed in Germany, at Duisburg- Essen University which has as main focus to demonstrate the feasibility of biological hydrogen production using sustainable waste and wastewater treatment concept. A pilot project is under the development and now it comprises and integrated process involving dark fermentation technology and anaerobic digestion of process effluents which presents higher yield of conversion [61] and [62].

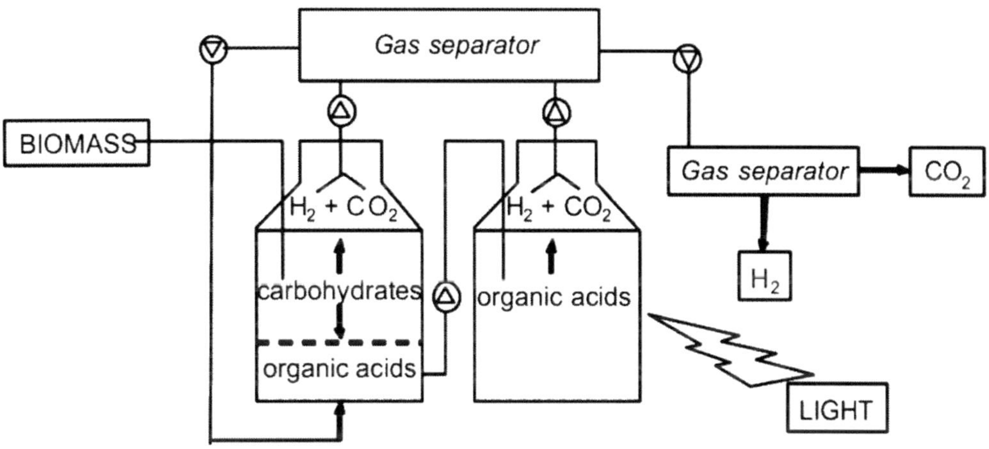


Fig. 6. Hyvolution biohydrogen integrated process

#### 4.4.3 Comparison of biohydrogen production processes

Using sugar as the main carbon source, Manish and Bannerjee [63] made an interesting analysis of four processes (dark fermentation, photo-fermentation, two-stage process and the biocatalyzed electrolysis) and compared with the steam methane reforming (SMR), taking into consideration, energy ratio, energy efficiency and greenhouse gas (GHG) emissions. Processes for biohydrogen production operate at ambient temperature and pressure. So it is expected that they are less intensive in energy than thermochemical processes to produce hydrogen. In this study it was used three different feedstocks: sugar cane bagasse, sugar cane juice, and potato processing wastewater and four processes were analyzed: photo-fermentation; dark- fermentation, two-stage process (integration of dark with photo-fermentation) and biocatalyzed electrolysis, considering the two cases; with by-products and without byproducts. The following definitions were used: **Net Energy Ratio (NER)**, which is the ratio between the hydrogen outputs to the **Non-Renewable Energy (NRE)** input; **Energy Efficiency (EE)** which is calculated by the ratio between hydrogen energy output and energy input. Using emissions factors, corresponding to **Greenhouse Gas (GHG)** emissions were obtained. Inventory results for **Steam Methane Reforming (SMR)** process were taken from Spath and Man [64] and the mass and energy balance were provided as input for the life cycle analysis using the software SimaPro 6 [65].

In order to show the importance of the biohydrogen processes studied by Manish and Bannerjee [63], the authors of this paper organized their data according to Table 7 for the case without by-products. This Table was built to discuss and compare the results of the biohydrogen processes. The following conclusions were driven from these comparisons:

- Efficiency of biohydrogen process increase significantly when by-products are taking into consideration, and all processes present higher efficiencies than the SMR one, according to item 4;
- Biohydrogen process use biomass. All processes considered in this analysis reduce GHG emissions and non-renewable energy use by 57-73% and 65-79%, respectively, as compared to the SMR process, according to items 2 and 3.
- NER are higher for all biohydrogen processes studied. The SMR process presented a non favorable results according item 1.

| **Process** | **SMR** | **Dark fermentation** | **Photo Fermentation** | **Two-stage Process** | **Biocatalyzed Electrolysis** |
|---|---|---|---|---|---|
| **Analysis for the case without by-products** | | | | | |
| 1- **NER** | 0.64 | 1.9 | 3.0 | 3.1 | 1.8 |
| 2- **GHG** (kg $CO_2$/kg $H_2$) | 12.8 | 5.5 | 3.5 | 3.4 | 5.3 |
| a) Reduction to SMR | - | 7.3 | 9.3 | 9.4 | 7.5 |
| b) % reduction | - | 57 | 73 | 73 | 58 |
| 3- **NRE** (MJ) | 188 | 61.7 | 40.1 | 39.3 | 64.8 |
| a) Reduction to SMR | - | 126.3 | 147.9 | 148.7 | 123.2 |
| b) % reduction | - | 67.2 | 78.7 | 79.1 | 65.5 |
| 4- **EE** (%) | - | - | - | - | - |
| a) With by-products | 64 | 89.1 | 82.3 | 81.6 | 76.8 |
| b) Without by-products | 64 | 9.6 | 25.6 | 27.2 | 25.7 |

Table 7. Results of the NER, GHG and NRE analysis for hydrogen processes compared with SMR

## 5. Conclusions

From the analysis of the material presented in this work it is possible to conclude that, in spite of biomass biodiversity and the renewable character of biofuels they are still gaining space at regional level and impacting land use. However, thermochemical conversion process to syngas of lignocellulosic resource –LCM and microalgae are gaining international importance due to the potentiality of these bioresources to produce biofuels.

This process also presents higher yields which greatly contribute for its economic aspects. Moreover this is a flexible technology that can produce also add value products, favoring the greenish of the chemical industry. So, thermal conversion products seem to be the most adequate to substitute some fossil derivates at the international market. Besides gasoline, green diesel and hydrogen, this technology enables the production of other biofuels to attend aviation and ships motors, as well as power generation. From this platform of applications, biomass thermal conversion is considered the promising technology to succeed fossil fuels, diminishing the $CO_2$-emissions and attending the IPPC and Millennium goals.

In spite of the research level, biohydrogen technology is another process that has a great contribution to achieve future green mobility, it means, to turn the transport sector free of CO2-emisions through the use of batteries and fuel cell driven vehicles. With a very innovative approach, biohydrogen technology will be emerging around the year 2020 as a powerful candidate to be the greenish fuel in global terms, as well as, to definitively implement the concept of green mobility on the roads. However, it is necessary to intensify the research to increase biohydrogen productivity process continuously, as well as, to diminish the cost of the fuel cell for the transport sector. This is the only decentralized technology that is in full accordance with the Millennium goals. It is important to observe the straight relationship of this technology with the anaerobic fermentation process which is the current practice to convert waste into biogas worldwide.

The ethanol productions in the USA, Brazil and in other countries are good examples of regional importance that are entering in the international market, but they will certainly have transport costs restrictions, as well as, internal impacts on land use. The actual and realistic opportunity to apply it as a real candidate to replace polyethylene or other biopolymers to produce green plastic is one important point to be considered. It is important to consider also the development of alcohol chemistry as add value products. At the moment it is possible to emphasize that, ethanol, biodiesel, and biogas are the current biofuels technologies in the market, attending automotive motor specifications. In our vision, biodiesel could not compete at international level in the long range, once the traditional oil seeds present low yields, which creates instability and will strongly impact negatively land use. However, vegetable oils cracking process is a possible technology to be applied to improve diesel specifications in the countries which do not have good oil to produce diesel according to automotive standards. Also this technology could use existing oil refineries infrastructure. A new patented on synthetic biology is emerging in Brazil from Amyris Company using specific microorganism which metabolize sugars to produce Biofene[R]. From this molecule synthetic diesel and jet fuel can be produced. Actually, this technology is being tested in demonstration phase.

Finally, it is possible to conclude that the development of biofuels technologies will definitively contribute to the increase of the renewable participation in the global primary energy supply, but also it will contribute to the achievement of the green economy goals, mainly those related to the arising of bioproducts which are strongly increasing in the market. These facts will also contribute for the achievement of the Millennium goals, ensuring environmental sustainability and helping to eradicate extreme poverty.

## 6. References


[1] United Nations Environment Programme. Environment for Development. Green Economy Report. Available at: http://www.unep.org/greeneconomy. Accessed in March, 2011.

[2] United Nations Environment Programme. Environment for Development. Green Economy Report. Available at: http://www.unep.org/greeneconomy/AboutGEI/tabid/1370.

[3] Carioca, J.O.B. "Biofuels Overview: Problems, Challenges and Perspectives". Published at Biotechnology Journal, reference biot.200900137. ISSN 1860-6768.

[4] Marques F.J., et all, "Indicadores de Sustentabilidade em Agroecosisstemas".Embrapa, Brasília, DF, (2003)

[5] United Nations Environment Programme - Environment and Development. Green Economy Report Available at: http://www.unep.org/greeneconomy/portals/88/documents/ger/GER_6_Renewable Energy.pdf.

[6] Goldemberg, J. "Ethanol for a Sustainable Energy Future". *Science,* 2007, 315, pp. 808-810.

[7] Ciamician, G. "The Photochemistry of the Future". *Science*, 1912, 36, pp. 385-394.

[8] International Energy Agency. World Energy Outlook 2006, OECD/IEA, 2006, Paris.

[9] Leite, A.L.S. and Leal, M.R.L.V. Biomass Future Based Supply. Brazilian Network on Green Chemistry: Awareness, Responsibility and Action. Edições UFC, Fortaleza, 2008, pp. 217-234.

[10] Amyris Company Overview. Seminário InformEx 2010 Latin America. realizado nos dia 23 e 24/08/10 no Centro de Convenções Amcham. São Paulo/SP.

[11]"Shell e Cosan anunciam fusão e criação da Raízen". Disponível em: http://www.ruralcentro.com.br/notícias/34825.

[12] Harada, J. "Plásticos Biodegradáveis e de Fontes Renováveis". BASF The Chemical Company. 12º Congresso de Atuação Responsável, ABIQUIM, São Paulo, 2008.

[13] Pereira, B.R. "Polietileno de Cana". DOW, 12º Congresso de Atuação Responsável, ABIQUIM, São Paulo, 2008.

[14] "Pessoa e Sustentabilidade no Centro da Estratégia". BRASKEM Petrochemical Company, 12º Congresso de Atuação Responsável, ABIQUIM, São Paulo, 2008.

[15] OECD-FAO Agricultural Outlook 2008-2017, OECD/FAO 2008.

[16] IPCC: Climate Change 2007: Mitigation of Climate Change. Contribution of Working Group III to the Fourth Assessment Report of the Intergovernmental Panel on Climate Change (IPCC). Edited by B. Metz, O.R. Davidson, P.R. Bosch, R. Dave and L.A. Meyer, Cambridge University Press, Cambridge, UK and New York, USA, 2007.

[17] IEA (2010d): World Energy Outlook 2010, International Energy Agency, OECD Publishing, Paris.

[18] IEA (2008b): World Energy Outlook 2008, International Energy Agency, OECD Publishing, Paris.

[19] AGECC: Energy for a Sustainable Future, Summary Report and Recommendations. The Secretary-General's Advisory Group on Energy and Climate Change (AGECC), 2010.

[20] IEA (2008c): IEA Energy Technology Perspectives 2008, International Energy Agency, OECD Publishing, Paris.

[21] GNESD (2004): Energy Access Theme Results - Summary for Policy Makers: Modern Energy: Impacts on Micro-enterprises Phase 1, Task 1.2 A Literature Review into the Linkages Between Modern Energy and Micro-Enterprise. For United Kingdom Department for International Development. By: K. Meadows, C. Riley, with G. Rao and P. Harris, Global Network on Energy for Sustainable Development (GNESD), July 2003.

[22] United Nations Environment Programme – Environment and Development. Green Economy Report Available at: http://www.unep.org/greeneconomy/portals/88/documents/ger/GER_6_Trasnport.pdf.

[23] ECMT: Assessment and decision making for sustainable transport, 2004.

[24] "Transport Technologies and Policy Scenarios to 2050". World Energy Council, 2007.

[25] Schmidt, P.; Zerta, M.; Stiller, C.; Landiger, H. "Alternative World Energy Outlook (AWEO) and the Role of Hydrogen in a Changing Energy Landscape". 2nd World Congress of Young Scientists on Hydrogen Energy Systems, Torino, Italy, 6-8 June 2007.

[26] Smith, J.M.; Van Ness, H.C.; Abbott, M.M. Introduction to Chemical Engineering Thermodynamics. Fifth Edition. McGraw-Hill International Editions, 1996.

[27] Srinivasan, S. "Fuell Cells for Electric Utility and Transportation Applications". Presented at National Seminar on Eletrochemical Systems: Batteries and Fuel Cells, Fortaleza, Brazil, 2-6 March 1980.

[28] VECTOR21 Vehicle Technology Scenario Model. German Aerospace Center (DLR), Institute of Vehicle Concepts, Stuttgart, Germany, 2010.

[29] Schmid, S. "The Future of Electromobility". Presentation on March, 24th 2011. German Aerospace Center, Institute of Vehicle Concepts, Stuttgart, Germany, 2011.

[30] Carioca, J.O.B. "Decentralized Integrated Systems for Biomass Production and Its Energy/Nonenergy Utilization". Presentation at the 1982 World's Fair International Energy Simposia Series-Symposium II. Knoxville,Tennesse, November 1981.

[31] "Ethanol fuel in Brazil". Available at: http://en.wikipedia.org/wiki/Ethanol_fuel_in_Brazil. Accessed in March 2011.

[32] "Etanol: Carros flex representam 92% das vendas de março, último mês de IPI reduzido", Disponível em http://www.unica.com.br/noticias/show.asp?nwsCode={9002676B-8B05-4401-8742-DD1AB14FB554}) (in Portuguese).UNICA, Brazil.2010-04-20.

[33] "Magneti Marelli apresenta a moto flexível em combustível" (http://www.webmotors.com.br/wmpublicador/Reportagens_Conteudo.vxlpub?hnid=38490 (in Portuguese). WebMotors.2007-11-07.

[34] Position Paper - Change in the Raw Materials Base. Produced and sponsored by: GDCh/Dechema/DGMK/VCI. Frankfurt, August 2010.
[35] Carioca, J.O.B. "The hard choice for alternative biofuels to diesel in Brazil". Published at *Biotechnology Advances*, reference JBA6230, DOI ion: 10.1016/j.biotechadv.2009.05.012. ISSN 0734-9750.
[36] Carioca, J.O.B.; Moraes, C.; Janssens, M.; Peixoto, M.C.S. "Biodiesel Constraints versus Synthetic Diesel-GTL, Dimethyl Eter-DME and Green Diesel Perspectives in Brazil". Presented at World Renewable Energy Congress - IX Programme, Florence, Italy, 19-25 August 2006.
[37] Adriano B. Kury et al. "Brazilian Biodiversity Research Programme - Diretrizes e estratégias para a modernização de coleções biológicas brasileiras e a consolidação de sistemas integrados de informação sobre a biodiversidade". Ministério da Ciência e Tecnologia do Brasil, Brasília, 2006.
[38] Wackett, L.P. "Microbial-based motor fuels: science and technology". *Microbial Biotechnology,* 2008, 1 (3), 211-225. doi: 10.111/j.1751-7915.2007.00020.x
[39] Carioca, J.O.B.; Moraes, C. "Brazilian Netwoork on Green Chemistry: Fundamentals". Brazilian Netwoork on Green Chemistry: Awareness, Responsibility and Action. Edições UFC, Fortaleza, 2008, pp. 35-73.
[40] Macedo, I.C. "A Energia da Cana-de-Açucar". Doze estudos sobre a agroindústria da cana-de-açucar no Brasil e a sua sustentabilidade". ÚNICA-União da Agroindústria Canavieira do Estado de São Paulo, São Paulo, 2005.
[41]"Etanol como combustível no Brasil". Disponível em: http://pt.wikipedia.org/wiki/Etanol_como_combust%C3%ADvel_no_Brasil#Compara.C3.A7.C3.A3o_entre_a_produ.C3.A7.C3.A3o_de_etanol_nos_Estados_Unidos_e_Brasil. Acessado em: 14/03/2011.
[42] Lima, J.A. Chemical and physicochemical characteristics of modified starches and its importance in productive food chain. Brazilian Netwoork on Green Chemistry: Awareness, Responsibility and Action. Edições UFC, Fortaleza, 2008, pp. 251-272.
[43] Huber, G.W. "Breaking the Chemical and Engineering Barriers to Lignocellulosic Biofuels: Next Generation Hydrocarbon Biorefineries". Based on the Workshop Washington, D.C. June 25-26, 2007.
[44] Pereira Jr., N. "Biomass of Lignocellulosic Compostion for Fuel Ethanol Production and the Context of Biorefinery". Brazilian Netwoork on Green Chemistry: Awareness, Responsibility and Action. Edições UFC, Fortaleza, 2008, pp. 591-627.
[45] Carioca, J.O.B.; Arora, H.L., Pannirselvam, P.V. and Horta, E.A. Lignocellulosic biomass fractionation: I-Solvent extraction in a novel reactor. Biotechnology Letters, 1985, 7, pp. 213-216.
[46]Carioca, J.O.B. "Potential for the Use of Cosolvency Diesel-Ethanol Technology". Trabalho apresentado no Encontro PROSUL 2007 - BIOENERGIA, 8 - 9 maio, São Paulo, Brazil, 2007.
[47] Brazilian Patent N° PI 0500591-4, submitted in February, 18, 2005 to Petróleo Brasileiro S.A., Author: Jefferson Roberto Gomes (Processo para hidroconversão de óleos vegetais).
[48] Doherty, W.; Reynolds,A.; Kennedy, D. "Simulation of a circulating fluidized bed biomass gasifier using ASPEN Plus:a performance analysis". In A.Ziebik,

Z.Kolenda and W.Stanek (eds). Proc. 21st International Conference on Efficiency, Cost, Optimization, Simulation and Environmental Impact of Energy Systems, Krakow,Poland, 2008, 2141-1248.

[49]Falabella, E.S.A.; Sousa, S.S.A. and Oliveira, F.B. BTL: A Solution to Brazilian Sugar Cane and other Biomass Residues. Brazilian Networok on Green Chemistry: Awareness, Responsibility and Action. Edições UFC, Fortaleza, 2008, pp. 439-448.

[50]National Renewable Energy Laboratory, Biomass Program, June 2006. http://www.nrel.gov/biomass/biorefinery.htm.

[51] Araújo, O. Q. F.; Carioca, J.O.B.; Macambira, F. S.; Medeiros, J. L. "Energy Evaluation of Technological Routes for Recycle of $CO_2$ Emissions from Natural Gas Power Plants". International Microalgae and Biofuels Workshop, 02-05 August 2010, Fortaleza, Brazil.

[52] Dermibas, A. "Use of algae as biofuel sources". Energy Conversion and Management 51 (2010) 2738-2749. Doi 10.1016/j.enconman.2010.06.010.

[53] "Hydrogen Production and Distribution". IEA Energy Technology Essentials. Available at: http://www.iea.org/techno/essentials5.pdf. April 2007.

[54] Benemann JR, Berenson JA, Kaplan NO, and Kamen MD (1973) Proc. Nat. Acad. Sci. USA, 70: 2317-2320.

[55] Benemann JR (1996) Nature Biotechnology, 14: 1101-1103.

[56] Melis, A. et al. Sustained photobiological hydrogen gas production upon reversible inactivation of oxygen evolution in the green alga *Chlamydomonas reinhartdtii*. Plant Physiol., 2000, 122, pp. 127-136.

[57] Polle P, Kanakagiri SD, Benemann JR and Melis A (2001) In, Biohydrogen II: An Approach to Environmentally Acceptable Technology (Edited by J Miyake), pp. 1-130; Pergamon Press, Oxford, UK.

[58] Krupp M., Brunstermann R., Widmann R. (2005), A Combined Bio-hydrogen/Methane Production Process compared with Single Stage Methane Mono-digestion, Proceedings of the 4th international Symposium: Anaerobic Digestion of Solid Waste, Copenhagen, 31.August-2.September 2005.

[59] Rechtenbach, D. (2009): Fermentative Erzeugung von Biowasserstoff aus biogenen Roh- und Reststoffen. Hamburger Berichte. Stuttgart, Abfall aktuell. Band 34.

[60] Claassen, P.A.M.; Vrije, T.; Friedl, A.; Wukovits, W.; Ahrer, W. "Hyvolution-Non-Thermal Production of Pure Hudrogen from Biomass". 1st Austrian Hydrogen Conference, Octuber 10-11, Messecenter Graz, Austria, 2005.

[61] Krupp, M. "Biohydrogen production from organic waste and wastewater by dark fermentation – a promising module for renewable energy production". Doktor Ingenieur Thesis, University Duisburg-Essen, 2007.

[62] Brunstermann, R. "Entwicklung eines zweistufigen anaeroben Verfahrens zur kontinuierlichen Wasserstoff- und Methanerzeugung aus organischen Abfällen und Abwässern". Doktor-Ingenieur Thesis, University Duisburg-Essen, 2010.

[63] Manish, S.; Banerjee, R. "Comparison of biohydrogen production processes". International Journal of Hydrogen Energy 33 (2008) 279-286. DOI.10.1016/j.ijhydene.2007.07.026.

[64] Spath, P.L.; Mann, M.K. "Life cycle assessment of hydrogen production via natural gas steam reforming". National Renewable Energy Laboratory (NREL) USA, NREL/TP-570-27637, February 2001.

[65] PRé consultants , SimaPro LCA software,
(http://www.pre.nl/simapro/http://www.pre.nl/simapro/default.htm); 2006.

21

# Bioproduction of Hydrogen with the Assistance of Electrochemical Technology

Soundarrajan Chandrasekaran[1] and Dachamir Hotza[2]
[1]*Central Electrochemical Research Institute (CECRI-CSIR), Karaikudi*
[2]*Federal University of Santa Catarina (UFSC), Florianópolis, SC*
[1]*India*
[2]*Brazil*

## 1. Introduction

The depletion of fossil fuel diverts us to the use of renewable resources as the supplement for fuel. Solar, hydroelectric power and microbial system are known to be abundant renewable resources for fuel production. Hydrogen has a energy yield of 122 kJ/g, which is 2.75 times greater than hydrocarbon fuels (Kapdan & Kargi, 2006). Hydrogen together with oxygen is the key element in the biological energy cycle on the earth. In all organic matter hydrogen atoms are bound to carbon, nitrogen, sulphur and other elements.

Biological processes for the production of hydrogen, which are environment-friendly and less energy intensive, may be categorized into bio-photolysis, photo-fermentation and dark fermentation. Bio-photolysis occurs in organisms such as green algae or cyanobacteria, which carry out plant-type photosynthesis, using captured solar energy to split water. Non-sulphur purple photosynthetic bacteria undergo photo-fermentation to perform an anaerobic photosynthesis. By dark fermentation, a variety of different microbes anaerobically breaks down carbohydrate rich substrates to hydrogen and by-products (Das & Veziroglu, 2001; Hallenbeck & Benemann 2002). Gaseous hydrogen is produced as well as consumed by living microorganisms in the presence or absence of oxygen (under both oxic and anoxic conditions). The anoxic condition is observed during dark fermentation of microbes.

Among the processes, dark fermentation presents a high rate of hydrogen production, using fermentative bacteria, such as *Enterobacter* species (Palazzi et al., 2000; Kumar & Das, 2000; Kumar & Das, 2001; Nakashimada et al., 2002; Kurokawa &Tanisho, 2005; Zhang et al., 2005; Shin et al., 2007), *Clostridium* species (Chin et al., 2003; Lee et al., 2004; Levin et al., 2006; Jo et al., 2008) and *Escherichia coli* (Yoshida et al., 2005). Hydrogen production through bacterial fermentation is currently limited to a maximum of 4 moles of hydrogen per mole of glucose, and under these conditions results in a fermentation end product (acetate; 2 mol/mol glucose) that bacteria were unable to further convert to hydrogen. Thermophiles produced up to 60–80% of the theoretical maximum, demonstrating that higher hydrogen yields can be reached by extremophiles than using mesophilic anaerobes (Chin et al., 2003). The oxidative pentose phosphate pathway as an alternative metabolic route exists for example in microalgae, which can produce stoichiometric amount of $H_2$ from glucose. However, this pathway is usually not functional for energetic reasons (Lee et al., 2004).

Several problems still remain for the commercial scale production of bio-hydrogen including low hydrogen yield. Alternatively the by-products are to be used by microorganisms or by bioelectrochemical technology, so that higher moles of hydrogen may be produced.

Electrolysis is a method of separating bonded elements and compounds by passing an electric current through them. One important use of electrolysis is to produce hydrogen, which has been suggested as an energy carrier for powering electric motors and internal combustion engines. All electrolysers work according to a principle of two electrodes separated by an electrolyte. A so-called half cell reaction resulting in the formation of hydrogen and oxygen respectively takes place at each electrode. The role of the electrolyte is to close the electrical circuit by allowing ions (but not electrons) to move between the electrodes.

Bioelectrochemically assisted microbial system has the potential to produce 8-9 mol $H_2$/mol glucose (Liu et al., 2005). Hence, the hybrid technology is an alternative for the production of hydrogen with higher efficiency.

## 2. Hydrogen production using microbial systems

Different microorganisms participate in the biological hydrogen generation by using photofermentation or dark fermentation such as green algae, microalgae and bacteria, as shown in Table 1.

### 2.1 Major enzymes for metabolizing and producing hydrogen

The enzymes catalyzing the formation and the oxidation of hydrogen are collectively called hydrogenases. The enzyme reaction is represented by Equation 1:

$$H_2 \leftrightarrow 2H^+ + 2e^- \quad (1)$$

In spite of many similarities between the hydrogenases their catalytic and physicochemical properties vary widely. There are three fundamentally different hydrogen producing and metabolizing enzymes found in algae and bacteria (Schlegel & Schneider, 1978):

- reversible or classical hydrogenases,
- membrane-bound hydrogenases, and
- nitrogenase enzymes

Hydrogenase can be differentiated with respect to their position in electron transport systems and their location in the cell. The natural electron donor/acceptor is known only for the soluble, cytoplasmic or loosely bound periplasmic enzymes. For the membrane-bound hydrogenases this information is incomplete or lacking. Details of types and properties of hydrogenases are presented in the literature (Schlegel & Schneider, 1978; Adams et al., 1981). A compilation of papers on function and structure of hydrogenases has been published (Yagi, 1981). Nitrogenase is also responsible for hydrogen evolution by many bacteria. Hence, hydrogenases and nitrogenases possessing microbes can produce hydrogen by their metabolic pathways (Schlegel & Schneider, 1978).

#### 2.1.1 Reversible hydrogenases

The reversible hydrogenase is located at the cytoplasmic membrane (Kentemich, 1991). It has the dual function of catalysing hydrogen evolution and hydrogen uptake (Lambert & Smith, 1981). It has been suggested that this enzyme functions as a valve for low potential electrons generated during the light reaction of photosynthesis, thus preventing the slowing

| Broad classification | Microorganisms | Enzymes Involved |
|---|---|---|
| Green algae | *Scenedesmus obliquus* | Hydrogenase |
| | *Chlamydomonas reinhardii* | |
| | *C. moewusii* | |
| Cyanobacteria Heterocystous | *Anabaena azollae* | Nitrogenase |
| | *Anabaena CA* | |
| | *A. variabilis* | |
| | *A. cylindrical* | |
| | *Nostoc muscorum* | |
| | *N. spongiaeforme* | |
| | *Westiellopsis prolifica* | |
| Cyanobacteria Nonheterocystous | *Plectonema boryanum* | Nitrogenase |
| | *Oscillotoria Miami BG7* | |
| | *O. limnetica* | Nitrogenase, Membrane-bound hydrogenase |
| | *Synechococcus sp.* | Nitrogenase |
| | *Aphanothece halophytico* | |
| | *Mastidocladus laminosus* | |
| | *Phormidium valderianum* | |
| Photosynthetic bacteria | *Rhodobater sphaeroides* | Nitrogenase, Membrane-bound hydrogenase |
| | *R. capsulatus* | |
| | *R. sulidophilus* | |
| | *Rhodopseudomonas sphaeroides* | |
| | *R. palustris* | |
| | *R. capsulate* | |
| | *Rhodospirillum rubnum* | |
| | *Chromatium sp. Miami PSB* | |
| | *Chlorobium limicola* | |
| | *Chloroexu aurantiacus* | |
| | *Thiocapsa roseopersicina* | |
| | *Halobacterium halobium* | |
| Fermentative bacteria | *Enterobacter aerogenes* | Hydrogenase |
| | *E. cloacae* | |
| | *Clostridium butyricum* | |
| | *C. pasteurianum* | |
| | *Desulfovibrio vulgaris* | |
| | *Magashaera elsdenii* | |
| | *Citrobacter intermedius* | |
| | *Escherichia coli* | |

Table 1. Microorganisms used for hydrogen generation (Gest, 1954; Das & Veziroglu, 2001)

down of the electron transport chain (Appel, 2000). It is available in the majority of the nitrogen-and non-nitrogen-fixing cyanobacteria (Eisbrenner, 1978). Reversible hydrogenase is a heterotetrameric, NAD-reducing enzyme, consisting of a hydrogenase (encoded by hoxY and hoxH genes) and a diaphorase part (encoded by hoxF and hoxU genes).

### 2.1.2 Uptake hydrogenases

Uptake hydrogenase is located at the cytoplasmic face of the cell membrane or thylakoid membrane, where it uses hydrogen evolved by nitrogenase. There is a considerable loss of energy through the production of hydrogen during nitrogen fixation. Some of this energy can be regained through the action of uptake hydrogenase. This enzyme splits the hydrogen and feeds the electrons back into the electron-transport chain. The reduction of a substrate with a relatively high redox potential like cytochrome through this hydrogenase seems to be a wasteful process. But since nitrogen-fixing cells maintain a highly reducing environment, it seems necessary to use part of the reductive power of hydrogen and saving reducing equivalents. Hydrogen-using uptake hydrogenase has several functions:

- It serves as one of the mechanisms to protect oxygen-sensitive nitrogenase (Robson & Postgate, 1980).
- It generates ATP in the hydrogen-dependent respiratory oxygen uptake (Knallgas or oxyhydrogen reaction) and
- It provides additional reducing equivalents to photosystem-I.

Uptake hydrogenase has been found in all heterocystous cyanobacteria and in some non-heterocystous cyanobacteria (Peschek, 1979). The structural genes encoding cyanobacterial uptake hydrogenases have been sequenced and characterized in only a few strains (Axelsson, 1999). The large subunit of the enzyme is encoded by hupL genes and small subunit is encoded by hupS genes. In the organisms studied so far, there is a high degree of homology in the gene sequence of hupSL (Tamagnini, 1997). However, the mode or rearrangement of the genes varies from one organism to another (Axelsson, 1999).

### 2.1.3 Nitrogenase

All nitrogenases studied so far are catalysts for $H_2$ production as they liberate $H_2$ during the reduction of nitrogen to ammonia. A minimum of 25% of the electron flux through nitrogenase is used in the reduction of protons to $H_2$.

$$8H^+ + 8e^- + N_2 + 16\ ATP \longrightarrow 2\ NH_3 + H_2 + 16ADP + 16Pi \qquad (2)$$

ATP, reductant and electrons are provided by photosynthesis or by degradation of sugars in cyanobacteria. Nitrogenase is a metalloenzyme complex consisting of dinitrogenase (MoFe protein: $\propto 2\beta 2$) and dinitrogenase reductase (Fe protein: $\gamma 2$). The Mo-Fe protein or component-I is a larger component is responsible for the catalytic reduction of substrate molecules. The Mo-Fe protein from all sources examined are $O_2$ labile, have molecular weights of approximately 220,000 daltons. Approximately 2 mol of molybdenum and 24±32 mol of iron and sulphide are found per mol of protein (Kim & Rees, 1994). The second protein dinitrogenase reductase or component II accepts electrons from donors such as ferredoxin or flavodoxin, or dithionite and transfers these electrons to dinitrogenase with the concomitant hydrolysis of two molecules of ATP per electron transferred. The six electron reduction of $N_2$ to $2NH_3$, therefore requires a minimum of 12 ATP molecules making nitrogen fixation an energetically expensive process. The Fe protein is also $O_2$ labile and has an average molecular weight of about 60,000 daltons. The protein consists of two subunits of equal weight (Kim & Rees, 1994). In addition to reducing nitrogen to ammonia, dinitrogenase can reduce a number of substrates such as protons, acetylene, cyanide, nitrous oxide and azide. Apart from the conventional molybdenum-based nitrogenase, an alternative vanadium-based nitrogenase has also been reported (Kentemich, 1988). *A. variabilis* can express a third nitrogenase when grown under vanadium and molybdenum deficiency (Kentemich, 1991). This nitrogenase contains vanadium in the prosthetic group. A novel mutant of Azotobacter which has a tungsten-based nitrogenase has also been isolated (Kajii, 1994). In photosynthetic bacteria and cyanobacteria, photohydrogen production is mainly associated with nitrogenase rather than hydrogenase and coupled with ferredoxin or flavodoxin (Kosaric & Lyng, 1988). It requires ATP and is inhibited by $N_2$ or $NH_4$. In this case, ferredoxin is reduced (1) directly by a light-driven reaction, (2) indirectly by ATP-driven reversed electron transport, or (3) by dehydrogenation or oxidative de carboxylation reactions of intermediary metabolism not involving electron transport chains (Kosaric & Lyng, 1988). Nitrogenase is an extremely common, if not universal, enzyme in photosynthetic bacteria (Stewart, 1973). It is difficult to ascertain its prevalence in cyanobacteria since oxygenic photosynthesis in these microbes is inherently incompatible with the nitrogenase protein. Cyanobacteria have evolved several mechanisms to overcome the $O_2$ incompatibility of nitrogenase.

### 2.2 Genetic engineering aspects of biohydrogen production

Genetic engineering is the transfer of genes of interest from one organism into other known organism for its ease of culturing and its efficient metabolic activity. Usually *E.coli* is considered as the universal host and it is consequently well characterized for harbouring the foreign genes. Especially for hydrogen, *E. coli* possesses different membrane-bound hydrogenases under specific conditions: the two enzymes are hydrogenase 3(Hyd-3) and hydrogenase 4(Hyd-4) responsible for hydrogen gas production as well as hydrogenase 1 (Hyd-1) and hydrogenase 2 (Hyd-2) responsible for hydrogen uptake. The entire gene regulation in *E.coli* for hydrogen production is shown in Fig. 1. *E. coli* cells convert glucose to various organic acids (such as succinate, pyruvate, lactate, formate, and acetate) to synthesize energy and hydrogen from formate by the formate hydrogen-lyase (FHL) system that consists of hydrogenase 3 and formate dehydrogenase-H. Bacterial strain, *E.cloacae* IIT-BT 08 was isolated and characterized shown enhancement in biohydrogen production (Kumar & Das, 2000). The gene [Fe]-hydrogenase encoding gene isolated from *E.cloacae* IIT-BT 08 has been over-expressed in fast growing non-hydrogen producing *E.coli* BL-21 using pGEX 4T-1 vector (Mishra, 2004). Hence genetic engineering helps in the effective production of hydrogen.

### 2.3 Biohydrogen production using phototrophic microorganisms

Photosynthetic bacteria can use small-chain organic acids as electron donors for the production of hydrogen at the expense of light energy. In such a system, anaerobic fermentation of carbohydrates (or organic wastes) produces intermediates such as low-molecular hydrogen by photosynthetic bacteria in the second step using a photobioreactor (Nath & Das, 2004). Complete degradation of glucose to hydrogen and carbon dioxide is impossible by anaerobic digestion. However, photosynthetic bacteria could use light energy to overcome the positive free energy of the reaction (bacteria can utilize organic acids for

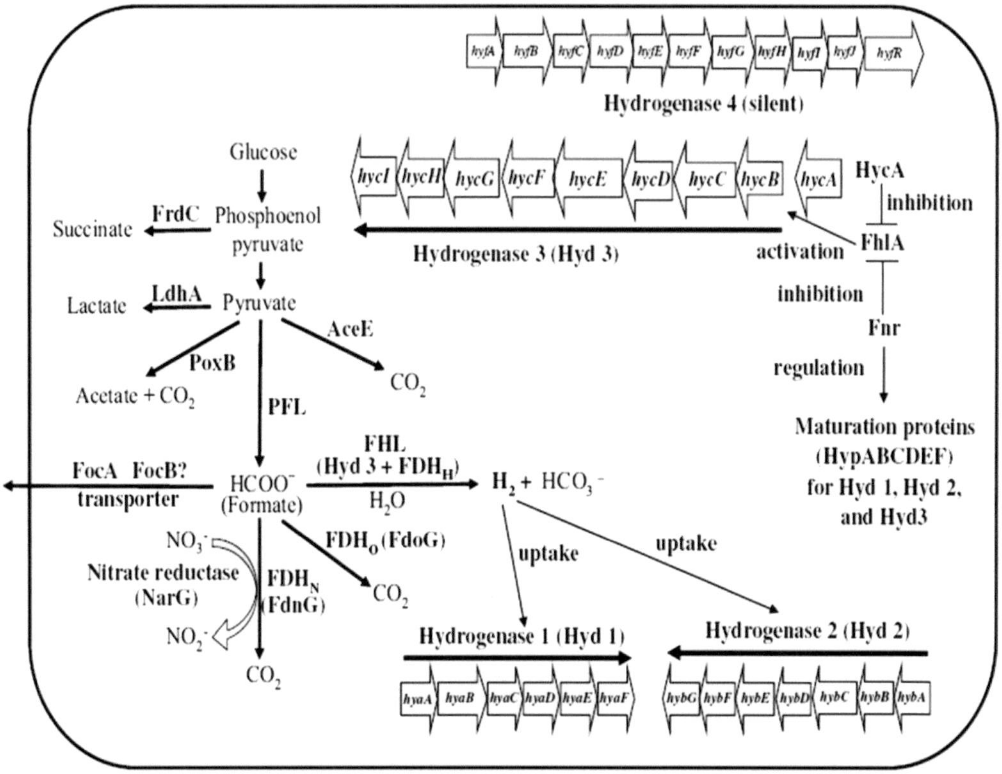


Fig. 1. Schematic of fermentative hydrogen production in E. coli (Vardar-Schara, 2008).

hydrogen production) (Das & Veziroglu, 2001). The conversion of malate and lactate to hydrogen by photosynthetic bacteria (mainly purple non-sulphur bacteria) has been documented (Koku, 2002; Kondo, 2002). Cyanobacteria are using two sets of enzymes to generate hydrogen gas (nitrogenase and hydrogenase). Hydrogen photo evolution catalyzed by nitrogenases or hydrogenases (Wünschiers et al., 2003) can only function under anaerobic conditions due to their extreme sensitivity to oxygen. Since oxygen is a by-product of photosynthesis, organisms have developed the following spatial and temporal strategies to protect the enzyme from inactivation by oxygen (Linus Pauling, 1970; Lopes Pinto, 2002). These factors can be arranged into two categories: environmental factors (light, temperature, atmosphere, nutrient availability) and intrinsic factors (genetic or certain sensitive proteins) (Beral & Zapan, 1977). Genetic engineering has made possible in cyanobacteria for effective hydrogen production (Theil, 1994). The strategies and regulatory studies of enzymes responsible for biohydrogen production in cyanobacteria was well characterized (Hansel & Lindblad, 1998). Cyanobacterial hydrogen production is not rapid which can be circumvented by combining electrochemical technology for higher efficiency of hydrogen production.

## 3. Hydrogen production by fermentative bacteria using acids

*Clostridium diolis* JPCC H-3 was obtained from soil and it is capable of producing hydrogen from slurry solution having acetic and lactic acid at higher rates compared with other isolated *clostridium* spp. Maximum hydrogen production by *C.diolis* JPCC H-3 of 6.03±0.15

ml from 5 ml of slurry solution was achieved at pH 6.8 and 40ºC (Matsumoto & Nishimura, 2007). *E. coli* produces hydrogen from formic acid with high productivity. Formic acid can be derived from biomass or carbon monoxide plus methanol. Bio-hydrogen production from formic acid by facultative anaerobe is catalyzed by formate hydrogen lyase (FHL) (Das & Veziroglu, 2001; Bagramyan & Trchounian, 2003; Sawers, 2005; Vardar-Schara, 2008). The direct decomposition of formic acid into $H_2$ and $CO_2$ by FHL would provide a high hydrogen production rate without the generation of by-products except $CO_2$. *Enterobacter* species have a higher potential for hydrogen production than *E. coli* (Das & Veziroglu, 2001). However, hydrogen production from formic acid by *Enterobacter* species has not been studied. It was reported that the hydrogen production by FHL-1 system in *E. coli* was also active only at acidic pH and high formic acid concentration (Bagramyan et al., 2002). Although acids are used by bacteria, hydrogen production is not same as that of stoichiometric yield. Hence, this biochemical barrier can be overcome by generating hydrogen gas from acids using electrochemical technology.

## 4. Electrochemical technology

### 4.1 Electrolysis

Many different types of electrolysis cells have been proposed and constructed. The different electrolysis cells can be divided into groups based on the electrolyte which capable of using $H_2O$ as reactant to produce $H_2$. However, only the solid oxide cell is capable of using $CO_2$ to produce CO (Table 2).

| Types | Alkaline | Acid | Polymer electrolyte | Solid oxide |
|---|---|---|---|---|
| Charge carrier | $OH^-$ | $H^+$ | $H^+$ | $O^{2-}$ |
| Reactant | Water | Water | Water | Water, $CO_2$ |
| Electrolyte | Sodium or Potassium hydroxide | Sulphuric or Phosphoric acid | Polymer | Ceramic |
| Electrodes | Nickel | Graphite with Pt, polymer | Graphite with Pt, polymer | Nickel, ceramics |

Table 2. Types of electrolysis cells (Vendt, 1990)

Generally, the electrolysis cell consists of two electrodes and an electrolyte. The electrolyte may be a liquid (alkaline or acid) or a solid (polymer electrolyte or solid oxide). It serves to conduct ions (the charge carrier) produced at one electrode to the other. There has been a great deal of research in splitting water to make hydrogen and oxygen; in fact its commercial uses date back to the 1890s (Norbeck et al., 1996). Water splitting in its simplest form uses an electrical current passing through two electrodes to break water into hydrogen and oxygen. Commercial low temperature electrolyzers have system efficiencies of 56–73% (70.1–53.4 kWh/kg $H_2$ at 1 atm and 25°C) (Turner et al., 2008). It is essentially the conversion of electrical energy to chemical energy in the form of hydrogen, with oxygen as a useful by-product using proton exchange membrane (PEM) (Grigoriev et al., 2006; Norbeck et., 1996; Pettersson et al., 2006). Currently, electrolysis is more expensive therefore if non-renewable power generation is used to make the electricity for electrolysis, and results in higher emissions compared to natural gas reforming (Bradley, 2000; Janssen et al., 2004). Several approaches have been addressed these shortcomings. These include using renewable sources of energy such as solar, wind, and hydro, to produce the electricity (Janssen et al., 2004; Koroneos et al., 2004) or excess power from existing generators to produce hydrogen during off-peak times (Yumurtaci & Bilgen, 2004). Since water needs high electrical energy for its electrolysis, use of weak acids or dilute acids which are obtained from wastes or by-products can be electrolyzed for supplementing hydrogen demands using low electrical appliances.

### 4.2 Electrohydrogenesis

Electrohydrogenesis is a recently developed electrolysis method for directly converting biodegradable material, organic acids into hydrogen using modified microbial fuel cells (MFCs) (Liu et al., 2005; Rozendal et al., 2006; Ditzig et al., 2007; Cheng & Logan, 2007; Rozendal et al., 2008; ). In fact, these types of cells are rather versatile and have been shown to be able to generate hydrogen from a variety of substrates, including some wastewaters (Ditzig et al., 2007). The open circuit potential of ~ -300mV is needed for the electrolysis of acetate, if hydrogen is produced at the cathode; the half reactions occurring at the anode and cathode are as follows:

Anode:

$$C_2H_4O_2 + 2H_2O \longrightarrow 2CO_2 + 8e^- + 8H^+ \quad (3)$$

Cathode:

$$8H^+ + 8e^- \longrightarrow 4H_2 \quad (4)$$

Producing hydrogen at the cathode requires a potential of at least E° = -410mV (NHE) at pH 7.0. This voltage is substantially lower than that needed for hydrogen derived from the electrolysis of water, which is theoretically 1210mV at neutral pH. In practice, 1800-2000mV is needed for water hydrolysis (under alkaline solution conditions) due to overpotential at the electrodes (Liu et al., 2005). Hence electrolysis of acids requires less electrical energy compared to electrolysis of water.

### 4.3 Types of ion exchange membranes

A thin sheet or film of ion-exchange material which may be used to separate ions by allowing the preferential transport or either cations (in the case of a cation-exchange membrane) or anions (in the case of an anion exchange membrane). If the membrane material is made from only ion exchanging material, it is called a homogeneous ion-exchange membrane. If the ion-exchange material is embedded in an inert binder, it is called a heterogeneous ion-exchange membrane. The difference between anion and cation exchange membrane are summarized in Table 3. The cation exchange membrane based on fluorinated polymer and sulfonic acid group is used as major membrane for PEMFC because of the excellent proton conductivity and durability. On the other hand, AEM based on quaternary ammonium group and hydrocarbon polymer backbone has been considered to have low thermal durability and low $OH^-$ conductivity under the condition of fuel cell (Gasteiger et al., 2008).

| Anion Exchange Membrane | Cation Exchange Membrane |
|---|---|
| $OH^-$ conductive | $H^+$ conductive |
| 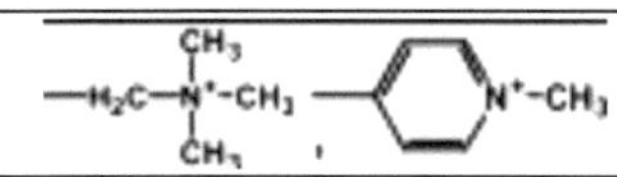 | $-SO_3^-$, ( $-PO_4^-$, $-CO_2^-$ ) |
| Pt free catalyst available<br>Advantage for cathode $O_2$ reduction | High ion conductivity<br>Excellent ionomer solution |
| Low ion conductivity<br>Low thermostability<br>Influence of $CO_2$ | High cost materials<br>Fuel crossover |

Table 3. Differences between ion exchange membranes

### 4.3.1 Cation exchange membrane electrolyser

PEM electrolyser is a recent advancement in PEM fuel cell technology. PEM-based electrolysers typically use platinum black, iridium, ruthenium, and rhodium for electrode catalysts and a Nafion membrane as the proton exchanger (Pettersson et al., 2006; Turner et al., 2008). The performance that is the hydrogen generation rate can be increased by using efficient electrodes, proton exchange membranes and by reducing electrode spacing (Liu et al., 2005). Proton exchange membranes (PEMs) are one of the most important components in microbial fuel cells (MFCs), since PEMs physically separate the anode and cathode compartments while allowing protons to transport to the cathode in order to sustain an electrical current. The Nafion 117 membrane used in this study is generally regarded as having excellent proton conductivity. Nafion, a sulfonated tetrafluorethylene, consists of a hydrophobic fluorocarbon backbone ($-CF_2-CF_2-$) to which hydrophilic sulfonate groups ($SO_3^-$) are attached. The presence of negatively charged sulfonate groups in the membrane explains the high level of proton conductivity of Nafion, while also showing a significant undesirable affinity for other cations rather than protons (Chae et al., 2008). Most MFCs are operated at a neutral pH in order to optimize bacterial growth in the anode chamber, while other cations ($Na^+$, $K^+$, $Ca^+$, $mg^{2+}$ and $NH_4^+$) contained in growth medium are typically present at a $10^5$ times higher concentration than protons (Rozendal et al., 2006). Consequently, these cations combine with the sulfonate groups of Nafion and inhibit the migration of protons produced during substrate degradation, causing a decrease in the MFC performance due to the pH reduction in the anode chamber. In addition, the frequent replacement of the buffer solution as a catholyte reduced the economic viability of MFCs. Nafion operated over a period of 50 days was contaminated with biofilm causing adverse effects on mass transport through the membrane (Chae et al., 2008).

### 4.3.2 Anion exchange membrane electrolyser

Anion exchange membrane fuel cells (AEMFCs) are a viable alternative to PEMFCs and are currently gaining renewed attention. In an AEMFC, an anion exchange membrane (AEM) conducts hydroxide (or carbonate) anions (as opposed to protons) during current flow, which results in several advantages: (1) The oxygen reduction reaction (ORR) is much more facile in alkaline environments than in acidic environments. This could potentially facilitate the use of less expensive non-PGM catalysts with high stability in alkaline environments. (2) The electro-oxidation kinetics for many liquid fuels (including non-conventional choices of importance to the military, such as sodium borohydride) is enhanced in an alkaline environment. (3) The electroosmotic drag associated with ion transport opposes the crossover of liquid fuel in AEMFCs, thereby permitting the use of more concentrated liquid fuels. This is an advantage for portable applications. (4) The flexibility in terms of fuel and ORR catalyst choice also expands the parameter space for the discovery of highly selective catalysts that are tolerant to crossover fuel. These potential advantages make AEMFCs an attractive future proposition (Christopher et al., 2010).

For a traditional AEMFC with hydrogen fuel and air/oxygen as the oxidant, the half cell and overall chemical reactions are as follows: (Varcoe & Slade, 2005)

$$H_2 + 2OH^- \rightarrow 2H_2O + 2e^- \text{ ; EO, anode} = 0.83 \text{ V} \tag{5}$$

$$\tfrac{1}{2} O_2 + H_2O + 2e^- \rightarrow 2OH^- \text{ ; EO, cathode} = 0.40 \text{ V} \tag{6}$$

Overall:

$$H_2 + \tfrac{1}{2} O_2 \rightarrow H_2O\text{; EO, cell} = 1.23 \text{ V} \tag{7}$$

In an AEMFC, hydroxide ions are generated during electrochemical oxygen reduction at the cathode. They are transported from the cathode to the anode through the anion conducting (but electronically insulating) polymer electrolyte, wherein they combine with hydrogen to form water. The electrons generated during $H_2$ oxidation pass through the external circuit to the cathode, where they participate in the electrochemical reduction of oxygen to produce -OH. Note that in practice, the ideal thermodynamic cell voltage of 1.23 V (at standard conditions) is not realized even at open circuit (zero current) due to myriad irreversibilities that arise during AEMFC operation. The phenomenological sources of irreversibility are very similar to those in PEMFCs and include oxygen and water activities that are less than unity, and gas crossover at open circuit leading to mixed potentials, and activation, ohmic, and mass transfer losses (overpotentials) during current flow. Hence, AEM may be a suitable membrane for electrolysis of acid wastes, waste waters and biomass.

## 5. Conclusion

In summary, biological hydrogen production may be the environmental pollutant free fuel for future energy needs. This could fulfil the demands of drastic fuel consumption. Some problems for the commercialization of biohydrogen as fuel can be overcome by the electrochemical technology. This review gives the details of the improvement of hydrogen production efficiently through electrochemical technology. The efficiency of hydrogen production from microbial system can be enhanced by the hybrid use of electrohydrogenesis cell. Application of this renewable hydrogen is mainly for transportation and industries. Electrohydrogenesis cell can contribute significantly to these hydrogen demands by producing large quantities of hydrogen from renewable resources and wastes such as biomass, wastewaters and acid wastes. Hence, commercialization of the biohydrogen technology can be possible with the electrochemical technology.

## 6. References

Adams, M.W.W.; Mortenson, L.E. & Chen, J.S. (1981). Hydrogenase. *Biochimica et Biophysica Acta*,Vol. 594, pp. 105-176

Appel, J.; Phunpruch, S.; Steinmuller, K. & Schulz, R. (2000). The bidirectional hydrogenase of *Synechocystis* sp. PCC 6803 works as an electron valve during photosynthesis. *Archives of Microbiology*,Vol.173, pp. 333-338

Axelsson, R.; Oxelfelt, F. & Lindblad, P. (1999). Transcriptional regulation of Nostoc uptake hydrogenase. *FEMS Microbiology Letters*, Vol. 170, pp. 77-81

Bagramyan, K.; Mnatsakanyan, N.; Poladian, A.; Vassilian, A. & Trchounian, A. (2002). The roles of hydrogenases 3 and 4, and the F0F1-ATPase, in $H_2$ production by *Escherichia coli* at alkaline and acidic pH. *FEBS Lett.* Vol. 516, pp. 172–178

Bagramyan, K. & Trchounian, A. (2003). Structural and functional features of formate hydrogen lyase, an enzyme of mixed-acid fermentation from *Escherichia coli*. *Biochemistry*, Vol. 68, pp. 1159–1170

Beral, E. & Zapan, M. (1977). Inorganic Chemistry, E.D.P. Bucharest

Bradley, M.J. (2000). Future Wheels Interviews with 44 Global Experts on the Future of Fuel Cells for Transportation and Fuel Cell Infrastructure and a Fuel Cell Primer, Northeast Advanced Vehicle Consortium, Boston, MA, p. 89

Chae, K.J.; Choi, M.; Ajayi, F.F.; Park, W.; Chang, S. & Kim, S. (2008). Mass transport through a proton exchange membrane (Nafion) in Microbial Fuel Cells. *Energy & Fuels*, Vol. 22, pp.169-176.

Cheng, S. & Logan, B. E. (2007). Sustainable and efficient biohydrogen production via electrohydrogenesis. *Proceedings of the National Academy of Science*, Vol. 104, pp. 18871–18873

Chin, H.L.; Chen, Z.S. & Chou, C.P. (2003). Fedbatch operation using *Clostridium acetobutylicum* suspension culture as biocatalyst for enhancing hydrogen production. *Biotechnology Progress*, Vol.19, pp. 383–388

Christopher G. Arges, Vijay Ramani, and Peter N. Pintauro, Anion Exchange membrane Fuel cells, The electrochemical society interface:summer 2010

Das, D. & Veziroglu, T.N. (2001). Hydrogen production by biological processes: a survey of literature. *International Journal of Hydrogen Energy*, Vol. 26, pp.13–28

Ditzig, J.; Liu, H. & Logan, B. E. (2007). Production of hydrogen from domestic wastewater using a bioelectrochemically assisted microbial reactor (BEAMR). *International Journal of Hydrogen Energy*, Vol. 32, pp. 2296–2304

Eisbrenner, G.; Distler, E; Floener, L. & Bothe, H. (1978). The occurence of hydrogenase in some blue-green algae. *Archives of Microbiology*, Vol. 118, pp. 177-184

Electrochemical Hydrogen Technologies, H. Vendt, Elsevier, Amsterdam (1990)

Gasteiger H.A. *et al.,(2008).* FC EXPO. Technical conference FC-10

Gest, H. (1954) Oxidation and Evolution of molecular hydrogen by microorganisms *Journal of Bacteriology*, Vol. 18, pp. 43-73

Grigoriev, S.A.; Porembsky, V.I. & Fateev, V.N. (2006). Pure hydrogen production by PEM electrolysis for hydrogen energy. *International Journal of Hydrogen Energy*, Vol. 31, pp. 171–175.

Hallenbeck, P.C. & Benemann, J.R. (2002). Biological hydrogen production; fundamentals and limiting processes. *International Journal of Hydrogen Energy*, Vol. 27, pp. 1185–1193.

Hansel, A. & Lindblad, P. (1998). Towards optimization of cyanobacteria as biotechnologically relevant producers of molecular hydrogen, a clean and renewable energy source. *Applied Microbiological Biotechnology*, Vol. 50; pp.153-160

Janssen, H.; Bringmann, J.C.; Emonts, B. & Schroeder, V. (2004). Safety- related studies on hydrogen production in high-pressure electrolysers. *International Journal of Hydrogen Energy*, Vol. 29, pp.759–770

Jo, J.H.; Lee, D.S. & Park, J.M. (2008). The effects of pH on carbon material and energy balances in hydrogen-producing Clostridium tyrobutyricum JM1. *Bioresource Technology*, Vol. 99, pp. 8485–8491

Kapdan, I.K. & Kargi, F. (2006). Bio-hydrogen production from waste materials. *Enzyme and Microbial Technology*, Vol. 38, pp. 569–582

Kajii, Y.; Kobayashi, M.; Takahashi, T. & Onodera, K. (1994). A novel type of mutant of Azotobacter vinelandii that fixes Nitrogen in the presence of tungsten. *Bioscience Biotechnology and Biochemistry*, Vol. 58, pp. 1179-1180

Kentemich, T.; Casper, M. & Bothe, H. (1991). The reversible hydrogenase in A. nidulans is a component of the cytoplasmic membrane. *Naturwissenschaften*, Vol. 78, pp.559-560

Kentemich, T.; Haverkamp, G. & Bothe, H. (1991). The expression of a third nitrogenase system in the cyanobacterium Anabena variabilis. *Zeitschrift fur Naturforschung*, Vol. 46, pp. 217-222

Kentemich, T.; Dannenberg, G.; Hundeshagen, B. & Bothe, H. (1988). Evidence for the occurring of the alternative vanadium-containing nitrogenase in the cyanobacterium Anabena variabilis. *FEMS Microbiology Letters*, Vol. 51, pp. 19-24

Kim, J. & Rees, D.C. (1994). Nitrogenase and biological nitrogen fixation. *Biochemistry*, Vol. 33, pp. 389-397

Koku, H.; Eroglu, I.; gunduz, U.; Yucel, M. & Turker, L. (2002). Aspects of the metabolism of hydrogen production by *Rhodobacter sphaeroides*. *International Journal of Hydrogen Energy*, Vol. 27, pp. 1315-1329

Kondo, T.; Arakawa, M.; Waakayama, T. & Miyake, J. (2002). Hydrogen production by combining two types of photosynthetic bacteria with different characteristics. *International Journal of Hydrogen Energy*, Vol. 27, pp. 1303-1308

Koroneos, A.; Dompros, G. & Roumbas, N. (2004). Moussiopoulos. *International Journal of Hydrogen Energy*,Vol. 29, pp. 1443–1450

Kosaric, N. & Lyng, R.P. (1988). Microbial production of hydrogen. In *Biotechnology*, Vol 6b, eds. H.J. Rehm & G. Reed. pp. 101-136. Weinheim: VCH Verlagsgesellschaft

Kurokawa, T. & Tanisho, S. (2005). Effects of formate on fermentative hydrogen production by Enterobacter aerogenes. *Marine Biotechnology*, Vol. 7, pp. 112–118

Kumar, N. & Das, D. (2000). Enhancement of hydrogen production by *Enterobacter cloacae* IIT-BT 08. *Process Biochemistry*, Vol. 35, pp. 589–593

Kumar, N. & Das, D., (2001). Continuous Hydrogen production by immobilized *Enterobacter cloacae* IIT-BT 08 using lignocellulosic material as solid matrices. *Enzyme Microbiological Technology*,Vol. 29, pp. 280–287

Lambert, G.R. & Smith, G.D. (1981). The hydrogen metabolism ofcyanobacteria (Blue-green algae). *Biological Reviews*,Vol. 56, pp. 589-660

Lee, K.H.; Lo, Y.S.; Lo, Y.C.; Lin, P.J. & Chang, J.S. (2004). Operation strategies for biohydrogen production with a high-rate anaerobic granular sludge bed bioreactor. *Enzyme Microbiological Technology*, Vol. 35, pp. 605–612

Levin, D.B.; Islam, R.; Cicek, N. & Sparling, R. (2006). Hydrogen production by *Clostridium thermocellum* 27405 from cellulosic biomass substrates. *International Journal of Hydrogen Energy*, Vol. 31, pp. 1496–1503

Linus Pauling. (1970). General Chemistry, Section 15-2. San Francisco

Liu, H.; Grot, S. & Logan, B. E. (2005). Electrochemically assisted microbial production of hydrogen from acetate. *Environmental Science and Technology,* Vol. 39, pp. 4317–4320

Liu, H.; Cheng, S. & Logan, B. E. (2005). Power generation in fed-batch microbial fuel cells as a function of ionic strength, temperature and reactor configuration. *Environmental Science and Technology,* Vol. 39, pp.5488–93

Liu, W.; Wang, A.; Ren, N.; Zhao, X.; Liu, L.; Yu, Z. & Lee, D. (2007). Electrochemically assisted biohydrogen production from acetate. *Energy Fuels,* Vol. *22,* pp. 159–163

Lopes Pinto, F. A.; Troshina, O. & Lindblad, P.(2002). A brief look at three decades of research on cyanobacterial hydrogen evolution. *International Journal of Hydrogen Energy,* Vol. 25 , pp. 1209-1215

Matsumoto, M. & Nishimura, Y. (2007). Hydrogen production by fermentation using acetic acid and lactic acid. *Journal of Bioscience and Bioengineering,* Vol. 103, pp. 236-241.

Mishra, J.; khurana, S.; Kumar, N.; Ghosh, A. K. & Das, D. (2004). Molecular cloning, characterization, and over expression of a novel [Fe]-hydrogenae isolated from a high rate of hydrogen producing *Enterobacter cloacae* IIT-BT 08. *Biochemical and Biophysical Research Communications,* Vol. 324, pp. 679-85

Nakashimada, Y.; Rachman, M.A.; Kakizono, T. & Nishio, N. (2002). Hydrogen production of Enterobacter aerogenes altered by extracellular and intracellular redox states. *Int. J. Hydrogen Energy,* Vol. 27, pp. 1399–1405

Nath, K. & Das, D. (2004). Improvement of fermentative hydrogen production: various approaches, *Appl Microbiol Biotechnol.,* Vol. 65, pp. 520-529

Norbeck, J.M.; Heffel, J.W.; Durbin, T.D.; Tabbara, B. Bowden,J,M. & Montani, M.C.(1996). Hydrogen Fuel for Surface Transportation, Society of Automotive Engineers Inc., Warrendale, PA, p. 548

Palazzi, E.; Fabiano, B. & Perego, P. (2000). Process development of continuous hydrogen production by Enterobacter aerogenes in a packed column reactor. *Bioprocess Eng,* Vol. 22, pp. 205–213

Peschek, G.A. (1979). Aerobic hydrogenase activiy in Anacystis nidulans. The oxyhydrogen reaction. *Biochimica et Biophysica Acta,* Vol. 548, pp. 203-215.

Pettersson, B.; Ramsey, D. & Harrison. (2006). A review of the latest developments in electrodes for unitised regenerative polymer electrolyte fuel cells. *Journal of Power Sources* Vol. 157, pp. 28–34

Robson, R.L. & Postgate, J.R. (1980). Oxygen and hydrogen in biological nitrogen fixation. *Annual Review of Microbiology,* Vol. 34, pp.183-207.

Rozendal, R. A.; Hamelers, H. V. M.; Euyerink, G.J.W. ; Metz, S.J. & Buisman, C. J. N.(2006). Principles and perspectives of hydrogen production through biocatalyzed electrolysis. *International Journal of Hydrogen Energy,* Vol. 31, pp. 1632-1640

Rozendal, R.; Jeremiasse, A.; Hamelers, H. & Buisman, C. (2008). Hydrogen production with a microbial biocathode. *Environmental Science and Technology,* Vol. 42, pp. 629–634

Sawers, R.G. (2005). Formate and its role in hydrogen production in *Escherichia coli. Biochemistry Society Transactions,* Vol. 33, pp. 42–46

Schlegel, H.G. & Schneider, K.(1978). Introductory report:distribution and physiological role of hydrogenases in microorganisms. In *Hydrogenases*: Their Catalytic Activity, Structure and Function. pp. 15-44. Goltze KG,Gottingen

Shin, J.H.; Yoon, J.H.; Sim, S.J.; Kim, M.S. & Park, T.H. (2007). Fermentative hydrogen production by the newly isolated Enterobacter asburiae SNU-1. *International Journal of Hydrogen Energy,* Vol. 32, pp.192–199

Stewart, W.D.P. (1973). Nitrogen fixation by photosynthetic microorganisms. *Annual Review of Microbiology,* Vol. 27, pp. 283-316

Tamagnini, P.; Troshina, P.; Oxelfelt, F.; Salema, R. & Lindblad, P. (1997) Hydrogenase in Nostoc sp. strain PCC 73120, a strain lacking a bi-directional enzyme. *Applied and Environmental Microbiology,* Vol. 63, pp. 1801-1807

Theil, T.(1994). Genetic analysis of cyanobacteria, In:Bryant DA(ed) The molecular biology of cyanobacteria. Kluwer academic Publishers, Dordrecht, The Netherlands. pp581-611

Turner, J. G.; Sverdrup, M.K.; Mann, P.C.; Maness, B.; Kroposki, M.; Ghirardi, R.J. & Evans, D. (2008). Renewable hydrogen production. *International Journal of Hydrogen Energy,* Vol. 32, pp. 379–407

Vardar-Schara, G.; Maeda, T. & Wood, T.K. (2008). Metabolically engineered bacteria for producing hydrogen via fermentation. *Microbial Biotechnology,*Vol. 1, pp. 107–125

Varcoe, J.R. & Slade, R.C.T.(2005). Prospects for alkaline anion-exchange membranes in low temperature fuel cells, *Fuel Cells,* Vol. 5, pp.187

Yagi, T. (1981). Function and structure of hydrogenases. *Natural Science,*Vol. 32, pp. 29-83

Yumurtaci, Z. & Bilgen, E. (2004). Hydrogen production from excess power in small hydroelectric Installations. *International Journal of Hydrogen Energy,*Vol. 29, pp. 687–693

Yoshida, A.; Nishimura, T.; Kawaguchi, H.; Inui, M. & Yukawa, H.(2005). Enhanced hydrogen production from formic acid by formate hydrogen lyase overexpressing *Escherichia coli* strain. *Applied Environmental Microbiology,*Vol. 71, pp. 6762– 6768

Wünschiers, R.; Batur. & Lindblad P (2003). Presence and expression of hydrogenase specific C-terminal endopeptidases in cyanobacteria. *BMC Microbiology* 3: 8 (on-line journal, 12 pages)

Zhang, C.; Xing, X.H. & Lou, K. (2005). Rapid detection of a gfp-marked Enterobacter aerogenes under anaerobic conditions by aerobic fluorescence recovery. *FEMS Microbiology Letters,* Vol. 249, pp. 211–281

# 22

# A Genetic-Fuzzy System for Modelling of Selected Processes in Diesel Engine Fuelled by Biofuels

Michał Kekez and Leszek Radziszewski
*Kielce University of Technology*
*Poland*

## 1. Introduction

The work of diesel engine fuelled by various fuels can be modelled by the methods based on the first law of thermodynamics (Rychter & Teodorczyk, 2006), or the CFD (computational fluid dynamics) methods (Amsden, 1997). However, all these methods have some deficiencies. The first group of methods can give results which have insufficient accuracy at some ranges of engine working cycle. The second group of methods produces models which cannot be easily applied to real-time control of a diesel engine. The concise and fast model of a diesel engine working cycle should be found. The methods which belong to the group of computational intelligence (CI) methods are also used for modelling selected aspects of engine work (Kalogirou, 2003) and many results are encouraging. However, the pressure course in cylinder or injection pipe was rarely modelled by CI methods. In the next sections we present the single model of cylinder pressure and injection pressure in a diesel engine fuelled by RME (rapeseed methyl esters) or its blends with diesel oil. The model was created by means of the GFSm system proposed by the authors. The methodology that allows the extension of the created model for other fuels is also presented. The obtained model can be applied to diagnostic and control of a diesel engine. One of the advantages of the proposed method is that the model is built for a one given engine and therefore describes its work with good accuracy.

## 2. Test bench

Experimental research and necessary measurements of cylinder pressure, injector pressure and injector needle lift curves were carried out on test bench, which consisted of a three cylinder diesel engine, water brake and control panel. There were four measurement chains (Fig. 1). The engine was fuelled by diesel oil, RME (methyl esters of rapeseed oil), and its blends. Cylinder pressure values and injection pressure values were recorded every 1.406° CA (exactly 512 measurements for one working cycle of four-stroke engine). At each measurement point, defined by crankshaft angle, values of parameters were recorded for 50 consecutive working cycles and later averaged. This methodology was used for all tested fuels. The experiments were conducted when engine operated at full-load and part-load conditions, with crankshaft rotational speeds of 1000, 1200, 1400, 1600, 1800, and 2000 rpm (revolutions per minute).

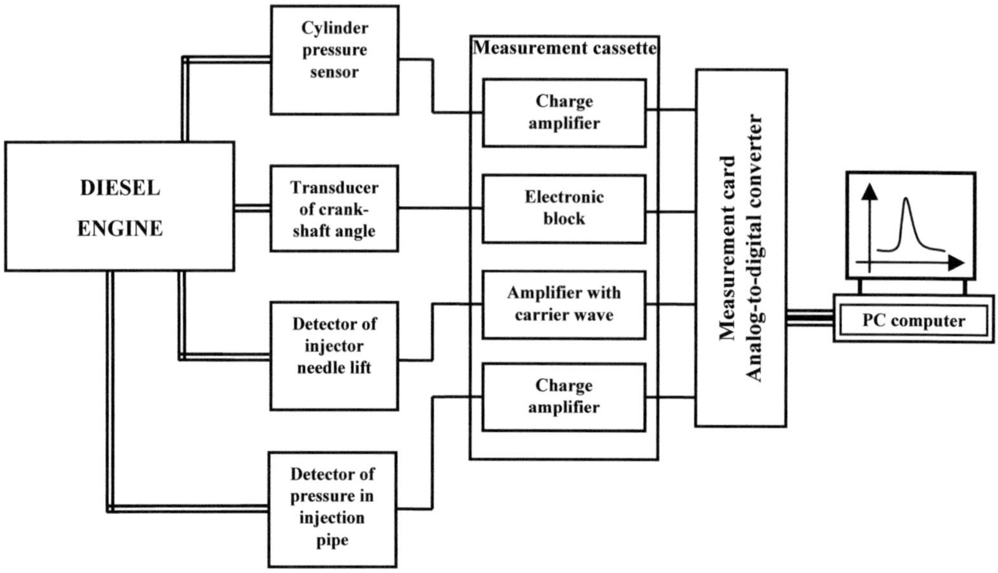


Fig. 1. The scheme of measurement chains (Ambrozik et al., 2010)

## 3. Results of experimental research

The measured values, when engine was fuelled by RME, are presented in Figure 2. Cylinder pressure (Fig.2a) in the range below 351 °CA is almost identical for all tested rotational speeds. Pressure starts to increase rapidly when it reaches 3 to 4 MPa, and when crankshaft rotation angle lies between 351° (for speed of 1000 rpm) and 357° CA (for speed of 2000 rpm). When rotational speed of crankshaft increases, the maximum value of pressure decreases from 8.5 to 8 MPa. The first derivative of pressure with respect to crankshaft angle decreases from 0.92 to 0.68 MPa/°CA with the increase of rotational speed from 1400 to 2000 rpm.

Injection pressure (Fig. 2b) in the range below 335 °CA is almost identical for all tested rotational speeds. Pressure increases rapidly in the range of about 335 to 340 °CA, but when reaches about 20 MPa (injector opening pressure), its increase becomes much slower or it starts to decrease. Maximum pressure value increases with the increase of rotational speed. The area below the pressure curve (which is related to amount of fuel) changes from 382 MPa ·°CA at 1000 rpm to 492 MPa ·°CA at 2000 rpm.

## 4. Genetic-fuzzy system GFSm

The proposed GFSm system (Kekez, 2008; Radziszewski & Kekez, 2010) belongs to the group of genetic-fuzzy rule-based systems (Cordon et al., 2001), i.e. fuzzy systems which use genetic algorithm to improve or create a set of rules. There are at least four approaches in these systems: Pittsburgh (Smith, 1983) and Michigan (Holland, 1976) approaches, iterative rule learning (Cordon et al., 2001), and co-evolutionary approach (Regattieri Delgado et al., 2004). GFSm uses genetic learning of the whole knowledge base, not only membership functions of fuzzy set, and uses modified Pittsburgh approach (with advanced encoding, or, in other words, special arrangement of data in a chromosome). The Pittsburgh approach

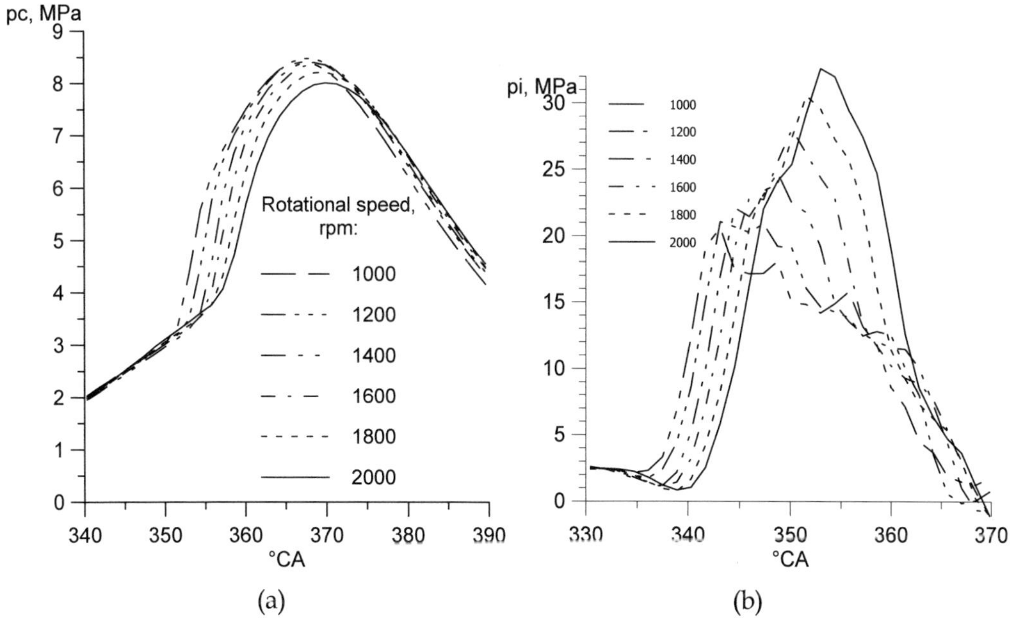


Fig. 2. The curves of $p_c$ cylinder pressure (a) and $p_i$ injection pressure (b) in function of crankshaft angle in the engine fuelled by RME at rotational speeds 1000, 1200, 1400, 1600, 1800 and 2000 rpm

offers wide exploration of the solution space, but has several drawbacks, which were avoided by using the above-mentioned encoding in GFSm system.

### 4.1 Fuzzy systems

For the fuzzy system having $m$ input variables, $x_i$, and one output variable, $y$, the fuzzy rule has at most $m$ conditions (or premises) of the form $x_i \text{ IS } A_i$, where $A_i$ denotes a fuzzy set. The consequent of the rule ($y \text{ IS } B_j$) is placed after "THEN" in the rule. The fuzzy rule has the form $\text{IF}\left[(x_1 \text{ IS } A_1)\text{AND}(x_2 \text{ IS } A_2)\text{AND}\ldots\text{AND}(x_m \text{ IS } A_m)\right]\text{THEN}\left(y \text{ IS } B_j\right)$ and is called the Mamdani-type rule. In the Takagi-Sugeno-type of fuzzy rule, the consequent has the form $y = w_0 + w_1 \cdot x_1 + w_2 \cdot x_2 + \ldots + w_m \cdot x_m$, where $w_i$ are constants. A set of fuzzy rules (also known as a rule base or a rule set) together with definitions of membership functions of fuzzy sets is called the knowledge base of a fuzzy system. Each fuzzy set is described by its membership function, which usually has two parameters, $a$ and $b$, called center and width, respectively; see e.g. equations (3) and (4) in Section 5. For reasoning in fuzzy systems, the defuzzification method must be also defined.

### 4.2 Genetic algorithm (GA) used in GFSm

Genetic algorithms (GAs) search for optimal solution. Solution parameters are encoded in a binary string called a chromosome. Specifically in GFSm, a solution refers to a fuzzy system. The number of chromosomes is called a population size $q$. Fragments of a chromosome are called genes (implemented as binary strings or single bits). The GA searches for optimal solution by exchanging fragments of chromosomes (using so-called crossover operator) and by randomly changing some bits in chromosome (using so-called mutation operator). The quality of each solution encoded by one chromosome is evaluated before the application of the above-mentioned operators. If GA uses the elitist strategy, the best chromosome at each stage is copied to the next stage without crossover or mutation. Initial set of chromosomes is called an initial population. The initial population in GFSm is built by the following routine:

1. The maximum number of fuzzy sets describing the *i*-th input of the fuzzy system ($i = 1,\ldots,m$) is set as a GFSm input parameter $j_{\max i}$.
2. For each input, create two fuzzy sets with sigmoidal membership functions (Eq. 2), and $j_{\max i} - 2$ fuzzy sets with Gaussian membership functions (Eq. 3); for each fuzzy set, create randomly two parameters (*a* and *b*) of these functions.
3. Create 10 fuzzy sets with triangular membership functions (Eq. 4), which uniformly divide the range of output values.
4. Given all these fuzzy sets, build a set of rules by using the (Wang & Mendel, 1992) method for Mamdani-type systems, or by using the (Abonyi et al., 2002) method for Takagi-Sugeno systems
5. Repeat steps 2-4 for all *q* fuzzy systems.

### 4.3 Encoding of information for GA used in GFSm

The idea of advanced encoding in GFSm system, proposed in (Kekez, 2008), means the fragmentation of rules. All information in the chromosome is grouped by the number of input to which it refers. In a chromosome, all pieces of rules regarding given input are located just after the definitions of fuzzy sets for this input (Fig. 3).

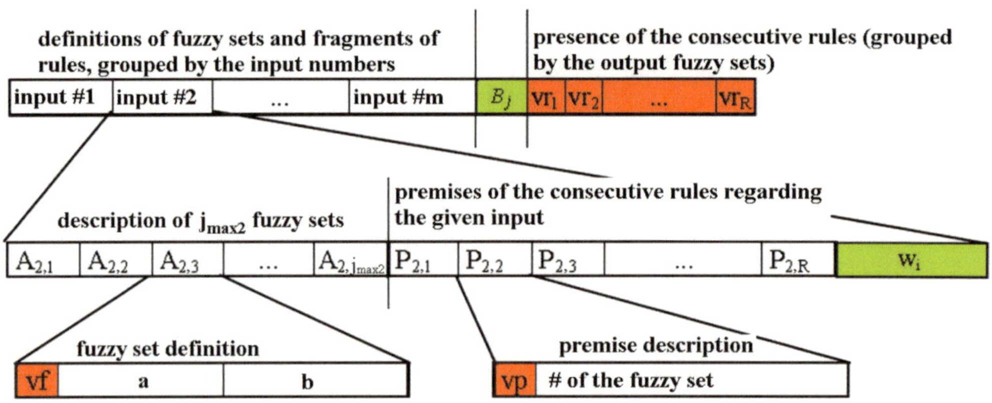


Fig. 3. Encoding of information in a chromosome, used in GFSm system (Kekez, 2008)

The last part of a chromosome stores the information about which rules are present ("active") in a rule base of a fuzzy system. The "validity" bits *vf*, *vp*, $vr_1$, $vr_2$, …, $vr_R$ (Fig. 3) allow to dynamically activate or deactivate a fuzzy set, a premise in a rule, and a whole rule, respectively. A chromosome has fixed length, but number of rules can vary in the range $\langle 0, R \rangle$, because each $vr_r$ bit ($r = 1 \ldots R$) activates or deactivates *r*-th rule. The maximum number of rules, *R*, is calculated during the creation of initial population. A rule need not to contain the premises ($x_i$ IS $A_{ij}$) regarding all inputs because each *vp* bit includes or excludes a premise in a rule. Similarly, the total number of fuzzy sets describing *i*-th input varies from 0 to $j_{\max i}$ because of *vf* bits. The $j_{\max i}$ value is set by the user before the start of the GFSm system, as described in Section 4.2. A chromosome also contains either the definitions

of fuzzy sets $B_j$, which describe system output (in Mamdani-type fuzzy systems), or $w_i$ parameters of the consequent of the rule (in Takagi-Sugeno-type systems).
The proposed idea of information encoding allows processing of high-dimensional data, and retains the connection between the definition of a given fuzzy set and its references in the rules. Like in Michigan approach, the crossover operator connects fragments of rules. These fragments come from rules which have the identical consequent.

## 5. Model of pressures in diesel engine

The results of experiments for two selected rotational speeds (1200 and 1800 rpm) when engine was fuelled by diesel oil were used as a training dataset for GFSm system. Each row in the training dataset consisted of ($x_1$, $x_2$, $x_3$, $y$) values. The $x_1$ was a crankshaft rotation angle $\alpha$, and $x_2$ was not a rotational speed $n$, but the time (in milliseconds) elapsed since the moment when crankshaft rotation angle was 335°CA:

$$x_2 = \begin{cases} 0 & \text{for } \alpha \le 335 \\ \dfrac{(\alpha - 335)\cdot 60}{n \cdot 360} \cdot 1000 & \text{for } \alpha > 335 \end{cases} \tag{1}$$

The $x_3$ denotes the type of pressure: it was equal 1 for injection pressure and 0 for cylinder pressure. The $y$ was the pressure value for a given $x_1$, $x_2$, and $x_3$.
The GFSm system was run twice with the following options:
- type of the fuzzy system: Mamdani,
- population size: 50,
- probability of crossover: 0.77.

In the first run we obtained 12 rules containing "$x_3$ is $A_{3,1}$" condition (regarding cylinder pressure), and in the second run we also obtained 12 rules with "$x_3$ is $A_{3,2}$" condition (regarding injection pressure).
By using GFSm system, we obtained one model of pressure courses consisting of 24 Mamdani-type fuzzy rules (Fig. 4).
In Fig. 4, the notations (S, $a$ , $b$), (M, $a$ , $b$), and (T, $a$ , $b$), where $a$ and $b$ are constants, identity fuzzy sets defined by a sigmoidal

$$\mu_A(x,a,b) = \frac{1}{1+e^{b\cdot(x-a)}}, b \neq 0 \tag{2}$$

Gaussian

$$\mu_A(x,a,b) = e^{-\left(\frac{x-a}{b}\right)^2} \tag{3}$$

or triangular

$$\mu_A(x,a,b) = \begin{cases} 0 & \text{for } x < a-b \\ \dfrac{x-a+b}{b} & \text{for } a-b \le x < a \\ \dfrac{a+b-x}{b} & \text{for } a \le x \le a+b \\ 0 & \text{for } x \ge a+b \end{cases} \tag{4}$$

membership functions, respectively. The $A_{3,1}$ and $A_{3,2}$ refer to singleton fuzzy sets which pertain to cylinder pressure ($p_c$) or injection pressure ($p_i$), respectively.
The operation of the model is depicted in Fig. 5, where "knowledge base" refers to membership functions of fuzzy sets and set of rules shown in Fig. 4.

if $x_1$ is (M,416.34,23.23) and $x_3$ is $A_{3,1}$ then y is (T,0.11,0.97)
if $x_1$ is (M,292.50,38.50) and $x_3$ is $A_{3,1}$ then y is (T,0.11,0.97)
if $x_1$ is (M,200.29,47.34) and $x_3$ is $A_{3,1}$ then y is (T,0.11,0.97)
if $x_1$ is (S,433.52,-0.07) and $x_3$ is $A_{3,1}$ then y is (T,0.11,0.97)
if $x_1$ is (M,412.02,93.24) and $x_2$ is (M,26.05,10.48) and $x_3$ is $A_{3,1}$ then y is (T,1.06,0.97)
if $x_1$ is (M,445.80,35.32) and $x_3$ is $A_{3,1}$ then y is (T,1.06,0.97)
if $x_1$ is (M,346.29,7.52) and $x_2$ is (M,6.77,4.43) and $x_3$ is $A_{3,1}$ then y is (T,1.06,0.97)
if $x_1$ is (M,180.37,137.34) and $x_2$ is (M,1.08,0.46) and $x_3$ is $A_{3,1}$ then y is (T,2.52,0.97)
if $x_1$ is (M,393.65,11.07) and $x_3$ is $A_{3,1}$ then y is (T,2.53,0.97)
if $x_1$ is (M,364.22,38.06) and $x_2$ is (M,26.05,10.48) and $x_3$ is $A_{3,1}$ then y is (T,7.89,0.97)
if $x_1$ is (M,364.22,38.06) and $x_2$ is (M,6.46,3.76) and $x_3$ is $A_{3,1}$ then y is (T,8.79,0.97)
if $x_1$ is (M,333.05,121.68) and $x_2$ is (M,3.91,1.49) and $x_3$ is $A_{3,1}$ then y is (T,8.79,0.97)
if $x_1$ is (S,363.69,-0.72) and $x_3$ is $A_{3,2}$ then y is (T,-0.90,3.34)
if $x_1$ is (M,352.12,15.31) and $x_2$ is (S,0.38,4.90) and $x_3$ is $A_{3,2}$ then y is (T,-0.90,3.34)
if $x_1$ is (S,335.39,0.21) and $x_3$ is $A_{3,2}$ then y is (T,3.28,3.34)
if $x_1$ is (M,362.23,2.46) and $x_2$ is (M,2.54,0.50) and $x_3$ is $A_{3,2}$ then y is (T,4.59,3.34)
if $x_1$ is (M,362.23,2.46) and $x_2$ is (S,3.01,-10.24) and $x_3$ is $A_{3,2}$ then y is (T,10.81,3.34)
if $x_1$ is (M,357.54,7.13) and $x_2$ is (M,2.94,0.66) and $x_3$ is $A_{3,2}$ then y is (T,12.69,3.34)
if $x_1$ is (M,350.37,11.20) and $x_2$ is (M,2.94,0.66) and $x_3$ is $A_{3,2}$ then y is (T,14.99,3.34)
if $x_1$ is (M,348.90,1.11) and $x_2$ is (M,2.12,0.92) and $x_3$ is $A_{3,2}$ then y is (T,14.99,3.34)
if $x_1$ is (M,344.00,4.15) and $x_2$ is (M,2.15,0.80) and $x_3$ is $A_{3,2}$ then y is (T,14.99,3.34)
if $x_1$ is (M,355.08,3.86) and $x_2$ is (M,2.12,0.30) and $x_3$ is $A_{3,2}$ then y is (T,24.18,3.34)
if $x_1$ is (M,337.24,7.01) and $x_2$ is (M,2.15,0.80) and $x_3$ is $A_{3,2}$ then y is (T,27.08,3.34)
if $x_1$ is (M,358.46,11.99) and $x_2$ is (M,1.61,0.39) and $x_3$ is $A_{3,2}$ then y is (T,30.84,3.34)

Fig. 4. Knowledge base of the model of pressure courses in diesel engine fuelled by diesel oil

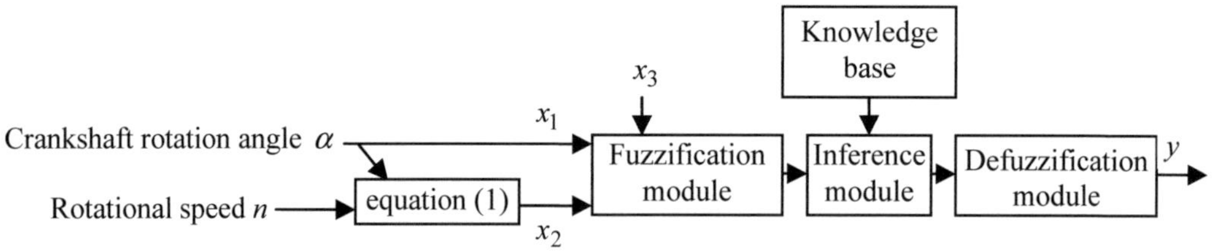


Fig. 5. Operation of the Mamdani-type fuzzy model of pressure courses, which has three input variables, $x_1$, $x_2$, and $x_3$, and one output variable, $y$

The time of computation of cylinder pressure curves and injection pressure curves for all tested rotational speeds (and time of writing of the results to disk) is less than 20 ms.

### 5.1 Extension of the model for other fuels

The model that was built for diesel oil (Fig. 5) can be extended for the prediction of pressure curves in the engine fuelled by other mineral fuels. In order to achieve this, two scaling functions were added to the existing model (fuzzy system), one for the inputs and another for the output of the system (Fig. 6).

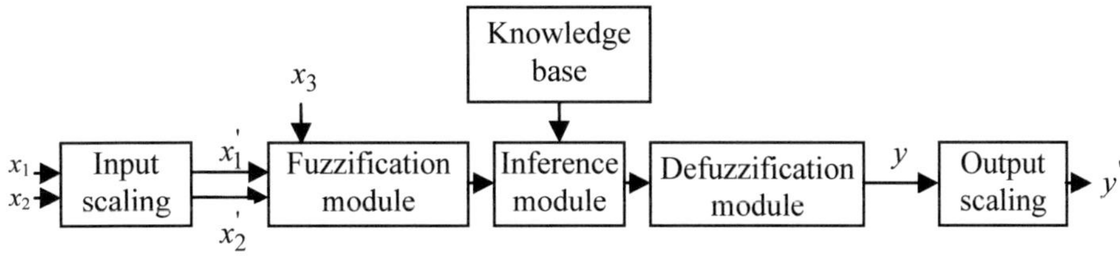


Fig. 6. Operation of the Mamdani-type fuzzy model of pressure courses, with scaling functions

The output scaling function has 4 parameters: $a_0$, $a_1$, $a_2$, and $a_3$:

$$y' = \begin{cases} y \cdot \dfrac{a_2}{a_0} & \text{for } y < a_0 \\ (y - a_0) \cdot \dfrac{a_3 - a_2}{a_1 - a_0} + a_2 & \text{for } y \geq a_0 \end{cases} \tag{5}$$

The input scaling function has only one parameter, $a_4$:

$$x_1' = x_1 - a_4 \tag{6}$$

The $x_2'$ value is calculated as in Eq. 1, but using the $\alpha = x_1'$ instead of $\alpha = x_1$. The values of parameters of output scaling function are calculated according to Table 1. Except for $a_0$ and $a_3$, parameter values are different for each fuel. In order to obtain these values for a new fuel, it is necessary to measure the pressure curve as a function of crankshaft angle for only one rotational speed, for example, 1800 rpm.

Let $\alpha_{p_{c\max}}$ denote $\alpha$ value for which $p_c$ reaches its maximum value, $p_{c\max}$. Let $\dfrac{d^2p_c}{d\alpha^2}$ denote the second derivative of $p_c$ with respect to $\alpha$. For a certain value of $\alpha$, the $\dfrac{d^2p_c}{d\alpha^2}$ reaches its maximum, denoted $\left(\dfrac{d^2p_c}{d\alpha^2}\right)_{\max}$. The $p_c$ value at this point is the value of $a_0$ parameter when we consider cylinder pressure for diesel oil (see Table 1).

Analogically, we define $p_{i\max}$ and $\left(\dfrac{d^2p_i}{d\alpha^2}\right)_{\min}$, but we does not search the global minimum, but only the minimum in the range where $\alpha \in \left\langle 330, \alpha_{p_{i\max}} \right\rangle$.

When $x_3$ is injection pressure, the $a_0 - a_2$, and these parameters can be considered as the estimation of injector opening pressure.

Once the output scaling function has been determined, and when $x_3$ is injection pressure, the $a_4$ value is chosen for a new fuel in such a way that the root mean square error (RMSE) between the experimental and computational $p_i$ curves for rotational speed of 1800 rpm is minimal

$$RMSE = \sqrt{\frac{\sum_{i=1}^{N}\left(y_{1,i} - y_{2,i}\right)^2}{N}} \tag{7}$$

where $y_{1,i}$ and $y_{2,i}$ are experimental and computational values of $p_i$, respectively.
When $x_3$ is cylinder pressure, we select $a_4$ in other way, by choosing a value for which the mean indicated pressure (Rychter & Teodorczyk, 2006) is calculated with the highest accuracy.

| Para-meter | Value (when $x_3$ is cylinder pressure) | | Value (when $x_3$ is injection pressure) | |
|---|---|---|---|---|
| $a_0$ | $p_c$ for $\alpha$ when $\left(\frac{d^2p_c}{d\alpha^2}\right)_{max}$ | for diesel oil | $p_i$ for $\alpha$ when $\left(\frac{d^2p_i}{d\alpha^2}\right)_{min}$ (for $\alpha \in \langle 330, \alpha_{p_{i\max}} \rangle$) | for diesel oil |
| $a_2$ | | for a new fuel (e.g. RME) | | for diesel oil |
| $a_1$ | $p_{c\max}$ | for diesel oil | $p_{i\max}$ | for diesel oil |
| $a_3$ | | for a new fuel (e.g. RME) | | for a new fuel (e.g. RME) |

Table 1. Values of parameters of the output scaling function

The Figs 7 and 8 present the way of calculation (7ab, 8ab) and application (7cd, 8cd) of output scaling function parameters.

### 5.2 Accuracy of the model

The accuracy of the model for RME fuel is presented in Table 2 (injection pressure) and Table 3 (cylinder pressure). The maximum pressure in injection pipe is predicted with error not exceeding 7.2%, and the area below this pressure curve – with the error not exceeding 8.2% for any tested rotational speed. The maximum cylinder pressure error does not exceed 3%, and mean indicated pressure is calculated with error not exceeding 8.1%.
The model allows prediction of the pressure curves (Fig. 9) with accuracy required in practical applications.

| Rotational speed, $n$, rpm | Area $S$ below the curve $p_i = f(\alpha)$, MPa·°CA | | | Maximum pressure $p_{i\max}$, MPa | | |
|---|---|---|---|---|---|---|
| | experimental | computed | $\delta S$, % | experimental | computed | $\delta p_{i\max}$, % |
| 1000 | 382 | 366 | -4.1 | 20.6 | 20.3 | -1.4 |
| 1200 | 394 | 386 | -2.1 | 21.1 | 21.8 | 3.3 |
| 1400 | 418 | 424 | 1.5 | 24.7 | 26.5 | 7.2 |
| 1600 | 440 | 442 | 0.3 | 28.0 | 28.4 | 1.2 |
| 1800 | 461 | 449 | -2.7 | 30.6 | 30.1 | -1.6 |
| 2000 | 492 | 452 | -8.2 | 32.6 | 31.0 | -4.9 |

Table 2. Accuracy of the pressure curves model of a diesel engine fuelled by RME; comparison of the area below the injection pressure curve and the maximum pressure in the injection pipe

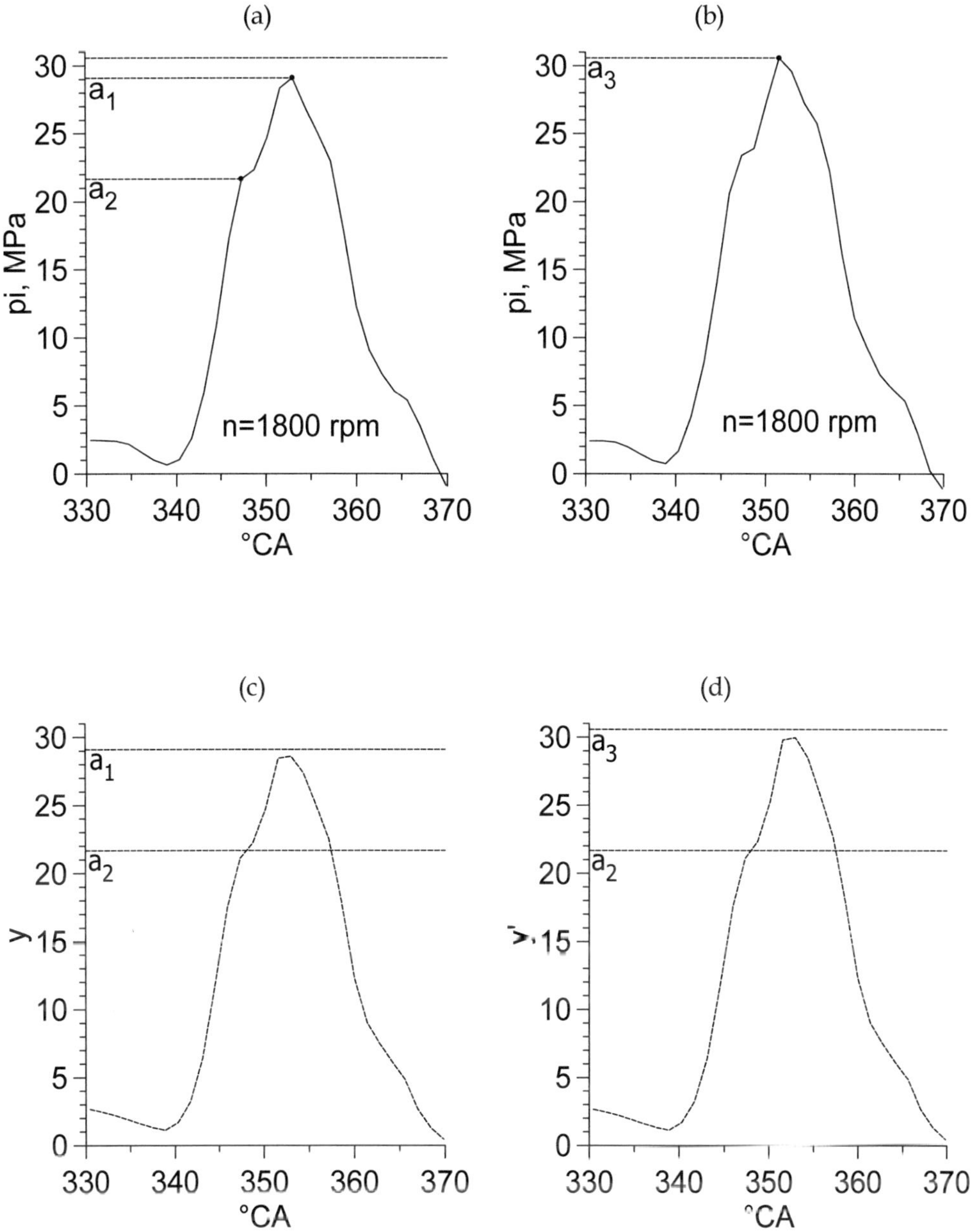


Fig. 7. Calculation and application of parameters of the output scaling function when $x_3$ is injection pressure: a) the experimental pressure curve $p_i$ and calculation of $a_1$, and $a_2=a_0$, when the engine is operated at a crankshaft speed of 1800 rpm and fuelled by diesel oil; b) the experimental pressure curve $p_i$ and calculation of $a_3$ when the engine is operated at 1800 rpm and fuelled by a new fuel (for instance, RME); c) the pressure curve $p_i$ modelled without output scaling (for diesel oil) for a given rotational speed; d) the pressure curve $p_i$ modelled with output scaling (for a new fuel, for instance, RME) for a given rotational speed

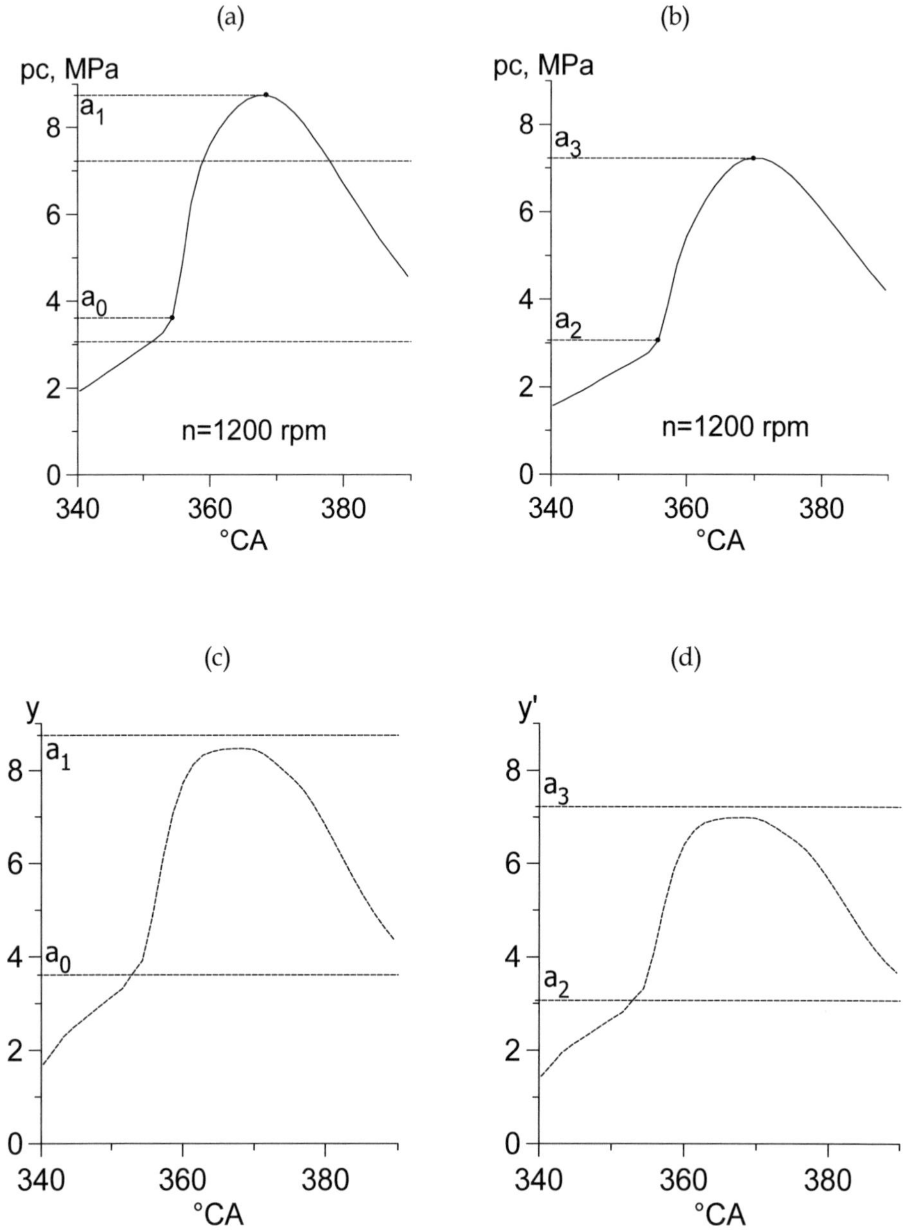


Fig. 8. Calculation and application of parameters of the output scaling function when $x_3$ is cylinder pressure: a) experimental pressure curve $p_c$ and calculation of $a_0$ and $a_1$ when the engine is operated at a crankshaft speed of 1200 rpm and fuelled by diesel oil;
b) experimental pressure curve $p_c$ and calculation of $a_2$ and $a_3$ when the engine is operated at 1200 rpm and fuelled by a new fuel (e.g., RME); c) the pressure curve $p_c$ modelled without output scaling (for diesel oil) for a given rotational speed; d) the pressure curve $p_c$ modelled with output scaling (for a new fuel, for instance, RME) for a given rotational speed

| Rotational speed, $n$, rpm | Mean indicated pressure $P$ | | | Maximum pressure $p_{c\max}$, MPa | | |
|---|---|---|---|---|---|---|
| | experimental | computed | $\delta P$, % | experimental | computed | $\delta p_{c\max}$, % |
| 1000 | 0.95 | 0.96 | 1,2 | 8.41 | 8.15 | -3.1 |
| 1200 | 1.03 | 1.03 | -0,2 | 8.47 | 8.20 | -3.2 |
| 1400 | 1.05 | 1.01 | -4,1 | 8.40 | 8.22 | -2.2 |
| 1600 | 1.07 | 1.00 | -7,1 | 8.39 | 8.15 | -2.9 |
| 1800 | 1.08 | 0.99 | -8,1 | 8.22 | 8.02 | -2.3 |
| 2000 | 1.08 | 1.01 | -6,8 | 8.01 | 7.85 | -2.0 |

Table 3. Accuracy of the pressure curves model of a diesel engine fuelled by RME; comparison of the mean indicated pressure (based on cylinder pressure curve) and the maximum cylinder pressure

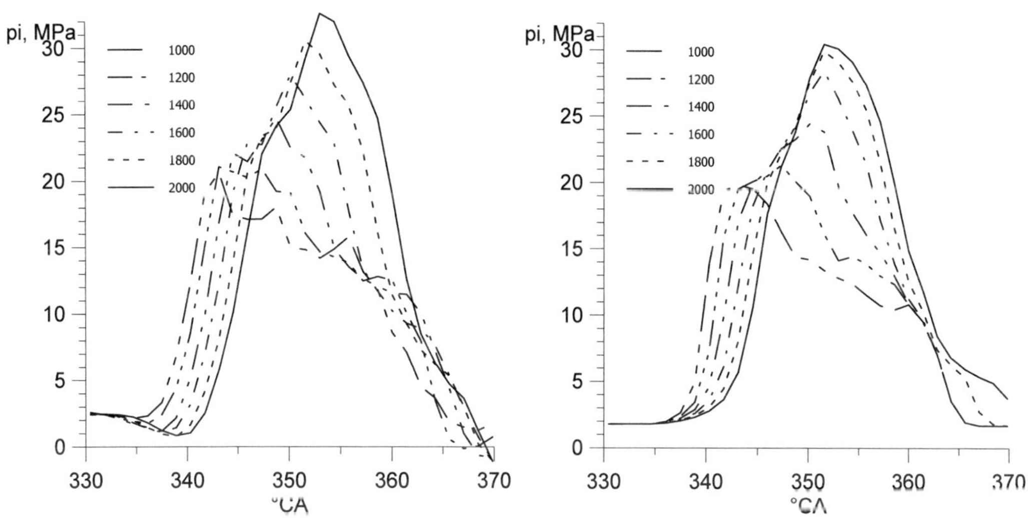


Fig. 9. The curves of injection pressure in function of crankshaft angle in the engine fuelled by RME at rotational speeds of 1000, 1200, 1400, 1600, 1800 and 2000 rpm: (a) experimental, (b) computed by the model obtained in the GFSm system

### 5.3 Comparison with models built by Fuzzy Logic Toolbox (FLT) and SANN

In this section we compare only the models regarding pressure in injection pipe when engine was fuelled by diesel oil. The GFSm model regarding cylinder pressure was discussed in (Radziszewski & Kekez, 2010). The model of injection pressure build by GFSm software was later transferred into Matlab Fuzzy Logic Toolbox (FLT) software in order to compare time of computation of the pressure curve by GFSm software and FLT software. Some new models of injection pressure were created using this Toolbox. Two of them are fuzzy models and one is a neuro-fuzzy model. We also used SANN (Statistica Automated Neural Networks) to automatically find the architecture of neural network and learning algorithm which produces the best model of pressure in injection pipe. The main disadvantage of the neural network model is the lack of its interpretability. The accuracy of modelling and time of computations for all models were compared.

### 5.3.1 Transfer of the model into FLT

It was necessary to write the code of the function that implements Center Average Defuzzification method with minor modifications used by the GFSm model (Kekez, 2008). The accuracy of the model transferred into FTL depends on the number of points, *k*, for which the value of the membership function of the aggregated output fuzzy set was calculated. For *k*=401 points, the accuracy of original model (GFSm) and the same model implemented in FLT was identical, but FLT made all calculations in 80 ms.

### 5.3.2 Fuzzy models in FLT

Attempts to build a new model of injection pressure using the FLT were carried out, using the same training dataset as used for GFSm (but without cylinder pressure data).The two models were built, using genfis3 function of the software: Takagi-Sugeno-type (Fig. 10a) and Mamdani-type (Fig. 10b). The first one has unsatisfactory accuracy, but the latter computed

(a)

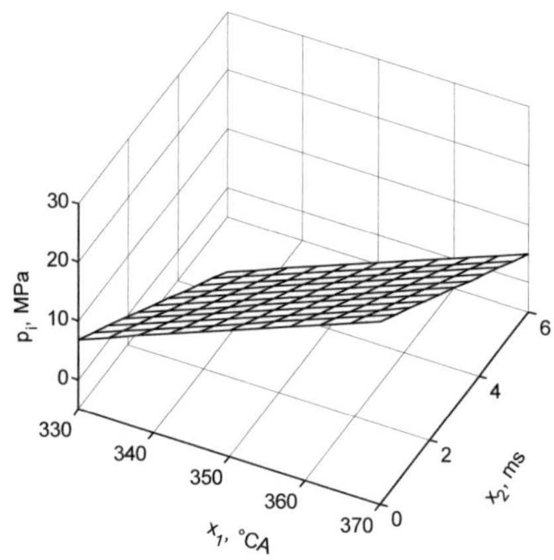


(b)

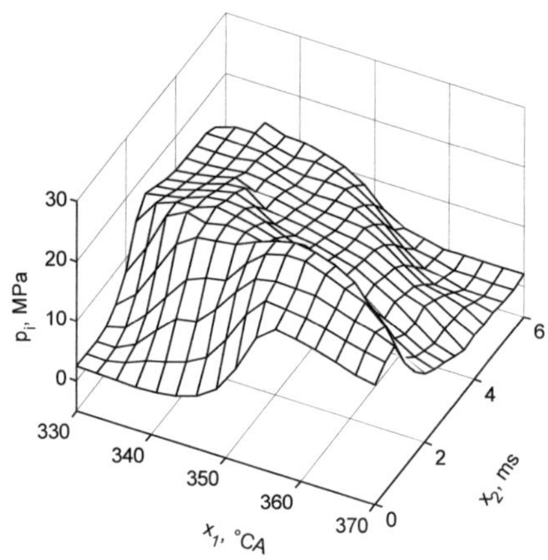


Fig. 10. Models of pressure in the injection pipe built by using the FLT function genfis3: a) Takagi-Sugeno-type, b) Mamdani-type

the area below the curve (for $n$=1200 rpm) with 4.3% error and the maximum pressure with 0.9% error. However, for other rotational speeds the accuracy of prediction of the maximum pressure was worse (errors up to 21.6% for 2000 rpm). Better accuracy could be achieved by checking other possible values of settings for the fuzzy c-means clustering procedure, or by changing the structure of training dataset.

### 5.3.3 Neuro-fuzzy model in FLT

We built the injection pressure model, using anfis function in FLT software, which implements neuro-fuzzy ANFIS method (Jang, 1993). In comparison with GFSm model (Table 4), the model built by ANFIS with default settings had slightly better accuracy of prediction of the area below the pressure curve for the speed of 1800 rpm, but over 60% error for 1000 rpm (as compared to 4% in case of GFSm). Other settings of ANFIS parameters (model no. 2 in Table 4) resulted in very high accuracy for the speeds that were represented in the training dataset (0.05% error) and very low accuracy for some rotational speeds (225% error for 1000 rpm). This model was over-fitted to the training data. Better results could be achieved by checking other possible values of ANFIS settings, or by changing the structure of training dataset.

| $n$, [rpm] | GFSm | ANFIS default model | ANFIS model no. 2 (64 rules) |
|---|---|---|---|
| | Error of calculation of the area below the pressure curve, % | | |
| 1000 | -4 | -61 | 225 |
| 1200 | -1 | 2 | 0 |
| 1400 | 4 | 18 | -10 |
| 1600 | 5 | 17 | -7 |
| 1800 | 1 | 2 | 0 |
| 2000 | -4 | 18 | -31 |
| | Error of calculation of the maximum pressure, % | | |
| 1000 | 6 | -19 | 763 |
| 1200 | -1 | -1 | 0 |
| 1800 | -2 | -4 | 0 |
| 2000 | -5 | -10 | 2 |

Table 4. Comparison of accuracy of GFSm model and two ANFIS models for a diesel engine fuelled by diesel oil

### 5.3.4 Neural-network models built by Statistica Automated Neural Network

Neural networks have ability to approximate any non-linear function (Korbicz et al., 1994). Because the value of pressure is non-linearly dependent on crankshaft angle and rotational speed, we used neural networks for modelling this relation. Our choice was the STATISTICA Automated Neural Networks (SANN) software package. SANN builds many neural networks, i.e. tests all possible combinations of the following elements: network structure (including the number of neurons in hidden layer), activation function, and learning algorithm. This allows to avoid a situation where one neural network, a priori chosen by the researcher, poorly describes given non-linear relation.

The SANN software was run with the following settings:

- analysis type: regression,
- sampling method: default (random selection, train 70%, test 15%, validation 15%),
- network type: multi-layer perceptron (MLP), number of neurons in a hidden layer: default (minimum 2, maximum 8),
- activation functions in hidden layer and output layer: linear, logistic, tanh (hyperbolic tangent), exponential,
- create 1000 different neural networks and save the 5 best networks.

The learning dataset was almost the same as for GFSm system, but contained injection pressure data for the rotational speeds of 1200 and 1800 rpm.

After the creation of the networks, the best one (in terms of prediction accuracy of maximum pressure) was chosen from five networks saved by SANN software. This network had: 9 neurons in a hidden layer, logistic activation function in hidden layer and output layer. It was trained using BFGS (Broyden-Fletcher-Goldfarb-Shanno) algorithm with SOS (Sum Of Squares) error function, and the best solution was found in 145th iteration. Accuracy of modeling of pressure course by this network is presented in Table 5 and Fig. 11.

| Rotational speed, $n$, rpm | Area $S$ below the curve $p_i = f(\alpha)$, MPa·°CA | | | Maximum pressure $p_{i\max}$, MPa | | |
|---|---|---|---|---|---|---|
| | experimental | computed | $\delta S$, % | experimental | computed | $\delta p_{i\max}$, % |
| 1000 | 381 | 358 | -6.0 | 21.3 | 21.0 | -1.5 |
| 1200 | 392 | 390 | -0.3 | 21.3 | 21.2 | -0.6 |
| 1400 | 406 | 415 | 2.2 | 24.2 | 24.5 | 1.1 |
| 1600 | 417 | 431 | 3.3 | 26.4 | 26.9 | 2.0 |
| 1800 | 439 | 441 | 0.4 | 29.1 | 28.1 | -3.6 |
| 2000 | 462 | 448 | -2.9 | 30.9 | 28.6 | -7.7 |

Table 5. Accuracy of injection pressure model, created by SANN

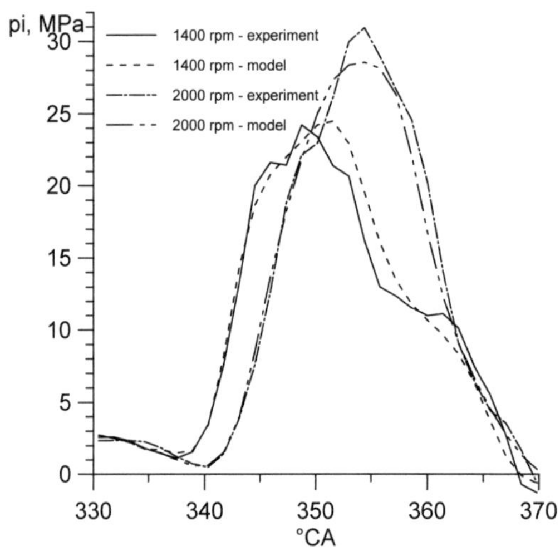


Fig. 11. Comparison of injection pressure model built by SANN software with experimental

## 6. Conclusion

The single model of pressures in a diesel engine (in cylinder and injection pipe), created by GFSm system was presented. The model was built for diesel oil and extended for biofuels, using scaling functions. The obtained model can be applied to diagnostic and control of a diesel engine. One of the advantages of the proposed method is that the model was built for a one given engine and therefore describes its work with good accuracy, sufficient for most technical applications. The time of computation of cylinder pressure curves and injection pressure curves for all tested rotational speeds (and time of writing of the results to disk) is less than 20 ms. Accuracy of the proposed model is better than accuracy of one neuro-fuzzy and one fuzzy model and slightly better than that of a neural network.

## 7. References

Abonyi, J., Babuska, R., & Seifert, F. (2002). Modified Gath-Geva Fuzzy Clustering for Identification of Takagi-Sugeno Fuzzy Models. *IEEE Trans. on Systems, Man and Cybernetics*, Vol. 32, (2002), pp. 612–621

Ambrozik, A., Ambrozik, T., Kurczyński, D. & Łagowski, P. (2010). Examination of fast-changing quantities in engine with common rail injection system. *J. of KONES Powertrain and Transport*, 2010, Vol. 17, No. 3, pp. 17-24, ISSN 1231-4005

Amsden, A. A. (1997). *KIVA-3V: A Block-Structured KIVA Program for Engines with Vertical or Canted Valves*. Los Alamos National Laboratory, LA-13313-MS

Cordon, O., Herrera, F., Hoffman, F. & Magdalena L. (2001). *Genetic Fuzzy Systems: Evolutionary Tuning and Learning of Fuzzy Knowledge Bases* (Advances in Fuzzy Systems - Applications and Theory 19), World Scientific, Singapore

Holland, J. H. (1976). Adaptation, In: *Progress in Theoretical Biology IV*, R. Rosen, F. M. Snell, (Ed.), pp. 263-293, Academic Press

Jang, J. S. R. (1993). ANFIS: Adaptive-Network-Based Fuzzy Inference Systems, *IEEE Transactions on Systems, Man, and Cybernetics*, Vol. 23, No. 3, (1993), pp. 665–685

Kalogirou, S. A. (2003) Artificial intelligence for the modeling and control of combustion processes: a review. *Progress in Energy and Combustion Science*, Vol. 29, No. 6, (2003), pp. 515–566, ISSN 0360-1285

Kekez, M. (2008). *Modelowanie pracy silnika o zapłonie samoczynnym z wykorzystaniem metod sztucznej inteligencji* (in Polish), Kielce University of Technology, Ph.D. thesis

Korbicz J., Obuchowicz A., Uciński D.: *Sztuczne sieci neuronowe. Podstawy i zastosowania* (in Polish), Akademicka Oficyna Wydawnicza PLJ, Warszawa, 1994

Radziszewski, L. & Kekez, M. (2010). Application of a genetic-fuzzy system to diesel engine pressure modeling. *The International Journal of Advanced Manufacturing Technology*, Vol. 46, No. 1-4, (2010), pp. 1-9, DOI 10.1007/s00170-009-2080-1

Regattieri Delgado, M., Von Zuben, F. & Gomide, F. (2004). Coevolutionary genetic fuzzy systems: a hierarchical collaborative approach, *Fuzzy Sets and Systems*, Vol. 141, No.1, (January 2004), pp. 89-106, ISSN 0165-0114

Rychter, T. & Teodorczyk, A. (2006). *Teoria silników tłokowych* (in Polish), Wydawnictwa Komunikacji i Łączności, ISBN 978-83-206-1630-1, Warszawa, Poland

Smith, S. F. (1983). Flexible learning of problem solving heuristics through adaptive search, *Proceedings of Eighth International Joint Conference on Artificial Intelligence*, pp. 422-425, Karlsruhe, Germany, August 1983, Morgan Kaufman

Wang, L. X. & Mendel, J.M. (1992). Generating fuzzy rules by learning from examples. *IEEE Trans. on Systems, Man and Cybernetics*, Vol. 22, No. 6, (1992), pp. 1414-1427

# 23

# Determination of the Impact of Biogas on the Engine Oil Condition Using a Sensor Based on Corrosiveness

C. Schneidhofer, S. Sen and N. Dörr
*AC²T research GmbH*
*Austria*

## 1. Introduction

Fuels, in particular low refined fuels and biofuels, may undergo considerable changes upon storage and circulation in the fuel supply system. (Bondioli et al., 2003) The knowledge of the quality of such fuels is crucial to maintain system reliability. Besides rubbing contact to parts in the fuel supply system, fuels and biofuels have also an impact on parts of the combustion section of an engine including the engine oil. (De Castro, 2007; Demirbas, 2009) Biofuels of interest are bioethanol, biodiesel and biogas. Biofuels usually differ significantly in chemistry from fossil fuels expressed by undesired interactions with oil additives accelerating oil alteration and hence significantly shortening oil intervals. (Luther, 2008)

The determination of the optimum oil change interval is increasingly important for both ecological and economical reasons. Therefore, different approaches exist to determine the right time for the oil change. For automotive applications, a common method is the forecast of the oil quality by algorithms based on indirect parameters such as the oil temperature, load profile or number of cold starts. (Fitch, 2004; Wang, 2001) For large engines or stationary machines, the current oil condition is determined by laboratory analysis of oil samples regularly taken. (Smolenski et al., 1994) The latter is the most accurate and reliable method for the determination of the current oil condition. But laboratory analysis suffers from the drawback of unacceptable time delays between sampling and analysis.

Biogas combusted in stationary gas engines is considered as one challenging renewable resource of energy. (Deublein, 2010) Besides methane and carbon monoxide as combustible compounds, undesired trace compounds can be found in biogas, sometimes at considerable levels (Harasimowicz et al., 2007). Among others, hydrogen sulphide, mercaptans, and halogen compounds are formed from processing of biomass at landfill sites, sewage plants, fermentation sites and wood gasification facilities.

These components in the biogas are able to form aggressive and potentially corrosive components during the combustion process (Abatzoglou et al., 2009; Agoston et al., 2005a). The uptake of these compounds in the lubricant results in a higher degree of oil acidification. Therefore, a higher potential for corrosion of engine parts is given and may result in engine breakdown. (Richard et al., 2010) The direct measurement of the resulting corrosiveness of the engine oil could be used as main indicator for an oil change.

Therefore, an oil condition sensor (Agoston et al., 2007) is proposed, able to measure the direct impact of the oil corrosiveness in form of a material loss of a sacrificial metal layer. Details about the principle of the proposed sensor can be found in chapter 1.3.
In this contribution, we report on the principle behaviour of the oil condition sensor for oil corrosiveness based on experiments at static conditions using defined model oils containing base oil and selected additives as well as acidic components. Furthermore, the sensor is tested with so-called artificial alteration method simulating the impact of biogases on engine oils to evaluate the online applicability of the proposed sensor concept. Results from field test gives emphases on the usefulness of the proposed sensor.

### 1.1 Biogas

Biogas as renewable source of energy is becoming increasingly important since biogas has some ecological advantages mainly being $CO_2$ neutral and hence reducing the formation of greenhouse gases. Furthermore, biogas represents a meaningful way of both waste use and waste disposal as agricultural, commercial and municipal waste from biogenic sources are used for the production of different types of biogas. The potential and possibilities for the successful application of renewable energy are illustrated by the "Modell Güssing" (EEE GmbH, 2008) - a small city in Austria which is independent from conventional energy providers as energy is entirely generated from renewable sources. A main part thereby is the gas engine driven with wood gas.

| component | concentration |
|---|---|
| methane | 40 – 75 % |
| carbon dioxide | 25 – 55 % |
| water (steam) | 0 – 10 % |
| nitrogen | 0 – 5 % |
| oxygen | 0 – 2 % |
| hydrogen | 0 – 1 % |
| hydrogen sulphide | 0 – 1 % |
| ammonia | 0 – 1 % |

Table 1. Typical components of a biogas (Schiffer, 2011)

The main components of biogas are methane ($CH_4$) and carbon dioxide ($CO_2$), but it also contains significant quantities of undesirable compounds. Table 1 shows typical compounds and their concentrations in a biogas. For the exploitation of energy gained from biogas, the amount of methane is essential: the higher the amount of methane the higher the output of energy from biogas. The other components are mostly useless for the energy production such as nitrogen or water. It is even possible that a negative impact on engine reliability can be provoked by trace components such as hydrogen sulphide ($H_2S$), ammonia ($NH_3$) and siloxanes. The trace components can be very harmful as the aggressive substances formed may destroy the engine, e.g., due to corrosion. (Harasimowicz et al, 2007; Schiffer, 2011)
The existence and amounts of these contaminants depend on the source of the biogas, i.e., landfill, anaerobic fermentation of manure or wood gasification. As they are aggressive to the engine, they should be removed before the gas is combusted in the engine using a biogas purification step. An important step during the purification process is to desulphurise the

gas in particular by removing hydrogen sulphide. Further steps are drying to remove water from the biogas or mechanical filtering to remove dust and other particles. (Abatzoglou et al., 2009; Besser et al., 2009; Harasimowicz et al, 2007)

Generally it can be stated that the better the purification steps the less the negative impact on the engine is to expect. But in turn, the installation and running costs are significantly higher. Consequently, an engine could be driven with an unpurified biogas including several harmful compounds directly impacting the lubricant condition. In this case, the oil reaches the end of the useful life time considerably earlier than with a purified biogas. For small engines, this method could be an economical alternative. But also due to ecological reasons, a gas purification should be applied resulting in less impact on the gas engine oil with the benefit of extended oil change intervals and lower costs of used oil disposal. The decision for or against a biogas purification system, respectively, requires a careful consideration of the economical and ecological influencing factors.

### 1.2 Oil condition monitoring

Independently of the type of fuel or lubricant, oil condition monitoring is getting increasingly important. For economical and ecological reasons, it is necessary to replace the used oil only when it is required. Therefore, the knowledge about the current oil condition is a prerequisite for the determination of the remaining useful life of the lubricant and the optimum time of an oil change, respectively. Different approaches of varying value are available to characterise the oil condition as described in more details in the following.

#### 1.2.1 Fixed oil change intervals

This is the easiest method to define the time for the oil change. The lubricant is changed after a defined driven distance, a time period or after a defined number of operating hours, e.g., 15 000 km or once per year for passenger cars (Exxon Mobil Corporation, 2011) or after 500 operating hours for some biogas engines. (Riemag, 2011) This method is mostly applied to older engines.

Here, the actual stress of the engine is not noticed but requires very constant operating conditions. The latter cannot be guaranteed especially in the field of biogas as remains from seasonal crops are often processed. Hence, the different strains on a lubricant can cause that the oil is changed too late which can lead to engine damage in the worst case. On the other hand, the oil can be replaced too early when the total oil stress is lower than expected. In that case, economical losses due to early purchase of fresh lubricant and disposal of still useful lubricant have to be faced. Due to these disadvantages, the oil change after a fixed interval is not recommended. But this method is currently applied for older and inexpensive engines.

#### 1.2.2 Algorithm based oil change intervals

Algorithm based methods for determining the optimum oil change interval are very common in the automotive sector. Based on the engines operating parameters like temperature, load profile or numbers of cold starts, the oil condition is determined with special algorithms. (Fitch, 2004; Wang, 2001)

An example of such a system is the General Motors GM Oil-Life™ system (Fitch, 2004). This system divides the load profile in four groups: easy freeway driving; high temperature, high load service; city driving and extreme short, cold start driving. Thereby, the oil deterioration mainly depends on the oil temperature regarding the first three groups. At cold starts driving, the lubricant is stressed by contaminations like water: the lower the temperature the higher the contamination with water. Based on these telemetric data, the oil change interval is estimated.

This method only records indirect parameters, hence the actual oil condition is not known at any point of the oil interval. Therefore, it can only react to variations of the operating conditions, but unforeseen influences like changes in the fuel or oil quality or a water intake cannot be detected. Consequently, such systems fail in the event of unforeseen incidents.

### 1.2.3 Laboratory analysis

This method is usually applied to large engines or stationary machines. Thereby, oil samples regularly taken are sent to a laboratory to analyse the oil condition using a large number of standardised methods. The set of analytical methods significantly depends on the application but mostly include viscosity, water content, acid content, base reserve and wear elements. Based on the analytical findings and if available including the history of the engine, a recommendation whether the oil should be change or still can be used will be done. (Newell, 1999; Smolenski et al., 1994; Toms et al., 2010)

This method is currently the most accurate means to determine the condition of the lubricant. But laboratory services suffer from the drawback of considerable time delay between sampling and analyses. Therefore, the actual condition of the lubricant is not immediately known. Consequently, sudden and unforeseen dramatical changes of the oil condition, e.g., caused by a water intake, cannot be detected which can result in rapid and severe engine breakdown.

### 1.2.4 Sensors for online oil condition monitoring

Sensors and sensor systems, respectively, directly installed in the lubricated system are proposed as the best way for condition monitoring as the current oil condition is immediately known and oil changes can be performed without delay especially in critical situations. Due to this crucial advantage over laboratory analysis, an extensive portfolio of different sensor systems exists on the market capable to measure various relevant oil parameters using a large number of sensor principles. But not every sensor principle is suitable for each oil application as the operating conditions as well as fuel and lubricant chemistry have to be considered. For this reason, the knowledge about the biofuel properties and critical oil parameters is indispensable for the selection of an appropriate sensor system.

Recently, a couple of sensor principles have been proposed aiming at the detection of physical or chemical properties of the oil. One typical parameter detected by sensors is the viscosity. (Agoston et al., 2004; Agoston et al., 2005b; Ronaldson et al., 2006) Here, it has to be pointed out that each sensor principle measures a different aspect of the viscosity. Electrical parameters of the oil such as permittivity or conductivity (Agoston et al., 2004; Basu et al., 2000; Gegner et al., 2010) are easy to measure and can be realized in robust devices. But the interpretation of the sensor signals is difficult as many factors influence permittivity or conductivity. The contribution of each factor cannot be clearly attributed without the knowledge of other oil parameters. Infrared sensors as proposed in (Agoston et al., 2008; Kudlaty et al., 2003) enable the measurement of several oil parameters, among others oxidation and additive depletion, but the constructional implementation of the interaction between infrared beam and oil is considered as largest difficulty.

In the case of online sensors, a similar thinking as for laboratory based analytical services should be considered: a single parameter is not sufficient to obtain reliable information about the oil condition. Therefore, the use of online systems equipped with multiple sensors able to measure several oil parameters should be implemented in the engine.
Particularly in stationary engines driven by gaseous biofuels, acidification is a crucial parameter for the evaluation of the condition of the engine oils (Felkel et al., 2008) as a correlation to corrosion of engine parts is expected. Here, elevated levels of acidification showed to be one of the main indicators for an oil change. For this reason, online monitoring of the engine oil corrosiveness by means of a chemical sensor is highly beneficial which is described in the following chapter.

### 1.3 Oil condition sensor based on corrosiveness

The proposed oil condition sensor is used to measure the corrosiveness of lubricants. For this purpose, a metal film acting as sacrificial layer is exposed to the lubricant. Due to the corrosive attack by the oil, a material loss of the metal film occurs which can be monitored by electrical means.

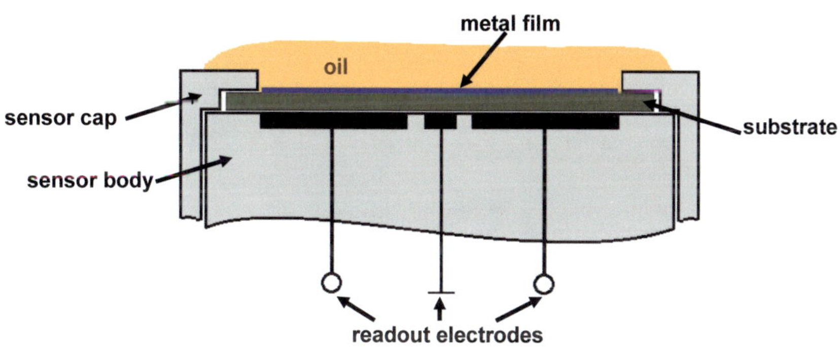


Fig. 1. Schematic setup of the proposed oil condition sensor for corrosiveness using capacitively coupled electrodes

Previous concepts measured the electrical resistance of a metal film (Agoston et al., 2005a; Agoston et al., 2007). This assembly requires a resistive connection between metal film and electronic measuring equipment. Experiments with this type of sensor setups clearly showed that corrosion occurred preferably at these contact points of the sacrificial layer. Furthermore, corrosion of the sacrificial layer was observed as even loss of material or as blotch type corrosion. In the latter case, the position of the blotches or spots on the sacrificial layer showed an influence on the sensor signal. (Agoston et al., 2006)
In order to overcome these shortcomings, a novel readout method was presented (Dörr et al., 2009) which is based on corrosion measurement without the need of an electrical contact using capacitively coupled planar electrodes. As shown in Figure 1, the novel sensor consists of an inert nonconductive substrate with a metal film as sacrificial layer on one side which is directly exposed to the corrosive medium. The material loss due to corrosion is monitored by measuring the capacitive coupling between the sacrificial layer and the readout electrodes on the backside of the substrate.
Using this setup, no resistive contact between the sacrificial layer and the electronic measurement equipment exists and hence no contact points with the danger of undesired corrosive attack exist, too.

## 2. Experimental setup

### 2.1 Sensor design and fabrication

The main part of the proposed sensor is the sacrificial element that consists of a metal film - the sacrificial layer - deposited on an inert substrate. For these experiments, a ceramic substrate from CeramTec AG was used. As metal film selected for these evaluations, lead was applied with a thickness of 600 nm. After this fabrication process, the metal films of the sensor were covered with a protective lacquer to avoid an undesired surface passivation by air during storage (prior to use).

For the implementation in corrosion experiments, a defined setup was fabricated consisting of the sensor body including the electrodes for the readout and a sensor cap for fixing the sacrificial element to the sensor body. The readout electrodes were mounted in a defined mould which was cast with epoxy resin afterwards. The sensor cap was also fabricated by casting of an other defined mould with epoxy resin. Figure 2 shows the setup of the manufactured sensor.

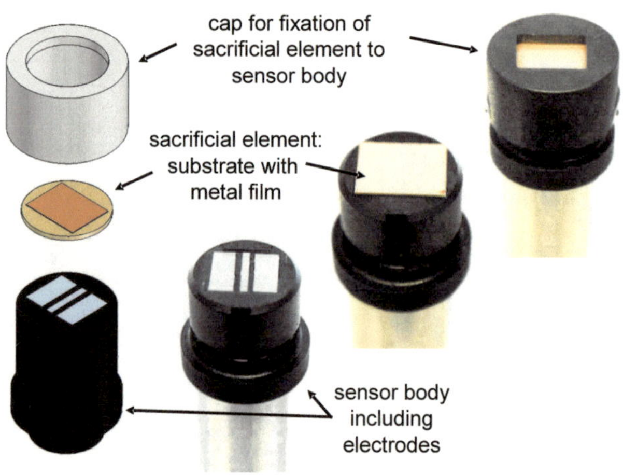


Fig. 2. Setup of the sensor for evaluation in the laboratory

An appropriate electronic measurement circuit which was directly connected to the readout electrodes evaluated the resonance frequency of the capacitive coupling between the electrodes and the metal film in combination with a constant resistor. For laboratory evaluations, the sensor was operated with a voltage output. This way, the resonance frequency was converted into a respective voltage. In the case of the field test also performed, the resonance frequency was transformed into an appropriate current.

As parameter for the corrosiveness of the lubricant monitored, the initial time (onset) for corrosion can be used representing the time when the sensor signal began to increase. Alternatively, the end time defined as the time when the sensor signal reached the end value can be determined. Hence, the end time refers to the moment when the entire metal film has been corroded away. Furthermore, an average corrosion rate can be easily calculated from the metal film thickness divided by the end time in (nm/h).

For the evaluation in laboratory, the sensor body was mounted on a tube for fixing the whole setup while immersing the sensor head with the sacrificial layer in the respective oil during the experiment. In this embodiment, the electronic measurement equipment was

separated from the sensor body because of the high temperature effected by artificial alteration (see 2.3). Accordingly, a higher stray capacity due to longer cables between sensor and electronics had to be accepted.
For the experiments in the field, the sensor body was built on a flange for easy integration in the oil circulation system of the observed gas engine (Figure 3). In this version, the electronic measurement equipment was also integrated in the same setup on the backside of the flange. So, a compact device could be integrated into the engine. Advantageously, the stray capacity was possible to be kept lower, too, because of short cables between readout electrodes and electronics.

Fig. 3. Setup of the sensor for the evaluation in the field without cap for fixing the sacrificial element and with electronics at the bottom (LEFT), and fully mounted with cap and sacrificial element (RIGHT)

**2.2 Experiments at static conditions**

For corrosion experiments at static conditions, the substrates with the sacrificial layer were cleaned by a solvent to remove the protective layer. Afterwards, the dried substrate was mounted on the sensor base body and fixed with the sensor cap. After these preparatory steps, the fully assembled sensor was immediately immersed into 40 mL of the oil sample in a hermetically sealed vessel. In such a vessel, a further chemical alteration of the oil was prevented and hence allowed constant (static) conditions. The assembled test units were kept at 100 °C and 120 °C, respectively, to monitor the corrosion process at an elevated rate. The chosen elevated temperature accelerated the corrosion process due to a higher rate of the chemical reactions initiated. At these conditions, the corrosion process of the metal film depends mainly on the current corrosiveness of the oil sample; therefore, the metal films corrode faster in more corrosive oils. In the event that after 100 hours no corrosion of the sacrificial layer could be observed, the experiment was aborted and consequently the oil investigated was classified as non-corrosive to the used metal for the sacrificial layer.
The sensor was tested at these conditions with a series of defined model oils containing acidic compounds and/or selected additives. Table 2 lists the oil samples used for these investigations as well as the chosen temperature and acid number (AN) according to (DIN 51558-1, 1979) of the selected oil samples. AN refers to the amount of acidic components in an oil given in (mg KOH/g oil).
The model oils M1 to M3 were used to evaluate the principle behaviour of the novel sensor setup. Therefore, a base oil was mixed with different amounts of acidic component (here naphthenic acid). The model oils containing mineral base oil and only one additional

component (oil additive or acid) were used to investigate the sensitivity of the used sacrificial layer to different components of the oil. The oil samples with both additive and acid were used to describe the interaction (competition) between the components and the impact on the corrosion process of the metal film, respectively.

| sample no. | oil type | experimental temperature [°C] | AN [mg KOH/g] |
|---|---|---|---|
| M1 | base oil + 3 mg/g naphthenic acid | 100 | 0.6 |
| M2 | base oil + 6 mg/g naphthenic acid | 100 | 1.3 |
| M3 | base oil + 10 mg/g naphthenic acid | 100 | 2.3 |
| BO | base oil | 120 | - |
| AO | base oil + 0.5% antioxidant | 120 | - |
| DET1 | base oil + 2.0% detergent A | 120 | - |
| DET2 | base oil + 2.0% detergent B | 120 | - |
| S1 | base oil + 1mg/g organic acid of middle chain length | 120 | 0.4 |
| S2 | base oil + 1mg/g organic acid of long chain length | 120 | 0.2 |
| DET1 + 0.1% S2 | base oil + 2.0% detergent A + 1 mg/g long chain length organic acid | 120 | - |
| DET1 + 0.5% S2 | base oil + 2.0% detergent A + 5 mg/g long chain length organic acid | 120 | - |
| AO + 0.1% S2 | base oil + 0.5% antioxidant + 1 mg/g long chain length organic acid | 120 | - |

Table 2. Defined model oils for corrosion experiments at static conditions; the amount of components is given in percent by weight

### 2.3 Experiments at dynamic conditions using artificial alteration

The online applicability was investigated by the integration of the oil condition sensor for corrosiveness in a novel artificial alteration apparatus. This lab-based apparatus is specifically designed for the simulation of the impact of biogases to lubricants during the application.

Artificial alteration means applying harsh conditions to the oil sample to accelerate oil degradation in a short-term scale. Nevertheless, an artificial alteration procedure should remain as close to reality as possible. This way, the quality of lubricants at special operating conditions and the usefulness of sensor systems for special applications can be investigated without the need to perform expensive field tests.

Currently used standardized methods for the artificial alteration of engine oils evaluate different parameters of the oil performance. Most standards examine the thermal-oxidative stability of oils by exposing the lubricant to elevated temperatures and oxygen from air. (ASTM D4871, 2006; DIN 51352, 1985)

| sample no. | alteration hours [h] | KV @ 100 °C [mm²/s] | TBN [mg KOH/g] | AN [mg KOH/g] |
|---|---|---|---|---|
| A-A-0 | 0 | 14.2 | 9.0 | 1.5 |
| A-A-1 | 16 | 14.1 | 8.5 | 1.3 |
| A-A-2 | 32 | 14.2 | 7.9 | 1.4 |
| A-A-3 | 48 | 14.3 | 7.9 | 1.6 |
| A-A-4 | 64 | 14.4 | 7.5 | 1.8 |
| A-A-5 | 80 | 14.5 | 7.1 | 2.1 |
| A-A-6 | 96 | 14.6 | 6.4 | 2.2 |
| A-A-7 | 112 | 14.8 | 5.9 | 2.4 |
| A-A-8 | 128 | 15.0 | 5.4 | 2.6 |
| A-A-9 | 144 | 15.0 | 5.1 | 3.0 |

Table 3. Oil samples taken during artificial alteration with "air" (A-A-0 to A-A-9) including alteration hours and oil properties determined by laboratory analysis

Based on the existing methods, an adapted alteration method was used to simulate oil life cycles in gas engines. Therefore, the lubricant was deteriorated at high temperature - in this case 160 °C - in a defined vessel able to blow gas into the oil. Additionally, a fixation was used to implement the proposed sensor setup in the artificial alteration device.

In order to simulate the impact of different fuel qualities on lubricant degradation, two different alteration methods were chosen. According to the first method, air was blown through the lubricant at a rate of 10 L/h. This method denoted with "air" should simulate mild conditions in an engine, e.g., the operation with natural gas almost free from contaminants forming aggressive compounds.

The conditions for the second method called "biogas" were based on the "air" method and adjusted to simulate the impact of biogas on accelerated oil degradation.

The lubricant was artificially altered at these conditions for 72 hours for "biogas" alteration and 144 hours for alteration "air", respectively. For oil condition monitoring and observation of the trend of oil degradation during the alteration procedure, oil samples were taken every 8 hours or 16 hours, respectively, and analysed in the laboratory. Table 3 for alteration "air" and Table 4 for alteration "biogas" summarise the most important properties of the oil samples: kinematic viscosity (KV) at 100 °C according to (ASTM D7042, 2010) using the so-called Stabinger viscosimeter, total base number (TBN) according to (DIN ISO 3771, 1985) and AN. TBN indicates the amount of bases in an oil possible to neutralise acidic components taken up or formed upon lubricant alteration. TBN is given in (mg KOH/g oil).

| sample no. | alteration hours [h] | KV @ 100 °C [mm²/s] | TBN [mg KOH/g] | AN [mg KOH/g] |
|---|---|---|---|---|
| **A-B-0** | 0 | 14.2 | 9.0 | 1.5 |
| **A-B-1** | 8 | 14.0 | 8.3 | 1.9 |
| **A-B-2** | 16 | 14.1 | 7.0 | 2.4 |
| **A-B-3** | 24 | 14.3 | 5.6 | 2.8 |
| **A-B-4** | 32 | 14.5 | 4.1 | 3.5 |
| **A-B-5** | 40 | 14.6 | 2.8 | 4.0 |
| **A-B-6** | 48 | 14.9 | 1.3 | 4.3 |
| **A-B-7** | 56 | 15.0 | 0.3 | 4.7 |
| **A-B-8** | 64 | 15.2 | 0.0 | 5.3 |
| **A-B-9** | 72 | 15.5 | 0.0 | 5.9 |

Table 4. Oil samples taken during artificial alteration "biogas" (A-B-0 to A-B-9) including alteration hours and oil properties determined by laboratory analysis

### 2.4 Evaluation in field operation

In order to evaluate the proposed sensor at real conditions, the sensor was integrated in the oil circulation system of a stationary gas engine driven with biogas from a wood gasification facility. The sensor signals were recorded every minute. The operating temperature of the engine oil at the sensor position was approximately 80 °C and hence was significantly lower than the temperature chosen in the laboratory tests (see 2.2 and 2.3).

During the observation period, oil samples were taken every 250 operating hours to carry out reference measurements in the laboratory. The laboratory analyses included KV at 100 °C, TBN and AN. The properties of the oil samples taken are listed in Table 5.

## 3. Results and discussion

### 3.1 Laboratory evaluation at static conditions

In laboratory tests, sensors were immersed in model oils (M1 to M3) containing an acidic component for fundamental evaluation of the behaviour of the novel setup with focus on sensitivity and reproducibility (Table 6). Thereby, a reasonable reproducibility was achieved (M1 - 1 and M1 - 2) as well as a higher corrosion rate and shorter time for complete corrosion, respectively, for higher amounts of acid could be demonstrated. (Dörr et al., 2009)

Based on these results, a series of model oils composed of base oil with selected lubricant additives and/or typical acids was tested under similar conditions. In Table 7, the results of corrosion experiments with model oil containing only one additional component (additive or acid) are listed. As shown, all additives selected (here antioxidant and detergents) as well as the mineral base oil used for these experiment showed no appreciable corrosive attack to the sacrificial layer of the sensor during the entire experiment.

However, model oils containing acidic compounds (here organic acids of different chain lengths) showed immediate corrosive attack to the metal film reflected by complete

| sample no. | operating hours [h] | KV @ 100 °C [mm²/s] | TBN [mg KOH/g] | AN [mg KOH/g] |
|---|---|---|---|---|
| F-0 | 0 | 13.7 | 8.9 | 1.5 |
| F-1 | 250 | 14.1 | 8.7 | 1.5 |
| F-2 | 500 | 14.5 | 8.0 | 1.7 |
| F-3 | 750 | 14.7 | 7.8 | 2.0 |
| F-4 | 1000 | 14.9 | 7.3 | 2.0 |
| F-5 | 1250 | 15.0 | 7.1 | 2.1 |
| F-6 | 1500 | 14.9 | 7.1 | 2.1 |
| F-7 | 1750 | 15.0 | 6.8 | 2.1 |
| F-8 | 2000 | 15.1 | 6.7 | 2.3 |
| F-9 | 2250 | 15.1 | 6.4 | 2.3 |
| F-10 | 2500 | 15.2 | 6.7 | 2.4 |
| F-11 | 2750 | 15.2 | 6.6 | 2.4 |
| F-12 | 3000 | 15.3 | 5.9 | 2.5 |

Table 5. Properties of used oil samples taken every 250 operating hours during the field test

| sample no. | corrosion rate [nm/h] | time for complete corrosion [h] |
|---|---|---|
| M1 - 1 | 350 | 1.7 |
| M1 - 2 | 375 | 1.0 |
| M2 | 550 | 1.1 |
| M3 | 750 | 0.8 |

Table 6. Corrosion rates of sensors immersed in model oils (M1 to M3) with different amounts of acid and time for complete corrosion of the sacrificial layer

consumption of the sacrificial layer within few hours. Furthermore, it can be seen that in the model oil containing the organic acid of middle chain length the metal film corrodes faster than that with long chain length. This can be explained by different amounts of effective acidic compound as both model oils contained the same percentage by weight of organic acid. But in the case of shorter chain lengths the number of reactive moieties is higher confirmed by determination of the acid number in Table 2. Eventually, the metal film started to corrode earlier.

For the investigation of the interaction of additives and acids, models oils containing both types of components were prepared (Figure 4). The results from static corrosion experiments clearly showed that the model oil containing antioxidant and acids corrodes similarly than the model oil containing only the acidic component (see Table 7). It can be concluded that the antioxidant does not influence the corrosion process which was only determined by the acidic component. Furthermore, the antioxidant is not able to prevent the metal film from corrosion and consequently it can not neutralise the acid in the oil.

| oil type | corrosive attack |
|---|---|
| BO | no corrosion |
| AO | no corrosion |
| DET1 | no corrosion |
| DET2 | no corrosion |
| S1 | corrosion after 2 h |
| S2 | corrosion after 4 h |

Table 7. Results of corrosion experiments at static conditions with selected model oils

The experiments with detergent and a low amount of acid (0.1 %) showed no significant corrosive attack to the metal film during the entire experiment. In this case, the detergent was possible to neutralise the acidic component and therefore prevent the corrosion of the metal film. The experiment with large amount of acid (0.5 %) showed a negligible corrosive attack to the metal film for about 50 hours then followed by a slow steady increase of the sensors signal. The observed behaviour showed that the detergent was not able to neutralise the total amount of acidic component and therefore a small quantity of reactive substances left induced slight corrosion of the sacrificial layer.

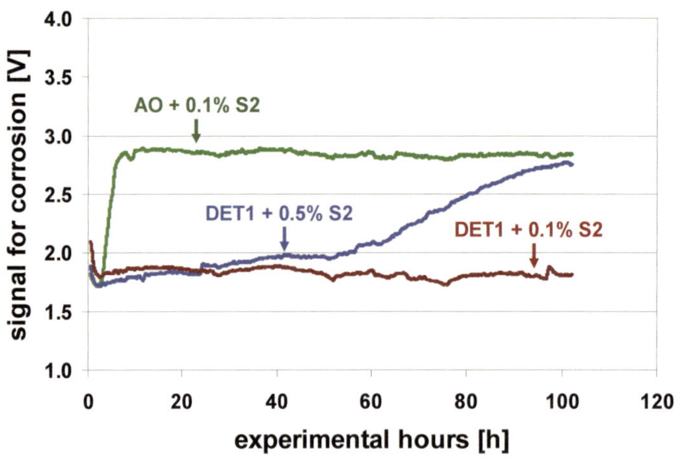


Fig. 4. Corrosion experiments performed at static conditions with model oils containing both additives and acidic compounds

The static conditions chosen - constant temperature and constant oil condition - showed good response and repeatability and hence allowed the pre-selection of elements of the proposed sensor. The results clearly showed that the metal film mainly corroded due to the acidic components dissolved in the model oils. Additives selected for these experiments (antioxidant and detergents) showed no significant corrosion. The knowledge about the ability of additives to neutralise acids and their interaction with surfaces is of essential importance for the understanding of the corrosion process and the behaviour of the sensor.

### 3.2 Laboratory evaluation at dynamic conditions – artificial alteration

As shown in Table 3 and Table 4, the lubricant deteriorated significantly faster when exposed to the alteration method "biogas". This is particularly expressed by significantly higher acidification determined as higher AN and lower TBN, respectively. Generally, an increase of viscosity can be observed when lubricants alter upon application. In the field of stationary gas engines, the oxidation stress results in the formation of acids (expressed as AN) from the blow-by gas, base oils but also from additives. Acidification by artificial alteration with "air" (simulating combustion with purified natural gas) increased up to 3.0 mg KOH/g after 144 hours. In comparison, artificial alteration with "biogas" (simulating combustion with unpurified biogas) reaches a value of 5.9 mg KOH/g within 72 hours. That is why a considerable base reserve (expressed as TBN) is indispensable to enhance the lifetime of the engine oil. As stated above, the stress is even higher when biogas is burnt as fuel: the base reserve is completely consumed after 64 hours of artificial alteration whereas an gas engine oil altered by "air" still has more than the half of the base reserve after 144 hours. So, it can be concluded that sensors giving information about the degree of acidification are more useful than sensors measuring viscosity as viscosity did not change considerably until the end of artificial alteration or the field test, respectively.

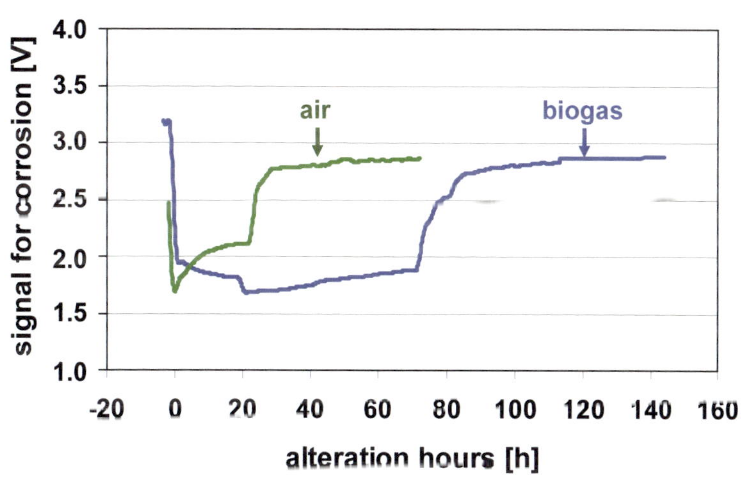


Fig. 5. Sensor signal during artificial alteration using both alteration methods

By looking at the sensor signals recorded during artificial alteration of the lubricant, an obvious shorter initial time for corrosion of about 20 hours with "biogas" can be seen in comparison to "air" causing an initial time for corrosion of about 70 hours (see Figure 5). The trend of the sensor signals shows a sharp decrease of the signal at the beginning of the alteration. This is due to the adaptation of temperatures to the alteration temperature of 160 °C. This period is followed by a relatively constant signal till the initial point for corrosion is observed. In this constant phase, no significant corrosion is noticeable. After the initial point of corrosion, the signal increases very fast due to corrosive attack of the sacrificial layer of the sensor till the whole metal film is corroded and the end signal of about 3 V is reached.

As it can be seen in Table 3 and Table 4, both oils had an acidification (AN) of 2 to 2.5 mg KOH/g and a base reserve (TBN) of about 7 mg KOH/g at the initial point of the corrosion. A similar initial point for corrosion was also reported in (Schneidhofer et al., 2009) using a resistive readout method. These results give evidence that the sensor is sensitive to a certain oil condition reached after a certain "amount" of stress independent from the time needed to achieve this condition. In other words, the sensor is capable of condition monitoring of engine oils run with different fuel qualities. An additional correlation with data from laboratory analysis is demonstrated in chapter 3.4.

### 3.3 Field operation

The usefulness of the proposed sensor concept is also illustrated by results determined by online condition monitoring in stationary gas engines. Figure 6 shows the trend of the sensor signal of a sensor system implemented into a gas engine driven with wood gas. The bright line signifies the moving average to smooth out the short-term fluctuations of the sensor signal. It is to note that the trend depicted includes only periods with running engine. Non-operating periods are not considered for better recognition.

After about 1300 operating hours, a significant increase in the sensor signal due to corrosion of the sacrificial layer was observed till the whole metal film was corroded after about 2200 operating hours followed by the constant end signal of the sensor. Before the onset of corrosion after 1300 operating hours, the sensor showed an almost constant signal (base signal) due to no significant corrosive attack to the sacrificial layer. As shown in Table 5, the oil was characterised by an acidification of about 2 mg KOH/g and a TBN of about 7 mg KOH/g at the initial point of corrosion. This critical oil condition was also detected during artificial alteration (see 3.2).

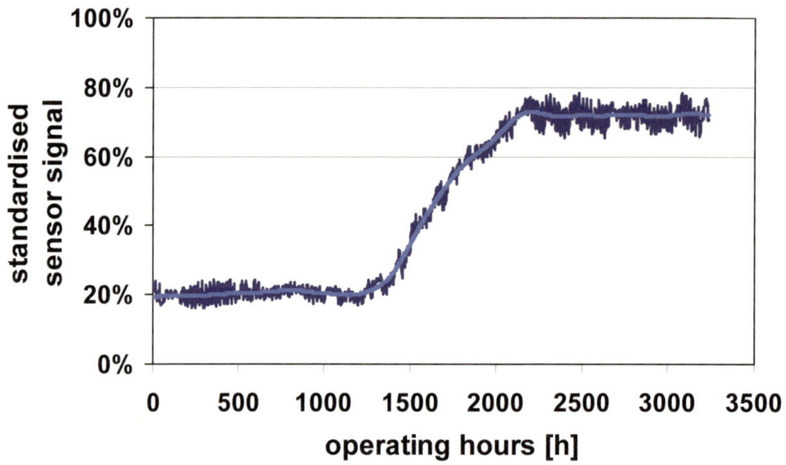


Fig. 6. Trend of the sensor signal during an entire oil interval of a stationary gas engine driven with wood gas; moving average of the sensor signal is highlighted

### 3.4 Correlation of sensor signal with data from laboratory analysis

The results from the experiments at static conditions (see 3.1) clearly showed that the metal film of the sensor is sensitive to the acidic components in the oil. If there are free reactive molecules which cannot be neutralised by the oil, the sacrificial layer is corroded according to the rule of thumb: the higher the amount of acid the faster the corrosion process.

The results obtained from the artificial alteration and field operation suggest that an initial point exists where a critical amount of acid is in the oil. For the correlation with data from laboratory analysis, the sensor signals at the time of sampling were extracted and compared with the conventional oil parameter AN determined in the laboratory. It can be seen in Figure 7 that the critical amount of acid equals AN of about 2 mg KOH/g for the used gas engine oil and the conditions applied. This value is equivalent to an increase of about 0.5 mg KOH/g in comparison to the acid number of the fresh oil (samples A-A-0 in Table 3, A-B-0 in Table 4 and sample F-0 in Table 5).

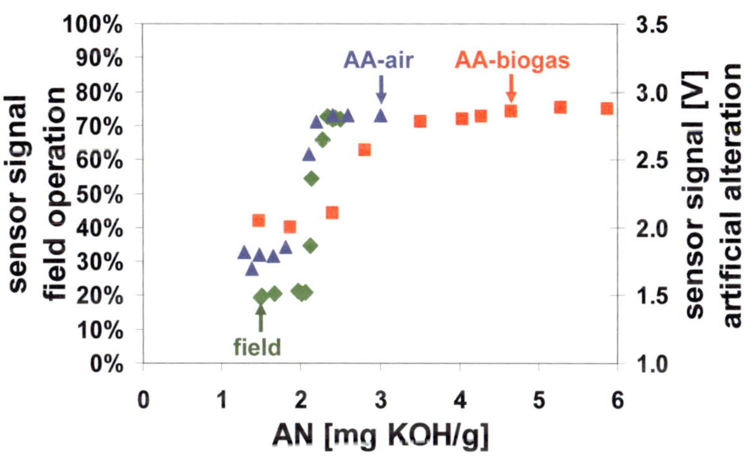


Fig. 7. Correlation of the sensor signal with AN of artificially altered oils series (AA-air and AA-biogas) as well as field samples (field)

The correlation of the sensor signal with the base reserve (Figure 8) showed a similar behaviour as identified for acidification: a critical value of about 7 mg KOH/g was observed for the gas engine oil selected which is equivalent to a decrease of TBN of approximately 2 mg KOH/g.

It has to be pointed out that a dependence of the critical values for acidification and base reserve on the composition of the engine oil (base oil and additives) may occur. That is why the initial point for the corrosion process of the sacrificial layer could vary. Consequently, it is important to know the critical initial point for a correct interpretation of the sensor signal for a proper decision on the optimal time for an oil change. The method of artificial alteration described has proven to be a valuable tool for a rapid but close-to-reality evaluation of the initial time as the same critical values for AN and TBN for both artificial alteration methods as well as at the field operation was observed.

The knowledge of the initial point of corrosion can be used as pre-warning for a critical amount of oil acidification reached. The period after initiation of corrosion can also be exploited for continuous oil condition monitoring till the whole metal film has been removed by corrosion. The time of the oil change can be adjusted to the end point of the corrosion of the metal film when following aspects are considered: proper adaptation of the thickness of the sacrificial layer, operating temperature and knowledge about engine behaviour.

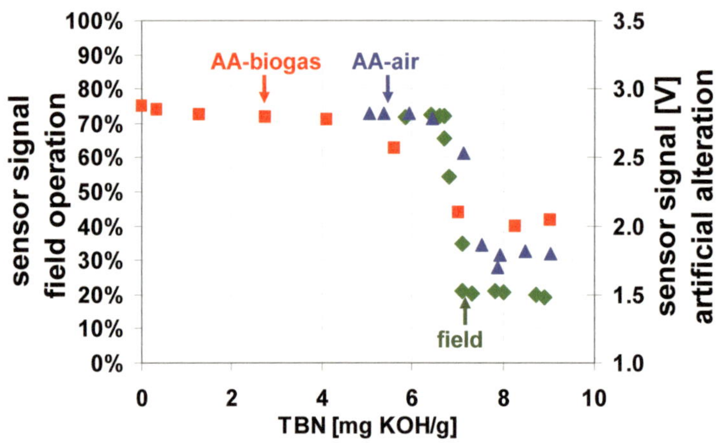


Fig. 8. Correlation of the sensor signal with TBN of artificially altered oils series (AA-air and AA-biogas) as well as field samples (field)

## 4. Summary and conclusions

Online oil condition sensors for the monitoring of the corrosiveness of lubricants are based on a metal film deposited on an inert substrate. Thereby, the material loss due to corrosion is recorded by measuring the capacitive coupling to readout electrodes. The focus of the application of the proposed sensor are gas engines driven with biogas because of the high amount of aggressive compounds formed during the combustion process which potentially lead to severe corrosion of engine parts.

The fundamental applicability of such a sensor system has been investigated with experiments at static conditions in the laboratory using defined model oils which were composed of base oil and selected additives and/or acidic components. The results clearly showed that the metal film of the sensor is sensitive to acidic compounds in the oil: the higher the amount of acid the faster the corrosion process. Selected additives (antioxidant and detergent) used for these experiments showed no significant attack to the sacrificial layer. In the case of model oils containing both additives and acids, the interaction between these components is essential for the corrosive attack. Here, it is essential whether the additive can neutralise the acidic compound or not.

Furthermore, the sensor was evaluated with different procedures of artificial alteration. In detail, two different methods were applied for the simulation of mild conditions ("air") in a gas engine (e.g., driven with natural gas) and harsh conditions caused by an unpurified biogas ("biogas"). The results from the sensor signals reported on higher corrosiveness (shorter initial time for the corrosion process) of the oil deteriorated with "biogas" in comparison to the artificial alteration "air". Artificial alteration is characterised by a significantly shorter experimental time enabling a rapid but close-to-reality evaluation of the sensor.

These results were confirmed by laboratory analyses of oil samples taken during artificial alteration. The correlation of the sensor signal with laboratory analytical data revealed that the initial point of corrosion is equivalent to an oil acidification (expressed as acid number) of about 2 mg KOH/g and a base reserve (expressed as total base number) of 7 mg KOH/g, respectively. In comparison to the values of the fresh oil, the onset of corrosion was observed at an increase of the acid number of 0.5 mg KOH/g and a consumption of the base reserve of 2 mg KOH/g, respectively. Below the initial point detected, the metal film exposed to corrosion remained almost unchanged. After the critical point, the metal film is corroded till the whole sacrificial layer is consumed.

In field operation, the sensor showed the usefulness and applicability of the proposed sensor concept. The results confirmed the initial point for the onset of corrosion already elucidated during artificial alteration.

This research work demonstrates that the proposed sensor for oil corrosiveness is characterized by a sufficient and reproducible sensitivity useful for online oil condition monitoring in engines as pre-warning system. The choice of the sacrificial material and thickness of the metal layer mainly depends on the application determined by fuel quality and the type of engine oil.

The oil condition sensor proposed is suitable for the indication of the impact of biofuel quality on the engine oil condition as successfully demonstrated on the example of a stationary gas engine run with biogas from wood gasification. Concluding, the sensor can be used for online oil condition monitoring in engines as pre-warning system for the indication of critical amounts of acids formed upon combustion of biofuel.

## 5. Acknowledgement

This work was funded by the Austrian "COMET Program" and has been carried out within the "Austrian Excellence Centre of Tribology".

## 6. References

Abatzoglou, N. & Boivin, S. (2009). A review of biogas purification processes, *Biofuels, Bioproducts and Biorefining* Volume 3, Issue 1, 42-71

Agoston, A.; Ötsch, C.; Novotny-Farkas, F.; Eisenschmid, H. & Jakoby, B. (2004). Sensors for a lubrication-monitoring-system, *Proc. Vol. III, 14th Int. Colloquium Tribology*, TAE, 13.-15.01.2004, pp. 1883-1889

Agoston, A.; Svasek, E. & Jakoby, B. (2005a). A novel sensor monitoring corrosion effects of lubrication oil in an integrating manner, *Proc. IEEE Sensors 2005*, 29.10.-03.11.2005, pp. 1120-1123

Agoston, A.; Keplinger, F. & Jakoby, B. (2005b). Evaluation of a vibrating micromachined cantilever sensor for measuring the viscosity of complex organic liquids, *Sensors and Actuators A*, Volume 123 - 124, 82 - 86

Agoston, A.; Dörr, N.; Schneidhofer, C. & Jakoby, B. (2006). Corrosion sensor for engine oils - evaluation and modified design, *Proc. Eurosensors 2006*, 9-2006, ISBN: 91-631-9280-2, Artikel W2C-O1, 4 S.

Agoston, A.; Doerr, N. & Jakoby, B. (2007). Corrosion sensors for engine oils - laboratory evaluation and field tests, *Sensors and Actuators B*: Chemical, Volume 127(1), pp. 15-21.

Agoston, A.; Schneidhofer, C.; Dörr, N. & Jakoby, B. (2008). A concept on an infrared sensor system for oil condition monitoring *E&I. Elektrotechnik und Informationstechnik*, Nr. 125/3, 71-75

ASTM Standard D4871 (2006). Standard guide for universal oxidation/thermal stability test apparatus, ASTM International, West Conshohocken, PA, 2006, DOI: 10.1520/D4871-06, www.astm.org

ASTM Standard D7042 (2010). Standard test method for dynamic viscosity and density of liquids by Stabinger viscometer (and the Calculation of Kinematic Viscosity), ASTM International, West Conshohocken, PA, 2006, DOI: 10.1520/D7042-10, www.astm.org

Basu, A.; Berndorfer, A.; Buelna, C.; Campbell, J.; Ismail, K.; Lin, Y.; Rodriguez, L. & Wang, S.S. (2000). Smart sensing of oil degradation and oil level measurements in gasoline engines, *SAE Technical Paper* 200-01-1366

Besser C.; Dörr N. & Grafl A. (2009). Performance von Ölen in stationären Gasmotoren (Performance of oils in stationary gas engines), Proceedings, CD-ROM, *ÖTG Symposium 2009*, ÖTG, Waidhofen/Ybbs (A), 26.11.2009, ISBN 978-3-901657-30-7, 2009.

Bondioli, P.; Gasparoli, A.; Della Bella, L.; Tagliabue, S. & Toso G. (2003). Biodiesel stability under commercial storage conditions over one year, *European Journal of Lipid Science and Technology*, Volume 105, Issue 12, 735-741

De Castro, J. (2007). Biofuels - An overview, 21.03.2011, Available from https://files.pbworks.com/download/VHPKm4xvxs/np-net/12638924/BZOS%20(2007)%20Biofuels%20in%20Africa_overview.pdf

Demirbas, A. (2009). Political, economic and environmental impacts of biofuels: A review, *Applied Energy* 86: S108–S117

Deublein, D. & Steinhauser, A. (2010). *Front Matter, in Biogas from Waste and Renewable Resources: An Introduction, Second Edition*, Wiley-VCH Verlag GmbH & Co. KGaA, Weinheim, Germany

DIN 51352 (1985). Testing of lubricants; determination of ageing characteristics of lubricating oils; increase of the Conradson carbon residue after ageing by passing air through the lubricating oil, Deutsches Institut für Normung, Berlin, Deutschland

DIN 51558-1 (1979). Determination of the Neutralization Number, colour-indicator titration, Deutsches Institut für Normung, Berlin, Deutschland

DIN ISO 3771 (1985). Petroleum products; total base number; perchloric acid potentiometric titration method, Deutsches Institut für Normung, Berlin, Deutschland

Dörr, N.; Schneidhofer, C. & Jakoby, B. (2009). Novel corrosion sensor based on capacitively coupled electrodes for the determination of the oil corrosiveness, Conference Proceedings, *STLE2009*, Orlando (USA), 17.-21.05.2009

EEE GmbH (2008). Modell Güssing, 21.03.2011, Available from http://www.eee-info.net/cms/

Exxon Mobil Corporation (2011). Mobil Super 3000 XE 5W-30, 21.03.2011, Available from http://www.mobil.com/Germany-German/Lubes/PDS/GLXXDEPVLMO Mobil_Super_3000_XE_5W-30.aspx

Felkel, J.; Dörr, N.; Varmuza, K.; Leimeter, T. & Novotny-Farkas, F. (2008). Implementation of chemometrics in oil condition monitoring - Trend Analysis Illustrated by Gas Engine Oils, *16th Int. Colloquium Tribology*, TAE, Ostfildern, 15.-17.01.2008, ISBN 3-924813-73-6

Fitch, J. (2004). Determining proper oil and filter change intervals: can onboard automotive sensors help?, *Practicing Oil Analysis* (1/2004), 21.03.2011, Available from http://www.machinerylubrication.com/Read/562/oil-change-filter-sensors

Gegner, J.; Kuipers, U. & Mauntz, M. (2010). Ölsensorsystem zur Echtzeit-Zustandsüberwachung von technischen Anlagen und Maschinen (Oil sensor system for online condition monitoring of technical equipment and machines), *tm - Technisches Messen*, Volume 77, No. 05, pp. 283-292

Harasimowicz, M.; Orluk, P.; Zakrzewska-Trznadel, G. & Chmielewski, A.G. (2007). Application of polyimide membranes for biogas purification and enrichment, *Journal of Hazard Materials*, Volume 144, Issue 3, 698–702

Kudlaty, K.; Purde, A. & Koch, A.W. (2003). Development of an infrared sensor for on-line analysis of lubricant deterioration, *Proc. IEEE Sensors 2003*, ISBN: 0-7803-8133-5, 903-908

Luther, R. (2008). Alternative fuels from the engine lubrication point of view, *MTZ worldwide Edition*, 2008-03

Newell, G.E. (1999). Oil analysis cost-effective machine condition monitoring technique, *Industrial Lubrication and Tribology*, Vol. 51 Iss: 3, pp.119 - 124

Richard, K. & McTavish, S. (2010). Impact of biodiesel on lubricant corrosion performance, SAE International Journal of Fuels and Lubricants March 2010 vol. 2 no. 2 66-71

Riemag (2011). Impressum, 21.03.2011, Available from http://www.riemag.de/impressum.html

Ronaldson, K.A.; Fitt, A.D.; Goodwin, A.R.H. & Wakeham, W.A. (2006). Transversely oscillating MEMS viscometer: The "Spider", *International Journal of Thermophysics*, Volume 27, Number 6, 1677-1695

Schiffer, F. (2011). Biogaszusammensetzung (Composition of biogas), 21.03.2011, Available from http://www.mifratis.de/biogaszusammensetzung.php

Schneidhofer C. & Dörr N. (2009). Online-Ölzustandsüberwachung mit chemischen Korrosionssensoren (Online oil condition monitoring using chemical corrosion sensors), *E&I. Elektrotechnik und Informationstechnik*, Nr. 1-2/2009, Vol 126, Springer-Verlag GmbH, 01.01.2009, ISSN 0932-383X, pp 31-36

Smolenski, D.J. & Schwartz, S.E. (1994). Automotive engine-oil condition monitoring, *Lubrication engineering*, Volume 50, Issue 9, 716-722

Toms A. & Toms L. (2010). Oil analysis and condition monitoring, *Chemistry and Technology of Lubricants 2010*, Part 4, 459-495

Wang, J.C.; Whitacre, S.D.; Schneider, M.L. & Dringenburg, D. H. (2001). System and method for determining oil change interval, US Patent 6,253,601 B1